Chemistry of Water Treatment

by
Samuel D. Faust
Osman M. Aly

BUTTERWORTHS
Boston • London
Sydney • Wellington • Durban • Toronto

An Ann Arbor Science Book

Ann Arbor Science is an imprint of Butterworth Publishers

Library of Congress Catalog Card Number 83-71823
ISBN 0-250-40388-9

Published by Butterworth Publishers
10 Tower Office Park
Woburn, MA 01801

Printed in the United States of America

To our wives,
Anne and Aida,
with affection.

ABOUT THE AUTHORS

Samuel D. Faust is Professor of Environmental Science at Rutgers University. He received his PhD in environmental sciences and chemistry from Rutgers University and a BA in chemistry from Gettysburg College. Dr. Faust was a Research Fellow in applied chemistry during the 1966–1967 term at Harvard University.

Dr. Faust is the editor of three books on the chemistry of aquatic environments and has published 95 papers in research journals. He has served as chairman and organizer of several conferences. Dr. Faust is a member of several honorary professional societies and committees, which include Standard Methods for the Examination of Water and Wastewater.

Osman M. Aly is Manager of Environmental Quality at the corporate headquarters of Campbell Soup Company. He received his PhD in environmental Sciences from Rutgers University, and held various research and teaching positions before joining the Campbell Soup Company in 1970. His major research interests include the chemistry of natural waters and environmental control.

Dr. Aly is the author of more than 30 publications on water and wastewater chemistry and treatment processes.

CONTENTS

PREFACE

The qualities of natural and treated water are areas of immense concern in our environment. One of the highest priority uses of water is, of course, its ingestion to sustain life. It is well established that water is a medium for transmission of enteric diseases and, in recent years, for transmission of alleged carcinogenic compounds. Consequently, we demand our water to be absolutely "safe" to drink, i.e., potable. On the other hand, the potability of water is not the only concern since water has other uses in the home and industry. As water is drawn from many ground and surface sources, it has diverse qualities which, in turn, dictate the types of treatment, the degree of treatment, and, to a large extent, the quality of the finished product.

In these days of widespread concern about environmental pollution, we are even more conscious of our drinking water quality. We expect it to be free from pathogenic organisms, free from scale-forming substances, free from toxic chemicals, free from corrosion and corrosion products, and free from aesthetically offensive substances. Thus, we speak of the six freedoms of water quality. This book intends to serve as a reference for professionals and others concerned with water quality and treatment.

The chemical and, to a lesser extent, microbiological aspects of water quality and treatment are emphasized. It is an application of these two sciences to these two interrelated parts of our immediate environment. This volume does not and is not intended to discuss the design of water treatment, per se. Rather, it is intended to demonstrate how chemistry influences the design of a water treatment plant, or, more appropriately, how it *should* influence the design. Historically, water treatment plants have been designed from hydraulic considerations with little regard to chemical aspects. The many chemical reactions used for removal of pollutants from water simply cannot be forced to occur within present designs. This traditional approach must be reexamined in view of the complexity of today's water quality and treatment.

Samuel D. Faust
Osman M. Aly

ACKNOWLEDGMENTS

A special acknowledgment must go to one of our previous professors, Dr. H. Heukelekian, now retired, who provided us with the opportunity to begin our professional careers. We are grateful and appreciative for this support, encouragement, and intellectual leadership from "Doc."

Once again, the authors wish to acknowledge the patience of our families who were neglected during the long hours of preparation. We are indebted to Mrs. Mildred McHose who typed the first chapter but had to retire due to an Achilles tendon operation. Ms. Carolyn Smith came to our rescue with her typing skills. We are appreciative of her long hours of labor well into the "wee" hours of the morning. We are grateful to our professional colleagues whose works are cited in this text. Again, we enjoyed the excellent cooperation of the staff of Ann Arbor Science.

CHAPTER 1

CRITERIA AND STANDARDS OF WATER QUALITY

HISTORY OF WATER STANDARDS

Man's quest for pure water is neither a recent nor a modern development. A compilation of medical lore in Sanskrit (circa 2000 B.C.) reads: "Impure water should be purified by being boiled over a fire, or being heated in the sun, or by dipping a heated iron into it, or it may be purified by filtration through sand and coarse gravel and then allowed to cool" [1]. This statement includes the essentials of modern water treatment, namely, disinfection and sand filtration. It also implies that water is a vector for transmission of impurities and diseases. Hippocrates, the father of medicine (460–354 B.C.), expressed concern about water quality when he said that "water contributed much to health," and asserted that rain water should be boiled and strained, for otherwise it would have a bad smell and cause hoarseness. Herodotus, an early Greek historian, reports that when warring, Cyrus the Great, King of Persia, took boiled water in silver flagons loaded on four-wheel carts drawn by mules [1]. This, too, implies the ancients' concern for water quality and suggests that the silver flagons were used for their disinfective property.

History also records some early epidemics of waterborne diseases: the infamous 1854 cholera outbreak in London, England, and later typhoid epidemics in the United States reported at Butler, Plymouth, New Haven, Nanticoke and Reading that involved 39,029 cases with 361 deaths [2]. By the early 1900s, it was fairly obvious that drinking water contained physical, chemical and biological impurities that impaired its quality. "Modern" water treatment probably started with the filtration of municipal supplies in Scotland in 1804 and in England in 1829 for the Chelsea Water Company of London [3]. By 1900, there were approximately 10 slow sand water filtration plants in the United States [1]. It was not until 1914 that any expression of water quality standards was noted with the introduction of testing for *Bacterium coli* as an indicator organism [4]. There was, however, previous concern about the levels of contaminants in drinking water with the 1899 publication [5] of the precursor to the present version of *Standard Methods for the Examination of Water and Wastewater.* Quality criteria and standards initially were confined to drinking water. Recently, however, attempts have been made to apply these concepts to other uses of water such as its recreational and esthetic uses, its use as a medium for fish and other aquatic life, its agricultural uses, and its value as an industrial resource. These additional water quality criteria were reported first in 1968 in the so-called "Green Book" entitled "Water Quality Criteria" [6].

LANGUAGE

Semantics

There is much confusion over the language used to establish water quality criteria and standards. This may be due to our federal bureaucracy's demand that several terms be defined when implementing water quality laws. Of course, there is the additional requirement that these definitions must change with each new water quality law! An attempt is made here to sort out the semantics of water quality criteria and standards in current use.

Perhaps the concept of distinguishing between criteria and standards began with the Water Quality Act of 1965, Public Law 89-234 [7]:

> In the absence of State action, such standards will be adopted by the Secretary of the Interior under procedures set forth in the Act.
>
> Standards adopted by a State will become the standards applicable if:
>
> 1. The Governor or water pollution control agency files by October 2, 1966, a letter of intent that the State, after public hearings, will, before June 30, 1967, adopt (a) water quality criteria applicable to interstate waters or portions thereof within the State, and (b) a plan for the implementation and enforcement of the criteria; and,
> 2. The State subsequently adopts such criteria and plan; and,
> 3. The Secretary determines that the State criteria and plan are consistent with the purposes of the Act, i.e., "...to enhance the quality and value of our water resources and to establish a national policy for the prevention, control, and abatement of water pollution."

Figure 1 [8] provides a conceptual framework for developing standards from criteria. The latter represents attempts to quantify water quality in terms of its physical, chemical, biological and esthetic characteristics. These guidelines apply in cases requiring scientifically determined measurements for judging the suitability of water quality for a designated use. The specific uses are cited in Figure 1. In this context, the definition of criteria as reported in the "Blue Book" [8] is "the scientific data evaluated to derive recommendations for characteristics of water for specific uses." These criteria represent the first step in the ultimate

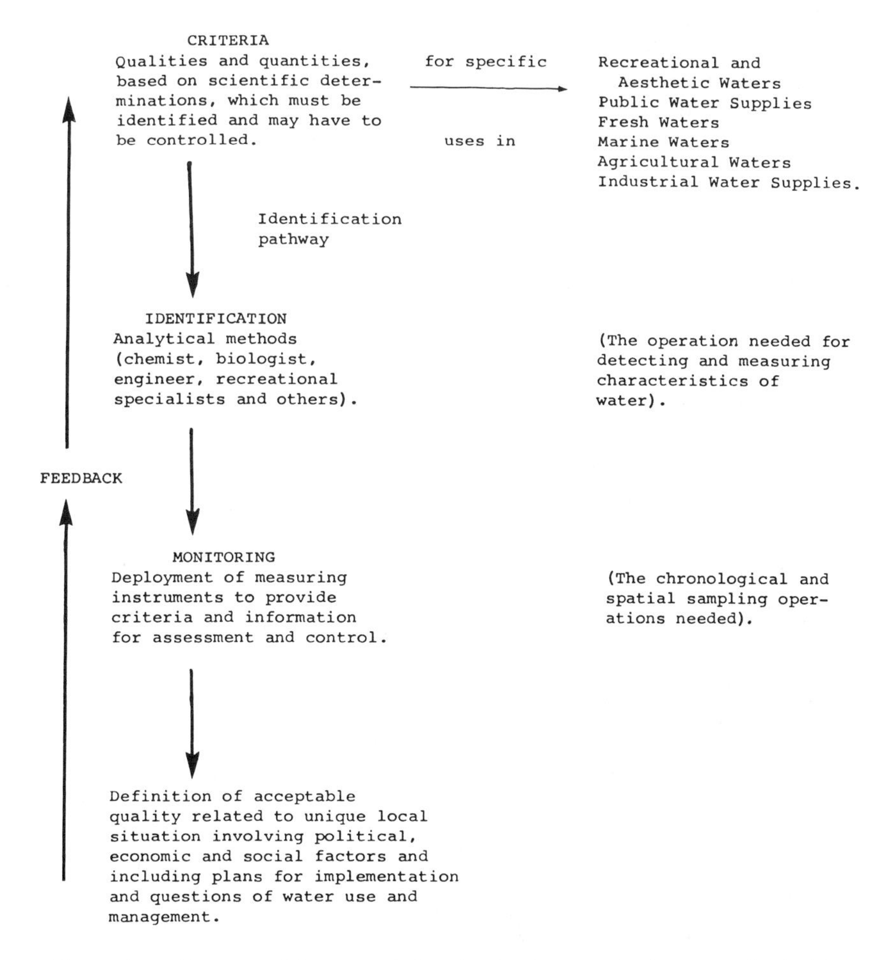

Figure 1. Conceptual framework for developing standards from criteria [8].

development of standards, as noted in Figure 1. The numbers assigned to criteria frequently are the same as those assigned to the standard. This is not intentional. One definition of water quality standards is noted in Figure 1. A second, similar definition of a standard is "a plan that is established by a governmental authority as a program for water pollution abatement and prevention" [7]. This latter definition is employed whenever enforcement and litigation activities are involved.

Public Water Supplies

Some additional terms are frequently confused. The 1962 U.S. Public Health Service (PHS) Drinking Water Standards were in effect in certain situations until the passage of the Safe Drinking Water Act of 1974 [9]. These 1962 standards employed two types of limits:

1. *Shall not exceed* limits, if exceeded, will be grounds for rejecting the supply. Substances in this category may have adverse effects on health when present in concentrations above the limit.
2. *Should not exceed* limits should not be exceeded whenever more suitable supplies are, or can be made, available at reasonable cost. When present in concentrations above the limit, substances in this category are either objectionable to an appreciable number of people or exceed the levels required by good water quality control practices.

The "shall not" type is a mandatory standard, whereas the "should not" is a recommended standard. However, the 1974 law introduced a new set of terms with a similar distinction. On December 16, 1974, the President of the United States signed into law the Safe Drinking Water Act recorded as Public Law 93–523 [10,11]. Among other things, this law provided for primary and secondary types of regulations:

1. Primary regulations set mandatory standards designed to protect public health.
2. Secondary regulations set mostly esthetic standards intended to provide suggestions that states may or may not implement.

To date, two versions of the primary standards have appeared [10,11]. Both versions have introduced another concept of a water quality standard; namely the "maximum contaminant level" (MCL), which is "the maximum permissible level of a contaminant in water which is delivered to the free-flowing outlet of the ultimate user of a public water system." (Details of the interim primary standards are given below.) The primary maximum contaminant level (PMCL) is perhaps somewhat similar to the "shall not" limit specified by the 1962 drinking water standards.

SAFE DRINKING WATER ACT

This act regulating drinking water quality was established in a three-step process. First, the U.S. Environmental Protection Agency (EPA) promulgated interim regulations based on the 1962 Public Health Service standards [9–11]. Also included were MCL and minimum monitoring frequencies. These regulations became effective on June 24, 1977 [12], and included limits for 10 inorganic and 6 organic chemicals, turbidity, coliform bacteria and radionuclides. The second step consisted of a two-year study conducted by the National Academy of Sciences (NAS) intended to gather basic information on all contaminants in drinking water that may have an adverse impact on mankind [12]. In the third step the EPA promulgated more comprehensive regulations, and, in addition, revised primary drinking water regulations following the results of the NAS study and other research activities.

Definitions

Public Law 93–523 provides some relevant definitions, some of which are [10,11]:

1. "Contaminant" means any physical, chemical, biological or radiological substance or matter in water.
2. "Maximum contaminant level" means the maximum permissible level of a contaminant in water that is delivered to the free-flowing outlet of the ultimate user of a public water system, except in the case of turbidity where the maximum permissible level is measured at the point of entry to the distribution system. Contaminants added to the water under circumstances controlled by the user, except those resulting from corrosion of piping and plumbing caused by water quality, are excluded from this definition.
3. "Person" means an individual, corporation, company, association, partnership, state, municipality or federal agency.
4. "Public water system" means a system for the provision to the public of piped water for human consumption, if such system has at least 15 service connections or regularly serves an average of at least 25 individuals daily at least 60 days out of the year. Such term includes (1) any collection, treatment, storage and distribution facilities under control of the operator of such system and used primarily in connection with

such system; and (2) any collection or pretreatment storage facilities not under such control that are used primarily in connection with such system. A public water system is either a community water system or a noncommunity water system.

The statement, "a system for the provision ... regularly serves an average of at least 25 individuals daily at least 60 days out of the year" was used in the definition of a public water system to include campgrounds, service areas and other public accommodations that are open for slightly less than three months each year.

5. "Community water system" means a public water system that serves at least 15 service connections used by year-round residents or regularly serves at least 25 year-round residents.
6. "Noncommunity water system" means a public water system that is not a community water system.

Public water systems were subdivided into community and noncommunity systems in order to relax the requirements of noncommunity systems in meeting the regulations. The possible health effects of a contaminant in drinking water consumed regularly over a long period of time are often quite different from those in water that is consumed occasionally. Thus, it is not imperative that the water consumed by transients from noncommunity systems be of the same quality as the water consumed by residents; that is, precautions should be taken to prevent the occurrence of contaminants (e.g., nitrate) which may pose a potential health hazard from short-term exposure.

Noncommunity systems include: hotels, motels, schools, service stations, highway rest areas, campgrounds and other public accommodations that have their own water systems with at least 15 service connections or that serve water to a daily average of at least 25 people. There are an estimated 200,000 noncommunity systems in the United States, but they are not the principal source of drinking water for their customers.

It is necessary to revert to the 1962 PHS drinking water standards for some definitions not covered by PL 93–523 [9]. These standards (1962) introduced new definitions and extended old definitions that are incorporated into the present EPA regulations. For example, in 1962, considerable attention was given to the "source and protection of the water supply":

> Mounting pollution problems indicate the need for increased attention to the quality of source waters. Abatement and control of pollution of sources will significantly aid in producing drinking water which will be in full compliance with the provisions of these Standards and will be esthetically acceptable to the consumer.

> Production of water supplies which poses no threat to the consumer's health depends upon continuous protection. Because of human frailties associated with this protection, priority should be given to selection of the purest source. Polluted sources should be used only when other sources are economically unavailable and then only when the provision of personnel, equipment, and operating procedures can be depended upon to purify and otherwise protect the drinking water supply continuously.
>
> Surface waters are subjected to increasing pollution and although some surface waters may be sufficiently protected to warrant their use as a supply without coagulation and filtration, they are becoming rare. Surface waters should never be used without being disinfected. Because of the increasing hazards of pollution, the use of surface waters without coagulation and filtration must be accompanied by intensive surveillance of the quality of the raw water and the disinfected supply in order to assure constant protection. This surveillance should include sanitary survey of the source and water handling, as well as biological, radiological, physical, and chemical examination of the supply.

In effect, these drinking water standards apply to the quality of the source of the water; they are, essentially, surface water standards. In the event that the quality of the source does not meet the standards, then the water must be treated. In turn, this leads to the following definition of "adequate protection by treatment":

> Any one or any combination of the controlled processes of coagulation, sedimentation, adsorption, filtration, disinfection, or other processes which produce a water consistently meeting the requirements of these Standards. This protection also includes processes which are appropriate to the source of supply; works which are of adequate capacity to meet maximum demands without creating health hazards, and which are located, designed, and constructed to eliminate or prevent pollution; and conscientious operation by well-trained and competent personnel.
>
> The degree of treatment should be determined by the health hazards involved and the quality of the raw water. During times of unavoidable and excessive pollution of a source already in use, it may become necessary to provide extraordinary treatment (e.g., exceptionally strong disinfection, improved coagulation, or special operation). If the pollution cannot be removed satisfactorily by treatment, use of the source should be discontinued until the pollution has been reduced or eliminated. When used, the source should be under continuous surveillance to assure adequacy of treatment in meeting the hazards of changing pollution conditions [9].

Another important point was introduced in the 1962 drinking water standards:

> these limits should apply to the water at the free-flowing outlet of the ultimate consumer. Delivery of a safe water supply depends upon the protection of the water in the distribution system as well as protection of the source and by treatment. Minimum protection in the distribution system should include programs which result in the provision of sufficient and safe materials and equipment to treat and distribute the water; disinfection of water mains, storage facilities, and other equipment after each installation, repair, or other modification which may have subjected them to possible contamination; prevention of health hazards, such as cross-connections or loss of pressure because of overdraft in excess of the system's capacity; and routine analysis of water samples and frequent survey of the water supply system to evaluate the adequacy of protection [9].

Two additional important definitions were introduced:

> a. Health hazards mean any conditions, devices, or practices in the water supply system and its operation which create or may create, a danger to the health and well-being of the water consumer. An example of a health hazard is a structural defect in the water supply system, whether of location, design, or construction which may regularly or occasionally prevent satisfactory purification of the water supply or cause it to be polluted from extraneous sources.
>
> b. Pollution, as used in these Standards, means the presence of any foreign substance (organic, inorganic, radiological, or biological) in water which tends to degrade its quality so as to constitute a hazard or impair the usefulness of the water [9].

Furthermore, intake from food and air was considered in establishing limits for toxic substances. Much of the above represented considerable extension of the authority of the PHS to regulate and protect U.S. public water supplies.

Primary Maximum Contaminant Levels (PMCL)

Inorganic Chemicals

The PMCL for inorganic chemicals are shown in Table I. The only PMCL that might be confusing are those for fluoride. There is a relatively narrow concentration range in which fluoride is effective in pre-

Table I. Contaminants and MCL in the National Interim Primary Drinking Water Regulations [10,11]

Contaminant	MCL
Inorganic	(mg/l)
Arsenic	0.05
Barium	1.0
Cadmium	0.01
Chromium	0.05
Lead	0.05
Mercury	.002
Nitrate (as N)	10.0
Selenium	0.01
Silver	0.05
Fluoride	
At ≤53.7°F (≤12.0°C)[a]	2.4
At 53.8–58.3°F (12.1–14.6°C)	2.2
At 58.4–63.8°F (14.7–17.6°C)	2.0
At 63.9–70.6°F (17.7–21.4°C)	1.8
At 70.7–79.2°F (21.5–26.2°C)	1.6
At 79.3–90.5°F (26.3–32.5°C)	1.4
Inorganic	(mg/l)
Endrin	0.0002
Lindane	0.004
Toxaphene	0.005
2,4-D	0.1
2,4,5-TP (Silvex)	0.01
Methoxychlor	0.1
TTHM	0.1
Radiological	(pCi/l)
Alpha Emitters	
Radium-226	5
Radium-228	5
Gross alpha Activity (excluding radon and uranium)	15
Beta and Photon Emitters[b]	
Tritium	20,000
Strontium-90	8
Turbidity (TU)	1[c]

[a] Average annual maximum daily air temperature.
[b] Based on a water intake of 2 liter/day. If gross beta particle activity exceeds 50 pCi/l, other nuclides should be identified and quantified on the basis of a 2-liter/day intake.
[c] One turbidity unit based on a monthly average. Up to 5 TU may be allowed for the monthly average if it can be demonstrated that no interference occurs with disinfection or microbiological determination.

venting tooth decay. Because the amount of fluoride increases as the amount of water consumed increases, and since people tend to drink more water when the weather is warm, different fluoride PMCL have been set for different air temperatures.

Organic Chemicals

The seven organic substances included in the Primary Drinking Water Standards are not naturally occurring. Four of these organics (endrin, lindane, toxaphene and methoxychlor) are chlorinated hydrocarbon insecticides. These synthetic organic insecticides are contributed to water supplies by industrial discharge during manufacture or runoff following use. The remaining two organics [2,4-D and 2,4,5-TP (Silvex)] are chlorophenoxy herbicides that are generally used for the control of aquatic vegetation. Contamination of water supplies occurs by manufacturing operation and/or use.

Recently incorporated into the standards is the regulation of total trihalomethanes (TTHM) [13]. These compounds (chloroform, bromodichloromethane, dibromochloromethane and tribromomethane) are not naturally occurring; rather, they are reaction by-products resulting from chlorination of water containing naturally occurring humic and fulvic compounds (see Chapter 2). Bromide and iodide ions are also reactants in the process.

Shortly after the preparation of this chapter, the American Water Works Association and the EPA agreed to settle a lawsuit challenging the TTHM standard of 100 μg/l [14]. Under the agreement, the EPA will establish a procedure for obtaining and maintaining a variance from the TTHM standard and will specify control measures for a specified water system. The treatment techniques of granular activated carbon (GAC) and biologically activated carbon (BAC) for TTHM control will not be required. Rather, the proposed new rule lists two groups of technologies for water utilities.

The first group is composed of widely recognized, relatively low cost technologies within the technical capability of most water systems. The second group is considered to be generally unavailable technologies that a system could be required to study and perhaps install under a compliance schedule.

Group one technologies include: use of chloramines as an alternative disinfectant; use of chlorine dioxide as an alternative disinfectant; improvement of existing clarification; alteration of the point of chlorination; and intermittent or seasonal use of powdered activated carbon at dosages not to exceed 10 mg/l on an average basis.

Group two technologies include: introduction of offline water storage facilities; aeration where geographically and environmentally appropriate; introduction of clarification where not currently practiced; and use of ozone.

The EPA or a state authority could require any group one item as a condition for granting and maintaining a variance to the TTHM MCL of

100 μg/l. A water system could challenge such a ruling by demonstrating that the treatment method was not technically appropriate and/or feasible, or that it would result in only a marginal reduction of TTHM for that system.

Group two technologies could be installed only if the EPA or a state demonstrated that, for the affected system, the treatment was technically feasible, economically reasonable and would achieve TTHM reduction at least equal in value to the cost of obtaining the treatment.

Turbidity

In the past, a high turbidity level was considered undesirable because it caused cloudy water. But more importantly, turbidity in water can be a health hazard. In general, it interferes with (1) disinfection, by creating a possible shield for disease-causing organisms, (2) maintaining an effective chlorine residual, and (c) bacteriological testing of the water.

The PMCL for turbidity (these apply only to surface water systems) dictate that: (1) the monthly average turbidity may not exceed 1 turbidity unit (TU) (states have the option to raise this PMCL to 5 TU); and (2) the two-day average turbidity may not exceed 5 TU.

Under the monthly average turbidity, the 5-TU PMCL was included as a state option in the regulations because there are certain kinds of turbidity that do not particularly interfere with the bacteriological quality of the water. This higher monthly turbidity level may be allowed in situations when iron or other mineral particles in an otherwise satisfactory water are the cause of the turbidity. In such cases, the state may authorize the 5-TU PMCL as applicable to the particular water system. The two-day turbidity limit of 5 TU is designed to protect against the presence of a particularly high turbidity level over a short period of time that would threaten the bacteriological quality of the water.

Bacteria

Coliform bacteria are used as indicator organisms to ascertain the bacterial quality of potable waters. The presence of coliform bacteria indicates that disease-causing bacteria are probably present. Appropriate PMCL have been established for each of the acceptable methods used in testing for coliform bacteria (membrane filtration method and most probable number method). These levels are given in Table II. In each case PMCL are based on the number of samples taken each month, which, in turn, depends on the size of the population served by the water system.

Table II. Maximum Contaminant Levels for Coliform Organisms

Detection Technique Used	Number of Samples Examined per Month[a]	Maximum Number of Coliform Bacteria
Membrane Filter		1/100 ml as arithmetic mean of all samples examined each month
	Fewer than 20	4/100 ml in no more than one sample
	20 or more	4/100 ml in no more than 5% of all samples examined each month
Fermentation Tube 10-ml Standard Portions		Coliforms shall not be present in more than 10% of the portions in any month
	Fewer than 20	Coliforms shall not be present in three or more portions in more than one sample
	20 or more	Coliforms shall not be present in three or more portions in more than 5% of the samples
100-ml Standard Portions		Coliforms shall not be present in more than 60% of the portions in any month
	Fewer than 5	Coliforms shall not be present in five portions in more than one sample
	5 or more	Coliforms shall not be present in five portions in more than 20% of the samples

[a] Based on population.

Radiochemical

Appropriate definitions for radiological contaminants include the following:

1. "Dose-equivalent" means the product of the absorbed dose from ionizing radiation and such factors as account for differences in biological effectiveness due to the type of radiation and its distribution in the body as specified by the International Commission on Radiological Units and Measurements (ICRU).
2. A "rem" means the unit of dose equivalent from ionizing radiation to

the total body or any internal organ or organ system. A "millirem" (mrem) is 1/1000 of a rem.

3. A "picocurie" (pCi) means that quantity of radioactive material producing 2.22 nuclear transformations per minute.
4. "Gross alpha particle activity" means the total radioactivity due to alpha particle emission as inferred from measurements on a dry sample.
5. "Manmade beta particle and photon emitters" means all radionuclides emitting beta particles and/or photons listed in "Maximum Permissible Body Burdens and Maximum Permissible Concentration of Occupational Exposure," NBS Handbook 69, except for the daughter products of thorium-232, uranium-235 and uranium-238.
6. "Gross beta particle activity" means the total radioactivity due to beta particle emission as inferred from measurements on a dry sample.

The radiological PMCL are divided into two categories of source: natural or manmade. Natural radioactivity can result from water passing over such radioactive elements as geologic uranium. The PMCL for natural radioactivity specify gross alpha activity and the radium isotopes 226 and 228. It is only necessary to test for radium if gross alpha levels exceed 5 pCi/l; if not, testing for natural radioactivity is not required for another four years. If gross alpha activity does exceed 5 pCi/l, then a test for radium-226 is mandatory. If radium 226 exceeds 3 pCi/l, then tests for radium-228 must be performed. If the total of 226 plus 228 exceeds 5 pCi/l, then use of the water is denied.

If it is necessary to perform all of the above tests, then the following PMCL apply to the drinking water: (1) If gross alpha activity is 15 pCi/l, then test for radon and uranium. (2) If alpha minus radon and uranium is 15 pCi/l, then water use is denied; and (3) if less, proceed with the individual tests for radium described above.

PMCL for manmade radiological contaminants apply to systems serving more than 100,000 people. Primarily through nuclear fallout, radioactive contaminants are present in all our surface waters. PMCL have been designed for gross beta (50 pCi/l), tritium (20,000 pCi/l), and strontium-90 (8 pCi/l).

Secondary MCL

The National Secondary Drinking Water Regulations (NSDWR) apply to public water systems and specify the secondary maximum contaminant levels (SMCL) necessary to protect public welfare. The contaminants covered by the secondary regulations are those which may create a nuisance to the consumer at elevated concentrations by adversely affecting the esthetic quality of drinking water, such as its taste, odor, color and appearance.

Secondary maximum contaminant levels have been set for chloride, color, copper, corrosivity, foaming agents, iron, manganese, odor, pH, sulfates, total dissolved solids and zinc (Table III). If these contaminants are present in drinking water at concentrations considerably higher than the SMCL, health implications may co-exist with esthetic degradation.

The NSDWR are only guidelines unenforceable by the EPA. They represent reasonable goals for drinking water quality, and states are encouraged to establish regulations based on these guidelines. The actual standards may be higher or lower depending on such local conditions as unavailability of alternative raw water sources or other compelling factors, provided that public health and welfare are adequately protected. Factors such as odor, color and other esthetic qualities are important to the public and should be considered when establishing regulations. Consequently, the public will not drink water that may have a potentially lower quality and greater risk to health.

The EPA's responsibility is limited under the NSDWR to notifying a state when it finds that a public water system is not meeting the secondary regulations and that a state is not taking reasonable action to ensure that the secondary regulations are being satisfied. Thus, the intent of these regulations is to maintain a federal/state alertness to the importance of the esthetic qualities of drinking water, rather than to empower the EPA to require states to adopt secondary regulations.

Table III. Secondary Maxium Contaminant Levels for Drinking Water

Contaminant	Level
Chloride	250 mg/1
Color	15 Color Units
Copper	1 mg/1
Corrosivity	Noncorrosive
Foaming Agents	0.5 mg/1
Iron	0.3 mg/1
Manganese	0.05 mg/l
Odor	3 TON
pH	6.5–8.5
Sulfate	250 mg/1
Total Dissolved Solids (TDS)	500 mg/1
Zinc	5 mg/1

Therefore, adoption of secondary regulations no less stringent than the federal regulations is not a requirement with which a state must comply in order to be granted primary enforcement responsibility under the Safe Drinking Water Act.

RATIONALE FOR SELECTION OF PMCL [12,16]

Inorganic Chemicals

Arsenic

Arsenic is a shiny, gray, brittle element possessing both metallic and nonmetallic properties. Compounds of arsenic are ubiquitous in nature, slightly soluble in water and occur mostly as arsenides and arsenopyrites. Samplings from 130 U.S. water stations have shown arsenic concentrations of 5–336 μg/l with a mean level of 64 μg/l. See Faust and Aly [17] for the aqueous chemistry of arsenic and arsenical compounds.

Arsenic exists in the trivalent and pentavalent states and its compounds may be either organic or inorganic. Trivalent inorganic arsenicals are more toxic than the pentavalent forms both to mammals and aquatic species. Though most forms of arsenic are toxic to humans, arsenicals have been used in the medical treatment of spirochetal infections, blood dyscrasias and dermatitis. Arsenic and arsenicals have many diversified industrial uses such as hardening of copper and lead alloys, pigmentation in paints and fireworks, and the manufacture of glass, cloth and electrical semiconductors. Arsenicals are used in the formulation of herbicides for forest management and agriculture.

Toxic Effects. Arsenic content of drinking water ranges from a trace in most U.S. supplies to approximately 0.1 mg/l. No adverse health effects have been reported from the ingestion of water containing 0.1 mg/l of arsenic.

The toxicity of arsenic is well known and the ingestion of as little as 100 mg can result in severe poisoning. In general, inorganic arsenicals are more toxic to man and experimental animals than the organic analogs; and arsenic in the pentavalent state is less toxic than the trivalent form.

Inorganic arsenic is absorbed readily from the gastrointestinal tract, the lungs and, to a lesser extent, the skin, and becomes distributed throughout the body tissues and fluids. Inorganic arsenicals appear to be oxidized slowly in vivo from the trivalent to the pentavalent state; however, there is no evidence that the reduction of pentavalent arsenic occurs within the body. Inorganic arsenicals are potent inhibitors of the intracellular sulfhydryl (SH) enzymes involved in cellular oxidations. Arsenic is excreted via urine, feces and the epithelium of the skin [16]. During chronic exposure, arsenic accumulates mainly in bone, muscle and skin, and, to a smaller degree, in liver and kidneys. After cessation of continuous exposure, arsenic excretion may last up to 70 days.

A number of chronic oral toxicity studies with inorganic arsenite and arsenate demonstrate the minimum-effect and no-effect levels in dogs, rats and mice. Three generations of breeding mice were exposed to 5 ppm of arsenite in the diet with no observable effects on reproduction. At high doses (i.e., 200 mg/l or greater), arsenic is a physiological antagonist of selenium and has been reported to counteract the toxicity of seleniferous foods when added to agriculture animals' feed water [16].

In man, subacute and chronic arsenic poisoning may be insidious and pernicious. In mild chronic poisoning, the only symptoms present are fatigue and loss of energy. The following symptoms may be observed in more severe intoxication: gastrointestinal catarrh, kidney degeneration, tendency to edema, polyneuritis, liver cirrhosis, bone marrow injury and exfoliate dermatitis. In 1962, 32 school-age children developed a dermatosis associated with cutaneous exposure to arsenic trioxide. It has been claimed that individuals become tolerant to arsenic. However, this apparent effect is probably due to the ingestion of the relatively insoluble, coarse powder, since no true tolerance has been demonstrated.

Since the early nineteenth century, arsenic was believed to be a carcinogen; however, evidence from animal experiments and human experience has accumulated to suggest strongly that arsenicals do not produce cancer [18]. One exception is a report from Taiwan showing a dose-response curve relating skin cancer incidence to the arsenic content of drinking water [19]. Other reports incriminated arsenic as a carcinogen, but it was later learned that agents other than the metalloids were responsible for such cancers. Later studies did show an increase in cancer of the skin when people were exposed to arsenic in drinking water and/or from an occupational exposure.

The acute fatal dose of arsenic trioxide for a human is between 70 and 180 mg [20]. This would be equivalent to a total arsenic dose of 53.2–136.8 mg, or 0.76–1.95 mg/kg body weight (for a 70-kg person). This dose is less than that reported based on animal tests.

Recommendation. In light of present knowledge concerning the potential health hazard from ingestion of arsenic, the arsenic content of drinking water shall not exceed 0.05 mg/l.

Barium

Barium is a yellowish-white metal of the alkaline earth group. It occurs in nature chiefly as barite ($BaSO_4$) and witherite ($BaCO_3$), both of which are highly insoluble salts. The metal is stable in dry air, but readily oxidized by humid air or water. Many salts of barium are soluble in both water and acid.

While barium is a malleable, ductile metal, its major commerical value lies in its compounds. Barium compounds are used in a variety of industrial applications, in the metallurgic, paint, glass and electronics industries, as well as for medicinal purposes.

The most commonly occurring natural form of barium is barite (barium sulfate), which has a low solubility, especially in waters containing sulfate. Soluble forms of barium are very toxic, whereas insoluble forms are considered nontoxic. Barite is used principally as a drilling mud in oil and gas well drilling, but other barium compounds are used in the production of glass, paint, rubber, ceramics and in the chemical industry. Faust and Aly [17] describe the aqueous chemistry of barium compounds.

Toxic Effects. Concentrations of barium in domestic drinking water supplies generally range from less than 0.6 μg/l to approximately 10 μg/l, with upper limits in a few Midwestern and Western states ranging from 100 to 3000 μg/l [17]. Barium enters the body primarily through air and water; no appreciable amounts are contained in foods.

The metabolism of barium is very similar to that of calcium as shown by radioactive tracing techniques. Barium is rapidly transferred from the digestive tract to the blood plasma from which it is cleared within 24 hr. Barium is excreted primarily through the feces rather than through urine. There is no known nutritional need for barium in humans.

The toxic form of barium is contained in its soluble salts which are often used in rodenticides. The fatal dose of barium chloride for man has been proven to be 0.8 g as recorded from accidental poisonings. Acute barium poisoning exerts a strong stimulation on all muscles, including the smooth muscle of the heart, gastrointestinal tract and urinary bladder [21]. In small doses it can produce an increase in blood pressure by causing blood vessels to constrict.

Recommendations. A drinking water guideline was derived from the 8-hr-weighted maximum allowable concentration in industrial air of 0.5 mg/m^3 set by the American Conference of Governmental Industrial

Hygienists [16]. It was assumed that, with an 8-hr inhalation of 10 m^3 of air, the daily intake would be 5 mg of barium, of which 75% was absorbed in the blood stream and 90% transferred across the gastrointestinal tract. Based on these assumptions, it was reasoned that a concentration of about 2 mg/l of water would be safe for adults. To provide additional safety for more susceptible members of the population, such as children, a level of 1 mg/l was recommended [16]. There have been no long-range feeding studies to confirm the safety of this barium intake. The limit set in the USSR is 4 mg/l of water. International and European standards do not list barium upper limits, because available information is insufficient.

It is rare to find sources of water that exceed a barium concentration of 1 mg/l, although a concentration of 1.55 mg/l has been recorded in drinking water [16]. The 1975 Analysis of Interstate Carrier Water Supply Systems showed none exceeding the 1 mg/l standard. Small numbers of people are known to be consuming well waters in Illinois, Kentucky, Pennsylvania and New Mexico that are at, or exceed by 10 times, the standard for barium.

Cadmium

Cadmium is a soft, white, easily fusible metal similar to zinc and lead in many properties and readily soluble in mineral acids. Cadmium is a biologically nonessential and nonbeneficial element recognized to be of high toxic potential. It is deposited and accumulated in various body tissues and is found in varying concentrations throughout all areas where man lives. Within the past two decades industrial production and use of the metal have increased. Concomitantly, there have been incidents of acute cases of clinically identifiable cadmiosis. Cadmium may function in or may be an etiological factor for various human pathological processes including testicular tumors, renal dysfunction, hypertension, arteriosclerosis, growth inhibition, chronic diseases of old age and cancer [16].

Cadmium occurs in nature chiefly as a sulfide salt, frequently in association with zinc and lead ores. Accumulations of cadmium in soils in the vicinity of mines and smelters may result in high local concentrations in nearby waters. The salts of the metal also may occur in wastes from electroplating plants, pigment works, textile and chemical industries. Seepage of cadmium from electroplating plants has resulted in groundwater concentrations of 0.01 to 3.2 mg/l. However, dissolved cadmium was found in less than 3% of 1577 water samples examined for the

United States, with a mean of slightly under 10 μg/l. Most freshwaters contain less than 1 μg/l [17].

Toxic Effects. The total daily intake of cadmium from air, water, food and cigarettes is estimated to range between 40μg/day (for nonsmoking rural residents who have negligible air exposure and consume a low-cadmium diet) and 190 μg/day (for smokers living in industrialized cities and consuming a high-cadmium diet) [16].

Absorption from the digestive tract is thought to average about 10%. However, a number of factors, including dietary calcium, protein and age, may have an important bearing on this. For the digestive tract route of assimilation, the major organs of cadmium storage are the liver and renal cortex. The renal cortex may contain one-third of the total cadmium body burden. The biologic half-life of cadmium in these organs is variously estimated at 13–38%/yr. Urinary excretion is low, 1–9 μg/day [16]. Because cadmium tends to accumulate, a more useful way of looking at the question is to consider the rate of accumulation. The human placenta is apparently highly impermeable to cadmium. The total body burden is estimated at 1 mg at birth and at 15–50 mg at the age of 50 years [22]. This is consistent with an average accumulation of 0.9–1.8 μg/day. There is a major need for a more reliable estimate of the rate of cadmium accumulation.

The renal cortex is considered to be the critical organ for accumulation of cadmium from low-level dietary exposures: the critical concentration for renal cortex is approximately 200 μg/g of of tissue (wet weight). At greater concentrations, irreversible renal injury may occur. In the outbreak of itai-itai disease on the Jintsu River, renal cortical cadmium concentration was estimated at 600–1000 μg/g of tissue (wet weight) in those most severely (and irreversibly) affected [23]. With an assumed water consumption of 1.5 l/day, the average cadmium intake from water was estimated at 5 μg/day, or less than 10% of the total intake.

For the general population, the major route of absorption is through the gastointestinal tract. The major effects are likely to be on the kidney. Experimental data indicate that the zinc:cadmium ratio in the organs is an important determinant of cadmium toxicity (in most foodstuffs, the dietary ratio of cadmium to zinc is 1.0—it is highest in meat products and lowest in dairy products), and there is some evidence that the intake of sodium may also influence cadmium toxicity. There are no dose-response data. Limited autopsy data suggest that average renal cortical concentrations of cadmium in American and European populations are generally less than 50 μg/g of tissue (wet weight)—less than the projected critical concentration by a factor of four or more [16].

In addition to the suspected interactions between cadmium, zinc and calcium, recent experimental studies indicate that cadmium at very high doses can interfere with the activation of vitamin D. There is also evidence from animal studies that cadmium is implicated in the etiology of hypertension; the thresholds and dose-response relationships are unknown. There is some evidence that cadmium is carcinogenic in the rat, but no substantial evidence to implicate it in human cancer. Cadmium is known to be teratogenic in the rat following rather high (2–13 mg/kg) doses on specific days of gestation.

Recommendation. Chronic kidney disease (renal tubular dysfunction) will begin to occur in an individual when the cadmium accumulation reaches a critical concentration. While the critical concentration varies from one individual to another, the threshold of observed effect is about 200 ppm of cadmium in the renal cortex. Individuals have been found with several times this level without evidence of kidney disease. At a 0.01-mg/l criterion, as recommended for drinking water, the maximum daily intake of cadmium would not exceed 20 μg from water, assuming a 2-liter daily consumption. The daily intake from other sources is up to 60 μg in the U.S. population [16].

Chromium

Although chromium has oxidation states ranging from Cr(II) to Cr(VI), the trivalent form is found more often in natural waters. Chromium is found rarely in natural waters. It ranks 27th or lower among the elements in seawater, and is generally found at concentrations well below 1 μg/l. For 1577 surface water samples collected at 130 U.S. sampling sites, 186 samples contained 1–112 μg/l; the mean was 9.7 μg/l chromium [24]. In a similar survey of 700 samples, none contained over 50 μg/l of hexavalent chromium, and 11 contained over 5 μg/l. Chromium is found in air, soil, some foods and most biological systems; it is recognized as an essential trace element for humans.

Toxic Effects. The earliest consequences of mild chromium deficiency in experimental animals is a reduced sensitivity of peripheral tissues to insulin; more severe deficiency in rats and mice results in fasting hyperglycemia, glycosuria and mild growth retardation that is probably due, at least in part, to reduced insulin activation. Glucose intolerance is a common human problem and one of its many possible causes is chromium deficiency [25].

Based on rat studies indicating that absorption of various Cr^{3+}

compounds can range from below 1 to 25% of a given dose, and assuming that the human case is similar, a dietary intake of 20–500 μg/day would balance urinary losses. It is estimated that daily chromium intakes in the United States are marginal and vary from 5 to 100 μg. It seems unlikely that any Cr^{3+} ingested via public drinking water would be appreciably assimilated [26].

No harmful effects were observed when food or water containing moderate amounts of Cr^{3+} was administered to laboratory animals, e.g., cats that were fed chromic phosphate or oxydicarbonate from 50 to 1000 mg/day for 80 days, or rats drinking water containing 25 mg/l for a year, or 5 mg/l throughout their lives [25].

Hexavalent chromium, on the other hand, is irritating and corrosive to the mucous membrances; it is absorbed via ingestion, through the skin and by inhalation, and is toxic when introduced systematically into laboratory animals [25]. Knowledge of the harmful human health effects of hexavalent chromium has been obtained almost entirely from occupational health effects. Lung cancer, ulceration and perforation of the nasal septum, and a variety of other respiratory complications and skin effects have been observed.

The PHS Drinking Water Standards [9] state that the presence of hexavalent chromium in excess of 0.05 mg/l shall constitute grounds for rejection of the supply, and no harmful human health effects have been reported at this level. The NAS Committee on Water Quality Criteria recommended that public water supply sources for drinking water contain no more than 0.05 mg/l total chromium, largely on the basis that lifetime tolerable levels of chromate ion are not known for man. There are insufficient data on the effect of the defined treatment process on chromium removal (see Chapter 8). Chromium as Cr^{3+} is not likely to be present in waters of pH 5 and above, because the hydrated oxide is very sparingly soluble. A family of four individuals drank water for a period of three years at a chromium level of 0.45 mg/l without known effects on their health [27].

Recommendation. The present drinking water standard of 0.05 mg/l is less than the no-observed-adverse-health-effect level. Consideration should be given to setting the chromium limit in terms of the hexavalent form. Extensive work is urgently needed to establish the role of dietary chromium with regard to atherosclerosis and glucose metabolism as well. It appears that a concentration of 0.05 mg/l of chromium in domestic water supply incorporates a reasonable safety factor to avoid any hazard to human health.

In addition, the possibility of dermal effects from bathing in water containing 0.05 mg/l chromium would likewise appear remote, although chromium is recognized as a potent skin sensitizer. Domestic water supplies should, therefore, contain no more than 0.05 mg/l total chromium.

Fluoride

Fluorine is the most electronegative of all elements, existing naturally in the form of fluoride. It is the 17th most abundant element in the earth's crust, occurring principally as fluorite (CaF_2) and fluoroapatite ($Ca_5(PO_4)_3F$). It is present in small amounts in most soils except those that have been strongly leached. The concentration of fluoride in natural waters depends principally on the solubility of fluoride-containing rocks in contact with water.

In 1969 the general Community Water Supply Survey of the PHS sampled 969 water supplies and found fluoride ranging from less than 0.2 up to 4.4 mg/l [28]. Fluoride concentrations greater than the recommended limits for this constituent were found in 52 systems. Most unfluoridated supplies have fluoride concentrations less than 0.3 mg/l.

A more extensive survey in the same year by the Dental Health Division of the PHS [28] showed that 8.1 million people in 2630 communities in 44 states were consuming water with more than 0.7 mg/l of naturally occurring fluoride. Most of the communities with more than 0.7 mg/l natural fluoride were in Arizona, Colorado, Illinois, Iowa, New Mexico, Ohio, Oklahoma, South Dakota and Texas. There were no reports of community water supplies with as much as 0.7 mg/l fluoride from Delaware, Hawaii, Massachusetts, Pennsylvania, Tennessee or Vermont [16].

For more than 30 years, fluoride has been added to drinking water in the United States to promote the reduction of dental caries. The principal chemicals used for this purpose are sodium fluoride, sodium silicofluoride, hydrofluorosilicic acid and ammonium silicofluoride. While these chemicals require care in handling and must be dispensed with properly designed chemical feeding systems, these systems are readily available.

When any one of these chemicals is dispersed in water at the 1-mg/l range, it dissociates almost completely. EPA endorses controlled fluoridation, although the Safe Drinking Water Act states that no primary drinking water regulations may require the addition of any substance for preventive health care purposes unrelated to the contamination of drinking water. Small amounts of fluoride on the order of 1 mg/l, depending on the environmental temperature in water and beverages, are generally conceded to have a beneficial effect on prevention of dental caries, particularly among children.

Toxic Effects. Two forms of chronic toxic effects are recognized generally as being caused by excess intake of fluoride over long periods of time. These are mottling of teeth enamel, or dental fluorosis, and skeletal fluorosis. In both cases, it is necessary to consider the severity, since the very mild forms are considered beneficial by some. The most sensitive of these effects is the mottling of tooth enamel, which, depending on the temperature, may occur to an objectionable degree with fluoride concentrations in drinking water of only 0.8–1.6 mg/l [29]. There has been little apparent systematic investigation of the degree to which consumers of drinking water with several milligrams per liter of fluoride regard the resultant mottling as an adverse health effect.

Skeletal fluorosis has been observed with use of water containing more than 3 mg/l [16]. It now appears that long-term consumption of water containing fluoride in excess of 1 mg/l runs into a fair probability of objectionable dental mottling and increased bone density in patients with long-standing renal disease or polydipsia. Increased bone density, however, has often been regarded as a beneficial rather than an adverse effect. Intake of fluoride for long periods in amounts greater than 20–40 mg/day may result in crippling skeletal fluorosis [16].

Other reported adverse health effects following the intake of specified milligram-per-liter levels of fluoride in drinking water, including individual physiological mongolism, cancer mortality, mutagenic or birth defects, and sensitivity, have either not been confirmed or have been found lacking in substance. Nor is there evidence of any difference between the effects of naturally occurring or intentionally added fluoride.

Epidemiological studies of water naturally high in fluoride have found no adverse effects, except in rare cases, until the concentration exceeds by many times that recommended for added fluoride. Controlled studies of fluoridation at the 1-mg/l level have reported no instances of adverse effects.

Recommendation. There is no generally accepted evidence that anyone has been harmed by drinking water with fluoride concentrations considered optimal for the annual mean temperatures in the temperate zones. It seems likely, however, that objectionable dental fluorosis occurred in two children with diabetes insipidus. Bone changes, possibly desirable, have been noted in patients being dialyzed against large volumes of fluoridated water. Similar changes can be expected in the rare renal patient with a long history of renal insufficiency and a high fluid intake that includes large amounts of tea. In this particular combination of circumstances, the lowest drinking-water concentration of fluoride associated with symptomatic skeletal fluorosis that has been reported to date is 3 mg/l. On the basis of previous studies, occasional objectionable mottling would be expected to occur in communities in the hotter regions of the United States with water that contains fluoride at 1 mg/l or higher and in any community with water that contains fluoride at 2 mg/l or higher [30].

Lead

In addition to their natural occurrence, lead and its compounds may enter and contaminate the global environment at any stage during mining, smelting, processing and use. The annual increase in lead consumption in the United States during the 10-year period 1962–1971 averaged 2.9%, largely due to increased demands for electrochemical batteries and gasoline additives [31]. Nonindustrial sources that may contribute to the human ingestion of lead include the in-house use of lead-bearing paints and plaster, improperly glazed earthenware, lead fumes or ashes produced in burning lead battery casings and exhaust from internal combustion engines.

Most lead salts are of low solubility. Lead exists in nature mainly as lead sulfide (galena); other common natural forms are lead carbonate (cerussite), lead sulfate (anglesite) and lead chlorophosphate (pyromorphite) [17]. Stable complexes result also from the interaction of lead with the sulfhydryl, carboxyl and amine coordination sites characteristically found in living matter. The toxicity of lead in water, like that of other heavy metals, is affected by pH, hardness, organic materials and the presence of other metals. The aqueous solubility of lead ranges from 500 μg/l in soft water to 3 μg/l in hard water. Lead enters the aquatic environment through precipitation, lead dust fallout, erosion and leaching of soil, municipal and industrial waste discharges, and the runoff of fallout deposits from streets and other surfaces. Extrapolations from recent studies indicate that nationally as much as 5000 ton/yr of lead may be added to the aquatic environment as a result of urban runoff [32]. It may be inferred from available data that the mean natural lead content of the world's lakes and rivers ranges from 1 to 10 μg/l; the lead content of rural U.S. soils is 10–15 μg/g, although many urban soil concentrations are much higher.

Available data generally indicate that the addition of lead to drinking water occurs chiefly in the distribution system, including household plumbing, and that this is most likely to occur in areas with soft "agressive" water (see Chapter 9).

Toxic Effects. As far as is known, lead has no beneficial or desirable nutritional effects. Lead is a toxic metal that tends to accumulate in the tissues of man and other animals. Although seldom seen in the adult population, irreversible damage to the brain is a frequent result of lead intoxication in children. Such lead intoxication most commonly results from the ingestion of lead-containing paint still found in older homes. The major toxic effects of lead include anemia, neurological disfunction and renal impairment. The most common symptoms of lead poisoning are anemia, severe intestinal cramps, paralysis of nerves (particularly of the arms and legs), loss of appetite and fatigue; the symptoms usually develop slowly. High levels of exposure produce severe neurologic damage, often manifested by encephalopathy and convulsions; such cases frequently are fatal. Lead is strongly suspected of producing subtle effects (i.e., effects due to low-level or long-term exposure insufficient to produce overt symptoms) such as impaired neurologic and motor development and renal damage in children. Subclinical lead effects are distinct from those associated with residual damage following lead intoxication [16].

Gastrointestinal absorption and retention of lead are greater in children than in adults (53% and 18% respectively) [33]. The average daily intake of lead from diet and air among young children probably amounts to up to two-thirds of the daily permissible intake, leaving a very narrow margin of safety [34]. Compared to adults, food and air intake by children is proportionally greater than their weight, e.g., a one-year-old child, with only about one-seventh of the body weight of an adult, had one-fourth to one-third of the daily adult air intake and 40–60% of the dietary intake of an adult, so that its lead intake is proportionally greater on a body-weight basis.

Recommendation. The lead content in U.S. public water supplies in 1962 ranged from traces to 62 μg/l [17]. Continuous monitoring of U.S. water supplies since 1962 has demonstrated that their lead content, in general, has not exceeded the PHS standard of 0.05 mg/l [9]. The range of lead concentrations in finished U.S. community water ranged from nondetectable to 0.64 μg/l [35]. Of 969 water supplies surveyed, 1.4% exceeded 0.05 mg/l of lead. In drinking water, lead should be kept to a minimum; a criterion of 0.05 mg/l is attainable and protective. Experience indicates that fewer than 4% of the water samples analyzed exceed the 0.05-mg/l limit and that the majority of these are due to corrosion problems rather than to naturally occurring lead content in raw waters.

Mercury

Mercury has three major oxidation states: (0), elemental mercury; (I), mercurous compounds; and (II), mercuric compounds. Widely distributed in the environment, mercury is a biologically nonessential or nonbeneficial element. It was historically recognized to possess a high toxic potential and was used as a germicidal or fungicidal agent for medical and agricultural purposes. Metallic mercury is highly insoluble in water.

Human poisoning by mercury or its compounds has been recognized clinically. Although its toxic properties are well known, dramatic instances of toxicosis in man and animals have occurred, e.g., the Minamata Bay poisonings [36]. In addition to the incidents in Japan, poisonings have also occurred in Iraq, Pakistan and Guatemala as a result of ingestion of flour and seed treated with methyl- and ethylmercury compounds [37]. Mercury intoxication may be acute or chronic, and the specific toxic effects will vary with the form of mercury and its mode of entry into the organism. The mercurous salts are less soluble and consequently less toxic than the mercuric. The fatal oral dose of mercuric salts for man ranges from 20 mg to 3.0 g [38]. Symptoms of acute inorganic mercury poisoning include pharyngitis, gastroenteritis, vomiting followed by ulcerative hemorrhagic colitis, nephritis, hepatitis and circulatory collapse. Chronic mercury poisoning results from exposure to small amounts of mercury over extended time periods. Chronic poisoning from inorganic mercurials most often has been associated with industrial exposure, whereas poisoning from the organic derivatives has been the result of accidents or environmental contamination. Alkyl compounds are the derivatives of mercury most toxic to man, producing illness, irreversible neurological damage or death from the ingestion of amounts in milligrams [39].

The mercury content of unpolluted U.S. rivers in 31 states where natural mercury deposits are unknown is less than 0.1 μg/l [40]. Most U.S. water contains less than 0.1 μg/l of mercury. The lower limit of detection in these studies totaled mercury values of 0.045 μg/l; this was recently determined in Connecticut River water using more sensitive detection methods [41]. Marine waters have been shown to contain concentrations of mercury from a low of 0.03 to a high of 0.2 mg/l, but most marine waters fall within the range of 0.05–0.19 mg/l mercury. Mining, agriculture and waste discharge contribute to the natural levels found.

Toxic Effects. Several forms of mercury, ranging from elemental to dissolved inorganic and organic species, are expected to occur in the environment. The discovery that certain microorganisms have the ability to convert inorganic and organic forms of mercury to the highly toxic methyl- or dimethylmercury has made any form of mercury potentially hazardous to the environment [42]. Studies of the biochemical kinetics of mercury methylation in water under naturally occurring conditions of pH and temperature have revealed that inorganic mercury can be converted readily to methylmercury [43].

It has been argued that whenever mercury in any form is added to the aquatic environment, a combination of microbially catalyzed reactions and chemical equilibrium systems is capable of leading to steady-state concentrations of dimethylmercury, methylmercuric ion, metallic mercury, mercuric ion and mercurous ion [44]. Thus, it is evident that the total mercury level should be the basis for a mercury criterion rather than any particular form in which it may be found within a sample.

Quantities of ingested mercury safe for man can be estimated from epidemiological evidence; the lowest whole-blood concentration of methylmercury associated with toxic symptoms is 0.2 μg/g. This blood concentration can be compared to 60 μg/g mercury in hair. These values, in turn, correspond to prolonged continuous exposure at approximately 0.3 mg/70-kg body weight/day. By using a safety factor of 10, the maximum dietary intake from all sources—air, water and food—should not exceed 30 μg/person/day mercury. Although the safety factor is computed for adults, limiting ingestion by children to 30 μg/day mercury is believed to afford some lesser degree of safety. If the exposure to mercury were from fish alone, the limit would allow for a maximum daily consumption of 60 g (420 g/week) of fish containing 0.5 mg/kg mercury.

Recommendation. The drinking water standard for total mercury is 0.002 mg/l. At this concentration, drinking 2 liters of water will contribute a total of 4 μg to the daily intake. Since only a small fraction of the total mercury in water is in an alkyl form, the contribution of methylmercury to the daily intake will be approximately 0.0004 μg or 0.01% of the 4 μg/day. At this level, the potential hazard to humans from mercury in U.S. water supplies is inconsequential compared with the contribution from food [16].

Nitrate

In oxygenated surface waters, all inorganic and organic nitrogen should be in the form of nitrate. Major sources of these two major forms of nitrogen are municipal and industrial wastewaters, feedlot discharges, fertilizers from runoff, leachates from landfills, atmospheric fallout, soil organic matter, etc. A significant source of nitrate in groundwaters is septic tank discharge. In severely polluted groundwaters, nitrate concentrations can become extremely high.

Toxic Effects. In quantities normally found in food or feed, nitrate becomes toxic only under conditions in which they are, or may be reduced to nitrites. Otherwise, at reasonable concentrations, nitrates are rapidly excreted in the urine. High intake of nitrates constitutes a hazard primarily to warmblooded animals under conditions that favor nitrate reduction to nitrite. Under certain circumstances, nitrate can be reduced to nitrite in the gastrointestinal tract, which then reaches the bloodstream and reacts directly with hemoglobin to produce methemoglobin, with consequent impairment of oxygen transport.

The reaction of nitrite with hemoglobin can be hazardous in infants under three months of age. Serious and occasionally fatal poisonings in infants have occurred following the ingestion of untreated well waters shown to contain nitrate at concentrations greater than 10 mg/l NO_3-N [8]. High nitrate concentrations frequently are found in shallow farm and rural community wells, often as the result of inadequate protection from barnyard drainage or from septic tanks [45]. Although many U.S. public water supplies contain levels that routinely exceed this amount, only one case of infant methemoglobinemia associated with a U.S. public water supply has ever been reported [46].

Because of the potential risk of methemoglobinemia to bottle-fed infants, and in view of the absence of substantial physiological effects at nitrate concentrations below 10 mg/l nitrate-nitrogen, this level is the criterion for domestic water supplies. Waters with nitrite-nitrogen concentrations over 1 mg/l should not be used for infant feeding [8]. Waters with a significant nitrite concentration usually would be heavily polluted and probably bacteriologically unacceptable.

Recommendation. Nitrate in water at concentrations less than 1000 mg/l is not of serious concern as a direct toxicant. It is a health hazard because of its conversion to nitrite. Nitrite is directly toxic on account of its reaction with hemoglobin to form methemoglobin and thereby cause methemoglobinemia. It also reacts readily under appropriate conditions with secondary amines and similar nitrogenous compounds to form N-nitroso compounds, many of which are potent carcinogens.

Epidemiological evidence of the occurrence of methemoglobinemia in infants tends to confirm a value near 10 mg/l nitrate as nitrogen as a MCL for water with no observed adverse health effects, but this value provides little margin of safety. However, the highly sporadic incidence of methemoglobinemia among users of drinking water that contains a concentration of nitrate much greater than the maximum suggests that factors other than nitrate intake are important in connection with development of the disease. On the basis of its high toxicity and more pronounced effect than nitrate, the NAS recommends that the nitrite-nitrogen concentration in public water supply sources should not exceed 1 mg/l.

Selenium

Selenium is a biologically essential, beneficial element recognized as a metabolic requirement in trace amounts for animals, but toxic to them when ingested in amounts ranging from about 0.1 to 10 mg/kg of food [16]. The national levels of selenium in water are proportional to its content in soil. In low-selenium areas, its content in water may be well below 1 μg/l [17]. In water from seleniferous areas, selenium levels of 50–300 μg/l have been reported [17]. Selenium appears in soil as basic ferric selenite, as calcium selenate and as elemental selenium. The latter must be oxidized to selenite or selenate before it has appreciable solubility in water.

Toxic Effects. Selenium is considered toxic to man. The symptoms of this toxicity are similar to those of arsenic poisoning [47,48]. Any consideration of the toxicity of selenium to man must take into consideration the dietary requirement for this element in amounts estimated to be 0.04–0.10 mg/kg of food. Considering this requirement in conjunction with evidence that ingestion of selenium in amounts as low as 0.07 mg/day has been shown to give rise to signs of toxicity, selenium concentrations above 0.01 mg/l should not be permitted in drinking water [49,50]. The PHS Drinking Water Standards recommend that drinking water supplies contain no more than 0.01 mg/l of selenium [9]. Chronic exposure of humans to selenium by inhalation or by ingestion results in central nervous system and gastrointestinal disturbances and dermatitis [51].

The high selenium content of diet and water in areas of seleniferous soils has been associated with alkali disease in cattle. Human populations living in these areas are not similarly affected. This is believed to be due to the wide geographical sources of food consumed and the loss of selenium during processing. Where feed is low in selenium, water containing 400–500 μg/l is too low to cause poisoning in cattle. The maximum recommended concentration for livestock water is 0.05 mg/l.

The only documentation of human toxicity from drinking water involved a family consuming well water that contained selenium at 9.0 mg/l. In 1942 the PHS Drinking Water Standard listed selenium for the first time along with fluoride and arsenic. The maximum level was 0.05 mg/l and was easily met by public water supplies.

Recommendation. The interim standard of 0.01 mg/l was recommended in 1962 in response to animal studies [9] that determined that selenium was carcinogenic. The current literature review of animal studies does not support this contention, nor is there any epidemiological evidence implicating a higher than normal daily intake of selenium.

The established requirement for selenium in most animal species indicates a need for more data on potential or real deficits or excesses in human populations. The concentration of selenium in U.S. waters varies widely and no evidence exists currently to suggest that it is causing health problems. The totality of evidence indicates that there is greater overall potential for selenium deficiency than for selenium toxicity given current intake levels of selenium. The maximum no-observed-adverse-health-effect level for selenium in water is not less than 0.1 mg/l and appears to be as great as 0.5 mg/l. A concentration of 0.02 mg/l just barely provides a minimum nutritional amount of selenium with a consumption of 2 liters/day.

Silver

Silver is a biologically nonessential, nonbeneficial element recognized as a cause of localized skin discoloration in humans and as a systemic toxin for aquatic life. Human ingestion of silver or silver salts results in deposition of silver in skin, eyes and mucous membranes that causes a blue-gray discoloration without apparent systemic reaction [52]. Because of its strong bactericidal action, silver has been considered for use as a water disinfectant (see Chapter 10). Dosages of 0.001–500 μg/l of silver have been reported sufficient to sterilize water. At these concentrations, the ingestion of silver has no obvious detrimental effect on humans. Occurrence data from 1577 samples collected across the United States showed concentrations ranging from 0.1 to 38 μg/l, with a median of 2.6 μg/l. Silver was present in 104 out of 130 locations [53].

Toxic Effects. The 1962 PHS Drinking Water Standards contained a limit for silver of 0.05 mg/l [9]. This limit was established because of evidence that silver, once absorbed, is held indefinitely in tissues, particularly the skin, without evident loss through usual channels of elimination or without reduction by transmigration to other body sites. An additional reason is the probable high absorbability of silver bound to sulfur components of food cooked in silver-containing water [54]. A study of the toxic effects of silver added to rats' drinking water showed pathologic changes in kidneys, liver and spleen at concentrations of 400, 700 and 1000 μg/l [55].

Recommendation. It is possible to derive a drinking water standard for silver by using silver deposited in excess of 1 g in the integument of the body as an end point that must not be exceeded [9]. Assuming that all silver ingested is deposited in the integument, it is calculated that 0.01 mg/l could be ingested for a lifetime before 1 g of silver would be attained from 2 liters of water intake per day; 0.05 mg/l silver could be ingested for approximately 27 years without exceeding a silver deposition of 1 g [9].

Organic Chemicals

Cyclodiene Insecticides

All of the cyclodienes—endrin, dieldrin, chlordane, heptachlor and heptachlor expoxide—are products of the diene reaction and are all derivatives of hexachlorocyclopentadiene. They date their development to Hyman's patented synthesis of chlordane in 1944 [16]. Since then, about 600 million pounds of these highly chlorinated, cyclic organic compounds have been dispersed in the ecosystem. Little is known about their fate in the environment, but traces of them and their stable epoxide oxidation products are ubiquitous. They are heavily bioconcentrated in the lipids of terrestrial and aquatic wildlife, humans and foods, especially in animal fats and milk. Degradation products of the cyclodienes are neurotoxic poisons whose metabolic pathways are largely unknown and of extreme complexity.

Standards Based on Chronic Toxicity. The federal MCL for chlorinated hydrocarbons in drinking waters have been calculated primarily on the basis of extrapolation to human beings of those concentrations causing minimal toxic effects in animals. Furthermore, safety factors are applied to these levels to reduce risk to a minimum. To reduce the risk of human consumption, a safety factor of 0.1 is applied to existing safe levels found in humans when such data exist. When animal data are corroborated by some human data, a factor of 0.01 is applied, and when animal data alone are available, a factor of 0.002 is employed. For such mathematical extrapolations, the dietary levels of animals must then be converted from concentrations of exposure into dosage concentrations based on the mass or weight of the experimental animal and then converted to levels meaningful to human consumption.

For example, 1 ppm was determined by bioassay to be the safe level for endrin exposure from food and water for dogs [56]. (In rats this level turned out to be 5 ppm, so the more sensitive organism, the dog, was chosen as representative of the safety level appropriate for human exposure.) Assuming the average dog weighs 10 kg, having a consumption of 0.2 kg of food per day, then (1 ppm $\cong$ 1 mg/kg):

$$\frac{1 \text{ mg/kg} \times 0.2 \text{ kg food/day}}{10 \text{ kg body wt}} = 0.02 \text{ mg/kg/day}$$

Multiplying this no-effect dosage in dogs by a safety factor (in this case 0.002) helps to ensure protection of exposure from all other sources:

$$(0.02 \text{ mg/kg/day})(0.002) = 0.00004 \text{ mg/kg/day}$$

Next, to calculate the allowable exposure to a human (assuming an average weight of 70 kg), multiply the above dosage by 70 kg:

$$(0.00004 \text{ mg/kg/day})(70 \text{ kg}) = 0.0028 \text{ mg/day}$$

The level thus arrived at, 0.0028 mg, is considered an acceptably safe level of endrin to which the average man may be exposed each day from all sources.

In order to calculate an acceptable level of exposure from drinking water, it is necessary to consider exposure from all media. In the case of chlorinated hydrocarbons, exposure is expected to occur mostly through diet, though some inhalation of pesticides from aerial spraying can occur. The Food and Drug Administration has estimated from samples taken from 1964 through 1970 that 20% of the total dietary intake of pesticide chemicals comes from drinking water. Since the sampling also revealed that the total dietary intake of endrin is 0.0035 mg/day, itself a good deal below the acceptable daily level for humans, 20% of that figure (the amount apportioned to pesticide exposure from water), results in only 0.0007 mg/day exposure from endrin in water under normal circumstances. The maximum acceptable concentration (MAC) allowable for endrin exposure from drinking water under any situation, then, is 20% of the maximum allowable daily dietary level for endrin (0.0028 mg) or

$$0.20 \times 0.0028 = 0.00056 \text{ mg/day}$$

Endrin

Endrin (1,2,3,4,10-hexachloro-6,7-epoxy-1,4,4a,5,6,7,8,8a-octahydro-endo-1,4-endo-5,8-dimethanonoaphthalene) is a member of the class of cyclodiene insecticides. Although not much is known about the metabolic path of endrin in living tissue, it is known that endrin can form isomers of ketoendrin on exposure to light, and that it is degraded to 9-ketoendrin, 9-hydroxyendrin and 5-hydroxyendrin in animals.

Toxic Effects. Human illness and death have been observed after poisoning during the manufacture, spraying or accidental ingestion of cyclodienes. Typical symptoms include stimulation of the central nervous system, blurred vision, insomnia, nausea, involuntary muscular movements and general weakness. Severe poisoning (0.2–2.5 mg/kg) causes convulsions in humans that can recur many months after cessation of exposure. Such seizures are marked by abnormal encephalographic patterns.

Epidemics of poisoning have occurred after eating bread contaminated with endrin. In Saudi Arabia, such an incident caused 26 deaths in 1956. Also, workers in a plant manufacturing cyclodienes, including endrin, had encephalograms suggesting brainstem injury, although normality seemed to return after six months.

The cyclodiene compounds are extremely toxic. Endrin is so toxic that it is officially registered as a rodenticide and its dermal toxicity (or toxicity via exposure through the skin) is roughly equivalent to its oral toxicity at acute doses (15 mg/kg LD_{50}).

The results of chronic feeding of the cyclodienes to laboratory animals are extremely severe, and there exists no adverse effect minimum dosages for some of these compounds (dieldrin, chlordane, heptachlor). Endrin fed to rats at 1 and 5 ppm in the diet produced no obvious effects over their life span, except for enlargement of the liver at 5 ppm. A dosage of 25 ppm caused degeneration of brain, liver, kidneys, adrenals and a shortened life span. When dogs were fed 2 and 4 ppm endrin, they exhibited convulsion and on autopsy revealed pathologic changes in the brain.

There are no data proving that endrin is mutagenic, though other cyclodienes are carcinogenic (chlordane, heptachlor, dieldrin and aldrin). Endrin did not produce tumors in rats fed up to 12 ppm for two years.

Endrin does seem to have some ill effects on reproduction. Endrin fed to mice at 5 ppm for 30 days produced significantly smaller litters than controls, but rats fed 2 ppm for three generations exhibited no effect. However, quails fed 1 ppm during their reproductive periods produced no eggs. Other studies, in hamsters and mice, also show teratogenic birth defects resulting from feeding pregnant animals endrin during their gestation periods.

Lindane

The commercial insecticide lindane is defined as a product containing at least 99% of the lowest melting point and most reactive isomer (the gamma isomer) of benzene hexachloride (BHC). BHC is a common name for 1,2,3,4,5,6-hexachlorocyclohexane, a mixture of many isomers. Production of the gamma isomer for lindane is achieved by selective crystallization. The relatively high water solubility and vapor pressure of lindane—10 mg/l and 0.14 torr—cause it to have relatively low persistence in the environment.

Toxic Effects. The highest level of lindane found to have minimal or no long-term effects in the most sensitive mammal tested, the dog, is 15 mg/kg in diet or 0.3 mg/kg of body weight per day [57]. Since no adequate human data are available to corroborate the animal studies, the total safe drinking water level is assumed to be 1/500 or 0.002, of the no-effect level reported for the most sensitive animal tested. Assuming that 20% of the total intake of lindane is from drinking water and that the average person weighs 70 kg and consumes 2 liters/day of water, then 0.3 $mg/kg \times 0.2 \times 70\ kg \times 1/500 \times 0.5 = 0.004$ mg, thus deriving the criterion level for domestic water supply of 4 $\mu g/l$.

Invertebrates have proven to be slightly more sensitive to lindane, with the midge, *Chironomus tentans,* being most sensitive at 2.2 $\mu g/l$, and *Daphnia magna* the least sensitive at 11 $\mu g/l$ in chronic bioassays. The brown trout (*Salmo gairdneri*) is apparently the fish most sensitive to lindane, with a 96-hr LC_{50} of 2 $\mu g/l$. Lindane apparently can be accumulated in fish and mollusk tissues in concentrations many hundred times greater than those in water, causing the FDA to recommend a 0.3-mg/kg guideline for lindane in edible fish tissue in this country.

Lindane is metabolized (as shown by experimentation in the rat) through various cyclic forms into chlorophenols. In humans, liver damage has been found on live biopsy of workers chronically exposed to BHC for many years and over 30 cases of anemia have been reported from exposure to BHC. No satisfactory mode of activity for this pesticide has ever been reported.

Carcinogenicity. The isomers of BHC have produced liver tumors in experimental rodents [58]. Certain levels of carcinogen are considered to be an acceptable risk, since there is no dosage without some effect being presumed. The risk estimates are expressed as a probability of cancer after a lifetime consumption of 1 liter/day of water containing a certain concentration of the carcinogen [12]. A risk is calculated from:

$$\text{risk} = \text{risk factor} \times Q \times \text{liters of water/day}$$

where Q = concentration of compound in μg/l and the liters of water ingested daily numbers 2.

For example, a risk of 1×10^{-6} (or 1 in a million) multiplied by the number of μg/l and the number of liters per day would result in $1 \times 10^{-6} \times 2$ or 2 in a million. This means that at a concentration of 10 μg/l during a lifetime of exposure this compound would be expected to produce one excess case of cancer for

$$2 \times 10^{-6} \times 10\ \mu\text{g/l} = 2 \times 10^{-5} = \frac{2}{100{,}000} = \frac{1}{50{,}000}$$

every 50,000 persons exposed. If the population of the United States is estimated at 220 million people, this translates into 220,000,000/50,000 = 4400 deaths from cancer over a lifetime of 70 years or 62.8 deaths per year.

For lindane (γ-BHC) at a concentration of 0.001 mg/l the estimated risk for man would be from 3.3×10^{-6} to 8.1×10^{-6} or from 6.6 to 16.1 excess cases of cancer for every million people.

Recommendation. Considering all of the above, the MCL for lindane is set at 0.004 mg/l.

Toxaphene

Toxaphene ($C_{10}H_{10}Cl_8$-technical chlorinated camphene, 67–69% chlorine) is a complex mixture of unspecified chlorinated camphene derivatives, the most heavily used and least understood organochlorine insecticide. The major reason that so little is known about its structure or metabolism is its complex nature: it is a mixture of at least 175 compounds, of which the structures of fewer than 10 are known.

Toxaphene is widely used as an insecticide on a variety of crops, and residues of the substance can be found in various grains and seeds and in the fat of domestic animals fed on contaminated feed. Like the other organochlorine pesticides, toxaphene is known to be somewhat persistent in the environment, especially in the soil.

Toxic Effects. Very little is known about the metabolism of toxaphene. Oral doses to rats were reportedly excreted 37% in feces and 15% in urine within nine days. Some rare cases of human poisoning by toxaphene have been reported both by accident (from agriculture exposure) and from voluntary studies. When people inhaled toxaphene, there was no evidence of toxicity at 0.4- and 250-ppb exposures. Chronic, low-level exposure (200 ppm) to toxaphene resulted in clinical signs of liver damage in rodents, dogs, cattle, sheep and rabbits. Kidney damage was noted also in rats and dogs [57].

Hughes [59] reported that lakes treated with toxaphene concentrations ranging from 40 to 150 μg/l remained toxic to fish for periods of a few months to five years. Terriere et al. [60] reported that a lake treated with toxaphene as a piscicide remained toxic to fish for at least five years. Bioconcentration factors of toxaphene were 500 for aquatic plants, 1000–2000 for aquatic animals other than fish and 10,000–20,000 for rainbow trout in the lake. Mayer et al. [61] observed accumulations of 5000–21,000 times the water concentration in brook trout exposed only through the water. Accumulation factors of 3400–17,000 from aqueous solution have been reported for bacteria, algae and fungi [62].

Recommendation. By application of the available data and based on the assumptions that (1) 20% of the total intake of toxaphene is from drinking water, and (2) the average person weighs 70 kg and consumes 2 liters/day of water, the calculation is:

$$1.7 \text{ mg/kg} \times 0.2 \times 70 \text{ kg} \times \frac{1}{500} \times 0.5 = 0.024 \text{ mg/l}$$

However, an organoleptic effect has been shown to occur at 0.005 mg/l [63]. Therefore, the MCL for toxaphene is 0.005 mg/l [53].

2,4-D and 2,4,5-TP

Since aquatic weeds have become undesirable in many U.S. waters, chemical control of this vegetation has won wide acceptance. However, waters receiving herbicides are sometimes employed as raw sources of drinking water. Consequently, a standard is needed for the more extensively used herbicides so as to protect the health of the water consumer.

Two widely used herbicides are 2,4-D (2,4-dichlorophenoxyacetic acid) and 2,4,5-TP (Silvex, 2,4,5-trichlorophenoxypropionic acid). Each of these compounds is formulated in a variety of salts and esters that may have a marked difference in herbicidal properties, but all are hydrolyzed rapidly to the corresponding acid in the body. A closely related compound, 2,4,5-trichlorophenoxyacetic acid, was used extensively at one time, but has been banned for major aquatic use.

Toxic Effects. The subacute oral toxicity of chlorophenoxy herbicides has been investigated in a number of species of experimental animals [57,64]. The dog was found to be sensitive and often displayed mild injury in response to doses of 20 mg/kg/day for 90 days. The no-effect level of 2,4-D is 0.5 mg/kg/day in the rat and 8.0 mg/kg/day in the dog [57].

Data are available on the toxicity of 2,4-D to man. A daily dosage of 500 mg (about 7 mg/kg) produced no apparent ill effects in a volunteer over a 21-day period [65]. When 2,4-D was investigated as a possible treatment for disseminated coccidiomycosis, the patient had no side effects from 18 intravenous doses during 33 days; each of the last 12 doses in the series was 900 mg (about 15 mg/kg) or more, the last being 2000 mg (about 37 mg/kg). A 19th and final dose of 3600 mg (67 mg/kg) produced mild symptoms [66].

Since 2,4,5-TP has been banned for all aquatic uses, there is considerable interest as to why this action was taken. Consequently, a discussion of the toxicity of this herbicide is included. In rats, the effects were evidenced by statistically increased proportions of litters affected and of abnormal fetuses within the litters (notably cleft palate and cystic kidneys). Effects were noted in both mice and rats, although the rat appeared to be more sensitive to this effect. (A dosage of 21.5 mg/kg produced no harmful effects in mice, while a level of 4.6 mg/kg caused minimal, but statistically significant, effects in the rat.) Research has indicated that a contaminant (2,3,7,8-tetrachlorodibenzo-*p*-dioxin) that was present at approximately 30 ppm in the 2,4,5-TP formulation originally tested was highly toxic to experimental animals and produced fetal and maternal toxicity at levels as low as 0.005 mg/kg. However, purified 2,4,5-TP has also produced teratogenic effects in both hamsters and rats at relatively high dosages. Current production samples of 2,4,5-TP that contain less than 1 ppm of dioxin did not produce embryo toxicity or teratogenesis in rats at levels as high as 24 mg/kg/day.

The acute oral dose of 2,4-D required to produce symptoms in man is probably 3000–4000 mg (or about 45–60 mg/kg). A comparison of other toxicity values of 2,4,5-TP indicates that the toxicity of these two agents is of the same order of magnitude. Thus, in the absence of any specific toxicologic data for 2,4,5-TP in man, it might be estimated that the acute oral dose of 2,4,5-TP required to produce symptoms in man would also be about 3000–4000 mg.

Recommendations. The no-effect level of 2,4-D in rats and dogs has been established as 62.5 and 10 mg/kg/day, respectively. Based on these data, an acceptable daily intake can be calculated as 0.0125 mg/kg/day.

Using an uncertainty factor of 1000, the no adverse effect in drinking water assumes an intake by a 70-kg person is 0.09 mg/l. This is rounded to 0.1 mg/l as the MCL for 2,4-D drinking water.

Based on a two-year feeding study of mice, the no-adverse-effect level for 2,4,5-TP is set at 5 mg/kg/day in dogs from which the daily intake levels were calculated to be 0.00075 mg/kg/day or 0.00525 mg/l from water. This is rounded to 0.01 mg/l as the MCL for 2,4,5-TP in drinking water [16].

Methoxychlor

Methoxychlor [1,1,1-trichloro-2.2-*bis*(*p*-methoxyphenyl)ethane] is an insecticide first introduced in 1945. It is used as a substitute for DDT on crops. No residues of this substance were detected in samples of drinking water from the Mississippi, New Orleans surveys of 1974 [16].

Toxic Effects. The highest level of methoxychlor to have minimal or no long-term effects in man is 2.0 mg/kg/day [57]. Since human data are available for corroborating the animal results, the total safe drinking water intake level is assumed to be 1/100 of the no-effect level for the most sensitive species tested, in this case, man. Methoxychlor gave negative results in mutagenic testing with bacteria, but can cause tumors in mice. This substance is considered a safer substance than DDT because of its low mammalian toxicity.

Recommendation. Applying the available data and assuming 20% of the total intake of methoxyclor is from drinking water, and that the average person weighs 70 kg and consumes 2 liters/day of water, then:

$$2 \text{ mg/kg} \times 0.2 \times 70 \text{ kg} \times 1/100 \times 0.5 = 0.14 \text{ mg/l.}$$

A level of 0.10 mg/l is recommended for domestic water supplies.

The criterion for aquatic life is set at 0.03 μg/l because sensitivity data on the freshwater fathead minnow, *Pimephales promelas* [67], indicate this to be a safe level for this sensitive species and also for the marine striped bass, *Morone saxatilis* [68].

Total Trihalomethanes

Total trihalomethanes (TTHM) are defined as the arithmetic sum of the concentrations of trichloromethane (chloroform), dibromochloromethane, bromodichloromethane and tribromomethane (bromoform).

The MCL applies to all community water systems serving 10,000 persons or more that add a disinfectant to their treatment process. See Chapter 2 for the details of occurrence and formation in drinking water.

Toxic Effects. Chloroform has been shown to be adsorbed rapidly on oral and intraperitoneal administration and subsequently metabolized to CO_2, Cl^-, phosgene and other unidentified metabolites [16]. The metabolic profile of $CHCl_3$ in such animal species as mice, rats and monkeys is qualitatively similar to man.

Mammalian responses to chloroform exposure include central nervous system depression, hepatotoxicity, nephrotoxicity, teratogenicity and carcinogenicity [16]. These responses are discernible in mammals after oral and inhalation exposure to high levels of chloroform ranging 30–350 mg/kg; the intensity of response is dependent on the dose. Although less toxicological information is available for the brominated THM, mutagenicity and carcinogenicity have been detected in some test systems. Physiological chemical activity should be greater for the brominated THM than for chloroform.

Although short-term toxic responses to THM in drinking water are not documented, the potential effects of chronic exposures to THM should be a matter of concern. Prolonged administration of chloroform at relatively high dose levels (100–138 mg/kg) to rats and mice manifested oncogenic effects (i.e., tumors). These effects were not observed at the lowest dose level (17 mg/kg) in three experiments. Since methods do not now exist to establish a threshold no-effect level of exposure to carcinogens, the preceding data do not imply that a safe level of exposure can be established for humans.

Human epidemiological evidence is inconclusive, although positive correlations with some sites have been found in several studies [16]. There have been 18 retrospective studies investigating some aspect of a relationship between cancer mortality or morbidity and drinking water variables. Due to various limitations in the epidemiological methods, in the water quality data, and problems with the individual studies, the present evidence cannot lead to a firm conclusion that there is an association between contaminants in drinking water and cancer mortality.

Causal relationships cannot be proven on the basis of results from epidemiological studies. The evidence from these studies thus far is incomplete, and the trends and patterns of association have not been developed fully. When viewed collectively, however, the epidemiological studies provide sufficient evidence for maintaining the hypotheses that there may be a potential health risk, and that the positive correlations may be reflecting a causal association between constituents of drinking water and cancer mortality.

Recommendation. Preliminary risk assessments made by the Science Advisory Board (SAB), NAS, and EPA's Carcinogen Assessment Group (CAG) using different models have estimated incremental risks associated with exposure from $CHCl_3$ in drinking water [16]. The exposure to TTHM for air and food have not been included in these computations. The risk estimates associated with the MCL at the 0.10-mg/l level are essentially the same in the NAS and CAS computations (3.4×10^{-4} and 4×10^{-4}) assuming 2 liters/day of water at 0.10 mg/l $CHCl_3$ for 70 years.

Radiological Contaminants

Background

Radioactivity refers to the ability of an element spontaneously to emit positively and negatively charged particles and electromagnetic waves. Alpha particles (α) are doubly charged ions of helium with a mass of four. Beta particles (β) are negatively charged particles (electrons) that move at almost the speed of light. Gamma rays (γ) are electromagnetic radiation or photons which travel at the speed of light. Release of an alpha particle results in the radioactive element decreasing its mass by four and its atomic number by two. Release of a beta particle results in an increase of one in the atomic number. Gamma radiation can accompany alpha and beta emission and is a result of energy level changes within the atom. The radiological standards are concerned with the amounts of these spontaneous radiations emitted by both natural and manmade substances [69].

The potency of radiation forms is measured in three ways: the *roentgen* gives the amount of radiation absorbed in air at a given point; the *rad* gives the amount of radioactive energy actually held in a gram of any material (doses of 600–700 rad are dangerous to man); the *rem* indicates the degree of potential danger to health, and is the rad multiplied by a given factor of potential danger. The radiation to which we are all exposed is made up mostly of the fast moving, highly penetrating X-rays, gamma rays and beta rays, where rem and rad are equal. Alpha particles are very harmful but have little penetrating force. On the other hand, they are quite dangerous if a radioactive substance is taken into the body.

Sources

Radiation has both natural and manmade origins. Natural radioactivity is found in groundwaters as a result of geological conditions.

Radium-226 has a wide natural distribution in the United States, particularly in the Midwest and Rocky Mountain regions. The natural background radiation dose varies geographically between 40 and 300 mrem/yr in the United States.

Manmade radioactivity was ubiquitous in surface waters because of fallout from nuclear weapons testing. For example, the presence of strontium-90 from atmospheric fallout has caused a background level of 1 pCi/l in public water supplies. In addition, there is contamination from nuclear power plants, science and industry. Water-cooled atomic piles require large quantities of water to dissipate the heat released by nuclear fission. The discharge of such cooling waters to rivers is of concern because of the induced radioactivity and the presence of fission products in the water. The use of atomic energy has made it necessary to establish limiting concentrations for radioactive substances because of their proven carcinogenic potential.

Many natural and artificial radionuclides have been found in water, but most of the radioactivity is due to a relatively small number of nuclides and their decay products. Among these are the following emitters of radiation of low linear energy transfer (LET), which refers to the average amount of energy lost by an ionizing particle or photon traversing a certain length of any medium but air: potassium-40, tritium, carbon-14 and rubidium-87. In addition, such high-LET alpha-emitting radionuclides, as radium-226, the daughters of radium-228, polonium-210, uranium, thorium, radon-220 and radon-222, may also be present in varying amounts.

Exposure from Drinking Water

Drinking water is not a major pathway for exposure from manmade radiations. Estimates are that the total radiation body dose from drinking water ranges from 0.00001 to 0.3 mrem/yr with 90% of the dose below 0.04 mrem/yr. Data on ambient levels in public water systems indicate that almost all radioactivity is due to residual activity from nuclear weapons testing.

The MCL for manmade radionuclides is expressed in terms of annual dose rate (mrem/yr) as a result of continuous ingestion. Concentration (pCi/l) is not used as a unit of MCL since concentration limits are appropriate for short-time but not lifetime exposure.

By far the largest contribution to radioactivity in drinking water comes from potassium-40, which is present as a constant percentage of total potassium-40 body burden. The total body dose from other possible radioactive contaminants of water constitutes a small percentage of the background radiation to which the population is exposed. Although the amounts of individual radioactive contaminants fluctuate from place to place, calculations made for a hypothetical water supply that might be typical for the United States have shown that a total soft-tissue dose of only 0.24 mrem/yr would be contributed by all the radionuclides found in the water. Even with rather wide fluctuations in the concentrations, the total contribution of the radionuclides will remain very small.

However, such bone-seeking radionuclides as strontium-90, radium-226 and radium-228 account for a somewhat larger proportion of the total bone dose. This is particularly true for the two isotopes of radium because they, or their daughters, emit high-LET radiation, and because certain restricted localities have been found to have rather high concentrations of radium in drinking water. Nevertheless, in the hypothetical typical water supply, less than 10% of the annual background dose comes from such radiation. It has also been estimated that the total population exposed to levels of radium greater than 3 pCi/l is about a million people. About 120,000 people are exposed to radium at levels greater than 9 pCi/l.

Levels of radiation in the environment may be measured directly, but the determination of resultant radiation doses to humans and their susceptible tissues is generally derived from pathway and metabolic models and through calculations of energy adsorbed. Relations between dose and effect are derived from human epidemiological studies and animal research. Although much is known about radiation dose-effect relationships at high levels of dose, a great deal of uncertainty exists when these high-level relationships are extrapolated to lower levels of dose, particularly when given at low dose rates. At present, EPA assumes a linear relationship between dose and effects and the number of cancers per rem from a high dose is assumed to be a practical predictor of the low-dose effects. Since the number of health effects caused by ionizing radiation at low doses is not known, the assumption is made that there is no harmless level of dose, i.e., no threshold level of radiation is considered safe.

Estimates of the radiation doses (rems) expected to be produced by radionuclide concentrations ingested in water were calculated according to formulas provided by the National Council of Radiation Protection and the International Commission on Radiation Protection [12]. Such formulas are complex and must take into account buildup, retention, decay and elimination of radionuclides in the body. Doses are computed for two types of exposure for 50 years: (1) what the dosage would be after a constant intake of 1 pCi/year, and (2) for 2 liters/day of water containing 1 pCi/l.

A separate mathematical formula was used to calculate radium and strontium doses, since these substances are bone-seeking radionuclides and are best represented by a special (power function) metabolic model to calculate the body burden imposed by the radionuclide. This formula results in burdens lower by a factor of 10 than the previous formula. The average body burden of radium in areas of normal radioactivity is about 50 pCi (40 pCi in the skeleton and 10 pCi in soft tissue). Intake of radium in food appears to be the main source at normal levels, because the daily intake in the United States is about 2 pCi in food and less than 0.1 pCi in water.

A quantitative relationship (the Lucas equation [15]) was developed by empirical observations of the body burden of radium-226 and the concentration of radium in drinking water from community water supplies. These observations tend to support the validity of mathematical models used to describe retention of radionuclides in the body. The EPA estimates that consumption of radium-226 at 5 pCi/l by 1 million people could result in 1.5 fatalities per year, while a 4-mrem annual total body exposure causes from 2 to 4 fatalities due to cancer per million people [12].

Maximum Contaminant Levels

In community water systems, the following are the maximum contaminant levels for radium-226, radium-228 and gross alpha particle radioactivity:

1. combined radium-226 and radium-228: 5 pCi/l, and
2. gross alpha particle activity (including radium-226, but excluding radon and uranium): 15 pCi/l.

Maximum contaminant levels for beta particle and photon radioactivity from manmade radionuclides in community water systems:

1. The average annual concentration of beta particle and photon radioactivity from manmade radionuclides in drinking water shall not produce an annual dose equivalent to the total body or any internal organ greater than 4 mrem/year.
2. Except for tritium and strontium-90 (Table IV), concentration of manmade radionuclides in drinking water shall not produce an annual dose equivalent of 4 mrem/year shall be calculated on the basis of a 2-liters/day drinking water intake using the 168-hr data listed in "Maximum Permissible Radionuclides in Air or Water for Occupational Exposure" [12]. If two or more radionuclides are present, the sum of their annual dose equivalent to the total body or to any organ shall not exceed 4 mrem/year.

Table IV. Types of Activities, Sensitive Target Tissues and MCL for Radiological Contaminants in Drinking Water [12]

Radionuclides	Type of Activity	MCL (pCi/1)	Health Effect
Natural Alpha particle activity $^{226}Ra + {}^{228}Ra$	Alpha	5	
Gross alpha activity (including ^{226}Ra)	Alpha	15	Bone cancer
Manmade radionuclides	Beta & photons	A concentration not to produce an annual dose equivalent greater than 4 mrem/yr to the body or any internal organ	Bone cancer

If gross beta particle activity exceeds 50 pCi/l, then the following major radioactive constituents must be analyzed for, and then dosages calculated:

Gross alpha activity	Strontium-90
Radium-226	Iodine-131
Radium-228	Cesium-134
Tritium	Gross beta
Strontium-89	

Average annual concentrations assumed to produce a total body or organ dose of 4 mrem/year include:

Radionuclide	Critical Organ	pCi/l
Tritium	Total body	20,000
Strontium 90	Bone marrow	8

3. The average annual concentration of tritium and strontium-90 estimated to produce a 4-mrem/year dosage at these levels (20,000 pCi/l and 8 pCi/l, respectively) are themselves designated as MCL.

Turbidity

Background

Materials suspended in drinking water include inorganic and organic solids as finely divided particles of sizes ranging from colloidal dimensions to more than 100 μm. Such particles may also have other substances and microorganisms attached to them. Small particles of some materials, such as the asbestos minerals, may have the potential to affect human health directly when they are ingested. There is widespread concern over the biological effects of such substances.

Many kinds of particles, though apparently harmless in themselves, may indirectly affect the quality of water by acting as vehicles for concentration, transport and release of other pollutants. Water treatment can often be effective in removing most of the suspended particles, but conventional methods of detecting the presence of particulate material by measurement of turbidity has serious deficiencies.

Particles that pass through conventional coagulation and filtration plants in significant amounts do so because of inadequate design, operation or control of these facilities. Some of these particles may be harmful in themselves. Most are innocuous, but have the potential to contain harmful substances. On the other hand, the absence of a detectable turbidity does not guarantee that a water is free from harmful particulates. A determination that a water has a high turbidity does indicate the presence of particulate materials and is a cause for concern.

Asbestos: Direct Effect on Health

Particles of asbestos and other fibrous minerals occur in raw water, usually in a range of sizes from fractions of a micrometer to a few micrometers. There are generally fewer than 10 million fibers per liter, but waters are found with from less than 10,000 to more than 100 million fibers per liter. Some of the highest counts have been found near some cities. Fibers in drinking water are typically less than 1 μm long; those longer than 2 μm are uncommon. The identification and counting of fibers is difficult and time-consuming, usually requiring a transmission electron microscope. The reported counts are highly variable, often differing from one count to the next by a factor of 10 or more [12].

Epidemiological studies of workers exposed to asbestos by inhalation have shown an increase in death rates from gastrointestinal cancer [16]. With respiratory exposure it is likely that more fibers are swallowed than remain in the lungs. The workers studied were exposed to asbestos with a large range of fiber lengths. It is not clear whether fiber length is pertinent to the development of cancer in the digestive tract in humans.

Epidemiological studies of cancer death rates in Duluth, Minnesota, where the water supply has been contaminated with mineral fiber, have so far not revealed any increase of gastrointestinal cancer with time, in comparison with death rates in other areas. The contamination of the

Duluth drinking water began less than 20 years ago, however, and since many cancers have long latency periods, these negative epidemiological findings do not exclude the possibility that an increase may appear within the next 5–15 years.

Animal deposition model studies have shown that fiber length and diameter affect the carcinogenic response seen; the long, thin fiber appears to be the active one. However, the relevance of these experimental models to the human experience is not clear. While some animal studies have shown penetration of the gastrointestinal epithelium, others have not.

Indirect Effects on Health

The concentration of inorganic, organic and biological pollutants is usually much higher in the suspended solids and sediments of streams and lakes than in water. Clay and organic particulates have large surface areas and strongly adsorb ions, polar and nonpolar molecules and biological agents. The occurrence of these materials in water is a consequence of natural events, as well as human activity, and is common in many waters that people drink. Although many of the clay or natural organic particulates, in themselves, may not have deleterious effects when ingested by humans, they may exert important health effects through adsorption, transport and release of inorganic and organic toxicants, bacteria and viruses. The clay or organic complex with a pollutant may be mobilized by erosion from the land, or complexes may form when eroded particulate matter enters a stream containing pollutants. The atmosphere is also an important pathway. In the adsorbed state, organic and inorganic toxicants may be less active; however, the possibility exists that the toxicants may be released from the particulate matter in the alimentary tract where they may exert toxic effects. It is not clear how complexes of particulate matter with viruses and bacteria behave in the gut. It is known, however, that some enzymes retain their activity when adsorbed on clay and that viral-clay particulates are infectious in tissue culture and in animal hosts.

Maximum Contaminant Levels

Based on the above and other studies not cited, the maximum contaminant levels for turbidity apply both to community water systems and noncommunity water systems using surface water sources in whole or in part. The maximum contaminant levels for turbidity in drinking water, measured at a representative entry point(s) to the distribution system, are:

1. one turbidity unit (TU), as determined by a monthly average, except that five or fewer turbidity units may be allowed if the supplier of water can demonstrate to the state that the higher turbidity does not interfere with disinfection, prevent maintenance of an effective disinfectant agent throughout the distribution system, or interfere with microbiological determinations.
2. five turbidity units based on an average for two consecutive days.

Microbiological Contaminants

Coliform Bacteria as Indicator Species

Procedures for the detection of specific disease-causing bacteria, viruses, protozoa, worms and fungi are complex, time-consuming and in need of further refinement to increase the levels of sensitivity and selectivity. Therefore, an indirect approach to microbial hazard measurement is required.

Coliform bacteria have been used as indicators of sanitary quality in water since 1880 when *Escherichia coli* and similar gram-negative bacteria were shown to be normal inhabitants of fecal discharges. Although the total coliform group as presently recognized in the PHS Drinking Water Standards includes organisms known to vary in characteristics, the total coliform concept merits consideration as an indicator of sanitary significance, because the organisms are normally present in large numbers in the intestinal tracts of humans and other warm-blooded animals.

Numerous stream pollution surveys over the years have used the total coliform measurement as an index of fecal contamination. However, occasional poor correlations to sanitary significance result from the inclusion of some strains on the total coliform group that have a wide distribution in the environment and are not specific to fecal material. Therefore, interpretation of total coliform data from sewage, polluted water and unpolluted waters is sometimes difficult. For example, *Enterobacter* (Aerobacter) *aerogenes* and *E. cloacae* can be found on various types of vegetation and in polluted water. Also included are plant pathogens and other organisms of uncertain taxonomy whose sanitary significance is questionable. All of these coliform subgroups may be found in sewage and in polluted water.

Fecal Coliform Measurements

The presence of any type of coliform organism in treated water suggests either inadequate treatment or contamination after postchlorination. It is true there are some differences between various coliform strains with regard to natural survival and their chlorination resistance, but these are minor biological variations that are more clearly demonstrated in the laboratory than in the water treatment system. The presence of any coliform bacteria, fecal or nonfecal, in treated water should not be tolerated.

Insofar as bacterial pathogens are concerned, the coliform group is considered a reliable indicator of the adequacy of treatment. As an indicator of pollution in drinking water supply systems and, indirectly, as an indicator of protection provided, the coliform group is preferred to fecal coliform organisms. Whether these considerations can be extended to include rickettsial and viral organisms has not been determined definitely.

Drinking Water Standards

The first U.S. national standards for bacteriological water quality were established in 1914 [9]. These standards were specifically applicable to water used on interstate carriers, but were adopted quite early (formally or informally) by many states. However, the standards are not precise and accurate indices of quality, but simply a convenient mode of analysis for comparative purposes that must be used with considerable caution.

It is obvious that when drinking water meets the standards there is no absolute assurance of the absence of pathogens, only a confidence that their presence is unlikely; hence, the probability of waterborne-disease transmission is decreased. The decline in morbidity and mortality from diseases such as cholera, typhoid fever, salmonellosis and shigellosis provides some evidence of the validity of this confidence, although some of the decline may be due to the generally better health of the population, making people less susceptible to infection.

The latest standards adopted by the PHS were those of 1962 [9]. Although none of the bacteriological numerical values was changed, a major procedural revision was made; the membrane-filter (MF) technique was accepted as an equivalent alternative to the multiple-tube-dilution technique that has been in use since 1914. The MF standard was set at one coliform/100 ml.

The EPA interim regulations [11] have broadened the applicability of the standards to all public water systems, specified a sample size of 100 ml when the MF technique is used and modified the required frequency of sampling. A monthly mean of less than one coliform/100 ml is still standard.

Recommendation

The MCL for coliform bacteria are given in Table II. The reader is referred to Chapter 10 for additional information, including the rationale for using coliform bacteria as indicator organisms. Many references to the scientific literature are given.

RATIONALE FOR SELECTING SMCL

Chloride

Chloride in reasonable concentrations is not harmful to humans, but it causes corrosion in concentrations above 250 mg/l and, above 400 mg/l, causes a salty taste in water that many people find objectionable. It is a constituent of total dissolved solids. Chloride can be removed from drinking water by distillation, reverse osmosis or electrodialysis, but in some cases the entry of chloride into a drinking water source can be minimized by proper aquifer selection and well construction [70].

Color

Color may indicate the presence of dissolved and colloidal organic material that may lead to the generation of trihalomethanes and other organohalogen compounds during chlorination. Color can also be caused by such inorganic species as manganese, cooper or iron for which SMCL exist. Color becomes objectionable and unaesthetic to most people at levels over 15 color units (CU). In some cases, color can be objectionable at the 5 CU level, and states, therefore, should also consider the regulation of color at levels below 15 CU. See Faust and Aly [17] for a discussion of the chemical nature of organic color.

Copper

Copper is an essential and beneficial element in human metabolism, but copper imparts an undesirable taste to drinking water. The 1 mg/l standard assures the absence of taste effects, but copper can stain porcelain as low as 0.5 mg/l. Small amounts of copper are generally regarded as nontoxic. Sufferers of Wilson's disease (an inherited liver ailment) cannot ingest even small doses of copper. At high concentrations (6.7

mg/l) over 14 months, copper caused infant mortality in one case [16]. In gram quantities this element is a gastrointestinal tract irritant.

Corrosivity

Corrosivity is a complex characteristic of water related to pH, alkalinity, dissolved oxygen and total dissolved solids plus other factors. A corrosive water, in addition to dissolving metals with which it comes in contact, also produces objectionable stains on plumbing fixtures. Corrosivity is controlled by pH adjustment, the use of chemical stabilizers or other means which are dependent on specific conditions of the water system.

The corrosivity of drinking water is a parameter which has not only esthetic significance, but health and economic significance as well. The products of corrosion having the greatest health significance, cadmium and lead, are addressed in primary regulations, but there is also a sufficient basis to include corrosivity in secondary regulations. The problem lies in the inability of existing numerical indices for assessing and limiting corrosivity. There are a number of indices in use, but no agreement on a single one which would, in all cases, definitively say whether or not a given water was corrosive. An attempt to circumvent the problem can be made by specifying, in lieu of an index, practical tests of corrosivity using pipe sections, metal coupons or water analyses to determine the corrosive properties of a water. Unfortunately, most of these tests, as well as most indices, are not universally applicable and require long periods of time to conduct. For a corrosivity test or index to be widely used and applied, the testing procedure must be rapid, simple and generally applicable. Chapter 9 gives additional information about the nature and treatment of corrosivity.

Foaming Agents

Foaming is a characteristic of water caused principally by the presence of detergents and similar substances. Water that foams is definitely unesthetic and considered unfit for consumption. The foamability of water is measured by the quantity of methylene blue active detergent substances (MBAS) present which indicates the levels of alkylbenzylsulfonate (ABS) and the biodegradable linear alkyl-benzyl sulfonate surfactants (LAS). LAS at 1 mg/l imparts an oily taste (because of LAS degradation products), while 10,000 ppm is a gastrointestinal irritant.

Foaming substances can be removed from drinking water by carbon adsorption, but it is preferable to prevent contamination of water by these substances.

Iron

Iron is a highly objectionable constituent of water supplies for either domestic or industrial use. Iron may impart brownish discolorations to laundered goods. The taste that it imparts to water may be described as bitter or astringent (0.1 mg/l), and iron may adversely affect the taste of other beverages made from water. The amount of iron causing objectionable taste or laundry staining constitutes only a small fraction of the amount normally consumed in the daily diet and thus does not have toxicologic significance. Iron can be removed from water by conventional water treatment processes or ion exchange and also by oxidation processes followed by filtering (see Chapter 8). If the iron comes from the corrosion of iron or steel piping, the problem can often be eliminated by practicing corrosion control.

Manganese

Manganese, like iron, produces discoloration in laundered goods and impairs the taste in drinking water and beverages, including tea and coffee. At concentrations in excess of 0.05 mg/l manganese can occasionally cause buildup of coatings in distribution piping which can slough off and cause brown spots in laundry items and unesthetic black precipitates. Manganese can usually be removed from water by the same process used for iron removal (see Chapter 8).

Odor

Odor is an important esthetic quality of water for domestic consumers and such process industries as food, beverage and pharmaceutical that require water essentially to be free of taste and odor. The threshold odor number (TON) is the dilution factor necessary to remove the odor. A TON of 1 implies odor-free water. It is usually impractical and often impossible to isolate and identify the odor-producing chemical, which can be organic or inorganic, or originate in waste discharges. The evaluation of odors and tastes is thus dependent on the individual senses of smell and taste. Panel members are used to rate this parameter. In many cases, sensations ascribed to the sense of taste are actually odors. Chlorophenols, which are the reaction product of the disinfectant, and phenol often give rise to a smell. Algae and fungi in water give rise to a distinct musty odor. Odors are usually removed by carbon adsorption or aeration (see Chapter 3).

pH

The range of pH values in public water systems may cause a variety of esthetic and health effects. Corrosion effects and the release of toxic metals are commonly associated with pH levels below 6.5. As pH levels are increased above 8.5, mineral incrustations and a bitter taste occur, the germicidal activity of chlorine is substantially reduced and the rate of formation of trihalomethanes is increased significantly. However, the impact of pH in any one water system will vary depending on the overall chemistry and composition of the water. As a result, a more or less restrictive range may be appropriate under specific circumstances.

Sulfate

Sulfate may cause detectable tastes at concentrations of 300–400 mg/l. At concentrations above 600 mg/l it may have a laxative effect. High concentrations of sulfate also contribute to formation of scale in boilers and heat exchangers. Sulfate can be removed from drinking water by distillation, reverse osmosis, ion exchange or electrodialysis. The laxative effect seldom affects regular users of the water, but transients are particularly susceptible. For this reason it is recommended that individual states institute monitoring programs for sulfate, and that transients be notified if the sulfate content of the water is high. Such notification should include an assessment of the possible physiological effects of consuming the water.

Total Dissolved Solids (TDS)

Total dissolved solids may have an influence on the acceptability of water in general, and, in addition, a high TDS value may indicate the presence of an excessive concentration of some specific substance that would be esthetically objectionable to the consumer. Excessive hardness, taste, mineral deposition or corrosion are common properties of highly mineralized water. Taste is usually noticed by nonregular users. Dissolved solids can be removed by chemical precipitation in some cases, but distillation, reverse osmosis, electrodialysis and ion exchange are more generally applicable.

Zinc

Zinc, like copper, is an essential and beneficial element in human metabolism. Zinc can also impart an undesirable taste to water. At high concentrations, zinc salts impart a milky appearance to water. Zinc can be removed from water by conventional water treatment processes or ion exchange, but since the source of zinc is often the coating of galvanized iron, corrosion control will minimize the introduction of zinc into drinking water. At the same time, corrosion control will minimize the introduction of lead and cadmium into the drinking water, since lead and cadmium are often contaminants of the zinc used in galvanizing.

CRITERIA FOR THE PRIORITY POLLUTANTS

In 1972, Congress altered the Federal Water Pollution Control Act of 1948 to its present form as the Clean Water Act (CWA) [71]. A major objective of the CWA is to control the chemical and industrial effluents that enter various ground and surface waters. It established guidelines for the employment of what it terms the "best practical technology" and the "best available technology economically achievable" in order to treat or remove these substances. EPA uses these guidelines as regulations when issuing effluent permits under the National Pollutants Discharge Elimination System (NPDES). EPA eventually identified 129 constituents, the so-called priority pollutants.

Since some of the drinking water is at one time effluent before it is cycled through different plants for treatment (i.e., along a river) these contaminants can enter raw sources of water. Section 304 of the Clean Water Act provides for the development of criteria for these 129 pollutants based only on data and scientific judgment about human health and not only on any economic or technological feasibility. Regulations based on other criteria (i.e., SDWA) may state different levels. The CWA also promulgates criteria based on the protection of aquatic life. The CWA used the same mathematical formulas for extrapolating risk from animal studies to humans except that their calculations for the intake of noncarcinogenic substances used intake from water, food and sometimes air. The acceptable daily intake is formulated using the exposure assumptions of 2 liters of water, 18.7 g of edible aquatic products and an average human body weight of 70 kg. The criteria for most of the priority pollutants became available in 1980 [72]. Table V shows the acceptable level and the carcinogenic risk (for one cancer in a population of one million) for noncarcinogens and carcinogenic compounds, respectively.

Table V. Priority Pollutants—Criteria as of November 1980 [72]

	Concentration (ppb)	
Contaminant	**Noncarcinogen Acceptable Level**	**Carcinogen[a] Risk/Population (10^{-6})**
Acenaphthene	20.0	
Acrolein	6.72	
Acrylonitrile		0.058
Aldrin		0.000071
Dieldrin		0.000074
Antimony	145.0	
Arsenic		0.002
Asbestos		30,000 fibers per liter
Benzene		0.66
Benzidine		0.00012
Benzo [a] pyrene		0.00097
Beryllium		0.0037
bis-(2-Chloroethyl) ether		0.03
bis-(2-Chloroisopropyl) ether		1.15
bis-(Chloromethyl) ether		0.000038
Bromodichloromethane	2.0	
Bromoethane (methyl bromide)	2.0	
Cadmium		10.0
Carbon Tetrachloride		0.4
Chlordane		0.00046
Chloroform		0.19
Chloromethane (methyl chloride)	2.0	
2-Chlorophenol	0.3	
Chromium		0.0008
Copper	1,000	
Cyanides	200	
DDT and metabolites		0.000024
Dibenzo [a,h] anthracene (DBA)		0.00043
Di-*n*-butyl phthalate	5,000	

Table V, continued

Contaminant	Concentration (ppb)	
	Noncarcinogen Acceptable Level	Carcinogen[a] Risk/Population (10^{-6})
Dichlorobenzenes	230	
Dichlorobenzidine		0.01
Dichlorodifluoromethane	3,000	
1,2-Dichloroethane		0.94
Dichloroethylene		0.033
Dichloromethane (methylene chloride)	2.0	
2,4-Dichlorophenol	0.5	
Dichloropropane	200.0	
Dichloropropene	0.63	
Di-2-ethylhexyl Phthalate	1,000	
Diethyl Phthalate	60,000	
2,4-Dimethylphenol		
Dimethyl Phthalate	160,000	
4,6-Dinitro-*o*-cresol	12.8	
2,3-Dinitrophenol	68.6	
2,4-Dinitrophenol	68.6	
2,5-Dinitrophenol	68.6	
2,6-Dinitrophenol	68.6	
3,4-Dinitrophenol	68.6	
3,5-Dinitrophenol	68.6	
Dinitrotoluenes		0.1
Diphenylhydrazines		0.04
Endosulfan	100	
Endrin	1.0	
Ethylbenzene	1,100	
Fluoranthene	200	
Heptachlor		0.00028
Hexachlorobenzene		0.000125
Hexachloroethane		1.9
Hexachlorobutadiene		0.45
Hexachlorocyclohexane (BHC)		
α-BHC		0.0092
β-BHC		0.0163
γ-BHC		0.0186
δ-BHC		0.0123
Hexachlorocyclopentadiene	1.0	
Isophorone	460	
Lead	50	
Mercury	0.2	
Monochlorobenzene	20	
Naphthalene	143	
Nickel	133	
Nitrobenzene	30	

Table V, continued

Contaminant	Concentration (ppb)	
	Noncarcinogen Acceptable Level	Carcinogen[a] Risk/Population (10^{-6})
N-nitrosodiethylamine		0.00092
N-nitrosodimethylamine		0.0014
N-nitrosodi-*n*-butylamine		0.0013
N-nitrosodiphenylamine		4.9
N-nitrosopryolidine		0.016
Pentachlorobenzene	0.5	
Pentachlorophenol	140	
Phenol	3,400	
Polychlorinated biphenyls (PCB)		0.000079
Polynuclear Aromatic Hydrocarbons (PAH) (total of 6 compounds together)		0.00028
Selenium	10	
Silver	10	
Tetrachlorobenzene	17	
Tetrachlorodibenzo-*p*-dioxin		0.000000046
1,1,2,2-Tetrachloroethane		0.17
Tetrachloroethylene		0.8
Thallium	4.0	
Toluene	12.4	
Toxaphene		0.00071
Tribromoethane (bromoform)	2	
Trichlorobenzene	13	
1,1,2,-Trichloroethane		0.06
Trichloroethylene		2.7
Trichlorofluoromethane	32,000	
2,4,5-Trichlorphenol	10	
2,4,6-Trichlorophenol	1.2	
2,3,5-Trinitrophenol	10	
2,3,6-Trinitrophenol	10	
2,4,5-Trinitrophenol	10	
2,4,6-Trinitrophenol (picric acid)	10	
Vinyl Chloride		2.0
Zinc	5,000	

[a]The designated concentration assumes one additional cancer case caused by the compound in a population of one million. Dividing this concentration by 10 gives the level causing cancer to one in ten million, and multiplying the concentration by 10 gives the concentration causing cancer in one in 100,000 people.

For many noncarcinogenic pollutants, the criterion was derived by multiplying the lowest concentration known to have an acute lethal effect on one-half of a test group of an aquatic species (LC_{50} value) by an application factor (10, 100, 1000), depending on the adequacy of the data, in order to protect against chronic effects. If the data showed a substance to be bioaccumulative, a factor was used to reduce concentrations to a presumed protective level.

Criteria for suspect or proven carcinogens are presented as concentrations in water associated with a range of incremental cancer risks to man. Criteria for noncarcinogens represent levels at which exposure to a single chemical is not anticipated to produce adverse effects in man.

REFERENCES

1. Baker, M. N. *The Quest for Pure Water* (New York: American Water Works Association, 1948).
2. Rosenau, M. J. *Preventative Medicine and Hygiene,* 5th ed. (New York: Appleton-Century, 1935).
3. Klein, L. *Aspects of Water Pollution* (New York: Academic Press, Inc., 1957).
4. *Public Health Reports* 29:2959 (1914).
5. Special Commission AAAS. "A Method, in Part, for the Sanitary Examination of Water, and for the Statement of Results, Offered for General Adoption," *J. Anal. Chem.* 3:398 (1889).
6. "Water Quality Criteria," Federal Water Pollution Control Administration, U.S. Department of the Interior, Washington, DC (1968).
7. "Guidelines for Establishing Water Quality Standards for Interstate Waters," U.S. Department of the Interior, Washington, DC (1966).
8. "Water Quality Criteria 1972," Report of Committee on Water Quality Criteria, National Academies of Sciences and Engineering, Washington, DC (1972).
9. "Public Health Service Drinking Water Standards, 1962," Public Health Service Pub. No. 956, U.S. Government Printing Office, Washington, DC (1962).
10. *Federal Register* 40(51):11990 (1975).
11. *Federal Register* 40(248):59566 (1975).
12. *Federal Register* 42(132):35764 (1977).
13. *Federal Register* 44(231):68624 (1979).
14. *Mainstream* 26(3), (1982).
15. *Federal Register* 44(140):42195 (1979).
16. *Drinking Water and Health, Vols. 1 and 3* (Washington, DC: National Academy of Sciences Press, 1977, 1980).
17. Faust, S. D., and O. M. Aly. *Chemistry of Natural Waters* (Ann Arbor, MI: Ann Arbor Science Publishers, Inc., 1981).
18. Fraumeni, J. F. *J. Nat. Cancer Inst.* 55:1039 (1975).

19. Tseng, W. P. *Environ. Health Pers.* 19:109 (1977).
20. Vallee, B. L. et al. *Arch. Ind. Health* 21:132 (1960).
21. Sollmann, T. H. *A Manual of Pharmacology and Its Applications to Therapeutics and Toxicology,* 8th ed. (Philadelphia: W. B. Saunders Co., 1957).
22. Schroeder, H. A. *J. Chronic Dis.* 18:647 (1965).
23. Kobayashi, J. Fifth International Water Pollution Research Conference I-25:1 (1970).
24. Kopp, J. F. Proceedings of the University of Missouri 3rd Annual Conference on Trace Substances in Environmental Health (Columbia, MO: University of Missouri, 1969), p. 59.
25. "Chromium, Medical and Biologic Effects of Environmental Pollutants," National Academy of Sciences, Washington, DC (1974).
26. "Trace Elements in Human Nutrition," World Health Organization Technology Report Series No. 532, Geneva, Switzerland (1973).
27. Davids, H. W., and M. Lieber. *Water Sew. Works* 98:528 (1951).
28. "Natural Fluoride Content of Community Water Supplies," National Institute of Health, U.S. Department of Health, Education, and Welfare, Washington, DC (1960).
29. Richards, L. F. et al. *J. Am. Dent. Assoc.* 74:399 (1967).
30. Hodge, H. C. *J. Am. Dent. Assoc.* 40:436 (1950).
31. Ryan, J. P. "Minerals Yearbook," U.S. Department of the Interior, Washington, DC (1971).
32. "Water Pollution Aspects of Street Surface Contaminants," EPA-R2-72-081, U.S. EPA, Washington, DC (1972).
33. Alexander, F. W. et al. "The Uptake and Excretion by Children of Lead and Other Contaminants," Proceedings of the International Symposium on Environmental Health Aspects of Lead, Amsterdam, Holland (1972), p. 319.
34. King, B. G. *J. Dis. Child.* 122:337 (1971).
35. McCabe, L. J. et al. *J. Am. Water Works Assoc.* 62:670 (1970).
36. Irukayama, K. *Adv. Water Poll. Res.* 3:153 (1967).
37. Baker, F. et al. *Science* 181:230 (1973).
38. Stokinger, H. E. "Mercury, Hg^{237}," *Ind. Hyg. Toxicol.* 2:1090 (1963).
39. Berglund, F., and M. Berlin. *Chemical Fallout* (Springfield, IL: Charles C. Thomas, Publishers, 1969), pp. 258–273.
40. Wershaw, R. L. "Mercury in the Environment," U.S. Geol. Survey Prof. Paper 713, Washington, DC (1970).
41. Fitzgerald, W. F., and W. B. Lyons. *Nature* 242:452 (1973).
42. Jensen, S., and A. Jernelov. *Nature* 223:753 (1969).
43. Bisogni, J. J., and A. W. Lawrence. "Methylation of Mercury in Aerobic and Anaerobic Environments," Technology Report 63, Cornell University Resources and Marine Sciences Center, Ithaca, NY (1973).
44. Wood, J. M. *Science* 183:1049 (1974).
45. Stewart, B. A. et al. *Environ. Sci. Technol.* 1:736 (1967).
46. Vigil, J. et al. *Public Health Rep.* 80:119 (1965).
47. Fairhill, L. T. *J. New England Water Works Assoc.* 55:400 (1941).
48. Kehoe, R. H. et al. *J. Am. Water Works Assoc.* 36:645 (1944).
49. Smith, M. I. et al. *Public Health Rep.* 51:1496 (1936).
50. Smith, M. I., and B. B. West. *Public Health Rep.* 52:1375 (1937).

51. Joffe, W. G. et al. *Arch. Latinoam Nutr.* 22:595 (1972).
52. Hill, W. B. et al. *Am. Silver Produc. Res. Proj. Rept.* (1957).
53. "Quality Criteria for Water," EPA 440/9-76-023, U.S. EPA, Washington, DC (1976).
54. Aub, J. C., and L. T. Fairhall. *J. Am. Med. Assoc.* 146:118 (1942).
55. Just, J., and A. Szniolis. *J. Am. Water Works Assoc.* 28:492 (1936).
56. Treon, A. et al. *J. Agric. Food Chem.* 8:842 (1955).
57. Lehman, A. J. "Summaries of Pesticide Toxicity," Association of Food and Drug Officials of the U.S., Topeka, KS (1965).
58. Thorpe, E., and A. I. T. Walker. *Food Cosmet. Toxicol.* 11:433 (1973).
59. Hughes, R. A. "Persistence of Toxaphene in Natural Waters," MS Thesis, University of Washington, Seattle, WA (1968).
60. Terriere, L. C. et al. *J. Agric. Food Chem.* 14:66 (1966).
61. Mayer, F. L. Jr. et al. EPA-IAG-0153, U.S. EPA, Washington, DC (1975).
62. Paris, D. F. et al. EPA-660/3-75-007, U.S. EPA, Washington, DC (1975).
63. Cohen, J. M. et al. *J. Am. Water Works Assoc.* 53:49 (1961).
64. Palmer, J. S., and R. D. Radcliff. *Ann. N.Y. Acad. Sci.* 111:729 (1964).
65. Kraus, A. cited by Mitchess, J. W. et al. *J. Animal Sci.* 5:226 (1946).
66. Seabury, J. H. *Arch. Environ. Health* 7:202 (1963).
67. Merna, J. W., and P. M. Eisele. EPA-R3-73-046, U.S. EPA, Washington, DC (1973).
68. Korn, S., and R. Earnest. *Calif. Fish Game* 60:128 (1974).
69. *Federal Register* 41(133):28402 (1976).
70. *Federal Register* 42(62):17143 (1977).
71. Public Law 92-500, October 18, 1972.
72. *Federal Register* 45(231):79318 (1980).

CHAPTER 2

ORGANIC COMPOUNDS IN RAW AND FINISHED WATERS

THE NATURE AND TYPES OF ORGANIC POLLUTANTS

Much concern has been expressed about the introduction and subsequent distribution of organic compounds throughout the environment, especially those from anthropogenic sources. This chapter examines the nature and types of organic pollutants in raw and finished waters. Some insight may be derived into the occurrence and distribution of organic compounds throughout various natural aquatic systems to determine whether they are hazardous or damaging to the environment as well as whether they need to be removed.

The synthetic organic compound and/or its derivative probably enters natural aquatic environments concurrently with the development of its initial manufacturing process. Waste disposal problems undoubtedly lead to its discharge into a river or some other body of water. There is some evidence in the scientific literature that potable, recreational, irrigational, fish and shellfish waters are contaminated with foreign organic compounds. Much of the early evidence was largely circumstantial as observed from physiological responses of various aquatic organisms. The advent of chromatographic separation procedures and of the confirmatory procedures, nuclear magnetic resonance and mass spectrometry, has led to positive identification of organic compounds in aquatic environments and in the attendant solid phases of bottom sediments.

Many 1974 newspaper headlines and television programs featured the "discovery" of low-molecular-weight halogenated hydrocarbons in drinking water [1]. It was charged that these organic compounds, especially trichloromethane (chloroform), were formed in the chlorination process for disinfection. Furthermore, it was suggested that some of these organic compounds may be physiologically harmful to humans because chloroform, for example, may be carcinogenic to laboratory animals, in this case, mice.

Numerous publications document, to some extent, the occurrence of organic compounds in aquatic environments [2]. These include the naturally occurring organics of color bodies [3], polynuclear aromatic hydrocarbons (PAH) [3] and synthetic compounds such as pesticides [4], detergents, polychlorinated biphenyls [5], petrochemicals and halogenated hydrocarbons. The sources and origins of these compounds are many, diverse and reasonably well known. Consideration is given here to some of the more significant compounds and their occurrences in potable water supplies.

OCCURRENCES

National Surveys

Prior to 1974, trace quantities of individual organic compounds had been isolated and identified from a finished municipal water supply on the Ohio River [6,7]. As seen in Table I, many of these compounds are the low-molecular-weight halogenated hydrocarbons. Also, a major organic contaminant was *bis*-(2-chloroisopropyl) ether. Its source was believed to be an industrial outfall located about 150 river miles upstream in the Ohio River.

In another survey, chloroform (94.0 μg/l), bromodichloromethane (20.8 μg/l), dibromochloromethane (2.0 μg/l) and ethyl alcohol were recovered and identified in a finished drinking water by gas-liquid chromatographic/mass spectrometric techniques [8]. Also, chloroform was found in the raw water source at a rather constant value (with time), 0.9 ± 0.2 μg/l. The difference in the chloroform content implies, of course, that it is formed in the water treatment plant from the chlorination process. In an effort to locate a source of these compounds, several grab samples of combined industrial and domestic wastewater were collected from a local treatment plant. These compounds were found: methylene chloride; chloroform; 1,1,1-trichloroethane; 1,1,2-trichloroethylene; 1,1,2,2-tetrachloroethylene; ethylene; and di- and trichlorobenzenes. Of course, some of these compounds could have industrial origins rather than being formed in the water. Again, bromo compounds were found for which no source was reported.

Similar results and compounds were reported for the drinking water of New Orleans, Louisiana [9,10]. In addition, five halogenated hydrocar-

Table I. Compounds Identifed from Finished Water [6,7]

Acetone	Dimethylsulfoxide
Acetophenone	Dinitrotoluene
Acetylenedichloride	Ethylbenzene
Benzene	Ethylenedichloride
Benzothiazole	Exo-2-camphanol
Bromobenzene	Hexachlorobenzene
Bromochlorobenzene	Hexachloroethane
Bromodichloromethane	Hydroxyadiponitrile
Bromoform	Isoborneol
Bromophenylphenylether	Isocyanic acid
Butylbenzene	Isopropanyl-isopropylbenzene
Camphanol	Isopropylbenzene
Camphor	*p*-menth-1-en-8-ol
Caprolactam	*o*-methoxyphenol
Carbon tetrachloride	2-Methoxybiphenyl
Clorobenzene	Methylbenzothiazole
Chlorodibromomethane	Methylbiphenyl
Chloroethoxyether	Methylchloride
Chloroethylether	Nitroanisole
Chloroform	Nitrobenzene
Chlorohydroxybenzophenone	Octane
bis-(2-chloroisopropyl) ether	Pentane
Chloromethylether	Propylbenzene
Chloronitrobenzene	Styrene
Chloropyridine	Tetrachloroethylene
Chloromethylethylether	Toluene
Dibromobenzene	Trichloroethane
Dichloroethane	Triglycodichloride
Dichloroethylether	Thiomethylbenzothiazole
Dimethoxybenzene	Vinylbenzene
Dimethylnaphthalene	Xlyene

bons were isolated from blood plasma pooled from eight humans: 1-chloropropene, chloroform, carbon tetrachloride, dichloroethane, trichloroethylene, dichloropropane, dichloropropene, bromodichloromethane, tetrachloroethylene and dibromochloromethane. Three isomers of dichlorobenzene were found also in the blood plasma. Speculation was offered that drinking water was the source of the halogenated hydrocarbons.

The reports cited above undoubtedly prompted the U.S. Environmental Protection Agency to conduct a nationwide survey of "the concentration and potential effects of certain organic chemicals in drinking water" [11]. This resulted in a National Organics Reconnaissance Survey (NORS) for chloroform, dibromochloromethane, bromodichloromethane and bromoform [the trihalomethanes (THM)] and carbon tetrachloride in 80 surface- and groundwater supplies throughout the United States. In general, these five compounds were not detected in all of the raw waters. However, many finished waters contained one or more of the four THM. Consequently, it was concluded that their occurrence was caused by chlorination practices and their concentrations were related to the organic content of the raw water. When detected, the THM were present at less than 4.0 μg/l in the raw water.

Results of the NORS study for finished waters are summarized below and in Figure 1 [12]:

Results of NORS Study for Finished Waters

Compound	Number of Cities with Positive Results	Concentration (μg/l)		
		Minimum	Median	Maximum
Chloroform	80	<0.1	21.0	311
Bromodichloromethane	78	0.3	6.0	116
Dibromochloromethane	72	<0.4	1.2	110
Bromoform	26	<0.8	[a]	92
Carbon Tetrachloride	10	<2.0		3

[a]93.8% of 80 cities had ≤5 μg/l bromoform.

where the frequency distributions are given for the four THM. Generally, the following conditions produce the highest concentrations of THM: surface water as the source, prechlorination and the presence of more than 0.4 μg/l of free chlorine. Obviously, this study focused on the relevant water treatment practices producing the THM.

As part of the NORS of drinking water supplies, there was also a detailed study of the occurrence of volatile organics in drinking waters of five cities: Miami, Seattle, Ottumwa (Iowa), Philadelphia and Cincinnati [13,14]. Altogether, there were 72 volatile compounds found in these finished waters with the following common to five cities [14]:

acetaldehyde
bromodichloromethane
2-butanone
chlorodibromomethane
chloromethane
dichloromethane
ethanol
3-methylbutanal
3-methyl-2-butanone
2-methylpropanal
propanal
2-propanone (acetone)
trichloromethane

Their concentrations and the standard deviations of the organics are seen in Table II [13]. It is doubtful that these compounds resulted from chlorination practices; rather, they may have originated from anthropogenic sources and were not removed by treatment.

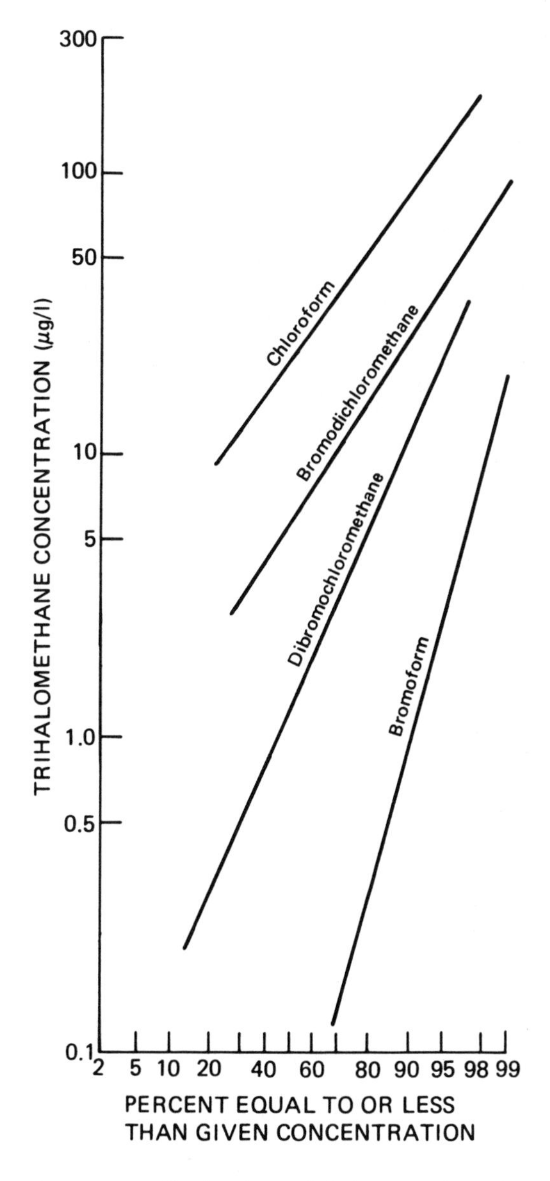

Figure 1. Frequency distribution of halogenated methane concentrations found in the NORS study of drinking water in 80 U.S. cities [12].

Trihalomethanes

Sources for the occurrence of trihalomethanes in raw and finished waters are many and diverse. For example, there is a relation between

Table II. Concentrations and Standard Deviations of Selected Organics from Five-City Survey[a]

Compound	Cincinnati, OH		Miami, FL		Ottumwa, IA		Philadelphia, PA		Seattle, WA
	Concentration by GCMS (μg/l)	Standard Deviation (μg/l)	Concentration by GCMS (μg/l)	Standard Deviation (μg/l)	Concentration by GCMS (μg/l)	Standard Deviation (μg/l)	Concentration by GCMS (μg/l)	Standard Deviation (μg/l)	Concentration by GCMS (μg/l)
Benzene	0.3	0.1	<0.1	—	<0.1	—	0.2	0.1	[b]
Chlorobenzene	0.1	0.04	1	0.5	[b]		<0.1	—	[b]
Chloroethene (vinyl chloride)	[b]	—	[c]	—	[b]	—	[c]	—	[b]
1,1-Dichloroethene (vinylidene chloride)	[b]	—	0.1	0.2	[b]	—	<0.1	—	[b]
cis-1,2-Dichloroethene	<0.1	—	12	2	[b]	—	0.1	0.1	[b]
trans-1,2-Dichloroethene	[b]	—	1	0.2	[b]	—	[b]	—	[b]
Nitrotrichloromethane (chloropicrin)	3	2.3	0.4	0.1	[b]	—	2	2	[b]
Tetrachloroethane	0.3	0.2	<0.1	—	0.2	0.2	0.4	0.3	[b]
Toluene	<0.1		[b]	—	[b]	—	0.7	0.4	[b]
Trichloroethene	0.1	0.1	0.3	0.1	<0.1	—	0.5	0.2	[b]

[a] Reproduced from Lingg et al. [13], courtesy of the American Water Works Association.
[b] Not detected.
[c] Not quantified by GCMS.

periods of high agricultural runoff and peaks in THM concentrations in finished waters from the Iowa River at Iowa City [15]. This is seen in Table III, where there is a direct correlation between rising river water turbidity and chloroform content. A peak occurred at 230 μg/l $CHCl_3$ and 450 JTU of turbidity. This observation indicates very strongly that organic materials from agricultural lands serve as precursors for formation of THM. Similar results were found in the Occoquan Watershed in Virginia [16].

Another positive relation occurs between total organic carbon (TOC) of raw and finished water supplies and the occurrence of the four THM [17]. Two recovery and separation techniques for THM in water were compared in this study; namely, liquid-liquid extraction (LLE) and the purge-and-trap method. It is readily apparent from the data in Table IV that, in general, the purge-and-trap method yielded higher TTHM and especially chloroform contents than the LLE. In addition, the finished

Table III. Halogenated Methanes in Iowa Finished Waters[a]

Turbidity Jtu	Haloforms (μg/1)			
	$CHCl_3$	CCl_4	$CHCl_2Br$	$CHClBr_2$
5	55	0.4	[b]	[b]
4	42	0.1	[b]	[b]
300	159	0.5	15	2
400	196	0.2	20	4
450	230	1.3	19	5
300	150	0.7	11	[c]
110	90	0.3	13	4
130	116	0.2	18	3
25	114	0.8	12	[c]
25	137	0.8	15	7
30	155	0.7	14	4
33	175	0.9	14	4
26	114	1.1	16	7
23	127	0.2	14	5
150	122	0.7	10	8
50	162	1.9	13	8
26	117	0.6	11	4
30	122	0.6	13	4
25	120	0.3	14	6

[a]Reproduced from Morris and Johnson [15], courtesy of the American Water Works Association.

[b]None detected.

[c]Not analyzed.

Table IV. Trihalomethane and Total Organic Carbon Levels in Raw and Finished East Texas Surface Waters[a]

System	Sample Water	Liquid-Liquid Extraction Method ($\mu g/l$)					Purge-and-Trap Method ($\mu g/l$)					Total Organic Carbon in Raw Water (mg/l)
		$CHCl_3$	$CHCl_2Br$	$CHClBr_2$	$CHBr_3$	TTHM	$CHCl_3$	$CHCl_2Br$	$CHClBr_2$	$CHBr_3$	TTHM	
12	Raw	1.2 ± 0.1				1.2 ± 0.1	1.1				1.1	7.20 ± 0
	Finished	129.0 ± 4.0	89.0 ± 2.0	39.8 ± 1.7	2.02 ± 0.25	259.0 ± 8.0	170.6	20.8	1.5		192.9	
13	Raw	1.2 ± 0.1				1.2 ± 0.1	1.3	0.5	3.5		5.3	11.95 ± 0.35
	Finished	215.3 ± 0.1	37.0 ± 0.2	4.2 ± 0		256.5 ± 0.3	270.8	36.7	3.1		310.6	
14	Raw	0.4 ± 0.1				0.4 ± 0.1	5.6	1.0			6.6	7.80 ± 0.10
	Finished	823.0 ± 19.0	71.0 ± 2.0	11.5 ± 0.6		905.0 ± 22.0	664.4	48.2	7.3		719.9	
15	Raw	1.6 ± 0.1				1.6 ± 0.1	1.3	0.3			1.6	11.75 ± 0.15
	Finished	225.0 ± 3.0	48.7 ± 0.7	6.0 ± 0.2		280.0 ± 4.0	269.9	48.3	6.5		324.7	
16	Raw	1.3 ± 0.4				1.3 ± 0.4	12.9				12.9	10.75 ± 0.05
	Finished	207.0 ± 6.0	63.0 ± 2.0	10.4 ± 0.2		280.0 ± 8.0	254.8	54.6	5.2		314.6	
17	Raw	1.6 ± 0.1				1.6 ± 0.1	2.3	2.9			6.2	2.90 ± 0
	Finished	68.8 ± 1.3	42.6 ± 0.3	31.2 ± 0.5	3.54 ± 0.25	146.1 ± 2.3	70.3	48.0	34.2		152.5	
18	Raw	1.7 ± 0				1.7 ± 0	2.3				2.3	4.65 ± 0.05
	Finished	19.6 ± 0.7	58.2 ± 1.6	125.0 ± 3.0	52.5 ± 1.5	255.2 ± 6.8	17.2	57.8	183.7	40.7	299.4	
19	Raw	1.3 ± 0.4				1.3 ± 0.4	0.7				0.7	5.65 ± 0.25
	Finished	61.2 ± 0.6	49.1 ± 0.5	41.2 ± 0.4	5.05 ± 0.25	156.5 ± 1.7	60.8	49.3	37.9		148.0	
20	Raw	1.9 ± 0.2				1.9 ± 0.2	26.4	57.3	6.9		90.6[c]	9.00 ± 0.10
	Finished	337.8 ± 1.1	44.1 ± 1.5	5.4 ± 0.2		387.3 ± 2.8	382.7	49.5	5.8		438.0	
21	Raw						5.4	0.7			6.1	9.85 ± 0.05
	Finished	81.2 ± 1.3	36.0 ± 0.7	12.9 ± 0.2		130.1 ± 2.2	109.8	35.9	10.0		155.7	
22[b]	Raw								not analyzed			7.90 ± 0

	Finished	264.0 ± 7.0	60.0 ± 1.3	9.0 ± 0.2		333.0 ± 8.0					
23	Raw	1.6 ± 0.5				1.6 ± 0.5	1.0	1.0		2.0	9.85 ± 0.05
	Finished	171.6 ± 0.6	30.8 ± 0	3.1 ± 0.2		205.5 ± 0.8	204.2	30.6	3.3	238.1	
24	Raw						3.6			3.6	15.80 ± 0.10
	Finished	882.0 ± 37.0	40.0 ± 2.0	trace		922.0 ± 39.0	814.8	32.9		847.7	
25[b]	Raw	0.5 ± 0				0.5 ± 0			not analyzed		11.12 ± 0.02
	Finished	144.3 ± 1.1	62.1 ± 0.8	27.1 ± 0	1.77 ± 0	235.3 ± 1.9					

[a]Reproduced from Glaze and Rawley [17], courtesy of the American Water Works Association.
[b]Sampled in May and June.
[c]The purge-and-trap results for system 20 are probably in error and indicate a contamination of the sample.

waters (Table IV) had higher THM values than the raw waters. This is due, of course, to the chlorination practices of the raw surface waters containing mg/l quantities of TOC. Similar data were obtained in an attempt to establish a seasonal variation in the THM contents of 14 finished waters in Iowa [18]. A one-year survey of THM contents in the distribution system of Southampton, England, provides additional evidence for the positive relation between chlorine residual and organic precursors for formation of THM [19].

That there are temporal variations in the THM contents of finished water is seen in Figure 2. There are daily fluctuations in THM values where the highest contents are observed in the morning portion of a 24-hr period [20]. These results are explained as a "real phenomenon" rather than an experimental variation. There is the probability that, during nocturnal hours, THM are formed due to lower water demands and are flushed out of the distribution system during the morning hours. Another explanation is that the organic content of the finished water varies considerably in the treatment process.

Seasonal variations in THM levels were reported as occurring from 50 to 335 $\mu g/l$ with a mean value of 173 $\mu g/l$ for the Iowa River [21]. This results from the variable nature of organic precursors and not from fluctuating pH values and temperatures. The average annual contribution of the individual THM to the total content were: $CHCl_3$, 73.1%, $CHBrCl_2$, 21.1%, $CHBr_2Cl$, 5.0% and $CHBr_3$, 0.8%. Figure 3 shows these trends for THM and chloroform.

A statewide survey of New York public water supply systems was conducted for the four THM as well as trichloromethylene, tetrachloroethylene, 1,1,1-trichloroethane and carbon tetrachloride [22]. The frequencies of detection for 235 grab samples were 89% for chloroform, 93% for bromodichloromethane, 57% for chlorodibromomethane and 0.47% for bromoform (Table V). These are somewhat different than the frequencies of occurrence for the Iowa River [21]. The mean and maximum for the four compounds were: 62.7 and 430 $\mu g/l$ for chloroform, 10.2 and 60 $\mu g/l$ for bromodichloromethane, 3.0 and 51 $\mu g/l$ for chlorodibromomethane, 0.3 and 38 $\mu g/l$ for bromoform and 76.2 and 443 $\mu g/l$ for THM. This survey also indicated that the four other low-molecular-weight halogenated hydrocarbons were detected in only two of the supplies tested. This and other surveys for THM in public water supplies leave little doubt about their ubiquitous distribution in raw and finished waters.

Anthropogenic Compounds

Early studies of the occurrence of synthetic organic compounds in raw waters were concerned with identifying the causes of tastes and odors

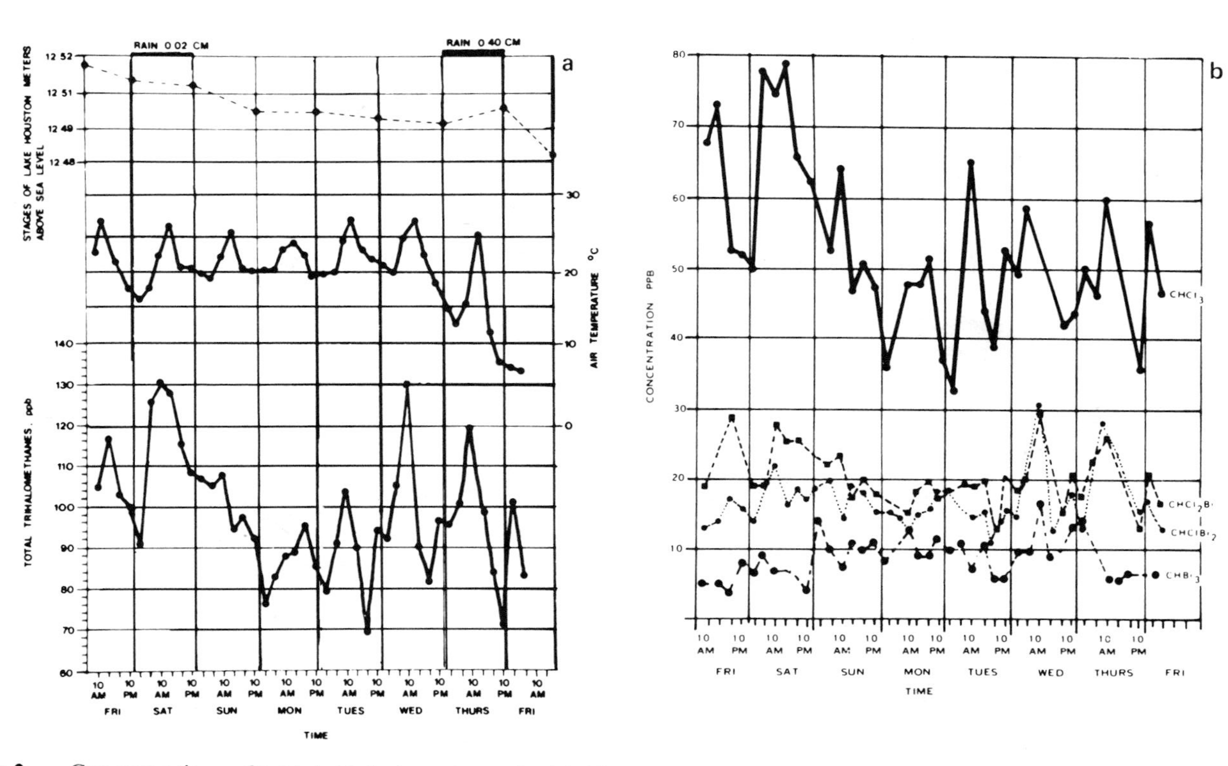

Figure 2. Concentrations of **(a)** total trihalomethanes in drinking water (and ambient conditions in the river basin of origin); and **(b)** $CHCl_3$, $CHCl_2Br$, $CHClBr_2$ and $CHBr_3$ in drinking water. Reproduced from Smith et al. [20], courtesy of the American Chemical Society.

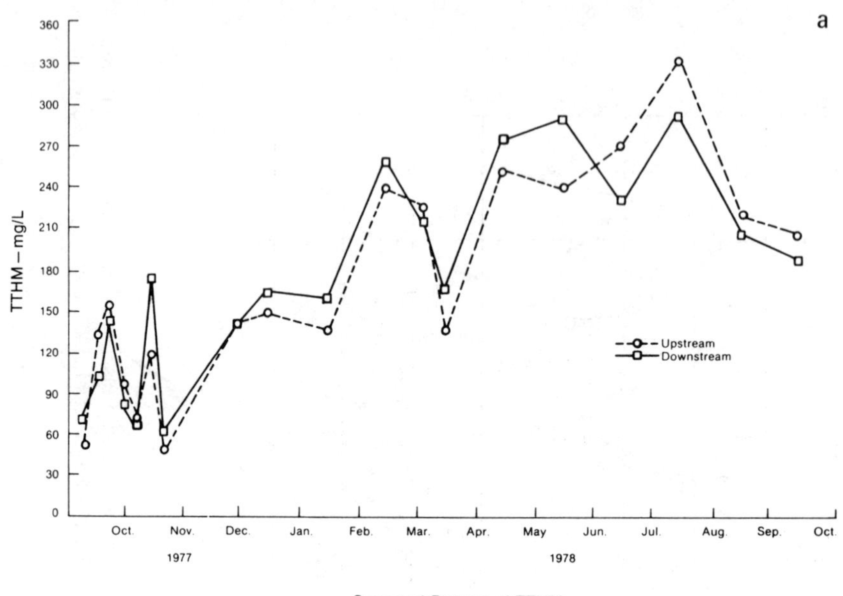

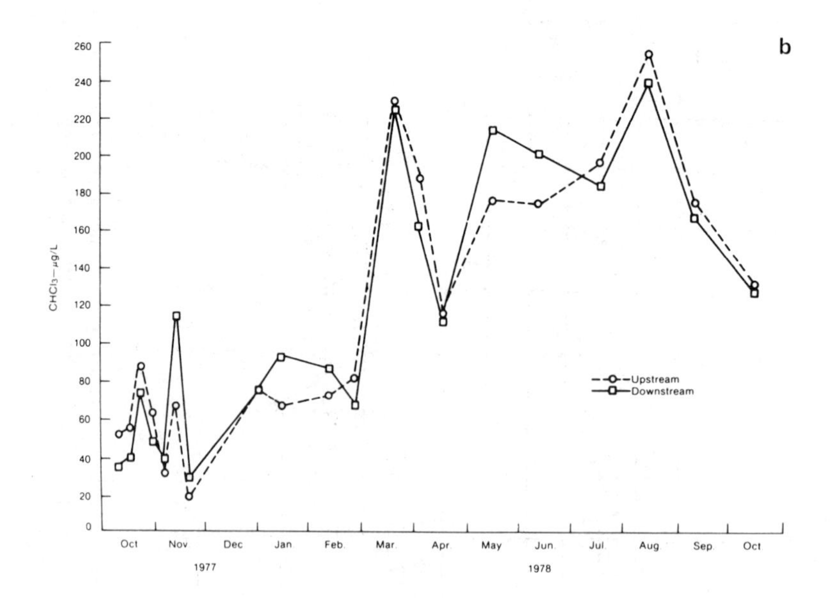

Figure 3. Seasonal patterns of **(a)** TTHM and **(b)** $CHCl_3$ in the Iowa River. Reproduced from Veenstre and Schnoor [21], courtesy of the American Water Works Association.

Table V. Summary of TTHM Data for the New York Water System (μg/l)[a,b]

Source Type	Number of Samples	Chloroform	Bromodichloro-methane	Chlorodibromo-methane	Bromoform	TTHM
Rivers						
Range	32	< 5–266.0	<2–60.0	<2–32.0		<14–266.0
Mean-Total		79.1	14.0	4.7	<5	97.8
Filtered[c]	25	80.2	14.3	5.0	<5	99.5
Unfiltered	7	75.0	13.1	3.9	<5	92.0
Reservoirs						
Range	35	15.0–349.0	<2–27.0	<2–5.0		18.0–369.0
Mean-Total		98.1	8.7	0.9	<5	107.7
Filtered	10	87.0	10.7	1.5	<5	99.2
Unfiltered	25	102.5	7.8	0.7	<5	111.0
Brooks and streams						
Range	58	<5–430.0	<2–28.0	<2–10.0		<14–443.0
Mean-Total		64.0	9.0	1.8	<5	74.8
Filtered	17	72.4	12.9	2.7	<5	103.3
Unfiltered	41	60.5	7.4	1.4	<5	63.0
Lakes						
Range	86	<5–176.0	<2–44.0	<2–51.0	<5–38.0	<14–221.0
Mean-Total		54.7	11.8	4.2	0.8	71.5
Filtered	41	49.0	13.0	5.5	1.6	69.2
Unfiltered	45	59.9	10.8	3.0	<5	73.7
Mixed						
Range	11	<5–109.0	<2–16.0	<2–5.0		<14–119.0
Mean-Total		31.8	6.6	1.6	<5	40.1
Filtered	4	31.8	8.5	3.0	<5	43.3
Unfiltered	7	31.9	5.5	0.8	<5	38.3

Table V. continued

Source Type	Number of Samples	Chloroform	Bromodichloro-methane	Chlorodibromo-methane	Bromoform	TTHM
Ground						
Range	13	<5–8.0	<2–8.0	<2–7.0		<14–15.0
Mean-Total		1.0	2.8	2.7	<5	6.5
Filtered	3	<5	3.3	2.0	<5	5.3
Unfiltered	10	1.3	2.6	2.9	<5	6.8
Total						
Range	235	<5–430.0	<2–60.0	<2–51.0	<5–38.0	<14–443.0
Mean-Total		62.7	10.2	3.0	0.3	76.2
Filtered	100	65.0	12.6	4.3	0.7	82.6
Unfiltered	135	61.0	8.4	2.1	<5	71.5

[a] All values less than the detection level were included as 0.0 for statistical purposes.
[b] Reproduced from Schreiber [22], courtesy of the American Water Works Association.
[c] Sand filtration.

[23-26] (see Chapter 3). Riverwaters were processed by the carbon adsorption method that, subsequently, was air-dried and extracted successively with chloroform and alcohol. Most attention was given to the chloroform extract that was separated into solubility groups: ether-insolubles, water-solubles, weak acids, and strong acids, bases and neutrals. The neutral group was separated by chromatographic procedures into aliphatic, aromatic and oxygenated compounds whereupon infrared spectra were run on selected groups. Some of the identified organics were: (a) DDT in the Mississippi, Missouri, Columbia and Detroit Rivers; (b) aldrin in the Snake River at Pullman, Washington; (c) *o*-nitrochlorobenzene in the Mississippi River at New Orleans in the concentration range of 1-2 μg/l; and (d) phenyl ether in the St. Clair, Kanawha and Ohio Rivers. These were, perhaps, the first reports in the United States that identified organic contaminants in natural waters.

Macroreticular resins can be used to extract organic compounds from raw waters [27,28]. These adsorbents are copolymers of polystryrene divinylbenzene (DVB) or polymethacrylate that are cross-linked with a suitable nonaromatic material [29]. These resins have considerably larger pore sizes than GAC. In this respect, they are better adsorbents than carbon.

Table VI shows a list of neutral organic compounds recovered from a water supply [15]. This was a unique situation in that this was groundwater contamination with organics of an industrial origin of many years' duration. At that time, an undesirable organoleptic water quality was the primary concern. The probable source of contamination was a pit used for the disposal of coal-tar residues from a coal-gas plant that operated in the 1920s.

Organic Pesticides

Much concern has been expressed about the occurrence of organic pesticides in drinking water supplies. The National Primary Drinking Water Regulations [30] have standards for four insecticides and two herbicides as seen in Table VII (see also Chapter 1). Maximum contaminant levels (MCL) were not promulgated for dieldrin, DDT, DDE, DDD, heptachlor, aldrin, organophosphates or carbamates because their registration labels have been banned, or they are relatively unstable in water. It is relevant, however, to review some historical data from the occurrence of organic pesticides in surface waters of the United States.

A National Pesticide Monitoring Program was initiated in the late 1960s to establish a network to survey major rivers in the United States

Table VI. Neutral Compounds in an Ames, Iowa, Contaminated Well [27]

Name of Component	Identification	Concentration (ppb)	Standard Deviation
Acenaphthylene	a,b,c,	19.3	1.4
1-Methylnaphthalene	a,b,d	11.0	0.6
Methylindenes (two isomers)	c,f	18.8	0.8
Indene	a,b	18.0	1.5
Acenaphthene	a,b	1.7	0.2
2-2-Benzothiophene	a,b	0.37	0.11
Isopropylbenzene	a		
Ethyl benzene	a		
Naphthalene	a		
2,3-Dihyroindene	e	15.	
Alkyl-2,3-dihyroindene	e,f		
Alkyl benzophiophenes	e,f		
Alkyl naphthalene	e,f		

[a] Identification was verified by comparison of retention time and mass spectrum with an authentic sample.
[b] Identification was verified by comparison of the ultraviolet spectrum with an authentic sample.
[c] Identification was verified by comparison of the proton magnetic spectrum with an authentic sample.
[d] Identification was verified by comparison of the infrared spectrum with an authentic sample.
[e] Identification based on mass spectral data alone.
[f] Knowledge of the exact positional isomer was not important for this work. This could be done by proton magnetic resonance if needed.

[31]. Initially, 39 rivers were selected for monthly sampling. In the early 1970s, the program was revised to quarterly samples of water as well as bed sediments of 161 sites in the conterminous United States, Alaska, Hawaii and Puerto Rico [32]. These pesticides were sought: 11 chlorinated hydrocarbons, 2 organophosphates and 3 phenoxyherbicides.

A five-year (1964–1968) summary of the pesticidal monitoring represents, perhaps, the "peak" contents in major U.S. rivers [33]. Several chlorinated hydrocarbons were found (in order of decreasing frequency): dieldrin, endrin, DDT, DDE, DDD, aldrin, heptachlor, heptachlor epoxide, lindane and BHC, and chlordane. The highest concentration was 0.407 μg/l of dieldrin in the Tombighee River at Columbia, Mississippi. Six organophosphorus pesticides were sought, but only two were found in one sample in the Snake River at Wawawai, Washington: parathion, 0.050 μg/l, and ethion, 0.380 μg/l. Table VII shows a comparison of the maximum pesticide concentrations with the primary

Table VII. Maximum Pesticide Concentration Found vs Primary Drinking Water Standards[a]

Pesticide	Proposed Standards (μg/l)	Maximum Concentration Found[e] (μg/l)
Dieldrin	[d]	0.407
Endrin	0.2	0.133
DDT	[d]	0.316
DDE		0.050
DDD		0.840
Heptachlor	[d]	0.048
Heptachlor Epoxide	[d]	0.067
Aldrin	[d]	0.085
Lindane (BHC)	4.0	0.112
Chlordane	[d]	0.169
Methoxychlor	100.0	[b]
Toxaphene	5.0	[c]
Oragnophosphates plus Carbamates	[d]	0.380
Herbicides		
2,4-D	100.0	[b]
2,4,5-TP	10.0	-

[a]After Reference 30.
[b]Not determined.
[c]Not detected.
[d]None recommended.
[e]After Reference 33.

drinking water standards found in this five-year survey. All concentrations are far below these permissible standards.

A survey of the Black Creek Watershed (Indiana) for pesticides and PCB residues was conducted in 1977–1978 for water and sediment [34]. Two of seven water samples contained PCB at 0.4 and 0.2 μg/l. Only the herbicide 2,4,5-T was found in surface water samples at contents of 0.2–7.7 μg/l. Atrazine, alachlor, carbofuran and malathion were not detected in any water or sediment sample. This survey undoubtedly reflects the current nationwide occurrence of organic pesticides in natural aquatic environments.

Polychlorinated Biphenyls

Some anxiety has been expressed about the occurrence and widespread distribution of the somewhat persistent polychlorinated biphenyl (PCB) compounds in aquatic environments. These contaminants are, perhaps, more ubiquitous than the chlorinated hydrocarbon pesticides. In fact, many PCB are positive interferences with gas-liquid chromatographic procedures for the analytical separation of chlorinated hydrocarbons [34]. Consequently, many of the early reports on the global distribution of pesticides may have been, in fact, referring to PCB.

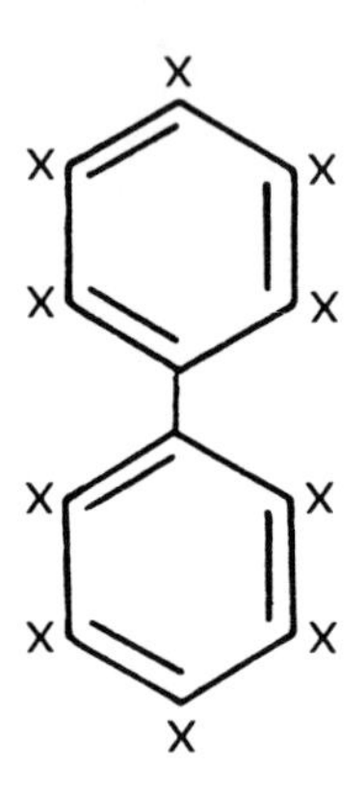

X = sites for Cl substituent

Surprisingly few reports are in the scientific literature, to date, about the occurrence of PCB in aquatic environments. The distribution of Aroclor® 1254 was investigated for the water, sediment and biota of Escambia Bay, Florida [36]. Most of this PCB resided in some sort of solid phase, i.e., fish, crustacean or sediment, since the contents in water were less than 0.1 μg/l. It was not detected at some sampling stations in the water phase. However, sediment samples taken near a wastewater outfall reached a high content of 486 ppm. In an effort to develop analytical methodology for PCB in water, contents in untreated and treated tap water were reported as: 0.50 and 0.33 ppt, respectively [37].

In the 1971–1972 National Pesticides Monitoring Program for the distribution of PCB in various aquatic environments of the United States, residues were detected in water samples from 19 states [5]. PCB concentrations ranged from 0.1 to 4.0 μg/l in unfiltered water samples and from 5.0 to 3200 μg/kg in bottom sediments. Residues were found also in fish and aquatic plants. This report provides strong evidence for the ubiquitous occurrence and distribution of PCB. In addition, PCB were found in eight sampling stations on Lake Ontario. Concentrations in water ranged from 35 to 97 μg/l [38].

Reverse osmosis was used to concentrate organic compounds from the drinking water of Cincinnati, Ohio [39]. Approximately 460 compounds were identified in a tap water extract that included 41 polynuclear aromatics (PNA), 15 PCB and a number of amines, amides and halogenated species.

In Other Environments

The occurrence and distribution of halogenated hydrocarbons in the "whole" environment from several researchers indicate that there was agreement on the universal distribution of fluorochloromethane, tri- and perchloroethylene, chloroform and carbon tetrachoride in the atmosphere at concentrations normally in the range between 1 and 10 ng/l [40]. The concentrations of $CHCl_3$ and CCl_4 were "surprising" because of their "limited losses" to the atmosphere [41]. It was suggested that these two compounds occur naturally by reactions between chlorine and methane in the atmosphere. Analyses of rainwater, rivers, municipal water supplies and seawater indicated that tri- and perchloroethylene, trichloroethane, chloroform and carbon tetrachloride are widely distributed at the 10^{-9} level or lower [40].

Some information is available also on the occurrence of chlorinated hydrocarbons in animal tissue, plant tissue, foodstuffs of man and human tissue. A summary of these surveys is presented in Table VIII. It appears that the above-mentioned compounds are distributed widely in the environment and at analytically significant concentrations. Two main conclusions were drawn from the background data for Table VIII: "The concentrations of DDT and PCB in fatty tissues are at least three orders of magnitude greater than those of the industrial solvents; and (b) Chloroform and carbon tetrachloride are very widely distributed and at unexpectedly high concentrations" [40].

Trichlorofluoromethane (CCl_3F) was detected in groundwaters at three locations in the United States [41]. CCl_3F is an entirely manmade compound, i.e., "Freon 11." It was produced commercially first in the United States in 1931 and has been used as a propellant in spray cans and in refrigeration processes. It was proposed that this compound could be used as a tracer and indicator of groundwater age because of its unique atmospheric history. The age relation comes from the fact that precipitation, exposed to CCl_3F in the atmosphere, will accumulate an amount proportional to the atmospheric concentration. A portion of this water infiltrates into the subsurface groundwater where it can be differentiated from older groundwater. This is all very well; however, this study does demonstrate that halogenated hydrocarbons may have been more widely

Table VIII. Occurrence of Chlorinated Hydrocarbons in the Environment: Typical Concentrations (w/w) of the Five Major Compounds (Chloroform, Carbon Tetrachloride, Trichloroethylene, Perchloroethylene, Trichloroethane)[a]

	Minimum	Maximum
Air	10^{-9}	10^{-8}
Rainwater	10^{-11}	10^{-9}
Surface water	10^{-11}	10^{-9}
Potable water	10^{-11}	10^{-9}
Seawater	10^{-10}	10^{-9}
Marine sediments	10^{-10}	10^{-9}
Marine invertebrates	10^{-9}	10^{-8}
Fish	10^{-9}	10^{-8}
Waterbirds	10^{-9}	$>10^{-7}$
Marine mammals	10^{-9}	$>10^{-8}$
Fatty foods	10^{-9}	10^{-8}
Nonfatty foods	10^{-9}	10^{-9}
Human organs	10^{-9}	10^{-9}
Human body fat	10^{-9}	10^{-8}

[a]Reproduced from McConnell [40], courtesy of *Endeavor*.

distributed and for a longer period of time than previously thought. The highest CCl_3F content reported was 36 ppt in a municipal well water in Texas.

FORMATION OF HALOGENATED HYDROCARBONS

General Reactions

One of the first reports for formation of THM in natural surface waters came from chlorination practices in the Rotterdam Waterworks [42]. Gas-liquid chromatographic and mass spectrometric techniques were employed to identify $CHCl_3$, CCl_4, $CHCl_2Br$, $CHClBr_2$ and $CHBr_3$ and traces of CH_2Cl_2, CH_2Br_2 and $C_2H_2Cl_2$. Some quantitative data are given in Table IX from laboratory experiments. The observation of these compounds led to two perplexing questions: (1) What is the precursor? and (2) What is the source of the bromine? For the latter question, no satisfactory explanation was given, but two possibilities exist: (1) contamination of the chlorine gas with elemental bromine, and (2) formation of Br_2 from naturally occurring Br^- in the surface water during chlorination.

Table IX. Haloforms Produced in Chlorination of Stored Surface Waters[a]

Ammonia Content (mg NH_4^+/l)	Chlorine Dose (mg/l)	pH During Reaction	Temperature (°C)	Inorganic Bromide (μg/l)	Color (mg Pt/l)	Concentration of Haloforms (μg/l)			
						$CHCl_3$	$CHCl_2Br$	$CHClBr_2$	$CHBr_3$
3.8	25	7.8–7.4	4.5	360	25	54.0	20.0	13.3	2.0
2.6	17	8.3–7.3	7	210	27	16.5	12.4	9.7	10.0
0.08	4	9.5–8.9	12	150	16	16.4	4.2	2.6	1.1
0.60	9	9.1–8.6	17	160	16	44.1	17.9	8.0	1.3
0.18	6	9.6–9.2	21	150	15	41.5	7.3	2.8	2.8
0.05	4	9.6–9.3	20	140	14	52.0	7.8	2.8	8.7
0.05	3	9.4–9.1	19	140	10	20.3	4.4	1.7	0.5
0.06	4	8.7–8.2	15.5	150	10	11.9	6.7	5.2	7.2
0.12	3	9.1–8.7	6	120	7	6.0	4.3	3.0	—
0.35	6.5	8.8–8.7	4.5	130	9	9.5	7.5	4.6	7.8
0.12	2	8.9–8.6	6	110	6	10.0	4.2	3.6	1.5

[a]Reproduced from Rook [43], courtesy of the Society for Water Treatment.

Rook's search for the precursor to the haloforms was more fruitful [43]. The possibility for acetone (0.04 mg/l) to serve in the reaction was confirmed by experiments in which $CHCl_3$, $CHCl_2Br$, $CHClBr_2$ and $CHBr_3$ were found after two-hour contact with 0.02 mg/l Cl_2 and 1.4 mg/l NaBr. Attempts to find acetone in the natural riverwaters at concentrations high enough (2–9 mg/l) to yield significant quantities of the haloforms were unsuccessful. Instead, compelling evidence was presented that the polyhydroxybenzene building blocks of natural color molecules are responsible for the haloform reaction, since there is a good correlation between $CHCl_3$ formation and color intensity. These four compounds were formed from the chlorination of several polyhydric phenols: pyrogallol; phloroglucinol; the *o*-, *m*- and *p*-dihydroxybenzenes and from two natural polyphenols—hesperidine and phlorizine.

Additional research into the supposition that organic color molecules are precursors in the formation of THM indicated that, under ordinary conditions of chlorination, quantitative yields of the haloform reaction were always very low [43,44]. For example, the maximum content of $CHCl_3$ was 50 and 20 μg/l for brominated compounds (0.6 μM of total haloforms). As a result, there was speculation that there should be a relatively large concentration of a naturally persistent organic compound from which 0.2% of its carbon reacts to produce THM. "The big molecules of the yellow acids or fulvic acids" provided this requirement. (See Reference 2 of Chapter 1 for discussion of the chemistry of natural organic color molecules.) Furthermore, it was reasoned that the haloform reaction would be site specific. That is, a number polyhydroxy aromatic rings and diketo alicyclic rings (degradation products of fulvic acids) was tested for $CHCl_3$ formation upon chlorination. Millimolar solutions of various organics were treated with 6–8 mM chlorine for four hours at 10°C. The $CHCl_3$ yields are given in Table X. "It became clear that an aromatic ring with two OH^- groups in the meta position, as well as cyclohexane rings with the configuration:

```
   O   H   O
   ‖   |   ‖
 — C — C — C —
       |
       H
```

(e.g., indandione and dimedon, 100% yields at pH 11.0) are well suited for the THM reaction via enolization to

Table X. Yields of Chloroform from Various Compounds (%) 4-hr Chlorination at 10°C[a]

Substance	pH 7.5	pH 11
Resorcinol[b]	75	80
Catechol[c]	Trace	5
Hydroquinone[d]	Trace	26
Pyrogallol	Trace	5
Phloroglucinol	–	1.5
1,3-Indandione	58	100
Dimedone[e]	85	100
1,3-Cyclohexanedione	50	71
1,3-Cyclohexanediol	0	0
1,2-Cyclohexanedione	0	0
1,4-Cyclohexanedione	0	14
Phenol	–	1
Acetone	Trace	6.6
Ethanol	0	0

[a]Reproduced from Rook [44], courtesy of the American Water Works Association.
[b]Meta-dihydroxybenzene.
[c]Ortho-dihydroxybenzene.
[d]Para-dihydroxybenzene.
[e]5,5-dimethylcyclo 1,3-hexanedione.

```
  H
  |
  O  H
  |  |
—C=C—
```

"whereupon the HOCl or HOBr would be added to the double bond."

One of the first studies under practical plant conditions reported the rate of formation of THM [44]. These results are seen in Figure 4. That low chlorine doses, as applied in postchlorination, produce measurable amounts of THM is seen in Figure 4a (precursor content expressed as TOC). These curves also indicate that the reaction is rapid enough to occur within the detention time of a treatment plant or in a distribution system. The pH value is certainly a factor in THM formation from fulvic acid. This is seen in Figure 4b where the $[CHCl_3]$ is increased as the $[H^+]$ is decreased, especially when the pH value is greater than 8. Formation of the phenoxide ion, $C_6H_5^-O$, at the alkaline pH values apparently

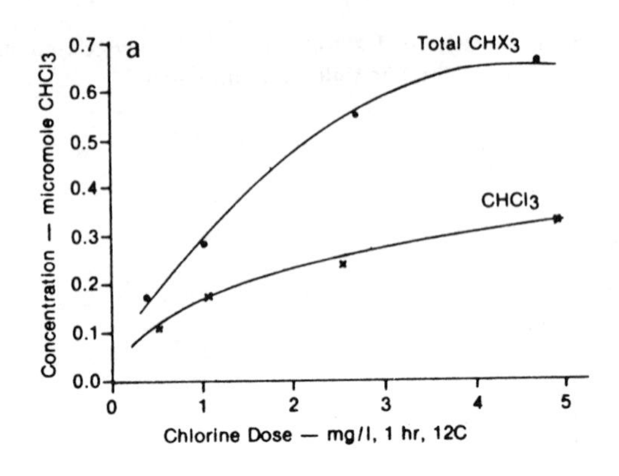

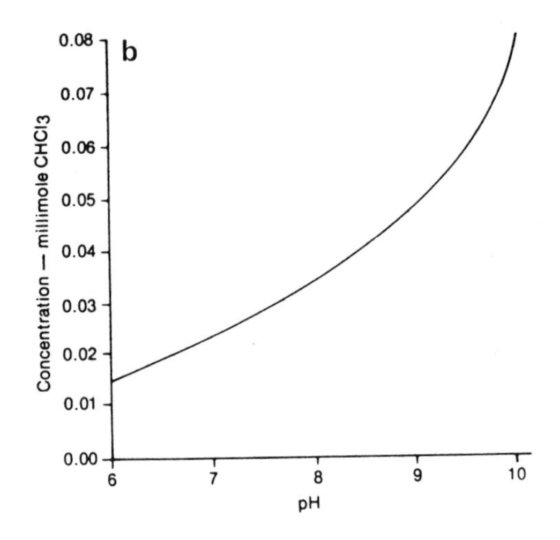

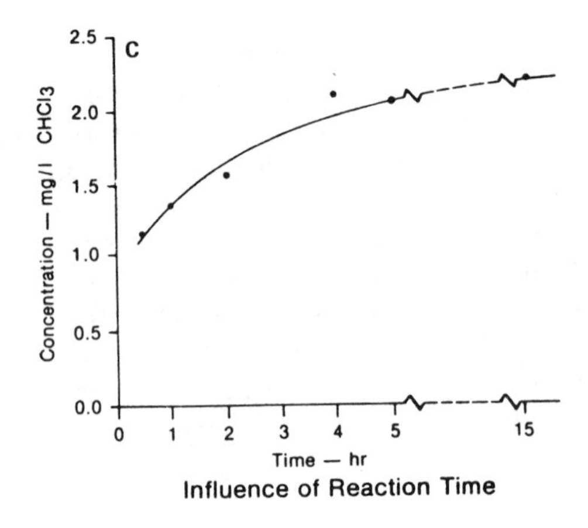

Influence of Reaction Time

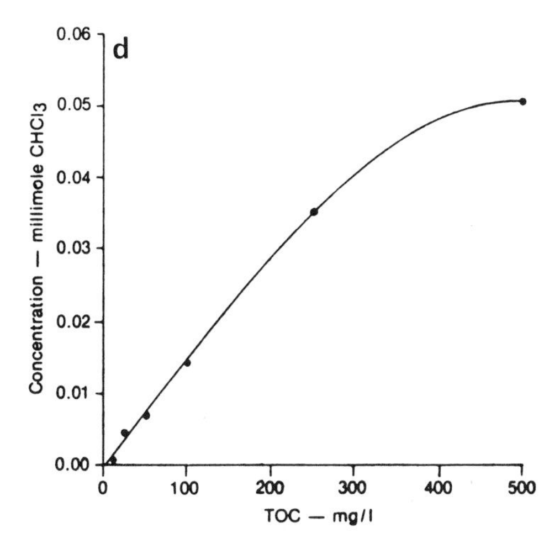

Figure 4. **(a)** Formation of haloforms in water with low organic matter contents. TOC = 3.6 mg/l. **(b)** Influence of pH on haloform reaction. Test conditions: 4 hr; 10°C; TOC, 250 mg/l; Cl_2, 870 mg/l. **(c)** Influence of reaction time. Test conditions: 10°C; Cl_2, 800 mg/l; fulvic acids, 500 mg/l. **(d)** Haloform reaction with increasing fulvic acid precursor content. Test conditions: Cl_2, 885 mg/l; 10°C; 4 hr. Reproduced from Rook [44], courtesy of the American Water Works Association.

increases the yield of $CHCl_3$ (see below and Rook [45] for explanation). Figure 4c shows that the THM reaction is time-dependent. Under these conditions the reaction was considered to be essentially complete within four hours. Using the TOC value as a measure of precursor content, $CHCl_3$ production from organic matter was linear up to 250 mg/l.

From Model Compounds

Additional research on chlorination reactions in natural waters emphasized model compounds that are degradation products of humic and fulvic acids [45]. For example, resorcinol, phlorogulcinol, pyrogallol, catcehol, orcinol, 2,6-dihydroxytoluene, *o*- and *m*-phthalic acids and 3,5-dihydroxybenzoic acids are reported frequently. On the basis of these compounds and proposed structures for humic and fulvic acids [46,47], "hydroxylated aromatic rings with two free meta-positioned OH groups are available active sites for haloform formation" [45]. To test this supposition, chlorination experiments were conducted with natural glycosides:

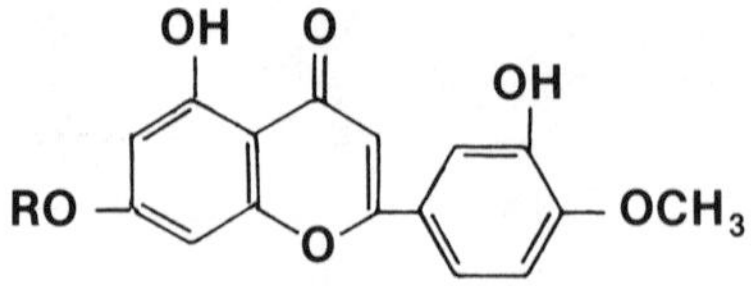

OH O OH RO OCH3 O

R = D-Glucose + L-Rhamnose
Hesperidin (30–32%)

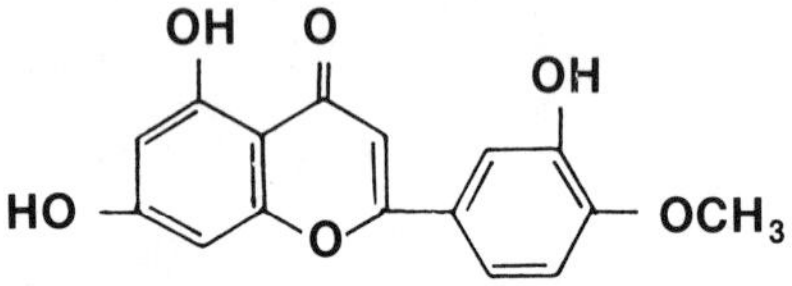

OH O OH HO OCH3 O

Hesperetin (50–56%)

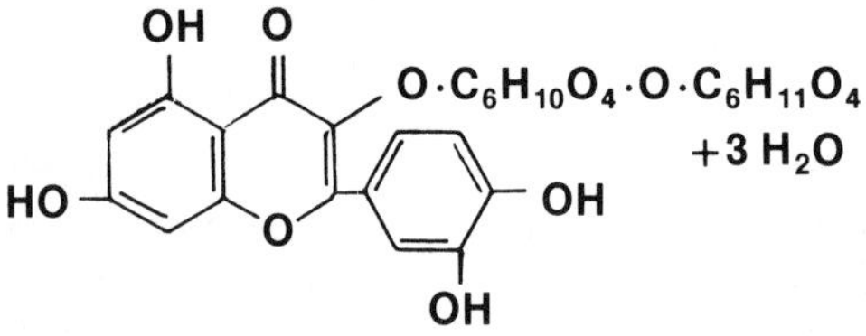

OH O O·C6H10O4·O·C6H11O4 +3 H2O HO OH O OH

Rutin (65–70%)

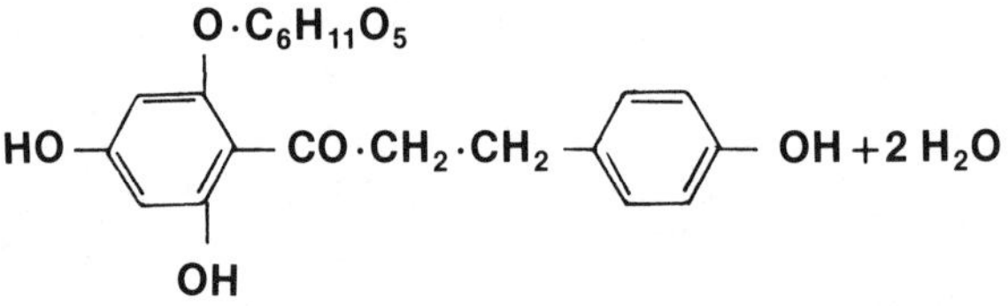

O·C6H11O5 HO CO·CH2·CH2 OH +2 H2O OH

Phlorizin (90–94%)

The reaction conditions, 5×10^{-6} *M* of model compound with a tenfold excess of chlorine, 2 hr, 20°C and a pH value of 7.2 yielded the (percentages) of $CHCl_3$. This is sufficient evidence that the most readily available site of reaction is the C between two meta-OH-groups [45]. There are several of these sites in the structures of fulvic acid. However, this must be the reason why yields of $CHCl_3$ are very low from natural organic carbon; that is, 0.5 m*M* of total carbon yields 0.001 m*M* $CHCl_3$. This means that only 1 of 500 carbon atoms is transformed into $CHCl_3$.

Additional model compounds with OH-substituted aromatic rings were tested for $CHCl_3$ yield, with pH as the variable condition [45]. The test conditions and $CHCl_3$ yields (% molar) are given in Table XI. Since the halogenation reaction is an electrophilic substitution, and since hydroxyl and *o*-alkyl groups are activating, and ortho- and para-directing substituents, the C atom between two meta-positioned OH or OR

Table XI. Comparison of $CHCl_3$ Yields (%) from Model Substances 0.001 *M*, Chlorine 0.012 *M*, 2hr, 15°C[a]

	pH 7	pH 11
1,2-Dihydroxybenzene	0.5	6
1,3-Dihydroxybenzene	85	100
1,3-Dihydroxynapthalene	42	100
1,4-Dihydroxybenzene	1.5	14
1,4-Quinone	2	13
3,5-Dihydroxytoluene	10	32
3,5-Dihydroxybenzoic acid	32	96
Pyrogallol	0	7
Phloroglucinol	60 (100)	1.5
Phloroglucinol monopentylether	15 (45)	89
Phenol	0	1
1-Hydroxy-3-methoxybenzene	10	4
1,3-Dimethoxybenzene	0	9
3-Hydroxybenzoic acid	6	1
1,3,5-Trimethoxybenzene	3	—

[a]Reproduced from Rook [45], courtesy of the American Chemical Society.

groups is activated from both sides. For example, 3,5-dihydroxytoluene (the OH groups are meta to the CH_3 substituent) yielded 32% of the theoretical amount of $CHCl_3$ (pH 11), whereas 1,3-dihydroxybenzene yielded 100% $CHCl_3$. Therefore, the C atom between two ortho hydroxyl groups is an active site for $CHCl_3$ formation. That a greater yield of $CHCl_3$ occurs at a pH value of 11 than at 7 with fulvic acid is explained by the greater formation of phenoxide ions. All of this is brought together in a proposed degradative pathway of fulvic acid and resorcinol shown in Figure 5. Here, the resorcinol type of moiety of fulvic acid may be substituted by R_1 (fulvic acid), R_2 (OH) and R_3 (H) which gives a phloroglucinol molecule in the fulvic acid structure. The proposed pathway is a rapid chlorination of the C atoms that are activated by the two ortho-OH substituents or phenoxide ions in an alkaline environment. An intermediate carbanion II is formed that is, in turn, rapidly halogenated to product IV or protonated to III. Both products, III and IV, undergo additional hydrolytic or oxidative fissions indicated by the dotted lines a, b or c in Figure 5. The final cleavage results in $CHCl_3$ or, in the presence of bromine, leads to the bromo-chloro-haloforms. Other cleavages may lead to other fission products. In any event, the proposed degradative pathway yielded many of the products observed by Rook [45].

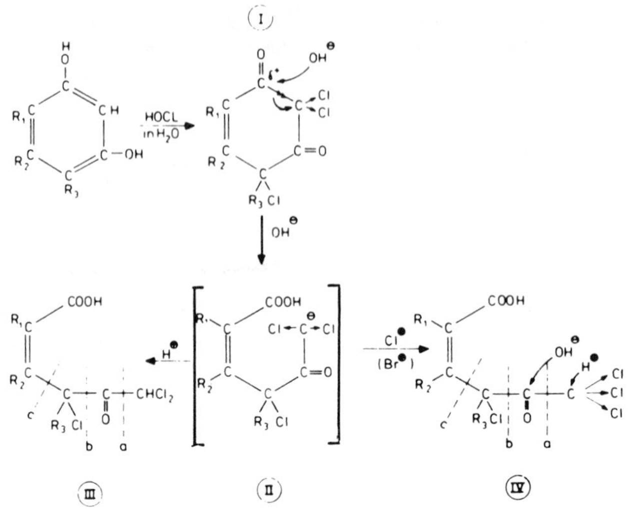

Figure 5. Proposed degradation pathways of fulvic acids and resocinol. Cl^+ represents in a simplified way any electrophilic halogenating species of series XOH_2^+, X_2, HOX, X_2O. Reproduced from Rook [45], courtesy of the American Chemical Society.

A more conventional reaction mechanism was proposed for the production of $CHCl_3$ from various precursors [48]. This is an OH catalyzed series of halogenation and hydrolysis reactions that occur with methyl ketones or compounds oxidizable to that structure. This "haloform reaction" is seen in Figure 6. There is one difficulty with this mechanism when acetone or other simple ketones are involved. They react too slowly to account for synthesis of $CHCl_3$ in the usual water chlorination. The rate-limiting step is the first one, namely, the ionization rate of the methyl ketone. For example, the rate of the ionization of acetone is known whose computed halflife for haloform formation, at pH 7 and room temperature, is 7500 hr or 0.856 yr. Compounds that ionize more rapidly to carbanions than the methyl ketones are required to account for $CHCl_3$ formation within contact times of treatment plants and distribution systems.

A functional group that ionizes rapidly to form carbanions is the β-diketone (B and C Figure 7a). Many examples of the rapid formation of $CHCl_3$ from this and similar structures (1,3-dihydroxy aromatics) have been observed [45]. Another structure which gives the prerequisite carbanion is the pyrrole ring (F in Figure 7a), where the hydrogens ortho to the nitrogen are activated and provide sites for chlorination.

$$R{-}\overset{O}{\overset{\|}{C}}{-}CH_3 \underset{slow}{\overset{OH^-}{\rightleftharpoons}} \left[R{-}\overset{O}{\overset{\|}{C}}{-}CH_2^{\ominus} \longleftrightarrow R{-}\overset{O^{\ominus}}{\overset{|}{C}}{=}CH_2\right]$$

$$\downarrow \text{fast} \quad (HOX \underset{fast}{\overset{H^+}{\rightleftharpoons}} H_2OX^+)$$

$$\left[R{-}\overset{O}{\overset{\|}{C}}{-}CHX^{\ominus} \longleftrightarrow R{-}\overset{O^{\ominus}}{\overset{|}{C}}{=}CHX\right] \underset{slow}{\overset{OH^-}{\rightleftharpoons}} R{-}\overset{O}{\overset{\|}{C}}{-}CH_2X$$

$$\downarrow \text{fast} \quad (HOX \underset{fast}{\overset{H^+}{\rightleftharpoons}} H_2OX^+)$$

$$R{-}\overset{O}{\overset{\|}{C}}{-}CHX_2 \underset{slow}{\overset{OH^-}{\rightleftharpoons}} \left[R{-}\overset{O}{\overset{\|}{C}}{-}CX_2^{\ominus} \longleftrightarrow R{-}\overset{O^{\ominus}}{\overset{|}{C}}{=}CX_2\right]$$

$$\downarrow \text{fast} \quad (HOX \underset{fast}{\overset{H^+}{\rightleftharpoons}} H_2OX^+)$$

$$\boxed{CHX_3} + R{-}\overset{O}{\overset{\|}{C}}{-}OH \xleftarrow{OH^-} H_2O + R{-}\overset{O}{\overset{\|}{C}}{-}CX_3$$

Figure 6. Reaction pathway of the haloform reaction. Reproduced from Morris and Baum [48], courtesy of Ann Arbor Science Publishers.

There is a widely occurring group of natural compounds whose structures are related to the β-ketonic link. These are the acetogenins occurring in nature as pigments [49]. They have the β-diketone as a prime characteristic of their molecules. Phloroacetophenone and orsellinic acid, Figure 7b, are the simplest representatives of two major families of these naturally occurring acetongenins.

Representative compounds of pyrrolic types and of acetogenins were researched for formation of $CHCl_3$ from water chlorination practices [48]. An excellent point was raised in this study through a concept of the chlorine demand. That is, Cl demand = Cl^- formed + C—Cl bonds formed. Consequently, a stoichiometry of chlorine demand may be calculated as it relates to $CHCl_3$ formation. Using resorcinol as an example,

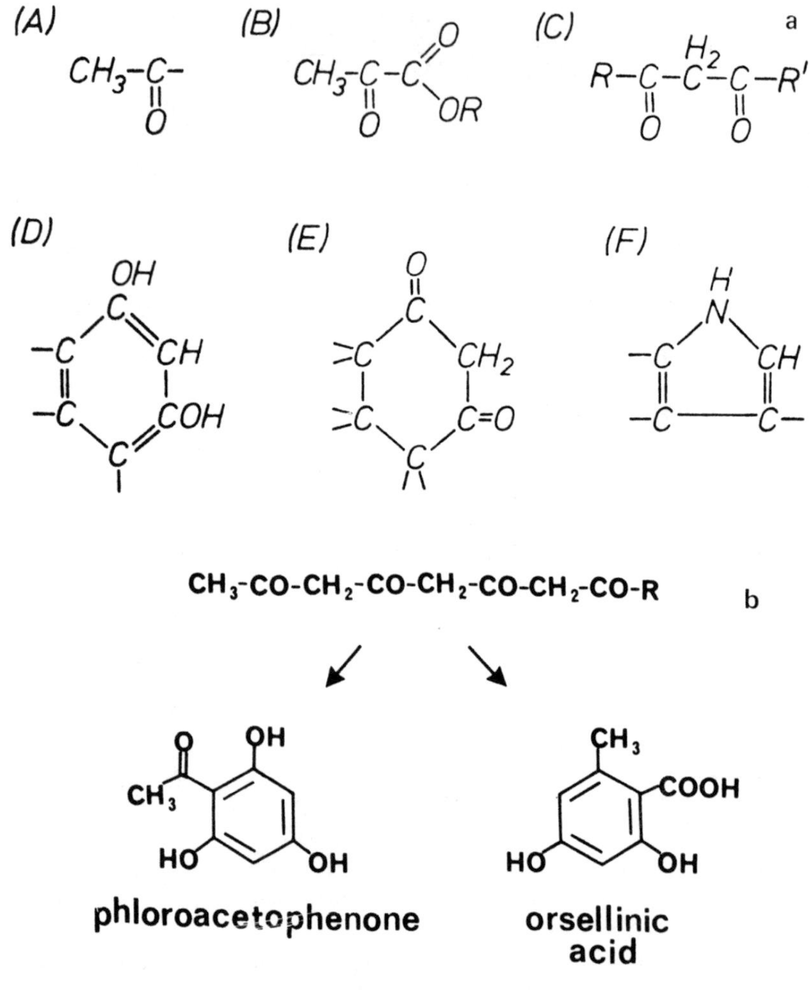

a

(A) $CH_3-C(=O)-$

(B) $CH_3-C(=O)-C(=O)-OR$

(C) $R-C(=O)-CH_2-C(=O)-R'$

(D)

(E)

(F)

b

CH_3-CO-CH_2-CO-CH_2-CO-CH_2-CO-R

phloroacetophenone

orsellinic acid

Figure 7. **(a)** Some molecular structures prospectively responsible for chloroform formation. **(b)** Biosynthetic pathways for acetogenins. Reproduced from Morris and Baum [47], courtesy of Ann Arbor Science Publishers.

a chlorine demand is calculated by assuming complete oxidation to CO_2 and H_2O by HOCl:

$$C_6H_6O_2 + 13HOCl = 6CO_2 + 3H_2O + 13H^+ + 13Cl^- \quad (1)$$

That is, the chlorine demand is 13 moles of HOCl to 1 mole of resorcinol. However, the stoichiometric demand is reduced to 12:1 when one mole of $CHCl_3$ is formed:

$$C_6H_6O_2 + 12HOCl = CHCl_3 + 5CO_2 + 4H_2O + 9H^+ + 9Cl^-$$

This would represent the maximum expected chlorine demand for a given compound. These figures are given in column 2 of Table XII. In the case of organic N compounds, breakpoint of oxidation to N_2 was assumed. For many compounds in Table XII, theoretical demands are reached after 24 hr when sufficiently high ratios of chlorine to organic compound are employed. Where the theoretical demand is not reached, this suggests formation of partially oxidized compounds, some of which may be chlorinated.

Two additional and important observations are made from this study: (1) the time dependence of $CHCl_3$ production and chlorine demand exertion, and (2) the effect of $[H^+]$ [48]. For the first point, the rate of formation and exertion may be nearly the same as shown in Figure 8a for phloroacetophenone. Here the chlorine demand is exerted at about the same rate as for the formation of $CHCl_3$. However, about 50% of the total expected yield of $CHCl_3$ is produced with satisfaction of 60% of the chlorine demand. These curves and the data in Table XII suggest that each molecule of phloroacetophenone is oxidized fully to $CHCl_3$, CO_2 and H_2O. In addition, the data indicate that the chlorination reaction proceeds more slowly at a pH value of 11 than at 7, "in spite of the expected base catalysis." Not all compounds react identically with phloroacetophenone. Inspection of the data in Table XII reveals that some reactions are considerably slower and the molar yield of $CHCl_3$ does not reach 100% within the reaction period. Vanillin, for example, does not have the 1,3-dihydroxy structure for rapid formation of $CHCl_3$. The second point, the effect of $[H^+]$, is seen in Figure 8b where yields of $CHCl_3$ are increased as the pH value is raised from 5.4 to 11. That is, an 18% yield is obtained at pH 5.4 and 100% at pH near 11. This is explained by chlorination of the tryptophan to intermediate products during the reaction period in slightly acidic or neutral solutions. At the higher alkaline pH values, $CHCl_3$ is liberated by hydrolytic reactions. A similar pattern was reported for uracil [48].

Several pragmatic studies have been reported for formation of $CHCl_3$ and other THM, all of which tend to support one another but frequently differ in some detail [50–56]. For example, several precursors, expressed as nonvolatile total organic carbon (NVTOC), "humic acid," acetone, acetaldehyde and aceteophenone were chlorinated under a variety of conditions simulating water treatment plant operation [50]. It was concluded that "the precursor to THM production during the chlorination process in drinking water treatment is probably a complex mixture of humic substances and simple low molecular weight compounds containing the acetyl moiety." This, of course, was confirmed [45,48].

Table XII. Some Potential Precursors of Chloroform in Water Chlorination[a]

Compound	Theoretical Molar Demand Ratio	Initial Molar Ratio Used	Theoretical Demand			Molar Yield $CHCl_3$ (%)
			Time (hr)	Ratio (%)	pH	
Orcinol	15	18	25	56	7.2	38
		18	25	56	9.2	70
Ferulic Acid	20	24	25	55	9.9	58
Phloroacetophenone	15	9	4	54	7	52
		18	48	63	6,9,11	96–100
Vanillin	16	28	26	75	9.2	22
		19	48	75	10	42
o-Vanillin	16	14	25	70	7	19
Syringaldehyde	18	18	24	48	9.2	22
		28	60	51	9.2	17
Pyrrole	9.5	13.6	24	74	10	30
Tryptophane	25	20	7	65	7.5	18
		25.2	47	100	6–11	100
Proline	11.5	16.5	25	130	6.6–11	30
Hydroxyproline	10.5	28	170	102	5.6–10	110
Uracil	7	8.2	25	117	6.5–11	100
Chlorophyll		23 (by wt)	100		9.2	15.5 (by wt)

[a]Reproduced from Morris and Baum [48], courtesy of Ann Arbor Science Publishers, Inc.

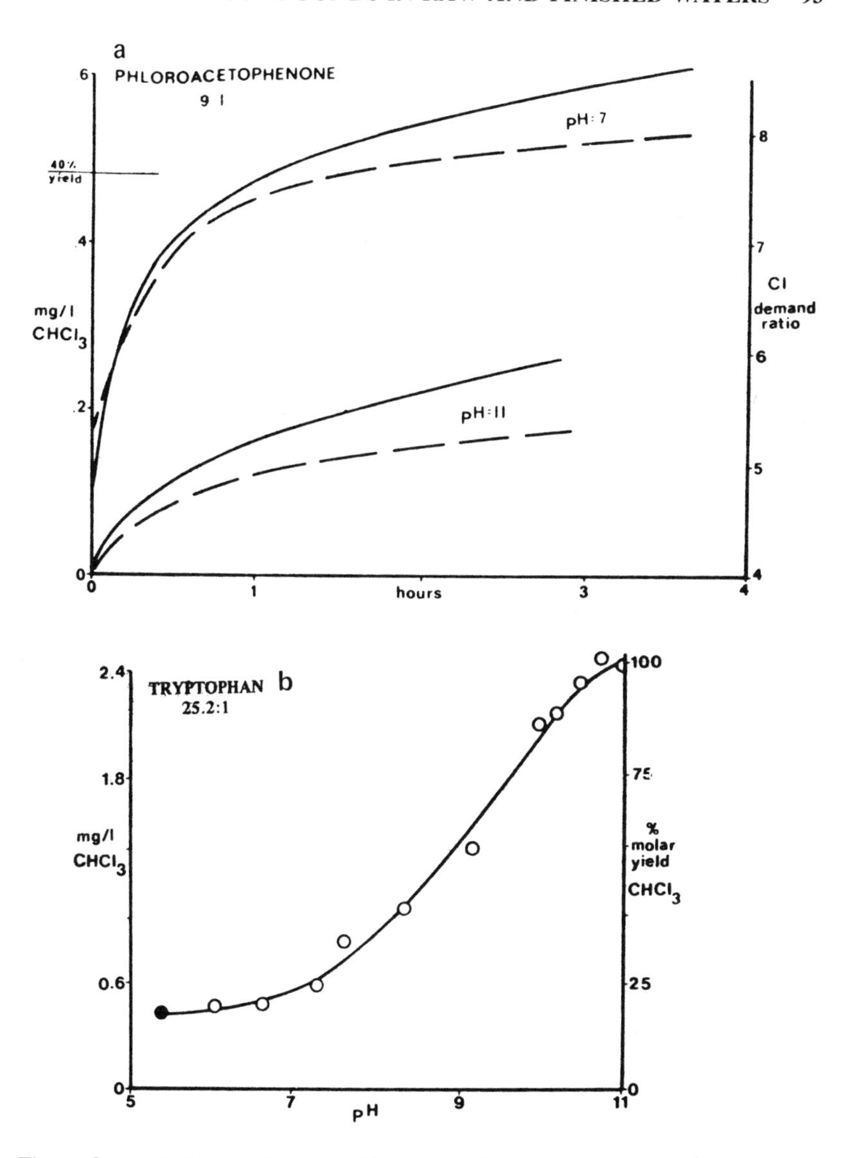

Figure 8. **(a)** Chlorination of phloroacetophenone. ——— = chloroform produced; --- = molar chlorine-demand ratio. Initial phloroacetophenone, 1.0×10^{-5} M; initial aqueous chlorine, 9×10^{-5} M (6.4 mg/l). **(b)** Yields of $CHCl_3$ from tryptophane with pH change. Initial tryptophane concentration = 2.0×10^{-5} M (36 mg/l). pH at 5.4 for 45 hr, then separate samples increased in pH in 0.5-unit steps with greatest pH near 11. Chloroform determinations at 47 hr. Reproduced from Morris and Baum [47], courtesy of Ann Arbor Science Publishers.

Again, the pH dependence of THM production is emphasized where THM and/or $CHCl_3$ contents are always higher at the alkaline values of 11, 11.5, 9.2, 10.2, etc. than at neutral or slightly acid values [50]. In terms of plant operation, where chlorination follows clarification and is conducted at nearly neutral pH values, effective coagulation and sedimentation may be sufficient to reduce the precursor concentration to levels where ultimate THM contents are low (paraphrasing the authors [50]). On the other hand, where chlorination is conducted at a "high" pH value (in a lime-soda softening plant, perhaps), the treatment for precursor removal is complicated. The chlorination process would have to occur at a different point, or the pH value would have to be lowered.

Experiments have been conducted also on the chlorination of acetone, 1 mg/l, 25°C, 10 mg/l Cl_2, at pH values of 6.7, 8.3 and 10.2 over an extended period of time, 95 hr [50]. At pH 6.7, there was a total THM content of 0.1 μM after 96 hr. However, at pH values of 8.3 and 10.2, there was a THM content of 0.8 μM after 25 hr. Similar effects were observed from chlorination of "humic acid." It was concluded that "acetone is not likely to be a significant precursor at pH 6.7," whereas it may be at pH 8.3 and 10.2 where the reaction rate was greater by a factor of 30. For acetone to be a precursor of THM, its concentration would have to be much greater than the experimental value of 1.0 mg/l. Rook had suggested a content of 9.0 mg/l or so [43,44].

An important operational variable is the yield of $CHCl_3$ (or THM) from naturally occurring organics. Several investigators have reported the ratio of $[CHCl_3]$ to [TOC] and the percent yield on a carbon basis. This is important data because of local or regional variations in the amount and chemical structure of the natural precursors. This information was compiled from several places (Table XIII) with a single parent source [21]. The yield of $CHCl_3$ is, of course, extremely small on the basis of carbon atoms (second column, Table XIII). This is caused mainly by the various sources of precursors, such as humic and fulvic acid extracts and river- and lakewaters. TOC values were mostly around 2.0 mg/l which indicates the "low" levels of organic matter from which the THM are formed [51]. Reaction times varied from 2 to 96 hr with the latter suggesting the occurrence of reactions in distribution systems. THM formation was observed also in raw waters, filtered (0.45-μm filter) water, and alum-treated and sand-filtered water [52]. This suggests either the occurrence of soluble, low-molecular-weight precursors or the inability of the treatment processes to remove the THM. Similar results were obtained in England where humic and fulvic fractions extracted from the River Thames were chlorinated [57]. A slightly different mechanism of $CHCl_3$ formation was proposed; that is, it was hypothesized

Table XIII. Comparative Yield Data for Chloroform Formation[a]

$CHCl_3$/TOC (m*M*/*M*)	Percent Yield, Carbon Basis (mg $CHCl_3$ as c/mg TOC) × 100	pH of Sample	Cl_2 Concentration Applied (mg/l)	TOC of Sample (mg/l)	Time of Reaction (hr)	Temperature (°C)	Material Studied	Reference
5.0	0.5	6.5	10	2.0	2	20	Humic acid extract	51
16.0	1.6	6.5	10	2.0	96	20	Humic acid extract	51
7.6	0.76	6.5	10	1.4	10	20	Humic acid extract	51
4.0–7.0	0.7–1.4	6.7–9.2	10	1.4	96	25	Aldrich humic acid	50
1.5–2.1	0.15–2.1	7.0–11	10	2.0	24	20	Humic acid	51
3.5	0.35	6.5	10	2.0	10	20	Fulvic acid extract	51
6.0	0.60	6.5	10	2.0	96	20	Fulvic acid extract	51
1.4–2.0	0.14–0.20	7.0–11	10	2.0	24	20	Fulvic acid extract	52
2.0	0.2	7.0	8	12			Fulvic acid extract	45
3–9	0.3–0.9						Fulvic acid extract	44
0.37	0.04[b]	8.2–8.3	2.2–2.9	2.93			River and lakes	53
2.5	0.25[c]		5.8	5.1[d]			Lake	54
2.7	0.27[c]		6.5	6.8[d]			Lake	54
0.7–2.5	0.07–0.25		10	1.4–11.0	24	20	River and lake	52
1.7	0.17[e]	7.25	1.6	0.9	0.5	8	River	55
1.1	0.11	7.5–7.9	7	5.9	2	12	River	44
0.1–0.5	0.01–0.05	10.8	3–8	4.4	10	3	Fractionated riverwater	56
0.3–1.7	0.03–0.17	10.8	6.4	1.8	10	3	Fractionated riverwater	21

[a]Reproduced from Veenstre and Schnoor [21], courtesy of the American Water Works Assocation.
[b]Based on average values over entire study.
[c]Based on average values over nine-month study.
[d]NVTOC.
[e]Average of six duplicate samples taken over nine days from Seattle tapwater.

that a "total" chloroform content was composed of two components: one produced by thermal decomposition of unidentified chlorinated intermediates, or "residual" $CHCl_3$, and the other one by "dissolved" $CHCl_3$. The "residual" $CHCl_3$ was determined by purging "dissolved" $CHCl_3$ from the water sample with nitrogen followed by injection into a gas chromatograph (200 °C). Whether or not there is a "residual" $CHCl_3$ is questionable, but this offers a possible explanation for the slow formation of THM.

Algae are also apparent precursors in the formation of THM [58, 59]. This follows the data of a previous study [48] where formation of $CHCl_3$ was reported from chlorination of soluble chlorophyll. Various amounts of $CHCl_3$ are produced from the chlorination of algal biomass of four species (two green and two blue-green in Table XIVa)[58]. Similar results were reported by Oliver and Shindler [59] for seven species of green, blue-green and diatom algae (Table XIVb).

Model compounds have been used in an effort to develop a mechanism of THM formation from their chlorination [44,45,48,52,60,61]. Many of the compounds are components of larger molecules, such as humic acid, fulvic acid and acetogenins. Reaction variables included reaction time, $[H^+]$, temperature and stoichiometry that were established to simulate water treatment plant operation. Results from some of these studies are given in Tables X (44), XI (45) and XII (48). There are, of course, some differences in the results. For example, formation of $CHCl_3$ from phenol (C_6H_5OH) was reported [52]:

Chloroform Produced from Chlorination of Selected Organics at pH 7 and 11[a]

Organic	Chloroform Concentration (μg/l)	
	pH 7	pH 11
Fulvic Acid	28	39
Humic Acid	29	42
Tannic Acid	11	61
Lignosulfonic Acid	3	9
Aspartic Acid	2	85
Phenol	24	31

[a]Reaction conditions: organic, 2 mg/l; chlorine, 10 mg/l; time, 24 hr; temperature, 20°C.

whereas it did not occur in other systems within 2–4 hr [44,45]. Perhaps, the longer contact time of 24 hr was responsible. Chloroform was observed from the chlorination (pH 7, 22°C, 2 hr) of: tannic acid,

Table XIVa. Chloroform Yields From Chlorinated Algal Biomass of Four Species at Various Stages of Growth[a,b]

Algal Species	Culture Age (days)	TOC Range (mg/l)	$CHCl_3$ Range ($\mu g/l$)	$CHCl_3$-C/TOC (percent)
C. pyrenoidosa	13–30	2.9–18.9	521–1500	0.3–2.8
S. quadricauda	9–35	20.7–75.8	622–1040	0.2–0.5
A. flos-aquae	9–35	10.9–30.5	479–1350	0.2–0.7
O. tenuis	6–37	2.1–23.5	470–1300	0.3–3.9

[a] Reproduced from Hoehn [58], courtesy of the American Water Works Association.
[b] Chlorine dosage = 50 mg/l.

Table XIVb. Chloroform Yields (μg $CHCl_3$/mg material) from the Chlorination of Algae[a]

	pH 7		pH 11	
Algae	Low Ratio[b]	High Ratio[c]	Low Ratio[b]	High Ratio[d]
Anabaena oscillarioides	3.4	7.3	11	47
Anacystis nidulans	2.8	4.4	16	24
Scenedesmus quadricauda	3.0	6.0	11	24
Scenedesmus basiliensis	2.5	4.8	12	27
Selanastrum capricornutum	3.4	4.2	17	31
Navicula minima	1.9	3.5	7.9	27
Navicula pelliculosa	2.4	5.0	15	38
Thomson Lake algae	3.2			
Lake Ontario *Cladophora*	1.4		5.3	
Water fulvic acid		17.0		29

[a] Reproduced from Oliver and Shindler [59], courtesy of the American Chemical Society.
[b] Low ratio: $[Cl_2]/[algae] = 1.5$.
[c] Reaction conditions: $[Cl_2] = 10$ mg/l; T = 20°C; reaction time = 24 hr.
[d] High ratio: $[Cl_2]/[algae] = 10$.

D-glucose, humic acid, vanillic acid, and gallic acid [60]. No chloroform was detected from α-pinene, acetic acid, propionic acid and butyric acid [60].

A series of similarly structured compounds was chlorinated in an effort to model the monomeric components of aquatic humic material [61]. Results are given in Table XVa for a reaction time of 40 min. Clearly, the *m*-dihydroxy substitution of resorcinol produces $CHCl_3$

very rapidly, which agrees with Rook [45]. The $CHCl_3$ yield is approximately 93% when the carbon atom is between two OH groups. The vanillic acid structure produces considerably less $CHCl_3$ despite a relatively high chlorine demand (5.38 moles of Cl_2 consumed per mole of compound). This is probably due to the lack of a doubly activated carbon atom between two free OH groups.

Quantitative aspects of the chlorine demand of model compounds are seen in Table XVa [61]. A portion of the chlorine demand is incorporated into non-$CHCl_3$ reaction products in addition to that demand required for $CHCl_3$ production and oxidation. Chlorinated products are

Table XVa. Results of Kinetics Studies[a]

Compound	Initial Concentration (Buffer Solution) (m*M*)	Cl_2/C	$CHCl_3$ Yield mol of $CHCl_3$/ mol of Compound[b]	Cl Consumption mol of Cl_2/ mol of Compound[b]
1,3-Dihydroxybenzene (resorcinol)	6.10	1.93	0.877	6.60
3,5-Dihydroxybenzoic Acid	4.36	1.95	0.450	7.06
3.5-Dimethoxybenzoic Acid	0.534	2.00	0.009 00	3.00
3.5-Dihydroxytoluene (orcinol)	0.570	1.82	0.852	6.26
3,5-Dimethoxy-4-hydroxybenzoic Acid (syringic acid)	3.35	1.76	0.005 02	5.18
3,5-Dimethoxy-4-hydroxycinnamic Acid	0.553	1.85	0.022 8	6.09
β-(3,5-Dimethoxy-4-hydroxyphenyl) propionic Acid	0.552	1.93	0.033 0	6.06
3-Methoxy-4-hydroxybenzoic Acid (vanillic acid)	0.573	1.73	0.037 5	5.38
3-Methoxy-4-hydroxycinnamic Acid	0.553	1.71	0.109	7.63
3-Methoxy-4-hydroxyhydro-cinnamic Acid	0.553	1.53	0.009 79	4.94

[a]Reproduced from Norwood [61], courtesy of the American Chemical Society.
[b]Forty minute reaction time.

Table XVb. Reaction Products of Model Compounds and HOCl at 25°C[a]

Reactant	Products Identified	
	at 0.5 Cl_2/C	at 2.0 Cl_2/C
1,3-Dihydroxybenzene 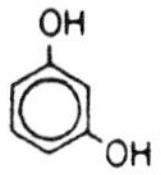 (OH, OH)	HO–(ring)–OH, Cl_x (x = 1-3); Cl, H, Cl, Cl, O, O (ring); $CHCl_3$	$CHCl_3$; 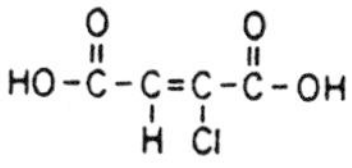 HO–C(=O)–C(H)=C(Cl)–C(=O)–OH; CCl_3COOH
3-Methoxy-4-Hydroxy-Cinnamic Acid 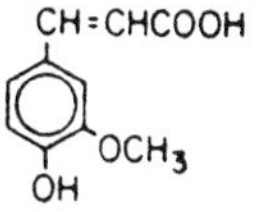 (CH=CHCOOH, OCH_3, OH)	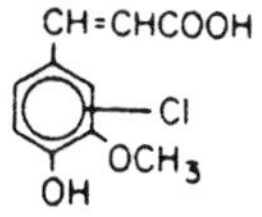 CH=CHCOOH, Cl, OCH_3, OH; 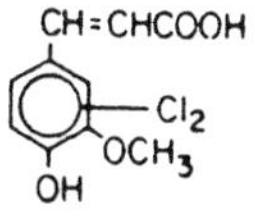 CH=CHCOOH, Cl_2, OCH_3, OH; 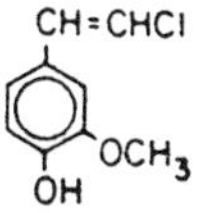 CH=CHCl, OCH_3, OH; 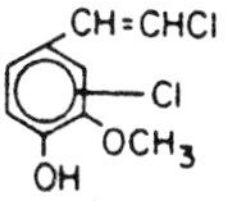 CH=CHCl, Cl, OCH_3, OH; $CHCl_3$	$CHCl_3$; $CHCl_2COOH$; CCl_3COOH

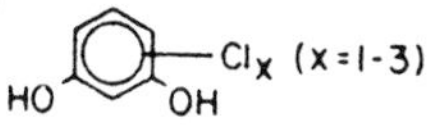

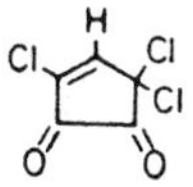

[a]Reproduced from Norwood [61], courtesy of the American Chemical Society.

given in Table XVb for two of the model compounds: resorcinol and 3-methoxy-4-hydroxycinnamic acid. "The distribution of specific forms of halogenated products is a function of the Cl_2/C dosage with higher Cl_2/C ratios favoring accumulation of chlorinated acids" [61].

THM formation and yield are also a function of precursor molecular weight [62,63]. Gel permeation chromatography with Sephadex G-75 was employed to separate naturally occurring organics from the Iowa River (December 1977 to January 1978) into apparent molecular weight fractions that were subsequently chlorinated to 2 mg/l of free residual and analyzed for THM. It was found that 96% of the TOC values occurred when the molecular weight was less than 3000, and 13.6% with a mol wt less than 1000 (Table XVI). The significance of this distribution of organics lies in the observation that 75% of the THM were formed from compounds with a mol wt less than 3000, and 20% from those with a mol wt less than 2000. Chloroform was the primary product of the chlorination of precursors with a mol wt less than 5000. There were also significant quantities of the brominated THM formed from the smaller organics with mol wt less than 1700. These observations are reasonably consistent with the observation that a majority of natural color molecules fall into the 700–10,000 mol wt range [64].

Aquatic humic material was fractionated into eight molecular weight ranges by ultrafiltration [63]. Each fraction was chlorinated: 1 mg/l as TOC, 15 mg/l Cl_2, pH 11, temperature at 20°C, and a reaction time of 72 hr. These conditions were designed to optimize $CHCl_3$ production. From the histograms in Figure 9, it is apparent that the most important fraction for formation of $CHCl_3$ from fulvic acids is in the 1000–10,000 range, whereas the range is 10,000–20,000 for lake humic acids and

Table XVI. THM Formation and Chloroform Yield as a Function of Precursor Molecular Weight Range for Iowa River, December 1977–January 1978[a]

Sample	App mol wt Range	TOC (mg/l)	$CHCl_3$ (ppb)	$CHCl_2Br$ (ppb)	$CHBr_2Cl$ (ppb)	$CHBr_3$ (ppb)	Yield (µg/l $CHCl_3$/mg/l TOC)
blank			3.4	<0.2	<0.2	<0.2	
I	>50 000	0.05	4.4	0.4	<0.2	<0.2	2.00
II	–33 000	0.30	5.4	0.3	<0.2	<0.2	6.67
III	–20 000	0.55	4.2	0.4	<0.2	<0.2	1.33
IV	–10 000	0.48	4.6	0.3	<0.2	<0.2	2.40
V	–6 000	0.60	4.2	0.3	<0.2	<0.2	1.33
VI	–3 000	2.23	15.0	1.4	<0.2	<0.2	5.27
VII	–1 700	13.3	43.0	8.0	0.8	0.2	2.98
VIII	–1 000	34.8	13.0	21.0	11.0	2.1	0.28
IX	<1 000	8.30	11.5	14.3	6.0	1.8	1.01
	total[b]	61.1	74.7	46.4	17.0	3.9	1.22

[a]Reproduced from Schnoor [62], courtesy of the American Chemical Society.
[b]Total excluding the blank; sample concentration factor ~2.5; G-75 Sephadex column. 10-hr Reaction Time.

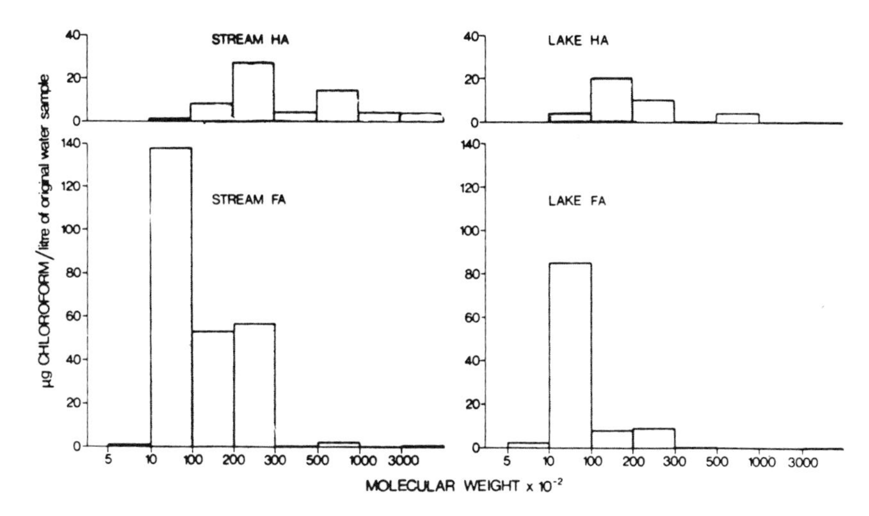

Figure 9. Effect of molecular weight on $CHCl_3$ production per liter of original sample from chlorination of lake and stream fulvic and humic acids. Reproduced from Oliver and Visser [63], courtesy of the International Association on Water Pollution Research.

20,000–30,000 for river humic acids. Greater $CHCl_3$ contents were obtained from the fulvic acids due, undoubtedly, to the lower molecular weight.

Kinetics of Formation

It has been stated that "the THM reaction is one of the slowest reactions of interest to the water utility practitioner" [65]. As illustrated in Figure 10a, the reaction occurs over several hours and there are systems where significant increases are observed after 24 hr of reaction. The kinetics of THM formation reaction is dependent on many factors such as $[H^+]$, $[Cl_2]$, [precursor] and temperature. Since no experimental conditions were given for Figure 10a, it is difficult to interpret this statement. Evidence is cited above showing that THM formation occurs within two hours [48,61]. On the other hand, it is difficult to follow kinetics with precision because there are competing reactions for the chlorine.

A kinetic equation for the THM reaction is:

$$\frac{d[Cl_2]}{dt} = -k_1[Cl_2][TOC] \qquad (3)$$

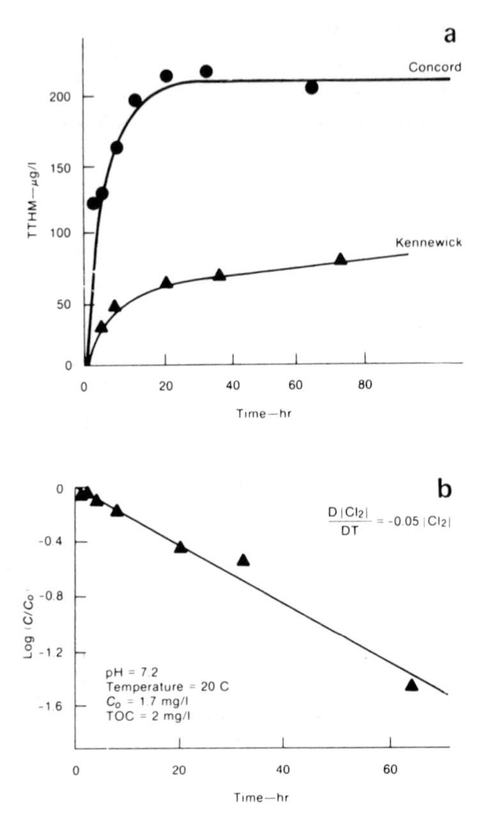

Figure 10. **(a)** THM formation. **(b)** First-order plot of chlorine consumption. Reproduced from Trussell and Umphres [65], courtesy of the American Water Works Association.

where it is assumed that the rate of chlorine consumption is dependent on its content and precursor content expressed as TOC [65]. Furthermore, if it is assumed that the [TOC] is not significantly reduced, then:

$$ln\frac{[Cl_2]_t}{[Cl_2]_o} = -k_1[TOC]t \qquad (4)$$

This equation represents a first-order reaction with respect to chlorine residual (Figure 10b). Another approach assumes that the rate of THM formation is related to residual chlorine to the first power:

$$\frac{d[THM]}{dt} = -\frac{dC}{dt} = k_2[Cl_2][C]^m \qquad (5)$$

where m is the order of reaction with respect to [C], the precursor concentration. When m is 3, a third-order reaction gives "a good fit" [65].

The kinetics of $CHCl_3$ and/or THM formation from humic acid (Aldrich Chemical Co.) and various natural precursors have been investigated [66]. Typical rate curves are seen in Figure 11a for $CHCl_3$ appearance from this humic acid. Initially, the chlorine disappears rapidly with concurrent appearance of $CHCl_3$. Both rates decrease with time. A log-log plot (Figure 11b) shows that the kinetics of THM formation from natural organics is somewhat slower and more complex than those in Figure 11a.

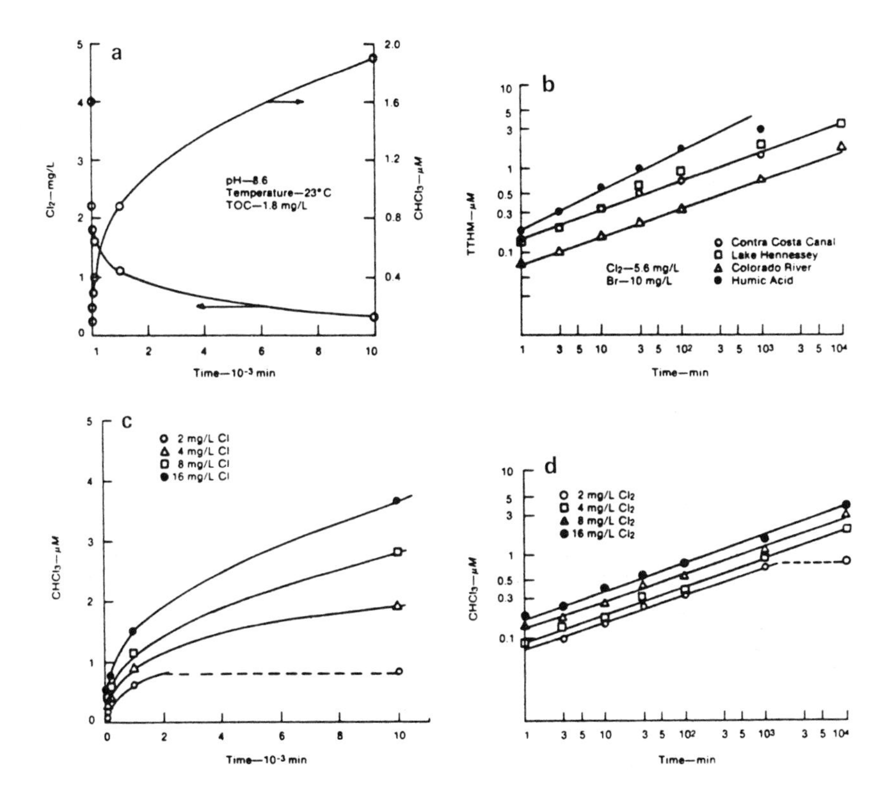

Figure 11. **(a)** Typical rate curves for chlorine consumption and THM formation. **(b)** Rate of THM formation for four samples from different sources. [TOC] = 2.8 mg/l, pH 7.3. **(c)** Effect of applied chlorine dose on rate of THM formation. Aldrich humic acid: TOC = 1.8 mg/l; pH 8.5; temperature 23°C; free chlorine residual >0.3 mg/l after 10^4 min for samples with $Cl_2 \geq 4$ mg/l. **(d)** Log-log plots of rate of THM formation as a function of applied chlorine dose. Reproduced from Kavanaugh et al. [66], courtesy of the American Water Works Association.

The well known haloform reaction in organic chemistry is zero order with respect to chlorine [67]. That is, the rate of formation of $CHCl_3$ should be independent of $[Cl_2]$ because the initial enolization step is rate limiting [48]. However, the experimental evidence, Figure 11c, shows the rate of THM formation to be strongly dependent on $[Cl_2]$. A log-log plot, Figure 11d, suggests "no shift in the reaction mechanism" [66]. TOC content is also a factor, but its precise nature is uncertain.

An empirical reaction kinetic model is [66]:

$$3A + B \xrightarrow{k_n} C \tag{6}$$

where A = HOCl, B = TOC, C = THM and k_n is the n-order overall rate constant. This states that three moles of HOCl react with one mole of carbon (i.e., organic precursor) to form one mole of THM. The rate expression for THM formation is:

$$\frac{dC}{dt} = k_n[B][A]^m \tag{7}$$

by assuming a first-order reaction with respect to [TOC] and m-order with respect to [HOCl]. This model was expanded to include an f factor, which is the yield of moles of THM formed per mole of Cl_2 consumed. The model was evaluated by assuming values of m = 1, 2 and 3 and using batch chlorination results from several waters with TOC contents 1–10 mg/l. Figure 12a shows a "good fit" for both second- and third-order reaction models (correlation coefficient, $r^2 = 0.97$). When m = 1, an r^2 value of 0.6 was obtained. When the experimental and computed results were compared (Figure 12b), the model fitted the experimental data when the reaction time was greater than 100 minutes. This suggests that the initial portion of the rate curve is more complex than assumed or, perhaps, that there is a two-stage reaction [45,68]. In any event, Equation 7 may have pragmatic use for evaluating the potential of THM formation in distribution systems [66].

Additional kinetic data are available from reactions of Cl_2 and ClO_2 with humic acid at pH values of 6.0, 7.0 and 9.5 at 25°C. [68]. A "fast" reaction was distinguished, less than 24 hr, from a "slow" reaction, 24–170 hr. Similar results were reported where "$CHCl_3$ production occurs in two phases, the first of which is rapid and strongly pH dependent whereas the second phase is slower and less dependent on pH" [57]. This observation is from chlorination of humic and fulvic acids isolated from the River Thames.

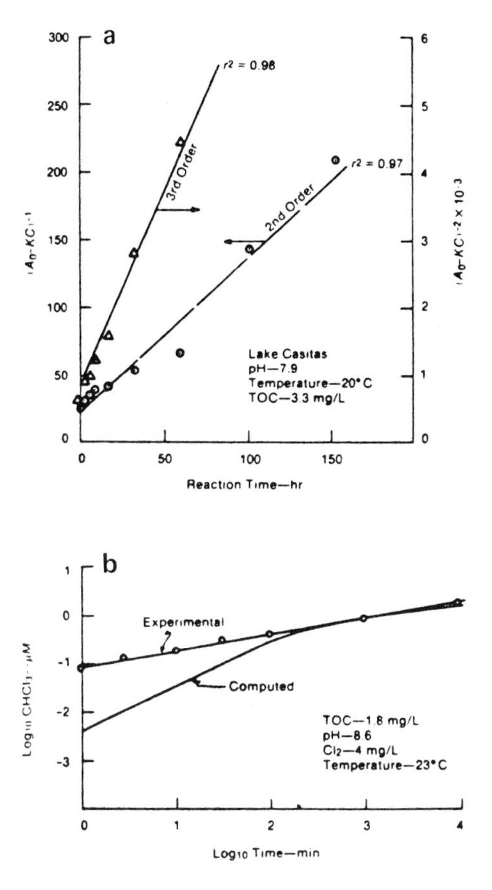

Figure 12. **(a)** Test of kinetic model—second and third order—with respect to chlorine. **(b)** Comparison of experimental and computed results for formation of $CHCl_3$. Aldrich humic acid, $k = 0.15 \text{ min}^{-1} \text{ m}M^{-3}$, $f = 0.1$. Reproduced from Kavanaugh et al. [66], courtesy of the American Water Works Association.

Case Histories

Granular activated carbon (GAC) is employed widely for removal, mainly, of organic compounds from drinking water supplies. That chlorination of GAC and adsorbed humic materials may produce chlorinated compounds has been reported [69,70]. Table XVII shows the compounds that were formed from chlorination of humic acid prior to adsorption by carbon; namely, the chlorophenols, chloromethoxyphenols and several "oxidized species." The humic acid blank solution had a surprisingly large number of compounds. Organics found in the influent and on the

Table XVII. Characterization of Organics from Humic Acid-Carbon Chlorine Systems[a,e]

Compound	Humic Blank Soln[b]	Chlorinated Influent Soln[b,c]	Carbon Profile Extract[d]	Effluent Extract after 310 L[b]
Acetophenone	+	+	+	+
Propylene Glycol	+	+		+
Methylbenzoate		+		+
2-Ethylhexanol	+	+	+	+
Propiophenone	+			
C_2 Benzaldehyde Deriv		+		+
C_2 Benzaldehyde Deriv		+		+
Naphthalene	+	+		+
Octanoic Acid	+	+	+	+
Benzothiazole	+			
Monochlorophenol		+ (79)	trace	+
p-Ethylacetophenone	+	+		+
3,5-Dimethylmethyl Benzoate	+	+		+
Hexachlorocyclopentadiene			+	
Lactic Acid	+	+	+	+
Dichlorophenol		+ (143)		+
Monomethoxychlorophenol		+		
Hydroquinone	trace	trace	+	trace
Trichlorophenol		+ (450)	trace	+
Methoxydichlorophenol		+		+
Dimethylphenol Deriv			+	
Methoxydihydroxybenzene			+	
Chlorodihydroxybenzene			+	
C_4 Dihydroxybenzene		+		+
2,6-Dimethylbenzoquinone			+	

Phthalate		+		+
Benzoate Deriv	+	+		
Tetramethylphenol		+		+
Benzoate Deriv	+	+	+	+
Benzophenone		+	+	+
Dimethoxydihydroxybenzene			+	
Chloromethoxydihydroxybenzene			+	
Anthracene (IS)	+	+	+	+

[a]Reproduced from McCreary and Snoeyink [69], courtesy of the American Chemical Society.
[b]XAD-2 extracts of aqueous solutions, eluted with MeOH/$MeCl_2$.
[c]Concentrations are in ng/l.
[d]Soxhlet extraction of 2 g of carbon.
[e]Determined by comparing the relative area ratios between the peak of interest and the internal standard in each chromatogram.

carbon are similar to the degradation products of humic acid [61]. These compounds are the various hydroxy- and methoxy-substituted benzoic acids formed from an alkaline $CuSO_4$ degradation of natural humic acid. It was concluded that there were no additional compounds present in the effluent of the experimental carbon column than had been in the influent [70]. Compounds may have been formed on the GAC, but more likely they originated from the humic acid. On the other hand, several chlorinated organics, including chloroform, bromodichloromethane, dichloroprene, trichloroethane and several chlorinated aromatics were formed from batch reactions of free chlorine and GAC, 2.5 g Cl_2/g of carbon (Table XVIII) [70].

An excellent case study of $CHCl_3$ formation in the public water supplies of Durham and Chapel Hill, North Carolina, was reported [71]. These cities have similar water supplies (surface impoundments) and

Table XVIII. Volatile Compounds in the Carbon-High Chlorine Dose Solutions[a,c]

No.	Compound	Chlorine Blank[b]	Carbon-Chlorine
1	Methylene Dichloride	+	+
2	1,1-Dichloroethylene	+	trace
3	Diethyl Ether	–	+
4	Chloroform	+	+
5	1,2-Dichloroethane	+	+
6	1,1,1-Trichloroethane	–	trace
7	Carbon Tetrachloride	+	+
8	Cycloalkane	—	+
9	Bromodichloromethane	—	+
10	Dichloropropene	—	+
11	Trichloroethylene	+	+
12	Benzene	+	+
13	1,1,2-Trichloroethane	+	+
14	Hexane	+	+
15	1,1,1,2-Tetrachloroethane	+	+
16	Tetrachloroethylene	+	+
17	Toluene	+	+
18	Chlorobenzene	—	+
19	Hexachloroethane	+	+
20	Chloroalkylbenzene	—	+
21	Chlorotoluene	—	+
22	Dichlorobenzene	—	+

[a] + indicates that the compound was present in the solution.

[b] The area ratio is the peak area for the chlorine-carbon solution divided by the area for the chlorinated blank (no carbon).

[c] Reproduced from Snoeyink et al. [70], courtesy of the American Chemical Society.

water treatment processes: chemical coagulation by alum, prechlorination, taste and odor control with $KMnO_4$ (Chapel Hill), rapid filtration and postchlorination. The raw waters average 5.1 mg/l of nonvolatile total organic carbon for Durham and 6.8 mg/l for Chapel Hill whereas their finished waters average 0.129 and 0.184 mg/l $CHCl_3$, respectively. Chlorine residuals in the finished waters were reasonably "normal": 2.0–2.5 mg/l for Durham and 1.0–1.5 mg/l for Chapel Hill. "That large chlorine additions must be employed to produce" these residuals was observed [72].

In order to obtain an insight into the extent of $CHCl_3$ formation during the various water treatments, the contents of $CHCl_3$, NVTOC and Cl_2 were sampled and measured at various points in the two systems [71]. Results are given in Figure 13a. It is obvious that the [$CHCl_3$] was increased as the water flowed through the plants and eventually into the distribution systems. There is the concurrent increase in [$CHCl_3$] with decrease in the [TOC]. It is interesting to note that there were 260 μg/l of $CHCl_3$ after 2.85 hr of detention in the Chapel Hill water, whereas there were 100 μg/l of $CHCl_3$ after 10.35 hr at Durham. Laboratory studies suggested that chlorination after coagulation reduced the $CHCl_3$ formation by more than 60%. This, in turn, indicated a modification of the point of chlorine addition in the flow diagram of the Durham plant. Prechlorination was moved to the collection channels following coagulation. Figure 14 shows that there were significant reductions in the [$CHCl_3$]. This was due, apparently, to reduction of the NVTOC from 5.8 to 3.7 mg/l.

This case study was followed by a detailed laboratory research of the chlorination of naturally occurring humic and fulvic acids before and after chemical coagulation [72]. That these two acids are precursors in the formation of THM is well established [44,45,50]. Confirmation was provided from a variety of reaction conditions simulating treatment plant operation [72]. Chemical coagulation of "colored" organic molecules in natural waters is, of course, effective for their removal. It is not surprising, therefore, that a reduction of [$CHCl_3$] is simultaneously observed. The quantitative aspects of this are seen in Figure 15 humic and fulvic acids. It was concluded that there is preferential removal of chloroform precursors from both acids by the coagulation. This may be a reaction rate phenomenon rather than a preferential removal.

Prevention of THM Formation

An approach for the elimination or reduction in the THM content in drinking waters is to employ a disinfectant other than chlorine. Obvi-

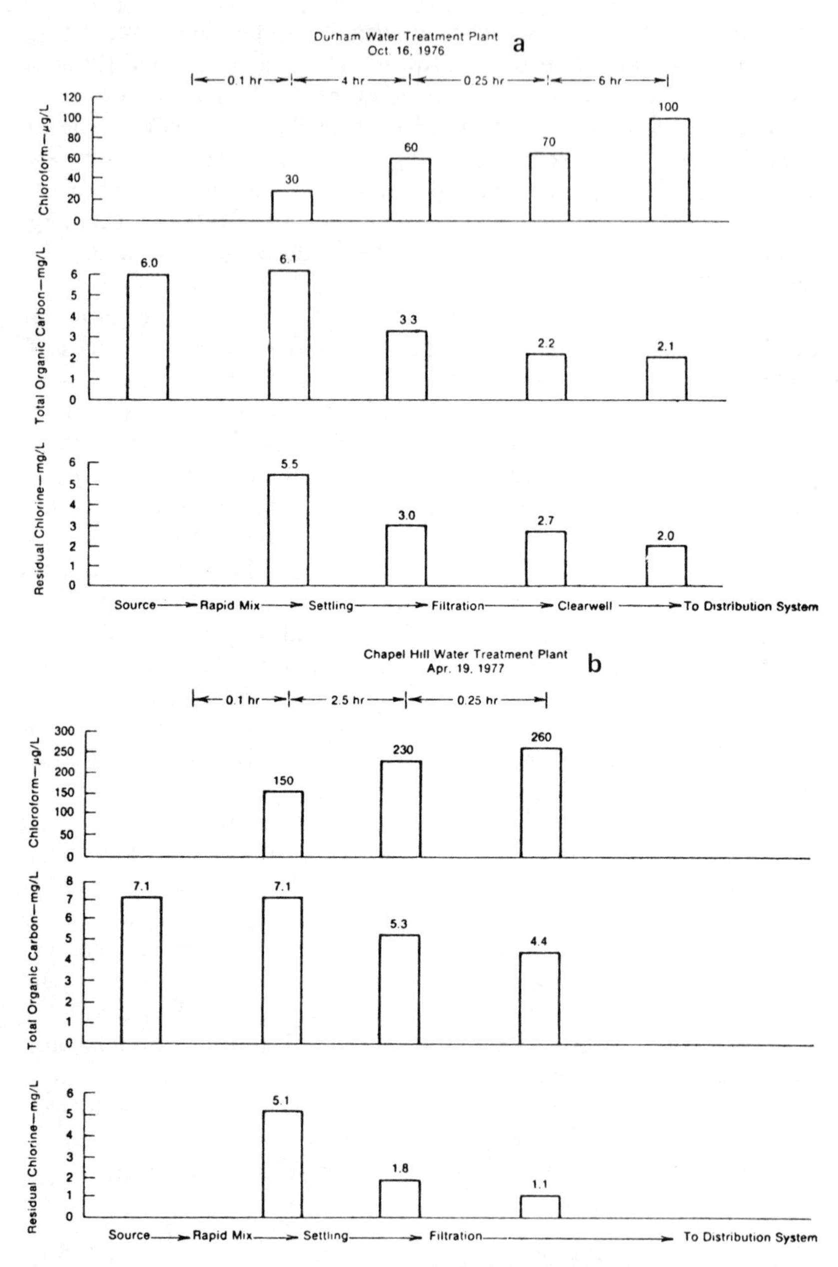

Figure 13. Extent of $CHCl_3$ formation at various stages of treatment at **(a)** Durham, North Carolina, and **(b)** Chapel Hill, North Carolina. Reproduced from Young and Singer [71], courtesy of the American Water Works Association.

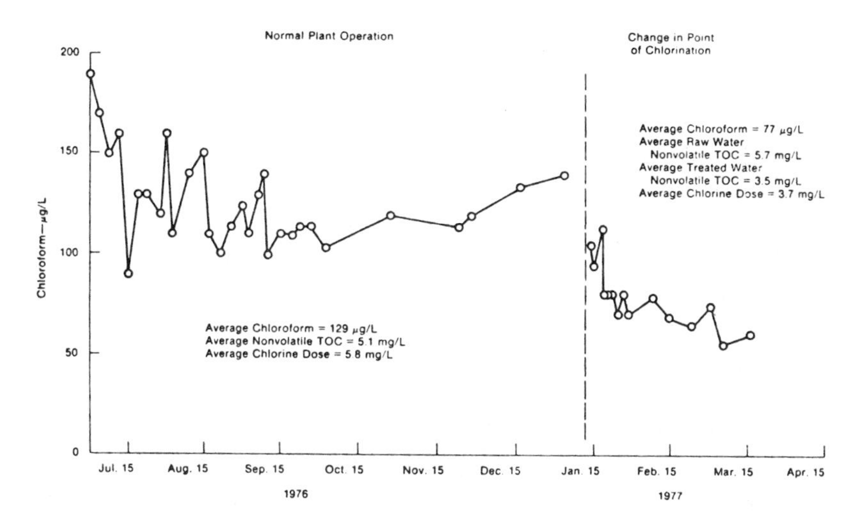

Figure 14. Sampling results before and after the change in point of chlorine addition at the Durham water treatment plant. Reproduced from Young and Singer [71], courtesy of the American Water Works Association.

ously, the alternative technique would have to be as effective as chlorine. The use of chloramine (NH_2Cl and/or $NHCl_2$, see Chapter 10) was proposed for the drinking water of Huron, South Dakota [73]. According to the NORS study [11], their water supply contained THM in excess of the proposed (at that time) MCL of 100 μg/l. Disinfection had been achieved by a combination of chlorine, $KMnO_4$ and lime-soda softening (i.e., $pH > 11.0$). The alternative disinfectant was chloramine treatment through addition of ammonium sulfate and chlorine. This was a reasonably good choice inasmuch as the softening process reduced the total coliform count some 68–100%. The appropriate laboratory and field studies were conducted to demonstrate the assurance of disinfection in the distribution system during and after the transition to the chloramine treatment. Figure 16 shows the [THM] during this period of time where percentage reductions ranged from 72 to 79, and the average of 37μg/l satisfied the EPA's maximum contaminant level. There was a subsequent discussion about the effectiveness of chloramine disinfection [74]. An extremely strong argument was presented for use of chloramines as a primary disinfectant from 30 years of experience: "when properly applied at effective dosages [1.5–1.8 mg/l, in their case], chloramine produces 100% kills of pathogenic bacteria species and also effectively reduces total bacterial populations to an acceptable range" [75].

Several options are available to control THM and achieve disinfection at the same time [76]:

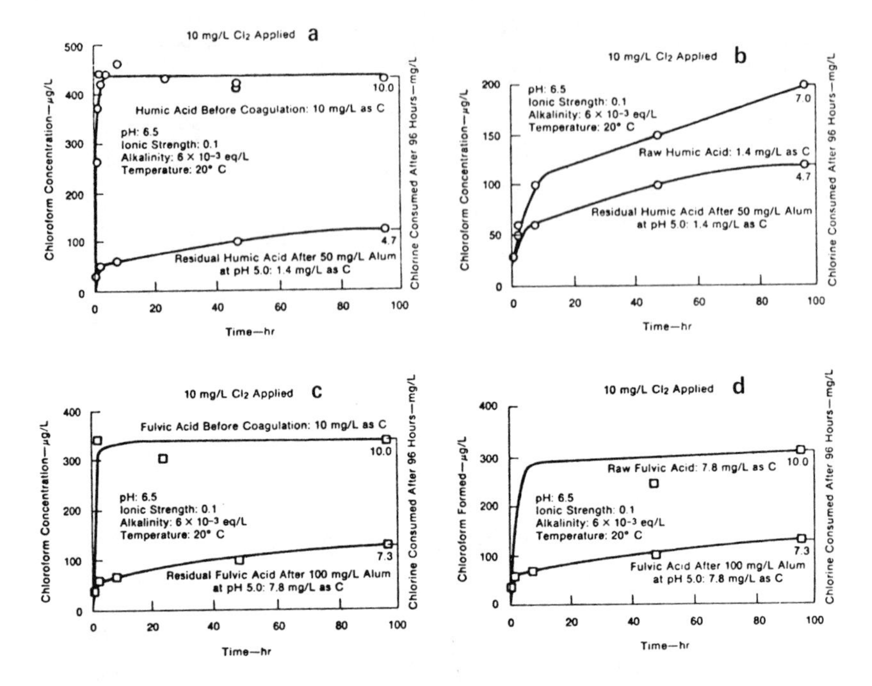

Figure 15. **(a)** Effect of coagulation of humic acid on subsequent chloroform formation. **(b)** Preferential removal of chloroform precursors by coagulation of humic acid with alum. **(c)** Effect of coagulation of fulvic acid on subsequent chloroform formation. **(d)** Preferential removal of chloroform precursors by coagulation of fulvic acid with alum. Reproduced from Babcock and Singer [72], courtesy of the American Water Works Association.

1. Use a disinfectant that does not produce THM.
2. Reduce the precursor concentration prior to chlorination.
3. Remove the THM after their formation.

In turn, there are several suboptions in each of the above. Also, there is the opportunity to employ a combination of all three.

Alternative disinfectants include chloramines (cited above), chlorine dioxide (ClO_2) and ozone (O_3). The last two oxidants can be used also to reduce precursor concentrations prior to disinfection. An argument against use of chloramine disinfection is the "100-fold increase in contact time to inactivate coliform bacteria and enteric pathogens as compared to free available chlorine" [76]. This is an "old" argument that has been rebutted many times. The experience of Brodtmann and Russo [75] has been repeated throughout the water treatment industry. Chlorine dioxide is not without problems. Chlorite ions (ClO_2^-) have a

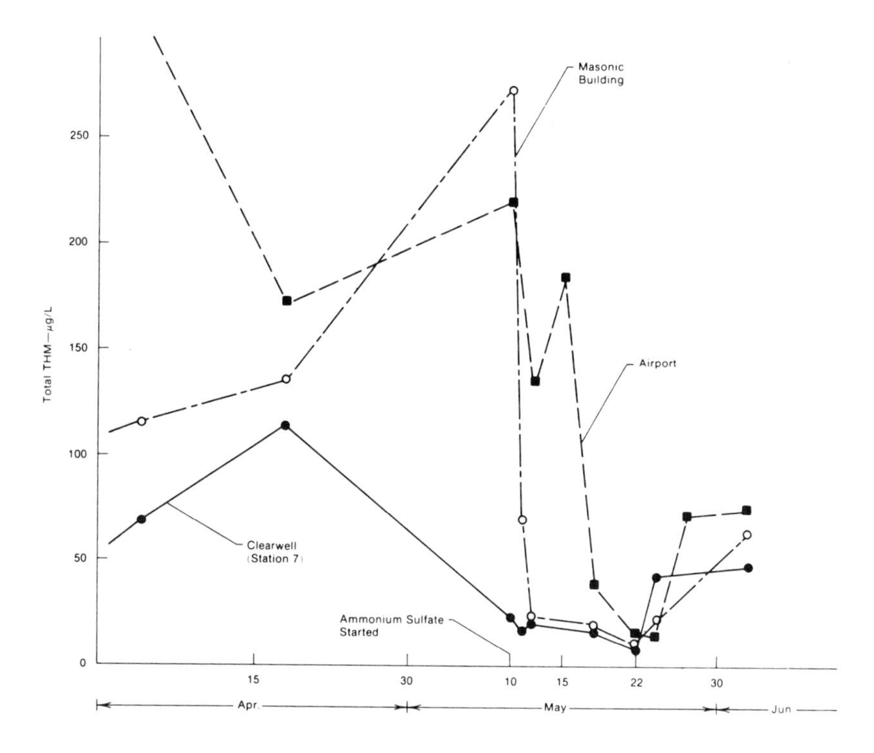

Figure 16. THM reduction using ammonium sulfate. Reproduced from Norman et al. [73], courtesy of the American Water Works Association.

deleterious effect on the red blood cell survival rate in cats when [ClO_2] exceeds 10 mg/l. Consequently, the EPA has recommended that the chlorite ion concentration should not exceed 0.5 mg/l. Commonly used water treatment alternatives are given in Figure 17 [76]. where the logic of "the least costly approach" was used. Any treatment modifications generally involve maximizing the efficiency of precursor removal during coagulation, flocculation, sedimentation and filtration as well as changing the chlorination point or dosage. The last recommendation has been successful at Cincinnati, Ohio [76] and Louisville, Kentucky [77] where reductions of THM were 75% and 40–50%, respectively. In the case of precursor removal, reductions can be increased by optimizing the coagulant dosage, pH and mixing conditions. Many laboratory studies have indicated that humic and fulvic acids are removed by coagulation to approximately 90 and 60%, respectively [78]. However, higher-than-normal coagulant dosages are required. Another modification is the installation of a second chlorination point in the distribution system. This was successful at the East Bay Municipal Utility District, Califor-

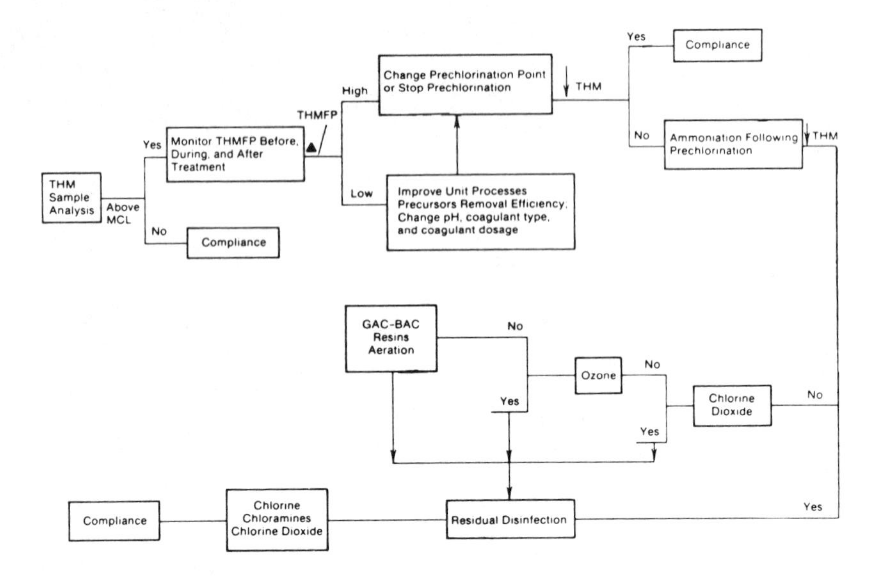

Figure 17. General approach to THM control. Reproduced from Vogt and Regli [76], courtesy of the American Water Works Association.

nia, where the initial chlorine dosage was decreased from 2.0 to 0.5 mg/l and an increase of the chlorine residual in the distribution system to 0.5 mg/l was effected. The [THM] was reduced by 50% at the consumer's tap [79]. More information on precursor removal and control of THM is given in Chapters 4 and 5.

An EPA publication (EPA 600/2-81-156), "Treatment Techniques for Controlling Trihalomethanes in Drinking Water," appeared in late 1981. It addresses the Groups 1 and 2 technologies for THM control that were cited in Chapter 1. In this publication, which is a compilation of several studies, three approaches were investigated for THM control: removal of THM, removal of THM precursors and the use of alternative disinfectants to chlorine. Of these three technologies, the applications of alternative disinfectants are the most effective and the least costly. ClO_2, O_3 and chloramines produce no significant concentrations of THM when used as disinfectants (see Chapter 10). However, the major disadvantage of this approach is that is does not remove precursors of the THM. Instead, ClO_2 and the chloramines may produce some other types of halogenated products from the organic matter in the water (see above). Also, there are some other disadvantages of these three disinfectants (summarized below).

For removal of the THM and their precursors, nine approaches are available: oxidation, aeration, adsorption, clarification, ion exchange, biodegradation, pH adjustment, source control and intense mixing during disinfection. Within the nine treatments, 19 techniques were researched. Table XIX summarizes the effectiveness of these techniques for control of THM. Estimated costs are available also in this EPA publication.

PHENOLIC COMPOUNDS

Hydroxy derivatives of benzene (i.e., phenol, C_6H_5OH) require special mention because of their possible effects on the taste and odor quality of potable water and potential toxicity to aquatic life in natural waters [80]. The role of phenol is these two effects on water quality is somewhat controversial, largely because of the analytical difficulties of quantitatively detecting and identifying these compounds in natural waters.

Naturally Occurring Sources

Many phenolic compounds may be derived from natural products. For example, seven hydroxy derivatives of benzene were found from degradative studies of the "color macromolecules" [81]. These compounds were: catechol, resorcinol, vanillin, vanillic acid, syringic acid, protocatechuic acid and 3,5-dihydroxybenzoic acid. In natural waters, there is the possibility that these compounds may arise from the microbially catalyzed degradation of color molecules. Many of these phenols have been recovered and identified in two rivers in Japan [82].

A rather unique source of naturally occurring phenolic compounds was reported [83]. The phenolic acids *m*-hydroxybenzoic acid, *m*-hydroxyphenylacetic acid and *m*-hydroxyphenylpropinoic acid were isolated from liquid manure (defined as urine from domestic animals). Average amounts of these three compounds were 35.8, 27.4 and 40.0 mg/l, respectively. In view of the rather substantial population of domestic farm animals in the world, liquid manure may represent a significant source of naturally occurring phenolic compounds.

Anthropogenic Sources

Many sources may be cited for industrially sponsored phenolic compounds: petroleum refineries, manufacturers of synthetic organic chemicals, ammoniacal liquors from coke ovens, domestic wastewater,

Table XIX. Effectiveness of THM-Removal Techniques

Treatment	Precursor	Trihalomethanes	Other Byproducts	Disinfection	Remarks
Ozone Oxidation	Good to very good destruction is technically feasible. The apparent concentration may increase at low doses. High doses and long contact times are required for good destruction, and complete destruction is difficult.	No effect by ozone, some incidental gas stripping.	Some are formed, but they will not contain chlorine, unless free chlorination or chlorine dioxide is employed. Bromine-containing THM may not be formed on later chlorination.	Excellent, but no residual is created. Organisms may regrow in the distribution system.	Slightly better at high pH.
Chlorine Dioxide Oxidation	Good destruction is technically feasible, but complete destruction was not achieved.	No effect.	Some are formed by the process and some will contain halogen.	Good and provides a residual. Slightly more effective at high pH.	Residual oxidant should be limited to 0.5 mg/l because of health effect.
Potassium Permanganate Oxidation	Fair destruction is technically feasible, but complete destruction was not achieved.	No effect.	Some are formed by the process and some will contain halogen, if free chlorine or chlorine dioxide is used.	Poor; a disinfectant must be used.	Pink water with overdose. Better at high pH.
Lowering pH	Fair decline of TermTHM concentration is technically feasible.	No effect.	None formed by the process, but some formed during final disinfection.	Free chlorine is more effective at lower pH.	May cause some corrosion problems.

	Affects the rate of reaction between free chlorine and precursor, thereby lowering resulting THM concentration.				
Diffused-Air Aeration	No effect and THM will form if free chlorine is used.	Good to very good removal is technically feasible, but bromine-containing THM are harder to remove than chloroform. High air to water ratios are difficult to achieve.	None are known to be formed by the process but some are still formed during disinfection. Byproducts will contain halogen if free chlorine or chlorine dioxide is used.	A disinfectant is required	Influent air can be cleaned. Possible air pollution problems. Removes regulated contaminant. Some removal of SOC[a] and T&O#[b] compounds.
Tower Aeration	No effect and THM will form if free chlorine is used.	Good to very good removal is feasible, but bromine-containing THM are harder to remove. High air to water ratios can be achieved.	None known to be formed by the process, but some are still formed during disinfection. Byproducts will contain halogen if free chlorine or chlorine dioxide is used.	A disinfectant is required.	Difficult to clean air, may entrain particulates. Possible air pollution problems. Removes regulated contaminant. May have to protect from freezing. Some removal of SOC and T&O# compounds.
Powdered Activated Carbon Adsorption	Good to very good removal is feasible. Removal is influenced by influent concentration	Good to very good removal is feasible. Bromine-containing THM are	None are formed by the process. Some removal of those coming to the process and less	Removes chlorine, so must post-disinfect. Some	Some removal of SOC and T&O# compounds. No desorption with decreasing concentration

Table XIX. continued

Treatment	Precursor	Trihalomethanes	Other Byproducts	Disinfection	Remarks
	and the loading is proportional to the influent concentration.	better adsorbed than chloroform. Removal is influenced by influent concentration and the loading is proportional to the influent concentration.	reformation as related to TOC removal. Will contain halogen if chlorine or chlorine dioxide is used.	reduction in disinfectant demand.	because PAC only used once. Sludge disposal a problem.
Granular Activated Carbon Adsorption	Good to very good removal technically feasible. Removal is nearly complete when adsorbent is fresh, then breakthrough toward exhaustion begins. Complete exhaustion generally does not occur, however. Loading is proportional to influent concentration and desorption may occur when the influent concentration declines.	Good to very good removal is technically feasible. Removal is nearly complete when adsorbent is fresh but then breakthrough to exhaustion occurs. Bromine-containing THM adsorbed better than chloroform. Loading is proportional to influent concentration and desorption will occur if the	None formed by the process and some can be removed. Because of good TOC removal, fewer are formed during disinfection.	Chlorine removed, so postdisinfection required. Disinfectant demand is lower than when TOC is removed.	SOC & T&O# compounds also removed. Requires reactivation or replacement. Complete removal does not last long. Possible corrosion problems if effluent TOC concentration near zero.

		influent concentration drops.			
Clarification by Coagulation, Sedimentation, Filtration	Good removal is feasible. If reaction with free chlorine is fast, delaying chlorination to after clarification will permit more removal. More removal will occur at lower pH but the reaction between free chlorine and precursor will be slower.	No effect.	None formed by the process and some may be removed. Because of TOC removal, fewer are formed later during disinfection. Some will contain halogen if free chlorine or chlorine dioxide is used.	Disinfectant demand lower if disinfection is delayed.	Sludge disposal problem. Iron salts may be somewhat better than alum.
By Precipitative Softening	Good removal is technically feasible. The faster reaction rate between free chlorine and precursor at higher pH should result in additional benefit by delaying chlorination.	No removal by process. High pH accelerates reaction to form THM.	None formed by the process. Because of TOC removal, fewer are formed during disinfection. Some will contain halogen if free chlorine or chlorine dioxide is used.	Effectiveness of free chlorine reduced at higher pH. Disinfectant demand will be lower if disinfection delayed.	Sludge disposal a problem.
By Direct Filtration	Good removal is technically feasible. THM concentrations will be lower if chlorination is delayed to after the process.	No effect.	None formed by the process. Because of TOC removal, fewer are formed during disinfection. Some will contain halogen if free chlorine or chlorine dioxide is used.	Disinfectant demand lower if disinfection follows clarification.	Little sludge produced. May require polymers.

[a]SOC = Synthetic Organic Compound.
[b]T&O# = Taste and Odor Number.
[c]TOC = Total Organic Carbon.

fungicides and pesticides, hydrolysis and photochemical oxidation of organophosphorus and carbamate pesticides, and microbial degradation of phenoxyalkyl acid herbicides. The first four sources may be, perhaps, the most significant and provide the greatest contents to the aquatic environments. Typical phenols and their contents in petroleum refinery effluents are given in Table XX [84] and in ammoniacal liquors from coke ovens in Table XXI [85]. The phenolic content of a typical domestic wastewater is in the 108–128 μg/l range [86].

Contents of Surface and Groundwaters

Few data surprisingly are available in the literature that document phenolic contents of surface waters of the United States. It would appear from the many sources of phenols cited above that they should have ubiquitous occurrence. Some indication may be obtained from surveys of the Delaware Estuary and several of its estuaries [87]. Table XXII shows these data were produced from the wet, colorimetric 4-aminoantipyrine [4-AAP] method for phenolic compounds [88,89]. This method does not yield a "total" phenol content largely because 4-AAP does not react with

Table XX. Determination of Phenols in Refinery and Petrochemical Plant Effluents[a]

	Concentration of Phenols Found in Various Streams by Chromatographic Method (ppm)		
Type of Phenol	**Biologically Treated Refinery Wastewater[b]**	**Once-through Refinery Cooling Water[c]**	**Petrochemical Plant Process Wastewater[d]**
Phenol	0.01	0.02	0.36
Cresols	0.22	0.22	0.04
Dimethyl Phenols	2.59	0.02	0.40
Trimethyl Phenols	0.71	0.03	0.17
Other Phenolics	0.31	0.07	0.21
Total Phenols	3.83	0.36	1.18
Total Phenols by Colorimetric Method	2.3	0.39	1.03

[a]Reproduced from Chriswell et al. [84], courtesy of the American Chemical Society.
[b]Major components: 2,5-dimethyl phenol (0.87 ppm), 3,5-dimethyl phenol (0.82 ppm).
[c]Major components: *m*- and *p*-cresol (0.20 ppm).
[d]Major components: phenol (0.36 ppm), 3,4-dimethyl phenol (0.32 ppm).

Table XXI. Concentration of the Major Phenols in Various Samples (mg/l)[a]

	A[b]	B	C	D	E
Phenol	2290	825	1375	1105	0.8
o-Cresol	358	100	386	272	0.7
m-Cresol	735	156	634	390	0.4
p-Cresol	376	109	485	212	0.2
o-Ethylphenol	3	1	20	25	0.1
m-Ethylphenol	35	11	71	17	0.2
p-Ethylphenol	14	3	112	2	0.3
2,3-Xylenol	26	5	117	6	0.4
2,4-Xylenol, 3,5-Xylenol	128	41	242	62	0.9
2,5-Xylenol	48	10	57	17	0.3
2,6-Xylenol	12	4	138	5	0.2
3,4-Xylenol	21	5	60	7	0.6
Total Monohydric Phenols	4046	1270	3697	2120	5.1
Catechol	28	8	3330	ND	0.7
3-Methylcatechol	ND	5	900	ND	ND
4-Methylcatechol	4	2	1200	ND	0.1
Resorcinol	22	7	150	ND	ND
Others	ND	2	160	ND	ND
Total Dihydric Phenols	54	24	5740	NIL	0.8
Total Phenols	4100	1294	9437	2120	5.9

[a] Reproduced from Cooper and Wheatstone [85], courtesy of the International Association on Water Pollution Research.

[b] A, coke-oven ammoniacal liquor; B, coke-oven ammoniacal liquor; C, low-temperature carbonization ammoniacal liquor; D, gas cooling water; E, drainage water; ND, not detected (i.e. less than 0.1 mg/l).

many parasubstituted phenols. As such, the concentrations given in Table XXII should be viewed as minimum values. How much greater is the actual phenol content is not known with any degree of certainty. The Delaware Estuary receives many industrial and several petroleum refinery effluents. These phenol contents reflect many, many sources.

The occurrence of "total" phenols was reported downstream from refinery outfalls in the region of Montreal [90]. Phenol concentrations seen in Table XXIII do, indeed, decrease with downstream distance from the source. Mass balance calculations at the many transverse sections indicated that this decrease in concentration was not attributed to dilution only. Microbially mediated degradation was a contributing factor.

On the other hand, phenolic compounds are quite persistent under the

Table XXII. Summary of the Phenolics Reconnaissance Survey of the Delaware Estuary, July 1974–June 1976 [87]

Sites	No. Sites	No. Samples	Phenolic Concentration (μg/l)	
			Average	Range
Delaware				
Water, All sampling	73	189	11	1–142
Water, Shore sampling[a]	27	64	17	1–142
Water, Boat sampling	46	125	8	1–32
Upper estuary[a]	24	85	10	1–32
Lower estuary[b]	22	40	5	1–29
Muds	58	64	227[c]	<10–917[c]
Tributaries, Water, Shore	14	17	16	1–36
Delaware, Above Trenton, Water	7	13	3	1–8
Storm Runoff, Pennypack Sewer[d]	1	25	16	1–32
Refinery Effluent, Arco	2	8	15	1–40

[a]Trenton to Artificial Island.
[b]Below Artificial Island.
[c]μg/kg, dry wt.
[d]Two storm events.

"proper conditions." These conditions are usually found in groundwater aquifers. For example, in 1962, the contamination of groundwaters by phenols after the end of World War II was observed at Alma, Michigan [91]. An accidental spill of commercial carbolic acid (95–100% phenol, C_6H_5OH) occurred in southeastern Wisconsin in July 1974 [92]. After 19 months of monitoring, the groundwater in the area surrounding the spill remained contaminated. Phenol contents ranged from 0.036 to 1130 mg/l, whereupon 200 mg/l remained at the conclusion of the monitoring period in the well nearest the spill (120 m).

Several specific phenolic compounds were identified in the Delaware Estuary survey [87]: *o*-chlorophenol, *o*-cresol, 6-chloro-*o*-cresol, *o*-nitrophenol and 3,5-xylenol were identified by gas-liquid chromatography. Such identification techniques can provide much information about sources of the phenols.

Environmental Hazards

Industrially sponsored phenols and their derivatives represent a class of undesirable organic compounds in aquatic environments. Their effect

Table XXIII. Phenol Concentrations in Transverse Sections at Various Distances from the North Shore of the St. Lawrence River (concentrations are mean values of measurements at different depths (summer) or single values (winter))[a]

	Distance from North Shore (m)	Phenol Concentration (ppb)	Mean Flow Velocity (m/sec)
Section 1: Sampling Date, June 6, 1972	90	2.6	
	180	1.8	
	300	2.8	
	650	1.3	
	850	0.8	
Section 2: Sampling Date, July 26, 1972	30	61.0	
	60	61.5	
	100	18.3	
	120	15.8	
	150	4.9	
	180	0.8	
Section 3: Sampling Date, August 3, 1972	30	52.0	
	60	45.0	
	100	22.3	
	120	29.5	
	180	11.0	
	250	1.3	
Section 4: Sampling Date, July 11, 1972	90	11.9	
	180	12.3	
	300	5.8	
	400	0.9	
	1000	1.8	
	1200	1.1	
	1400	1.0	
	1500	0.5	
Section 6: Sampling Date, July 19, 1982	90	8.2	
	180	4.0	
	300	1.4	
Section 4: Sampling Date, January 31, 1973	60	95.5	0.26
	130	93.5	0.42
	180	25.0	0.40
	250	16.0	0.36
	400	1.2	0.40
	600		0.42
Section 5: Sampling Date, January 30, 1973	30	87.2	0.24
	90	76.0	0.39
	150	30.1	0.40
	240	1.0	0.32

Table XXIII. continued

	Distance from North Shore (m)	Phenol Concentration (ppb)	Mean Flow Velocity (m/sec)
	300	2.0	0.16
	450	<0.5	0.18
Section 7: Sampling Date, January 30, 1973	30	68.9	0.47
	60	84.2	0.50
	90	57.8	
	150	11.0	0.54
	200	5.3	0.45
	250	<0.5	0.37

[a]Reproduced from Polisois et al. [90], courtesy of the Fisheries Research Board of Canada.

Table XXIV. Concentrations (in $\mu g/l = ppb$) of Chlorophenols in the River Rhine at Kilometer 865 (Lobith) in 1976 and 1977[a]

	1976			1977		
Phenol	Frequency (%)	Max.	Med.	Frequency (%)	Max.	Med.
2-Chlorophenol	2	2.3	—[b]	0	—	—
3-Chlorophenol	4	6.0	—	0	—	—
4-Chlorophenol	16	3.9	—	0	—	—
2,3-Dichlorophenol	10	0.72	—	25	0.86	—
2,4-Dichlorophenol	47	0.59	—	48	0.35	—
2,5-Dichlorophenol	37	0.29	—	35	0.26	—
2,6-Dichlorophenol	75	0.45	0.15	64	0.43	0.10
3,4-Dichlorophenol	0	—	—	2	0.23	—
3,5-Dichlorophenol	37	0.52	—	12	0.77	—
2,3,4-Trichlorophenol	0	—	—	4	0.21	—
2,3,5-Trichlorophenol	0	—	—	12	0.18	—
2,3,6-Trichlorophenol	4	0.05	—	23	0.11	—
2,4,5-Trichlorophenol	80	0.61	0.15	46	0.66	—
2,4,6-Trichlorophenol	94	2.5	0.19	87	0.51	0.18
3,4,5-Trichlorophenol	6	0.07	—	29	0.37	—
2,3,4,5-Tetrachlorophenol	0	—	—	4	0.14	—
2,3,4,6-Tetrachlorophenol	88	0.38	0.14	90	0.56	0.08
2,3,5,6-Tetrachlorophenol	12	0.06	—	65	0.17	0.03
Pentachlorophenol	100	2.4	0.73	100	11	1.1

[a]Reproduced from Wegman and Hofster [100], courtesy of the International Association on Water Pollution Research.
[b]— = Nondetectable.

on the organoleptic quality of potable water is reasonably well documented and is considered in more detail in Chapter 3. Insofar as aquatic life is concerned, phenolics are acutely toxic to fish [80]. For example, acute toxicity of phenol (C_6H_5OH) to minnows is 0.07 mg/l in 30 min [93]. There exists no adequate documentation, to date, of chronic effects of phenolics on fish and other aquatic organisms, and little knowledge exists of sublethal effects on various trophic levels. Also, the extent of biomagnification of phenolics is essentially unknown [94].

The tainting of fish flesh occurs at much lower phenolic concentrations than does toxic activity. In fact, *o*-chlorophenol has caused tainting at a level of 0.001 mg/l [95]. Taste of certain fish has been affected by 2,4-dichlorophenol at the 0.001-mg/l level and by 2-methyl-6-chlorophenol at a concentration of 0.003 mg/l [92]. An EPA-sponsored study of the effect of water pollutants on flavor of fish showed that of 27 organic compounds, a chlorophenol had the greatest effect, with a threshold concentration of only 0.0004 mg/l [95]. It has been recommended that receiving waters which support usable fish populations should, at no time or at any place, contain phenolic concentrations greater than 0.1 mg/l [93]. A study of fish taken from the Rhine and Elbe Rivers considered the low levels of phenol present (0.02–0.07 mg/l) to be damaging to fish [96]. Considerable pathological lesions were found in the fish.

Halogenated Compounds

Halogenated phenolic compounds frequently occur from the chlorination of waters, such as natural waters or wastewaters, that contain precursors or a parent molecule. Several examples are given here and in Chapter 3.

Hypochlorous acid (HOCl) was added to aqueous solutions of 2,6-dichloro-, 2,4-dichloro-, 2,4,5-trichloro-, 2,4,6-trichloro-, 2,3,4,6-tetrachloro-, penta-, and *o*-nitrophenol at pH values of 3.5 and 6.0 [97–99]. Several products were identified: chlorinated phenols, chloro-*p*-benzoquinones, chlorinated 2,5-cyclohexadienes and derivatives of cyclohexenone. Reaction periods of 20–24 hr were employed. These chlorinated products may very well occur in drinking waters in the process of superchlorination.

That chlorophenols occur in contaminated surface waters in substantial quantities and frequency was reported for several rivers in the Netherlands. Typical data are seen in Table XXIV for the Rhine River [100]. 2,6-Dichlorophenol, 2,4,5-trichlorophenol, 2,4,6-trichlorophenol, 2,3,4,6-tetrachlorophenol and pentachlorophenol occur with the greatest frequencies and have the highest concentrations. Industrial and/or municipal wastewater discharges were undoubtedly the sources of these chlorophenols. Pentachlorophenol is used extensively as a wood preservative. Its occurrence in natural aquatic environments is almost ubiquitous [101].

Chlorophenols are formed also from the chlorination of naturally occurring organic compounds [102]. Hypochlorous acid was reacted with two natural compounds: *p*-hydroxybenzoic acid and vanillic acid under a variety of conditions. A mixture of 4-chlorophenol, 2,4-dichlorophenol and 2,4,6-trichlorophenol was produced when *p*-hydroxybenzoic acid was chlorinated. These kinds of chlorination reactions presumably account for the occurrence of chlorophenols in such natural waters as the Rhine River, Delaware Estuary and Weser Estuary [103].

Bromophenols may be formed during chlorination of phenol (C_6H_5OH) in the presence of bromide ions. The formation of 2,4,6-tribromophenol was reported under these conditions [104]. A brominated phenol was observed also in the process of chlorinating humic acid with Br^- in the system [105].

A special notation should be given to the distribution of pentachlorophenol (PCP) in aquatic environments. Many of the reports [97–105] cite the occurrence of this chlorinated phenol in natural waters and, frequently, in the attendant bottom sediments. In addition, significant quantities of PCP were detected in sewage influent and effluent, Willamette River water and subsequently the finished drinking water of Corvallis, Oregon [106]. "Potentially dangerous levels" of PCP were found in a tributary of the Delaware River [107]. An industrial wastewater discharge was responsible for PCP concentrations as high as 10.5 ppm. Likewise, significant quantities of PCP were found in storm water runoff into the Delaware Estuary [101] and from industrial discharges into two urban rivers near Tokyo, Japan [82]. Additional sources of chlorinated phenols, in general, are human urine [108] and PCP, in particular [109].

REFERENCES

1. *Willing Water,* 18(12):4 (1974).
2. Faust, S. D. In: *Fate of Pollutants in the Air and Water Environments* I. H. Suffet, Ed. (New York: Wiley-Interscience, 1976).
3. Faust, S. D., and O. M. Aly. In: *Chemistry of Natural Waters* (Ann Arbor, MI: Ann Arbor Science Publishers, Inc., 1981).
4. Lichtenberg, J. J. et al. *Pestic. Monit. J.* 4(2):71 (1970).
5. Crump-Wiesner, H. J. et al. *Pestic. Monit. J.* 8(3):157 (1974).

6. Kleopfer, R. D., and B. J. Fairless. *Environ. Sci. Technol.* 6:1036 (1972).
7. Tardiff, R. G., and M. Deinzer. *Proc. 15th Water Quality Conf.,* University of Illinois, Urbana, IL (1973), pg. 23.
8. Bellar, T. A. et al. *J. Am. Water Works Assoc.* 66:703 (1974).
9. Dowty, B. et al. *Science* 187:75 (1975).
10. Dowty, B. et al. *Environ. Sci. Technol.* 9:762 (1975).
11. Symons, J. M. et al. *J. Am. Water Works Assoc.* 67:634 (1975).
12. "Chloroform, Carbon Tetrachloride, and Other Halomethanes: An Environmental Assessment," National Academy of Sciencies, Washington, DC (1978).
13. Lingg, R. D. et al. *J. Am. Water Works Assoc.* 69:605 (1977).
14. Coleman, W. E. et al. In: *Identification and Analysis of Organic Pollutants in Water,* L. H. Keith, Ed. (Ann Arbor, MI: Ann Arbor Science Publishers, Inc., 1981).
15. Morris, R. L., and L. G. Johnson. *J. Am. Water Works Assoc.* 68:492 (1976).
16. Hoehn, R. C. et al. *J. Environ. Eng. Div., ASCE* EE5:803 (1977).
17. Glaze, W. H., and R. Rawley. *J. Am. Water Works Assoc.* 71:509 (1979).
18. Argnello, M. D. et al. *J. Am. Water Works Assoc.* 71:504 (1979).
19. Brett, R. W., and R. A. Calverley. *J. Am. Water Works Assoc.* 71:516 (1979).
20. Smith, V. L. et al. *Environ. Sci. Technol.* 14:190 (1980).
21. Veenstre, J. N., and J. L. Schnoor. *J. Am. Water Works Assoc.* 72:583 (1980).
22. Schreiber, J. S. *J. Am. Water Works Assoc.* 73:154 (1981).
23. Middleton, F. M. et al. *J. Am. Water Works Assoc.* 44:538 (1952).
24. Middleton, F. M., and A. A. Rosen. *Pub. Health Reports* 71:1125 (1956).
25. Middleton, F. M., and J. J. Lichtenberg. *Ind. Eng. Chem.* 52:99A (1960).
26. Middleton, F. M. *Chem. Eng. Prog. Symp. Ser.* 59(45):26 (1963).
27. Burnham, A. K. et al. *Anal. Chem.* 44:139 (1972).
28. Burnham, A. K. et al. *J. Am. Water Works Assoc.* 65:722 (1973).
29. Gustafson, R. L., and J. Paleos. In: *Organic Compounds in Aquatic Environments,* S. D. Faust and J. V. Hunter, Eds. (New York: Marcel Dekker, Inc., 1971).
30. "Interim Primary Drinking Water Regulations," *Federal Register* 40(248):59566 (1975).
31. Green, R. S., and S. K. Love. *Pestic. Monit. J.* 1(1):13 (1967).
32. Feltz, H. R. et al. *Pestic. Monit. J.* 5(1):54 (1971).
33. Lichtenberg, J. J. et al. *Pestic. Monit. J.* 4(2):71 (1970).
34. Dudley, D. R., and J. R. Karr. *Pestic. Monit. J.* 13:155 (1980).
35. Reynolds, L. M. *Residue Reviews* 34:27 (1971).
36. Duke, T. W. et al. *Bull. Environ. Contam. Toxicol.* 5(2):171 (1970).
37. Ahling, B., and S. Jensen. *Anal. Chem.* 42(13):1483 (1970).
38. Haile, C. L. et al. "Chlorinated Hydrocarbons in the Lake Ontario Ecosystem (IFYGL)," EPA-660/3-75-022, U.S. EPA, Corvallis, OR (1975).
39. Coleman, W. E. et al. *Environ. Sci. Technol.* 14:576 (1980).
40. McConnell, G. et al. *Endeavor* 34(121):13 (1975).
41. Lovelock, J. A. et al. *Nature* 241:194 (1973).

42. Thompson, G. M., and J. M. Hayes. *Water Resources Res.* 15:546 (1979).
43. Rook, J. J. *Proc. Soc. Water Treat. Exam.* 23Pt.(2):234 (1974).
44. Rook, J. J. *J. Am. Water Works Assoc.* 68:168 (1976).
45. Rook, J. J. *Environ. Sci. Technol.* 11:478 (1977).
46. Dragunor, S. In: *Soil Organic Matter,* M. M. Kononova, Ed. (New York: Pergamon Press, Inc., 1961), pg. 45.
47. Kleinhempel, D. *Albrecht Thaer Arch.* 14:3 (1970).
48. Morris, J. C., and B. Baum. In: *Water Chlorination, Vol. 2,* R. L. Jolley et al., Eds. (Ann Arbor, MI: Ann Arbor Science Publishers, Inc., 1978).
49. Hendrickson, J. B. *The Molecules of Nature* (Menlo Park, CA: W. A. Benjamin, Inc., 1965).
50. Stevens, A. A. et al. *J. Am. Water Works Assoc.* 68:615 (1976).
51. Babcock, D. S., and P. C. Singer. *J. Am. Water Works Assoc.* 71:149 (1979).
52. Oliver, B. G., and J. Lawrence. *J. Am. Water Works Assoc.* 71:161 (1979).
53. Cohen, R. S. et al. *J. Am. Water Works Assoc.* 70:647 (1978).
54. Young, J. S., and P. C. Singer. *J. Am. Water Works Assoc.* 71:87 (1979).
55. Simmler, J. J. PhD Thesis, University of Washington, Seattle, WA (1978).
56. Nitzschke, J. L. MS Thesis, University of Iowa, Iowa City, IA (1978).
57. Peters, C. J. et al. *Environ. Sci. Technol.* 14:1391 (1980).
58. Hoehn, R. C. et al. *J. Am. Water Works Assoc.* 72:344 (1980).
59. Oliver, B. G., and D. B. Shindler. *Environ. Sci. Technol.* 14:1502 (1980).
60. Youssefi, M. et al. *J. Environ. Sci. Health* A13:629 (1978).
61. Norwood, D. L. et al. *Environ. Sci. Technol.* 14:187 (1980).
62. Schnoor, J. L. et al. *Environ. Sci. Technol.* 13:1134 (1979).
63. Oliver, B. G., and S. A. Visser. *Water Research* 14:1137 (1980).
64. Ghassemi, J., and R. F. Christman. *Limnol. Oceanog.* 13:583 (1968).
65. Trussell, R. R., and M. D. Umphres. *J. Am. Water Works Assoc.* 70:604 (1978).
66. Kavanaugh, M. C. et al. *J. Am. Water Works Assoc.* 72:578 (1980).
67. Morrison, R., and R. Boyd. *Organic Chemistry* (Boston, MA: Allyn & Bacon, Inc., 1973).
68. Noack, M. G., and R. L. Doerr. In: *Water Chlorination, Vol. 2,* R. L. Jolley et al., Eds. (Ann Arbor, MI: Ann Arbor Science Publishers, Inc., 1978).
69. McCreary, J. J., and V. L. Snoeyink. *Environ. Sci. Technol.* 15:193 (1981).
70. Snoeyink, V. L. et al. *Environ. Sci. Technol.* 15:188 (1981).
71. Young, J. S., and P. C. Singer. *J. Am. Water Works Assoc.* 71:87 (1979).
72. Babcock, D. B., and P. C. Singer, *J. Am. Water Works Assoc.* 71:149 (1979).
73. Norman, T. S. et al. *J. Am. Water Works Assoc.* 72:176 (1980).
74. White, G. C., and T. S. Norman. *J. Am. Water Works Assoc.* 73:63 (1981).

75. Brodtmann, N. V., and P. J. Russo. *J. Am. Water Works Assoc.* 71:40 (1979).
76. Vogt, C., and S. Regli. *J. Am. Water Works Assoc.* 73:33 (1981).
77. Hubbs, S. A. et al. In: *Water Chlorination, Vol. 2,* R. L. Jolley et al., Eds. (Ann Arbor, MI: Ann Arbor Science Publishers, Inc., 1978).
78. Kavanaugh, M. *J. Am. Water Works Assoc.* 70:613 (1978).
79. Carns, K. E., and K. B. Stinson. *J. Am. Water Works Assoc.* 70:637 (1978).
80. "Report on Monohydric Phenols and Inland Fisheries," *Water Res.* 7:929 (1973).
81. Christman, R. F., and M. Ghassemi. *J. Am. Water Works Assoc.* 58:723 (1966).
82. Matsumoto, G. et al. *Water Res.* 11:693 (1977).
83. Kump, O. *Water Res.* 8:899 (1974).
84. Chriswell, C. D. et al. *Anal. Chem.* 47:1325 (1975).
85. Cooper, R. L., and K. C. Wheatstone. *Water Res.* 7:1375 (1973).
86. Hunter, J. V. In: *Organic Compounds in Aquatic Environments,* S. D. Faust and J. V. Hunter, Eds. (New York: Marcel Dekker, Inc., 1971).
87. Hunt, G. et al. MS Thesis, Rutgers—The State University, New Brunswick, NJ (1976).
88. Faust, S. D., and O. M. Aly. *J. Am. Water Works Assoc.* 54:235 (1962).
89. Faust, S. D., and E. W. Mikulewicz. *Water Res.* 1:509 (1967).
90. Polisois, G. et al. *J. Fish. Res. Bd. Can.* 32:2125 (1975).
91. Deutsch, M. *Proc. Soc. Water Treat. Exam.* 11:94 (1962).
92. Delfino, J. J., and D. J. Dube. *J. Environ. Sci. Health* A11:345 (1976).
93. "Water Quality Criteria Blue Book 1972," Environmental Studies Board, National Academies of Science and Engineering, Washington, DC (1974).
94. Burkema, A. L. et al. *Mar. Environ. Res.* 2:87 (1979).
95. Fetterolf, C. M. *Proc. Ind. Waste Conf. Purdue Univ.* 115:174 (1964).
96. Reichenbach-Klinke, H. D. *Arch. Fisch. Wiss.* 16:1 (1965).
97. Smith, J. G. et al. *Environ. Lett.* 10:47 (1975).
98. Smith, J. G. et al. *Water Res.* 10:985 (1976).
99. Smith, J. G., and S. F. Lee. *J. Environ. Sci. Health* A13:61 (1978).
100. Wegman, R. C. C., and A. W. M. Hofster. *Water Res.* 13:651 (1979).
101. Fillmore, L. et al. *J. Environ. Sci. Health,* A17(6):797 (1982).
102. Larson, R. A., and A. L. Rockwell. *Environ. Sci. Technol.* 13:325 (1979).
103. Eder, G., and K. Weber. *Chemosphere* 9:111 (1980).
104. Sweetman, J. A., and M. S. Simmons. *Water Res.* 14:287 (1980).
105. Quimby, B. D. et al. *Anal. Chem.* 52:259 (1980).
106. Buhler, D. R. et al. *Envion. Sci. Technol.* 7:929 (1973).
107. Fountaine, J. E. et al. *Water Res.* 10:185 (1976).
108. Edgerton, T. R. et al. *Anal. Chem.* 52:1774 (1980).
109. Edgerton, T. R., and R. F. Moseman. *J. Agric. Food Chem.* 27:197 (1979).

CHAPTER 3

TASTES AND ODORS IN DRINKING WATER

SOURCES OF TASTE AND ODOR

Consumers often judge the potability of drinking water by its organoleptic qualities, i.e., taste and odor. An off-tasting and/or odorous water is thought to be unsafe to drink. This association is frequently based on a fallacy, since odorous water may be free of pathogenic bacteria and may meet the appropriate bacteriological standards. Nevertheless, when consumers detect a change in water quality through taste and odor, they frequently will register a complaint with the water company. Complaints of taste and odor are those most frequently made about water quality.

Sources of tastes and odors are many and varied. They range from such biological sources as algae and actinomycetes to individual organic compounds, such as chlorophenols. Tastes and odors may occur also within the water distribution system through chemical reactions. The reaction between chlorine and phenol to form chlorinated phenols is cited frequently. Metallic tastes may arise from corrosion of iron and copper pipes. This chapter attempts to document the many sources of tastes and odors in drinking water. Emphasis is placed on the odor quality because *Standard Methods* [1] warns that "taste intensity measurements are more difficult than odor threshold tests, because the physical contact of the sample with the observer produces after effects which are more difficult to eliminate." Furthermore, there is always the concern about hazardous substances being in the water that may endanger the taste tester. *Standard Methods* [1] does provide a taste test, but suggests that it should be performed "only on samples known to be safe for ingestion." Also, taste and odor sensations are most likely to be associated together.

MEASUREMENTS OF TASTE AND ODOR

Weber-Fechner "Law"

It is extremely difficult to quantify and systematize the highly complex taste and odor sensations. Of the several attempts made to organize and describe these sensations, perhaps the most often cited relationship is the Weber-Fechner "law" [2]:

$$S = K \log R \tag{1}$$

where S = sensation
R = stimulus
K = constant peculiar to the odorous substance and the observer's sensitivity

This "law" states that the stimulus must increase geometrically if the sensation is to increase by an equal amount in an arithmetic proportion. This relationship is seen below in the threshold odor test. However, the Weber-Fechner "law" is subjective at best and does not have universal application.

Threshold Odor Number and Odor Intensity Index

Laboratory measurement of an odorous sensation of water may be reported by either the Threshold Odor Number (TON) or by the Odor Intensity Index (OII) [1,3]. The TON is the ratio by which the odorous sample is diluted with odor-free water to be "just detectable" to the observer. Mathematically, the TON is:

$$TON = \frac{A + B}{A} \qquad (2))$$

where A = amount (ml) of original sample,
B = amount (ml) of odor-free water.

TON is dimensionless. OII is the number of times the original sample must be diluted in half with odor-free water for the odor to be "just detectable." OII is mathematically expressed as:

$$OII = 3.3 \log \frac{200}{A} + 3D \tag{3}$$

where A = amount (ml) of original sample,
D = number of 25:175 primary dilutions required to reach the "just detectable" odor level.

The relationship between TON and OII is:

$$TON = 2^{OII} \quad (4)$$

In conducting the actual measurement, a panel of five or more testers is selected. The measurements ideally should be taken in an odor-free atmosphere and in a special room ventilated with air that is filtered through activated carbon and maintained at a constant temperature and humidity. The test consists essentially of preparing the sample and dilution water at 40°C. A higher temperature of 60°C may be employed also. A series of dilutions of the sample is prepared in accord with column 2 in Table I. Note that the original sample volume is 200 ml and that all subsequent dilutions are made to this volume. The observer may be presented with seven flasks in this order [1]:

ml sample diluted to 200 ml	12	0	17	25	0	35	50
Response	–	–	–	+	–	+	+

The odorous samples are presented to the observer in an increasing order of concentration. At least two blanks are inserted into the series. The appropriate TON value is recorded for each observer. The results are averaged arithmetically if the spread of values is not too great. The OII test is basically the same as the TON test with the exception that the observer is presented with three clean, coded and odor-free flasks for each test trial, of which two contain odor-free water. The reader should consult the latest edition of *Standard Methods* [1] for recent innovations in odor testing.

BIOLOGIC SOURCES OF TASTES AND ODORS

Surface water supplies, especially those contained in reservoirs, frequently have a taste and odor problem due to such biologic sources as actinomycetes and algae. Descriptions of these tastes and odors range from fishy to earthy, woody, musty, haylike, manurelike and geraniumlike. These descriptors are not standard or official terms, but they do

Table I. Dilution of Sample and Reporting of Results [1,3]

	Volume Transferred to Odor Flask (ml)[a]	Threshold Odor Number (Dilution Factor)	Odor Intensity Index
Original Sample	200	1	0
	100	2	1
	50	4	2
	25	8	3
	12.5	16	4
Dilution A (25 ml	50	32	5
of original sample	25	64	6
diluted to 200 ml)	12.5	128	7
Dilution B (25 ml	50	256	8
of dilution A	25	512	9
diluted to 200 ml)	12.5	1,024	10
Dilution C (25 ml	50	2,050	11
of dilution B	25	4,100	12
diluted to 200 ml)	12.5	8,200	13
Dilution D (25 ml	50	16,400	14
of dilution C	25	32,800	15
diluted to 200 ml)	12.5	65,500	16
Dilution E (25 ml	50	131,000	17
of dilution D	25	262,000	18
diluted to 200 ml)	12.5	524,000	19
	6.25	1,050,000	20

[a]Volume in odor flask made up to 200 ml with odor-free water.

represent an attempt to describe the organoleptic effects of microscopic aquatic organisms.

Actinomycetes

It was found in the early 1950s that tastes and odors in reservoirs of the southwestern United States were attributable to actinomycetes [4,5]. Much of the evidence came from laboratory cultures of various aquatic organisms from different lakes in the Southwest. These organisms, in turn, gave byproducts that had tastes and odors similar to those found in the lakes. This early research concentrated on two reservoirs, Lake

Overholser and Lake Hefner, that supply Oklahoma City. An important point came from these observations that, at a water temperature of 3°C in February, a TON of 22 was recorded despite low plankton counts [5]. Furthermore, the types of odors produced by actinomycetes were very offensive when the TON reached a value of five.

Another important and fundamental observation is concerned with the nitrogen requirements of the actinomycetes. Substantial quantities of nitrogen apparently are required for their growth and maintenance. A symbiotic association with algae is essential. A heavy growth of the green algae, *Cladophora,* was observed in the reservoirs. One hypothesis held that "certain bacteria were attacking the *Cladophora*" and slowly decomposing them with the subsequent release of nitrogen. Another possible source of nitrogen would be atmospheric nitrogen fixation by such blue-green algae as *Anabaena* and *Nostoc* [6]. In any event, "thousands of colonies of actinomycetes" were observed to be associated with the *Cladophora* [5].

Some additional and empirical information comes from Morris' [7] and Erdei's [8] observations of tastes and odors arising from Midwestern rivers. This organoleptic quality problem chronically occurs at runoff periods during late winter thaws. These tastes and odors are very intense but persist usually for only short periods of time and decrease with subsidence of the river flood stage. Furthermore, the taste was intensified by chlorination of the water which produced bitter "medicinal tastes and odors." Early chemical analysis indicated that the taste-producing compounds were primarily phenolic in nature. In one specific instance, the actinomycetes' density was correlated with TON (Figure 1) [8]. In the springs of 1961, 1962 and 1963, a taste and odor of a distinct musty or earthy character developed in the water supply of Cedar Rapids, Iowa [7], and Omaha, Nebraska [8]. Two genera of actinomycetes subsequently were isolated and identified: *Micromonospora* and *Streptomyces.* The earthy, musty odor rising from the laboratory plates was characteristic of the odor in the Iowa River at Cedar Rapids. Figure 1 shows the empirical relation between the numbers of actinomycetes per milliliter and the TON in the Missouri River at Omaha, Nebraska [8].

Actinomycetes can be a problem in water distribution systems since these microorganisms, under certain environmental conditions, can produce the typical earthy, moldy and musty tastes in pipes inside buildings [9]. In England, these tastes and odors have arisen in cold water pipes adjacent to hot water pipes and where there was little use of water at night or on weekends. These citations were made when the temperature was 20°C. This is consistent with the knowledge concerning the effect of temperature on the growth rate of actinomycetes. Growth nor-

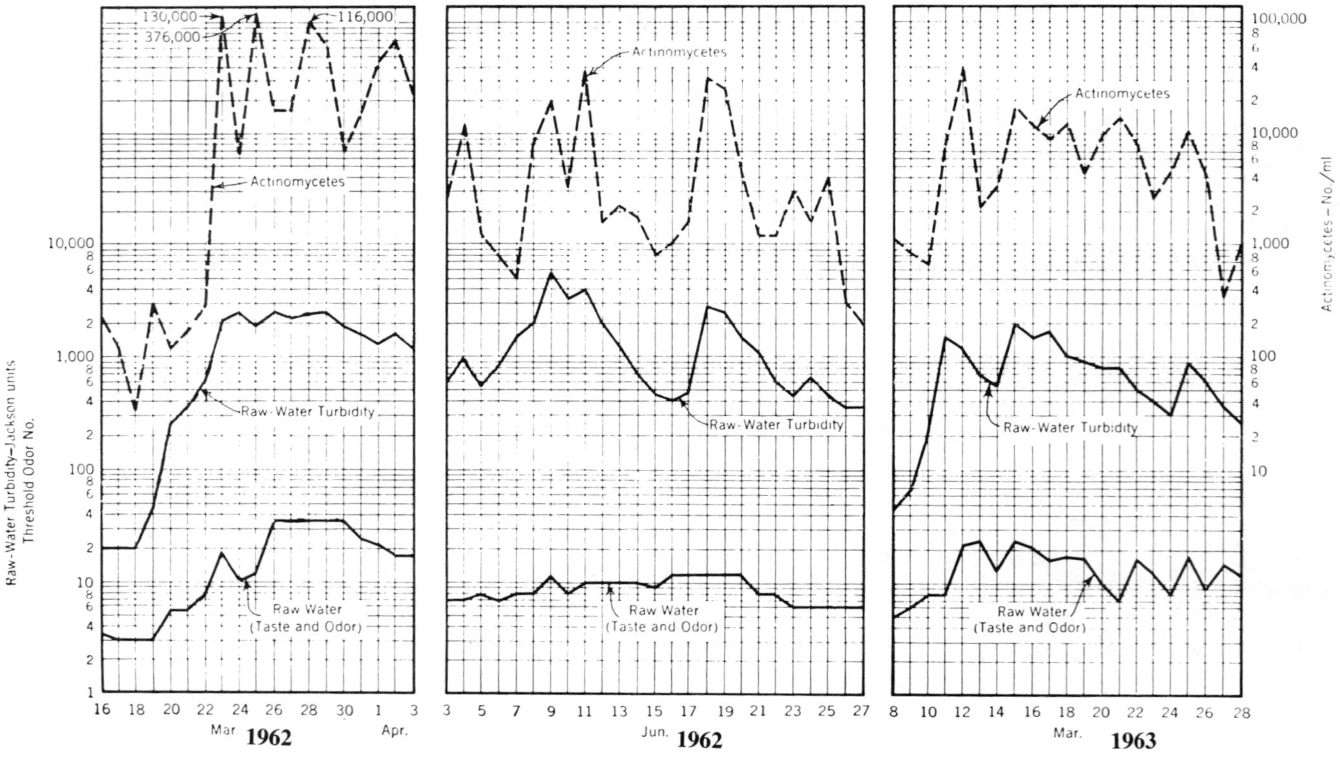

Figure 1. Densities of actinomycetes in raw river water. Reproduced from Erdei [8], courtesy of the American Water Works Association.

mally begins at approximately 7–8°C. However, rapid growth does not begin until after 20°C is reached. From this point, the growth rate is very rapid to 33–34°C whereupon it begins to diminish slightly. When 41°C is reached, growth diminishes abruptly. The other important growth factor, nutrients, may be provided by adsorption of the necessary organic matter on scales and encrustations on the internal side of the pipes. Some support for this contention may be seen in Table II that shows "typical" data for the occurrence of actinomycetes in a distribution system [9]. Another problem area may be found in the dead ends of distribution systems where the water does not circulate and where there is infrequent flushing. Scale, etc., tends to accumulate in these dead ends which then provide the necessary nutrients for the growth of microorganisms.

Algae

Many individuals involved in the production of potable and palatable water strongly believe that algae are the most important cause of tastes and odors in raw supplies [10]. Algae are ubiquitous and contribute heavily to the organoleptic quality of drinking water. A few algae are reasonably well known for producing specific distinctive tastes and odors, whereas a greater number of others contribute to the organoleptic quality according to local conditions. Certain of the diatoms, blue-green algae and flagellates are the principal offenders, but certain of the green algae, including desmids, are involved also. Most of the attention, however, has been focused on the blue-green algae. Palmer [10] has tabulated many of the algae associated with tastes and odors in drinking water.

Table II. Occurrence of Actinomycetes in a Distribution System [9]

	TON	Actinomycetes (org./ml)	Temperature (°C)
Raw Water	1	700	26
Finished Water	2	11	27
Clear Well	2	12	27
Distance from Plant			
2 miles	10	44	27.6
4 miles	16	106	28
8 miles	24	450	28

Several representative species of these various algae are seen in Table III with the taste and odor descriptor and, in some cases, the tongue sensation given in Table IV. Again, these descriptors are not "standard" terms, but they do represent an attempt to describe the type of taste and odor associated with these algae.

Some excellent field data have been developed for the relationship between blue-green algae and actinomycetes with the resultant odors [11,12]. This is seen in Figure 2 where an annual cycle of blue-green algae and actinomycetes is suggested. It is hypothesized that blue-green algae are a nutrition source for the actinomycetes whose population peaks after the blue-green population begins to decline. The TON peaks upon decline of both biologic organisms. Bacterial decomposition of the algae and actinomycetes residues apparently releases odorous material to the water.

Table III. Taste and Odor Algae, Representative Species [10]

Group and Algae
Blue-Green Algae (Myxophyceae):
Anabaena circinalis
Anabaena planctonica
Anacystis cyanea
Aphanizomenon flos-aquae
Cylindrospermum musicola
Gomphosphaeria lacustris, kuetzingianum type
Oscillatoria curviceps
Rivularia haematites
Green Algae (nonmotile Chlorophyceae, etc.):
Chara vulgaris
Cladophora insignis
Cosmarium portianum
Dictyosphaerium ehrenbergianum
Gloeocystis planctonica
Hydrodictyon reticulatum
Nitella gracilis
Pediastrum tetras
Scenedesmus abundans
Spirogyra majuscula
Staurastrum paradoxum
Diatoms (Bacillariophyceae):
Asterionella gracillima
Cyclotella compta

Table III. continued

Diatoma vulgare
Fragilaria construens
Stephanodiscus niagarae
Synedra ulna
Tabellaria fenestrata

Flagellates (Chrysophyceae, Euglenophyceae, etc.):
Ceratium hirundinella
Chlamydomonas globosa
Chrysosphaerella longispina
Cryptomonas erosa
Dinobryon divergens
Euglena sanguinea
Glenodinium palustre
Mallomonas caudata
Pandorina morum
Peridinium cinctum
Synura uvella
Uroglenopsis americana
Volvox aureus

Chemistry of Odors from Biologic Sources

Several attempts have been made to characterize chemically the compounds responsible for tastes and odors from biologic sources [13]. Several techniques have been employed, most of which utilize the classical separation scheme for organic compounds according to solubility. For example, a series of ethyl ether extractions of a steam distillate was used to isolate odorous compounds from laboratory cultures of actinomycetes [14]. After steam distillation of the culture, a "neutral" fraction is eventually obtained that has the characteristics of a pale yellow oil. It was felt that this neutral fraction was responsible for the typical mustiness of the actinomycetes. Subsequent work was devoted to chemical characterization of compounds in this fraction. Initial tests with 2,4-dinitrophenylhydrazine gave a weak positive test for a carbonyl group. Unsaturation was indicated by the Baeyer's test. Analyses revealed the percentages of C, H and O to be 74.98, 8.93 and 16.69, respectively, with an apparent mol wt of 103. Infrared analysis confirmed the presence of the carboxyl OH and a carbonyl group, but did not yield any additional information about the compound's identity. Later, the mol wt was revised to 194 with an empirical formula of $C_{12}H_{18}O_2$ [15]. A compound called "mucidone" (from the Latin *mucid* meaning musty and *one*

Table IV. Odors, Tastes and Tongue Sensations Associated with Algae in Water [10]

Algal Genus	Algal Group	Odor When Algae Are:		Taste	Tongue Sensation
		Moderate	Abundant		
Actinastrum	Green		Grassy, musty		
Anabaena	Blue-green	Grassy, nasturtium, musty	Septic		
Anabaenopsis	Blue-green		Grassy		
Anacystis	Blue-green	Grassy	Septic	Sweet	
Aphanizomenon	Blue-green	Grassy, nasturtium, musty	Septic	Sweet	Dry
Asterionella	Diatom	Geranium, spicy	Fishy		
Ceratium	Flagellate	Fishy	Septic	Bitter	
Chara	Green	Skunk, garlic	Spoiled, garlic		
Chlamydomonas	Flagellate	Musty, grassy	Fishy, septic	Sweet	Slick
Chlorella	Green		Musty		
Chrysosphaerella	Flagellate		Fishy		
Cladophora	Green		Septic		
(*Clathrocystis*)	See *Anacystis*				
Closterium	Green		Grassy		
(*Coelosphaerium*)	See *Gomphosphaeria*				
Cosmarium	Green		Grassy		
Cryptomonas	Flagellate	Violet	Violet	Sweet	
Cyclotella	Diatom	Geranium	Fishy		
Cylindrospermum	Blue-green	Grassy	Septic		
Diatoma	Diatom		Aromatic		
Dictyosphaerium	Green	Grassy, nasturtium	Fishy		
Dinobryon	Flagellate	Violet	Fishy		Slick
Eudorina	Flagellate		Fishy		
Euglena	Flagellate		Fishy	Sweet	
Fragilaria	Diatom	Geranium	Musty		
Glenodinium	Flagellate		Fishy		Slick

(Gloeocapsa)	See *Anacystis*				
Gloeocystis	Green		Septic		
Gloeotrichia	Blue-green		Grassy		
Gomphosphaeria	Blue-green	Grassy	Grassy	Sweet	
Gonium	Flagellate		Fishy		
Hydrodictyon	Green		Septic		
Mallomonas	Flagellate	Violet	Fishy		
Melosira	Diatom	Geranium	Musty		Slick
Meridion	Diatom		Spicy		
(Microcystis)	See *Anacystis*				
Nitella	Green	Grassy	Grassy, septic	Bitter	
Nostoc	Blue-green	Musty	Septic		
Oscillatoria	Blue-green	Grassy	Musty, spicy		
Pandorina	Flagellate		Fishy		
Pediastrum	Green		Grassy		
Peridinium	Flagellate	Cucumber	Fishy		
Pleurosigma	Diatom		Fishy		
Rivularia	Blue-green	Grassy	Musty		
Scenedesmus	Green		Grassy		
Spirogyra	Green		Grassy		
Staurastrum	Green		Grassy		
Stephanodiscus	Diatom	Geranium	Fishy		Slick
Synedra	Diatom	Grassy	Musty		Slick
Synura	Flagellate	Cucumber, muskmelon, spicy	Fishy	Bitter	Dry, metallic slick
Tabellaria	Diatom	Geranium	Fishy		
Tribonema	Green		Fishy		
(Uroglena)	See *Uroglenopsis*				
Uroglenopsis	Flagellate	Cucumber	Fishy		Slick
Ulothrix	Green		Grassy		
Volvox	Flagellate	Fishy	Fishy		

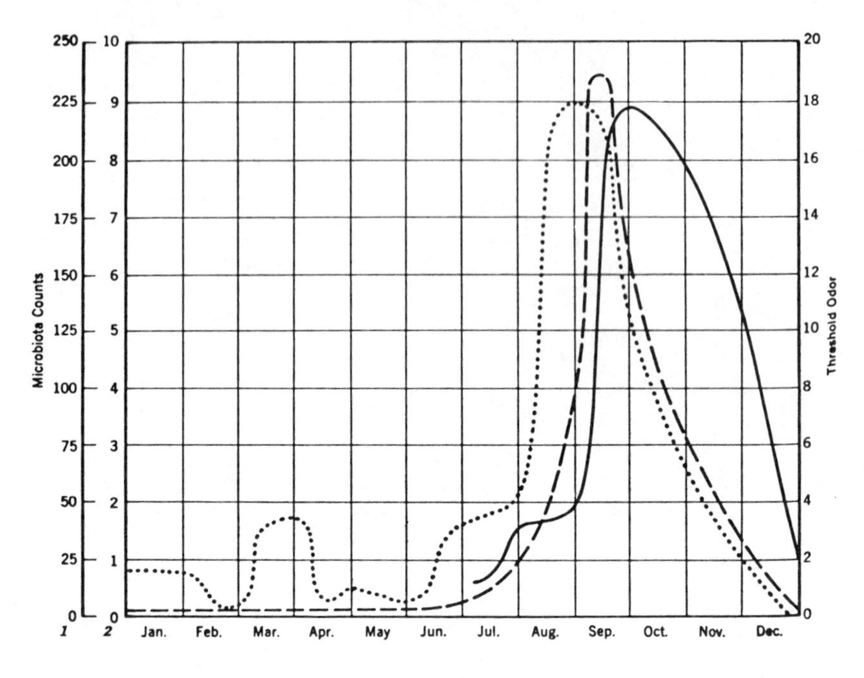

Figure 2. Average annual cycle of blue-green algae and actinomycetes with resulting threshold odors. Reproduced from Silvey and Roach [11], courtesy of the American Water Works Association.

denoting a ketone) was proposed to be responsible for the musty odor from actinomycetes. From nuclear magnetic resonance techniques (NMR) and mass spectroscopic measurements, a structure was suggested:

O
||
C
HC $C(C_5H_9)CH(CH_3)_2$
|| |
HC——O

mucidone

The volatile constituents responsible for odor produced by *Streptomyces odorifer* (an actinomycete) were divided into neutral, acidic and basic fractions after steam distillation [16]. The major odor-producing substances identified were acetic acid, acetaldehyde, ethanol, isobutyl alcohol, isobutyl acetate and ammonia. These compounds were considered to be contributors to the earthy odor produced by *Streptomyces odorifer*. Other constituents awaited identification.

Gerber and Lechevalier [17–19] may have provided the correct solution to the earthy, musty odor from actinomycetes through their "discovery" of geosmin (from the Greek *ge* meaning earth and *osme* meaning odor). This compound is a neutral oil with an approximate boiling point of 270°C. It contains C, H and O, but no nitrogen. This compound was isolated from a number of cultures of *Streptomyces* by extraction, distillation and gas-liquid chromatographic techniques. High-resolution mass spectroscopy and NMR techniques [18,19] indicated molecular formulas of $C_{12}H_{22}O$ for geosmin and $C_{12}H_{20}$ for argosmin. Their structural formulas are:

OH CH_3 CH_3

CH_3 CH_3

geosmin **argosmin C**

Argosmin C is obtained from geosmin by treatment with 10% HCl standing at room temperature for four days. Argosmin has no odor.

Several other investigators have identified geosmin from other microorganisms. An earthy odor during the routine transfer of a culture of the filamentous blue-green algae *Symploca muscorum* was observed [20]. This odor subsequently was identified as geosmin. The separation of geosmin from a pure culture of *Streptomyces griseolutens* was reported [21]. Preparative gas-liquid chromatography, high-resolution mass spectroscopy, infrared and proton magnetic resonance were employed to confirm the identification of geosmin from several cultures of aquatic actinomycetes and from two blue-green algae: *Symploca muscorum* and *Oscillatoria tenuis* [22].

A camphor-smelling compound, 2-exo-hydroxy-2-methyl bornane (2-methylisoborneol), was isolated and identified from the culture broth of three actinomycetes: *Streptomyces antibioticus, S. praecox* and *S. griseus* [23]. The structural formula of this compound is:

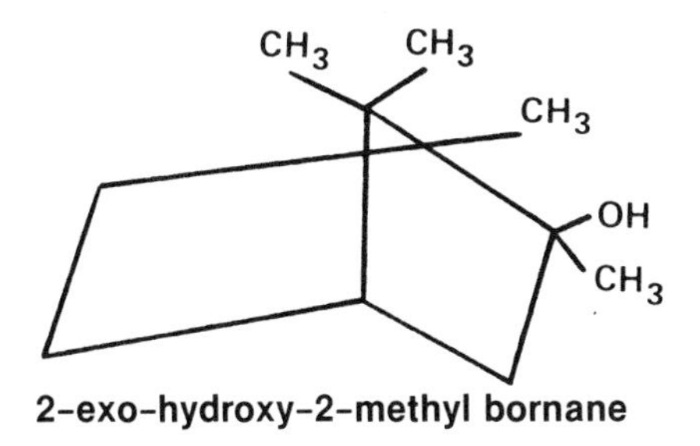

2-exo-hydroxy-2-methyl bornane

The same compound was identified independently from a culture of *S. lavendulae* [24]. An odor threshold concentration (OTC) ot approximately 0.1 μg/l was reported. A quantitative analytical method for 2-methylisobornel in water using gas-liquid chromatographic techniques has been developed [25].

Odorous sulfur compounds from aquatic organisms have been reported [26,27]. Hydrogen sulfide (H_2S) was produced by *S. odorifer* [26]. Several mercaptans and sulfides were extracted from three species of blue-green algae [27]. Tentatively identified were: isopropyl mercaptan, isobutyl mercaptan, *n*-butyl mercaptan, dimethyl disulfide, methyl mercaptan, dimethyl sulfide and dimethyl disulfide. The odor of dimethyl sulfide at low concentrations strongly resembles the "fishy" odor associated with such low-molecular-weight amines as methylamine and ethylamine.

CHEMICAL SOURCES OF TASTES AND ODORS

Organic Compounds

Many tastes and odors in drinking waters arise from various organic compounds reacting as individual molecules or reacting in an additive, synergistic or antagonistic manner. Sources of these organics are many and diverse, but the majority originate from a manufacturing wastewater that has been discharged in an improper manner. The practice of chlorination in a water treatment plant frequently will enhance the organoleptic quality. For example, the chlorination of water containing phenol (C_6H_5OH) produces a series of chlorophenols that have OTC at the μg/l level. This reaction is examined below in greater detail. Descriptors for the types of odors originating with organic compounds are many and varied. The American Society for Testing Materials (ASTM) D-19 Committee on Water classified the odors by chemical types [3]. This classification is seen in Table V where most of the chemical types are organic compounds.

Odor Threshold Concentrations of Individual Compounds

Several investigators have reported the OTC of many organic compounds that may be present in raw water supplies. Most of these values were determined in the laboratory. Table VI gives the odor characteristics and OTC values for several organic sulfur compounds [28]. The OTC values for several industrial chemicals found in the Kanawha River are given in Table VII [29].

Table V. Odors Classified by Chemical Types[a]

Odor Characteristics[b]						
Sweetness	Pungency	Smokiness	Rottenness	Odor Class	Chemical Types	Examples
100	50	0–50	50	Estery	Esters, ethers, lower ketones	Lacquer, solvents, most fruits, many flowers
100	50–100	0–100	50	Alcoholic	Phenols, cresols, alcohols, hydrocarbons	Creosote, tars, smokes, alcohol, liquor, rose and spicy flowers, spices and herbs
50	50	0–50	50	Carbonyl	Aldehydes, higher ketones	Rancid fats, butter, stone fruits and nuts, violets, grasses and vegetables
50	100	0–50	50	Acidic	Acid anhydrides, organic acids, sulfur dioxide	Vinegar, perspiration, rancid oils, resins, body odor, garbage
100	50–100	50–100	0–100	Halide	Quinones, oxides, ozone, halides, nitrogen compounds	Insecticides, weed killers, musty and moldy odors, husks, medicinal odors, earth, peat
50	50	100	100	Sulfury	Selenium compounds, arsenicals, mercaptans, sulfides	Skunks, bears, foxes, rotting fish and meat, cabbage, onion, sewage

Table V, continued

Odor Characteristics[b]						
Sweetness	Pungency	Smokiness	Rottenness	Odor Class	Chemical Types	Examples
100	50	50	100	Unsaturated	Acetylene derivatives, butadiene, isoprene	Paint thinners, varnish, kerosene, turpentine, essential oils, cucumber
100	50	0–50	100	Basic	Vinyl monomers, amines, alkaloids, ammonia	Fecal odors, manure, fish and shellfish, stale flowers such as lilac, lily, jasmine and honeysuckle

[a]Reproduced from *Annual Book of ASTM Standards* [3], courtesy of the American Society for Testing Materials.

[b]The degree of odor characteristic perceived is designated as follows: 100 indicates a high level of perception, 50 indicates a medium level of perception, and 0 indicates a low level of perception.

Table VI. Odor Characteristics and Threshold Concentrations[a]

Substance	Formula	Threshold Odor (mg/l)	Remarks
Allyl Mercaptan	$CH_2{:}CH \cdot CH_2 \cdot SH$	0.00005	Very disagreeable, garliclike odor
Ammonia	NH_3	0.037	Sharp, pungent odor
Benzyl Mercaptan	$C_6H_5CH_2 \cdot SH$	0.00019	Unpleasant odor
Chlorine	Cl_2	0.010	Pungent, irritating odor
Chlorophenol	$Cl \cdot C_6H_4 \cdot OH$	0.00018	Medicinal odor
Crotyl Mercaptan	$CH_3 \cdot CH{:}CH \cdot CH_2 \cdot SH$	0.000029	Skunk odor
Diphenyl Sulfide	$(C_6H_5)_2S$	0.000048	Unpleasant odor
Ethyl Mercaptan	$CH_3CH_2 \cdot SH$	0.00019	Odor of decayed cabbage
Ethyl Sulfide	$(C_2H_5)_2S$	0.00025	Nauseating odor
Hydrogen Sulfide	H_2S	0.0011	Rotten egg odor
Methyl Mercaptan	CH_3SH	0.0011	Odor of decayed cabbage
Methyl Sulfide	$(CH_3)_2S$	0.0011	Odor of decayed vegetables
Pyridine	C_6H_5N	0.0037	Disagreeable, irritating odor
Skatole	C_9H_9N	0.0012	Fecal odor, nauseating
Sulfur Dioxide	SO_2	0.009	Pungent, irritating odor
Thiocresol	$CH_3C_6H_4 \cdot SH$	0.0001	Rancid, skunklike odor
Thiophenol	C_6H_5SH	0.000062	Putrid, nauseating odor

[a]Reproduced from Dague [28], courtesy of the Water Pollution Control Federation.

Table VII. Threshold Odors of Industrial Chemicals in the Kanawha River [29]

Chemical	Concentration at Which Odor Is Just Detectable (mg/1)
Naphthalene	0.0068
Tetralin	0.018
Styrene	0.037
Acetophenone	0.065
Ethyl Benzene	0.14
Bis-(2-Chloroisopropyl) Ether	0.20
2-Ethyl Hexanol	0.27
Bis-(2-Chloroethyl) Ether	0.36
Diisobutylcarbinol	1.30
Phenyl Methylcarbinol	1.45
2-Methyl-5-ethyl Pyridine	0.019

The OTC values of several organic chemicals were reported by Baker [30] (Table VIII). These compounds (pure substances) were selected on the basis of frequency of use as an industrial chemical that may eventually find its way into a wastewater.

The detection and occurrence of light halogenated hydrocarbons (LHH) and other organic compounds in drinking water affect the organoleptic qualities as well as the physiological qualities (see also Chapter 2). This initiated a study of the organic compounds that are probably involved in the taste impairment of drinking water in 20 cities of the Netherlands [31]. Data are rare for the taste quality of drinking water. As a result of the 20-city study, some 20 organic compounds from a total of 180 identified by gas-liquid chromatography/mass spectrometry (GC-MS) were identified as impairing the taste of drinking water. These compounds are given in Table IX along with their OTC. The C:OTC ratio values in column 4 represent the odor potential for a given compound. All of the compounds had a C:OTC ratio of at least 0.1 (1%) suggesting that these compounds would affect the water's taste at these values. The threshold taste values apparently are less than the odor threshold values.

Table IX lists the most frequently detected organic compounds in the 20 drinking waters of the Netherlands. In turn, these compounds were grouped together into similar families (e.g., total haloforms: chloroform and bromoform) in order to evaluate their effect on taste quality. A multiple regression technique was employed for taste rating and the presence of representative compounds of the haloform group, chlorobenzene group, chlorinated ether group, etc. In turn, multiple regression equations were calculated where "high correlation coefficients" indicated that a particular group of compounds accounted for the variance of the taste rating. Actual taste tests were performed with individual compounds by a national panel of 50 adults. Table IX gives the correlation coefficients for the seven groups of organics. These data indicated that water taste is associated with all parameters except the polycyclic aromatics and phthalates.

OTC of Mixtures

Rarely does a natural water contain an individual odorous compound. More likely, a natural water receives a mixture of compounds from a variety of sources. In this case, the odorous compounds may react in one of four ways: (1) independence, (2) subtraction, (3) addition, or (4) synergism. These interactive effects of odor threshold have been examined through laboratory studies and statistical analysis of the data [32,33].

Table VIII. Odor Threshold Concentrations for Various Chemicals[a]

Chemical	No. of Panelists	No. of Observations	Threshold Odor Level (ppm)[b]	
			Avg.	Range
Acetic Acid	9	9	24.3	5.07–81.2
Acetone	12	17	40.9	1.29–330
Acetophenone	17	154	0.17	0.0039–2.02
Acrylonitrile	16	104	18.6	0.0031–50.4
Allyl Chloride	10	10	14,700	3660–29,300
n-Amyl Acetate	18	139	0.08	0.0017–0.86
Aniline	8	8	70.1	2.0–128
Benzene	13	18	31.3	0.84–53.6
n-Butanol	32	167	2.5	0.012–25.3
p-Chlorophenol	16	24	1.24	0.02–20.4
o-Cresol	13	21	0.65	0.016–4.1
m-Cresol	29	147	0.68	0.016–4.0
Dichloroisopropylether	8	8	0.32	0.017–1.1
2,4-Dichlorophenol	10	94	0.21	0.02–1.35
Dimethylamine	12	29	23.2	0.01–42.5
Ethylacrylate	9	9	0.0067	0.0018–0.0141
Formaldehyde	10	11	49.9	0.8–102
2-Mercaptoethanol	9	9	0.64	0.07–1.1
Mesitylene	13	19	0.027	0.00024–0.062
Methylamine	10	10	3.33	0.65–5.23
Methyl Ethyl Pyridine	16	20	0.05	0.0017–0.225
Methyl Vinyl Pyridine	8	8	0.04	0.015–0.12
β-Naphthol	14	20	1.29	0.01–11.4
Octyl Alcohol	10	10	0.13	0.0087–0.56
Phenol	12	20	5.9	0.016–16.7

Table VIII. continued

Chemical	No. of Panelists	No. of Observations	Threshold Odor Level (ppm)[b]	
			Avg.	Range
Pyridine	13	130	0.82	0.007–7.7
Quinoline	11	17	0.71	0.016–4.3
Styrene	16	23	0.73	0.02–2.6
Thiophenol	10	10	13.5	2.05–32.8
Trimethylamine	10	10	1.7	0.04–5.17
Xylene	10	21	2.21	0.26–4.13
n-Butyl Mercaptan	8	94	0.006	0.001–0.06

[a]Reproduced from Baker [30], courtesy of the American Water Works Association.
[b]Temperature = 40 ± 1 °C.

Table IXa. Organic Compounds Probably Involved in Taste Impairment of Drinking Water in 20 Cities of the Netherlands[a]

Compound	Highest Concentration Detected ($\mu g/l$)	OTC in Water ($\mu g/l$)	C:OTC Ratio
Chlorination Products			
Tribromomethane	10	300	0.03
Trichloromethane	60	100	0.6
Compounds of Biological Nature			
2-Ethyl-1-hexanol	3.0	300	0.01
Heptan-3-on	0.1	7.5	0.01
Heptanol	0.1	3.0	0.03
Octanol	0.03	0.7	0.04
Octene-1	0.03	0.5	0.06
Nanonal	0.1	1.0	0.1
Decanal	0.1	0.1	1.0
2-Methylisoborneol	0.03	0.002	1.5
Geosmin	0.03	0.02	1.5
Compounds of Industrial Nature			
bis-(2-Chloroisopropyl)ether	3.0	300	0.01
3,4-Dichloroaniline	0.03	3	0.01
o-Dichlorobenzene	0.1	10	0.01
Hexachlorobutadiene	0.1	6	0.02
Naphthalene	0.1	5	0.02
γ-Hexachlorocyclohexane	0.1	0.3	0.03
5-Chloro-*o*-toluidine	0.3	5	0.06
1,2,4-Trichlorobenzene	0.3	5	0.06
2-Chloroaniline	0.3	3	0.1
Biphenyl	0.1	0.5	0.2
1,3,5-Trimethylbenzene	1.0	3	0.3
p-Dichlorobenzene	0.3	0.3	1.0

Table IXb. Organic Compounds Detected Most Frequently in 20 Types of Drinking Water in the Netherlands[a]

Types of Compounds	Detection Frequency[b]	Maximum Concentration ($\mu g/l$)
Hydrocarbons		
Toluene	20	0.3
Xylenes	19	0.1
C-Benzenes	19	1.0
Decanes	18	0.3
Ethylbenzene	17	0.03

Table IXb. Organic Compounds Detected Most Frequently in 20 Types of Drinking Water in the Netherlands[a] (continued)

Types of Compounds	Detection Frequency[b]	Maximum Concentration (μg/l)
Fluroanthene	16	0.05
Nonanes	15	0.3
Naphthalene	14	0.1
Oxygen Compounds		
Dibutyl Phthalate	17	0.1
1,1-Dimethoxyisobutane	13	0.3
Methyl Isobutryate	13	1.0
Halogen Compounds		
Chloroform	16	60
Tetrachloromethane	15	0.7

Table IXc. Correlation Between Average Water Taste Rating and the Presence of Groups of Organic Micropollutants in Tapwater[a]

Group of Organic Compounds	Correlation Coefficient (r)
Total Haloforms	0.45
Total Alkylbenzenes	0.47
Total Polycyclic Aromatics	0.17
Total Phthalates	0.27
Total Chlorobenzenes	0.56
Total Chlorinated Ethers	0.63
Total Anilines	0.55

[a]Reproduced from Zoeteman et al. [31], courtesy of the American Water Works Association.
[b]Detection frequency is number of tapwaters, among the 20 types, in which the compound was detected.

For example, the odor stimulus of a mixture of odorants, A and B, is represented by R_{A+B} (see Equation 1) [32]. For these four cases,

$$\text{independence:} \quad R_{A+B} = R_A \text{ or } R_B \tag{5}$$

$$\text{subtraction:} \quad R_{A+B} < R_A \text{ or } R_B \tag{6}$$

$$\text{addition:} \qquad R_{A+B} = R_A + R_B \tag{7}$$

$$\text{synergism:} \qquad R_{A+B} > R_A + R_B \tag{8}$$

Mixtures of 1-butanol, *p*-cresol and pyridine were employed to investigate the interactive effects of odorous compounds [31]. Panels of 16–20 judges determined the OTC of these compounds individually or in mixtures of varying proportions. The consistent series method was employed [1]. The investigators had to consider the nonlinear property of odor sensation expressed by the Weber-Fechner expression (Equation 1) and the wide range of OT values reported by individual panel members. A geometric mean (GM) of the OT values was calculated, as well as the range and standard deviation of this value in accord with a statistical analysis suggested by Thomas [34]. The OTC values of the individual odorants are given in Table Xa.

In order to detect addition or synergism of the various combinations of butanol, *p*-cresol and pyridine, the data were expressed in olfactie units [32,35]. An olfactie unit is defined as the odor stimulus provided by exactly one OTC of any odorant. This definition permits computation of the ratio:

$$\text{olfactie units} = \frac{\text{concentration of odorant in water}}{\text{OTC}} \tag{9}$$

For example, if compound A has an OTC of 1 mg/l, and if it was detected in a natural water at 10 mg/l, compound A would give 10 olfactie units. The data in Table Xb indicate the fractions of individual olfacties that, in turn, produced 1.00 olfactie unit of odor in their mixture. The "total" column indicates the additive olfactie unit for the seven mixtures. If this sum were $\simeq 1.00$, then the odors would have acted

Table Xa. OT Concentrations of Single Odorants[a]

Compound	Geometric Mean OTC (ppm)	Range of Geometric Mean, antilog $(M - \sigma M)$ to antilog $(M + \sigma M)$
1-Butanol	1.00	0.78–1.3
p-Cresol	0.055	0.044–0.068
Pyridine	0.82	0.66–1.01

Table Xb. Odor Addition in Mixes[a]

Test No.	Olfacties Each Added Compound at Threshold (1.00 olfactie found in each case)			
	Butanol	*p*-Cresol	Pyridine	Total
2A	0.46	—	0.53	0.99
2B	0.35	0.40	—	0.75
3A	0.18	0.21	0.21	0.60
3B	0.15	0.09	0.26	0.50
3C	0.12	0.21	0.07	0.40
3D	0.37	0.14	0.42	0.93
3E	0.24	0.19	0.29	0.72

Table Xc. Apparent Odor Synergism in Mixtures[a]

Test No.	Observed TO/ Calculated TO	Total Olfacties Added, Mean + σ	Total Olfacties Found, Mean − σ
2A	1.0	1.17	0.81
2B	1.3	0.88	0.82
3A	1.6	0.70	0.74
3B	2.0	0.59	0.80
3C	2.5	0.46	0.79
3D	1.1	1.07	0.80
3E	1.4	0.83	0.82

[a] Reproduced from Rosen et al. [32], courtesy of the Water Pollution Control Federation.

in an additive manner as suggested by Test 2A. If this sum were <1.00 then the odors would have acted in a synergistic manner as suggested by Tests 3B and 3C. If this sum were >1.00, then the odors would have acted in a subtractive manner (none was reported).

Some of the data in Tables Xb and Xc suggest odor synergism. Tests 3B and 3C indicate that butanol, *p*-cresol and pyridine react to enhance the odor. In order to make certain that synergism had occurred, a rigorous statistical criterion was applied [32]. Each of the computed sums of olfactie units in Table Xc, as well as the OT values, is an inexact number, and each has its own standard deviation. A minimum criterion for synergism was the following: the GM $\pm\sigma$ (σ = standard deviation) of the olfactie units of the mixture must lie completely above the calculated

GM $\pm \sigma$. This may be seen in Figure 3 where Tests 3A, 3B and 3C are clearly synergistic and the others may be additive. Synergistic effects were observed also from the OT of binary mixtures of *m*-cresol, *n*-butanol, pyridine, *m*-butyl mercaptan, *n*-amyl acetate, acrylonitrile, 2,4-dichlorophenol and acetophenone [33]. Tests 2A, 2B, 3D and 3E suggest that "addition" of the odors had occurred. Rosen et al. [32] did not report any cases of independence or antagonism. However, Baker [33] did report an antagonistic odor response of a mixture of 75% *m*-cresol and 25% acetophenone.

Chlorophenols

Drinking water frequently may have an "iodine" or a "medicinal" taste. This condition is thought to be the result of chlorination of pheno-

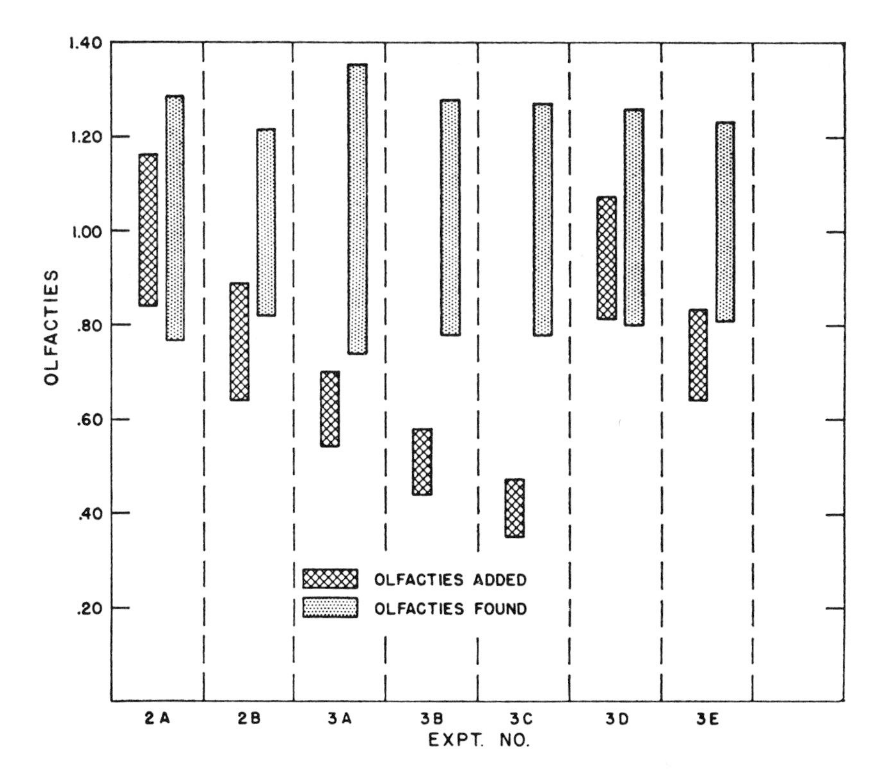

Figure 3. Odor addition and synergism. Reproduced from Rosen et al. [32], courtesy of the Water Pollution Control Federation.

lic compounds (Figure 4) [36]. These compounds have extremely "low" OTC (see Table VIII), whereas phenol (C_6H_5OH) has a relatively "high" OTC value—5.9 mg/l [30]. Consequently, a finished water may leave the treatment plant with acceptable taste and odor qualities. If this water should contain a "trace" concentration (<1.0 mg/l) of phenol, then chlorophenolic compounds may be formed in the distribution system. Development of this condition becomes a kinetic problem; namely, it is essential to know the rate at which the chlorophenols are formed.

Soper and Smith [37] were, perhaps, the first to investigate the chlorination of phenols through kinetic experiments. These investigators hypothesized that the interaction may occur either between the phenol molecule and the hypochlorite ion, or between hypochlorous acid and the phenoxide ion. The velocity equations are:

$$v = k_2' \ [OCl^-] \ [PhOH] \tag{10}$$

$$v = k_2 \ [HOCl] \ [PhO^-] \tag{11}$$

where v = velocity
k_2' and k_2 = proportionality constants

Kinetic data were obtained to suggest that Equation 10 describes the mechanism of the reaction between phenol and hypochlorous acid.

The intensity of tastes and odors produced by stepwise chlorination of phenolic compounds was determined [38,39]. Table XI shows the effects of adding chlorine (0.0–10.0 mg/l) on the threshold taste number of a 1.0-mg/l solution of phenol (C_6H_5OH). Taste observations were made after reaction times of 2 and 24 hr. These data reveal the extremely significant fact that the "medicinal" taste of chlorophenols develops slowly. This "slow" reaction would permit formation of chlorophenols

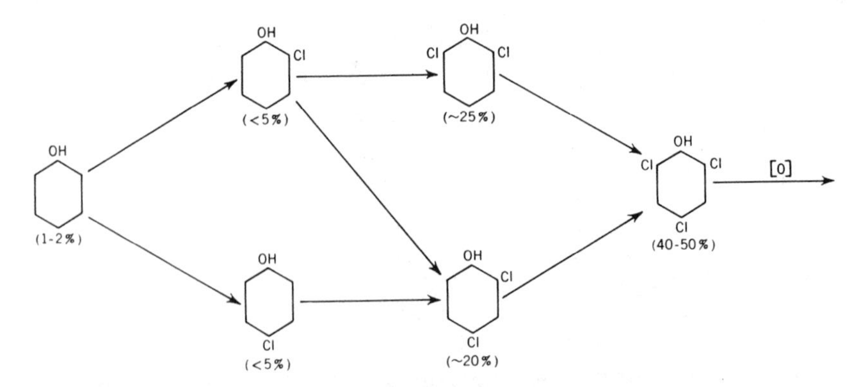

Figure 4. Course of chlorination of phenol. Reproduced from Burttschell et al. [36], courtesy of the American Water Works Association.

Table XI. Effects of Stepwise Chlorination of 1 ppm Phenol on Taste Intensity[a]

Chlorine Added (*ppm*)	2-hr Observations		24-hr Observations			
	pH	Flash[b]	pH	Flash[b]	Residual (o-t.)	Dilutions to Threshold
0	7.8	–	8.0	–	0.00	0
0.5	7.9	–	8.2	–	0.00	100
1	8.1	–	8.2	–	0.00	200
1.5	8.2	–	8.3	–	0.00	200
2	8.3	–	8.3	–	0.00	400
3	8.2	–	8.2	–	0.00	100
4	8.1	–	8.2	–	0.00	0
5	7.9	–	7.9	–	0.00	0
[illegible]	[illegible]		[illegible]			

p157

[a]Reproduced from Ettinger and Ruchhoft [38], courtesy of the American Water Works Association.
[b]*p*-Aminodimethylaniline flash test.

in a water distribution system provided the ratios (on a basis of ppm) of Cl_2:phenol were in the 0.5:1 to 3:1 range. These data in Table XI also establish another significant fact about the operational aspects of water treatment. Phenolic compounds may be oxidized completely by chlorine when the reaction conditions are appropriate; namely when the ratio of Cl_2:phenol exceeds 4:1 and when a sufficient reaction time is provided. Table XII shows some additional chlorination studies of several phenols.

Additional efforts were made to identify the chlorophenolic compounds responsible for the off taste (see Figure 4) [56]. Using infrared spectroscopy and paper chromatographic techniques, the products obtained by chlorination of phenol (C_6H_5OH) to maximum taste were:

Component	Percent of Product
Phenol	1–2
2-CP	2–5
4-CP	2–5
2,4-DCP	20
2,6-DCP	25
2,4,6-TCP	40–50
Carbonyl compound (oxidation product)	Clearly present

Table XII. Stepwise Free Residual Chlorination Studies of Some Phenolic Materials[a,b]

Material	Dilution to Threshold, Unchlorinated	Chlorine at Maximum Taste (ppm)	Dilution to Threshold at Taste Maximum	Chlorine Added to Produce Free Residual (ppm)	Chlorine Required to Eliminate Taste (ppm)
Phenol	0[c]	2	400	7	4
o-Cresol	10	2	800	5	5
m-Cresol	4	1	1000	5	5
p-Cresol	50	⅔	2000	4	3
1-Naphthol	30	0	30	5	4
2-Chlorophenol	100	½	400	5	3
4-Chlorophenol	2	1	8	6	3
2,4-Dichlorophenol	4	0	4	6	2
2,4,6-Trichlorophenol	0[c]	–	0	3	–
2,4,5-Trichlorophenol	0[c]	–	0	2	–
2,3,4,6-Tetrachlorophenol	0[c]	–	0	1.5	–
Pentachlorophenol	0[c]	–	0	1.0	–

[a]Reproduced from Ettinger and Ruchhoft [38], courtesy of the American Water Works Association.
[b]Contact time, 24 hr; 1 ppm solutions studied.
[c]Could not be tasted by observer.

There was no trace of these chlorophenols: 3-CP, 2,5-DCP, 3,4-DCP, 2,4,5-TCP, 2,3,4,5-tetra CP and PCP. Apparently, 2,4-DCP and 2,6-DCP were responsible for the maximum taste intensity. That 2,4-DCP, 2,6-DCP and 2-CP have the greatest tastes and odors may be seen in Table XIII. The threshold OTC [36] differ widely from the values reported in Table VIII [30]. Nevertheless, the above observations have an extremely useful and practical value. For example, two operational parameters were investigated. The effect of the pH value is given in Table XIV [36]. Maximum taste intensities were observed at a pH value of 8.0, whereas lesser tastes were observed at pH values of 6.0, 9.0 and 10.0. This is due to formation of dichlorophenols at a pH of 8.0. Another operational parameter was the combined effect of ammonia and pH value on the rate of chlorination of phenol. The appearance of 2,6-DCP was employed to follow this reaction rate. Three appropriately buffered solutions were prepared, each containing 1 ppm phenol, 7 ppm chlorine and 3 ppm ammonia. Aliquots were withdrawn periodically and checked for 2,6-DCP by paper chromatography. Phenol was present in all cases, but 2,6-DCP formed slowly at pH 9, slightly faster at pH 8 and not at all at pH 6. After five days, residual chlorine was still present in all three solutions.

These results indicate that the course of chlorination (Figure 4) resulting in products of intense taste also occurs in the presence of ammonia, but much more slowly. Under suitable conditions, this phenomenon may result in a "medicinal" taste that will be apparent to the water consumer, even if it is not detectable at the water plant.

Another intensive investigation of the kinetics of the chlorination of phenol and several chlorophenols was reported [40]. Figure 5 shows the

Table XIII. Taste and Odor Threshold Concentrations[a,b]

Component	Geometric Mean Thresholds (ppb)	
	Taste	Odor
Phenol	>1,000	>1,000
2-CP	4	2
4-CP	>1,000	250
2,4-DCP	8	2
2,6-DCP	2	3
2,4,6-TCP	>1,000	>1,000

[a]All tests were made at room temperature, about 25°C.
[b]Reproduced from Burtschell et al. [36], courtesy of the American Water Works Association.

Table XIV. Effect of pH on Chlorophenolic Products[a]

pH	Chlorine Residual (ppm) 2 hr	Chlorine Residual (ppm) 20 hr	Components Detected
6.0	<0.05	<0.05	Phenol; traces of 2-CP and 4-CP
8.0	<0.05	<0.05	2,4,6-TCP; much 2,4 and 2,6-DCP; traces of phenol, 2-CP, and 4-CP
9.0	0.15–0.20	<0.05	Same as for pH 8.0, except less 2,4,6-TCP
10.0	0.50	0.10–0.15	Same as for pH 9.0

[a]Reproduced from Burtschell et al. [36], courtesy of the American Water Works Association.

observed rates (k_{ob} is the second-order rate constant in liter/mol-min) of chlorination of several phenolic compounds within the pH range of 4.0–12.0. For 2-CP, phenol and 4-CP, maximum rates were observed at pH values of 8.0–9.0, whereas a range of 7.0–8.0 was found for the other compounds. These observations were consistent with those of Soper and Smith [37] and partially confirm the mechanism that, in neutral or alkaline pH ranges, the reaction proceeds between HOCl and the phenolate anion. The rate of formation of OT values was computed for simulated conditions of water treatment practices [40]. These curves are given in Figure 6 for three concentration ratios of chlorine and phenol and are based on the OT values for chlorophenols reported by Burttschell et al. [36]. The curves in Figure 6c are especially significant and are representative of conditions that may exist in a water distribution system. Another significant point is that these kinetic studies were conducted at 25°C. Natural water temperatures are somewhat lower (5–15°C). Consequently, these reaction rates are slower at the lower temperatures. The overall reaction rate would be lower by a factor of 2 for each 10°C decrease.

Chapter 2 contains additional information concerning the formation of halogenated phenols from a variety of precursors.

Inorganic Compounds

An unacceptable taste frequently may be imparted to drinking water from dissolved inorganic substances. These compounds may give the water a "mineral" taste or a "bitter metallic" taste. Some available research indicates that consumer acceptance of mineralized waters

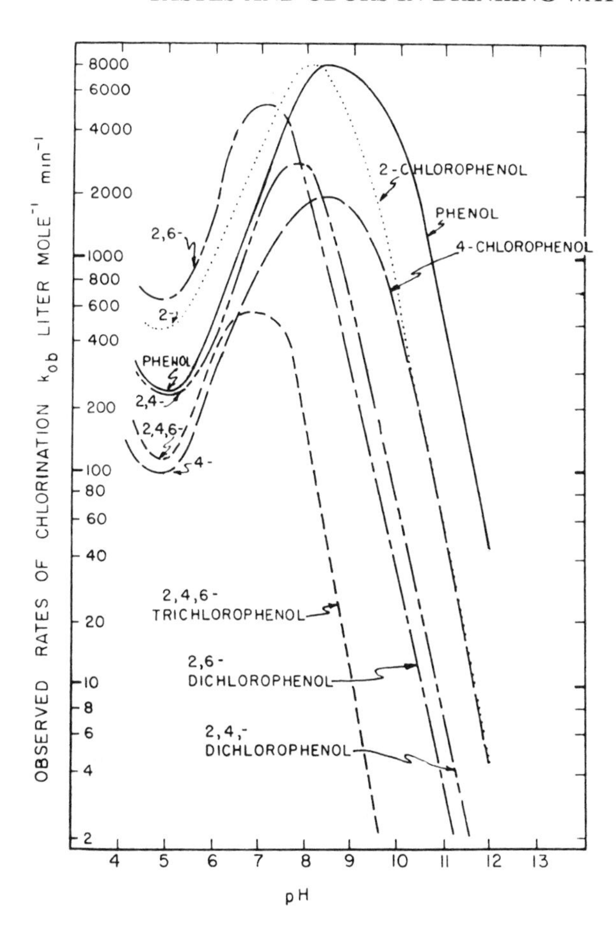

Figure 5. Observed rates of chlorination. Reproduced from Lee and Morris [40], courtesy of the International Association of Water Pollution Research.

decreased as the total dissolved solids (TDS) increased [41,42]. Also, the corrosion products of iron and copper pipes impart a "bitter metallic" taste to drinking water. Again, the point may be made that the consumer judges the potability of drinking water by its taste characteristics.

"Mineralized" Water

Several laboratory studies and surveys have reported the consumer's assessment of the mineral taste in California public waters [41–43]. The consumer survey was composed of an interview schedule containing questions and attitude scales and a taste scale rating procedure that was

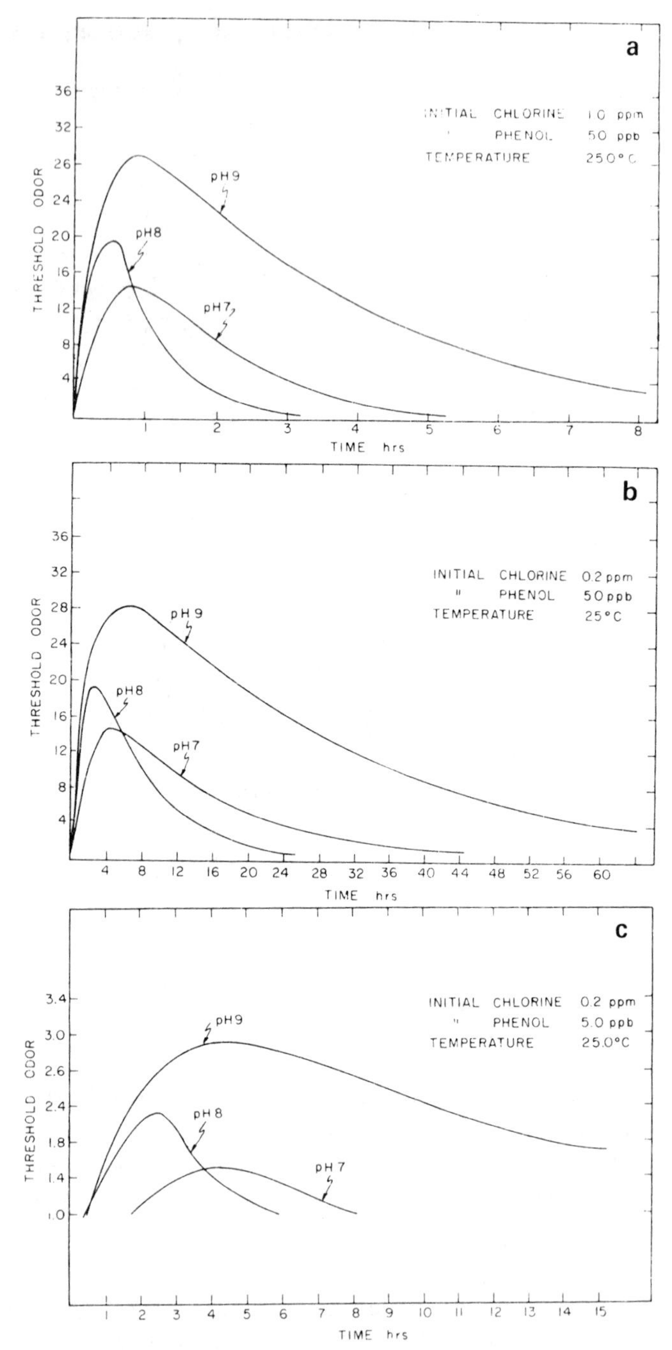

Figure 6. Threshold odor from chlorination of phenol at 25°C. **(a)** 50 μg/l phenol, 1.0 mg/l chlorine; **(b)** 50 μg/l phenol, 0.2 mg/l chlorine; **(c)** 5.0 μg/l phenol, 0.2 mg/l chlorine. Reproduced from Lee and Morris [40], courtesy of the International Association on Water Pollution Research.

administered to respondents in their homes [43]. Scale values for the attitude taste scale and the taste scale rating procedure are given in Table XV. To complete the interview, the respondent tasted a sample of tapwater under direction of the interviewer. The interviewee's response to this taste test was reported on the TRS scale. These surveys were performed for 29 water systems in California. Each system was selected carefully so that there were no tastes interfering with the TDS. That there was an inverse relationship between TDS values and consumer acceptability of the water is seen in Figure 7 for the two attitude scales and the one taste scale. There are, of course, many variables that affect consumer assessment of mineralized water. Nonetheless, there is qualitative value in the data presented in Figure 7.

The results of a laboratory study were reported in which a panel of 15 trained subjects rated the taste intensity of 125- to 2000-mg/l solutions of eight selected inorganic compounds: $NaCl$, $NaHCO_3$, Na_2SO_4, $CaSO_4$, $CaCl_2$, Na_2CO_3, $MgSO_4$ and $MgCl_2$ [44]. For each compound, 3-ml coded samples of eight concentrations from 125–2000 mg/l were pre-

Table XV. Samples Items from the Attitude Taste Scale (ATS), Attitude Adjective Taste Scale (AATS) and the Taste Scale Rating (TSR) Procedure[a]

Item	Scale Value
ATS	
Perfect	10.57
Good	7.67
Neither good nor bad	6.00
A little bad	4.33
Bad	2.16
AATS	
Delicious	10.57
Fine	8.04
Average	6.09
Inferior	3.54
Awful	1.94
TSR	
Excellent taste	10.67
Good taste	8.45
Neutral taste	6.00
Bad taste	2.95
Horrible taste	1.16

[a]Reproduced from Bruvold et al. [43], courtesy of the American Water Works Association.

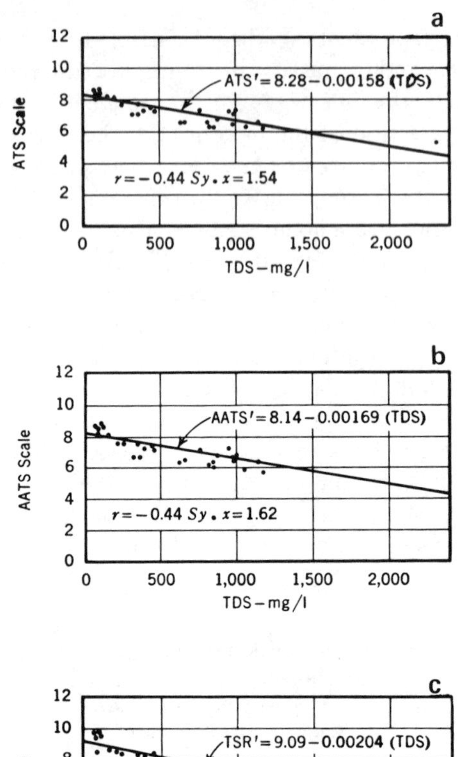

Figure 7. Linear regression on TDS values of **(a)** ATS, **(b)** AATS, and **(c)** TSR scores. Reproduced from Bruvold et al. [43], courtesy of the American Water Works Association.

sented to the judge at room temperature (22 ± 1°C) in a random order. The judges were instructed to rate the total taste intensity on a 13-point scale where 0 = none and 12 = extremely intense. A total of 45 replications (15 subjects × 3 test days) was obtained per sample. The average taste intensities of the eight mineralized solutions are shown in Figure 8. At concentrations between 750 and 2000 mg/l, the carbonates and chlorides were more intense than were the sulfates. The judges were required to report a descriptive term for each mineral: $CaCl_2$ is "bitter,"

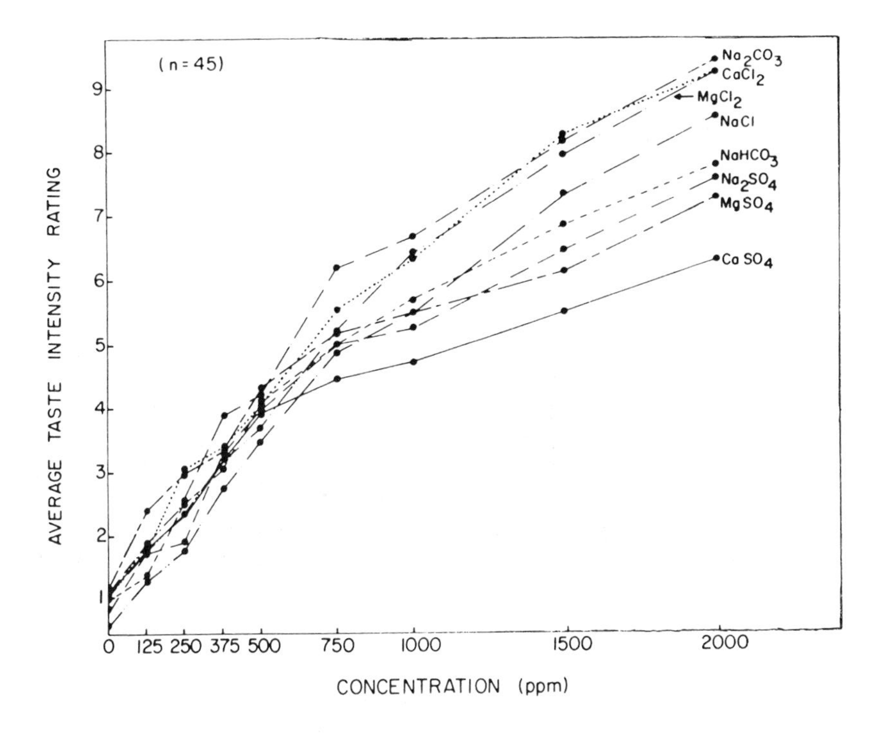

Figure 8. Rated taste intensity of distilled and mineralized waters at eight concentrations. Reproduced from Pangborn et al. [44], courtesy of the American Water Works Association.

$MgCl_2$ is slightly "bitter" and "sweet," $NaHCO_3$ is "very sweet," Na_2SO_4 is "salty," Na_2CO_3 is "bitter" and "sweet," and NaCl is unquestionably "salty." No distinguishing taste emerged for $CaSO_4$ and $MgSO_4$. These two salts had the lowest taste intensities of the eight minerals.

In communities where the drinking water contains a high percentage of unpalatable minerals, consumers remark that refrigeration of water greatly improves its taste. It is uncertain whether coolness of the water alone makes the water more acceptable, or whether perception of the undesirable tastes is diminished by the cold temperature. These considerations led to a study of the influence of temperature on taste intensity and the degree of consumer acceptance of the water [45]. Figure 9 shows some typical findings, namely, that taste intensities were lowest at 0°C and highest at 22°C. Also, intensity values were related to the TDS in the water. Consequently, the characteristic taste intensities of mineralized

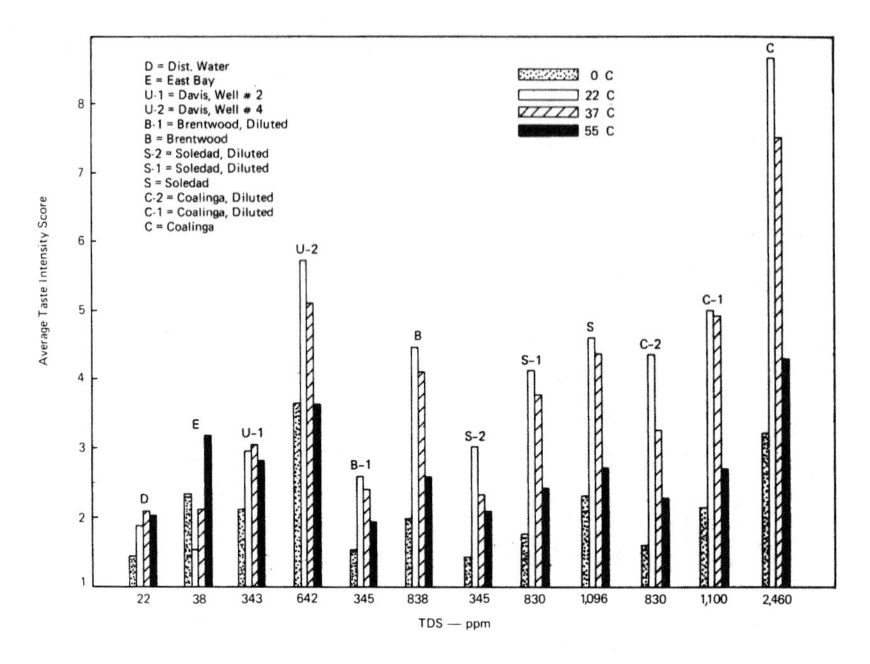

Figure 9. Taste intensity of distilled water, six natural waters and mixtures of the two at 0, 22, 37 and 55°C. Reproduced from Pangborn and Bertolero [45], courtesy of the American Water Works Assocation.

water were significantly reduced by lowering the solution temperature. The less the taste intensity, the greater is the degree of consumer acceptance. Perhaps this can be explained by the preference of consumers in a warm climate for drinking cold rather than hot beverages.

Additional references exist on the taste intensity of mineralized water [46–50].

Chlorinated and Dechlorinated Water

An undesirable taste frequently occurs in waters that are chlorinated for disinfection purposes. The effect of 0.8 mg/l Cl_2 was reported for the taste intensities of several mineralized waters at 750 mg/l at 1 and 24°C [44]. As seen in Figure 10, chlorine was perceived with considerable difficulty in the Na_2CO_3 solutions which averaged 63% correct compared to 88% correct for $NaHCO_3$ and over 96% correct for the other compounds. The right half of Figure 10 gives the taste intensity of the chlorine solutions. $CaSO_4$ appears to enhance the taste, whereas $NaHCO_3$, $CaCl_2$ and Na_2CO_3 appear to suppress the taste.

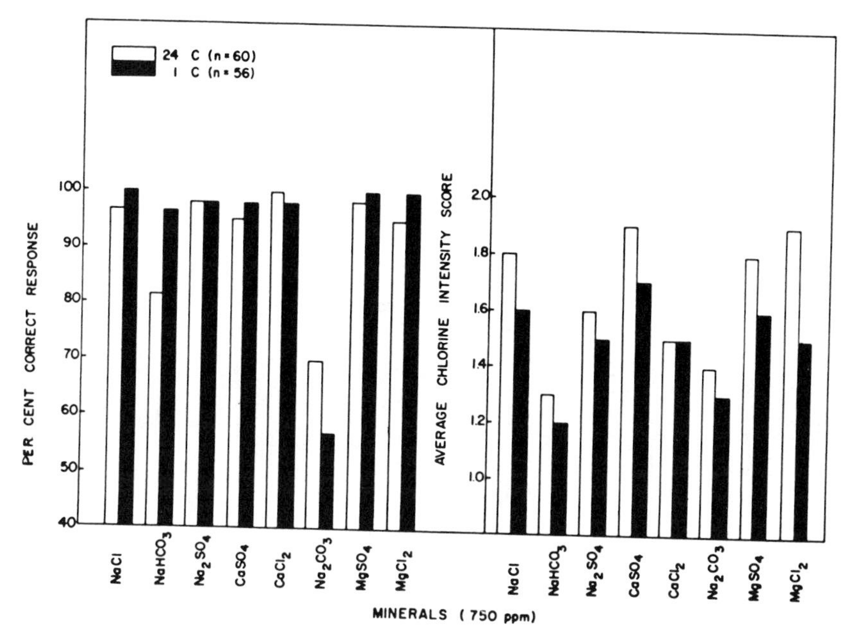

Figure 10. Sensory response to solutions of 0.8 ppm chlorine in mineralized waters at 24 and 1°C. Reproduced from Pangborn et al. [44], courtesy of the American Water Works Association.

The threshold taste values of chlorine, bromine and iodine in water were determined [51]. These halogens are, of course, employed as disinfectants in drinking and swimming pool waters. The triangle taste procedure [6] was used by a panel of 11–12 people. All samples were served in reagent water at 40 ± 1°C. The threshold taste concentrations and ranges are given in Table XVI for the three halogens at pH values of 5.0, 7.0 and 9.0. At a pH value of 5.0, chlorine gave the lowest threshold concentration, 0.075 mg/l. HOCl is apparently more sensitive to taste than is OCl^- which is the predominant chlorine species at a pH value of 9.0. The threshold taste values for bromine did not vary significantly over the pH range that was studied. There was no apparent or detectable effect of bromine species: Br_2, HOBr and OBr^-. Nor was the pH value a factor in the threshold taste concentration of iodine and the four species: I_2, HOI, HIO_3 and I^-.

There are occasions when a peculiar taste and/or odor results from the practice of dechlorinating waters. Sometimes it is necessary to effect superchlorination (i.e., dosages of $Cl_2 > 10$ mg/l) at water treatment plants for various purposes. If the water is not dechlorinated, a "chlo-

Table XVI. Panel Test Thresholds of Free-Halogen Residuals in Aqueous Solution (mg/l)[a]

pH	Chlorine		Bromine		Iodine	
	Threshold Concentration	Threshold Range	Threshold Concentration	Threshold Range	Threshold Concentration	Threshold Range
5.0 ± 0.1	0.075	0.013–0.136	0.226	0.111–0.340	0.204	0.118–0.290
7.0 ± 0.1	0.156	0.020–0.290	0.212	0.078–0.346	0.190	0.001–0.394
$9.0 + 0.1$	0.450	0.144–0.760	0.215	0.118–0.313	0.147	0.001–0.420
Unadjusted						
5.3 ± 0.1	0.050	0.001–0.220	0.168	0.043–0.291		
6.2 ± 0.1					0.155	0.056–0.252

[a]Reproduced from Bryan et al. [51], courtesy of the American Water Works Association.

rine" taste may result. On the other hand, the dechlorinating reagents, sodium thiosulfate and sodium arsenite, may also impart an off taste or off odor. Baker [52] examined this alleged problem through laboratory studies of chlorinated solutions of *m*-cresol, a phenolic compound of industrial origin. Solutions of 4.0 mg/l *m*-cresol with an OII value of 18 were prepared in odor-free and Delaware River water. The latter had an OII value of 3. The *m*-cresol solution prepared with riverwater, in turn, had a determined OII value of 22 against a predicted value of 21 (Table XVII). The difference of ±1.0 OII unit suggests no synergism or antagonism of odor effects. Then a series of *m*-cresol solutions were dosed with 8.5 mg/l Cl_2 and were permitted to stand for 24 and 48 hr before dechlorination with sodium thiosulfate. That the dechlorination reagent intensifies the OT is seen from an increase in the OII values from 21 to 29 (odor-free water) and from 21 to 26 (riverwater).

Dissolved Oxygen

It is commonly believed that the flat, insipid taste of water is caused by a low content of dissolved oxygen (DO) or by mineral salts. Some labora-

Table XVII. Odor Effects with *m*-Cresol Before and After Chlorination-Dechlorination[a]

Sample	Panelist A	Panelist B	Panelist I
	Threshold Odor Values—OII[b]		
Riverwater[c]	3	3	5
Riverwater + 4 ppm *m*-cresol			
Predicted	21	21	23
Actual	22	22	22
m-Cresol in odor-free water	22	21	22
After chlorination at 8.5 ppm			
m-Cresol in odor-free water			
24 hr	29	29	29
48 hr	29	29	29
m-Cresol in riverwater			
24 hr	26	26	26
48 hr	26	26	26

[a]Reproduced from Baker [52], courtesy of the American Water Works Association.
[b]TON is defined by the relation $TON = 2^{OII}$.
[c]Delaware River water (pH 7.1).

tory data were obtained for the taste of water with various dissolved oxygen contents [45,53]. Triangle tests [6] were employed to determine whether there was a significant difference between two extreme DO levels [3.25 and 18.8 mg/l (averages)]. The individual judge's ability to differentiate between the two contents of DO in distilled water and in 1000 mg/l NaCl at temperatures of 0 and 22°C is seen in Table XVIII. No significant differences were observed in any of the comparisons, as 18 correct separations in 36 trials are required for the chi-square to be significant at the $p = 0.05$ level. These data led to the conclusion that DO content had little, if any, effect on the taste of drinking water. This was confirmed later by Pangborn and Bertolero [45].

Metals in Drinking Water

Water occasionally acquires a "bitter," "metallic" taste due to such metals as copper, iron, manganese and zinc. Three of these metals, Cu, Fe and Zn, may arise from the corrosion of pipes through which water is conveyed. This is especially true of waters that are soft and acidic in nature. The threshold taste concentrations of these four metals as well as a review of the literature on this subject prior to 1960 were reported [54].

Table XVIII. Triangle Test Differentiation by Judges (A–F) Between Two Levels of DO in Distilled Water and in 1,000 mg/l Sodium Chloride Solutions[a]

	A	B	C	D	E	F	Totals
Study 1							
Dist. H_2O at 22°C	4	0	2	1	1	3	11/36
Dist. H_2O at 0°C	4	2	3	4	3	0	16/36
NaCl at 22°C	1	4	0	2	2	1	10/36
NaCl at 0°C	3	2	1	2	2	3	13/36
	12	8	6	9	8	7	
Study II							
Dist. H_2O at 22°C	2	2	1	3	1	1	10/36
NaCL at 22°C	2	1	3	3	3	1	13/36
	4	3	4	6	4	2	Y

[a]Reproduced from Bruvold and Pangborn [53], courtesy of the American Water Works Association.

A duo-trio test [55] with a panel of 18 or more judges was employed to provide threshold taste values of the four metals in distilled water and spring water. A summary of these values is given in Table XIX. Zinc was detected at the 5% frequency level (i.e., 5% of the panel) at 4.3 mg/l. The taste generally was characterized as bitter and, on occasion, it was detectable also by an astringent sensation which persisted for some time after tasting. Copper was detected at 2.6 mg/l at the 5% frequency level, and its taste persisted after the testing period. Of the four metals, iron was determined at the lowest concentration, 0.04 mg/l, at the 5% level. Manganese was detected at 0.9 mg/l. Another indication of metallic taste sensitivity may be seen in the range of concentrations from the most acute level (5% of panel) to the level at which 95% of the panel tasted the metal. These ranges are seen also in Table XIX. Copper appears to be the metal most universally tasted. Regulatory agencies take into consideration these threshold taste values of metals when they establish drinking water quality standards.

TREATMENTS FOR TASTES AND ODORS

Many treatments and/or combinations thereof are available to reduce tastes and odors in drinking water. Much of this originates with plant practice obtained by trial and error. Successful treatment often lies with years of experience by plant operators. Powdered and granular activated carbon and chemical oxidants (Cl_2, $KMnO_4$, ClO_2) generally have been the major treatments. They are applied frequently in conjunction with

Table XIX. Range of Concentrations in Distilled Water Detectable by Panel Members[a]

Metal	Threshold Distribution	
	5%	95%
	Concentration (ppm)	
Zinc	4.3	62
Copper	2.6	15.8
Iron	0.04	256
Manganese	0.9	487

[a]Reproduced from Cohen et al. [54], courtesy of the American Water Works Association.

other treatment such as chemical coagulation. The operating experiences from several locations are given below.

Case Histories

The taste and odor qualities of Missouri River water historically have been extremely difficult treatment problems [8]. The causative agents are many and diverse: silt, decaying vegetation, algae, actinomycetes, and wastewaters from packing plants, stockyards and oil refineries. The taste and odor problem is especially apparent during heavy runoff in the early spring. In addition, there is usually a severe turbidity content due to silts and suspended sediments.

A schematic flow diagram of the Metropolitan Utilities Districts, Omaha, Nebraska, water treatment plant is seen in Figure 11 [8]. There are long contact times for the carbon (presumed to be powdered) and alum treatments. Chlorine is added intermittently for taste and odor control. Applying it after the presedimentation basin was abandoned eventually in favor of applying it after the upflow sludge contact. This move reduced the chlorine demand and also avoided a chlorinous or fishy odor in the finished water. Potassium permanganate had been applied unsuccessfully for taste and odor control. Table XX shows operational data for odor control in the Missouri River water. That this is an extremely difficult treatment problem is seen in the odor values in the tapwater. These data were obtained in March, when low water temperatures retard the rate of chemical reactions for odor removal.

A flow diagram for the water treatment plant at Kansas City, Missouri, is seen in Figure 12 [56]. Here, the taste and odor were treated with potassium permanganate, chlorine and powdered carbon. Turbidity was removed by the coagulants, ferric sulfate and alum, and hardness was reduced by lime and soda ash (see Chapter 7). In so far as taste and odor control was concerned, "threshold odor values were reduced by the permanganate accompanied by a characteristic change. The chlorine did not reduce odor values and on two occasions increased them substantially."

Potassium permanganate and activated carbon were successfully applied for odor control to the waters of the Racoon River at Des Moines, Iowa [57]. Carbon was fed during all periods of river intake, and when the OT exceeded 25, $KMnO_4$ was fed. An interesting comment was made: "The bland taste resulting from the activated carbon-potassium permanganate treatment has practically eliminated consumer complaints."

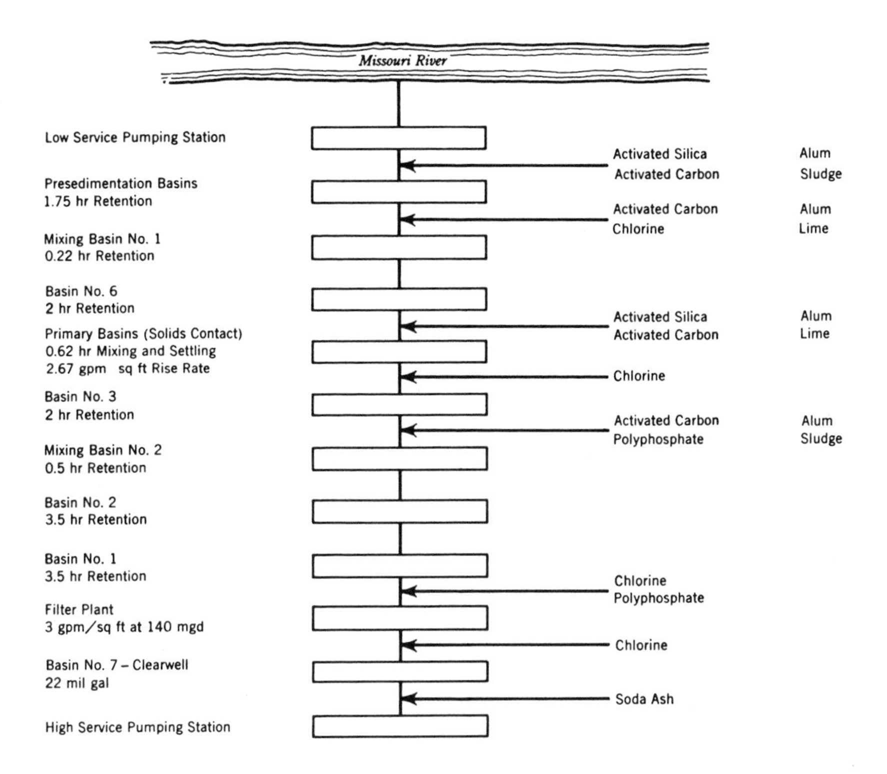

Figure 11. Schematic diagram of treatment at metropolitan utilities. All retention times are based on the maximum plant capacity of 140 mgd. Reproduced from Erdei [8], courtesy of the American Water Works Association.

Taste and odor problems with the drinking water from the Kanawha River, Nitro, West Virginia, may have ushered in the application of granular activated carbon (GAC) for their treatment [58,59]. The tastes and odors were due to a variety of organic compounds being discharged from a petrochemical complex upstream from Nitro (see Table VII). Figure 13 shows the flow diagram of the original system, an experimental system and the eventual new system. In the latter system, the water first is double-aerated and then directed to a 24 hr sedimentation basin where alum is added to effect coagulation, and chlorine is used for disinfection. After sedimentation, the water passes through GAC beds where filtration and adsorption occur. It is important to note the dual use of the GAC where the adsorptive efficiency is decreased by filtered materials. Nonetheless, the treatment scheme was reasonably effective in reducing the OT value to below the PHS standard of 3 (Figure 14).

Table XX. Operational Data for Odor Control of Missouri River Water[a]

Day of Month	Odor[b]	Odor Numbers at Sampling Points: Raw	Primary Influent	Primary Effluent	Lab. Tap	Tap Turbidity (ppm)	Carbon (ppm)[c]	$KMnO_4$ (ppm)[d]	Cl_2 (ppm)[e]
8	M	5	5	2.5	2.5	0.10	6–4	1.5–1.0	
9	M	6		3.0	2.0	0.10	4		
10	E	8			2.0 M	0.10	4–8		
11	E&M	6 M	6	3.0	3.0 S&M	0.20	8–12	3–1–2	
12	B	22	6 B	5.0 M	4.0 M	0.30	12	2–3	
	O	24							
13	E	24	7 M	5.0	3.0	0.60	12	3–1–2	
14	E&M	13	6 E&M	12.0 OM	3.0	0.80	12	1–0	3
15	E&M	21	8 OM	6.0 OM	8.0 M	0.75	14	2–4	3–5
16	M	21	6	10.0 OM	6.0 M	0.45	14	3–2–3	5–3
17	M	17	8	10.0	5.0	0.25	14	3–2	3–4
18	M	18	12	7.0	4.0	0.25	14	2–0	4–5
19	M	17 M	10	8.0	4.0	0.25	14–10–8		5–3
		21 M							
20	M	10	7	8	5	0.25	10–8		3–2.5
	M	12							
21	M	7	8	14 G	5 M	0.20	8–10		2.5
	M	12							
22	M	17	8	7	4 S&M	0.20	10		2.5–2
23	E	12	8 M	6	4	0.15	10		2.0
24	E	8	6 M	6	4	0.15	10–8		2.0
25	M	18	8	5	3.5	0.15	8–10		2.0
	M	17							

26	M	9	6	6	3.5		0.10	10–4		2.0
	M	12								
27	M	15	8	5	3.0		0.10	4		2.0

[a]Reproduced from Erdei [8], courtesy of the American Water Works Association.
[b]Odor description: B, barnyard; E, earthy; G, grassy; M, musty; O, offensive; S, sweet.
[c]Applied at presedimentation and Mixing Basin No. 2.
[d]Applied at clarification basin.
[e]Applied at primary effluent.

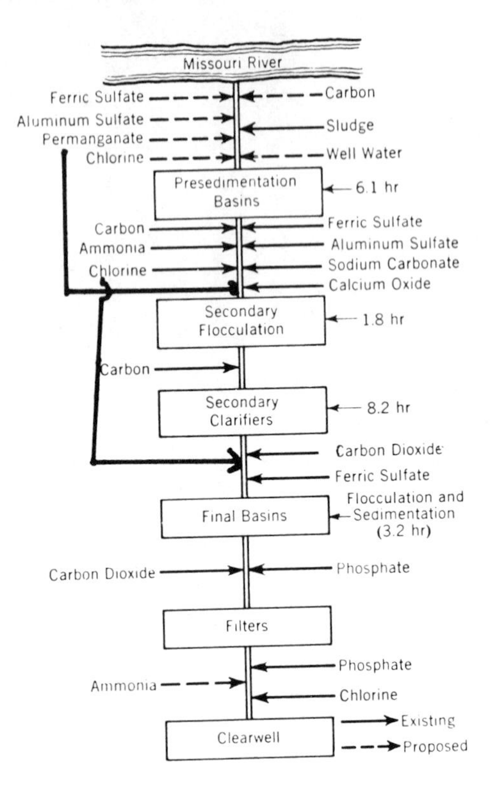

Figure 12. Flow diagram of the Kansas City, Missouri, water treatment plant. Reproduced from Popalisky and Pogge [56], courtesy of the American Water Works Association.

Chlorination

In 1940–1960, chlorination was a "popular" technique for taste and odor control. It was replaced subsequently by powdered and granular carbon. However, the various chlorination practices of combined residual (chloramine treatment), free-residual chlorination and chlorine-chlorine dioxide (see Chapter 10 for definitions) are still viable treatments provided, of course, that no THM or other halogenated products are formed. The ideal would be complete oxidation of the organoleptic compound(s) to inert products. Here it is assumed that organic compounds are responsible for the off-flavored water. Also, emphasis is given to the application of ClO_2.

At room temperature, ClO_2 is a greenish-yellow gas with an odor similar to $Cl_2(g)$, but is more irritating and more toxic. It is soluble in

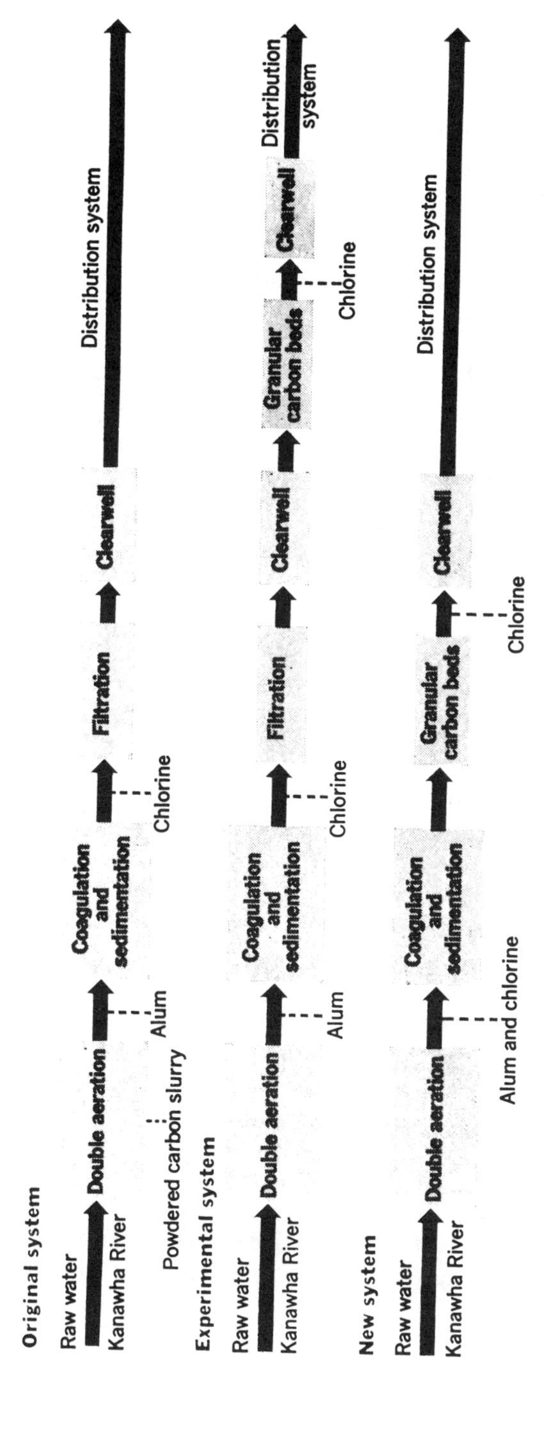

Figure 13. Flow diagram of Nitro, West Virginia, water purification plant. Reproduced from Hager [58], courtesy of the American Water Works Association.

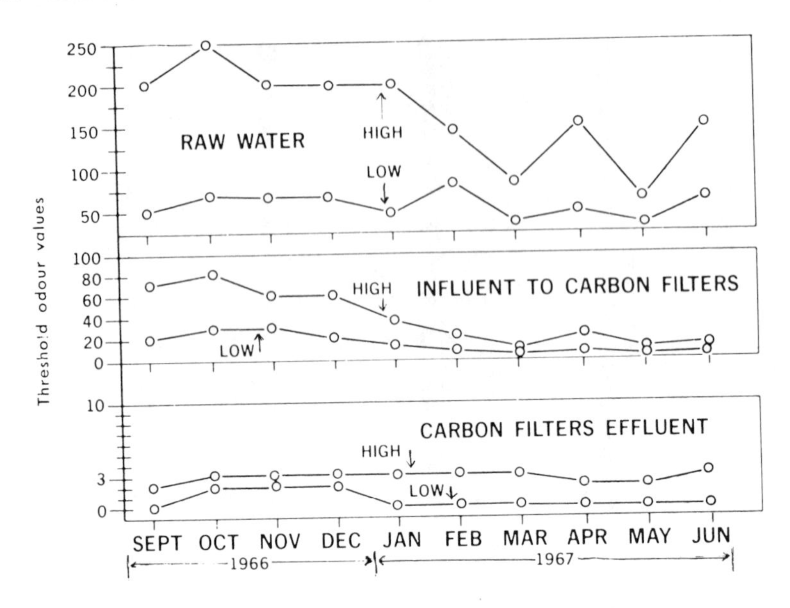

Figure 14. Threshold odor removal, Nitro, West Virginia. Reproduced from Hager and Fulker [59], courtesy of the Society for Water Treatment and Examination.

water at room temperature to the extent of 2.9 g ClO_2/l at 30 mm Hg partial pressure. It is compressed to a liquid with a density of 2.4, has a boiling point of 11°C and a melting point of −59°C.

One of the basic reactions for preparation of ClO_2 is the acidification (HCl or H_2SO_4) of a solution of sodium chlorate:

$$NaClO_3 + 2HCl = ClO_{2(g)} + 0.5Cl_2 + NaCl + H_2O \tag{12}$$

Another method for generating $ClO_{2(g)}$ in waterworks practice is the chlorination of sodium chlorite:

$$2NaClO_2 + Cl_2 = 2ClO_{2(g)} + 2NaCl \tag{13}$$

Therefore, 1.34 lb of pure $NaClO_2$ reacts with 0.5 lb of Cl_2 to produce 1 lb of ClO_2. The point of reactions 12 and 13 is that ClO_2 is prepared onsite. Bottled $ClO_{2(g)}$ is available for small-scale and emergency uses.

The oxidation state of Cl in ClO_2 is (IV), which makes it unstable in water (see Figure 6, Chapter 10). Some reduction reactions are:

$$ClO_2 + 4H^+ + 5e^- = Cl^- + 2H_2O, \qquad V^0 = +1.511 \text{ v} \tag{14}$$

$$ClO_2 + 2H^+ + 3e^- = ClO^- + H_2O, \qquad V^0 = +1.374\ v \qquad (15)$$

Reaction 14 is cited frequently as evidence that ClO_2 is 2.5× as powerful an oxidant as Cl_2. Reaction 15 is the most probable reduction reaction in waterworks practice. White [60] has recorded additional physical and chemical properties of ClO_2.

Perhaps the first use of ClO_2 for taste and odor control of phenolic compounds occurred at Niagara Falls, New York, in 1944 [61]. Other installations were made throughout the states of New York and South Carolina and in Philadelphia [62]. All of these water supplies experienced problems from phenolic tastes and odors as a result of wastewater discharges during and after World War II. Some success was achieved with control of tastes and odors from decaying algae and other aquatic vegetation. However, the greatest use of ClO_2 may be for the specific oxidation of phenols and chlorophenols [63]. According to White [60], there are four reasons why ClO_2 is the chemical of choice for oxidation of phenols:

1. ClO_2 will not react with ammonia or nitrogenous compounds, including the simple amino acids as will chlorine... It is always available, as would be HOCl, if the latter did not react to form chloramines from ammonia and other nitrogenous compounds.
2. ClO_2 reacts completely to destroy the taste-producing phenolic compounds many times faster than does free available chlorine. This is undoubtedly a result of its oxidizing capacity of two and a half times that of HOCl.
3. In all cases studied so far, it has been reported that ClO_2 will always destroy any chlorophenol taste caused by prechlorination.
4. The efficiency of ClO_2 is not impaired (as is HOCl) by a high pH environment.

Chlorine dioxide was applied successfully for control of phenolic tastes and odors in the various water supplies of Philadelphia in the late 1940s and early 1950s [62]. According to the operational experiences in this water supply, phenolic contents in the raw water were greatly affected by seasonal changes in temperature. Warmer temperatures undoubtedly increase the natural biological degradation of the phenols [64]. Correlation of the variables of temperature, phenol content and ClO_2 dosages is seen in Table XXI [62]. It follows from the temperature effect that the highest phenol contents and ClO_2 applications occurred in the coldest waters. Some success was experienced with algae control by ClO_2 added to the inlet of a storage reservoir at Philadelphia.

Table XXI. Relation of Seasonal Temperature Changes, Presence of Phenols and Chlorine Dioxide Use

Month	Average[c] Water Temperature (°F)	Average[d] Phenolic Odors	Average Presence of Phenols (ppb)		Sodium Chlorite Applied	
			Nov 1954–Dec 1955	Nov 1955–Oct 1956	Percentage of Yearly Use	Yearly Use (lb)
Jan	38	51	3,508	25	23.1	21,768
Feb	38	63	1,825	22	19.0	17,949
Mar	43	42	190	28	14.8	13,928
Apr	53	20	61	42	7.1	6,680
May	64	19	33	35	2.7	2,527
Jun	73	19	21	49	3.4	3,235
Jul	80	10	17	4	2.0	1,850
Aug	78	13	18	14	0.2	219
Sep	73	9	28	18	0.8	726
Oct	64	51	50	28	0.4	340
Nov	49	43	[e]	49[f]	8.5	8,026
Dec	40	73	4,225[e]	63[f]	18.0	17,021

Extent of Period	Temperature for Period	Percentage of Total of Phenolic Odor Averages 1948–1956	Percentage of Total Phenol Detected		Percentage of Total Chlorite Used 1949–1956
			Nov 1954–Oct 1955	Nov 1955–Oct 1956	
Six Months	>55°F	29	2	39	9.5
Six Months	<55°F	71	98	61	90.5

[a]Reproduced from Bean [62], courtesy of the American Water Works Association.
[b]Chlorine dioxide use is stated in terms of sodium chlorite applied. Note that data shown are not all for the same period.
[c]For period 1950–1955.
[d]Product of percent time and average TON. For phenolics at water intake: period October 1948–September 1956.
[e]For November and December 1954 only.
[f]For November and December 1955 only.

A kinetic study has been conducted for the phenol (C_6H_5OH) and ClO_2 reaction [65]. The stoichiometry of the reaction is:

$$\text{phenol} + 2ClO_2 \xrightarrow{k2} \text{products} \tag{16}$$

The reaction is first-order with respect to each reactant on the basis of initial rates. The appropriate rate expression is:

$$-d[ClO_2]/dt = 2k_2[P_t][ClO_2] \tag{17}$$

where $[P_t]$ = total phenol content
k_2 = observed second-order rate constant.

When the phenol content was held in excess (at least fivefold), Equation 17 becomes:

$$-d[ClO_2]/dt = 2k_{ob}[ClO_2] \tag{18}$$

where

$$k_{ob} = k_2[P_t] \tag{19}$$

Table XXII. Second-Order Rate Constants at 25.0°C for the Disappearance of ClO_2[a]

pH	$[ClO_2]_0/$ 10^{-6}, M	$[\text{phenol}]_0/$ 10^{-5}, M	k_{obsd},[b] (sec^{-1})	$k_2/10^4$, (M^{-1}-sec^{-1})
7.05	2.28	5.17	1.14	2.22
7.05	2.28	5.17		2.46
7.01	12.5	5.17		2.56
7.01	12.5	5.17		2.65
6.92	2.14	50.0	13.4	2.69
6.92	2.14	50.0	13.0	2.61
6.99	12.5	50.0	13.2	2.65
6.99	12.5	50.0		2.84
6.89	2.14	126.0	34.2	2.70
6.89	2.14	126.0	34.6	2.74
6.92	12.8	126.0	28.6	2.26
6.92	12.8	126.0		2.34
6.91	23.0	126.0	32.0	2.54
6.91	23.0	126.0		2.81

[a]Reproduced from Wajon et al. [65], courtesy of the American Chemical Society.
[b]When no value is listed for k_{obsd}, k_2 was obtained directly by fitting a second-order kinetic expression to the data.

and k_{ob} is the pseudo-first-order rate constant. Subsequent experimentation determined k_{ob} from the appropriate first-order plots, whereupon k_2 values were calculated by Equation 19. These k_2 values are given in Table XXII for ClO_2 oxidation of phenol at 25°C and for pH values 6.91–7.05 [65]. (The k_2 values are also available in the pH range of 0.17–7.99). Similar data were obtained for the ClO_2 oxidation of hydroquinone.

Under the appropriate conditions,

> the rapidity of the reaction of ClO_2 with phenol and the complete absence of chlorinated products are advantageous for rapid, efficient removal of taste and odor at significantly lower doses (6 mg/l ClO_2) than with the use of aqueous chlorine. A further advantage is that, although HOCl is a product of the ClO_2 treatment of phenols and probably also of guaiacol units of humic materials, trihalomethanes have not been found, and ClO_2-treated waters do not appear to be carcinogenic [65].

REFERENCES

1. *Standard Methods for the Examination of Water and Wastewater* 15th ed. (New York: American Public Health Association, 1980).
2. Fechner, G. T. *Elemente der Psychophysik* (Leipzig: Breitkipf and Hartel, 1889).
3. *Annual Book of ASTM Standards* Pt 31 (Philadelphia, PA: American Society for Testing and Materials, 1975).
4. Silvey, J. K. G. et al. *J. Am. Water Works Assoc.* 42:1018 (1950).
5. Silvey, J. K. G., and A. W. Roach. *J. Am. Water Works Assoc.* 45:409 (1953).
6. Fogg, G. E., and W. D. Steward. *Sci. Progr.* 53:191 (1965).
7. Morris, R. L. *Water Sew. Works* 109:76 (1962).
8. Erdei, J. *J. Am. Water Works Assoc.* 55:1506 (1963).
9. Silvey, J. K. G. *Proceedings of the 5th Sanitary Engineering Conference*, University of Illinois, Engineering Experimental Station, Circular No. 81 (1963).
10. Palmer, C. M. "Algae in Water Supplies," Public Health Service, Publication No. 657 (1959).
11. Silvey, J. K. G., and A. W. Roach *J. Am. Water Works Assoc.* 56:60 (1964).
12. Silvey, J. K. G. et al. *J. Am. Water Works Assoc.* 64:35 (1972).
13. Silvey, J. K. G. et al. *J. Am. Water Works Assoc.* 60:440 (1968).
14. Morris, R. L. et al. *J. Am. Water Works Assoc.* 55:1380 (1963).
15. Dogherty, J. D. et al. *Science* 152:1372 (1966).
16. Gaines, H. D., and R. P. Collins. *Lloydia* 26:247 (1963).
17. Gerber, N. N., and A. A. Lechevalier. *Appl. Microbiol.* 13:935 (1965).
18. Gerber, N. N., *Biotechnol. Bioeng.* 9:321 (1967).
19. Gerber, N. N. *Tetrahed. Lett.* 25:2971 (1968).

20. Safferman, R. S. et al. *Environ. Sci. Technol.* 1:429 (1967).
21. Rosen, A. A. et al. *Appl. Microbiol.* 16:178 (1968).
22. Medsker, L. L. et al. *Environ. Sci. Technol.* 2:461 (1968).
23. Medsker, L. L. et al. *Environ. Sci. Technol.* 3:476 (1969).
24. Rosen, A. A. et al. *Water Treatment Exam.* 19:106 (1970).
25. Wood, N. F., and V. L. Snoeynik. *J. Chromato.* 132:405 (1977).
26. Collins, R. P., and H. D. Gaines. *Appl. Microbiol.* 12:335 (1964).
27. Jenkins, D. et al. *Environ. Sci. Technol.* 1:731 (1967).
28. Dague, R. R. *J. Water Poll. Control Fed.* 44:583 (1972).
29. Rosen, A. A. *Proceedings of the 11th Sanitary Engineering Conference* (Urbana, IL: University of Illinois, 1969), p. 59.
30. Baker, R. A. *J. Am. Water Works Assoc.* 55:913 (1963).
31. Zoeteman, B. C. J. et al. *J. Am. Water Works Assoc.* 72:537 (1980).
32. Rosen, A. A. et al. *J. Water Poll. Control Fed.* 34:7 (1962).
33. Baker, R. A. *J. Water Poll. Control Fed.* 35:728 (1963).
34. Thomas, H. A. *J. Am. Water Works Assoc.* 35:751 (1943).
35. Rosen, A. A. et al. *J. Water Poll. Control Fed.* 35:777 (1963).
36. Burttschell, R. H. et al. *J. Am. Water Works Assoc.* 51:205 (1959).
37. Soper, F. G., and G. F. Smith. *J. Chem. Soc.* (*London*) p. 1582 (1926).
38. Ettinger, M. B., and C. C. Ruchhoft. *J. Am. Water Works Assoc.* 43:561 (1951).
39. Adams, B. A. *Water Water Eng.* 33:387 (1931).
40. Lee, G. F., and J. C. Morris. *Int. J. Air Water Poll.* 6:419 (1962).
41. Bruvold, W. H. et al. *J. Am. Water Works Assoc.* 59:547 (1967).
42. Bruvold, W. H., and H. J. Ongerth. *J. Am. Water Works Assoc.* 61:170 (1969).
43. Bruvold, W. H. et al. *J. Am. Water Works Assoc.* 61:575 (1969).
44. Pangborn, R. M. et al. *J. Am. Water Works Assoc.* 62:572 (1970).
45. Pangborn, R. M., and L. L. Bertolero. *J. Am. Water Works Assoc.* 64:511 (1972).
46. Pangborn, R. M. et al. *Percep. Psychophys.* 8:69 (1970).
47. Bruvold, W. H., and R. M. Pangborn. *J. Appl. Psychol.* 50:22 (1966).
48. Bruvold, W. H., and W. R. Gaffey. *J. Exp. Psychol.* 69:369 (1965).
49. Bruvold, W. H., and W. R. Gaffey. *J. Appl. Psychol.* 53:517 (1969).
50. Bruvold, W. H. *J. Am. Water Works Assoc.* 69:562 (1977).
51. Bryan, P. E. et al. *J. Am. Water Works Assoc.* 65:363 (1973).
52. Baker, R. A. *J. Am. Water Works Assoc.* 56:1578 (1964).
53. Bruvold, W. H., and R. M. Pangborn. *J. Am. Water Works Assoc.* 62:721 (1970).
54. Cohen, J. M. et al. *J. Am. Water Works Assoc.* 52:660 (1960).
55. Peryam, D., and V. Swartz. *Food Technol.* 4:390 (1950).
56. Popalisky, J. R., and F. Pogge. *J. Am. Water Works Assoc.* 59:1121 (1967).
57. Maloney, J. R. *J. Am. Water Works Assoc.* 60:1195 (1968).
58. Hager, D. G. *Environ. Sci. Technol.* 1:287 (1967).
59. Hager, D. G. and R. D. Fulker. *J. Soc. Water Treat. Exam.* 17:41 (1968).
60. White, G. C. *Handbook of Chlorination* (New York: Van Nostrand Reinhold Company, 1972).
61. Aston, R. N. *J. Am. Water Works Assoc.* 39:687 (1947).

62. Bean, E. L. *J. Am. Water Works Assoc.* 49:205 (1957).
63. Ingols, R. S., and G. M. Ridenour. *Water Sew. Works* 95:187 (1948).
64. Fillmore, L. et al. *J. Environ. Sci. Health,* A17(6):797 (1982).
65. Wajon, J. E. et al. *Environ. Sci. Technol.* 16:396 (1982).

CHAPTER 4

REMOVAL OF ORGANICS BY ACTIVATED CARBON

HISTORY OF ACTIVATED CARBON USE

Activated carbon has been long recognized as one of the most versatile adsorbents to be used for the effective removal of low concentrations of organic substances from solution. Charcoal, the forerunner of modern activated carbon whose ability to purify water dates back to 2000 B.C., was used for medicinal purposes by ancient Egyptians. Its adsorptive properties for gases were first reported by Scheele in 1773 and later by Fontanna in 1777 [1]. In 1785 Lowitz observed the ability of charcoal to decolorize many liquids. This led to the widespread application of wood and bone chars in the refinery of cane and beet sugars [2]. Several attempts were made during the nineteenth century to prepare activated carbons from other sources, such as blood, coconut, flour and paper mill wastes. The development of modern commercial activated carbon is attributed to Ostrejka, who described a basic process for producing activated charcoal from vegetable materials [2]. This led to the manufacture of powdered activated carbon in Europe as a substitute for bone char in established operations. The real development of activated carbon began during World War I, when hard granular activated carbon (GAC) was developed for use in gas masks. These earlier carbons had relatively less adsorptive capacity than modern carbons. Further developments in the manufacture of activated carbon with sufficient hardness to resist abrasion and the ability to be regenerated for reuse paved the way for utilization of activated carbon in many industrial applications [1,2]. During the past two decades, increased awareness of the occurrence of many organic substances in natural waters and the need for their control, led to the emergence of activated carbon adsorption as one of the most effective methods for removing these compounds from drinking waters.

PORE STRUCTURE AND SURFACE AREA

The properties of activated carbon are attributed mainly to its highly porous structure and relatively large surface area. A particle of activated carbon is composed of a complex network of pores that can be divided into two distinct classes with respect to size: macropores and micropores. The macropores are arbitrarily defined as pores greater than 500 Å in diameter. These pores are large capillaries that extend all through the interior of the particle. The micropores have diameters in the range of 10–500 Å and mostly branch off the macropores. Since the surface area of cylindrical or irregularly shaped pores increases considerably with decreasing diameter, the walls of the micropore system contribute to the major part of activated carbon's inner surface area that may reach 500–1500 m^2/g of activated carbon [3]. This large surface area is one of the main reasons for the high adsorptive capacity of activated carbon. The pore size distribution depends on the type of starting material and the method of manufacturing the activated carbon. The macropores serve as conduits for transport of the adsorbate from the outer particle surface to the interior micropore surfaces where most of the adsorption occurs. A proper distribution of both types of pores is therefore required for the efficient utilization of activated carbon.

Determining Specific Surface Area

The most common method for determining the surface area of activated carbon and other porous adsorbents is the Brunauer, Emmett and Teller (BET) method [4–6]. This method utilizes gas adsorption for the evaluation of monolayer capacity, Vm. This is defined as the volume of gas in cubic centimeters at standard temperature and pressure (STP) that would be required to cover the adsorbent with a monolayer. The specific surface area, Σ (m^2/g), and Vm are related by:

$$\Sigma = \frac{Vm}{22{,}400} \times 6.03 \times 10^{23} \times 10^{-20} \times \sigma_m \quad (1)$$

$$\Sigma = 0.269 Vm \sigma_m \quad (2)$$

where σ_m = area ($Å^2$) that one adsorbed molecule would occupy in a completed monolayer

Assuming an adsorbed molecule would have the same packing that a molecule of the condensed phase would have in their plane of closest packing, σ_m can be evaluated from:

$$\sigma_m = 3.646\left(\frac{M}{4 \cdot 2^{1/2} \cdot N \cdot \sigma}\right) \quad (3)$$

where M = molecular weight
N = Avogadro's number
σ = density of the condensed phase (solid or liquid) at the temperature of the isotherm

The monolayer capacity Vm is determined by performing an adsorption isotherm of nitrogen gas on the activated carbon at liquid nitrogen temperature (195.8°C). The adsorption isotherm is plotted according to the BET equation, which may be written in a linear form:

$$\frac{P}{V(P_o - P)} = \frac{1}{VmC} + \left(\frac{C-1}{VmC} \cdot \frac{P}{P_o}\right) \quad (4)$$

where P = equilibrium pressure
P_o = saturation pressure of nitrogen at the temperature of the experiment
V = volume of nitrogen adsorbed
C = constant characteristic of the gas-solid pair

A straight line is obtained by plotting $P/V(P_o - P)$ against P/P_o and the value of Vm can be calculated from the slope $(C-1)/VmC$.

Determining Pore-Size Distribution

The pore size and pore-size distribution of activated carbon are determined by an analysis of gas adsorption isotherms in the range where capillary condensation accompanies physical adsorption. The Kelvin equation [7] generally is applied to the desorption portion of the adsorption isotherm. The Kelvin equation relates the equilibrium vapor pressure of a curved surface, such as that of a liquid in a capillary or pore, to the equilibrium pressure of the same liquid on a plane surface. Equation 5 is a convenient form of the Kelvin equation:

$$ln\frac{P}{P_o} = -\frac{2\gamma\bar{V}}{rRT}\cos\theta \quad (5)$$

where P = equilibrium vapor pressure of the liquid contained in a narrow pore of radius r
P_o = saturation pressure or the equilibrium pressure of the same liquid exhibiting a plane surface
γ = surface tension

$\bar{V}$ = liquid molar volume
θ = contact angle at which the liquid meets the pore wall

The pore volume at each partial pressure (P/P_o) is given by [8]:

$$V_p = \frac{W}{\rho_L} \tag{6}$$

where V_p = the pore volume or the volume of adsorbate
W = the weight of adsorbate
ρ_L = the liquid density

A total pore volume can be obtained by determining the weight of the adsorbate at saturation pressure (i.e., $P/P_o \approx 1$). With each pore volume there is an associated surface area. If the pores are regarded as cylindrical segments, then for segments of radius r, the pore volume V and the surface area A are related by:

$$r = 2k\,V/A \tag{7}$$

where k = conversion factor for units of measurement

If the adsorbate is desorbed or adsorbed in small increments, and if the surface area is determined, a curve may be plotted showing the variation of surface area with the change in occupied pore space. The reciprocal of the slope of this curve at any pore volume is dV/dA and may be substituted into Equation 4 for V/A. With this information, pore-size distribution curves (r vs V or dV/dA vs r) may be constructed.

Juhola and Wiig [9] applied the Kelvin equation to a water isotherm to determine the micropore-size distribution in several activated carbons. This is the method most commonly used by carbon manufacturers. However, the adsorption isotherms of other adsorbates, such as benzene or nitrogen, can be used. The latter is used quite extensively for pore-size analysis of porous solids.

The useful range of the Kelvin equation is limited to the narrow micropore diameter range of 15–1000 Å. Above 1000 Å, the equation is not reliable due to the rapid change of the pore diameter with relative pressure. At partial pressure P/P_o above 0.99, *ln* P/P_o approaches 1, and r rapidly approaches infinite size and the accuracy of the pore diameter measurement decreases.

Mercury porosimetry techniques are used to study macropore distribution above r = 500 Å [8]. These techniques are based on capillary depression. Because of its high surface tension, mercury tends not to wet most solid surfaces and must be forced to enter a pore. When forced under pressure into a pore of radius r and length l the amount of work W required is proportional to the increased surface exposed by the mercury at the pore wall. Therefore, assuming cylindrical pores:

$$W = 2\pi r l \gamma \tag{8}$$

where γ = surface tension of mercury

Since the wetting angle θ of mercury is greater than 90° and less than 180° on all surfaces with which it does amalgamate, the work required is reduced by $\cos\theta$, and Equation 8 becomes:

$$W_1 = 2\pi r l \gamma \cos\theta \tag{9}$$

When a volume of mercury ΔV is forced into a pore under external pressure P the amount of work W_2 performed is:

$$W_2 = P\Delta V = P\pi r^2 l \tag{10}$$

At equilibrium, Equations 9 and 10 are combined:

$$Pr = 2\gamma \cos\theta \tag{11}$$

Because the product, Pr, is constant, and assuming constant γ and θ, Equation 11 indicates that, as the pressure increases, mercury will intrude into progressively narrower pores [8]. Consequently, pore-size distribution can be measured by forcing mercury into a porous solid, e.g., activated carbon, and measuring the penetrating quantity V as a function of the pressure applied. From Equation 11, assuming the pores have a circular cross section, the pore radius r associated with P may be calculated. By plotting V as a function of r, the pore-volume distribution curve, dV/dr as a function of r, can be obtained by graphical differentiation [8].

The volume of mercury forced into the pores is usually monitored in a penetrometer, which is a calibrated stem of a glass cell containing the sample and filled with mercury. As intrusion occurs, the level in the capillary stem decreases. Modern porosimeters can achieve maximum pressures in the range of 50,000–60,000 psi, which corresponds to pore radii ranging from 10.7×10^3 Å at 1 psi to 18 Å at 60,000 psi. Figure 1 shows a pore-size distribution curve for two carbons obtained by porosimetery [10].

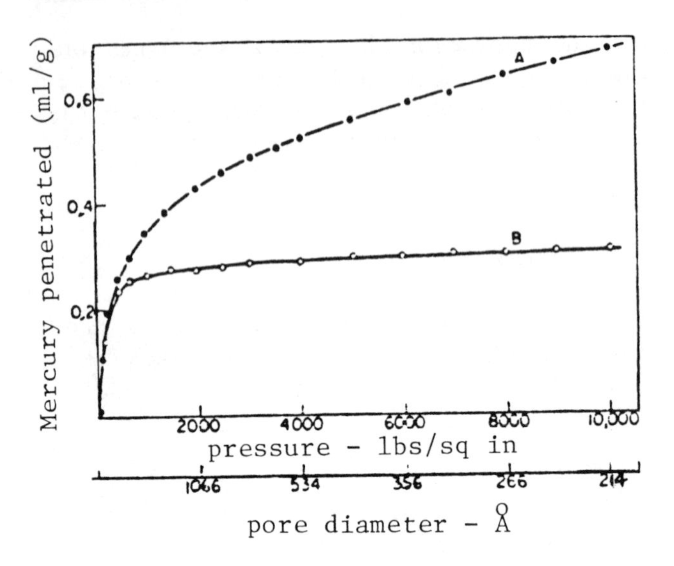

Figure 1. Mercury penetration into activated carbon. A = Darco carbon, B = Columbia carbon. Reproduced from Drake [10], courtesy of the American Chemical Society.

MANUFACTURING ACTIVATED CARBON

Activated carbons have been produced from every conceivable form of carbonaceous substance [2]. The most common materials are wood, coal, peat, lignin, nut shells, sawdust, bone and petroleum coke. The nature of the starting materials does not have a significant effect on the properties of the resulting activated carbon, since different characteristics can be imparted by the selective introduction of additives and by controlling the production processes.

The production of activated carbon involves two processes: carbonization followed by activation. Carbonization (pyrolysis) is sometimes called charring, which consists of slowly heating the starting material in the absence of air to temperatures usually below 600°C [2]. The carbonization of carbohydrate- or cellulose-containing materials (e.g., wood) can be controlled by use of dehydrating additives such as zinc chloride or phosphoric acid where a considerably greater pore system is obtained [2]. The dehydrating agents are usually recovered for reuse. During carbonization, volatile pyrolysis products are removed from the starting material which leaves a residue called char or coke. Chars, frequently called amorphous carbon, have been shown by X-ray studies to be composed of crystals of submicroscopic dimensions termed microcrystallites [11,12].

Activation consists of treating the chars with oxidizing agents such as steam, carbon dioxide, or oxygen at elevated temperatures, 800–900°C. The oxidizing agents selectively attack portions of the char, transforming it into gaseous products. The resulting hollow pores are systematically developed and enlarged. As a result of studies of the kinetics and mechanisms of activation [3,13], the reaction now can be carefully controlled, enabling the production of a well defined pore system in commercial activated carbons.

Before carbonization and activation, the starting materials must be brought to uniform size. Such physical properties of the final product as the granule size, shape, roughness and hardness are influenced by production techniques [3]. Smooth uniform cylinders with particle sizes between 1 and 9 mm are prepared by extrusion, at high pressures, of a blend of the pulverized material and a binder (sugar, tar, pitch and lignin) [2] or a dehydrating agent such as zinc. Pelletization is used to produce relatively smooth spherical particles in the size range of 4–9 mm. Agglomeration with subsequent crushing and screening is used for shaping small particle base material. Starting materials with natural granularity can be shaped by crushing. The last two methods result in granules of irregular rough surface with a wide range of grain sizes, 1–10 mm. Pelletization and crushing result in final products with acceptable hardness, while extrusion or agglomeration results in products with very good hardness [3].

Jüntgen [3] reported studies on changes in the properties of GAC during manufacturing. The shaping process of the granules seems to determine the macropore system. Most of the macropores are formed during the pyrolysis process in the void volume filled by the binder. The macropore volume depends on the forming pressure, particle size and particle size distribution of the starting material in the granule. Large particles with a wide range of sizes result in a macropore system with a large volume and a wide range of pore radii. On the other hand, smaller particle sizes and a narrow range of particle-size distribution results in a low macropore volume and a very narrow range of pore radii.

The activation process has a decisive effect on the most important properties of activated carbon as illustrated in Figure 2 [3]. The degree of activation has a profound effect on the micropore system. The micropore volume increases almost linearly with the degree of activation. The sharp increase is due to the gasification reaction occurring predominantly on the inner surface of the coke; thus, the majority of pores formed are fine. The macropore volume undergoes a relatively moderate increase in the course of activation. In addition, pore-size distribution of the macropore volume is not significantly influenced by the activation process. The average adsorption pore diameter increases sharply with

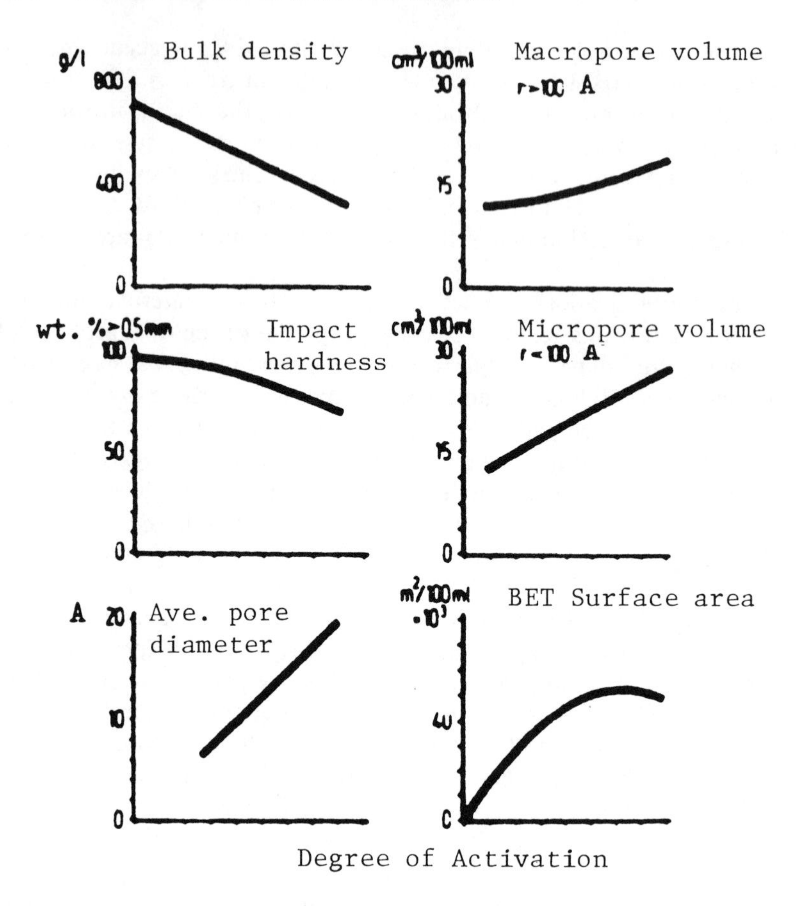

Figure 2. Manufacture of activated carbon, influence of degree of activation [3].

activation, i.e., the carbon becomes increasingly wide pored with activation. The inner surface area does not increase uniformly over the whole activation process. A sharp initial increase is followed by a slight decrease as activation is continued beyond a maximum value. The initial increase in the surface area is attributed to the formation of predominantly narrow pores which contribute largely to the inner surface area. Further activation causes a growth of smaller pores. Hence, the increase in the specific inner surfaces is slowed and eventually is decreased. The bulk density decreases almost linearly with increasing degree of activation. The decrease of bulk density is due to the pore system becoming larger with increasing activation, due to a decrease in apparent density and the constant real density of the carbon. With increasing micropore and macropore volume, the carbon structure becomes more fragile and the impact hardness decreases, first fairly slightly, then at a more pronounced rate beyond medium values of degree of activation.

STRUCTURE OF ACTIVATED CARBON

The structure of activated carbon was shown, from X-ray diffraction studies [11,12], to be composed of microcrystallites consisting of fused hexagonal rings of carbon atoms with a structure similar to graphite. The structure of graphite, as shown in Figure 3, is composed of infinite layers in which the carbon atoms are arranged in a hexagonal lattice [14]. Each graphitic layer is analogous to a very large polynuclear aromatic molecule (fused hexagons). Within each layer, three of each carbon atom's four valence electrons form covalent bonds (σ bonds) with three neighboring carbon atoms, while the fourth electron resonates between several valence bond structures (π bond), giving each carbon-carbon bond a one-third double-bond character. The layers are stacked with a separation distance of 3.35 Å and are held together by weak van der Waals forces. The carbon layers form an ABAB stacking sequence in which one-half of the carbon atoms in any one plane lie above the center of the hexagons in the layers immediately below it [14,15].

In microcrystalline carbon, the graphite-like layers are stacked parallel to each other in packet of 5–15 layers about 20–50 Å in height [12]. The diameter of each layer has been estimated to be 20–50 Å. The structure of microcrystallites, however, differs from that of graphite in many ways. Interior vacancies exist in the carbon microcrystallite, and their formation depends on the presence of impurities and the method of preparation. In addition, such foreign atoms as oxygen, sulfur and hydrogen are always present in different amounts in microcrystallites, which may be bound at the edge of the planes to form heterocyclic ring systems or functional groups [14]. Heterocyclic groups tend to affect the distance of separation of adjacent planes, which is slightly higher in microcrystallites, about 3.6 Å, then in graphite [15]. Also, the stacking sequence in microcrystallite varies from that found in graphite to a completely random orientation in turbostratic carbons [14]. The functional groups terminating the planes interconnect the microcrystallites and are, at least, partially responsible for the turbostratic character, in that they prevent orientation of the planes with respect to each other.

The size of microcrystallites is influenced by the temperature of carbonization and the structure of raw material [2]. Microcrystallites in

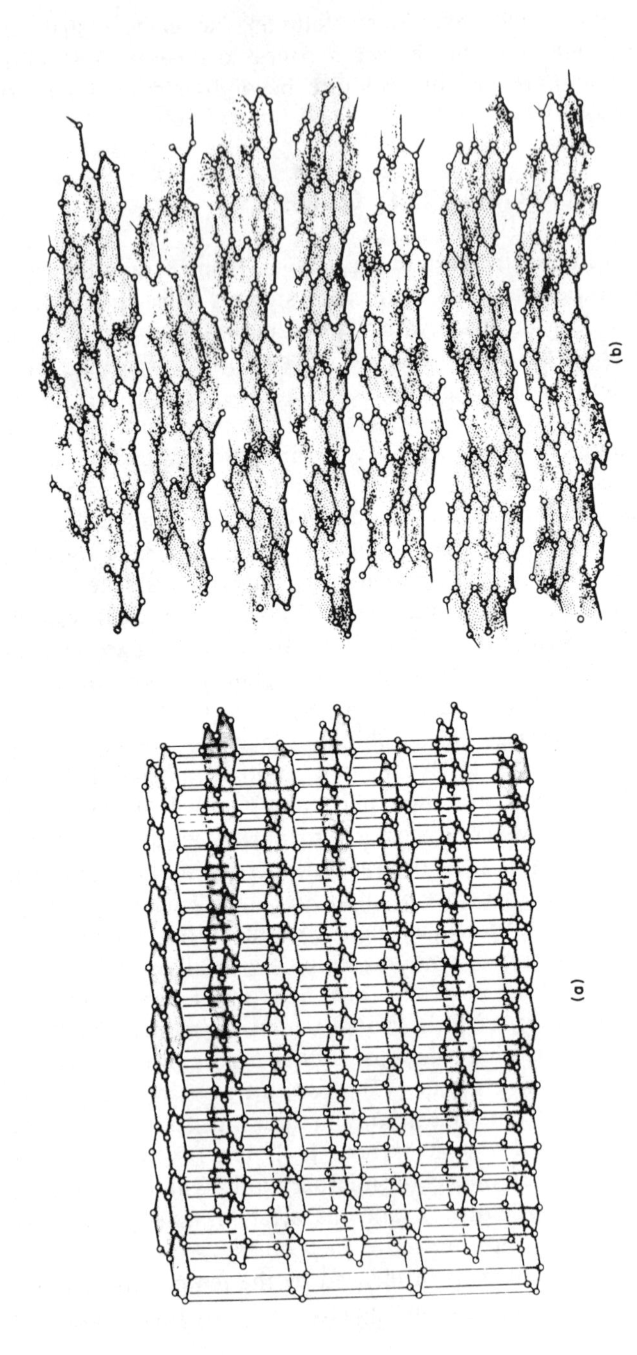

Figure 3. Schematic diagram comparing **(a)** a three-dimensional graphite lattice with **(b)** a turbostratic structure. Reproduced from Walker [14], courtesy of *American Scientist.*

chars prepared from carbohydrates (such as cellulose) increase in size with increasing temperature of carbonization. The macropore structure is developed by burnout of the reactive materials in the interstices between the microcrystallites and through the graphitic regions. The micropores, on the other hand, are formed mainly by burnout of the microcrystallite basal planes during the process of activation [16]. As indicated above, the basal planes are composed mainly of fused hexagonal rings, and are thus expected to be fairly uniform. The involvement of carbon atoms in these planes in the formation of sigma bonds with the neighboring carbon atoms precludes the presence of any functional groups. Therefore, planar surfaces of the micropores, which contribute most of the surface area, are responsible for most of the adsorptive properties of activated carbon. Adsorption on these surfaces would be mostly physical due to weak van der Waal forces, although electrons are available for bonding if the structure of the adsorbate molecule is conducive to such bonding [17,18].

The edges of microcrystallites are more reactive and contain a wide variety of functional groups. During the process of carbonization, a large number of repaired electrons, which are resonance stabilized, are trapped in the microcrystallite due to bond breakage at the edge of the planar structures [17,19]. Foreign atoms such as oxygen and hydrogen interact with these electrons to form surface complexes or functional groups [20]. Surface sites associated with the functional groups constitute a small portion of the total surface area of activated carbon. However, they can influence the characteristics of carbon by participating in chemisorptive interactions or forming "specific adsorption" sites [17,21]. For example, presence of hydrogen on the surface of carbon imparts specific characteristics such that adsorption of iodine vapor on certain carbons is reported to be exclusively associated with that part of the surface associated with hydrogen sites [22]. Hydrogen exists at edges of the basal planes in the form of terminal groups or part of hydrocarbon functional groups.

The chemistry of carbon surfaces has been studied extensively by several investigators [17,23–26]. Several reviews have been presented by Weber and co-workers [17,27]. The relevant aspects of surface functional groups pertaining to activated carbons used in water purification are discussed below.

The type of starting material and the method of activation used in producing activated carbon determines, to a large extent, the nature of surface functional groups. The most important of these are the oxygen-containing surface groups because of their probable effect on the adsorption process. The oxygen content of activated carbon ranges between 1 and 25% and has been shown to vary considerably with the activation temperature [28]. The amount of oxygen decreases with an increase in the activation temperature and a very low oxygen content is obtained at activation temperatures higher than 1000°C [26]. The major source of oxygen in activated carbon is the interaction of the oxidizing gases used in the process (oxygen, carbon dioxide or steam). Here, the oxygen is chemisorbed and bound as surface oxides on the edges of the layer planes [17,28].

Chemisorbed oxygen can be removed from activated carbon by heat, which it evolves as carbon dioxide, $CO_{2(g)}$, or carbon monoxide, $CO_{(g)}$, depending on the temperature. CO_2 is evolved at temperatures below 600°C, whereas CO is evolved between 500 and 800°C. The complete removal of all bound oxygen requires a temperature of about 1200°C. Under vacuum, the oxygen is completely removed as CO_2, CO and H_2O [30,31]. The activation temperature has a significant effect on the type of surface oxide formed in activated carbon. Activation temperatures from about 200 to 500°C result in the formation of surface functional groups that are thermally removed as CO_2, while at the higher temperature range, functional groups form that are thermally removed as CO [32,33]. The nature of the surface oxides formed at different activation temperatures has been studied extensively by several investigators using acid-base titrations [24,29,30,32,34]. Carbons activated at low temperatures, 200–400°C, termed L-carbons, generally will develop acidic surface oxides and will lower the pH value of neutral or basic solutions. They sorb primarily bases from solutions, they are hydrophilic, and they exhibit a negative zeta potential. Those carbons activated at higher temperatures, 800–1000°C, termed H-carbons, will develop basic surface oxides and will raise the pH value of neutral or acidic solutions. They sorb acids and exhibit a positive zeta potential. However, cooling H-carbons in contact with air will change the zeta potential to negative apparently due to formation of acidic surface oxides. Commercial activated carbons used in water treatment are activated at temperatures ranging between 500 and 900°C and exhibit amphoretic properties. Carbons activated by chemical treatment in aqueous solutions with such oxidizing agents as, chlorine, permanganate, persulfate, H_2O_2 and nitric acid, develop the same characteristics as L-carbons. It has been suggested [30,34] that surface oxides thermally removed as CO_2 are responsible for the physicochemical properties of L-carbons and that the oxides evolved as CO are responsible for H-carbons' characteristics.

Many techniques have been used to identify the surface oxides of activated carbon. An acid-base titration is the method most extensively used [24,25,30–34]. For example, the acidic surface oxides can be identified by titration with a base of different strength. Other methods include the typical identification reactions of organic chemistry for such specific groups as carboxylic, phenolic and ketones, etc.: polarography, infrared and internal reflectance infrared spectroscopy [26]. The groups most frequently suggested for the acidic surface oxides include carboxylic groups, phenolic hydroxyl groups and quinone-type carbonyl groups. There are further suggestions of normal lactones, fluorescein type lactones, carboxylic acid anhydrides, cyclic peroxides and the enol form of 1, 3-diketone. Figure 4 shows schematic structural representations of these groups. The presence of acidic surface oxides on activated carbon has been shown to influence the extent of adsorption of aromatic compounds from aqueous solutions. Carboxylic surface oxides, resulting from wet chemical oxidation of carbon, have a negative effect on the adsorption of phenol from dilute solutions [25]. These groups impart a polar character to the surface of carbon which could result in preferential adsorption of water molecules on these sites. These water-oxygen functional group complexes could prevent the migration of organic molecules to a large portion of the surface area [35]. The presence of surface-carbonyl groups or quinonic structures, on the other hand, enhances adsorption of such aromatic compounds as phenol and *p*-nitrophenol due to the formation of a donor-acceptor complex involving the benzene ring's π electrons with the partial positive charge on the carbonyl groups [26].

The basic surface oxides on H-carbons that sorb acids from aqueous solutions were not as thoroughly investigated as acidic surface oxides. Garten and Weiss [36] proposed a chromene surface structure responsible for chemisorption of acids. Chromene groups contain active methylene groups, $>CH_2$ or $>CHR$, that can react with strong acids in the presence of oxygen to form carbonium ions and hydrogen peroxide (Figure 4). The evolution of hydrogen peroxide was observed earlier by Kolthoff [37] during the adsorption of acid by carbon. The carbonium ion tends to associate strongly with a negative anion, which accounts for the fact that sorbed acids are very difficult to wash from activated carbon.

ADSORPTION

Adsorption is defined as the increase in concentration of a particular component at the surface or interface between two phases. In any solid or liquid, atoms at a surface are subject to unbalanced forces of attraction normal to the surface plane. These forces are merely extensions of the forces acting within the body of the material. A molecule in the

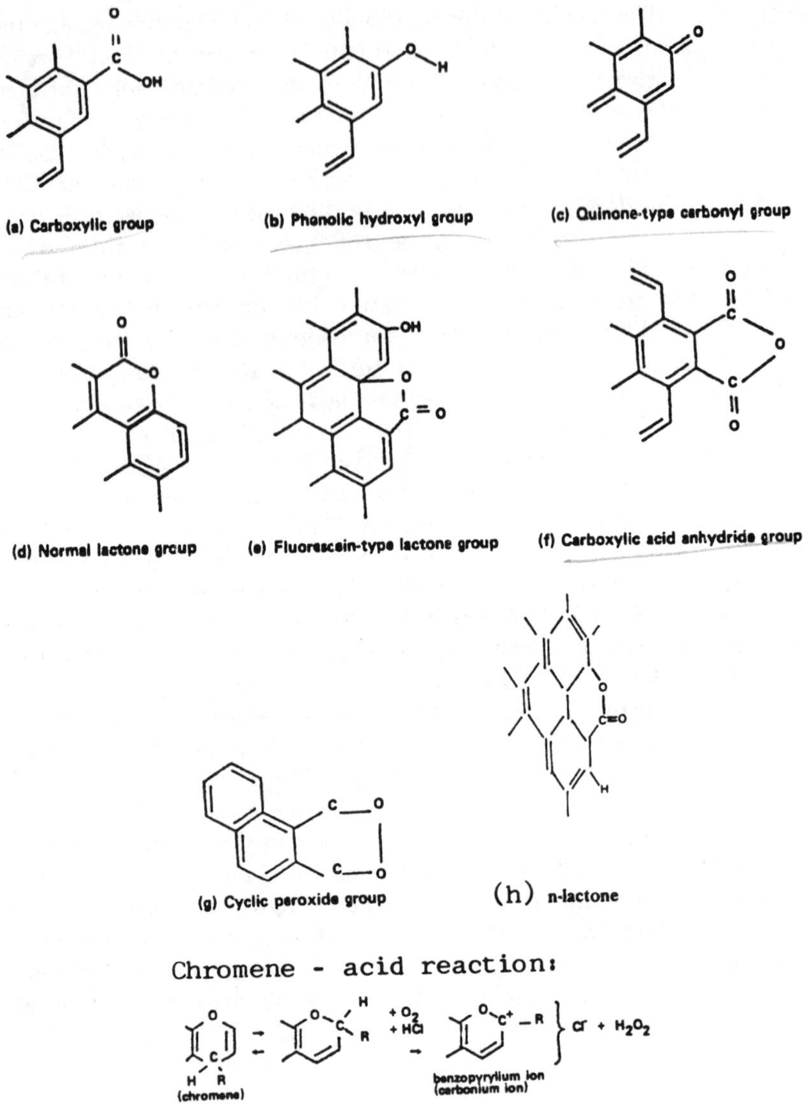

(a) Carboxylic group (b) Phenolic hydroxyl group (c) Quinone-type carbonyl group

(d) Normal lactone group (e) Fluorescein-type lactone group (f) Carboxylic acid anhydride group

(g) Cyclic peroxide group (h) n-lactone

Chromene - acid reaction:

(chromene) benzopyrylium ion (carbonium ion)

Figure 4. Oxygen surface functional groups.

center of a liquid drop is attracted equally from all sides, while at the surface, the attractive forces acting between adjacent molecules result in a net attraction into the bulk phase in a direction normal to the surface. Because of the unbalanced attraction at the surface, there is a tendency for these molecules to be pulled from the surface into the interior, and for the surface to shrink to the smallest area that can enclose the liquid. The work required to expand a surface by 1 cm^2 in opposition to these attractive forces is called the surface tension.

The above concept applies equally well to solids. Molecules in a solid surface are also in an unbalanced attractive field and possess a surface tension or a surface free energy. Whereas the surface tension of a liquid can be measured easily, that of a solid cannot, since, to increase the surface, extraneous work must be done to deform the solid [38]. Adsorption partially restores the balance of forces and is accompanied by a decrease in the system's free energy. Adsorption processes may be classified as physical or chemical, depending on the nature of the forces involved.

Physical adsorption on solids is attributed to forces of interactions between the solid surface and adsorbate molecules that are similar to the van der Waals forces between molecules. These forces that involve the electrons and nuclei of the system are electrostatic in origin, and are termed dispersion forces. The dispersion forces exist in all types of matter and always act as an attractive force between adjacent atoms and molecules no matter how dissimilar. They are always present regardless of the nature of other interactions, and often account for the major part of the adsorbate-adsorbent potential [38,39]. The nature of the dispersion forces was first recognized in the 1930s by London [40]. Using quantum mechanical calculations, it was postulated that the electron motion in an atom or molecule would lead to a rapidly oscillating dipole moment. At any instant, the lack of symmetry of the electron distribution about the nuclei imparts a transient dipole moment to an atom or molecule that would average zero over a longer time interval. When in close proximity to a solid surface, each instantaneous dipole of an approaching molecule induces an appropriately oriented dipole moment in a surface molecule. These moments interact to produce an instantaneous attraction. These forces are known as dispersion forces because of their relationship, noted by London [40], to optical dispersion. The dipole-dispersion interaction energy can be determined by [40]:

$$E = -C/r^6 \tag{12}$$

where E = dispersion energy or potential
C = a constant
r = distance of separation between the interacting molecules

In addition to dipole-dipole interactions, other possible dispersion interactions contributing to physical adsorption include dipole-quadrapole and quadrapole-quadrapole interactions. If these two are included, the total dispersion energy becomes [39,40]:

$$E = -C/r^6 - C'/r^8 - C''/r^{10} \quad (13)$$

where C' = a constant for dipole-quadrapole interactions
C'' = a constant for quadrapole-quadrapole interactions

The contribution to E from the terms in Equation 13 clearly depends on the separation r between the molecules; therefore, the dipole-dipole interactions will be most significant. Quadrapole interactions involve symmetrical molecules with atoms of different electronegativities such as CO_2. This molecule has no dipole moment, but does have a quadrupole ($^-O—^+C^+—O^-$) that can lead to interactions with polar surfaces.

When an adsorbate molecule comes very close to a solid surface molecule to allow interpenetration of the electron clouds, a repulsive interaction will arise. The overall interaction energy is expressed as:

$$E = -C/r^6 + B/r^{12} \quad (14)$$

where B = repulsive interaction constant

A more detailed discussion of interaction forces in physical adsorption can be found elsewhere [39].

The second type of adsorptive interaction is chemisorption. This is characterized mainly by large interaction potentials that lead to high heats of adsorption approaching the value of chemical bonds. This fact, coupled with other spectroscopic, electron spin resonance and magnetic susceptibility measurements, confirms that chemisorption involves the transfer of electrons and the formation of true chemical bonding between the adsorbate and the solid surface [39]. Because chemisorption involves chemical bonding, it often occurs at high temperatures and is usually associated with activation energy. Also, the adsorbed molecules are localized on specific sites and, therefore, are not free to migrate about the surface.

Physical adsorption can be distinguished from chemisorption according to one or more of the following criteria:

1. Physical adsorption does not involve the sharing or transfer of electrons, and thus always maintains the individuality of interacting species. The interactions are fully reversible, enabling desorption to occur at the same temperature, although the process may be slow because of diffusion effects. Chemisorption involves chemical bonding and is irreversible.

2. Physical adsorption is not site-specific; the adsorbed molecules are free to cover the entire surface. This enables surface area measurements of solid adsorbents. In contrast, chemisorption is site specific; chemisorbed molecules are fixed at specific sites.
3. The heat of physical adsorption is generally low compared to chemisorption; however, heat of adsorption is not usually a definite criterion. The upper limit for physical adsorption may be higher than 20 kcal/mol for adsorption on adsorbents with very narrow pores. The heat of chemisorption ranges from over 100 kcal/mol to less than 20 kcal/mol. Therefore, only very high or very low heats of adsorption can be used as a criterion for this type of adsorption process [8].

FACTORS AFFECTING ADSORPTION

Nature of the Adsorbent

Surface Area and Pore Structure

Since the adsorption process results in a concentration of solutes at the surface, it is obvious that the surface area is one of the principal characteristics affecting the adsorptive capacity of an adsorbent. The adsorptive capacity of solid adsorbents generally is proportional to the specific surface area; that is, the adsorption of certain solutes increases with an increase of surface area. However, the specific surface area alone is frequently inadequate to explain the adsorptive capacity of porous solids, such as activated carbon, for different solutes. As discussed above, the pore size distribution in activated carbon indicates that micropores contribute a major portion of the specific surface area. Many of the micropores possess molecular dimensions. It is, therefore, reasonable to expect that a solute molecule will readily penetrate into a pore having a certain critical diameter and will be excluded from pores smaller than this size. In other words, the surface area accessible to the sorbate will be influenced by its molecular size, and only those pores that are accessible to the sorbate will contribute to the effective or measured surface area. Not only has such a molecular sieve effect been well established but also a linear relationship between the specific surface area and sorbate molecular size has been demonstrated [41].

Particle Size

The surface area of nonporous adsorbents increases considerably with a decrease in particle size. Consequently, the adsorptive capacity should increase with reduction in particle diameter. However, for such highly porous adsorbents as activated carbon, most of the surface area resides in the internal pore structure, and the adsorptive capacity is expected to be independent of the particle size. Zogorski [42] showed that, crushing Columbia LCK carbon to particle sizes ranging between 0.66 and 1.4 mm had no effect on the N_2-BET surface area, and that the adsorptive capacity for 2,4-dichlorophenol and 2,4-dinitrophenol was independent of the particle size. This is in contrast to earlier findings by Weber and Morris [43] that a fourfold decrease in particle size of Columbia LCK activated carbon (0.502–0.126 mm) resulted in almost doubling the adsorptive capacity for 3-dodecylbenzenesulfonate. It was suggested by these authors that breaking up large particles served to open some tiny sealed channels in the carbon that might then become available for adsorption. This yielded the dependence of equilibrium capacity on particle size. The particle size of activated carbon has no apparent effect on the adsorptive capacity for such solutes of small molecular sizes as phenols (area per molecule is about 52 $Å^2$) that would have access to most of the small diameter micropores [35]. On the other hand, the capacity for solutes of larger molecular size such as the benzenesulfonates (area per molecule ranges between 100 and 280 $Å^2$) [44] may show dependence on particle size probably due to the exposure of more accessible surface area by crushing of the larger particles.

Chemistry of the Surface

The presence of specific functional groups on the surface of the adsorbent imparts certain characteristics that affect the adsorption process. The formation of polar surface groups of chemisorbed oxygen during the activation process of activated carbon affects the adsorptive capacity for many solutes. Surface oxides consisting of acidic functional groups reduce the capacity of carbon for adsorption of many organic solutes such as oxalic and succinic acids [35] and trihalomethanes from aqueous solutions [45]. This was attributed to the preferential adsorption of water and hence blockage of a part of the surface [35]. On the other hand, the abundance of surface oxides consisting of carbonyl groups enhances the adsorption of such aromatic solutes as phenol, *p*-nitrophenol and naphthalene. This enhanced adsorption was attributed to the interaction of aromatic ring π electrons with the carbonyl groups by a donor-acceptor mechanism involving the carbonyl oxygen as the electron donor and the aromatic ring as the acceptor [26].

Nature of the Adsorbate

Adsorption by activated carbon in aqueous solutions is influenced by several physicochemical properties of the organic solutes. Solubility is, perhaps, the most significant property affecting the adsorptive capacity. In general, a higher solubility indicates a strong solute-solvent interaction or affinity, and the extent of adsorption is expected to be low because of the necessity of breaking the solute-solvent bond before adsorption can occur. A general rule for adsorption of organic solutes, known as Traube's rule [46], states that the adsorption of organic substances from aqueous solutions increases strongly and regularly as the homologous series is ascended. The adsorption of a series of aliphatic acids, formic through butyric, on activated carbon increases with an increase in the chain length. The increase in adsorption is due to the increase in hydrophobicity and the decrease in solubility with the increase in the chain length. The reverse order of adsorption of these species is observed for their adsorption from nonpolar solvents. Adsorption isotherms for the fatty acids—acetic, propionic, *n*-butyric, *n*-valeric, *n*-caproic and *n*-heptylic—on carbon were shown by Hansen and Craig [47] to be superimposable when plotted as the amount adsorbed vs the reduced concentration. The latter is the equilibrium concentration divided by the solubility at the temperature of measurement, and is equivalent to the partial pressure P/P_0 in gas adsorption. Dividing the solute concentration by its solubility normalizes the solute-solvent interactions that are considered to be the same for each member of the homologous series.

The molecular weight and size of the adsorbate molecule also affects the adsorptive capacity. Weber [27] and Weber and Morris [43] showed that the adsorptive capacity of activated carbon increases with an increase in mol wt of a series of sulfonated alkylbenzenes ranging from the unsubstituted and up to the sulfonated tetradecylbenzene. The increase in the side chain length increases the hydrophobicity of the molecule, which results in greater adsorption.

The properties of solute molecules are discussed better in terms of polarity, which has a profound effect on adsorbability from water. As indicated above, a major portion of the carbon surface is nonpolar or hydrophobic; therefore, hydrophobic interactions become important. These interactions originate from a net repulsion between the water and nonpolar regions of the carbon surface as well as the nonpolar moieties of the solute or adsorbate. As a consequence of these repulsive interactions and the high cohesive energy of water, the nonpolar molecules and nonpolar regions of the surface can associate when they are in close proximity to each other [48,49]. In other words, in aqueous systems, nonpolar solutes are more strongly adsorbed on activated carbon than polar solutes. Polarity of organic compounds results when the centers of opposite charges in a molecule do not coincide. These charges (positive and negative) in a molecule originate from unequal distribution of electron density between two atoms of different electronegativities (electron-attracting ability). The introduction of functional groups imparts polar characteristics to the molecule. Polarity of the functional groups is thought to follow the order [50,51]: $-COOH > -OH > -C{=}O > -C-O-C$. Since solubility of organic solutes in water involves formation of hydrogen bonding between the partially positive hydrogen atoms of water and the partially negative atoms of the organic molecule, increasing the polarity is expected to increase solubility and in turn decrease adsorbability.

Several investigators evaluated the effects of functionality, mol wt, branching, solubility and polarity of solutes on carbon adsorption [52,54]. Giusti et al. [52] reported that functionality has a substantial effect on adsorption, which is interrelated with solubility and polarity. For straight-chain aliphatic compounds, the relative adsorbability on carbon for compounds of less than four carbon atoms are: undissociated organic acids > aldehydes > esters > ketones > alcohols > glycols. An increase in mol wt due to an increase in chain length in any series results in a decrease in solubility; therefore, there is an increase in adsorptive capacity [27,43,52]. Branching in the chain generally results in reduction in adsorbability. For example, adsorption decreases in the alcohol series in the following order: normal > iso > tertiary [48]. Branching presumably results in changes in the geometry of molecules, which become more spherical, and, hence, have less surface area for interaction with the carbon surface.

The adsorptive capacity of carbon for aromatic compounds is influenced by substitution in the ring structure. Introduction of substituent groups to phenols results in increased absorbability in the following order [48,53]: $-OCH_3 > -CH_3 > -Cl > -NO_2$. The position of the substituents seems to have no effect on adsorbability [27,53,55,56]. Thus, the adsorptive capacity for ortho-, meta- or para-cresol, or ortho- or para-chlorophenol is the same for each series. Adsorption is enhanced when the number of substituents on the phenol molecule is increased. The disubstituted phenols: 2,3- 3,4- and 2,6-dimethylphenol, are more strongly adsorbed than *o*-, *m*- or *p*-cresol [55]. For chlorinated phenols, adsorbability increases in the following order: 2,4,6-trichlorophenol > 2,4- or 2,6-dichlorophenol > 2- or 4-chlorophenol > phenol [56]. Length of the side chain also increases adsorption according to the following order: 2-isopropylphenol > dimethylphenol > ethylphenol > methylphenol > phenol [55].

Effect of $[H_3^+O]$

The adsorption of nonelectroytes by activated carbon from aqueous systems is generally not affected by the solution pH, although some exceptions can occur. However, the effect of $[H_3^+O]$ on the adsorption of weak electrolytes, both acids and bases, is quite pronounced. Extensive studies have been reported for the adsorption, at different pH values, of phenolic compounds [26,56–60] and organic acids [26,43,61–63] that may occur as contaminants in natural waters. The adsorption characteristics of these compounds follow more or less a general pattern. Both the undissociated and ionized forms of a species can be adsorbed on activated carbon with undissociated forms being more strongly adsorbed than ionized forms. The adsorption behavior of a weakly acidic compound is illustrated in Figure 5. This shows the adsorption versus pH curve for 2,4-dinitrophenol ($pK_a = 4.09$) developed from the adsorption data reported by Zogorski [42]. This adsorption curve follows the dissociation curve of 2,4-dinitrophenol in a general way, although it is shifted by a few pH units. This behavior is observed for adsorption of most organic acids on activated carbon. At alkaline pH values greater than the pK_a of a weak organic acid, the adsorptive capacity is greatly reduced, and the anions are adsorbed when no undissociated molecules are present in bulk solution. The adsorptive capacity increases with a decrease in pH and a maximum occurs in the range where pH is numerically equal to pK_a, i.e., where both anions and undissociated molecules exist in equal concentrations. At pH values more acidic than the pK_a value, the adsorptivae capacity usually decreases.

The dependence of activated carbon's adsorptive capacity for weakly acidic compounds on the solution pH has been attributed to changes in the characteristics of the adsorbent and the solute molecules. Activated carbon can adsorb both acids (H_3^+O) or bases (OH^-) [23,24,26,27] from aqueous solution that could result in changes of the surface characteristics. The decrease in capacity at pH values higher than pK_a values has been attributed to the development of repulsive forces between the anions and the carbon surface or between the anions themselves [57]. The negative surface charge of the carbon may be increased by adsorption of hydroxyl ions on the surface or by ionization of the very weak acidic functional groups on the surface or both [57]. The increased

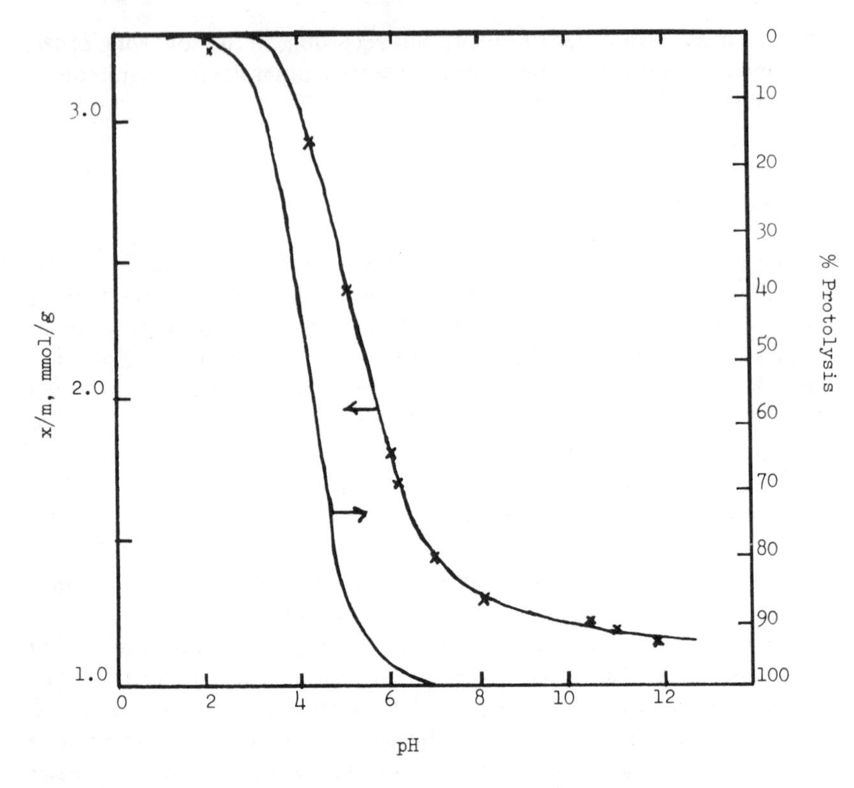

Figure 5. Effect of pH on adsorption of 2,4-dinitrophenol [42].

adsorptive capacity for carboxylic acids as the pH value is decreased to the range of pK_a was attributed by Getzen and Ward [62] to the adsorption of hydronium ions as well as anions. Since the concentration of H_3^+O ions rapidly exceeds that of the weak electrolyte anions, the former adsorb on the carbon far in excess of the anions, and subsequently enhance anion adsorption, i.e., a synergistic effect occurs. The decrease in adsorptive capacity at pH values much more acidic than pK_a was suggested to be due to the competitive interaction of the hydronium ions with the carbonyl oxygen groups on the carbon surface [62].

Manes [64] considered the effect of pH on adsorption of weak acids to be due to competitive adsorption between the acid and its anion where the ratio of both is controlled by the pH. An anion is much more weakly adsorbed than its conjugate acid at the same concentration because of the greater solubility of the anion over the acid and the increased negative entropy of concentration of the anion. The latter arises from the necessity to concentrate two particles (the ions) rather than one (the neutral molecule) in the course of adsorption. As a result, the net free energy driving force for the anion will be much smaller, and, therefore, can be readily displaced by the strongly adsorbed neutral species.

Effect of Temperature

Since the process of adsorption is spontaneous, it is accompanied by a decrease in the system's free energy. There is always a decrease in entropy due to loss of degrees of freedom of the solute in passing from the dissolved state to the adsorbed state [39]. It follows from the relationship:

$$\Delta G = \Delta H - T\Delta S \tag{15}$$

that the adsorption process must always be exothermic, i.e., ΔH must have a negative value regardless of the nature of the interaction forces. The enthalpy change, ΔH, for physical adsorption is in the range of 2–15 kcal/mol [8]. An increase in temperature, therefore, will result in a reduction of the equilibrium adsorptive capacity, whereas lower temperatures will favor an increased capacity. Morris and Weber [44] reported a reduction in the capacity of activated carbon for adsorption of benzenesulfonate by increasing the temperature from 9.5 to 48.5°C. Similar results were obtained by Snoeyink et al. [57] for the adsorption of *p*-nitrophenol.

Zogorski [42] showed that the effect of temperature on adsorption of phenol is more pronounced at low surface coverage or at low initial concentrations as shown in Table I. Thus, at a phenol concentration of 30 μmol/l, an increase in temperature from 8 to 29°C results in a 42% decrease in the adsorptive capacity of activated carbon. At initial concentrations higher than 200 μmol/l, the temperature effect is relatively small.

The temperature of natural surface waters in northern states fluctuates widely during the year, ranging from 34°F ($\approx$0.5°C) during the winter months to 95°F ($\approx$32°C) during the summer months. Such a wide range of temperature has a significant effect on the adsorptive capacity of activated carbon used to remove the relatively low concentrations of organic contaminants encountered in surface waters used for drinking water.

The heat of adsorption ΔH in gas-solid adsorption is measured by introduction of a definite quantity of gas to the adsorbent sample in a thermally insulated calorimeter. The adsorption process produces a temperature rise in the entire system, and the heat of adsorption is obtained

Table I. Adsorptive Capacity of Columbia LCK Carbon for Phenol as a Function of Temperature and Initial Adsorbate Concentration [42]

Concentration of Phenol (μmol/1)	$X_{8°C}$[a] (μM/g)	$X_{29°C}$ (μM/g)	Decrease in Adsorptive Capacity of Adsorbent (%)
30	895	520	42
50	1040	780	25
100	1250	1050	16
200	1480	1320	11
300	1620	1500	7
400	1710	1625	5
500	1800	1725	4

[a]Amount of phenol adsorbed per unit weight of carbon at the temperature indicated.

by multiplying the temperature rise by the heat capacity of the system. Direct measurement of the heat of adsorption from solutions, however, is not possible by this method because of the interaction of the solvent with the solid surface. Morris and Weber [44] measured the differential heat of adsorption by computing the monolayer capacity X_m at two different temperatures from the Langmuir equation and by using the van't Hoff equation in the form:

$$\Delta H = 2.303R \frac{(T_1 T_2)}{(T_2 - T_1)} (\log X_{m,2} - \log X_{m,1}) \tag{16}$$

The differential heat of adsorption for benzenesulfonate was reported to be $\Delta H = -1.4$ kcal/mol [44]. The negative sign of ΔH confirms that the adsorption is an exothermic process and the value 1.4 kcal/mol is within the expected range for physical adsorption.

ADSORPTION EQUILIBRIA

Adsorption from aqueous solutions involves concentration of the solute on the solid surface. As the adsorption process proceeds, the sorbed solute tends to desorb into the solution. Equal amounts of solute eventually are being adsorbed and desorbed simultaneously. Consequently, the rates of adsorption and desorption will attain an equilibrium state, called adsorption equilibrium. At equilibrium, no change can be observed in the concentration of the solute on the solid surface or in the bulk solution. The position of equilibrium is characteristic of the entire system, the solute, adsorbent, solvent, temperature, pH and so on. Adsorbed quantities at equilibrium usually increase with an increase in the solute concentration. The presentation of the amount of solute adsorbed per unit of adsorbent as a function of the equilibrium concentration in bulk solution, at constant temperature, is termed the adsorption isotherm. Typical adsorption isotherms for adsorption from water systems are shown in Figure 6 [56]. Isotherms of this type are typical for

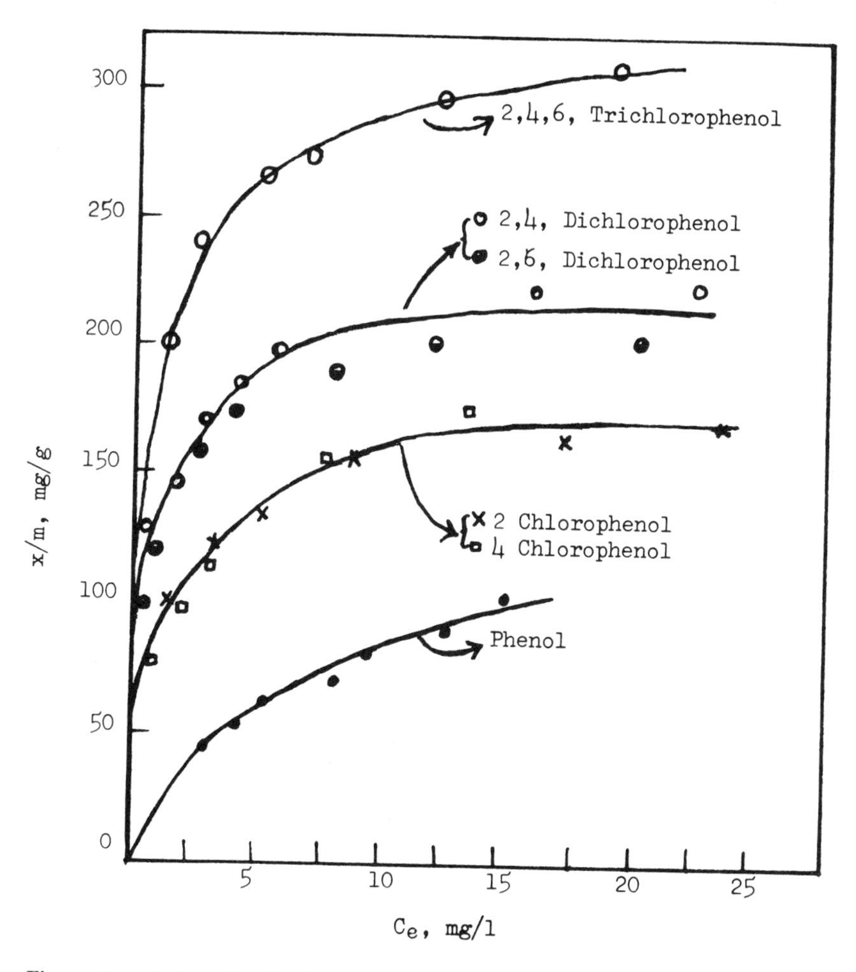

Figure 6. Adsorption isotherms of phenolic compounds on activated carbon [56].

adsorption by activated carbon from aqueous solutions when adsorption does not proceed beyond a monomolecular layer [6], whereas multilayer adsorption in these systems is not usually encountered.

The surface of activated carbon is heterogenous not only in surface structure but also in the distribution of surface energy. During the course of adsorption, the heat of adsorption is not constant for each incremental increase in adsorption. Usually the initial portions of adsorbed solute have greater differential heats of adsorption than subsequent ones. Thus, a steep initial drop of the heat of adsorption with an increase of the amount adsorbed, indicates that the first molecules to arrive at the bare surface are preferentially adsorbed on the most attractive sites or on positions on the surface where their potential energy will be a minimum [39,41]. As adsorption proceeds, the less active sites become occupied. Therefore, adsorption occurs on sites of progressively decreasing activity. Smooth adsorption isotherms are usually obtained because of the presence of sufficiently large number of sites which may occur in patches of equal energy or randomly distributed sites of unequal energy [39]. Several models can be used for the description of the adsorption data with Langmuir's adsorption isotherm and Freundlich's adsorption isotherm as the ones most commonly used.

Langmuir Adsorption Isotherm

The basic assumptions underlying Langmuir's model, which is also called the ideal localized monolayer model, are: (1) the molecules are adsorbed on definite sites on the surface of the adsorbent; (2) each site can accommodate only one molecule (monolayer); (3) the area of each site is a fixed quantity determined solely by the geometry of the surface; and (4) the adsorption energy is the same at all sites. In addition, the adsorbed molecules cannot migrate across the surface or interact with neighboring molecules. The Langmuir equation was originally derived from kinetic considerations [65]. Later, it was derived on the basis of statistical mechanics, thermodynamics, the law of mass action, theory of absolute reaction rates and the Maxwell-Boltzmann distribution law [39]. The Langmuir adsorption isotherm is expressed as:

$$X = \frac{X_m b C_e}{1 + bC_e} \tag{17}$$

where X = x/m, the amount of solute adsorbed x per unit weight of adsorbent m

C_e = equilibrium concentration of the solute
X_m = amount of solute adsorbed per unit weight of adsorbent required for monolayer coverage of the surface, also called monolayer capacity
b = a constant related to the heat of adsorption Q [$b \propto \exp(-\Delta H/RT)$] [39]

Equation 17 indicates that X approaches X_m asymptotically as C_e approaches infinity. For linearization of the data, Equation 17 can be written in the form:

$$\frac{C_e}{X} = \frac{1}{bX_m} + \frac{C_e}{X_m} \tag{18}$$

When C_e/X is plotted vs C_e, a straight line, should result, having a slope $1/X_m$ and an intercept $1/bX_m$. Another linear form can be obtained by dividing Equation 18 by C_e:

$$\frac{1}{X} = \frac{1}{X_m} + \left(\frac{1}{C_e}\right)\left(\frac{1}{bX_m}\right) \tag{19}$$

Plotting $1/X$ against $1/C_e$, a straight line is obtained having a slope $1/bX_m$ and an intercept $1/X_m$.

The monolayer capacity X_m determined from the Langmuir isotherm, defines the total capacity of the adsorbent for a specific adsorbate. Also, it may be used to determine the specific surface area of the adsorbent by utilizing a solute of known molecular area. Reliable X_m values can be obtained only for systems exhibiting type I isotherms of the Brunauer's classification shown in Figure 7 [6,39].

It must be indicated that conformity to the algebraic form of Langmuir's equation does not constitute conformity to the ideal localized monolayer model, even if reasonable values of b and X_m are obtained. Therefore, a constant value of b [$b \propto \exp(-\Delta H/RT)$] may be due to cancellation of variations in ΔH. In turn, constancy of ΔH may be due to internal compensation of opposing effects, such as attractive lateral interactions and surface nonuniformity [39]. In addition, the assumption that the area of the sites and, in turn X_m, is determined solely by the nature of the solid and is independent of the nature of the solute is contrary to what is encountered in adsorption systems. However, orientation of solute molecules on the surface of activated carbon has been shown to affect the monolayer capacity X_m as determined from the area occupied by the molecules [6,35,44]. Therefore, nonconformity with the physical

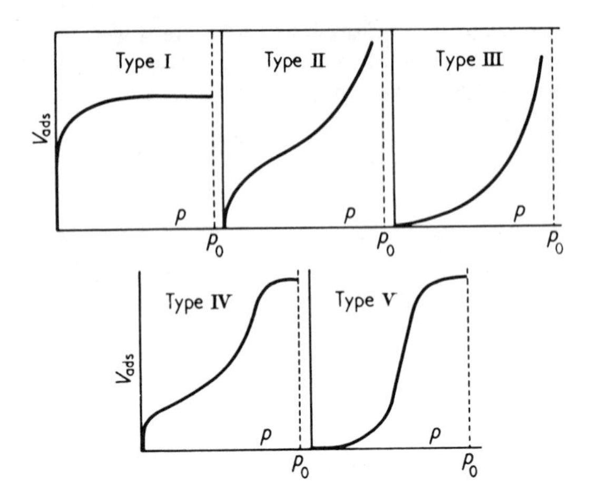

Figure 7. The five typical shapes of isotherms for physical adsorption. Reproduced from Young and Crowell [39], courtesy of Butterworths, Ltd.

model should not detract from the usefulness of the Langmuir isotherm for analytical description of adsorption systems that do not proceed beyond monomolecular layers and conform to type I isotherms [6].

Freundlich Adsorption Isotherm

The Freundlich adsorption equation is perhaps the most widely used mathematical description of adsorption in aqueous systems. The Freundlich equation is expressed as [66]:

$$\frac{x}{m} = KC_e^{1/n} \tag{20}$$

where
- x = the amount of solute adsorbed
- m = the weight of adsorbent
- C_e = the solute equilibrium concentration
- $K, 1/n$ = constants characteristic of the system

The Freundlich equation is an empirical expression that encompasses the heterogeneity of the surface and the exponential distribution of sites and their energies [39,67]. For linearization of the data, the Freundlich equation is written in logarithmic form:

$$\log \frac{x}{m} = \log K + \frac{1}{n} \log C_e \tag{21}$$

Plotting log x/m vs log C_e, a straight line is obtained with a slope of 1/n, and log K is the intercept of log x/m at log $C_e = 0$ ($C_e = 1$). The linear form of the isotherm can be obtained conveniently by plotting the data on log-log paper (Figure 8). The value of 1/n obtained for adsorption of most organic compounds by activated carbon is <1. Steep slopes, i.e., 1/n close to 1, indicate high adsorptive capacity at high equilibrium concentrations that rapidly diminishes at lower equilibrium concentrations covered by the isotherm. Relatively flat slopes, i.e., $1/n \ll 1$, indicate that the adsorptive capacity is only slightly reduced at the lower equi-

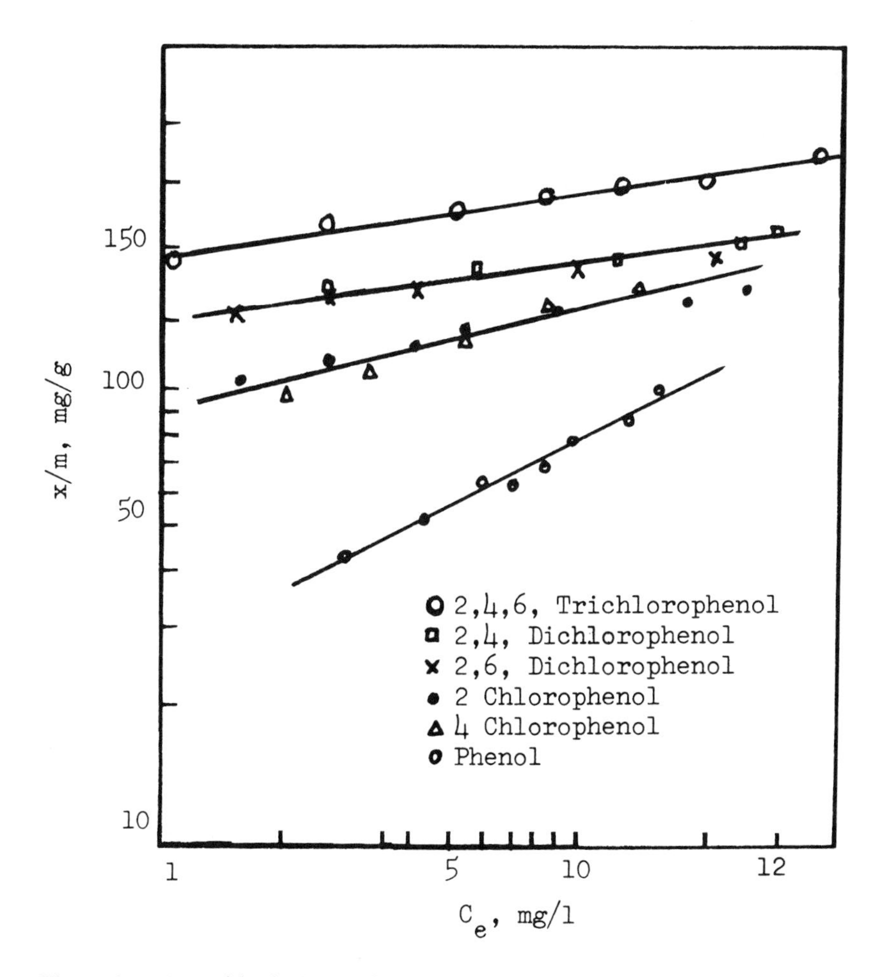

Figure 8. Logarithmic form of Freundlich adsorption isotherms for phenolic compounds on activated carbon [56].

librium concentrations. As the Freundlich equation indicates, the adsorptive capacity or loading factor on the carbon, x/m, is a function of the equilibrium concentration of the solute. Therefore, higher capacities are obtained at higher equilibrium concentrations.

The Freundlich equation can be used for calculating the amount of activated carbon required to reduce any initial concentration to a predetermined final concentration. By substituting x in Equation 21 for $C_o - C_e$, where C_o is the initial concentration:

$$\log\left(\frac{C_o - C_e}{m}\right) = \log K + \frac{1}{n}\log C_e \tag{22}$$

Equation 22 is useful for comparing different activated carbons in removal of different compounds or removal by the same carbon.

Most of the adsorption studies reported in the literature have been conducted in distilled water systems. However, inorganic salts have been shown to affect the adsorptive capacity of activated carbon for certain solutes. Snoeyink et al. [57] and Zogorski [42] reported on the enhancement of adsorptive capacity of activated carbon for some phenolic compounds at high pH values (anionic species) in the presence of inorganic salts. This effect was suggested to be due possibly to a reduction of the repulsive forces between adsorbed molecules and the carbon surface or between anions adsorbed on the surface. Although the concentrations of the inorganic salts used in these studies were too high to be encountered in drinking water supplies, the implication of the potential effects of inorganic ions on the carbon's adsorptive capacity should not be overlooked. Indeed, Weber et al. [68] showed that the presence of low concentrations of calcium and magnesium salts enhances the adsorption of humic acids on activated carbon possibly due to the formation of an ion-humate-carbon complex. Adsorption isotherms of humic acid in tap water systems showed higher carbon adsorptive capacities than those in distilled water systems. Therefore, adsorption studies for application of activated carbon in water treatment plants should be conducted using the natural water or a synthetic medium of equivalent composition. In addition, the isotherms should be conducted within the concentration range corresponding to the levels likely to be encountered for the compound of interest, since extrapolation of the isothermal data can lead to erroneous results.

ADSORPTION KINETICS

Evaluating the performance of unit processes utilizing adsorption requires an understanding of the kinetics of uptake or the time dependence of the concentration distribution of the organic solute in both bulk solution and solid adsorbent and identification of the rate-determining step. Adsorption of organics from aqueous solution by a porous adsorbent such as activated carbon can be described by three consecutive steps [27,69,70].

The first step is the transport of the adsorbate from bulk solution to the outer surface of the adsorbent granule by molecular diffusion. This is called external or film diffusion. There is no actual film surrounding the granule, but the term is used generally to describe the resistance to mass transfer at the surface of the particle. The mass transfer coefficient K_f in film theory, is related to the free liquid diffusivity of the solute, D_e, and the thickness of the diffusional sublayer δ as follows:

$$K_f = D_e/\delta \tag{23}$$

The concentration gradient in the liquid film around the granule is the driving force in film diffusion.

The second step, termed internal diffusion, involves the transport of the adsorbate from the particle surface into interior sites by diffusion within the pore-filled liquid and migration along the solid surface of the pore (surface diffusion) [27,70–73]. Because these two transport processes act in parallel, the more rapid one will control the overall rate of transport. Keinath and Weber [72] found that, for a typical water or wastewater sorption system, the rate parameter for surface diffusion was two to four orders of magnitude greater than the liquid phase pore diffusion.

The third step is adsorption of the solute on the active sites on the interior surfaces of the pores. Since the adsorption step is very rapid, it does not influence the overall kinetics [27,69]. The overall rate of the adsorption process, therefore, will be controlled by the slowest step, which would be either film diffusion or internal diffusion. However, control might also be distributed between intraparticle and external mechanisms in some systems [27,70,73,74].

The nature of the rate-limiting step in batch systems can be determined, in a general way, from the properties of the solute and the adsorbent. Rates of adsorption are usually measured by determining the change in concentration of the solute in contact with the carbon as a function of time. Linearization of the data is obtained by plotting the amount adsorbed per unit weight of adsorbent, x/m, vs $t^{0.5}$ for the initial fraction of the reaction [42,44].

The adsorption rate, determined from the slope of the line with units of $mg/g/hr^{0.5}$ or $mmol/g/hr^{0.5}$ or equivalent, are not true reaction rates but relative rates useful for comparative purposes. For processes controlled by film diffusion, the adsorption rate is expected to be proportional to the first power of concentration. Zogorski [42,59] showed that the adsorption of 2,4-dichlorophenol by activated carbon is film-diffusion controlled at concentrations less than 400 μmol/l, and that a direct linear relationship existed between the initial concentration and the adsorption rate. At concentrations of 2,4-dichlorophenol greater than 500 μmol/l, the reaction rate was shown to be intraparticle-diffusion controlled.

Variation of the rate with concentration was not linear, which is expected from the diffusion theory [27]. Therefore, the concentration dependence of the adsorption rate can be used to identify the rate-limiting step. It is also recognized that the adsorption rate is film-diffusion controlled during initial stages of the adsorption process in batch reactors and initial breakthrough in fixed-bed reactors. As the carbon becomes loaded with the adsorbate, the reaction rate becomes controlled by intraparticle-diffusion [73,75].

For processes controlled by intraparticle diffusion, the size and configuration of the adsorbate molecule should affect the overall rate of adsorption. The larger the molecule or the more branched, the lower should be the rate at which it diffuses. Hence, the adsorption rate should be lower. Morris and Weber [44] reported that a linear relationship existed between observed adsorption rates and mol wt for a series of alkylbenzenesulfonates ranging from 2-hexylbenzenesulfonate to 2-tetradecylbenzenesulfonate. An increase in the chain length by a $-CH_2-$ group resulted in reduction of the adsorption rate. An increase in the chain length apparently reduces mobility of the molecule and also increases its affinity to the carbon surfaces in accord with Traube's rule. Both effects result in a decrease of the surface diffusion inside the pores and, hence, an overall decrease in the observed adsorption rate. Adsorbates that exhibit high affinity to activated carbon usually result in intraparticle-diffusion control of the adsorptive process [73]. This is attributed to a decrease of surface diffusion of adsorbates that is tightly held, and thus, is less flexible to migrate on the internal surfaces of the carbon. Branching of the adsorbate molecule also results in a decrease of the observed adsorption rate due to a decrease in surface diffusion rate. Morris and Weber [44] showed that 6-dodecylbenzenesulfonate, which contains a bifurcated chain, and the extensively branched technical benzenesulfonate exhibited lower adsorption rates than the less branched 2-dodecylbenzenesulfonate.

Characterization of the rate-limiting step in adsorption by activated carbon can be obtained by determining variation of the rate with particle size of the adsorbent. For a film-diffusion controlled process, the rate of adsorption is expected to vary as the reciprocal of the diameter of the particles for a given total weight of adsorbent. This is because the rate, in this case, is a first-order function of exterior surface area which, in turn, is inversely proportional to particle diameter [42,44]. Zogorski [42,59] showed that the adsorption rate of 2,4-dichlorophenol, at concentrations less than 400 μmol/l, exhibited a linear relationship with the reciprocal of the activated carbon particle's diameter in the range of 0.16–1.41 mm. This supports the conclusion that the adsorption rate was film-diffusion-controlled. On the other hand, for internal-diffusion-controlled processes, the rate of adsorption is expected to vary as the reciprocal of some higher power of the particle's diameter. This relationship was demonstrated for the adsorption rates of benzenesulfonates on activated carbon which are internal-diffusion-controlled. The adsorption rates were shown to vary as a function of the inverse of the diameter squared [42,44].

The nature of the rate-controlling step can also be obtained by determining the activation energy of the process. This can be achieved by studying the effect of temperature on the rates of adsorption that are expected to increase with increase in temperature. This relationship was shown for adsorption of 2,4-dichlorophenol [42] and alkylbenzenesulfonates [44]. The activation energy can be computed from:

$$ln \frac{k_1}{k_2} = \frac{Ea(T_1 - T_2)}{RT_1T_2} \tag{24}$$

where k_1,k_2 = observed adsorption rates at temperatures T_1 and T_2 in °K
R = ideal gas law constant (1.987 cal/mol/°C)

The activation energy for internal diffusion-controlled processes is expected to be in the range of 3–5 kcal/mol. The activation energy for adsorption of 3-dodecylbenzenesulfonate was reported to be 4.3 kcal/mol which indicates a diffusion-controlled process [44].

Application of adsorption kinetics has been in the predictive modeling and design of fixed-bed GAC absorbers. Several mathematical models have been developed that deal with the relative contribution of film diffusion and intraparticle diffusion and methods for calculating the film diffusion coefficient and the intraparticle diffusion coefficient [27,70,73–75]. Discussion of development of the mathematical equations is beyond the scope of this book. However, the mathematical equations developed by Liu and Weber [75] are presented here. The model accounts for the film-transfer coefficient, intraparticle-diffusion coefficient and isotherm parameters. The equations are derived from mass balances for the solid and fluid phases involved in fixed-bed absorber or ion-exchange columns. In addition to the traditional assumptions made for fixed-bed models—such as spherical adsorbent particles and plug flow—effective surface diffusivity is used to describe the rate-controlling component of intraparticle diffusion.

Solid-phase mass balance:

$$\frac{\partial q}{\partial t} = D_s\left(\frac{\partial^2 q}{\partial r^2} + \frac{2\partial q}{r\partial r}\right) \tag{25}$$

Liquid-phase mass balance:

$$-D_p\frac{\partial^2 C}{\partial z^2} + v_i\frac{\partial C}{\partial z} + \left(\rho\frac{1-\epsilon}{\epsilon}\right)\frac{\partial q}{\partial t} + \frac{\partial C}{\partial t} = 0 \tag{26}$$

Freundlich adsorption isotherm (similar to Equation 20):

$$q_s = K_F C_s^n \tag{27}$$

Initial and boundary conditions:

$$C = C_o + \left(\frac{D_p}{v_i}\right)\frac{\partial C}{\partial z} \quad \text{at} \quad z = 0 \tag{28}$$

$$\frac{\partial C}{\partial z} = 0 \quad \text{at} \quad z = L \tag{29}$$

$$\frac{\partial q}{\partial r} = 0 \quad \text{at} \quad r = 0 \tag{30}$$

$$\frac{\partial q}{\partial r} = \frac{k_f}{D_s\rho}(C - C_s) \quad \text{at} \quad r = R \tag{31}$$

$$q = 0 \quad \text{at} \quad t = 0 \tag{32}$$

$$C = 0 \quad \text{at} \quad t = 0 \tag{33}$$

where
- q = solid phase concentration ($mM/g/°C$)
- D_s = effective surface diffusivity (cm^2/sec)
- r = radial distance (cm)
- z = longitudinal distance in column (cm)
- t = time (sec)
- C = fluid phase concentration (mM/l)
- C_s = fluid phase concentration near the particle surface (mM/l)

q_s = solid phase concentration at the particle surface (mM/g-°C)
K_F = constant for the Freundlich isotherm
n = exponent for the Freundlich isotherm (this n-value is not the same as the one in Equation 20)
k_f = film-transfer coefficient (cm/sec)
D_p = dispersion coefficient (cm^2/sec)
v_i = interstitial fluid velocity (cm/sec)
R = radius of particles (cm)
ϵ = bed void fraction (dimensionless)
ρ = particle density (g/cm^3)

The following dimensionless transforms may be performed to simplify the above equations:

$$\bar{r} = r/R \tag{34}$$

$$\bar{q} = q/q_o \tag{35}$$

$$\bar{c} = C/C_o \tag{36}$$

$$\bar{t} = t/(\tau D_g) \tag{37}$$

$$D_g = \frac{q_o \rho (1 - \epsilon)}{C_o \epsilon} = \text{solute distribution parameter} \tag{38}$$

$$N_d = \frac{D_s D_g \tau}{R^2} = \text{surface diffusion modulus} \tag{39}$$

$$Pe = \frac{v_i L}{D_p} = \text{Péclet number} \tag{40}$$

$$Sh = \frac{k_f R}{D_g D_s} \left(\frac{1 - \epsilon}{\epsilon} \right) = \text{modified Sherwood number} \tag{41}$$

where C_o = initial fluid phase concentration (mM/l)
q_o = solid phase concentration in equilibrium to C (mM/g/°C)
τ = retention time for column (s)
L = column length

Equation 25 then becomes:

$$\frac{\partial \bar{q}}{\partial \bar{t}} = N_d \frac{1}{\bar{r}^2} \frac{\partial}{\partial \bar{r}} \left(\bar{r}^2 \frac{\partial \bar{q}}{\partial \bar{r}} \right) \tag{42}$$

Equation 26 becomes:

$$-\frac{1}{Pe}\frac{\partial^2 \bar{c}}{\partial \bar{z}^2} + \frac{\partial \bar{c}}{\partial \bar{z}} + \frac{\partial \bar{q}}{\partial \bar{t}} + \frac{1}{D_g}\frac{\partial \bar{c}}{\partial \bar{t}} = 0 \tag{43}$$

and Equation 27 becomes:

$$\bar{q}_s = \bar{c}_s{}^n \tag{44}$$

The transformed initial and boundary conditions are:

$$\bar{c} = 1 + \frac{1}{Pe}\frac{\partial \bar{c}}{\partial \bar{z}} \quad \text{at} \quad \bar{z} = 0 \tag{45}$$

$$\frac{\partial \bar{c}}{\partial \bar{z}} = 0 \quad \text{at} \quad \bar{z} = 1 \tag{46}$$

$$\frac{\partial \bar{q}}{\partial \bar{r}} = 0 \quad \text{at} \quad \bar{r} = 0 \tag{47}$$

$$\frac{\partial \bar{q}}{\partial \bar{r}} = Sh(\bar{c} - \bar{c}_s) \quad \text{at} \quad \bar{r} = 1 \tag{48}$$

$$\bar{q}(\bar{r}, \bar{z}) = 0 \quad \text{at} \quad \bar{t} = 0 \tag{49}$$

$$\bar{c}(\bar{z}) = 0 \quad \text{at} \quad \bar{t} = 0 \tag{50}$$

Equations 47–50, along with the Freundlich exponent n and the dimensionless groups D_g, Pe, Sh and N_d represent the dynamics of breakthrough curves for fixed-bed adsorbers. The parameters n and D_g can be determined experimentally from isotherm studies. Pe is negligible for adsorbers of reasonable length.

The mass transport parameters, k_f and D_s, can be determined by the microcolumn method described by Liu and Weber [75]. Values of k_f, the external film-transfer coefficient, can be determined from the initial portion of the microcolumn breakthrough profile. The internal surface diffusion coefficient, D_s, can be determined from the entire breakthrough profile using regression analysis. These equations are solved numerically by the use of a computer.

MULTISOLUTE ADSORPTION

The occurrence of a wide variety of organic substances in natural waters results in simultaneous adsorption of different species by activated carbon used for treatment. Since adsorption from solution is usually restricted to a monolayer, multisolute adsorption involves competition between different compounds for the available sites on the carbon surface. Mutual reduction in the adsorptive capacity of each of the competing species is usually encountered in aqueous systems. The extent of such reduction depends on the affinity of each solute for the carbon surface, relative concentrations, molecular sizes and possible mutual interactions during adsorption. Competitive adsorption is illustrated in Figure 9 which shows carbon adsorption equilibria of the

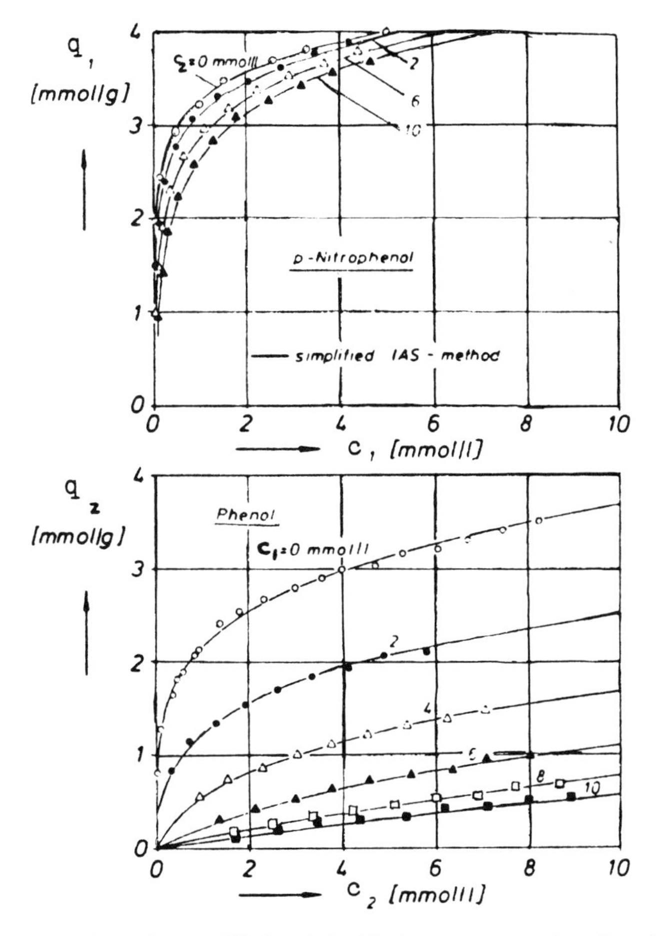

Figure 9. Adsorption equilibria of the bisolute system *p*-nitrophenol/phenol. Activated carbon: B 10 I (apparent particle density $\rho_K = 0.48$ g/cm^3, equivalent particle radius R = 0.62 mm). Reproduced from Fritz et al. [74], courtesy of Ann Arbor Science Publishers.

bisolute system phenol-p-nitrophenol [74]. The upper family of curves shows adsorption isotherms for p-nitrophenol in the presence of different initial concentrations of phenol, whereas the lower graph shows the adsorption of phenol in the presence of p-nitrophenol.

Mutual inhibition of the adsorptive capacity of the two solutes is apparent, but the extent of the interference is different. There is only a slight reduction in the adsorptive capacity of p-nitrophenol in the presence of phenol, whereas the uptake of phenol is very much reduced by the strongly adsorbed p-nitrophenol. Several similar studies have been reported on the kinetics and equilibria of adsorption of simple bisolute mixtures on activated carbon in batch systems which indicate that competitive adsorption generally results in mutual inhibition of the adsorptive capacity of each solute [27,42,55,60,76].

In fixed-bed adsorbers, which are used in water treatment, competitive adsorption results in displacement of weakly adsorbed species by more strongly adsorbed ones resulting in the "chromatographic effect." This means that under certain conditions, the effluent of the carbon bed contains concentrations of certain compounds higher than in the influent. Figure 10 shows the breakthrough curves of the bisolute system phenol-p-nitrophenol [74] at different bed depths Z. The influent concentration of each solute is about 5.0 $\mu M/l$. As can be seen, early breakthrough of phenol always occurs at any bed depth, and its concentration in the effluent increases sharply up to 70% at a bed depth of 20 cm above the influent concentration. After the initial phenol breakthrough, the effluent does not contain any p-nitrophenol for some time, until it starts to appear in the effluent. The increase of p-nitrophenol concentration in the effluent is generally gradual, i.e., the breakthrough curve is not as steep as the one for phenol. The increase of phenol concentration in the effluent can be explained by the fact that as the influent is introduced into the column, the more strongly adsorbed p-nitrophenol is adsorbed on the top carbon layers. The weakly adsorbed phenol is subsequently adsorbed at lower bed depths essentially as a single solute without competition. As the influent continues to flow into the column, additional p-nitrophenol is introduced, which displaces the previously adsorbed phenol into lower bed depths and eventually into the effluent at a higher concentration. This behavior is well recognized in chemical separations by chromatographic techniques. The relative concentrations of each solute in the binary mixture determine the positions of the breakthrough curves [73,74]. A higher relative concentration of the strongly adsorbed species results in earlier breakthrough of the weakly adsorbed one.

Several mathematical models have been developed for predicting the adsorptive behavior of mixed solutes from knowledge of the adsorptive properties of the individual components. The Langmuir competitive

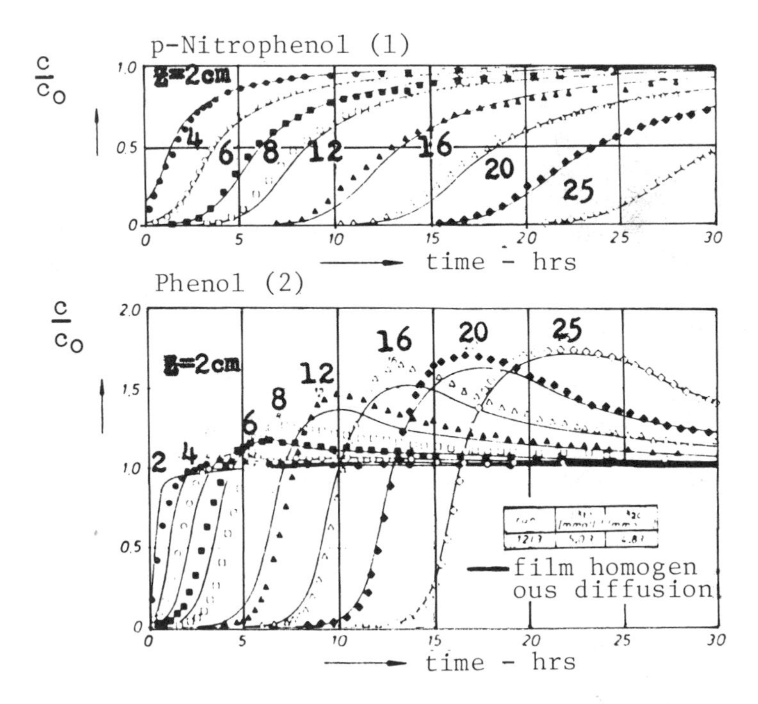

Figure 10. Calculated and experimental breakthrough curves of the bisolute system *p*-nitrophenol/phenol. Reproduced from Fritz et al. [74], courtesy of Ann Arbor Science Publishers.

model developed by Butler and Ockrent [77] has been used by several investigators for predicting equilibrium concentrations of a mixture in batch systems. This model allows calculation of X_i, which is the amount of species i adsorbed per unit weight of adsorbent at an equilibrium concentration C_e in the presence of a j-solute mixture.

$$X_i = \frac{X_{m,i} b_i C_i}{1 + \sum_{j=1}^{n} (b_j C_j)} \tag{51}$$

The constants X_m and b are obtained from single-solute systems. Equation 51 or modifications of it, have been applied by several investigators [27,55,60,76] for predicting equilibrium concentrations in bisolute mixtures. Langmuir's competitive model, despite its limitations, provides a useful analytical description of bisolute systems where the Langmuir isotherm is obtained for pure components [39,55]. This model, however, has not been tested for a mixture of several solutes.

Another approach for predicting multisolute adsorption in dilute solution from single-solute adsorption isotherms is the ideal adsorbed solution (IAS) model. This model was originally developed by Myers and Prausnitz [78] to describe competitive adsorption of gases, and was extended by Radke and Prausnitz [79] to multisolute adsorption from aqueous solutions. The predictive model utilizes the thermodynamic concept of spreading pressure of a solute, π_i, which is defined as the difference between the interfacial tension of the pure solvent-solid interface and that of the solution-solid interface for that solute. The spreading pressure is calculated from the expression:

$$\pi_i = \frac{RT}{A} \int_o^{C_i^*} \frac{X_i}{C_i} \, dC_i \tag{52}$$

Kidnay and Myers [80] modified Equation 52 to:

$$\pi_i = \frac{RT}{A} \int_o^{X_i^*} \frac{d \log C_i}{d \log X_i} \, dX_i \tag{53}$$

where R = ideal gas law constant
T = absolute temperature
A = surface area of the carbon per unit weight
C_i, X_i = the solution phase and solid phase equilibrium concentration of species i in the multisolute system
C_i^* = the equilibrium concentration of species in single-solute systems, which gives the same spreading pressure as that of a mixture
X_i^* = the corresponding amount of solute adsorbed

For single solute systems, C_i and X_i can be computed from Freundlich or Langmuir adsorption isotherm equations or from experimental data where:

$$X_i = f(C_i) \tag{54}$$

Details of calculation of multisolute equilibria using the IAS model are described elsewhere [55,79]. Good agreement between predicted and experimental results has been reported for two- or three-component systems.

Predictive modeling for fixed-bed continuous-flow adsorbers developed for single solutes, as discussed previously, has been modified for multisolute adsorption by incorporation of appropriate relationships [70,81], including the Langmuir competitive model [83] or the IAS model [73,74,81]. The models developed so far have been successful in predicting breakthrough curves for two or three components in pure systems. However, application of these models to natural water systems that contain a large number of organic compounds of unknown nature or adsorbability on activated carbon is limited at this time.

POWDERED ACTIVATED CARBON

The use of powdered activated carbon in most modern water treatment plants processing surface water is a routine practice. The first application in American waterworks dates back to 1929 for taste and odor control [84]; its use in water treatment subsequently became widespread. Almost 90% of water treatment plants worldwide that use activated carbon, use it in the powdered form [85]. The particle size distribution is such that at least 80% of the carbon should pass No. 325 mesh, or measure less than 44 μm.

Powdered carbon is usually applied to raw water in the form of a slurry at any treatment stage before filtration (in the batch-type system). The optimum point of application should allow adequate dispersion of the carbon and sufficient contact time to ensure maximum adsorption. The carbon slurry may be added to the mixing tank or a flocculation tank ahead of the settling tanks. This application point allows adequate contact time and the carbon can be removed easily with the settled sludge. However, the carbon particles may become embedded in the coagulant flocs leading to a reduction of adsorption efficiency. In addition, many organic compounds are adsorbed that would be removed by coagulation alone, therefore requiring more activated carbon. The cost of increased carbon use is offset in many instances by savings in the coagulant dose due to improvement in the flocculation process as a result of a seeding effect of the carbon particles. Another point of application of powdered activated carbon would be to the filter's influent. A small dose of a polymer usually is added as well to improve the filter's ability to retain carbon particles. Addition of carbon to the sand filter's influent should not normally exceed 4 mg/l; however, multimedia beds possess a high capacity for storage of solids. Therefore, carbon doses up to 10 mg/l can be tolerated for short durations.

Simultaneous addition of chlorine and activated carbon should be avoided because of carbon's dechlorination capability. In such cases where chlorine addition is required, the activated carbon should be applied at a point upstream or ahead of the chlorine. This allows the use of lower carbon doses, reduces the potential of formation of trihalomethanes and other chlorination products and avoids the potential loss of the carbon's adsorptive capacity due to formation of oxidized surface groups. The effect of other treatment processes should be considered also in the application of powdered activated carbon. For example, the addition of softening chemicals produces high pH values that are not favorable for adsorption of most organic compounds, whereupon high carbon doses would be required.

In situations where large quantities of activated carbon are required to reduce the concentration of organic compounds, adding the carbon in two steps (split dose) usually results in reducing the total amount of carbon. This can be readily determined from the Freundlich adsorption isotherm parameters and computation of the amount of carbon needed to reduce organic compounds to a desired level for a single dose and a split dose using Equation 22. The first dose can be applied ahead of the settling tank and the second dose ahead of the filter.

The use of powdered activated carbon in water treatment is more adaptable to applications where relatively low contaminant levels require less carbon. Also, when applications are periodical or seasonal in nature, such as taste and odor problems in some surface water supplies, the capital investment in equipment is relatively small, and the carbon addition can be adjusted to changes in water quality. However, if continuous application of activated carbon is required, or if the contaminants have low adsorbability, the powdered form may be more costly and may generate sludge disposal problems. In addition, batch systems do not allow efficient utilization of the adsorptive capacity, since the carbon will be in contact with a continuously decreasing solute concentration.

GRANULAR ACTIVATED CARBON

GAC is used in water treatment in the form of fixed beds where the water is contacted with the carbon by gravity flow downward through the bed. The carbon beds may be placed in structures similar to conventional filters in a water treatment plant or in the sand filters after replacing the medium with the carbon. In the latter case, carbon will function both as a filtering medium and adsorber. Pressure-type cylindrical filters are also widely used in the food processing and bottling industries. After the carbon is exhausted, it is either replaced with fresh carbon or removed and reactivated for further reuse. Considerable amount of research has been conducted in recent years on the use of GAC in water treatment in response to a proposed amendment by the EPA to the Interim Primary Drinking Water Regulations. The use of GAC filtration would be required as a treatment technique for removing a broad spectrum of organic contaminants from drinking water supplies [86]. Commercially available particle sizes of GAC expressed in limiting U.S. Standard Sieve sizes include 8×16, 8×30, 10×30, 12×40, 14×40 and 20×40 with effective sizes ranging from 0.55 mm to 1.35 mm [81].

The fixed-bed continuous flow operation results in maximum utilization of the carbon's adsorptive capacity. The carbon in the column acts like a series of layers with each layer in contact with fresh solution of constant solute concentration. This results in maximum loading of the carbon (x/m) from the solute's adsorption isotherm) at constant solute concentration in contrast to continuously declining solute concentration in batch processes.

Figure 11 illustrates the concentration of adsorbed species on the surface of the adsorbent (x/m) with bed depth [88]. Under operational conditions, adsorbed material accumulates at the top of the bed until the amount adsorbed is in equilibrium with the influent contaminant concentration. At this time, the adsorbent is loaded to capacity and that portion of the bed is exhausted. Below this zone is a second zone where dynamic adsorption is occurring, i.e., the contaminant is being transferred from the liquid solute to the adsorbed phase. This zone is called the "mass transfer zone," and its depth is controlled by many factors, depending on the contaminant being adsorbed, characteristics of the adsorbent, hydraulic factors and others. The depth of the mass transfer zone is a measure of physical/chemical resistance to mass transfer. Once formed, the mass transfer zone moves down through the adsorbent bed until it reaches the bottom, whereupon the effluent concentration of the contaminant in the aqueous phase begins to rise (Figure 12). Figure 12A shows the concentration gradient of adsorbed material (x/m) in an adsorber as the mass transfer zone moves down the column with time. As the mass transfer zone reaches the bottom of the column, "breakthrough" of the contaminant occurs as noted by a detectable increase in effluent concentration (Figure 12B). When the adsorber is operated to exhaustion (at equilibrium, $C_{in} = C_{out}$), the breakthrough profile (plot of effluent concentration with time) takes on the classical "S" shape—a shape controlled by the shape and length of the mass transfer zone. Steeper slopes for the breakthrough curves generally are obtained for systems that exhibit high film transfer coefficients, high internal diffusion coefficients or flat Freundlich adsorption isotherms, i.e., smaller $1/n$ values [71]. The areas A and B in Figure 12B show deviation of the behavior of the adsorbers from ideal plug flow projection, i.e., leakage of the solute appears long before exhaustion of the carbon bed. When an adsorber is removing all of a contaminant, the mass transfer zone in Figure 11 may also be called the "critical depth" because this is the

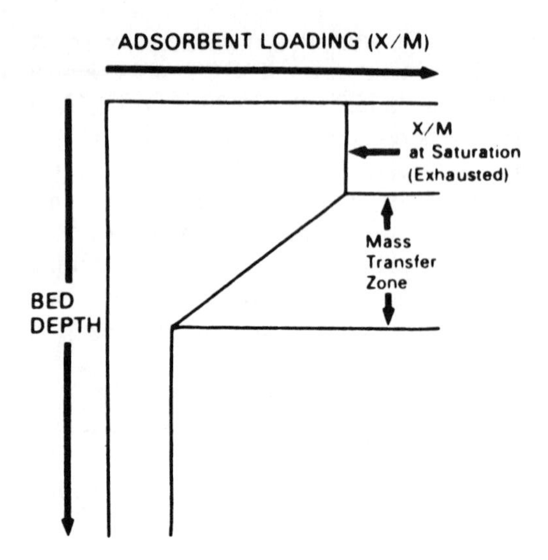

Figure 11. Mass transfer zone in a GAC adsorber [88].

minimum design depth for an adsorber that will allow it to remove all of a contaminant.

The time or volume of water treated to the breakthrough point generally is decreased by: (1) increased particle size, (2) increased solute concentration in the influent, (3) increased pH value of the water in the case of adsorption of weak electrolytes, (4) increased flowrate—usually expressed as gal/min/ft^2, and (5) decreased bed depth [27,42,44,58]. The typical smooth S-shaped breakthrough curve described above is obtained for single-solute systems where the influent concentration to the carbon column is assumed to be constant. However, in natural water systems, there is a broad spectrum of organic contaminants and a wide variation in the concentration of any single compound is not unusual. Competitive adsorption, displacement of weakly adsorbed compounds by strongly adsorbed ones (chromatographic effect), and concentration variations complicate the prediction of the shape and position of the breakthrough curves. Therefore, pilot-plant tests must be conducted at the treatment plant site employing the carbon to be used and the water to be treated in order to select the appropriate design parameters for the carbon beds.

The proposal by the U.S. Environmental Protection Agency (EPA) for use of GAC for control of organics in drinking water describes two adsorption schemes [89]. The first is to retrofit an existing water treatment plant by replacing the filtration media with GAC, and the second, called postfilter adsorption, requires separate contactors

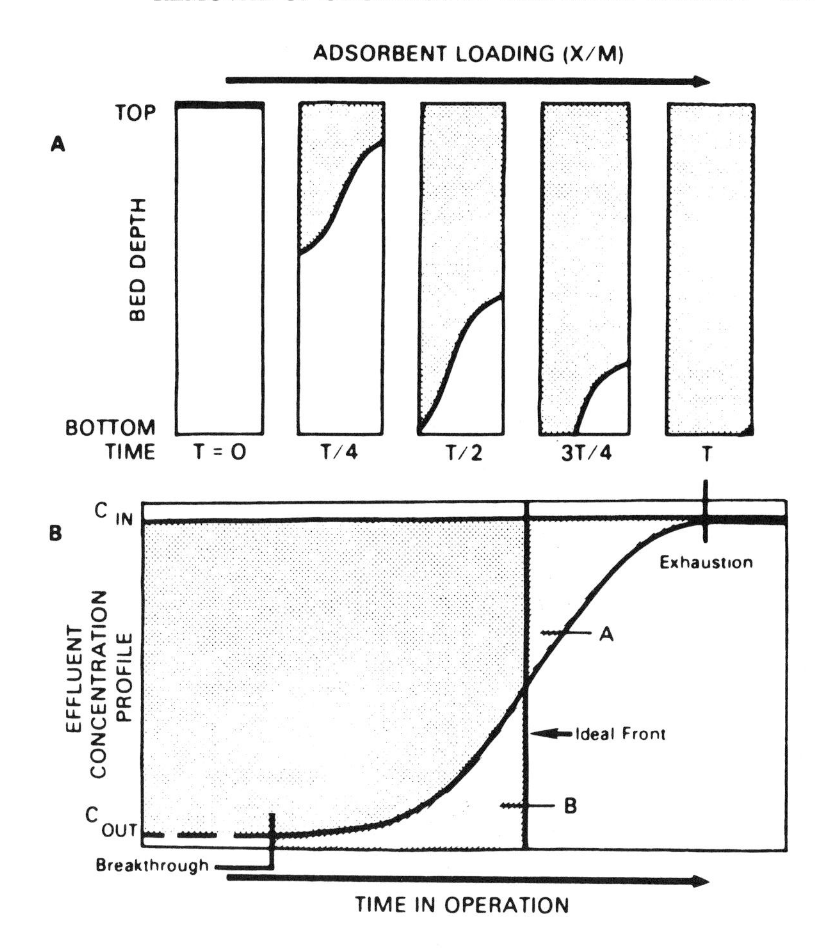

Figure 12. Typical adsorption column performance [88].

following the filtration step. Pilot studies conducted at the site to optimize the design should incorporate several "empty bed contact times," (EBCT) [89]. The choice of this criterion is based on the assumption that adsorption is dependent on contact time, i.e., true equilibrium is not achieved in column operation. EBCT is not a true measure of the contact time between the water and the carbon, but an "apparent contact time" can be obtained by multiplying EBCT by the porosity of the carbon. EBCT is related to the different operating parameters as follows:

$$\text{EBCT} = \frac{\text{bed volume (ft}^3) \times 7.48}{\text{hydraulic surface loading (gal/min/ft}^2) \times \text{bed surface area (ft}^2)} \quad (55)$$

$$\text{EBCT} = \frac{\text{bed volume (ft}^3) \times 7.48}{\text{flowrate (gal/min)}} \quad (56)$$

$$\text{EBCT} = \frac{\text{bed depth (ft)}}{\text{linear velocity (ft/min)}} \quad (57)$$

$$\text{EBCT} = \frac{\text{bed depth (ft)} \times 7.48}{\text{hydraulic surface loading (gal/min/ft}^2)} \quad (58)$$

The above relationships indicate that EBCT, for the same adsorber, is increased by increasing the bed depth, hence the bed volume, or by decreasing the surface loading rate. The usual range for the surface loading rate is 2–8 gal/min/ft^2. Figure 13 shows the relationship of length of carbon service to the bed depths or contact times used in a pilot study at a water treatment plant in Miami, Florida [88].

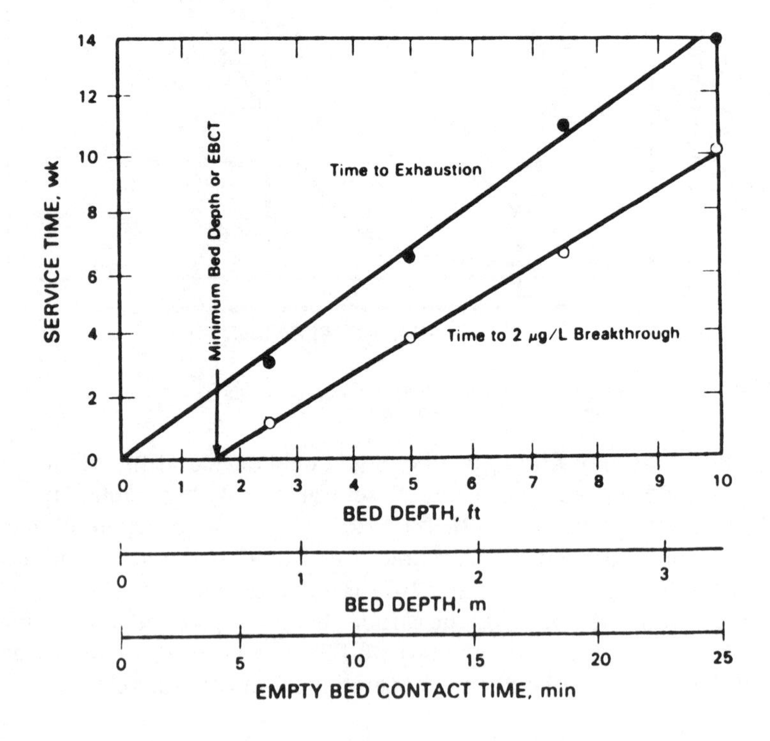

Figure 13. Bed depth/service times for removal of chloroform in Miami, Florida, water by GAC [88].

Another useful parameter suggested for evaluating performance of carbon filters is the "adsorbent use rate" [89]. This is the weight of activated carbon per unit volume of water needed to meet the controlling criterion for any empty bed contact time expressed in units such as pounds per million gallon or grams per liter. Adsorbent use rate is calculated by dividing the dry weight of adsorbent for a given empty bed contact time by the total volume of water passing through the carbon bed(s) until a performance criterion is exceeded. If the adsorbent use rate decreases with increasing empty bed contact time, then a more than proportional improvement in performance is gained by increasing EBCT. The decrease in use rate with increasing EBCT varies considerably among contaminants. At some point, however, the use rate becomes constant, implying a point at which increasing EBCT no longer increases efficiency of removal (Figure 14) [89,91]. When the use rate becomes constant with increasing bed depth, a shallower bed would be more cost-effective. Multiple beds in series offer advantages in better carbon usage.

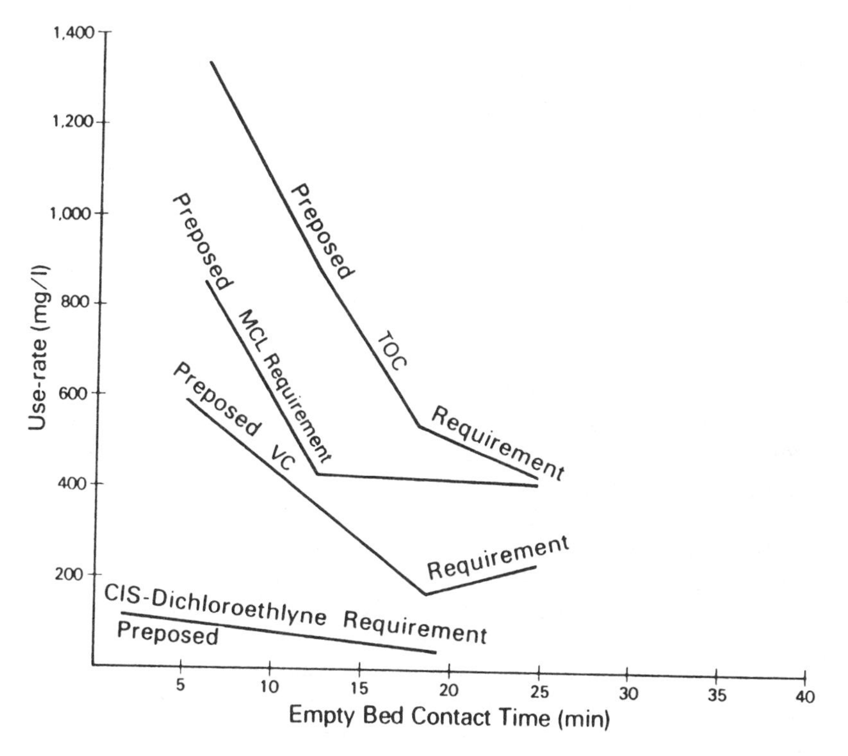

Figure 14. Use rate vs contact time for removal of contaminants [91].

General practice is to operate the series system until the last bed in a series reaches the effluent criteria. The first bed is then replaced with regenerated carbon and becomes the last bed in the series arrangement. The new lead bed is the one that was previously second in the series [92]. However, an economic analysis must be conducted that includes the influence of EBCT, adsorbent use rate, and frequency of carbon replacement or reactivation in order to decide on the most cost effective system. The number of contactors generally should be minimized and the volume per contactor maximized.

APPLICATIONS OF ACTIVATED CARBON

Taste and Odor Removal

The use of activated carbon for taste and odor control represents the earliest successful application of carbon adsorption in water treatment. Most modern water treatment plants use powdered activated carbon for control of tastes and odors of biological or industrial origin in batch processes as discussed earlier in this chapter [82]. The use of GAC was revived in the early 1960s, first at Hopewell, Virginia, in 1962, and next at the Nitro, West Virginia, water treatment plants in 1966 for taste and odor control [92,93]. Replacement of the sand with carbon allowed the filters to serve a dual adsorption and filtration function. The success of GAC beds for controlling taste and odor at these plants led to the acceptance of the practice in other water utilities. A 1977 survey indicated that about 35 plants are using GAC, either alone or on top of some sand medium as both a filter medium for particulate removal and as an adsorptive medium for taste and odor control [89]. The effective service life of these filters ranged between 2 and 4 years, and the odor breakthrough occurred much later than the breakthrough of other organics [89]. The exhausted carbon is usually replaced by virgin carbon and reactivation is not practiced.

Geosmin and 2-methylisoborneol (MIB), the causative agents of the earthy-musty odor in water supplies, were shown to adsorb strongly by activated carbon [60,94,95]. Snoeyink et al. [60] reported that humic substances, when present significantly reduce the adsorptive capacity of carbon for these compounds; they do so more markedly before equilibrium is achieved than at equilibrium. Commercial humic acids and humic substances from wellwater each had different competitive effects on MIB. The capacity of carbon for geosmin was reduced to a greater extent than was observed for MIB. Application of distilled water to a partially saturated carbon bed resulted in almost no elusion of MIB, indicating it was strongly adsorbed. Assuming complete saturation of the carbon, no leakage, and no biological activity, predicted bed life for reduction of MIB or geosmin from 10 μg/l to its threshold odor (TO) level of 0.1 μg/l in a 2-ft-deep bed is much greater (several months to years) than for reduction of humic substances from 5 to 1 mg/l (1–2 months), for example. When both MIB or geosmin and humic substances must be removed, the latter will control the life of the bed.

Chlorophenolic compounds that impart medicinal or chemical taste and odor to drinking water are also strongly adsorbed by activated carbon. These compounds may be produced as a result of chlorination of phenolic compounds in the water supply. The adsorption of chlorinated phenols increases with an increase in substitution and the position of substituents does not influence the adsorptive capacity.

As expected, adsorption of chlorophenols is affected by the solution pH [42,56–58,60] being substantially reduced at alkaline values greater than their pKa values (see above). Substantial reductions in adsorptive capacity of one phenol is affected by the presence of another chlorophenol. Thus, the adsorption of phenol was shown to be reduced by 30% in the presence of 2,4-dichlorophenol; the presence of 2,4,6-trichlorophenol reduced the adsorption of 2,4-dichlorophenol by 50% [60]. Evaluation of the competitive effects of commercial humic acid, soil fulvic acid and leaf fulvic acid showed that the presence of these materials decreased the capacity of carbon for chlorophenols and that each of the materials competed somewhat differently. However, even in the presence of humic substances and another chlorophenol species, it appears that the carbon adsorptive capacity is even greater for chlorophenol than it is for MIB and that bed life for chlorophenol will be greater than for MIB and much greater than for humic substances [60].

Removal of Organics

The ability of granular activated carbon to remove a broad spectrum of organic compounds from water is well documented. Table II shows adsorption isotherm data developed by Dobbs and Cohen [96] for some of the compounds studied as single solutes in aqueous systems. These data illustrate clearly the wide range of organic compounds of different structures, sizes, functionality, etc. that can be adsorbed by GAC (ground to a powdered form). It can also be seen that these compounds exhibit different adsorption characteristics, i.e., some are strongly adsorbed, whereas others are weakly adsorbed. Their adsorption

Table II. Freundlich Adsorption Isotherm Constants for Toxic Organic Compounds [96]

Compound	K (mg/g)	1/n
bis(2-Ethylhexyl Phthalate	11,300	1.5
Butylbenzyl Phthalate	1,520	1.26
Heptachlor	1,220	0.95
Heptachlor Epoxide	1,038	0.70
Endosulfan Sulfate	686	0.81
Endrin	666	0.80
Fluoranthene	664	0.61
Aldrin	651	0.92
PCB-1232	630	0.73
beta-Endosulfan	615	0.83
Dieldrin	606	0.51
Hexachlorobenzene	450	0.60
Anthracene	376	0.70
4-Nitrobiphenyl	370	0.27
Fluorene	330	0.28
DDT	322	0.50
2-Acetylaminofluorene	318	0.12
alpha-BHC	303	0.43
Anethole	300	0.42
3,3-Dichlorobenzidine	300	0.20
2-Chloronaphthalene	280	0.46
Phenylmercuric Acetate	270	0.44
Hexachlorobutadiene	258	0.45
gamma-BHC (Lindane)	256	0.49
p-Nonylphenol	250	0.37
4-Dimethylaminoazobenzene	249	0.24
Chlordane	245	0.38
PCB-1221	242	0.70
DDE	232	0.37
Acridine Yellow	230	0.12
Benzidine Dihydrochloride	220	0.37
beta-BHC	220	0.49
n-Butylphthalate	220	0.45
n-Nitrosodiphenylamine	220	0.37
Phenanthrene	215	0.44
Dimethylphenylcarbinol	210	0.34
4-Aminobiphenyl	200	0.26
beta-Naphthol	200	0.26
alpha-Endosulfan	194	0.50
Acenaphthene	190	0.36
4,4′-Methylene-*bis*-(2-chloroaniline)	190	0.64
Benzo [k] fluoranthene	181	0.57

Table II. continued

Compound	K(mg/g)	l/n
Acridine Orange	180	0.29
alpha-Naphthol	180	0.32
4,6-Dinitro-*o*-cresol	169	0.27
alpha-Naphthylamine	160	0.34
2,4-Dichlorophenol	157	0.15
1,2,4-Trichlorobenzene	157	0.31
2,4,6-Trichlorophenol	155	0.40
beta-Naphthylamine	150	0.30
Pentachlorophenol	150	0.42
2,4-Dinitrotoluene	146	0.31
2,6-Dinitrotoluene	145	0.32
4-Bromophenyl Phenyl Ether	144	0.68
p-Nitroaniline	140	0.27
1,1-Diphenylhydrazine	135	0.16
Naphthalene	132	0.42
1-Chloro-2-nitrobenzene	130	0.46
1,2-Dichlorobenzene	129	0.43
p-Chlorometacresol	124	0.16
1,4-Dichlorobenzene	121	0.47
Benzothiazole	120	0.27
Diphenylamine	120	0.31
Guanine	120	0.40
Styrene	120	0.56
1,3-Dichlorobenzene	118	0.45
Acenaphthylene	115	0.37
4-Chlorophenyl Phenyl Ether	111	0.26
Diethyl Phthalate	110	0.27
2-Nitrophenol	99	0.34
Dimethyl Phthalate	97	0.41
Hexachloroethane	97	0.38
Chlorobenzene	91	0.99
p-Xylene	85	0.19
2,4-Dimethylphenol	78	0.44
4-Nitrophenol	76	0.25
Acetophenone	74	0.44
1,2,3,4-Tetrahydronaphthalene	74	0.81
Adenine	71	0.38
Dibenzo [a,h] anthracene	69	0.75
Nitrobenzene	68	0.43
3,4-Benzofluoranthene	57	0.37
1,2-Dibromo-3-chloropropane	53	0.47
Ethylbenzene	53	0.79
2-Chlorophenol	51	0.41

Table II. continued

Compound	K (mg/g)	l/n
Tetrachloroethene	51	0.56
o-Anisidine	50	0.34
5-Bromouracil	44	0.47
Benzo [a] pyrene	34	0.44
2,4-Dinitrophenol	33	0.61
Isophorone	32	0.39
Trichloroethene	28	0.62
Thymine	27	0.51
Toluene	26	0.44
5-Chlorouracil	25	0.58
N-Nitrosodi-n-propylamine	24	0.26
bis (2-Chloroisopropyl) ether	24	0.57
Phenol	21	0.54
Bromoform	20	0.52
Carbon Tetrachloride	11	0.83
bis (2-Chloroethoxy) Methane	11	0.65
Uracil	11	0.63
Benzo [g,h,i] perylene	11	0.37
1,1,2,2-Tetrachloroethane	11	0.37
1,2-Dichloropropene	8.2	0.46
Dichlorobromomethane	7.9	0.61
Cyclohexanone	6.2	0.75
1,2-Dichloropropane	5.9	0.60
1,1,2-Trichloroethane	5.8	0.60
Trichlorofluoromethane	5.6	0.24
5-Fluorouracil	5.5	1.0
1,1-Dichloroethylene	4.9	0.54
Dibromochloromethane	4.8	0.34
2-Chloroethyl Vinyl Ether	3.9	0.80
1,2-Dichloroethane	3.6	0.83
1,2-*trans*-Dichloroethene	3.1	0.51
Chloroform	2.6	0.73
1,1,1-Trichloroethane	2.5	0.34
1,1-Dichloroethane	1.8	0.53
Acrylonitrile	1.4	0.51
Methylene Chloride	1.3	1.16
Acrolein	1.2	0.65
Cytosine	1.1	1.6
Benzene	1.0	1.6
Ethylenediaminetetraacetic Acid	0.86	1.5
Benzoic Acid	0.76	1.8
Chloroethane	0.59	0.95
N-Dimethylnitrosamine	6.8×10^{-5}	6.6

behavior is expected to be affected by the different factors discussed so far in this chapter, either in batch processes or continuous-flow systems.

In natural surface waters, however, contaminants do not exist as single solutes but rather as mixtures comprising, perhaps, hundreds of compounds. In removing these compounds by GAC beds, competitive effects of the different species for the available sites and displacement effects where the more strongly adsorbed compounds displace the weakly adsorbed ones will affect the shape and position of the breakthrough curves. In addition, the concentration of many contaminants in surface water is quite variable and may result in the establishment of different equilibrium conditions in the carbon beds. These effects may result, for certain compounds, in early breakthrough or effluent concentrations exceeding those in the influent to carbon beds. Suffet et al. [45, 97] used computer-reconstructed gas chromatographic profiles that demonstrated the displacement effects (chromatographic effects) in GAC beds treating surface waters.

Several light halogenated hydrocarbons (other than trihalomethanes) have been found in many surface and groundwaters (see Chapter 2). The effectiveness of GAC for removal of these compounds has been reported by several investigators [89,90,98,99]. Figure 15 shows Freundlich adsorption isotherms for trichloroethylene and tetrachloroethylene in single-solute systems [99]. These data indicate that tetrachloroethylene is more strongly adsorbed by GAC than is trichloroethylene and that effective removal of these compounds from dilute solutions can be achieved by GAC treatment. Pilot-plant studies for evaluation of the effectiveness of GAC columns for removal of a mixture of these compounds found in a groundwater supply (Table III) were conducted in a water treatment plant in southern Florida [89,90,100]. Figure 16 shows the breakthrough curves for all the halogenated hydrocarbons, expressed as total volatile organic halogens (TOX) from four columns in series 2.5 ft each [90]. GAC was effective in removing most of the compounds, and the breakthrough of individual compounds was shown to be dependent on the initial concentration, contact time and competitive effects of other species [90].

Polynuclear aromatic hydrocarbons (PAH), some of which are known carcinogens, have been detected in many surface and groundwaters at low concentrations [101,103]. Effective removal of these compounds in natural waters can be achieved by activated carbon treatment [102,103]. Borneff [102] showed that several types of activated carbon removed 99–99.9% of solubilized PAH under laboratory conditions. Use of activated carbon for removal of PAH from natural waters is shown in Table IV where a two-stage treatment consisting of powdered activated carbon

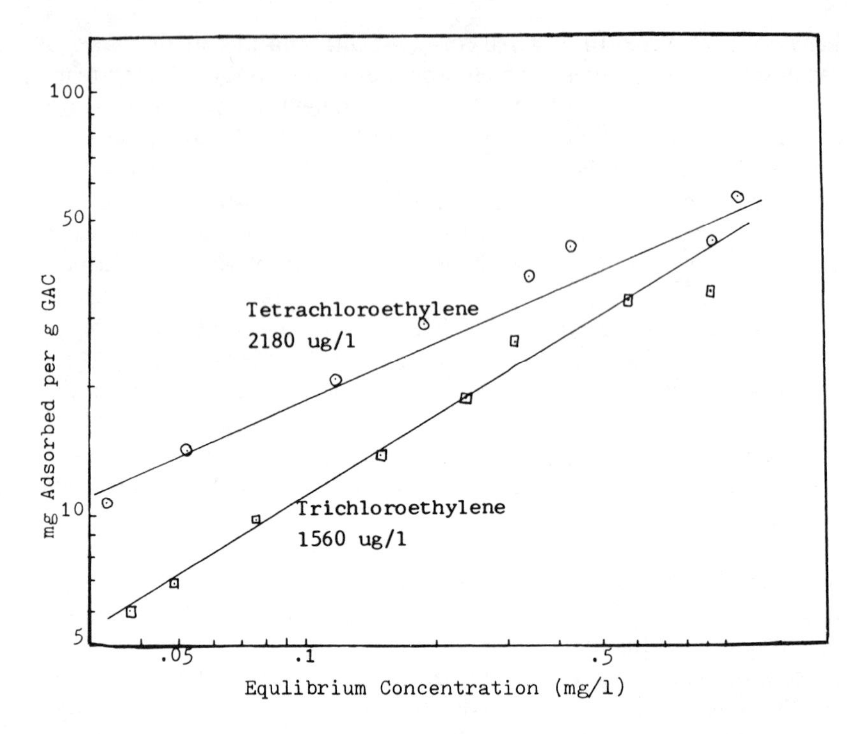

Figure 15. Logarithmic Freundlich isotherm for tetrachloroethylene and trichloroethylene in single-compound studies [99].

followed by GAC filtration reduced the concentration to extremely low levels [103]. The presence of naturally occurring humic acids in waters which are poorly adsorbed on activated carbon was shown not to interfere with the rate of adsorption of a model PAH, anthracene [62].

Nitrosamines, which are known carcinogens, also have been detected in natural waters. However, concern was raised about their potential occurrence in industrial wastewater discharges. Borneff [102] showed that filtration of several nitrosamines at 1-mg/l concentrations through GAC was effective in their removal. The percentage removals were: dimethylnitrosamine, 99.7%; diethylnitrosamine, 99.3%; dipropylnitrosamine, 99.4%; and dibutylnitrosamine, 96.8%.

The National Interim Primary Drinking Water Regulations [104] established maximum contaminant levels for six organic pesticides; endrin, lindane, methoxychlor and toxaphene, which are chlorinated hydrocarbons, and two phenoxy acids, 2,4-dichlorophenoxyacetic acid (2,4-D), and 2,4,5-trichlorophenoxypropionic acid (Silvex) (see Chapter 1). Effective removal of the chlorinated hydrocarbons and silvex by acti-

Table III. Average Concentrations of Purgeable Halogenated Organic Compounds in Raw, Lime-Softened and Finished Waters at the Preston Plant, Hialeah, Florida[a]

Compound	Average Concentration (μg/l)		
	Raw Water	Lime-Softened Water	Finished Water
Vinyl Chloride	12.8	9.7	6.2
Methylene Chloride	0.45	b	b
trans-1,2-Dichloroethene	1.5	0.95	0.77
1,1-Dichloroethane	0.58	0.45	0.4
cis-1,2-Dichloroethene	25.6	22.3	19.9
Chloroform	0.08	1.2	67.3
1,1,1-Trichloroethane, 1,2-Dichloroethane, Carbon Tetrachloride	0.1	0.12	7.7
Trichloroethylene	0.34	0.33	0.68
Bromodichloromethane	c	0.35	47
Tetrachloroethylene	0.003	0.004	0.003
Chlorodibromomethane	c	0.13	33.6
Chlorobenzene	1.1	0.72	0.86
Bromoform	c	0.007	2.6
p-Chlorotoluene	0.02	0.03	0.1
m-Dichlorobenzene	c	c	c
p-Dichlorobenzene	0.51	0.28	0.21
o-Dichlorobenzene	0.17	0.16	0.14
Total	43.25	36.73	187.36

[a]Reproduced from Wood and De Marco [100], courtesy of the American Water Works Association.
[b]Not determined.
[c]Nil.

vated carbon was reported by Robeck et al. [105]. Aly and Faust [106] showed that activated carbon treatment was the most effective treatment method for the removal of 2,4-D and its derivatives from natural waters. Table V lists additional organic pesticides that have been reported to be reduced in concentration by activated carbon treatment. In general, activated carbon adsorption has proven to be the most effective treatment process for reducing the concentrations of these contaminants [117].

The ability of GAC beds to remove the total organic content in surface waters has been monitored by measurement of such surrogate parameters as COD, total organic carbon (TOC), UV absorption or fluorescence measurement. TOC removal has received more attention in recent

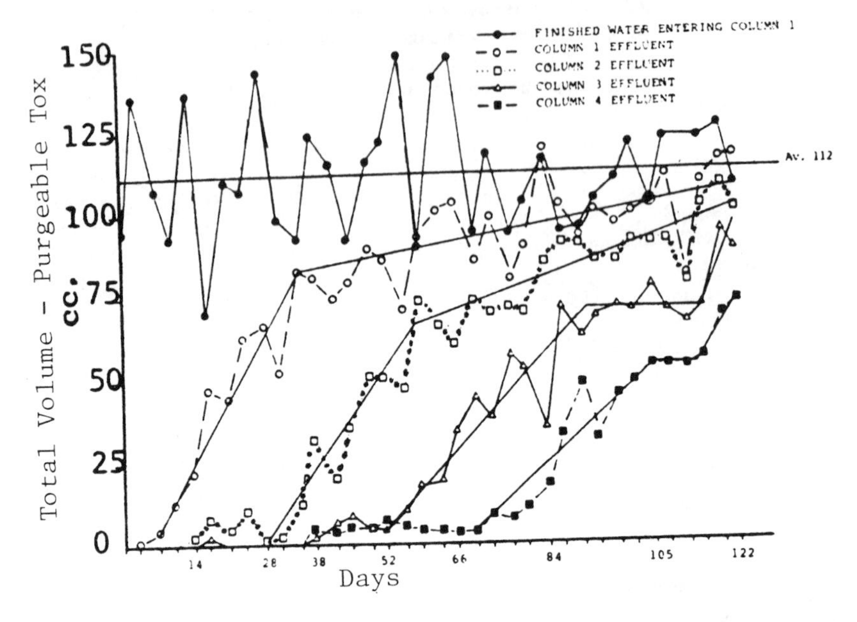

Figure 16. Volume adsorbed (cm³) by each GAC (F-400) column for all TOX added together [90].

Table IV. Removal of PAH by Activated Carbon Treatment (Pittsburgh, PA) [103]

Compound	Concentration (ng/l)[a]	
	Raw	Finished
Fluoranthene	408.3	nd[b]
Benzo[j]fluoranthene	35.7	0.3
Benzo[k]fluoranthene	19.1	0.2
Benzo[a]pyrene	42.1	0.4
Indeno[1,2,3-c,d]pyrene	60.4	1.2
Benzo[g,h,i]perylene	34.4	0.7

[a]Two-stage activated carbon treatment; powdered Activated Carbon, then GAC filtration (≈30–40 min of EBCT).
[b]Nd-Not detected; limits ranging 0.1–4.6 ng/l.

years because it represents a measurement of the total organic content in water and was suggested as one of the criteria for design of GAC beds [89]. Several pilot- and full-scale studies [45,89,97,118,119] have demonstrated that TOC concentration can be reduced effectively by fresh or

Table V. Additional Organic Pesticides Adsorbable on Activated Carbon [89]

	Reference
Aldrin	104
Baygon	105
alpha-BHC	104
2,4-D	106
Dibrom	107
Dieldrin	104
2,4-Dinitro-*o*-cyclohexyl Phenol	106
Diuron	108
Diquat	109
Dimethoate	108
DDT	104
Endosulfan	110
Endrin	108
Heptachlor	111
Heptachlor Epoxide	111
Hexachlorobenzene	111
Juglone	107
Lindane	111
Linuron	111
Malathion	111
Methoxychlor	111
Paraquat	109
Parathion	108
PCB	111
Rotenone	107
Sevin	105
Simazine	108
Strychnine	112
TFM	107
2,4,5-T	104,106
Toxaphene	108,113
Telodrin	111

virgin carbon beds. However, the removal efficiency decreases relatively rapidly resulting in early breakthrough and short bed lives. The breakthrough of TOC does not correlate with the breakthrough of many other compounds present in natural waters [118], and a fraction of TOC has always resisted removal by carbon beds (Figure 17) [89]. In addition, TOC measurement gives excessive weight to naturally occurring high-mol-wt organic compounds which are not hazardous and are poorly adsorbed by activated carbon. Therefore, surrogate monitoring seems to be very restrictive and is not a feasible criterion for performance evaluation or design of GAC beds.

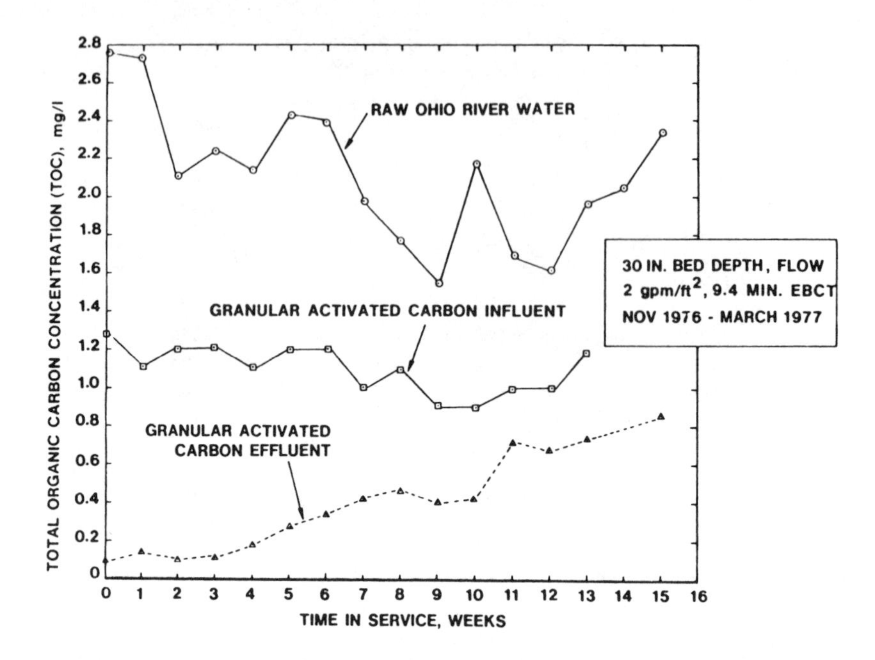

Figure 17. Organic carbon breakthrough curve for pilot water treatment plant [89].

Removal of Trihalomethanes

The discovery that disinfection of drinking waters containing low concentrations of naturally occurring organic matter with chlorine results in formation of trihalomethanes led the EPA to amend the National Interim Primary Drinking Water Regulations to include a maximum contaminant level of 0.10 mg/l for total trihalomethanes (see Chapters 1 and 2 for amendments). Compounds covered by the regulation are: chloroform, bromodichloromethane, dibromochloromethane and bromoform. Consequently, removal of these compounds by activated carbon adsorption was studied extensively in both pure systems and pilot and full-scale studies. Figure 18 shows Freundlich adsorption isotherms for the four trihalomethanes on GAC [120]. These data indicate that the brominated methanes are more strongly adsorbed than chloroform. The adsorptive capacities are in this order: bromoform > dibromochloromethane > dichlorobromomethane > chloroform. The breakthrough behavior of these compounds in column operation follows the same order observed in the isotherm studies, i.e., chloroform, the least adsorbed, exhibits an early breakthrough compared to the more strongly adsorbed

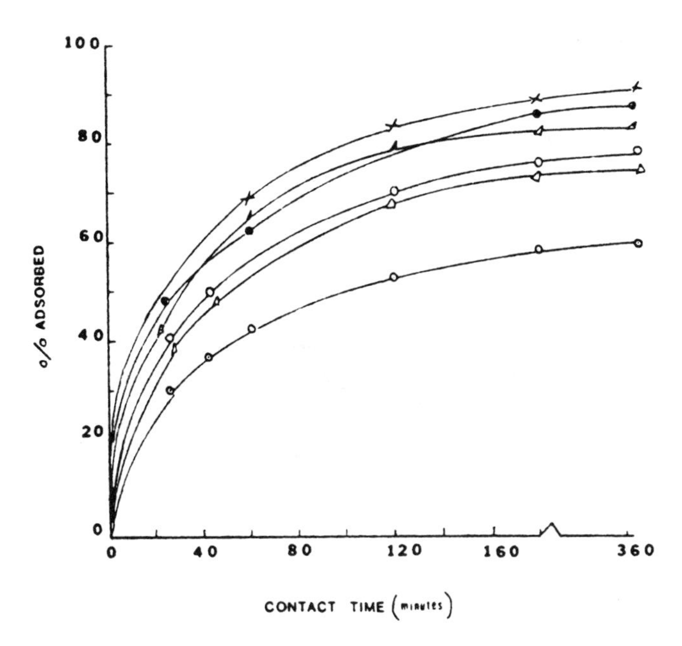

Figure 18. Removal of light halogenated hydrocarbon mixture by GAC. Carbon = 0.8 g/l, pH 7.0 (phosphate buffer); ⊙ = $CHCl_3$, 13.54 μg/l; X = $CHBr_3$, 37.21 μg/l; Δ = $CHCl_2Br$, 11.42 μg/l; Δ = $CHClBr_2$, 23.05 μg/l; O = CCl_4, 16.62 μg/l; ⊗ = $Cl_2C=CCl_2$, 17.48 μg/l. Reproduced from Youssefi and Faust [120], courtesy of Ann Arbor Science Publishers.

bromoform. Pilot- and full-scale studies using GAC beds were shown to follow the same general pattern discussed above with the expected variation in the shape and position of the breakthrough curves due to competitive effects of other organics, variations in the influent concentrations, and the operating parameters such as the EBCT [45,88,90,97]. Figure 19 shows typical breakthrough curves obtained in a pilot study using drinking water from a sample water supply [88]. Review of the data obtained from most of the pilot- and full-scale studies indicated that the carbon bed life or time to breakthrough of the trihalomethanes ranged from 4 to 26 weeks depending on influent concentrations, types of compounds removed or empty bed contact times [45,88,90,97]. The short service life of carbon beds used for trihalomethane removal of several weeks compared to several years obtained for taste and odor removal will require frequent replacement or reactivation of the exhausted carbon. Therefore, the use of GAC treatment should be economically evaluated against other treatment alternatives for trihalomethane reduction to the maximum contaminant level of 0.1 mg/l before a final choice is made.

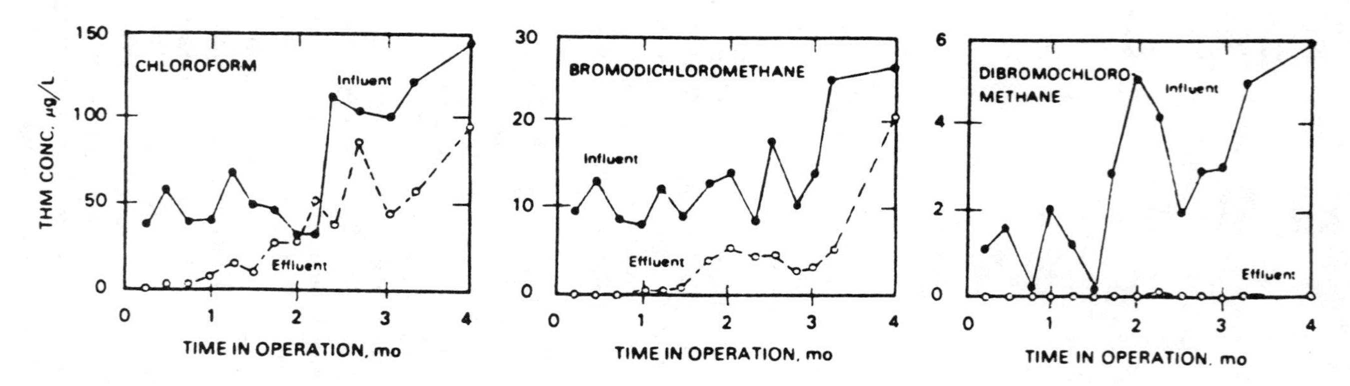

Figure 19. Removal of THM from Cincinnati, Ohio, tapwater by lignite-based GAC, type HD 10 × 30; bed depth, 76 cm (30 in.); hydraulic loading, 5 m/hr (2 gpm/ft^2); EBCT, 9 min. Bromoform was not found [88].

The use of powdered activated carbon may prove to be insufficient because of the high doses required for reducing the concentration of the poorly adsorbed chloroform that frequently constitutes the major portion of the total trihalomethane concentration [88].

Removal of Trihalomethane Precursors

Removal of trihalomethane precursors before the chlorination process offers a logical alternative to the control of trihalomethane levels in drinking water. A major portion of trihalomethane precursors consists of naturally occurring humic substances that are complex mixtures of organic compounds with varying characteristics (see Chapter 2). Youssefi and Faust [120] and Snoeyink et al. [60] showed that chloroform formation potential of humic and other naturally occurring organic substances varied widely depending on source of the material. The adsorptive capacity of activated carbon for humic substances is generally low and is variable depending on the source and nature of the humic compound being adsorbed [88,66,120].

The effectiveness of powdered activated carbon for removing trihalomethane precursors in full-scale and pilot-plant studies indicated a wide variability at any location [88,121,122]. In general, doses of powdered activated carbon much higher than conventional use in existing water treatment plants are required to obtain significant removal of trihalomethane precursors. Individual studies onsite are required to determine the actual effectiveness of this treatment method.

Removal of THM precursors by GAC beds has been studied at several locations [88,89,118,123]. The removal by virgin carbon is initially very effective but a rapid breakthrough occurs after a short operating period. A steady state develops during which a rather constant percentage of precursor material continues to be removed (Figure 20) possibly because of biodegradation [88]. The time to steady-state conditions in most of the studies varied between 2 and 24 weeks. Because of the variability of waters being treated, continuous flow studies must be conducted onsite at the treatment plant in order to determine breakthrough patterns and potential long-term removal at steady-state conditions.

The above discussion indicates that use of activated carbon adsorption for precursor control should be compared also with such other treatment alternatives as coagulation and filtration (see Chapter 6) or the use of other oxidants (ozone, ClO_2 or chloramines) in order to select the most economically effective treatment method.

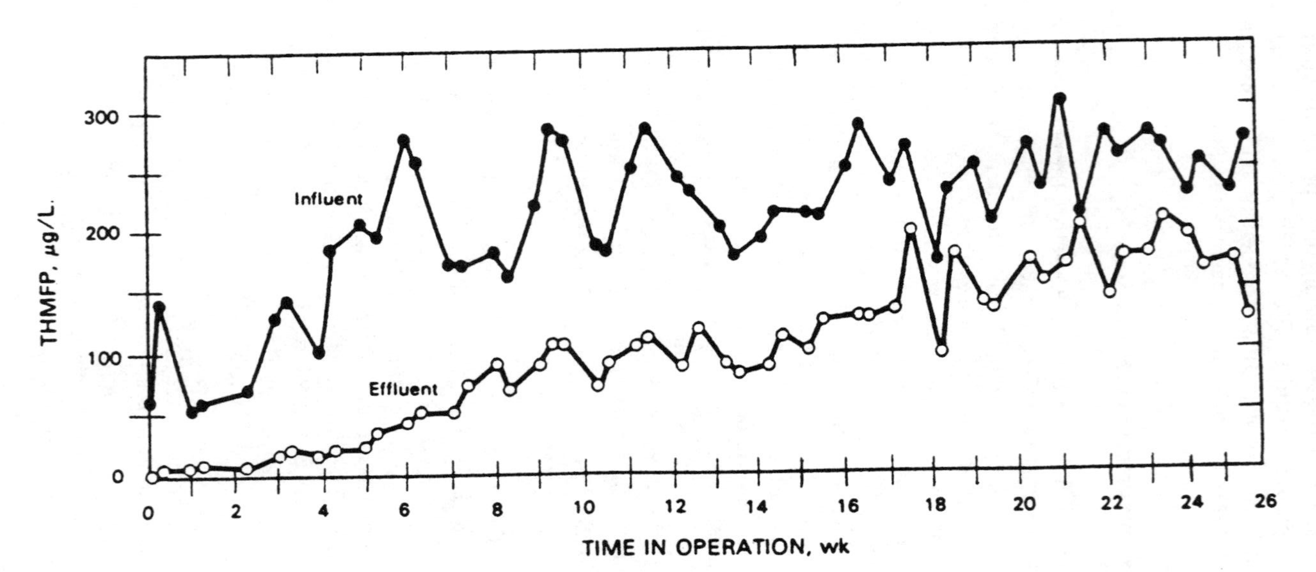

Figure 20. Removal of THM precursors by postfilter GAC adsorber at Jefferson Parish, Louisiana. GAC type, WVG 12 × 40; bed depth, 71 cm (28 in.); hydraulic loading, 1.9 m/hr (0.75 gpm/ft^2); EBCT, 23 min; THM formation potential conditions: pH 10, 23°C (70°F); storage time, 5 days [88].

Removal of Free and Combined Chlorine

The principal use of chlorination in water treatment is disinfection (see Chapter 10). However, chlorination is also practiced for taste and odor control (superchlorination), control of odor-causing algae and microorganisms and control of ammonia by breakpoint chlorination. The chlorine doses required to achieve any of the above goals, whether using free residual chlorination or combined residual chlorination, can be sufficiently high that dechlorination may be required. GAC adsorption has proven to be an effective dechlorination process. It is used widely in the food processing and bottling industries.

The reaction between free chlorine and activated carbon has been reported to proceed as [124–126]:

$$C^* + HOCl = CO^* + H^+ + Cl^- \tag{59}$$

where C^* = an activated carbon site
CO^* = a surface oxide on the carbon.

If OCl^- is the free chlorine species, then no H^+ would be expected as an end product unless the oxide is acidic and liberates a proton [127]. At pH values below 6.6, HOCl predominates, while at pH values higher than 8.6, the hypochlorite ion (OCl^-) is the predominant species. Both forms of free chlorine are converted to the chloride ion.

The reaction of chloramines with activated carbon has been studied by Bauer and Snoeyink [128] who proposed the following stoichiometric relationship for the interaction of dichloramine with carbon:

$$C^* + 2NHCl_2 + H_2O = CO^* + 4H^+ + 4Cl^- + N_2 \tag{60}$$

The nitrogen in the dichloramine is converted to nitrogen gas. Also, two dichloramine molecules react with carbon to produce one surface oxide, while each molecule of free chlorine is capable of forming such an oxide. Therefore, four times as much dichloramine (expressed as Cl_2) are needed to react with active carbon to produce an equivalent amount of surface oxides compared to free chlorine [126]. The capacity of activated carbon for dechlorination of dichloramine is, therefore, much higher than for free chlorine.

The reaction between monochloramine and activated carbon is proposed [128] to proceed according to:

$$C^* + NH_2Cl + H_2O = CO^* + H^+ + Cl^- + NH_3 \tag{61}$$

The monochloramine is reduced to ammonia and a surface oxide group is formed. However, in the presence of an already existing surface oxide group on the carbon, the following reaction occurs [128]:

$$CO^* + 2NH_2Cl = C^* + 2H^+ + 2Cl^- + H_2O + N_2 \quad (62)$$

In this case, the monochloramine is oxidized to nitrogen and the carbon is regenerated by removing the surface oxide. Therefore, in a continuous-flow operation, surface oxides formed according to Reaction 61 are regenerated according to Reaction 62, and a steady state is reached whereby the amount of ammonia produced would be equal to the net reduction in chloramine-N (NH_3-N plus NH_2Cl-N) [126,128].

The dechlorination reactions are affected by the pH value of the solution since it affects the distribution of the different species. The kinetics of the reactions have been studied, and mathematical models have been developed for the design or prediction of performance of the dechlorination process [126–130]. Intraparticle diffusion of the chlorine species was shown to be the rate-limiting step. Therefore, in continuous-flow operations, all factors that affect mass transfer such as flowrate, particle size and influent concentration change the shape and position of the breakthrough curve.

Biological Growth in Carbon Beds

GAC beds used in water treatment develop substantial microbial growth after a short period of operation, both in the carbon granules and the effluent water [45,88,89,131–135]. This increase in bacterial population is encountered irrespective of the pretreatment method including prechlorination. Residual chlorine in the influent is adsorbed in the top carbon layers and is rendered ineffective in controlling the bacterial population. The main source of the microorganisms is the influent water population that is readily adsorbed by the carbon. Also, virgin carbon may contain a significant number of bacteria. Several batches of GAC were analyzed in the author's laboratory and were found to contain an average of 2000 cells/g of carbon (standard plate count at 35°C). Establishment of the bacterial population is attributed to an increased concentration of adsorbed organics on the carbon surfaces that serve as nutrients and the large surface area of the carbon granules that serve as an attachment medium. Attachment of the microorganisms to the carbon surface has been shown by scanning electron microscopy to be in sparsely scattered single bacterial clumps [133,134,136]. A small fraction of the surface appears to be occupied by the bacteria and no uniform biofilm around the carbon granules was observed. The highest bacterial density is reported to be in the macropores. Individual bacterial cells are small enough to fit into the macropores (50–100 μm in size) of the GAC where they are protected from shear forces. On the other hand, bacteria are too large (1–16μm) to fit into the micropores, some of which are less than 0.5 μm in diameter. Since the micropores are about 99% of the available carbon surface area, bacteria would occupy only 1% of the total GAC surface area and, therefore, would not interfere with the adsorption process [135,136]. Bacterial growth is attributed to the utilization of adsorbed organics by one of two mechanisms: the dissolved organics adsorbed on the external surface and the macropores are utilized rapidly, or the organics adsorbed in the micropores are dissolved and are utilized slowly. Desorption can occur by displacement of the loosely adsorbed, biodegradable compounds by more strongly adsorbed organics or by enzymes secreted by bacteria present in the outer macropores. Neither of these proposed mechanisms has been demonstrated conclusively to date [137]. The mechanism of surface attachment of the bacteria has been discussed recently by Marshall [138]. Attachment apparently can be either reversible, with detachment produced by shearing action of fluid mixing and by mobility of the sorbed bacteria, or it can be irreversible, mediated by extracellular enzymes.

The bacterial density in carbon beds and effluent waters has been discussed by several investigators. Conflicting results have been reported due to variation in operating conditions, sample locations and nonuniformity in the bacterial measurement techniques (type of medium, incubation temperature and incubator period). However, the general trend is that the bacterial population increases rapidly in the carbon granules almost immediately after startup. It may reach values in the range of 10^3–10^5 cells/g of wet carbon and continues to rise until a pseudo-state cell level in the range of 10^5–$10^{7/8}$ of wet carbon is reached. The bacterial level in the filter effluent also rises sharply after startup from 10^1 to 10^5 cells/ml of water [88,131,134]. Size of the population within the filter and in the effluent varies with changes in the organic content of the water and with seasonal changes in temperature, usually increasing during warmer weather (Table VI). Distribution of the bacteria in the carbon filter varies with the adsorber bed age and perhaps the hydraulic flow rate or approach velocity. Flow rate is probably critical because it affects nutrient transport to the microorganism on the carbon granules [88]. Table VII shows the bacterial distribution in a GAC bed and the filter's effluent [88]. These data indicate that the bacterial density near the bottom does not correlate with the bacterial density of the effluent. Also,

Table VI. GAC Study Using Beaver River Source Water at the Beaver Falls, Pennsylvania, Municipal Authority [88]

	Source Water			GAC Influent					GAC Effluent[a]			
				Residual Chlorine (mg/l)					Residual Chlorine (mg/l)			
Week	Temperature (°C)	Turbidity (NTU)	Total Coliforms/ 100 ml	Free	Total	Turbidity (NTU)	Total Coliforms/ 100 ml	Standard Plate Count/ ml	Free	Total	Total Coliforms 100 ml	Standard Plate Count/ ml
1	21	44	98,000	2.0	[b]	5.6	<1	[b]	0	0	64	[b]
2	21	28	71,000	1.7	1.7	4.8	<1	[b]	0	0	75	[b]
3	15	22	140,000	1.3	1.4	2.3	<1	[b]	0	0	98	1,000
4	11	9.5	150,000	1.1	1.3	2.9	<1	100	0	0	45	1,400
5	16	7.5	39,000	1.2	1.4	2.5	<1	800	0	0	34	25,000
6	16	9	190,000	1.4	1.6	3.3	<1	350	[c]	<0.1	42	2,000
7	16	10	80,000	1.1	1.2	3.6	<1	10	[c]	<0.1	28	20,000
8	10	9	98,000	1.0	1.0	3.2	<1	42	[c]	<0.1	22	29,000
9	10	16	220,000	1.3	1.6	4.6	2	110	[c]	[c]	13	6,500
10	8	10	120,000	1.0	1.3	4.5	<1	33	[c]	<0.1	12	1,600
11	6	14	120,000	1.4	1.6	3.7	<1	95	[c]	<0.1	2	960
12	3	10	69,000	1.0	1.1	5.9	1	360	[c]	<0.1	1	270
13	4	22	89,000	1.2	1.7	4.6	<1	660	[c]	<0.1	<1	480
14	2	10	75,000	1.3	1.5	6.6	1	200	[c]	[c]	1	440
15	1	10	65,000	1.0	1.2	4.8	<1	120	[c]	<0.1	<1	44
17	1	12	48,000	1.4	1.7	5.9	<1	150	[c]	<0.1	<1	50
18	1	8	27,000	1.0	1.1	5.5	<1	33	[c]	<0.1	<1	21
22	1	14	6,000	0.4	1.6	6.4	<1	30	[c]	0.3	<1	30

23	4	10	23,000	0.3	1.6	5.8	<1	24	c	0.4	<1	13
25	4	150	84,000	c	1.4	6.6	<1	38	c	0.2	<1	21
27	7	12	13,000	b	1.4	6.3	<1	58	b	0.1	<1	41
29	10	8	24,000	0.2	1.1	1.7	<1	33	b	0.1	<1	31
32	11	6	8,400	0.2	1.6	1.9	<1	17	b	0.3	<1	69

[a]Turbidity was below 1 TU.
[b]Not run.
[c]Trace.

Table VII. Bacterial Counts[a] from Top, Mid-point and Bottom of an Activated Carbon Bed and from Its Effluent [88]

Column Age[b] (days)	Organisms/0.5 g dry wt[c]			Effluent (counts/ml)
	Top	Mid-point	Bottom	
6	[d]	58,000	55,000	250,000
11	550,000	45,000	28,000	135,000
17	130,000	4,400,000	2,700,000	30,000
20	2,790,000	460,000	320,000	44,000
25	7,700	90,000,000	50,000,000	520,000

[a]R2-A pour plates (35°C) incubation for 6 days.
[b]Ambient room temperature.
[c]Activated carbon particles sonicated for 4 min.
[d]Not run.

the bacteria appear to become established in the lower part of an adsorber bed. Furthermore, these populations may pulse widely in densities because they reflect numerous variables in the adsorber column ecosystem [88].

The microbial population in GAC beds and in the effluent has been identified by several investigators [88,134,137,139] and a list of the species identified is shown in Table VIII. It is noted that all the identified species are nonpathogenic in nature and are found in naturally occurring soils and waters. Concern has been expressed regarding the possible formation of endotoxins in GAC adsorbers because of bacteriological activity [89]. Endotoxins are pyrogenic lipopolysaccharide-protein complexes produced in the cell walls of gram-negative bacteria. The EPA Health Effects Research Laboratory has monitored bacterial endotoxin concentration in untreated water and effluent from GAC units. There was no increase in pyrogenic activity in the effluent from the GAC beds [89].

Attempts to reduce the bacterial population in the carbon beds by frequent backwashing with finished water or highly chlorinated water results only in an initial reduction in the population density followed by a gradual buildup to the higher levels in few days [31]. Therefore, it is expected that carbon filtered waters will contain high total plate counts with the levels depending on the quality of the influent water to the carbon beds. In addition, Symons et al. [88] have reported some cases where coliform bacteria penetrated the carbon bed and appeared in the filter effluent as shown in the example in Table VII. Both the potential penetration of coliforms in the carbon beds, which may not be a common occurrence, and the high standard plate counts make it essential to dis-

Table VIII. Microbial Species Found in Effluents of GAC [88,134,139]

Pseudomonas alcaligenes
Pseudomonas cepacia
Pseudomonas facilis
Pseudomonas fluorescens
Pseudomonas lemoignei
Pseudomonas mendocina
Pseudomonas ruhlandii
Pseudomonas stutzeri
Pseudomonas sp.
Gluconobacter oxidans
Azomonas agilis
Azomonas insignis
Azomonas macrocytogenes
Chromobacterium violaceum
Neisseria sicca
Acinetobacter calcoaceticum
Micrococcus luteus
Staphylococcus saprophyticus
Bacillus cereus
Bacillus circulans
Bacillus licheniformis
Bacillus megaterium
Bacillus pumulis
Bacillus thuringensis
Corynebacterium sp.
Micromonospora sp.

Pseudomonas putida
Pseudomonas maltophilia
Pseudomonas pseudoalcaligenes
Achromobacter sp.
Alcaligenes odorans
Flavobacterium odoratum

Citrobacter freundii
Enterobacter cloacae
Klebsiella pneumonia
Klebsiella oxytoca
Moraxella sp.

Filamentous Fungi
- *Phialophora hoffmannii*
- *Phialophora mutabilis*
- *Taphrina* sp.

Yeasts
- *Rhodotorula minuta* var. *texensis*
- *Cryptococcus uniguttulatus*
- *Candida guillermondii* var. *guillermondii*
- *Hansenula anomala* var. *anomala*

infect the carbon filtered water before final distribution. It is fortunately the common practice of water treatment plants in the United States to maintain a high disinfectant residual in the distribution system, thus assuring the safety of the delivered water.

Biological Activated Carbon

Biological activated carbon (BAC) is a process being used increasingly in European water treatment works where biological activity is deliberately enhanced in the GAC beds. The process, therefore, allows removal of soluble organic matter by utilization of the adsorptive capacity of the carbon and the biological oxidation provided by the microorganisms. The benefits claimed for the BAC process include: biological oxidation of organic compounds, biological nitrification of ammonia and partial biological regeneration of the GAC [137].

Aerobic biological growth in the activated carbon filters is promoted by introduction of sufficient dissolved oxygen into the water ahead of the contactor. If the water to be treated contains easily biodegradable organics, then simple aeration or perhaps oxygenation would provide adequate dissolved oxygen. Drinking water treatment plants along the upper Rhine River in Germany are utilizing this approach where pollution is mostly from municipal sewage [137].

However, the presence of organics that are not readily biodegradable in water requires the addition of an oxidant ahead of the carbon filters. To date, this has involved the use of ozone. Preozonation converts larger, less biodegradable organic compounds into smaller, more biodegradable molecules [89,132,137]. The extent of such conversion depends on the ease of oxidation of the specific organics present and on the amount of ozone employed. At the same time, ozonation introduces a large quantity of oxygen into the water that promotes aerobic bacterial growth. Slowly biodegradable organic molecules can be sorbed on the activated carbon where the microorganisms can metabolize them. This reactivates the formerly occupied carbon sites [135,137], which leads to an extended life of the activated carbon before it is replaced. High-mol-wt naturally occurring humic acids are not readily adsorbed by activated carbon. Preozonation of water containing these compounds results in their conversion into lower-mol-wt and more readily biodegradable organic compounds [137,140,141] that can be easily removed in biologically activated carbon beds. The activated carbon, therefore, will not become saturated rapidly with biorefractory materials, and the bed service life will be extended. Additional information on the role of microbial activity in BAC and carbon beds can be obtained in a publication by the AWWA Research and Technical Practice Committee [135].

Miller et al. [137] have reviewed in detail the development and current practices and operation of BAC systems where an estimated 46 full-scale drinking water treatment plants are presently in operation in Europe. The BAC process currently being utilized in the water treatment plant in Mulheim, Germany [131,137,142], is discussed here briefly for illustrative purposes. The raw water for this plant is the River Ruhr which, until mid-April 1977, was treated by breakpoint chlorination for ammonia removal, flocculation and sedimentation, and by GAC for dechlorination, and finally by ground and/or sand filtration and disinfection. The chlorine dose required for ammonia removal was as high as 10–50 mg/l. This resulted in the formation of chlorinated organics that were not completely removed by the activated carbon beds. Regeneration of the activated carbon columns was required every 4–8 weeks. A two-year pilot-plant study was conducted on the use of preozonation followed by activated carbon adsorption for removing ammonia and organics. The results indicated that breakpoint chlorination could be eliminated completely and the BAC was reliable for removing ammonia. At the same time, the dissolved organic carbon (DOC) was reduced to the desired levels. This process subsequently involves ozonation ahead of sand filtration and GAC adsorption. It was installed in the plant and began operating in mid-April 1977. The newly installed process involves preoxidation with about 1 mg/l ozone for manganese oxidation. After flocculation and sedimentation, 2 mg/l of ozone is added to oxidize dissolved organics where a retention time of 15–30 min is provided. The water is then filtered through sand filters and finally through BAC where the bulk of DOC and ammonia is removed. The treated water is subsequently chlorinated before discharge into the distribution system. Table IX lists the old process and the new process and compares their performance. The life of the full-scale carbon columns is estimated to be 2 years and the filters are backwashed every 10 days.

Reactivating GAC

Exhaustion of the adsoprtive capacity of GAC requires its removal from the adsorber so that it can be regenerated. Regeneration involves removing the adsorbate and simultaneously reestablishing the original adsorptive capacity of the virgin carbon. The most commonly used method is thermal regeneration where the spent carbon is heated in an inert atmosphere at high temperatures (800°C) in the presence of oxidiz-

Table IX. Process Parameters and DOC Data at the Dohne Waterworks (Mulheim, Germany) Before and After Treatment Changes[a]

Treatment Step	Old Treatment (Before 1977)		New Treatment[b] (April–July 1977)	New Treatment (November 1977–June 1978)
Preoxidation Dosing Power Input	10–50 mg/l Cl_2 4–6 mg/l Al^{3+} 0.1 kW/m^3 Ret. time—0.5 min		1 mg/l O_3 4–6 mg/l Al^{3+} 2.5 kW/m^3 Ret. time—0.5 min	1 mg/l O_3 4–6 mg/l Al^{3+} 2.5 kW/m^3 Ret. time—0.5 min
Flocculation Sedimentation	5–15 mg/l $Ca(OH)_2$ Ret. time—1.5 hr		5–15 mg/l $Ca(OH)_2$ Ret. time—1.5 hr	5–15 mg/l $Ca(OH)_2$ Ret. time—1.5 hr
Ozonation			2 mg/l O_3 Ret. time—5 min	2 mg/l O_3 Ret. time—5 min
Filtration with Preflocculation	v^c = 10.7 m/hr		v = 9 m/hr 0.2 mg/l Al^{3+} 0.1 mg/l polyelectrolyte	10 mg/l pure oxygen
Activated Carbon Filter	v = 22 m/hr h^d = 2 m		v = 18 m/hr v = 2 m	v = 18 m/hr h = 4 m
Ground Passage	Ret. time—12–50 hr		Ret. time—12–50 hr	Ret. time—12–50 hr
Safety Chlorination	0.4–0.8 mg/l Cl_2		0.2–0.3 mg/l Cl_2	0.2–0.3 mg/l Cl_2
DOC Data (mg/l)	**1975**	**1976**		
Raw Water (Ruhr)	3.9	5.0	3.6	2.4–3.7
After Flocculation and Sedimentation	3.2	4.0	2.9	1.8–3.0
After Filtration	3.2	3.8	2.6	1.7–3.1

After Activated Carbon	3.0	3.7	2.3	1.0–2.6
After Ground Passage	1.8	2.1	0.9	—

[a]Reproduced from Miller et al. [137], courtesy of Ann Arbor Science Publishers.
[b]Filters filled with fully exhausted GAC, used during pre-1977 treatment.
[c]Velocity.
[d]Bed height.

ing gases such as steam, carbon dioxide and oxygen [89,143,144]. The regeneration process proceeds in three steps [143,144]: (1) a drying step where the water in the wet carbon is driven off at temperatures above 100°C; (2) a pyrolysis step where the adsorbed materials are volatilized at high temperatures (during this step a fraction of the organic matter undergoes polymerization or condensation and carbonization); and (3) an activation step where residues in the micropores are oxidized by activation gases. Some loss of the activated carbon generally occurs during the regeneration due to burnoff and mechanical attrition.

The types of furnaces currently in use for reactivating GAC are: multiple-hearth, rotary kiln, fluidized bed furnace and infrared tunnel [89]. Figure 21 shows a schematic diagram of a multiple-hearth furnace. The total regeneration time ranges 20–30 min.

The properties of the regenerated carbon depend on the proper control of reactivation conditions. Several investigators reported changes in the pore size distribution [89] and specific surface area [145] of the regenerated carbon. However, Juntgen [143] showed that it is possible to determine the conditions of desorption and activation enabling the best possible restoration of the original surface area and pore size distribution with minimum loss of the activated carbon during regeneration. Figure 22 shows the effect of activation residence time on the changes in activity and weight of activated carbon during regeneration [143] and indicates, in this case, that an activation temperature at 800°C for 20 min does not significantly change the properties of activated carbon. Studies by Van Vliet and Weber [70] illustrate that the regeneration process may not result in significant changes in the pore size distribution of the activated carbon.

Onsite regeneration at drinking water treatment plants is not practiced in the United States. The general practice is to replace the spent carbon with virgin carbon. Most of the GAC filters in these plants are used for taste and odor control where the service life of the carbon can extend to several years, making onsite regeneration economically unfeasible.

SYNTHETIC ADSORBENTS FOR REMOVAL OF ORGANICS

The potential use of a wide range of synthetic adsorbents as alternatives to GAC for removing organics from drinking water has received considerable attention in recent years. The types that have been considered for potential use in water treatment are: polymeric adsorbents, carbonaceous adsorbents and ion-exchange resins. All of these adsorbents

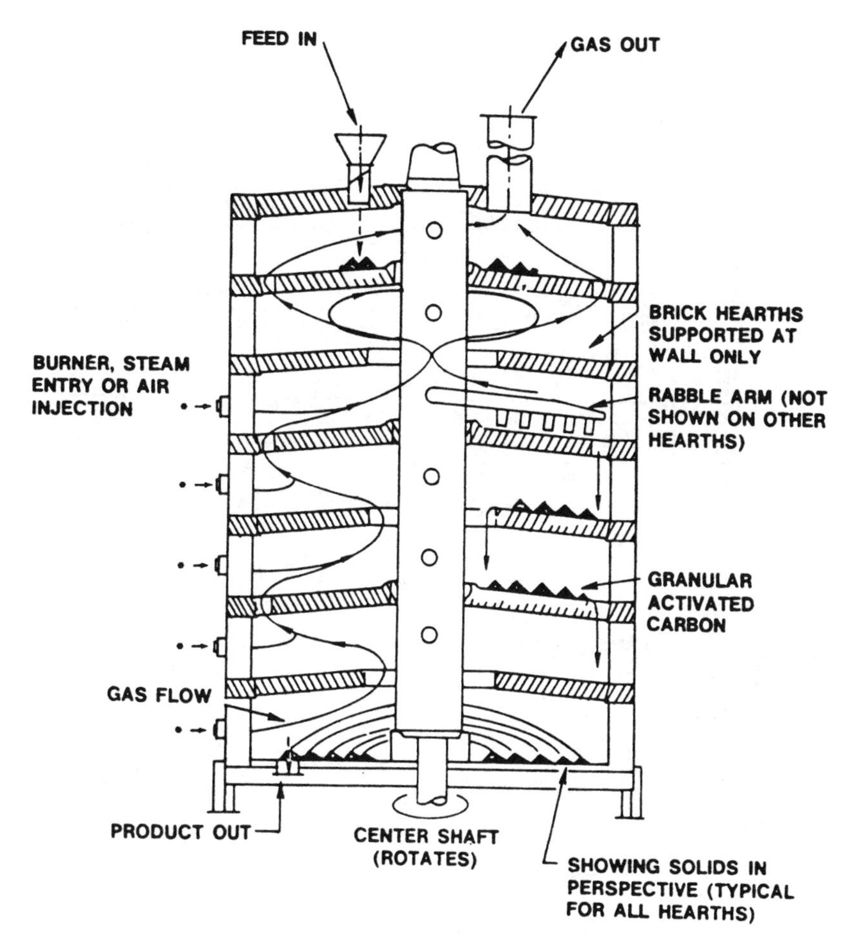

Figure 21. Cross section of multihearth furnace [89].

are based on cross-linked polymers having polystyrene, phenol-formaldehyde or acrylate structures (matrices) as shown in Figure 23 [146]. A major advantage of using these adsorbents is claimed to be the ease of regeneration in place with suitable regenerants [146,147].

Polymeric Adsorbents

Most of the commercially available adsorbents that have been studied are copolymers of polystyrene-divinylbenzene or polymethacrylates cross-linked with a suitable nonaromatic material. These adsorbents,

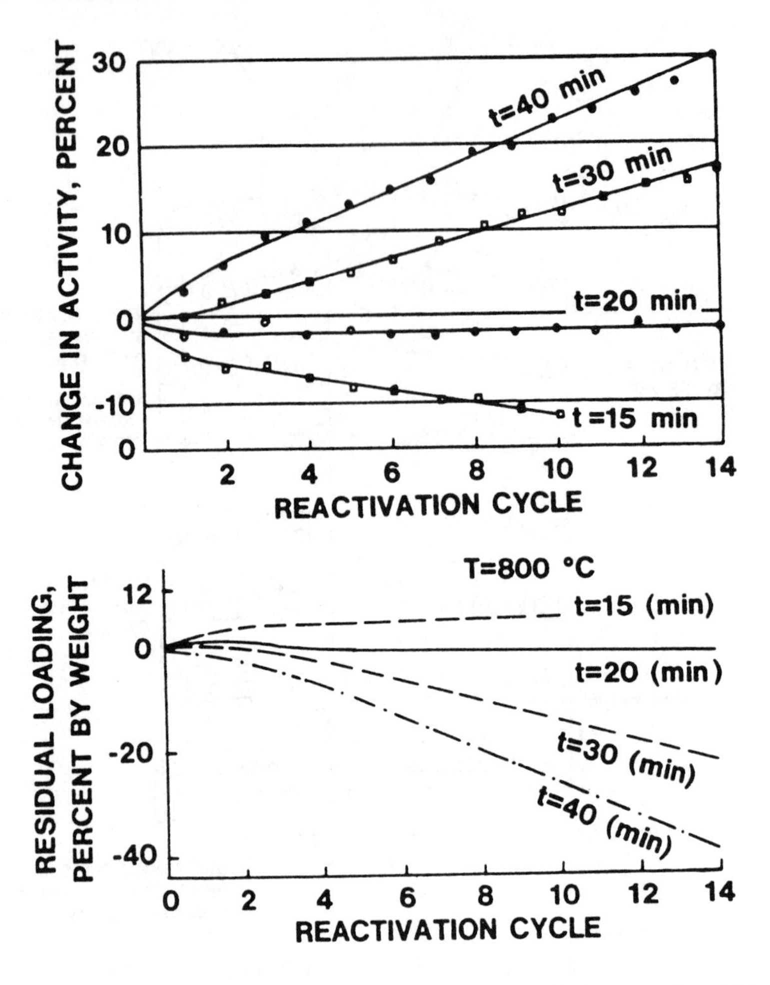

Figure 22. Effect of residence time on activity and change of weight of activated carbon [143].

commonly termed macroreticular resins, are characterized by being highly porous and having a high surface area. The pore size distribution is controlled during manufacturing by varying the level of cross linking. Another important feature is the absence of any ionic functional groups in their structure. Therefore, they can serve as hydrophobic adsorbents capable of removing hydrophobic solutes from solution via van der Waals interactions [147–149]. The characteristics of some of these adsorbents are shown in Table X.

Polymeric adsorbents have been used successfully in concentrating

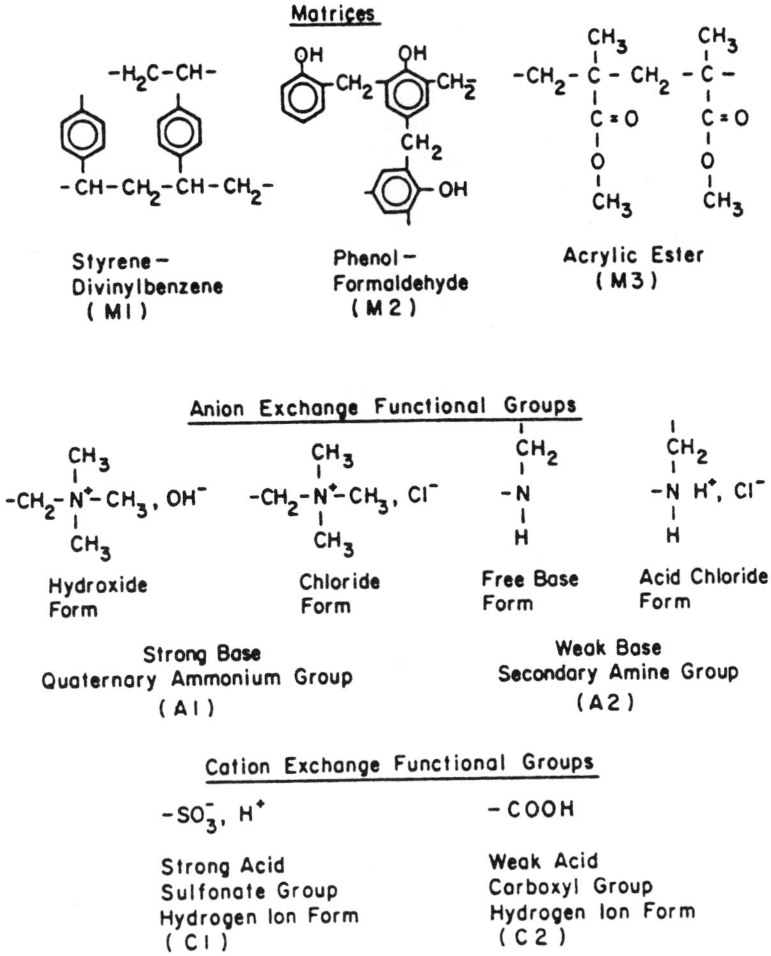

Matrices

Styrene–Divinylbenzene (M1)

Phenol–Formaldehyde (M2)

Acrylic Ester (M3)

Anion Exchange Functional Groups

Hydroxide Form

Chloride Form

Free Base Form

Acid Chloride Form

Strong Base Quaternary Ammonium Group (A1)

Weak Base Secondary Amine Group (A2)

Cation Exchange Functional Groups

$-SO_3^-, H^+$

$-COOH$

Strong Acid Sulfonate Group Hydrogen Ion Form (C1)

Weak Acid Carboxyl Group Hydrogen Ion Form (C2)

Figure 23. Matrices and functional groups of resins commonly used for water purification. Reproduced from Kim et al. [146], courtesy of the Water Pollution Control Federation.

organic contaminants in raw and finished waters for analytical purposes [150–152]. High recoveries of the sorbed organic by desorption in an organic solvent for subsequent analysis contributes to the high efficiency of the analytical procedures. They are used also for removing chlorinated pesticides [153] and dyes [154] from industrial wastes. Regeneration is accomplished with acetone, isopropanol or methanol, and the spent regenerate can be reclaimed for reuse. Adsorption studies from solutions

Table X. Properties of Macroreticular Resins[a]

Adsorbent	BET Surface Area (m^2/g)	Particle Density (g/cm^3 particles)	Pore Volume (cm^3/g)	Average Pore Diameter[b] (Å)
XAD-2[c]	300.0	0.582	0.7750	103.0
XAD-4[c]	725.0	0.548	0.9000	49.7
XAD-8[c]	140.0	0.445	0.4340	410.0
ES-863[d]	400.0	0.676	0.6150	61.5

[a]Reproduced from Weber and Van Vliet [158], courtesy of the American Water Works Association.
[b]Qualitative pore size, based on cylindrical homogeneous pores = 4×10^4 (pore volume/(BET surface area).
[c]Rohm and Haas Co., Philadelphia, Pennsylvania.
[d]Diamond Shamrock Chemical Co., Dallas, Texas.

of relatively high concentrations of phenols, carboxylic acids and alkylbenzenesulfonates indicate that the styrene-divinylbenzene and polymethacrylate polymeric adsorbents can sorb these compounds, but do so less effectively than activated carbon [149].

These adsorbents, however, proved to be inefficient for removing contaminants of concern in water treatment. The styrene-divinylbenzene adsorbent showed a very low capacity for removal of TOC from riverwater [155] and an insignificant capacity for humic substances [95,156,157] and chloroform [95]. Capacity of the same adsorbent for 2-methylisoborneol, which causes taste and odor of biologic origin in natural water, was also low, whereupon the resin could not be effectively regenerated by steam and ethanol [95]. Polymeric adsorbents also exhibited steep adsorption isotherms for phenolic compounds, alkylbenzenesulfonate, and carbon tetrachloride such that their capacity at the low μg/l range of interest in water treatment is very low compared to activated carbon [156–159]. Overall, polymeric adsorbents have not shown promise for successful application in water treatment.

Synthetic Carbonaceous Adsorbents

Synthetic carbonaceous adsorbents are manufactured by partial pyrolysis of beads of macroreticular copolymers under such conditions that the polymer chars during pyrolysis and retains the macroporous structure [160]. The average size of the macropores in the copolymers can be varied over a wide range, 50–100,000 Å, depending on the method of preparing the beads. Size of the micropores is controlled by adjusting the temperature during pyrolysis. The material has the appearance of activated carbon and reportedly was prepared for the removal of low-mol-wt halogenated hydrocarbons from drinking water [160]. The carbonaceous adsorbent most commonly studied is Ambersorb XE-340 (Rohm and Haas Company). It is prepared so that the size of the macropores is small enough to allow rapid internal diffusion of low-molecular-weight compounds and to exclude larger size molecules (molecular sieving). Size of the micropores is designed for maximum adsorption of chloroform and ranges between 15 and 30 Å [160]. The specific surface area is about 400 m^2/g and is encountered mostly in pore sizes up to 200 μm. Surfaces of the carbonaceous resins are highly hydrophobic and fairly homogeneous. Consequently, the adsorptive interactions are mainly due to van der Waals forces. The kinetics of adsorption and desorption are rapid, and it is claimed that regeneration can be accomplished with steam [160].

The carbonaceous adsorbent XE-340 has a demonstrated high capacity for removing chloroform and other volatile halogenated hydrocarbons (including trihalomethanes) in both pure-solute systems [95,160] and pilot-plant studies [88,90]. Wood and DeMarco [90] reported that XE-340 had four times the capacity of activated carbon for removing chloroform that was calculated on weight basis per 100 g of adsorbent. The adsorptive capacity of XE-340 for removing individual halogenated hydrocarbons from raw and lime-softened and settled water was approximately three times that of activated carbon [90]. Similar results were obtained in other pilot-plant studies at several locations [88,98]. Removals generally were higher in waters containing low levels of TOC than those higher in TOC and other contaminants because of competitive adsorption [90]. Table XI shows a list of halogenated hydrocarbons that have been removed by XE-340 [90,98].

The ability of XE-340 for removal of trihalomethane precursors is much less than that of activated carbon. Results obtained in a pilot-plant study indicated that the precursors were only partially removed from raw water and more was removed from lime-softened and settled water [118]. Activated carbon was generally and relatively more efficient for removal of THM precursors in raw and finished waters than XE-340. Humic substances, a major source of THM precursors, are not adsorbed by the carbonaceous resins [95,157], which is probably due to molecular screening.

The efficiency of carbonaceous resins for adsorption of 2-methylisoborneol [95], phenolic compounds and benzenesulfonates [158,159] was shown to be less than that of activated carbon. A reported advantage of carbonaceous adsorbents is their ability to be regenerated with steam. However, attempts to regenerate XE-340 loaded with light halogenated hydrocarbons with steam in a pilot plant study was not completely suc-

Table XI. List of Compounds Removed by XE-340 [90,98]

Chloroform
Bromodichloromethane
Dibromochloromethane
Bromoform
Tetrachloroethylene
1,1-Dichloroethane
1,1-Dichloroethylene
Methyl Chloroform
cis-1,2,Dichloroethylene
1,2-Dichloroethane
Trichloroethylene
Carbon Tetrachloride

cessful [98]. Similarly, 2-methylisoborneol was not completely desorbed with steam or ethanol [95] and restoration of the original capacity of the resin has not, so far, been reported.

Overall, carbonaceous adsorbents are more selective in their adsorptive characteristics and could be applied in drinking water treatment only in situations where chloroform or other volatile halogenated hydrocarbons must be removed from waters low in organic content.

Ion Exchange Resins

Ion-exchange resins are based on cross-linked polystyrene, phenol-formaldehyde or acrylic ester polymers converted into cation or anion exchangers by the introduction of appropriate functional groups (Figure 23). These resins can be produced as macroporous or gel types. The macroporous resins offer the advantage of large pores that allow high-molecular-weight ions to be adsorbed from solution and more completely eluted from the resin upon regneration [147]. The gel type of resin of the styrene divinylbenzene or acrylic type does not contain any true porosity. Ions to be exchanged must diffuse through the gel structure to the exchange sites. Ion-exchange resins are used in drinking water treatment for softening and sometimes for deionization and demineratlization (see Chapter 7).

Anion exchange resins, strongly or weakly basic, are commonly studied for removal of organics from drinking waters [88,146,147,156, 157–159]. This is attributed to the fact that most naturally occurring organic matter in waters is acidic in nature [95,147]. Humic substances, which are composed mainly of humic and fulvic acids, are examples of these compounds. Both strong base and weak base anion exchangers were effective in removing humic substances [147, 157]. The adsorptive capacity is generally a function of the sources of the organics. Boening et al. [157] reported that both types of anion exchangers removed soil and leaf fulvic acid more efficiently than GAC. Evans and Maalmen [161] obtained a 65% removal of humic acids from the Rhine River water using a weak base resin.

Strong base anion exchange resins were found to perform better than weak base resins for removal of TOC from Mississippi River water [162]. Kölle [163] reported a 58% removal of dissolved organic matter from groundwater containing natural color by a strong base anion exchange resin. On the other hand, the same resin used by Kölle was much less effective than activated carbon for removal of TOC from the Thames River. However, it was more effective than weak base resins [155]. Jayes and Abrams [164] found that a weak base resin was quite effective for removing a natural color at Lawrence, Massachusetts.

Removal of trihalomethane precursors by anion exchange resins is dependent on the quality of the water to be treated and the type of resin used. Wood and DeMarco [118] found the strong base anion exchanger (Amberlite IR-904) to be more effective than activated carbin in a 2.5-ft bed for removing THM precursors from raw groundwater. However, no significant removals were obtained from lime-softened and finished water. Rook [165] also obtained good THM precursor removal using a strong base anion exchange resin at the Rotterdam Waterworks. Rook and Evans [166] reported THM precursor removals in the range of 52–68% from Meuse River water by two different weak base anion exchange resins. These resins were regenerated by lime followed by HCl after about six days of operation.

The removal of specific organic compounds by anion exchange resins indicates that nonionic molecules have a very low affinity for adsorption by the resins. Weak electrolytes, on the other hand, are more strongly adsorbed depending on the type and functionality of the resin. Thus, a strong base styrene-divinylbenzene type anion exchange resin, as expected, did not adsorb trihalomethanes or other halogenated hydrocarbons from raw or softened water in Miami, Florida [90]. Insignificant removal efficiencies for THM were observed also in studies at Kansas City, Missouri, using a weak base resin [88]. A phenol-formaldehyde weak base anion exchange resin was shown to have essentially no capacity for adsorption of 2-methylisoborneol [95]. On the other hand, strong base resins (polystyrene matrix) in the chloride form had a high capacity for alkylbenzenesulfonates [167] and 2,4-dichlorophenoxyacetic acid [168]. Weak base anion exchange resins (phenol-formaldehyde) in the acid form also had a significant capacity for ABS [146]. The mechanism of adsorption of these compounds seemed to be by ion exchange [167,168]. Phenol [169,170], *p*-nitrophenol [146] and 2,4-dichlorophenol [168] are strongly adsorbed by anion exchange resins of different matrices and functionalities. Strong base anion exchange resins in the free base form generally had a higher capacity for phenols than the chloride form of that resin as well as weak base resins. These latter resins, in the free base form, however, had an appreciable capacity for adsorption of phenols [170]. The adsorption mechanism, in the case of weak base resins (free base), was attributed to hydrogen bonding with the amino-nitrogen or to physical adsorption on the resin matrix [146,158].

Regeneration of the anion exchange resins is usually accomplished by a strong solution of alkali or brine. An alternative regeneration process for the weak base ion exchangers has been suggested, using lime and HCl in place of NaCl and base [166]. However, disposal of the spent regenerant could pose a problem. Strong base exchangers are generally not regenerated as efficiently as weak base anion exchangers. Fouling of strong base resins, particularly those with small pores, can occur due to the irreversible adsorption of humic substances. Kunin [147] suggests that strong base acrylic resins are less selective for organic matter than the styrene based resins, and, therefore, can be regenerated more effectively.

In addition to the difficulties in regenerating strong base anion exchange resins, another problem has emerged from recent studies that warrants careful consideration of the use of these resins in drinking water treatment. The resin seems to catalyze a reaction between free chlorine, if present in the water, and the adsorbed organics resulting in the formation of trihalomethanes and other non-THM halogenated organics [90]. Other studies indicate that nitrosamines may be generated by the passage of water over the resin. The exact mechanism is not yet known, but the resin concentrates the potential reactants and catalysts (iron and copper) for formation of nitrosamines [147].

A comment on the potential use of synthetic adsorbents in drinking water treatment is warranted here. In general, these materials exhibit selective adsorption characteristics and none has been demonstrated to be a broad spectrum adsorbent like activated carbon. Studies by Weber and Van Vliet [158,159] indicated that polymeric and resinous adsorbents show adsorption isotherms for several organic compounds that are considerably steeper than those for activated carbon or carbonaceous adsorbents. These steep slopes indicate that these adsorbents would not be effective for removing organics in the μg/l range usually encountered in drinking water treatment. Carbonaceous adsorbents seem to perform effectively in situations for which they are specifically designed, that is, high trihalomethane or volatile halogenated hydrocarbon content in waters low in organic content.

REFERENCES

1. *J. Am. Water Works Assoc.* 73:391 (1981).
2. Hassler, J. W. *Activated Carbon* (New York: Chemical Publishing, 1974).
3. Jüntgen, H. "Translation of Reports on Special Problems of Water Technology Adsorption," EPA-600/9-76-030, U.S. EPA Cincinnati, OH [1975], p. 16.
4. Brunauer, S. et al. *J. Am. Chem. Soc.* 60:309 (1938).
5. Emmett, P. H., and S. Brunauer. *J. Am. Chem. Soc.* 59:1553 (1937).
6. Brunauer, S., and P. H. Emmett. *J. Am. Chem. Soc.* 59:2682 (1937).
7. Thompson, A. *Phil. Mag.* 42(4):448 (1871).
8. Lowell, S. *Introduction to Powder Surface Area* (New York: John Wiley & Sons, Inc., 1979).
9. Juhola, A. J., and E. O. Wiig. *J. Am. Chem. Soc.* 71:1069 (1949).
10. Drake, L. C. *Ind. Eng. Chem.* 41:780 (1949).
11. Hofmann, U., and D. Wilm. *Z. Electrochem. Angew. Physik. Chem.* 42:504 (1936).
12. Wolff, W. F. *J. Phys. Chem.* 62:653 (1959).
13. Hedden, K. In: *Ullmann's Enzyklopadie der Technische Chemie* 10:362 (1958).
14. Walker, P. L. Jr. *Am. Scientist* 50:259 (1962).
15. Boehm, H. P., and R. W. Coughlin. *Carbon* 2:1 (1964).
16. Dubinin, M. M. et al. *Carbon* 2:261 (1964).
17. Snoeyink, V. L., and W. J. Weber, Jr. *Environ. Sci. Technol.* 1:228 (1967).
18. Gasser, C. G., and J. J. Kipling. *Proceedings of the 4th Conference on Carbon* (New York: Pergamon Press, 1960), p. 59.
19. Ingram, D. J. E. *Proceedings of the 3rd Conference on Carbon* (New York: Pergamon Press, 1959), p. 93.
20. Harker, H. et al. *Proc. Roy. Soc.* A262:328 (1961).
21. Zettlemoyer, A. C. *Chem. Rev.* 59:937 (1959).
22. Kipling, J. J., and P. V. Shooter. *J. Colloid. Interface Sci.* 21:238 (1966).
23. Boehm, H. P. In: *Advances in Catalysis, 16,* D. D. Eley et al, Eds. (New York: Academic Press, 1966).
24. Garten, V. A., and D. E. Weiss. *Rev. Pure Appl. Chem.* 7:69 (1957).
25. Coughlin, R. W., and F. S. Ezra. *Environ. Sci. Technol.* 2:291 (1968).
26. Mattson, J. S., and H. B. Mark, Jr. *Activated Carbon—Surface Chemistry and Adsorption from Solution* (New York: Marcel Dekker, Inc., 1971).
27. Weber, W. J. Jr. *Physicochemical Processes for Water Quality Control* (New York: Wiley-Interscience, 1972).
28. Wolff, W. F. *J. Phys. Chem.* 63:653 (1959).
29. Garten, V. A., and D. E. Weiss. *Aust. J. Chem.* 8:68 (1955).
30. Puri, B. R. *Proceedings of the 5th Conference on Carbon*, 1 (New York: Pergamon, Press, 1962), p. 125.

31. Frumkin, A. *Kolloid Zh.* 47:44 (1929).
32. Puri, B. R., and R. C. Bansal. *Carbon* 1:451 (1964).
33. Fedorov, G. G. et al. *Zh. Fiz. Khim.* 37:2344 (1963).
34. Puri, B. R. *Carbon* 4:391 (1966).
35. Puri, R. P. In: *Activated Carbon Adsorption of Organics from the Aqueous Phase, 1,* I. H. Suffett and M. J. McGuire, Eds. (Ann Arbor, MI: Ann Arbor Science Publishers, Inc., 1980), p. 353.
36. Garten, V. A., and D. E. Weiss. *Aust. J. Chem.* 10:309 (1957).
37. Kolthoff, I. M. *J. Am. Chem. Soc.* 54:4473 (1932).
38. Osipow, L. I. In: *Principles and Applications of Water Chemistry—Proceedings of the 4th Rudolfs Conference,* S. D. Faust and J. V. Hunter, Eds. (New York: John Wiley & Sons, Inc., 1967), p. 75.
39. Young, D. M., and A. D. Crowell. *Physical Adsorption of Gases* (London: Butterworths, 1962).
40. London, F. *Trans. Faraday Soc.* 33:8 (1937).
41. Calgon Corporation. "Basic Concepts of Adsorption on Activated Carbon," Pittsburgh, PA.
42. Zogorski, J. S. "The Adsorption of Phenols onto Granular Activated Carbon from Aqueous Solution," PhD Thesis, Rutgers University, New Brunswick, NJ (1975).
43. Weber, W. J. Jr., and J. C. Morris. *J. San. Eng. Div.* (*ASCE*) 90(SA3):79 (1964).
44. Morris, J. C., and W. J. Weber, Jr. "Adsorption of Biochemically Resistant Materials from Solution, 1" Environmental Health Series, AWTR-9, U.S. Department of Health, Education, and Welfare, Washington, DC (1964).
45. Yohe, T. L. et al. In: *Activated Carbon Adsorption of Organics from the Aqueous Phase, Vol. 2,* M. J. McGuire and I. H. Suffet, Eds. (Ann Arbor, MI: Ann Arbor Science Publishers, Inc., 1980), p. 27.
46. Adamson, A. W. *Physical Chemistry of Surfaces,* 2nd ed. (New York: Academic Press, Inc., 1967).
47. Hansen, R. S., and R. P. Craig. *J. Phys. Chem.* 58:211 (1954).
48. Belfort, G. *Environ. Sci. Technol.* 13:939 (1979).
49. Franks, H. S., and M. W. Evans. *J. Chem. Phys.* 13:507 (1945).
50. Pinsky, J. *Mod. Plast.* 34:145 (1957).
51. Diamond, J. M., and E. M. Wright. *Ann. Rev. Physiol.* 31:581 (1969).
52. Giusti, D. M. et al. *J. Water Pol. Control Fed.* 46:947 (1974).
53. Al-Bahrani, K. S., and R. J. Martin. *Water Res.* 10:731 (1976).
54. Martin, R. J., and K. S. Al-Bahrani. *Water Res.* 11:991 (1977).
55. Singer, P. C., and C. Yen. In: *Activated Carbon Adsorption of Organics from Aqueous Phase, Vol. 1,* I. H. Suffet and M. J. McGuire, Eds. (Ann Arbor, MI: Ann Arbor Science Publishers, Inc., 1980), p. 167.
56. Aly, O. M., and S. D. Faust. "Sorption of Phenolic Compounds from Aqueous Solutions," paper presented at the Kendall Award Symposium, American Chemical Society, 163rd National Meeting, Boston, MA, April, 9, 1972.
57. Snoeyink, V. L. et al. *Environ. Sci. Technol.* 3:918 (1969).
58. Aly, O. M., and S. D. Faust. "The Adsorption of Chlorinated Phenols by Activated Carbon," Research Report, Department of Environmental Sciences, Rutgers University, New Brunswick, NJ (1970).

59. Zogorski, J. S., and S. D. Faust. *J. Colloid Interface Sci.* 55:329 (1976).
60. Snoeyink, V. L. et al. "Activated Carbon Adsorption of Trace Organic Compounds," EPA-600/2-77-223, U.S. EPA, Cincinnati, OH (1977).
61. Schwartz, H. G. Jr. *Envrion. Sci. Technol.* 1:332 (1967).
62. Getzen, F. W. and T. M. Ward. *J. Colloid Interface Sci.* 31:441 (1969).
63. Ward, T. M., and F. W. Getzen. *Environ. Sci. Technol.* 4:64 (1970).
64. Manes, M. In: *Activated Carbon Adsorption of Organics from the Aqueous Phase, Vol. 1,* I. H. Suffet and M. J. McGuire, Eds. (Ann Arbor, MI: Ann Arbor Science Publishers, Inc., 1980), p. 43.
65. Langmuir, I. *J. Am. Chem. Soc.* 37:1139 (1915).
66. Freundlich, H. *Colloid and Capillary Chemistry* (London: Methuen and Co., 1926).
67. Sips, R. *J. Chem. Phys.* 16:490 (1948).
68. Weber, W. J. Jr., et al. In: *Activated Carbon Adsorption of Organics from the Aqueous Phase, Vol. 1,* I. H. Suffet and M. J. McGuire, Eds. (Ann Arbor, MI: Ann Arbor Science Publishers, Inc., 1980), p. 317.
69. Smith, J. M. In: *Adsorption from Aqueous Solution,* Advances in Chemistry Series 79, Robert F. Gould, Ed. (Washington, DC: American Chemical Society, 1968), p. 8.
70. Weber, W. J. Jr., and A. B. Van Vliet. *Activated Carbon Adsorption of Organics from the Aqueous Phase, Vol. 1,* I. H. Suffet and M. J. McGuire, Eds. (Ann Arbor, MI: Ann Arbor Science Publishers, Inc., 1980), p. 15.
71. Weber, W. J. Jr., and R. R. Rumer, Jr. *Water Resources Res.* 1:361 (1965).
72. Keinath, T. M., and W. J. Weber, Jr. *J. Water Poll. Control Fed.* 40:743 (1968).
73. Sontheimer, H. "Translation of Reports on Special Problems of Water Technology, 9, Adsorption," EPA-600/9-76-030, U.S. EPA, Cincinnati, OH (1976), p. 29.
74. Fritz, W. et al. In: *Activated Carbon Adsorption of Organics from the Aqueous Phase, Vol. 1,* I. H. Suffet and M. J. McGuire, Eds. (Ann Arbor, MI: Ann Arbor Science Publishers, Inc., 1980), p. 193.
75. Liu, K. T., and W. J. Weber, Jr. *J. Water Poll. Control Fed.* 53:1541 (1981).
76. Jain, J. S., and V. L. Snoeyink. *J. Water Poll. Control Fed.* 54:2463 (1973).
77. Butler, J. A. V., and C. Ockrent. *J. Phys. Chem.* 34:2841 (1930).
78. Myers, A. L., and J. M. Prausnitz. *Am. Inst. Chem. Eng. J.* 11:121 (1965).
79. Radke, C. J., and J. M. Prausnitz. *Am. Inst. Chem. Eng. J.* 18:761 (1972).
80. Kidnay, A. J., and A. Myers. *Am. Inst. Chem. Eng. J.* 12:981 (1966).
81. DiGiano, F. A. et al. In: *Activated Carbon Adsorption of Organics from the Aqueous Phase, Vol. 1,* I. H. Suffet and M. J. McGuire, Eds. (Ann Arbor, MI: Ann Arbor Science Publishers, Inc., 1980), p. 213.
82. *Crittenden, J. C., and W. J. Weber, Jr. J. Environ. Eng. Div.* (*ASCE*) 104(EE6):1175 (1978).
83. Wilde, K. A. In: *Activated Carbon Adsorption of Organics from the Aqueous Phase, Vol. 1,* I. H. Suffet and M. J. McGuire, Eds. (Ann Arbor, MI: Ann Arbor Science Publishers, Inc., 1980), p. 251.

84. *The Quest for Pure Water Vol. II* (Denver, Co: American Water Works Association, 1981), p. 147.
85. Sontheimer, H. "Translation of Reports on Special Problems of Water Technology, 9, Adsorption," EPA-600/9-76-030, U.S. EPA (1976), p. 67.
86. *Federal Register* 43(28):2756–5780 (1978).
87. "AWWA Standard for Granular Activated Carbon," *J. Am. Water Works Assoc.* 66:672 (1974).
88. Symons, J. M. et al. "Treatment Techniques for Controlling Trihalomethanes in Drinking Water," EPA-600/2-81-156, U.S. EPA, Cincinnati, OH (1981).
89. "Interim Treatment Guide for Controlling Organic Contaminants in Drinking Water Using Granular Activated Carbon," J. M. Symons, Ed., U.S. EPA, Cincinnati, OH (1978).
90. Wood, P. R., and J. DeMarco. In: *Activated Carbon Adsorption of Organics from the Aqueous Phase, Vol. 2,* M. J. McGuire and I. H. Suffet, Eds. (Ann Arbor, MI: Ann Arbor Science Publishers, Inc., 1980), p. 85.
91. Clark, R. M., and P. Dorsey. "Influence of Operating Variables on the Cost of Granular Activated Carbon Adsorption Treatment," U.S. EPA, Cincinnati, OH, May 2, 1979.
92. Dostal, K. A. et al. *J. Am. Water Works Assoc.* 57:663 (1965).
93. Hager, D. G. *Environ. Sci. Technol.* 1:287 (1967).
94. Herzing, D. R. et al. *J. Am. Water Works Assoc.* 69:223 (1977).
95. Chudyk, N. A. et al. *J. Am. Water Works Assoc.* 71:529 (1979).
96. Dobbs, R. A., and J. M. Cohen. "Carbon Adsorption Isotherms for Toxic Organics," EPA-600/8-80-023, U.S. EPA, Cincinnati, OH (1980).
97. Cairo, P. R. et al. In: *Activated Carbon Adsorption of Organics from the Aqueous Phase, Vol. 2,* M. J. McGuire and I. H. Suffet, Eds. (Ann Arbor, MI: Ann Arbor Science Publishers, Inc., 1980), p. 3.
98. Ruggiero, D. D. et al. "Removal of Organic Contaminants from Drinking Water Supply at Glen Grove, N.Y.," EPA-600/8-80-030, U.S. EPA, Cincinnati, OH (1980).
99. Benedict, A. L. "The Adsorption-Desorption of Trichloroethylene, Tetrachloroethylene, and Carbontetrachloride from Granular Activated Carbon," MS Thesis, Rutgers University, New Brunswick, NJ (1982).
100. Wood, P. R., and J. DeMarco. *J. Am. Water Works Assoc.* 71:675 (1979).
101. Borneff, J., and H. Kunte. *Arch. Hyg.* (*Berlin*) 148:585 (1964).
102. Borneff, J. In: *Activated Carbon Adsorption of Organics from the Aqueous Phase, Vol. 1,* I. H. Suffet and M. J. McGuire, Eds. (Ann Arbor, MI: Ann Arbor Science Publishers, Inc., 1980), p. 145.
103. Sorrell, R. K. et al. "A Review of Occurrences and Treatment of Polynuclear Aromatic Hydrocarbons," EPA 600/D-81-066, U.S. EPA, Cincinnati, OH (1981).
104. "National Interim Primary Drinking Water Regulations," EPA-570/9-76-003, Office of Water Supply, U.S. EPA, Washington, DC (1976).
105. Robeck, G. G. et al. *J. Am. Water Works Assoc.* 57:181 (1965).
106. Aly, O. M., and S. D. Faust. *J. Am. Water Works Assoc.* 57:221 (1965).

107. Whitehouse, J. D. "Study of the Removal of Pesticides from Water," University of Kentucky Water Resources Institute, Research Report No. 8, Lexington, KY (1967).
108. El-Dib, M. A. et al. *Water Res.* 9:795 (1975).
109. Weber, W. J. Jr., and J. P. Gould. *Organic Pesticides in the Environment,* Advances in Chemistry Series 60, R. F. Gould, Ed. (Washington, DC, American Chemical Society, 1966), p. 280.
110. Dawson, V. K. et al. *Trans. Am. Fish. Soc.* 105:119 (1976).
111. Vrochinskiy, K. K. *Gig. I. Sanit.* (*USSR*) 38:76 (1973).
112. Parkash, S. *Carbon* 12:483 (1974).
113. Greve, P. A., and S. L. Wit. *J. Water Poll. Control Fed.* 43:2338 (1971).
114. Schmidt, K. *Gas Wasser-Abwasser* 115:72 (1974).
115. Stoltenberg, D. H. *Public Works* 103:59 (1972).
116. Cohen, J. M. et al. *J. Am. Water Works Assoc.* 52:1551 (1960).
117. "Manual of Treatment Techniques for Meeting the Interim Primary Drinking Water Regulations," EPA-600/8-77-005, U.S. EPA, Cincinnati, OH (1978).
118. Wood, P. R., and J. DeMarco, In: *Activated Carbon Adsorption of Organics from the Aqueous Phase, Vol. 2,* M. J. McGuire and I. H. Suffet, Eds. (Ann Arbor, MI: Ann Arbor Science Publishers, Inc., 1980), p. 115.
119. DeMarco, J., and A. A. Stevens "Application of Organic Analysis for Evaluation of Granular Activated Carbon Performance in Drinking Water Treatment," EPA-600/D-80-037, U.S. EPA, Cincinnati, OH (1981).
120. Youssefi, M., and S. D. Faust. In: *Activated Carbon Adsorption of Organics from the Aqueous Phase, Vol. 1,* I. H. Suffet and M. J. McGuire, Eds. (Ann Arbor, MI: Ann Arbor Science Publishers, Inc., 1980), p. 133.
121. Lykins, B. W. Jr., and J. DeMarco. "An Overview of the Use of Powdered Activated Carbon for Removal of Trace Organics in Drinking Water," U.S. EPA, Cincinnati, OH (1980).
122. Zogorski, J. S. et al. "Removal of Chloroform from Drinking Water," Research Report No. 111, University of Kentucky Water Resources Research Institute, Lexington, KY (1978).
123. Brodtmann, N. V. Jr. et al. In: *Activated Carbon Adsorption of Organics from the Aqueous Phase, Vol. 2,* M. J. McGuire and I. H. Suffet, Eds. (Ann Arbor, MI: Ann Arbor Science Publishers, Inc., 1980), p. 179.
124. Magee, V. *Proc. Soc. Water Treat. Exam.* 5:17 (1956).
125. Puri, B. R. *Chemistry and Physics of Carbon, Vol. 2,* P. L. Walker, Ed. (New York: Marcel Dekker, Inc., 1970), p. 191.
126. Suidan, M. T. et al. In: *Activated Carbon Adsorption of Organics from the Aqueous Phase, Vol. 1,* I. H. Suffet and M. J. McGuire, Eds. (Ann Arbor, MI: Ann Arbor Science Publishers, Inc., 1980), p. 397.
127. Suidan, M. T. et al. *Environ. Sci. Technol.* 11:785 (1977).
128. Bauer, R. C., and V. L. Snoeyink. *J. Water Poll. Control Fed.* 45:2290 (1973).
129. Suidan, M. T. et al. *J. Environ. Eng. Div.* (*ASCE*) 103:667 (1977).
130. Kim, B. R. "Analysis of Batch and Packed Bed Reactor Models for the Carbon Chloramine Reactions," PhD Thesis, University of Illinois, Urbana, IL (1977).

131. Klotz, M. P. et al. "Translation of Reports on Special Problems of Water Technology, 9, Adsorption," EPA-600/9-76-030, U.S. EPA, Cincinnati, OH (1976), p. 312.
132. Sontheimer, H. *J. Am. Water Works Assoc.* 71:618 (1979).
133. Schalekamp, M. *J. Am. Water Works Assoc.* 71:638 (1979).
134. Cairo, P. R. et al. *J. Am. Water Works Assoc.* 71:660 (1979).
135. "An Assessment of Microbial Activity on GAC," *J. Am. Water Works Assoc.* 73:447 (1981).
136. Weber, W. J. Jr. *Environ. Sci. Technol.* 12:817 (1979).
137. Miller, G. W., et al. In: *Activated Carbon Adsorption of Organics from the Aqueous Phase, Vol. 2,* M. J. McGuire and I. H. Suffet, Eds. (Ann Arbor, MI: Ann Arbor Science Publishers, Inc., 1980), p. 323.
138. Marshall, K. C. *Interfaces in Microbial Ecology* (Cambridge, MA: Harvard University Press, 1976).
139. Warner, P. et al. "Microbiological Studies of Activated Carbon Filtration," International Conference on Oxidation Techniques in Drinking Water Treatment, Federal Republic of Germany, Engler-Bunte Inst. der Univ. Karlsruhe (September 1978).
140. Kuhn, W. et al. "Ozone-Chlorine Dioxide Oxidation Products of Organic Materials," R. G. Rice and J. A. Cotruvo, Eds. (Cleveland, OH, Ozone Press International, 1978), p. 426.
141. Guirguis, W. A. et al. In: *Activated Carbon Adsorption of Organics from the Aqueous Phase, Vol. 2,* M. J. McGuire and I. H. Suffet, Eds. (Ann Arbor, MI: Ann Arbor Science Publishers, Inc., 1980), p. 371.
142. Sontheimer, H. et al. *J. Am. Water Works Assoc.* 70:393 (1978).
143. Juntgen, H. "Translation of Reports on Special Problems on Water Technology, 9, Adsorption," EPA-600/9-76-030, U.S. EPA, Cincinnati, OH (1976), p. 269.
144. Juhola, A. J., and F. Tepper. "Regeneration of Spent Granular Activated Carbon," TWRC-7, U.S. Department of the Interior, Federal Water Pollution Control Administration, Ohio Basin Region, Cincinnati, OH (1969).
145. DeJohn, P. B., and R. A. Hutchins. *Textile Chemist Colorist* 8:34 (1976).
146. Kim, B. R. et al. *J. Water Poll. Control Fed.* 48:120 (1976).
147. Kunin, R., and I. H. Suffet. In: *Activated Carbon Adsorption of Organics from the Aqueous Phase, Vol. 2,* M. J. McGuire and I. H. Suffet, Eds. (Ann Arbor, MI: Ann Arbor Science Publishers, Inc., 1980), p. 425.
148. Kunin, R. "Two Decades of Macroreticular Resins as Adsorbents," Bulletin 161, Rohm and Haas Co., Philadelphia, PA (1977).
149. Gustafson, R. L., and J. Paleos. In: *Organic Compounds in the Aquatic Environment,* S. D. Faust and J. V. Hunter, Eds. (New York: Marcel Dekker, Inc., 1971).
150. Junk, G. A. et al. *J. Chromatog.* 99:745 (1974).
151. Junk, G. A. et al. *J. Am. Water Works Assoc.* 68:218 (1976).
152. Van Rossum, P., and R. G. Webb. *J. Chromatog.* 150:381 (1978).
153. Kennedy, C. *Envrion. Sci. Technol.* 7:138 (1975).
154. Kennedy, C. "A New Adsorption/Ion Exchange Process for Treating Dye Waste Effluents," Pollution Control Research Department, Rohm and Haas Co., Philadelphia, PA (1973).

155. Gauntlett, R. B. "A Comparison Between Ion-Exchange Resins and Activated Carbon for the Removal of Organics from Waters," Water Research Center Technology Report TR-10, Medmenham, England (1975).
156. Snoeyink, V. L. et al. "Bench Scale Evaluation of Resins on Activated Carbon for Water Purification," EPA-R-804433, U.S. EPA, Cincinnati, OH (1978).
157. Boening, P. H. et al. *J. Am. Water Works Assoc.* 72:54 (1980).
158. Weber, W. J. Jr., and B. M. Van Vliet. *J. Am. Water Works Assoc.* 73:420 (1981).
159. Weber, W. J. Jr., and B. M. Van Vliet. *J. Am. Water Works Assoc.* 73:426 (1981).
160. Neely, J. W. In: *Activated Carbon Adsorption of Organics from the Aqueous Phase, Vol. 2,* M. J. McGuire and I. H. Suffet, Eds. (Ann Arbor, MI: Ann Arbor Science Publishers, Inc., 1980), p. 417.
161. Evans, S., and F. J. Maalman. *Environ. Sci. Technol.* 13:741 (1978).
162. Anderson, C. T., and W. J. Maier. *J. Am. Water Works Assoc.* 71:278 (1979).
163. Kölle, W. "Translation of Reports on Special Problems of Water Technology, 9, Adsorption," EPA-600/9-76-030, U.S. EPA, Cincinnati, OH (1976).
164. Jayes, D. A., and I. M. Abrams. *New England Water Works Assoc.* 82:12 (1958).
165. Rook, J. J. *J. Am. Water Works Assoc.* 68:168 (1976).
166. Rook, J. J., and S. Evans. *J. Am. Water Works Assoc.*71:520 (1979).
167. Abrams, I., and S. M. Lewon. *J. Am. Water Works Assoc.* 54:537 (1962).
168. Aly, O. M. "Studies on Fate and Persistence of 2,4-D in Natural Waters," PhD Thesis, Rutgers, University, New Brunswick, NJ (1963).
169. Anderson, R. E., and R. D. Hansen. *Ind. Eng. Chem.* 47:72 (1955).
170. Pollio,. F. X., and R. Kunin. *Environ. Sci. Technol.* 1:160 (1967).

CHAPTER 5

REMOVAL OF PARTICULATE MATTER BY COAGULATION

Almost all natural surface waters contain particulate matter and/or colloidal substances that are not removed rapidly by sedimentation. Consequently, it is necessary to catalyze this sedimentation by physical and/or chemical coagulation. The term flocculation occasionally is used interchangeably with coagulation. La Mer and Healey [1] have defined these two terms in a mechanistic fashion:

> We propose that coagulation be used for the general kinetic process obeying the simple Smoluchowski equation independent of θ, whereby colloidal particles are united (L. *coagulare*—to be driven together) as typified by the effects of electrolytes upon gold sols. Coagulation is brought about primarily by a reduction of the repulsive potential of the electrical double layer in accordance with the ideas advanced by Derjaguin, Landau, Verwey, and Overbeek.
>
> We propose that the term flocculation should be restricted more in accordance with original usage corresponding to the Latin meaning of "floc" (L. *flocculus*—a small tuft of wool or a loosely fibrous structure).
>
> Flocculation is usually brought about by the action of high molecular-weight materials (potato starch and polyelectrolytes in general) acting as linear polymers of the dispersion into a random structure which is three dimensional, loose, and porous.

Defined operationally, coagulation includes the addition of the chemicals, floc formation, coagulation and sedimentation prior to sand filtration. This chapter will use both the mechanistic and operational definitions of the term coagulation.

HISTORICAL ASPECTS

Clarification of water as an aid to sedimentation of particulate matter has been practiced from ancient times. The predominant chemical agent has been, since these times, aluminum sulfate ($Al_2(SO_4)_3$) whose common name is alum. Lime ($Ca(OH)_2$) has been employed also, either alone, or with alum, and with such iron salts as ferric sulfate ($Fe_2(SO_4)_3$) or ferric chloride ($FeCl_3$). Some other interesting substances have been improvised for use as coagulants: almonds, beans, nuts, toasted biscuits and Indian meal [2].

The Egyptians practiced a form of coagulation as early as 2000 B.C. After they collected water from streams and transported it in camel-skin bags, they placed it in wide, round-belly and oblong vessels. Then they smeared the mouth of the vessel with five sweet almonds, after which a person would plunge his arm into the water up to the elbow and twist it with a vigorous motion for an appropriate period of time. The arm was withdrawn and the crushed almonds were left in the jar. After about three hours of sedimentation the water would be clear enough to store in little earthen jars where it would become clearer and cooler [3]. Reports claim that this technique was applied also to the waters from the Nile River.

Alum was also known to the early Romans of 2000 B.C. [4]. Pliny (ca. 77 A.D.) was, perhaps, the first to mention use of alum as a coagulant when he described the use of both lime (chalk of Rhodes) and alum (argilla of Italy) in making bitter water potable. The Egyptians apparently used alum as an important item of trade long before its usefulness for water treatment was known. The purification of alum eventually reached proportions large enough for its manufacture. An attempt in 1461 by Pope Pius II to develop a monopoly on the production of alum led, eventually, to a manufacturing force of approximately 8000 workmen. By 1757 muddy water in England was treated with 2–3 grains of alum per qt, followed by flocculation and filtration of the supernatant [2]. This led eventually to the coagulation of municipal water supplies at Balton, England, in 1881. Alum was added by a dosage of 1.5 grains per imperial gallon at the intake of the water treatment plant. From about 1885, coagulation was employed ahead of sand filtration.

In 1884 the first patent was granted to Isaiah Smith Hyatt for the perchloride of iron coagulation of the New Orleans Water Company's supply. A year later, the Sommerville and Raritan Water Company of New Jersey installed Hyatt's coagulation-filtration system that, in turn, established these processes as full-scale treatments.

The first scientific investigation into the use of alum for coagulation was conducted and reported by Austen and Wilbur of Rutgers University in 1885 [5]. They concluded that, "by the addition of two grains of alum to the gallon (~34 ppm), or half an ounce to 100 gallons, water can be clarified by standing, and that neither, taste nor physiological properties will be impaired to it by this treatment." Later, a series of experiments was conducted on turbid Ohio River water at Louisville, Kentucky, in the years 1895–1897 [6]. Of the several compounds tried—alum, potash alum and lime—alum was found to be the most suitable. These experiments led to widespread use of alum coagulation as an adjunct to rapid sand filtration in the United States. There was concurrent research on the use of such iron compounds as ferrous sulfate, ferric chloride, chlorinated copperas (Cl_2 added to ferrous sulfate) and ferric sulfate as coagulants. These compounds are reasonably effective in certain situations (see below), but alum remains the most widely used coagulant today.

NECESSITY

It is inferred from the above passages that chemical coagulation is required to clarify turbid waters. In turn, turbidity is composed of various kinds of biotic species, i.e., bacteria, plankton, etc., and of abiotic substances such as inorganic and organic colloids. Table I [7] gives the kinds and particle sizes of particulate matters found in water. Plain sedimentation removes gravel, coarse and fine sands, and, perhaps, some of the silt fraction. The finer particles—bacteria, plankton and

Table I. Effect of Decreasing Size of Spheres[a]

Diam. of Particle (mm)	Order of Size	Total Surface Area[b]	Time Required to Settle[c]
10	Gravel	0.487 in^2	0.3 sec
1	Coarse sand	4.87 in^2	3 sec
0.1	Fine sand	48.7 in^2	38 sec
0.01	Silt	3.38 ft^2	33 min
0.001	Bacteria	33.8 ft^2	55 hr
0.0001	Colloidal particles	3.8 yd^2	230 days
0.00001	Colloidal particles	0.7 ac	6.3 yr
0.000001	Colloidal particles	7.0 ac	63 yr (minimum)

[a]Reproduced from Powell [7], courtesy of the McGraw-Hill Book Company.
[b]Area for particles of indicated size produced from a particle 10 mm in diameter with a specific gravity of 2.65.
[c]Calculations based on sphere with a specific gravity of 2.65 to settle 1 ft.

colloids—require coagulation to larger particle sizes that can settle within one to two hours. The large surface areas of colloids play an important role in their coagulation by a variety of coagulants. The interim primary drinking water standard is one turbidity unit (TU) based on a monthly average (see Chapter 1). Even high-quality treated water meeting this standard will contain a number of particles [9]. Assuming a suspended solids content of 0.1 mg/l and spheres with a 1-μm diameter and a specific gravity of 1.01, a liter of treated water will contain about 2×10^8 particles.

Several inorganic salts of iron and aluminum are available commercially for coagulation. These are seen in Table II [9], where their availability is in dry or liquid forms (frequently alum appears with 18 H_2O). As stated above, alum finds almost exclusive application in drinking water treatment, whereas the iron compounds are used predominantly in treating domestic and industrial waste waters. When these coagulants are added to water, their hydrous oxides are precipitated. For alum, the stoichiometric coagulation reaction is:

$$Al_2(SO_4)_3 \cdot 18\,H_2O + 3\,Ca(HCO_3)_2 = 2\,Al(OH)_{3(s)} + 3\,CaSO_4 + 6\,CO_{2(g)} + 18\,H_2O \quad (1)$$

In this reaction, calcium bicarbonate represents the natural or added alkalinity needed to form $Al(OH)_{3(s)}$. It is important to note that 1 mg/l

Table II. Properties of Common Coagulants[a]

Common Name	Formula	Equiv. Weight	pH at 1%	Availability (%)
Alum	$Al_2(SO_4)_3 \cdot 14\,H_2O$	100	3.4	Lump—17 Al_2O_3 Liquid—8.5 Al_2O_3
Lime	$Ca(OH)_2$	40	12	Lump—as CaO Powder—93-95 Slurry—15-20
Ferric Chloride	$FeCl_3 \cdot 6\,H_2O$	91	3-4	Lump—20 Fe Liquid—20 Fe
Ferric Sulfate	$Fe_2(SO_4)_3 \cdot 3\,H_2O$	51.5	3-4	Granular—18.5 Fe
Copperas	$FeSO_4 \cdot 7\,H_2O$	139	3-4	Granular—20 Fe
Sodium Aluminate	$Na_2Al_2O_4$	100	11-12	Flake—46 Al_2O_3 Liquid—2.6 Al_2O_3

[a]Reproduced from Kemmer [9], courtesy of the McGraw-Hill Book Company.

of alum with 18 H_2O requires 0.45 mg/l of alkalinity as $CaCO_3$ and releases 0.9 mg/l of $CO_{2(g)}$ as $CaCO_3$ if all of the alkalinity is bicarbonate. A representative iron coagulation reaction is:

$$Fe_2(SO_4)_3 + 3\,Ca(HCO_3)_2 = 2\,Fe(OH)_{3(s)} + 3\,CaSO_4 + 6\,CO_{2(g)} \quad (2)$$

CHEMISTRY OF COAGULATION

The chemical coagulation of turbid and/or naturally colored surface waters involves the interaction of particulate and/or colloids with a destabilizing agent. The essential purpose of coagulation is to aggregate these particles into larger sizes that will settle quickly within an hour or two and/or will be filtered by sand or other media. This aggregation process is also called destabilization of colloidal systems.

Colloids are characterized by their size and by the mechanism by which they are stabilized in water. Figure 1 shows a size spectrum of waterborne particles and of filter pores [10]. Colloids have diameters usually less than 10 μm that remain suspended in water because their sedimentation by gravity is less than 10^{-2} cm/sec. It should also be noted from Figure 1 that bacteria and algae are sized in the upper colloid range, whereas viruses are at the lower range. These microorganisms should be removed, in part, by coagulation.

The time periods over which colloidal systems are stable can range from a few seconds to several years. This is noted in Table I, as is the increase in total surface area as the particle size decreases. The latter is also an important factor in the stability of colloidal systems. The interface of large surface areas represents a substantial free energy that, by agglomeration, tends to lower values. Thus, the coagulation of many small particles into fewer large particles is favored in a thermodynamic sense.

Another characteristic of colloids is their affinity for the solvent in which stabilization occurs. This is the process of "solvation." Lyophilic is the general term given to colloids "loving" the solvent. In water, this becomes hydrophilic, and such colloids are stabilized by the formation of adherent thick layers of oriented water molecules around the particle. Lyophobic is the general term given to colloids "hating" the solvent which, in water becomes hydrophobic; such colloids are stabilized by an electrostatic repulsion between particles arising from ions that are attracted to the surface from bulk solution or dissolved out of the solid's surface. These stabilizing forces are seen in Figure 2 [11].

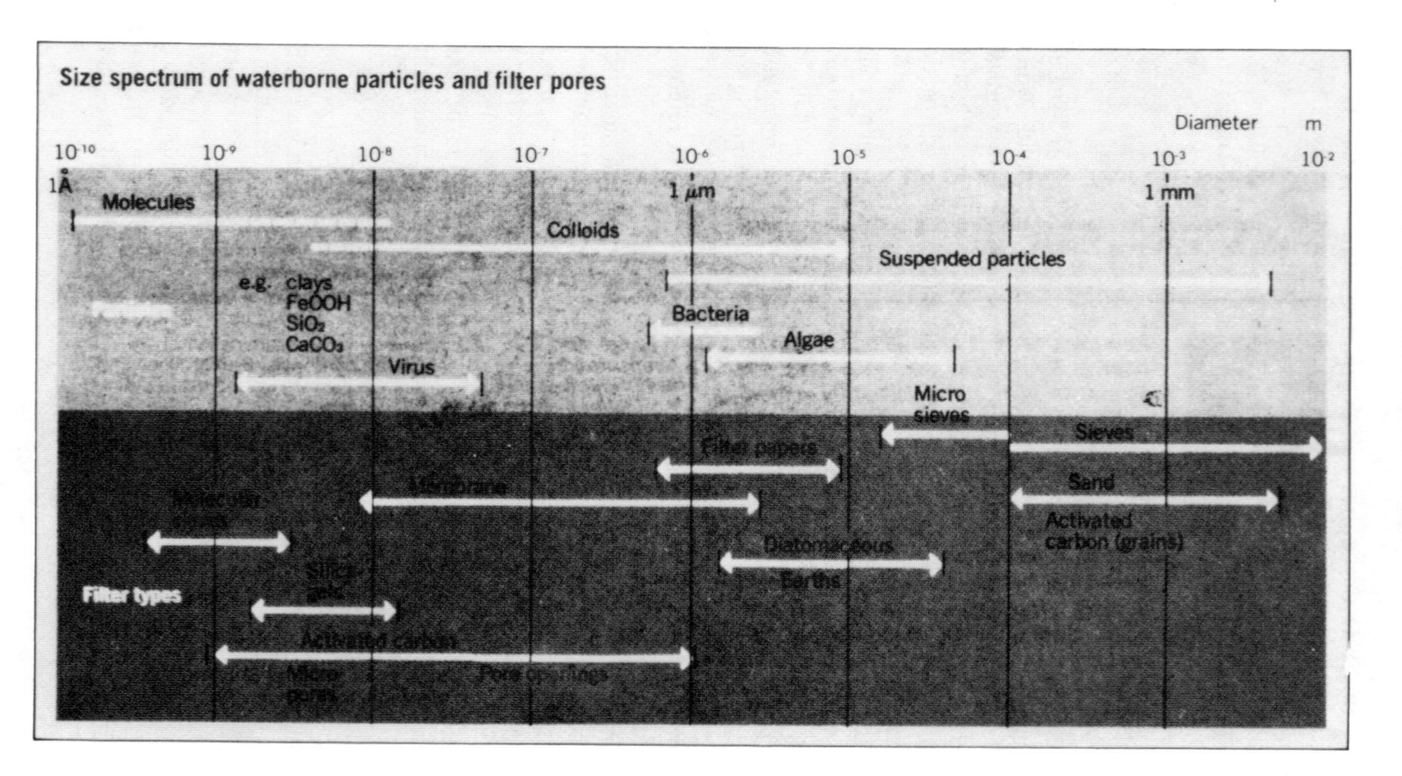

Figure 1. Suspended particles in natural and wastewaters varying in diameter from 0.005 to about 100 μm (5×10^{-9} to 10^{-4} m). Reproduced from Stumm [10], courtesy of the American Chemical Society.

Stability of Colloids

Colloids are stable in aqueous systems, by virtue of the hydration and/or electrostatic charge on their surfaces. Both phenomena depend primarily on the chemical structure and composition of the particle at the water-solid interface (Figure 2). The particle presents to the water an electronic or electrostatic capacity which, in turn, represents the attractive forces for molecules of water and/or various cations and anions. These stabilizing water molecules and ions are held by forming bonds with the particle. In accord with their energies (kcal/mol), these bonds are: ionic crystal bonds (150–200), covalent (50–100), hydrogen (1–10) and polar (<5). There may be some induced polarization (London-van der Waals forces) also in molecules and atoms that are neutral electrically under normal conditions. These forces of attraction are weak and have bond energies similar to the polar bonds. It is, however, the electrostatic stability of hydrophobic colloids that are of concern in coagulation of natural waters.

In natural waters, colloids are predominantly negatively charged due to a variety of negative functional groups on the surface of the particle (Figure 2). This negative charge was reported in 1929 from an early study of chemical coagulation [12]. Therefore, if the particle's surface has a negative charge, and the bulk of the solution is neutral electrically, a potential difference exists between this surface and a point at some distance into the bulk solution [13]. This distance can be calculated exactly from a derivation of the Boltzmann-Poisson distribution equation for the probability of finding a single electronic charge at any distance from an attracting plane (in this case, the particle's surface). In the presentation of electrostatic models of colloids in water, it is convenient to use a spherical shape. The Boltzmann-Poisson equation is:

$$\kappa^2 = \frac{1}{\delta^2} = \frac{8\pi n \epsilon^2 z^2}{DkT} \tag{3}$$

where κ = inverse of the thickness of the double layer δ (Figure 3)
n = number of ions per cm^3
z = their valence
ϵ = their electronic charge (esu)
D = dielectric constant
k = Boltzmann constant
T = absolute temperature

Equation 3 is defined for the case of equal valence of counterions (ions opposite in charge to the colloid's surface charge) and similions (ions

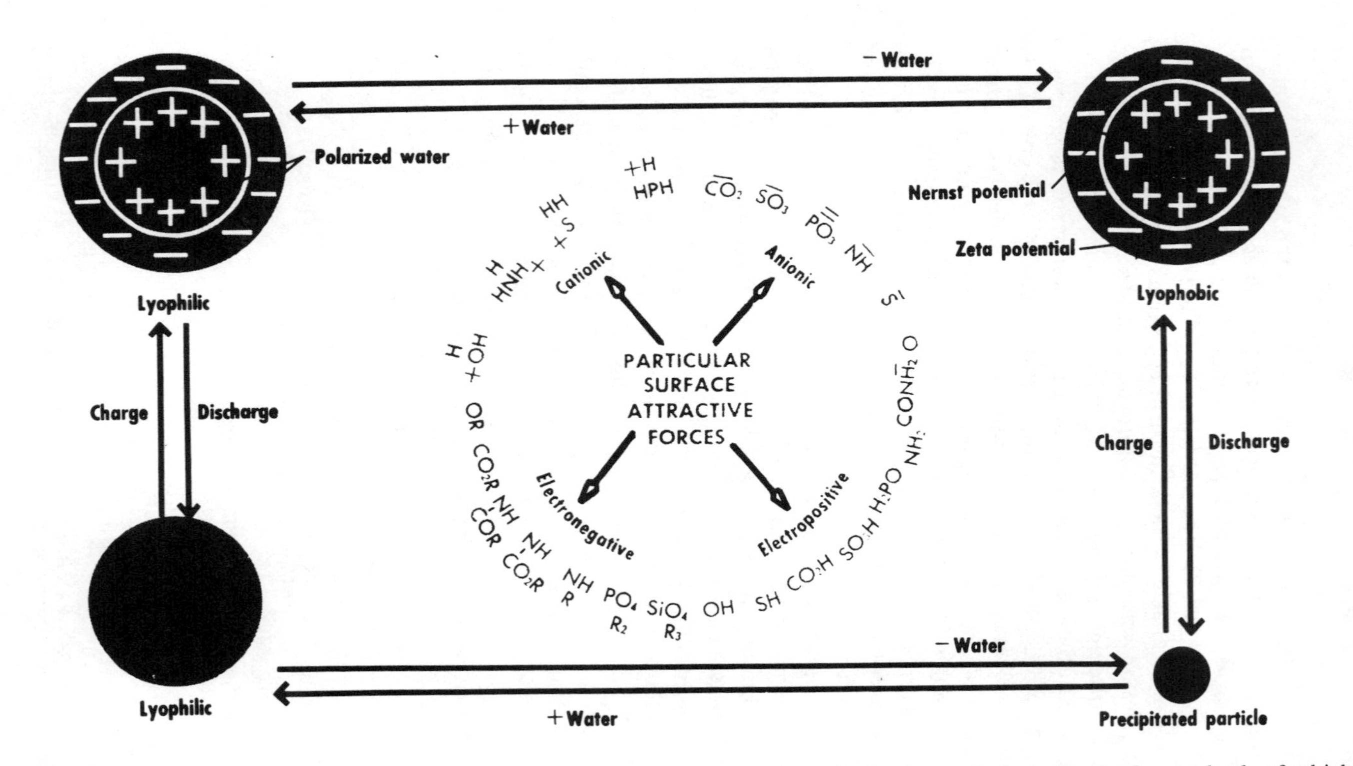

Figure 2. Stability of colloidal particles in aqueous suspensions depends on hydration and electrostatic charge, both of which depend on chemical composition and structure of substrate at the liquid-solid interface. Reproduced from Preising [11], courtesy of the American Chemical Society.

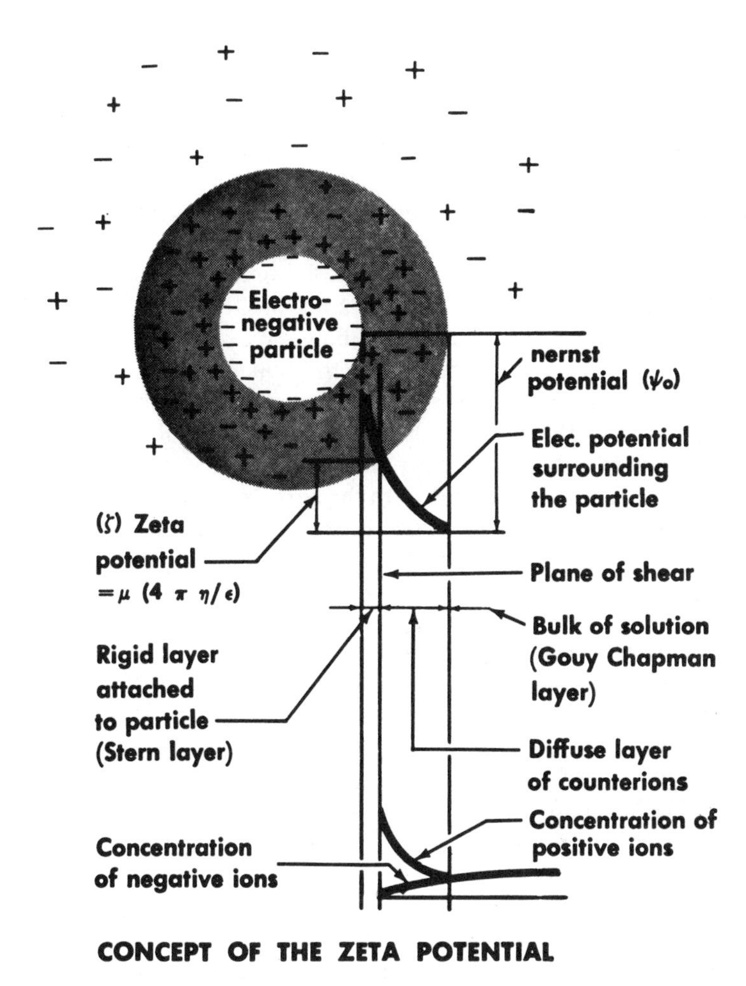

Figure 3. Zeta potential is the potential gradient across the diffuse double layer, which is the region between the Stern layer and the bulk of the solution. Reproduced from Preising [11], courtesy of the American Chemical Society.

similar in charge) [13]. For example, the thickness δ of the double layer is, in the case of monovalent ions in a 0.1 M solution, approximately 10^{-7} cm or 10 Å (10^{-8} cm).

Figure 3 illustrates the concept of the distribution and position of counterions and similions around a spherical particle. There are two layers of ions whose potential and potential decay is of interest in colloidal stability. One layer is at the surface of the particle itself. This is the so-called Stern layer where the positively charged counterions may have originated from the particle itself or by preferential adsorption from solution. This compact "double" layer that measures the total potential of that layer is given the symbol psi (ψ) which is sometimes distinguished as ψ_o, the Nernst potential. In the absence of thermal agitation (Brownian motion), the Stern layer would be formed simply by a very compact layer of counterions. However, the Brownian motion overcomes, in part, the electrostatic attractions and causes the diffuseness of the double layer out into bulk solution. Electroneutrality, therefore, is established at some finite distance into the solution.

The second layer is formed by the boundary between the Stern layer, the so-called "plane of shear," and the bulk solution. This region is called the diffuse double layer, or the Gouy-Chapman layer. The potential gradient over this region is the zeta potential Z. When a particle is put into motion by a direct current potential gradient (electrophoresis), the Stern layer apparently migrates with the particle in a rigid manner so that the dividing line between these two layers is called the "plane of shear." There is, of course, disagreement as to the precise location of this "plane of shear." For example, Mysels [13] gives a model of the distribution of counterions and similions around a particle that has these layers. It is tempting to label this model "triple layer." Mysels has a layer of bound solvent (water molecules) between the Stern and Gouy-Chapman diffuse layer. This apparently accommodates the water of hydration of the counterions. Mysels's "plane of shear" begins with these bound water molecules.

Concentration and valence of the counterions does affect, of course, the thickness and decay of the Nernst and zeta potentials. An increase of concentration in bulk solution leads to a corresponding and proportional increase in the concentration of counterions near the particle's surface. This gives a screening effect which, in turn, causes the potential to decay more quickly in terms of distance (Figure 4) [14]. Thus, an increase in the concentration of simple electrolytes gives a compaction of the double layer.

If the charge of the counterions is increased to di- and trivalent cations, there is a drastic effect on the double layer. The effect on the electric field is doubled and tripled, whereupon the counterions are brought closer to the particle's surface. Consequently, the thickness of the double layer is decreased with the concomitant quicker decay of potential. Figure 4d illustrates this point where a trivalent cation has entered the fixed double layer. This lowers the zeta potential to such a low value that coagulation might occur.

Another consideration in determining the double-layer thickness and potential decay is the charge density σ on the particle's surface. This is usually expressed as the electronic charge per 100 $Å^2$, i.e., per 1 $m\mu^2$.

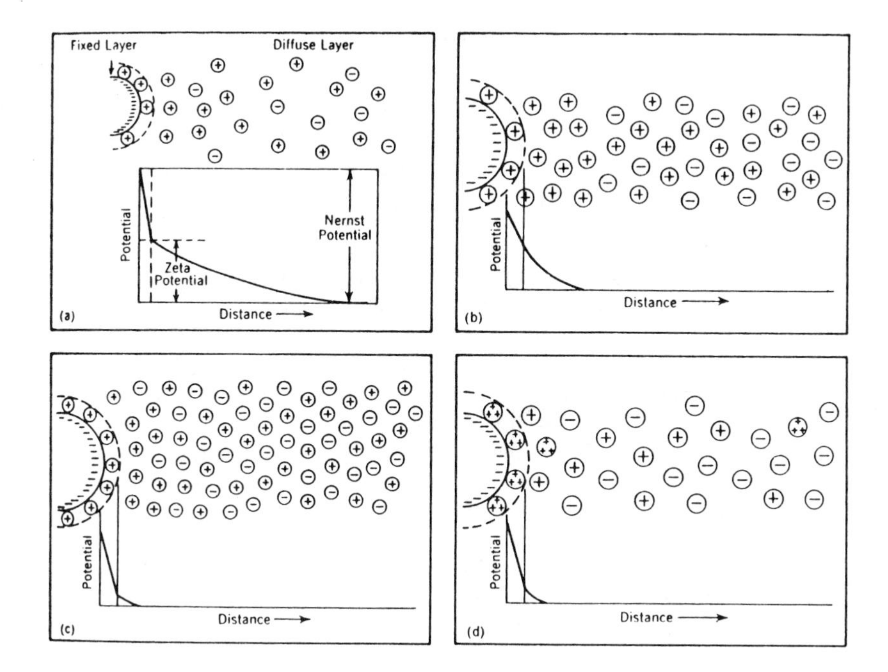

Figure 4. Source of zeta potential and effect of ions of opposite charge. Figure 4a represents a negative colloidal particle in a low concentration of monovalent ions. As the ion concentration increases from that in Figure 4b to that in Figure 4c, more positive ions enter the fixed double layer, and the zeta potential is reduced. In Figure 4d, a trivalent positive ion has entered the fixed double layer and so reduced the zeta potential that coagulation probably would result. Reproduced from Black [14], courtesy of the American Water Works Association.

According to Mysels [13], a monolayer of average-sized ions corresponds to a charge density on the order of five such units. All things being equal, the potential increases as charge density is increased. Doubling the charge density does not necessarily result in a doubling of the potential. In fact, the potential is less than doubled because the double layer becomes compressed due to the attraction of more counterions. It should be noted that the charged surface and its double layer are electrically neutral when considered together. Consequently, the particle's charge density is also the charge density of the diffuse double layer.

Electrokinetic Phenomena

Electrokinetic phenomena are based on the fact that a major part of the ions in the double layer is within the free solvent and is capable of moving along the surface. The diffuse double layer is, of course, bound to the charged surface by electric forces which prevent the ions from escaping to bulk solution. The motion of ions along the surface may be caused either by electrical forces acting on the charges or by mechanical forces acting on the liquid and the surface. There are three principal electrokinetic phenomena:

1. Streaming potential: The surface of the particle is held rigid and the liquid flows along the surface. The liquid then captures the ions of the double layer which causes a bulk motion of charges, hence a potential. A neutralizing current is carried by the ions in solution, and in the surface, as seen in Figure 5a. An alternative manner of producing relative motion is to drag the particles through a stationary solution by gravity or centrifugation, as seen in Figure 6a.
2. Electroosmosis: This is the reverse of streaming potential. When an electric field is applied along a fixed surface, it exerts a force on both parts of the double layer. Only the ions of the diffuse part move which, in turn, tends to carry the solvent along, causing a flow. This is seen in Figure 5c where net flow is prevented by application of proper hydrostatic pressure; thus, the electroosmotic pressure is obtained.
3. Electrophoresis: When there is a motion of the surface of a particle with respect to an immobile solution, an applied electric field pulls the charged particle in one direction and the counterions in the opposite direction. The result is that both move with respect to the observer. This is seen in Figure 6b, where the observed velocities measure the electrophoretic mobilities of the particles.

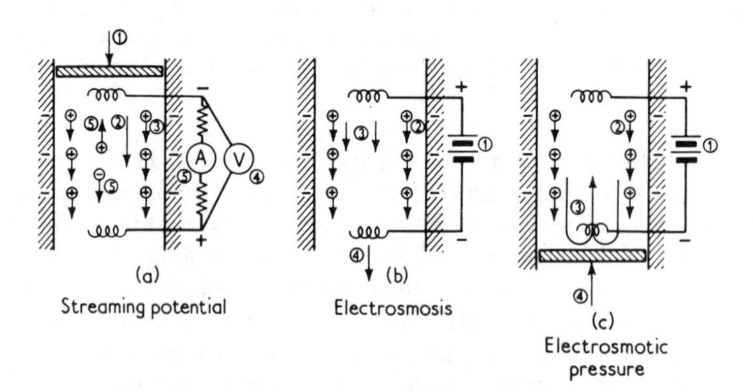

Figure 5. Electrokinetic phenomena at a fixed plane. 1 = applied stress; 2 = resulting motion of the double layer or of the solution; 3 = consequent entrainment of the solution or of the double layer; 4 = measured effect of the latter; 5 = additional path of dissipation of potential; V = voltmeter; A = ammeter. Reproduced from Mysels [13], courtesy of John Wiley & Sons, Inc.

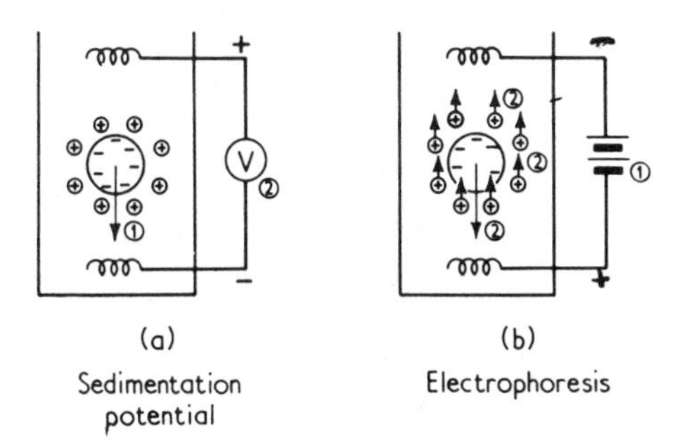

Figure 6. Electrokinetic phenomena of particles. 1 = applied stress; 2 = observed effect. Reproduced from Mysels [13], courtesy John Wiley & Sons, Inc.

These three electrokinetic phenomena all depend on the relative motion of the surface and of the diffuse double layer. The "plane of shear" delimits the flow of ions in the double layer with respect to the particle's surface. The potential at this plane is the zeta potential Z. Electrokinetic measurements yield information only about the zeta potential and nothing about the Nernst potential ψ_0, which is, in general, only slightly larger than the zeta potential [13].

Exact theories of electrokinetic effects are extremely complex and involve the theory of flow and various models of the double layer. However, the Helmboltz approximation of the structure of the double layer is cited frequently [13]. This double layer is analogous to a condenser of two equal and opposite charges separated by the thickness, δ (see Equation 3 and Figure 3):

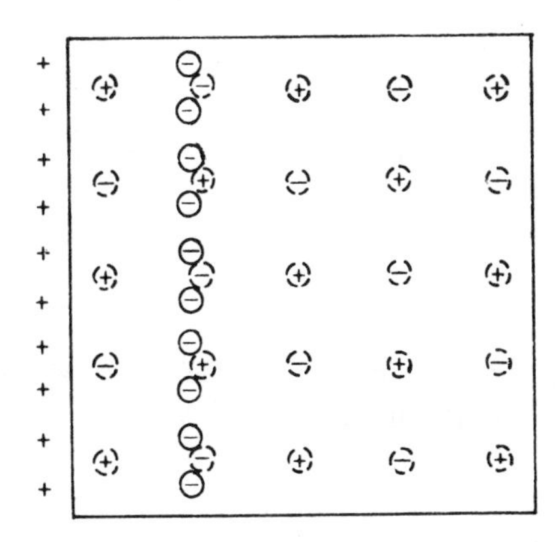

In turn, the charges are given by the product of the area and the charge density σ. In this model, the potential of this condenser is the zeta potential:

$$Z = \frac{4\pi\sigma\delta}{D}\left(\frac{\text{esu-cm}}{\text{dyn}}\right) \tag{4}$$

where the zeta potential varies inversely with the dielectric constant of the medium, which, in this case, is water.

In electrophoresis, two opposing forces act on the particle. One is a driving force, σ (E/L), where E is the applied electrical potential (V), and L is the distance (cm) over which the potential is applied. E/L is the applied potential gradient in V/cm. The resisting force is $\eta\mu/\delta$ where η is the viscosity, μ is the electrophoretic velocity in cm or μm/sec, and δ is the double-layer thickness. Combining these two forces with Equation 4 yields:

$$\mu = (1/4)\left(\frac{ZD}{\pi}\right)\left(\frac{E}{L}\right)\left(\frac{1}{\eta}\right) \tag{5}$$

Another version of Equation 5 uses the Debye-Hückel fraction of 1/6 for spherical particles [15]. In any event, μ is proportional to (E/L), the applied potential gradient. Consequently, it is more convenient to express electrophoretic results in terms of electrophoretic mobility:

$$\upsilon = \frac{\mu}{(E/L)} \tag{6}$$

which is the velocity in a potential gradient of 1 V/cm. Combining Equations 5 and 6:

$$\upsilon = \frac{ZD}{4\pi\eta} \tag{7}$$

or,

$$Z = \left(\frac{4\pi\eta}{D}\right)\upsilon \tag{8}$$

Expressing υ, the electrophoretic mobility in μ/sec/V/cm and Z in mV, Equations 7 and 8 become for water at 25°C:

$$\upsilon = 7.8 \times 10^{-2} Z \tag{9}$$

and

$$Z = 13v \tag{10}$$

according to Mysels [13].

The electrophoretic mobility of colloidal particles may be determined directly by a microscopic technique that views the motion in a small horizontal flat cell connected to two electrodes. Figure 7 [13] shows a schematic version of the Briggs cell [16]. The rate of a particle's motion is determined by timing its passage across a few micrometers on an ocular scale. The electric field is computed from the measured current intensity, the known dimensions of the cell and the (separately) measured conductivity of the solution. These measurements are brought together:

$$v = \frac{dX}{tIR_s} \tag{11}$$

where
v = electrophoretic mobility (μ/sec/V/cm)
d = distance (μm) over which the particle is timed
t = time (sec)
X = cross-sectional area of the cell (cm^2)
I = current (A)
R_s = specific resistance of the solution (ohm-cm)
X/IR_s = applied field strength

The Briggs cell was used in several extensive studies of chemical coagulation [17–19].

Destabilization

Figure 2 depicts the destabilization of colloidal systems by removal of the solvent molecules and "neutralization" of the charged particle. In

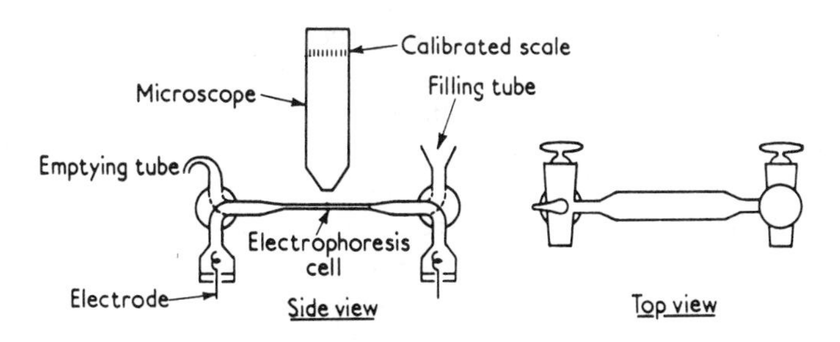

Figure 7. Principle of a microscopic method of measuring electrophoretic mobilities. Reproduced from Mysels [13], courtesy of John Wiley & Sons, Inc.

water treatment practice and research, colloidal destabilization is the aggregation of these particles for their rapid sedimentation and filtration. This is in contrast to the aggregation of colloids in research systems where destabilization is measured by an increase rather than a decrease of turbidity. This is confined usually to silver halide hydrosols, where useful information is obtained frequently that may be applied to water treatment. In any event, at least four distinct mechanisms of destabilization exist: (1) compression of the double layer, (2) adsorption for neutralization of charges, (3) entrapment in a precipitate and (4) adsorption for interparticle bridging [20].

Double-Layer Compression

When the addition of an electrolyte causes no specific interaction between it and the particle, these counterions enter into the double layer and determine its nature. Figure 8a shows that, under these conditions, the particle's charge remains constant, but the double-layer thickness will decrease, and the Nernst and zeta potentials will decrease. If a sufficient amount of an "indifferent" electrolyte is added, coagulation occurs. In addition, this effect increases greatly as the valence of the counterion is raised from one to two, and from two to three, etc. That is, a trivalent ion will require a lesser concentration than a divalent ion and a monovalent ion. This valence effect of the counterions on coagulation is called the Schulze-Hardy rule [21,22].

Figure 8b shows the situation where the Nernst potential remains constant despite the addition of an electrolyte. In this case, the so-called potential determining ions (whose ionization or adsorption is responsible for the charge of the particle) overcome the effects of concentration and valence of the added counterions. While the zeta potential decreases, there is a considerable effect on coagulation of these systems.

Figure 8c shows where counterions penetrate into the Stern layer and overcome the stabilizing effect of the potential determining ions. As the concentration of counterions increases, more are adsorbed in the Stern layer. The Nernst potential remains constant, but the double layer is compressed, and the zeta potential is decreased. The valence of the counterion is significant here in the same way it is significant in Figure 8a. This is the situation in water treatment practice.

Several coagulation curves are seen in Figure 9, where residual turbidity (%) is plotted against coagulant dose. Figure 9a represents the empirical Schulze-Hardy rule and the effects of indifferent electrolytes on colloidal stability. It does not, however, account for all of the mechanisms by which colloids are destabilized in the coagulation of

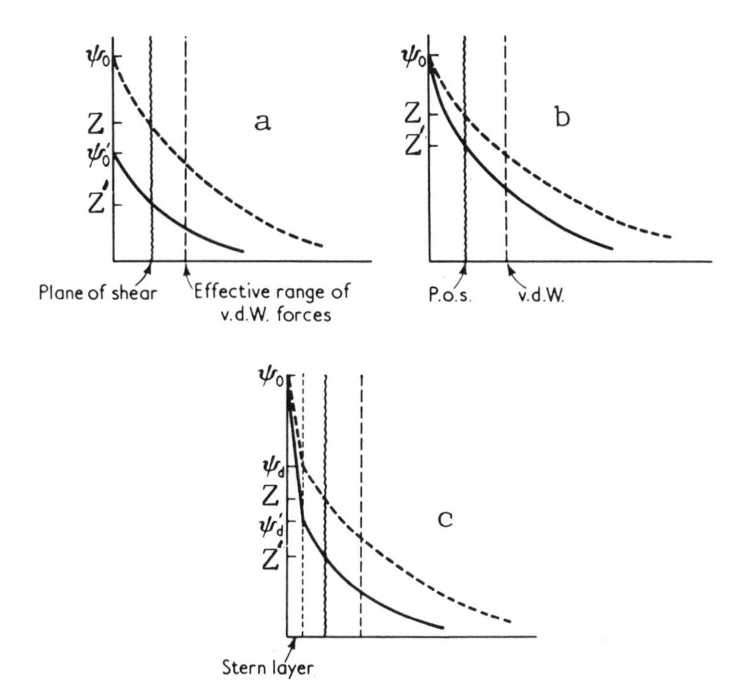

Figure 8. **(a)** In the absence of any interaction, the double layer is compressed and both ψ_o and Z are lowered. **(b)** In the diffuse double layer, if ψ_o is maintained constant, the double layer is compressed and, therefore, Z is lowered. **(c)** In the presence of a Stern layer, and with constant ψ_o, Z is lowered by both a lower ψ_d and a compression of the double layer. Reproduced from Mysels [13], courtesy of John Wiley & Sons, Inc.

water and wastewaters. There are many interactions of the coagulant with the particles in addition to the electrostatic effects.

Adsorption and Charge Neutralization

Researchers recently have compiled a considerable amount of information from various coagulant-colloidal systems indicating that interactions other than electrostatic are responsible for destabilization. For example, Figure 9b shows the reaction of dodecylammonium ions ($C_{12}H_{25}NH_3^+$) with negative silver iodide sols [23]. On the basis of electrostatic models, this monovalent cation should induce coagulation in a manner similar to Na^+ ions. Figures 9a and b show that a $[Na^+]$ of about 10^{-1} *M* is needed for coagulation, whereas approximately 6×10^{-5} *M* of the organic amine was effective. This suggests a coagulation

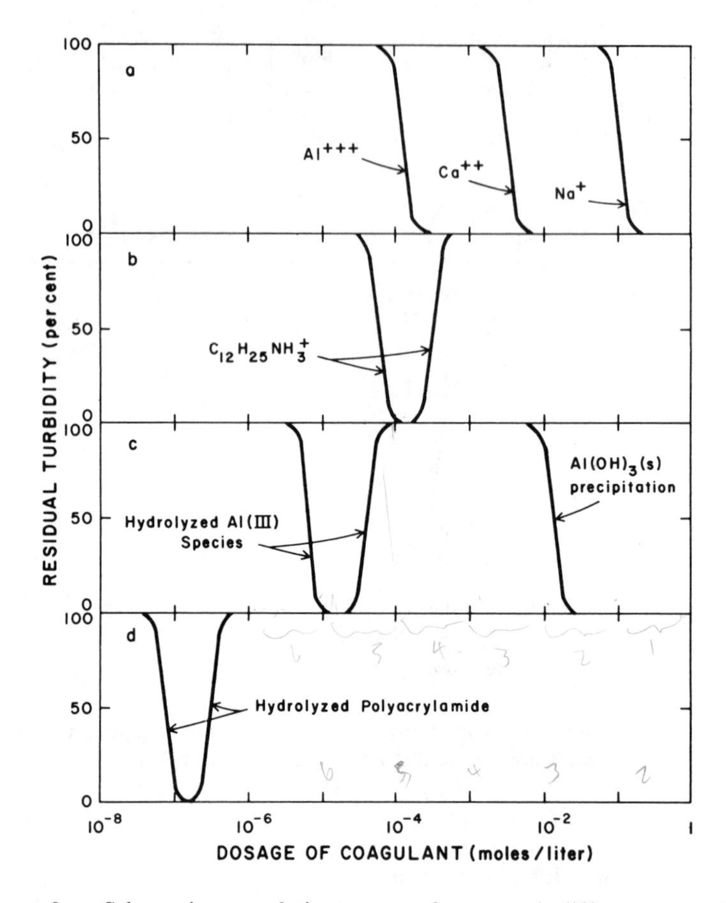

Figure 9. Schematic coagulation curves for several different coagulants. Reproduced from O'Melia [20], courtesy of John Wiley & Sons, Inc.

mechanism in addition to an electrostatic interaction. An additional effect can be seen in Figure 9b that shows a restabilizing of the AgI sols above the coagulation concentration of the organic amine. This is accompanied by a charge reversal of the sols from negative to positive that is explained by adsorption of excess counterions.

Another example of the effects of adsorption on colloidal stability is seen in Figure 9c where residual turbidity after settling is plotted against aluminum nitrate concentration at pH 5.0 [24]. This curve shows: (1) coagulation with aluminum compounds is effective at "low" dosages ($6 \times 10^{-6}\ M$); (2) the colloidal system was restabilized and the charge was reversed at a higher dose of aluminum nitrate ($4 \times 10^{-5}\ M$); and (3) there was another zone of coagulation at higher coagulant doses ($\sim 10^{-2}$ *M*) where $Al(OH)_{3(s)}$ is precipitated. This second area of coagulation is entrapment of the colloidal particles by precipitation of $Al(OH)_{3(s)}$. The colloids in this case were AgI and AgBr.

Entrapment

In this situation, the coagulant is precipitated rapidly, floc is formed, and an interaction occurs with the colloid. This is called frequently "orthokinetic" coagulation [25]. It is the entrapment of a "low" zeta potential colloid with a precipitate with, perhaps, a net charge of zero. In this case, the coagulant ($Al(OH)_{3(s)}$, $Fe(OH)_{3(s)}$, $Ca(OH)_{2(s)}$, etc.) physically destabilizes the colloidal system.

Adsorption and Interparticle Binding

There has been recently a large increase in the utilization of synthetic organic polymers in the treatment of water and wastewaters as coagulants or aids to coagulation. Optimum treatment frequently is obtained with anionic and polymeric destabilization of negatively charged particles. It is obvious that an electrostatic mechanism is not the only means of destabilization. A "bridging" theory was proposed by La Mer and Healy [26] and others [20] to account for the destabilization of colloidal systems by high-molecular-weight organic polymers. Adsorption of the polymer on specific sites of the colloid plays an important role in the "bridging" theory. This theory resembles the "binder" theory proposed for alum precipitation and coagulation [25].

In order for a polymer molecule to be an effective destabilizer, it must contain constituents that can interact with sites on the colloidal particle. This is depicted in Reaction 1, Figure 10 [20], which results in coagulation. In Reaction 2, the unattached portion of the polymer can interact with a second particle, which leads also to coagulation. Thus a "bridge" occurs between the two particles. Reaction 3 shows the possibility that bridging may not occur, resulting in little or no coagulation.

An example of anionic polymer coagulation is seen in Figure 9d where the curve describes aggregation of a negatively charged kaolinite suspension [27]. A comparison of Figures 9d and 9a shows that coagulation occurs at a much lower dosage than with Al^{3+}, and that restabilization occurs with excess doses. This type of restabilization cannot be due to charge reversal, since the colloid and polymer have the same apparent negative charge. Reaction 4 (Figure 10) may describe the restabilization

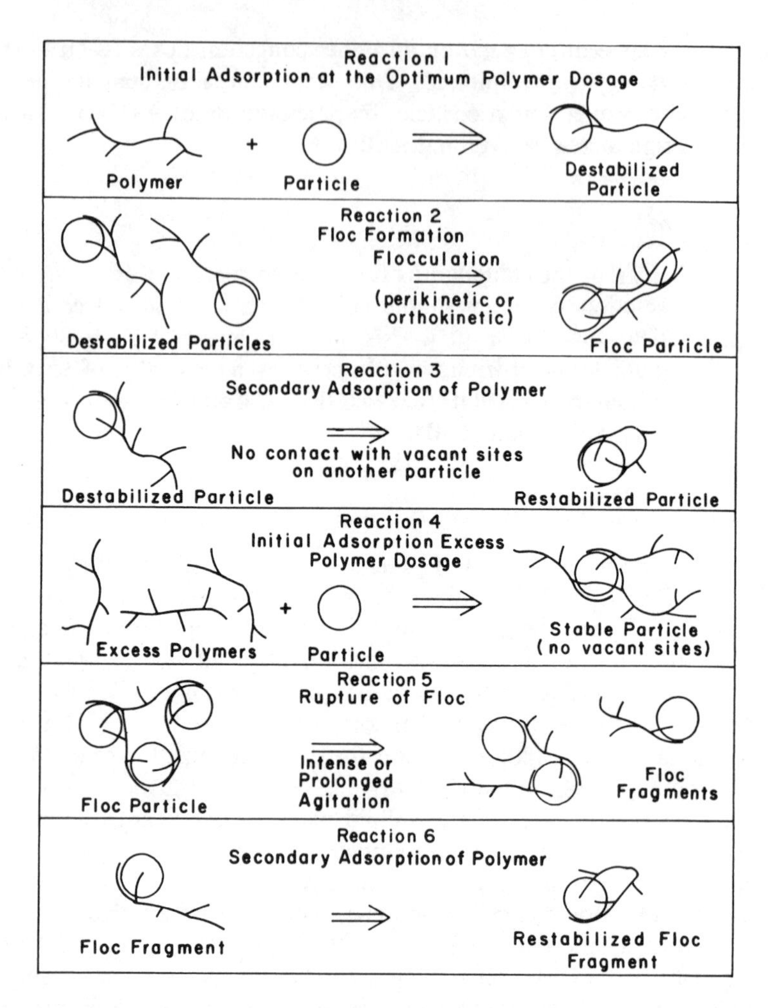

Figure 10. Schematic representation of the bridging model for the destabilization of colloids by polymers. Reproduced from O'Melia [20], courtesy of John Wiley & Sons, Inc.

by saturation of the particle's surface with excess polymer. Reactions 5 and 6 describe situations where systems that had been aggregated are restabilized by extended agitation.

Hydrolytic Chemistry of Aluminum and Iron

Aluminum and iron cations undergo hydration reactions in aqueous systems to an extent governed by the ligand properties of the cation, the metal ion concentration and the $[H^+]$ concentration. That is, "naked" metal ions do not exist in water, but occur as aquocomplexes of varying degrees. The addition of Al(III) and Fe(III) salts to water at concentrations lower than the solubility product constants of their amorphous hydroxides leads to the formation of soluble monomeric, dimeric and polymeric hydroxometal complexes. Some of these reactions and their equilibrium constants for Al and Fe (Equations 12–29) are given in Table III at 25°C [28]. It should be noted that both Al and Fe are amphoteric, that is, cationic and anionic complexes are formed. Experimental evidence confirms the existence of all complexes in Table III; however, there is doubt about the significance of such "giant" cations as $Al_{13}O_4(OH)_{24}^{7+}$ in the coagulation of waters in plant practices.

Whenever the solubility product constants are exceeded for $Al(OH)_{3(s)}$ (Reaction 21) and for $Fe(OH)_{3(s)}$ (Reaction 29), a series of hydrolytic reactions occurs (Table III). Amorphous forms of the metal hydroxides eventually precipitate. As depicted in Figure 11, $[H^+]$ affects the solubility of these precipitates. The quantities of Al and Fe salts employed in treatment plant practice of coagulation are sufficient to exceed the Ks

Table III. Hydrolytic Constants for Aluminum and Iron Equilibria [28]

Reaction No.	Reaction	Log Keg[a]
12	$Al^{3+} + H_2O = AlOH^{2+} + H^+$	−4.97
13	$Al^{3+} + 2\ H_2O = Al(OH)_2^+ + 2\ H^+$	−9.3
14	$Al^{3+} + 3\ H_2O = Al(OH)_{3(aq)} + 3\ H^+$	−15.0
15	$Al^{3+} + 4\ H_2O = Al(OH)_4^- + 4\ H^+$	−23.0
16	$2\ Al^{3+} + 2\ H_2O = Al_2(OH)_2^{4+} + 2\ H^+$	−7.7
17	$3\ Al^{3+} + 4\ H_2O = Al_3(OH)_4^{5+} + 4\ H^+$	−13.9
18	$13\ Al^{3+} + 28\ H_2O = Al_{13}O_4(OH)_{24}^{7+} + 32\ H^+$	−98.7
19	$\alpha\text{-}Al(OH)_{3(s)} + 3\ H^+ = Al^{3+} + 3\ H_2O$	8.5
20	$Al(OH)_3 + 3\ H^+ = Al^{3+} + 3\ H_2O$ amorph.	10.5
21	$Al^{3+} + 3\ OH^- = Al(OH)^{3(s)}$	33.0
22	$Fe^{3+} + H_2O = FeOH^{2+} + H^+$	−2.19
23	$Fe^{3+} + 2\ H_2O = Fe(OH)_2^+ + 2\ H^+$	−5.67
24	$Fe^{3+} + 3\ H_2O = Fe(OH)_{3(aq)} + 3\ H^+$	< -12
25	$Fe^{3+} + 4\ H_2O = Fe(OH)_4^- + 4\ H^+$	−21.6
26	$2\ Fe^{3+} + 2\ H_2O = Fe_2(OH)_2^{4+} + 2\ H^+$	−2.95
27	$\alpha\text{-}FeOOH(s) + 3\ H^+ = Fe^{3+} + 2\ H_2O$	0.5
28	$(am)FeOOH(s) + 3\ H^+ = Fe^{3+} + 2\ H_2O$	2.5
29	$Fe^{3+} + 3\ OH^- = Fe(OH)_{3(s)}$	38.0

[a]At 25°C and I = 0.0.

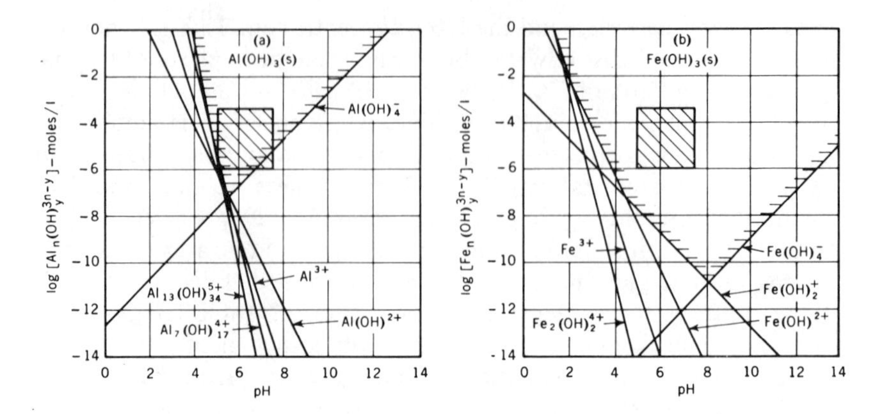

Figure 11. Solubility equilibria of amorphous $Al(OH)_{3(s)}$ and $Fe(OH)_{3(s)}$. Reproduced from O'Melia [20], courtesy of John Wiley & Sons, Inc.

values of their hydroxides. For example, a 30-mg/l dosage of alum as $Al_2(SO_4)_3$ is approximately 10^{-4} *M*, and is well within the solid region of $Al(OH)_{3(s)}$ at pH values of 6–8.

It is argued that the dosages of Al(III) and Fe(III) are always high enough to exceed the solubility product constants of their hydroxides in plant practice [20]. The significance of this is that the polymeric forms, i.e., hydrometal complexes, are adsorbed onto the colloidal particles and cause their destabilization. The quantity required for coagulation depends, of course, on the quantity of colloids in the water. This relation describes the "stoichiometry" of the coagulation reaction [29].

A series of studies was conducted for the hydrolysis products of Al(III) and Fe(III) [30–32]. Potentiometric titrations of these two metals with NaOH were performed with subsequent calculation of $\bar{n}$ values ([OH] bound/[Me]) for the various hydroxylated species. It is interesting to note that such complex ions as $Al_2(OH)_2^{4+}$, $Al_3(OH)_4^{5+}$, etc. (Table III) were not computed in these studies.

There has been a considerable amount of information on the effects of $[H^+]$ and "foreign" ions on the simple precipitation of $Al(OH)_3$ [33,34]. It was reported, for example, that the pH range for "rapid" precipitation of the hydroxide from $AlCl_3$ in the absence of electrolytes was 7.3–7.4. Addition of Na_2SO_4 and $CaSO_4$ progressively widens the zone of precipitation on the acid side of the pH range, and reduces the minimum time for appearance of the alum flocs (Table IV, Figure 12). The similar effect of $CaSO_4$ and Na_2SO_4 suggests that the SO_4^{2-} anion was responsible for the rapid precipitation of $Al(OH)_{3(s)}$. Similar experiments were repeated by Packham [35] using $Al_2(SO_4)_3 \cdot 18\ H_2O$ where a pH

Table IV. Effect of Sulfate Content on Formation of Aluminum Hydroxide

	Salt Conc.	pH range	Min. Time	pH at Min. Time
Na_2SO_4	25	7.4	9	7.4
	50	7.1–7.4	8	7.4
	100	6.1–7.4	8.0	6.75
	150	5.7–7.5	7.5	6.2
	200	5.45–7.75	6.5	6.0
	500	4.8–7.6	4.5	5.5
$CaSO_4$	25	7.5	12.5	7.5
	50	6.65–7.4	8.0	7.4
	100	5.8–7.4	7.0	6.2
	200	5.3–7.65	5.5	6.1
	300	5.2–7.4	5.5	5.9
	500	5.1–7.6	5.0	5.5

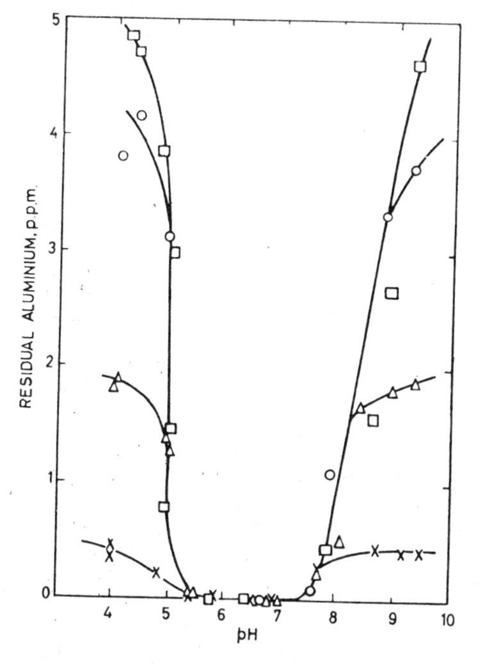

Figure 12. Hydrolysis of aluminum sulfate. Effect of pH on residual aluminum. X = 5 ppm $Al_2(SO_4)_3 \cdot 18$ $H_2O = 0.4$ ppm Al; Δ = 25 ppm $Al_2(SO_4)_3 \cdot 18$ $H_2O = 2.0$ ppm Al; O = 75 ppm $Al_2(SO_4)_3 \cdot 18$ $H_2O = 6.1$ ppm Al; □ = 200 ppm $Al_2(SO_4)_3 \cdot 18$ $H_2O = 16.2$ ppm Al. Reproduced from Packham [35], courtesy of the *Journal of Applied Chemistry*.

zone of $Al(OH)_{3(s)}$ precipitation, 5.5–7.2, independent of alum dose was defined. These types of experiments led to the universal use of aluminum sulfate for coagulation of drinking water supplies.

It is interesting to note that few data exist on the electrophoretic characteristics of $Al(OH)_{3(s)}$ and $Fe(OH)_{3(s)}$. While a multitude of data exists for systems containing $Al(OH)_{3(s)}$/clay mineral/turbidity/suspended solids/organic color, etc. (see discussion below). A qualitative report reads:

> When aluminum chloride was used, if the initial pH on addition of the coagulant was greater than 8.2, the colloidal particles were negatively charged and the pH decreased with time. If the initial pH was less than 8.2, the particles were of positive charge and the pH increased with time. When aluminum sulfate was used, the behavior was similar to that of aluminum chloride when the initial pH was greater than 8.2. When the initial pH was less than 8.2, an increase with stirring was noted, but if the final pH became as great as 7.6 to 8.2, flocculation occurred. Colloidal particles were positively charged below 7.6 and negatively charged above 8.2 [36].

There are, however, some electrokinetic data for oxides of aluminum from mineralogical research [37,38]. Figure 13 shows the zeta potential of corundum (Al_2O_3) as a function of pH and ionic strength. These data generally agree with Larson and Buswell [36], except that the positive charge was extended to a pH value of 9.45. The charged surface of Al_2O_3 in water is [38]:

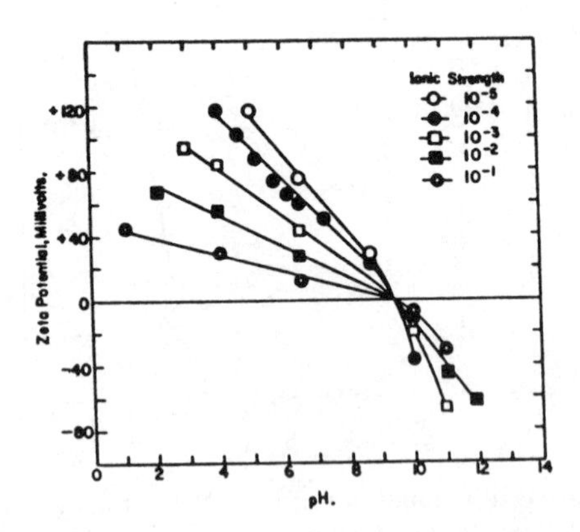

Figure 13. Zeta potential of corundum as a function of pH for different total ionic strengths. Reproduced from Modi and Fuerstenau [37], courtesy of the American Chemical Society.

$$\geq Al - OH^{+}_{2(surf)} \rightleftharpoons \geq Al - OH_{(surf)} + H^{+}_{(aq)} \rightleftharpoons \geq Al - O^{-}_{(surf)} + 2H^{+}_{(aq)} \tag{30}$$

Positive surface — Zero point of charge — Negative charge

Figure 14 shows the electrophoretic mobility values of α-alumina in aqueous systems as a function of pH and ionic strength. A summary of the zero point of charge (ZPC) for aluminum oxides is seen in Table V. A positive charge occurs at pH values below the ZPC, whereas a negative

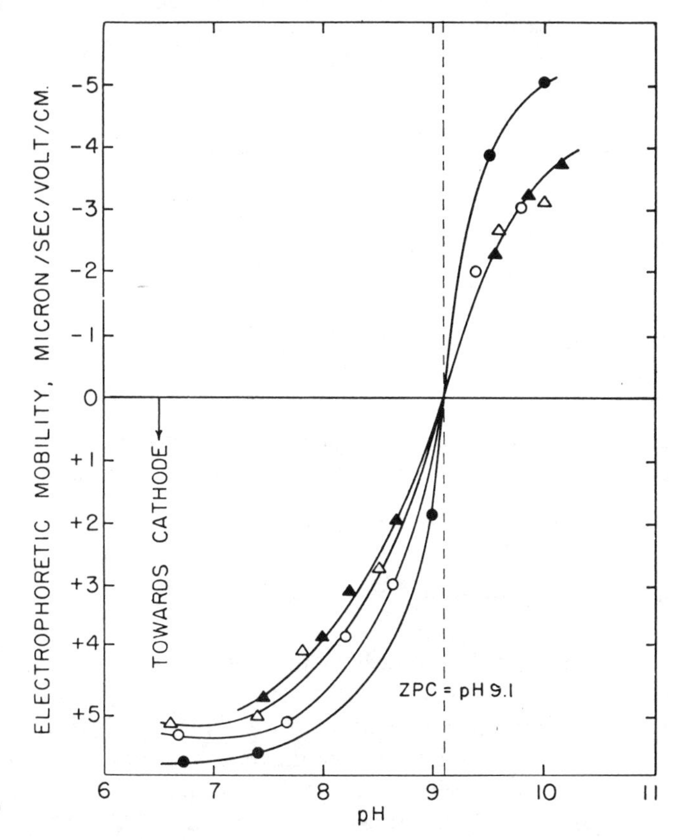

Figure 14. Electrophoretic mobility of alumina in aqueous systems as a function of pH. ● = H_2O; ○ = 10^{-3} N $KClO_4$; ▲ = 10^{-3} N KNO_3; △ = 10^{-3} N KCl. Reproduced from Yopps and Fuerstenau [38], courtesy of the *Journal of Colloid Science.*

Table V. Summary of the Zero Point of Charge of Aluminum Oxides and Hydrated Aluminum Oxides[a]

Material	zpc, pH	Experimental Method	Investigators	Remarks
α-Al_2O_3	8.4	Electrophoresis	Johansen and Buchanan	Natural corundum
	8.6–8.8	Streaming potential	Johansen and Buchanan	Nat. corundum, 0.94% SiO_2
	9.4	Streaming potential	Modi and Fuerstenau	Artificial sapphire
	9.1	Streaming potential	Robinson	Artificial sapphire
	6.7	Electroosmosis	Dobiáš, Spurný, and Freudlová	Artificial sapphire
γ-Al_2O_3	8.0	Electrophoresis	Fricke and Keefer	Artificial
γ-AlOOH	9.4	Electrophoresis	Fricke and Keefer, Fricke and Leonhardt	Artificial boehmite
$Al(OH)_3$	9.4	Electrophoresis	Fricke and Keefer	Amorphous, artificial
	9.2	Electrophoresis	Fricke and Keefer Fricke and Leonhardt	γ form, artificial
	7.7	Minimum solubility	Gayer, Thompson, and Zajicek	β form, artificial
	6.6–7.3	Electrokinetic	Tewari and Ghosh	Artificial

[a]Reproduced from Yopps and Fuerstenau [38], courtesy of the *Journal of Colloid Science.*

charge occurs above. It is not unreasonable to assume that $Al(OH)_{3(s)}$ precipitated in situ in plant practice would have the same electrokinetic behavior.

Mechanisms of Coagulation

A schematic representation of residual turbidity as a function of coagulant dosage when the pH value is held constant is seen in Figure 15a [20,29]. These four curves represent natural waters treated with Al(III) and Fe(III). Each water contains a different content of colloidal material where this is expressed as colloidal surface/unit volume of suspension, $\bar{S}$ (e.g., m^2/l). Four "zones" of coagulant dosages react with the colloidal matter. In zone 1, no destabilization occurs, since the coagulant dosages are "low." In zone 2, destabilization occurs because of higher dosages of the coagulant. Here, aggregation occurs and the residual turbidity is low

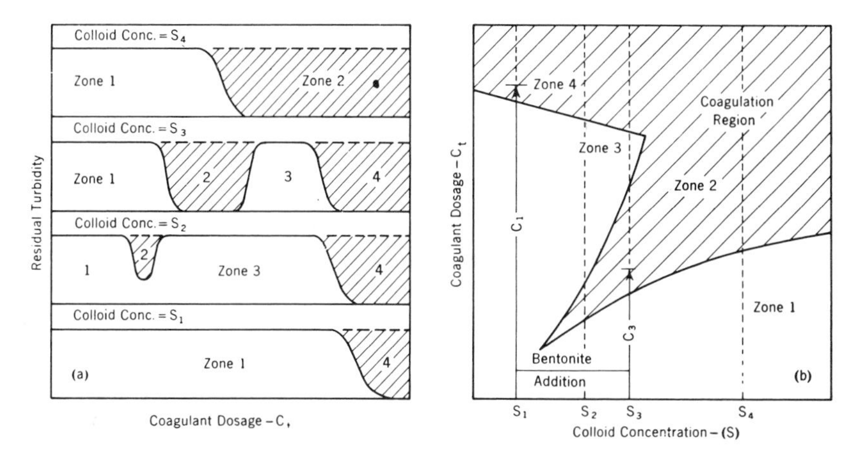

Figure 15. Schematic representation of coagulation observed in jar-tests using Al(III) or Fe(III) slats at constant pH. Reproduced from O'Melia [20], courtesy of John Wiley & Sons, Inc.

or nil. In zone 3, restabilization of the colloidal materials occurs due to the higher coagulant contents when the pH value is appropriate. In zone 4, coagulation occurs as the result of a "large" quantity of the precipitates of Al and Fe. This is the so-called zone of "sweep floc" where the colloids are swept from solution by the coagulants. In plant practice, this type of coagulation predominates.

Figure 15b shows a graphic representation of the relation between coagulant dosage and colloidal surface concentration at a constant pH value. When the colloid concentration is "low," $\bar{S}_1$, it is necessary to have an excess of coagulant in solid form. This is typical of most natural surface waters that contain "low" turbidities, less than 50 JTU. It is necessary to overwhelm the widely dispersed particulates with a "large" quantity of the coagulant in the "sweep floc" zone. When colloid concentrations are higher, $\bar{S}_2$ and $\bar{S}_3$, lower coagulant doses are needed for floc formation and destabilization. There is a greater number of colloidal particles at $\bar{S}_2$ and $\bar{S}_3$, which affords a greater contact with the floc. This occurs in zone 2 and where there is a "stoichiometric" relation between the colloid concentration and the coagulant dosage [29]. When the colloid concentration is very "high," a "large" quantity of coagulant is needed. This is represented by $\bar{S}_4$ in Figure 15b, where coagulation occurs in zone 2. An example is the destabilization of sludges accumulated at wastewater treatment plants, or the conditioning of water treatment sludges for ultimate disposal.

In general, there are two mechanisms of coagulation, both depicted in Figures 15a and b. One is neutralization of surface charges on the colloid by positively charged cations of Al and Fe or their hydrous oxides. In this case, it is not necessary to achieve electroneutrality. Rather, a reduction in the zeta potential frequently results in destabilization of the colloids. The second mechanism is physical entanglement of the colloid with floc. In this case, the zeta potential of the colloids is "low" or, at least, very close to the point of destabilization.

Clay Minerals and Inorganic Turbidity

There have been numerous laboratory studies of Al and Fe coagulation of turbid waters created by suspension of clay minerals. For the most part, microelectrophoresis techniques were used to measure the electrophoretic mobilities of the particulates and coagulant flocs [16,39]. In an early study, hydrolysis products of $AlCl_3$ were used to "neutralize" the electronegative particles of Sharkey clay [39]. It was noted that, at a ratio of 1.5 meq $AlCl_3$ to 1.5 meq NaOH, "rapid" flocculation occurred with a concurrent reduction of the positively charged particles. In 1958 one of the first electrophoretic studies of water coagulation was published [17] which, for the most part, confirmed previous conclusions [39] and, at the same time, initiated a "modern" era of water treatment research.

Representative clay minerals of the three major classes of clay minerals are employed frequently in coagulation research:

Type	Base Exchange Capacity (BEC) (meq/100 g)
Kaolinite	8.7
Fuller's Earth (illite)	26.5
Montmorillonite	115.

These clays are suspended in water with 50 mg/l NaCl or $NaHCO_3$. Electrophoretic mobilities of the three clays are seen in Figure 16a over the pH range of 3.3–10. The three clays remain negatively charged over this pH range and have about the same electrophoretic mobility value. Figure 16b shows the effect of 5 mg/l of alum on residual turbidity and the electrophoretic mobility and as a function of $[H^+]$. These data indicate that a "small" quantity of alum is sufficient to reverse the charge of kaolinite at a pH of about 5 whereas the floc particles formed with the other two clay minerals vary slightly in their negative charge. In Figure 16c, a dosage of 15 mg/l alum reverses the charge of kaolinite and Fuller's

Earth, but not montmorillonite. This suggests that the BEC is, indeed, a factor in destabilizing these clay mineral suspensions. The curves in Figures 16b and 16c also show the pH zones within which destabilization occurs. All three clays are coagulated somewhat within the pH range of 3.5–4.5 where the particles are weakly and negatively charged. As the pH value is increased, there is a charge reversal of kaolinite and Fuller's Earth over a pH range which defines a zone of no floc formation. Finally, the data indicate that all three clays are flocculated best in the pH range of 7.4–8.2 where the particles have a slight negative charge. The curves in Figure 16d show that a "sweep floc" dosage of alum, 100 mg/l, neutralizes the charge with destabilization at pH values 3.5–4.5. As the pH value is increased, the charge is reversed first to positive values with restabilization occurring within 4.5–5.5. Finally, the three clay minerals are destabilized as the pH value is increased beyond 5.5 with the charge slowly becoming negative.

These studies do not explain that charge neutralization alone effects coagulation of the three clay minerals by alum. This is evident where the best turbidity removal occurs at pH values of 7.5–8.5 in a region of negative particle mobility rather than at the point of charge reversal. Also, there is not a well defined stoichiometry between the alum dosage and the BEC of the clay minerals. Here again, a "large" alum dosage, 100 mg/l, is required to reverse the charge on montmorillonite (Figure 16d), but only a small amount, 15 mg/l, is needed for "good" coagulation (Figure 16c). Since the isoelectric point (i.e., zero net charge) of $Al(OH)_{3(s)}$ is somewhere in the pH range of 7.5–8.5 (see Table V), this suggests that coagulation is occurring with a neutral floc. In turn, this is explained by the "sweep" floc mechanism, or by an adsorption mechanism. Also, the microelectrophoresis technique is unable to distinguish uncoagulated clay particles from coagulated clay particles. Consequently, it is difficult to interpret the montmorillonite curve in Figure 16c where the charge is negative between pH 7.5 and 8.0, and where optimum turbidity removal is observed.

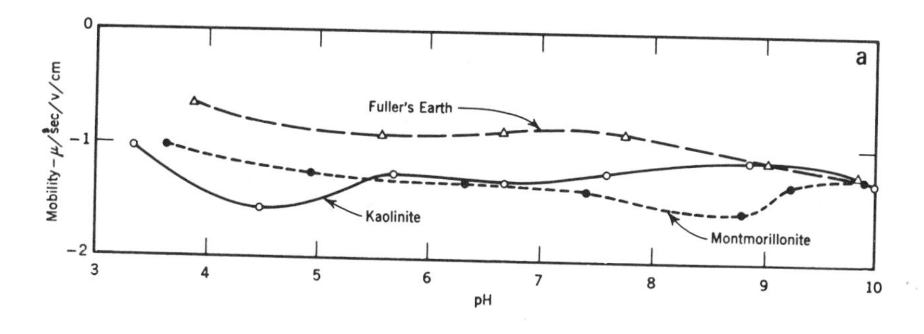

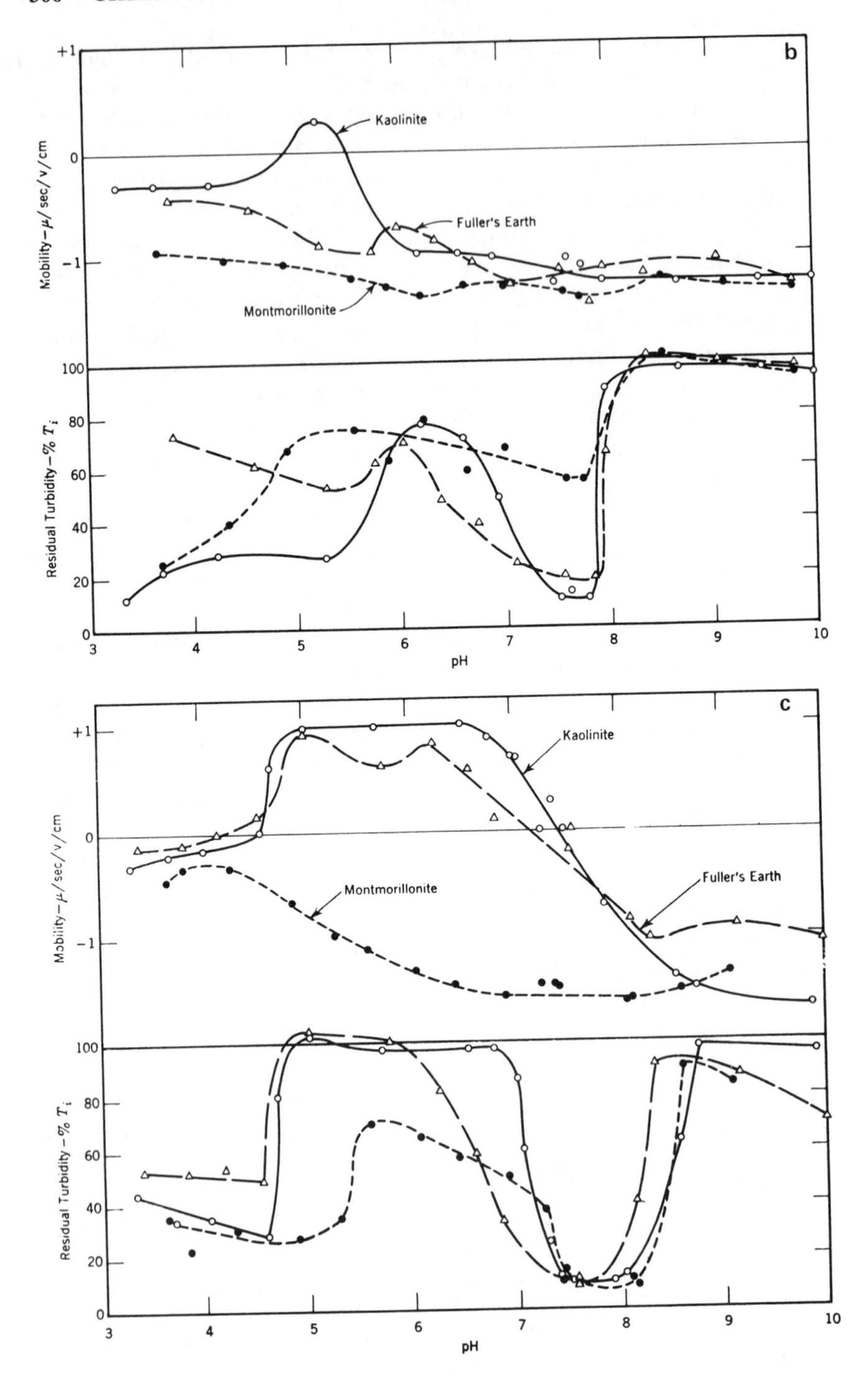
b
+1
0
-1
Mobility—μ/sec/v/cm
Kaolinite
Fuller's Earth
Montmorillonite
100
80
60
40
20
0
Residual Turbidity—% T_i
3
4
5
6
7
8
9
10
pH
c
+1
0
-1
Mobility—μ/sec/v/cm
Kaolinite
Montmorillonite
Fuller's Earth
100
80
60
40
20
0
Residual Turbidity—% T_i
3
4
5
6
7
8
9
10
pH

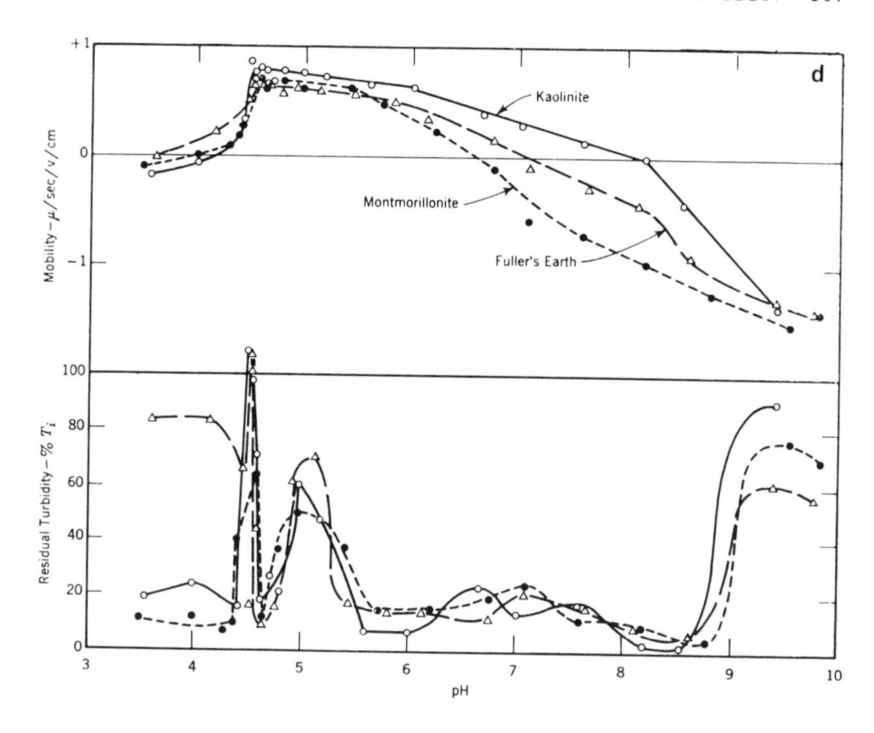

Figure 16. **(a)** Mobilities of clays without alum. **(b)** Coagulation of three clays with a dosage of 5 ppm alum. **(c)** Coagulation of three clays with a dosage of 15 ppm alum. **(d)** Coagulation of three clays with a dosage of 100 ppm alum. Reproduced from Black and Hannah [40], courtesy of the American Water Works Association.

An investigation was made of the destabilization by alum of natural clay sediments obtained from three American rivers: the Sacramento, Colorado and Apalachicola [41]. Typical results are seen in Figure 17 where optimum turbidity removal occurs in the pH range of 7–8 and where the particles are negatively charged. These results are similar to those obtained from the synthetic suspensions. It was concluded that coagulation is strongly controlled by the properties of the coagulant. Furthermore, "The characteristics of clay particles; namely, the base exchange capacity, size, and charge may influence the coagulant dosage, but not the basic mechanisms of coagulation for a particular coagulant" [42].

In another electrophoresis study, destabilization of the three clay minerals with ferric sulfate was investigated over the pH range of 3–10 [43]. In general, these clay suspensions are coagulated, and exhibit turbidity removals by ferric sulfate in a manner similar to alum. There are,

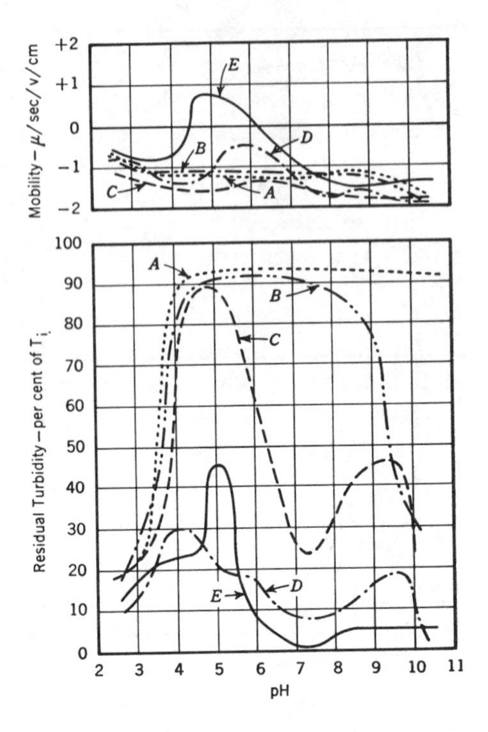

Figure 17. Effects of pH and alum dosage on mobility and turbidity of Apalachicola River sediment suspension. A = 0 ppm alum; B = 5 ppm alum; C = 10 ppm alum; D = 20 ppm alum; E = 50 ppm alum. Reproduced from Black and Chen [41], courtesy of the American Water Works Association.

however, two major differences: (1) substantially lower dosages of ferric sulfate are required for equivalent turbidity removal, ca. 5 mg/l, and (2) optimum flocculation for the three clays occurs at pH values lower (4.0–5.0) than with alum. Furthermore, the zero point of charge was reached at about pH 6.2. Maximum turbidity removal is observed at pH values greater than 6.5 where the particles are negatively charged. Here again, the latter is explained by the "sweep" floc mechanism. In this respect, hydrous ferric oxide is a better coagulant than alum, but it would leave an undesirable residual of iron in the water.

In an effort to elucidate the mechanism of alum coagulation of kaolinite, experiments were conducted so that the roles of Al^{3+}, polymeric multivalent aluminum ions and $Al(OH)_{3(s)}$ were evaluated side by side [44]. This was accomplished by conducting the coagulation at pH values of 3.0, 5.0 and 8.0, respectively (see Figure 11a). It was concluded that:

(1) Al^{3+} (pH 3.0) destabilizes the clay mineral by compression of its electrical double layer, which lowers the zeta potential (calculated values of these two colloidal properties were reported); (2) that at pH 5.0, coagulation of the kaolinite "can be attributed to a reduction of the zeta potential through the specific adsorption of the polymeric multivalent aluminum ions" [see Table III [44] (Freundlich adsorption isotherms were presented to support this conclusion)]; and (2) physical enmeshment, at pH 8.0, is offered as the explanation for destabilization of the kaolinite clay suspension by $Al(OH)_{3(s)}$ when the $[Al]_t$ was higher than 1.2×10^{-4} M. When the alum content is less than this value, and the pH is 8.0, "mutual flocculation between positively charged aluminum hydroxide colloids and negatively charged clay particles better describes the mechanism producing destabilization" [44].

Figure 18 shows that there is a relation between CEC and alum dosage

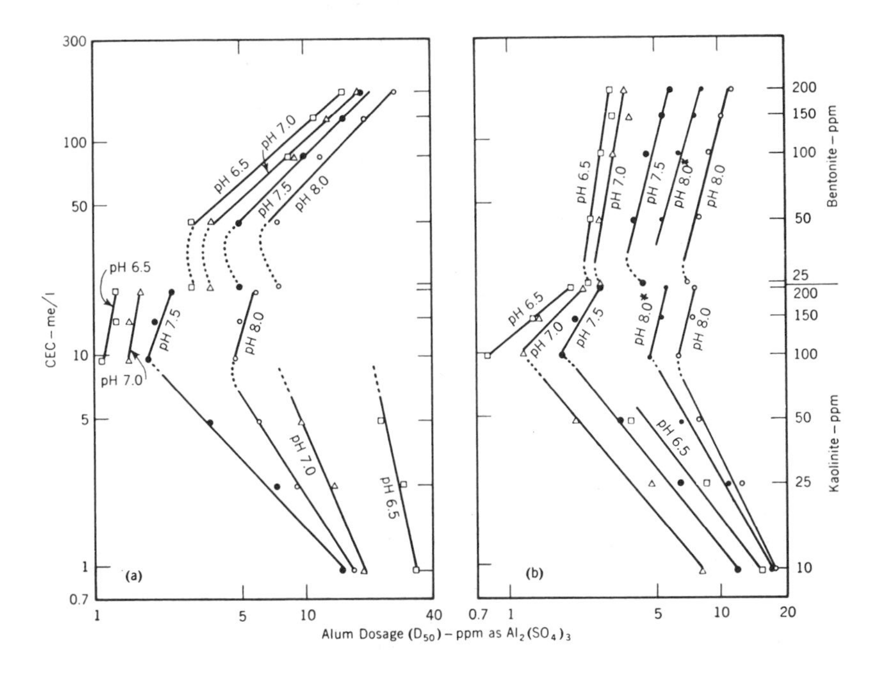

Figure 18. Relation of CEC to alum dosage. **(a)** alkalinity = 1 meq/l, Na^+ concentration = 1.0 meq/l; **(b)** alkalinity = 4 meq/l, Na^+ concentration = 4 meq/l. The two asterisked pH values have an alkalinity of 1 meq/l and a Na^+ concentration of 4 meq/l. Both charts are log-log scale. Reproduced from Kim et al. [45], courtesy of the American Water Works Association.

needed for a 50% reduction of turbidity (D_{50}) [45]. However, it is not linear and in direct proportion. At "low" quantities of kaolinite (less than 100 ppm) there is an inverse relation between the alum D_{50} dosage and CEC (less than 10 meq/l) for a given pH value. Better coagulation is obtained when the pH value is near the isoelectric point of alum (see the pH 7.5 and 8.0 lines when CEC <10 meq/l). The isoelectric point is defined as the pH value at which "the net charge of the aluminum hydroxide complex was zero and corresponded to least solubility." This suggests that the "sweep" floc mechanism (orthokinetic coagulation) predominates for coagulation of "weakly" charged clay minerals. However, when the CEC values of the clay mineral exceed 10 meq/l, there is a direct relation to the alum dosage. This suggests that the mechanism of perikinetic coagulation is operative. That is, the zeta potential of the colloid is lowered by Al^{3+} ions or positively charged alum floc. Destabilization occurs, whereupon alum floc "binds" the particles for rapid sedimentation. Support for this observation is seen in Figures 18a and b where better coagulation is obtained at pH values 6.5 and 7.0, when the CEC exceeds 10 meq/l. In general, these results [45] agree with Black et al. [40,41,44].

That there is a relation of the zeta potential of aluminum hydroxide to coagulation of kaolinite particles is seen in Table VI where electrophoretic mobility measurements are given [46]. It was concluded:

> On this basis a particular value of zeta potential appears to be an arbitrary standard of floc-forming ability valid only for one partic-

Table VI. Relationship of Z (mV) to the Flocculation of Kaolinite by Various Doses of Aluminum Sulfate[a,b]

Dose of Aluminum Sulfate ($10^{-4}M$)	pH											
	4.0	4.5	5.0	5.2	5.5	6.0	6.5	7.0	7.5	8.0	8.5	9.0
6.0	+1	+16	+19	+20	+20	+18	+13	+9	+4	0	−14	−28
1.5	−2	+5	+17	+18	+19	+18	+13	+9	+4	0	−14	−28
0.60				+16	+13	+16	+13	+9	+3	−7		−21
0.15	−5	−5	−2	−1	+3	+13	+10	+3	−9			

[a]Reproduced from Hall [46], courtesy of the *Journal of Applied Chemistry*.

[b]Values of Z given in the area bounded by thick lines were given by flocs after thorough flocculation. The values enclosed in thin lines were measured when the optical density caused by the suspended solids was reduced to a half of its initial value. Those in the lower area relate to conditions under which flocculation did not occur.

ular pH value. This would explain why some workers have advocated the attainment of zero potential as a condition for flocculation. However, the principle does not appear to be sufficiently well-founded to serve as a general rule'' [46].

A systematic investigation was made of the aggregation of montmorillonite with alum, from which an excellent perspective of the mechanisms of coagulation is provided [47]. Laboratory experiments were conducted using the famous ''jar-test'' procedure of Cohen [48], and graphical procedures were used to analyze the data. When $[Al]_t$ was held constant, the sedimentation curves were plotted as absorbance (i.e., residual turbidity) against pH value. The critical values for coagulation (pHc) and stabilization (pHs) were determined by extrapolation of the steepest portions of the curves to zero absorbance. A typical example is seen in Figure 19a. The pHc value is defined as a limit after which a slightly higher pH results in aggregation and sedimentation. The pHs value is the limit after

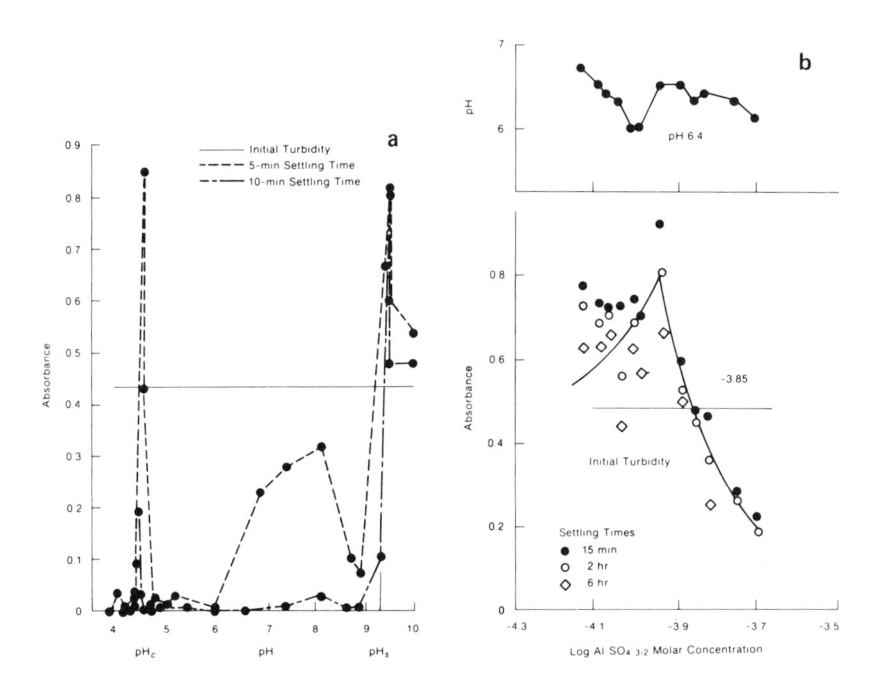

Figure 19. Coagulation of montmorillonite **(a)** by 1×10^{-3} *M* $Al_2(SO_4)_3$ and **(b)** as a function of alum concentration at constant pH. Reproduced from Rubin and Blocksidge [47], courtesy of the American Water Works Association.

which a slight increase produces a stable colloid. Coagulation occurs between pHc and pHs. When the pH value is held constant, the data are plotted as absorbance against log $[Al]_t$, for which a typical result is seen in Figure 19b. The point (solid line) where the plot intersects the initial turbidity line is selected as the critical coagulation concentration (CCC), which is defined as the minimum content of coagulant needed to aggregate a colloidal system at a specified pH value.

Figure 20 shows a "stability limit diagram" for the coagulation of montmorillonite by alum. Such a diagram defines the $[Al]_t$-pH boundaries for destabilization, restabilization and "sweep" floc formation. It is constructed from the $[Al]_t$-pHc values (dark circles), -pHs values (open circles), and -CCC values (dark squares and squares). The domain between the boundaries formed by the pHc and pHs points, and above the lowest CCC values, is the area of $Al(OH)_{3(s)}$ precipitation. This is fre-

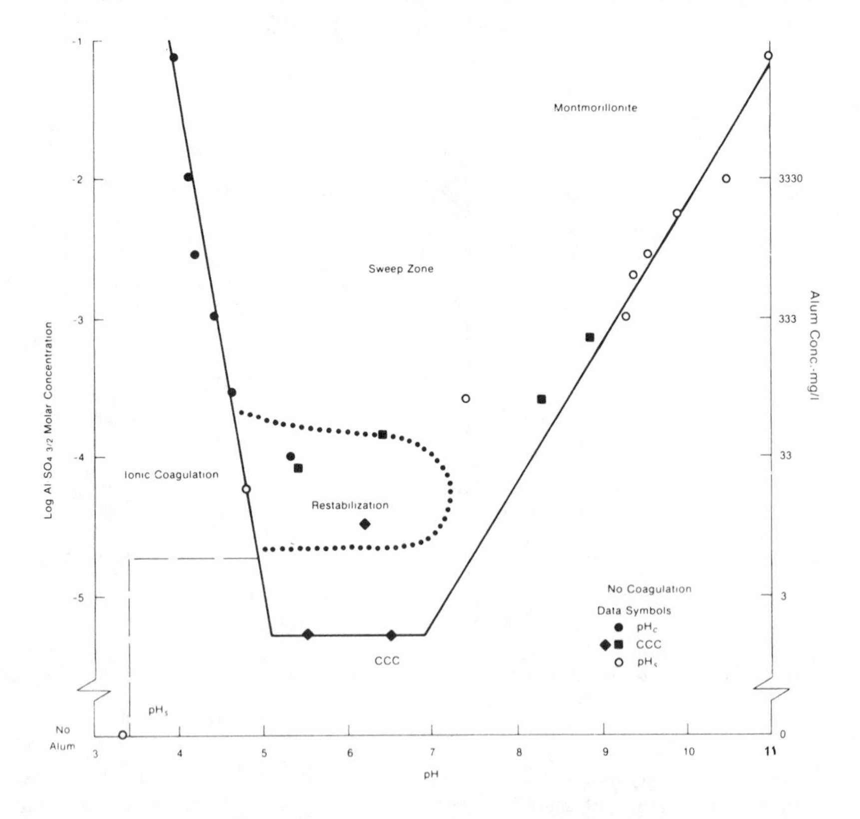

Figure 20. Aluminum sulfate concentration-pH stability limit diagram for montmorillonite. Reproduced from Rubin and Blocksidge [47], courtesy of the American Water Works Association.

quently called the "sweep" zone of coagulation. That is, colloidal particles in this region are entrapped in the alum floc, which is removed then by sedimentation [49]. The region labeled "ionic coagulation" is used to denote a pH region below the limit of $Al(OH)_{3(s)}$ precipitation. Aggregation in this region is believed to be due to Al^{3+} [47]. The area of the diagram labeled "no coagulation" is where montmorillonite remains stable toward coagulation for a "long period of time." Finally, the lowest $[Al]_t$ capable of coagulating montmorillonite is about 5×10^{-6} *M*. This CCC forms the lowest boundary of the stability limit diagram.

A chemical relationship exists between the zone of ionic coagulation and the sweep zone in Figure 20. It has been demonstrated that, in addition to Al^{3+}, the cations, $Al_8(OH)_{20}^{4+}$ and $AlOH^{2+}$ are the significant hydrolyzed species in equilibrium with $Al(OH)_{3(s)}$ at low pH values [50]. These cationic species undoubtedly are responsible for the ionic coagulation noted in Figure 20. The zone of no coagulation is due to insufficient $[Al]_t$ and/or to the presence of $Al(OH)_4^-$ at the higher pH values.

Some very pragmatic information comes from such stability limit diagrams as seen in Figure 20. An idealized version of the stability limit diagram (Figure 21) was constructed from which the problem of restabilization of colloidal systems can be addressed [51]. Some of the treatment alternatives for destabilization are: (1) in zone A, simply increase the coagulant dosage; (2) in zone B, decrease the coagulant dosage to get below the critical stabilization concentration (CSC); (3) in zone C, raise the pH value of the water by the addition of alkalinity usually in the form of $Ca(OH)_2$ or NaOH; and (4) in zone D, add such coagulant aids as polyelectrolytes (see below) to flocculate restabilized colloids.

Additional sources of information are listed here: (1) the coagulation of kaolinite suspensions with alum yielded a stability limit diagram drawn in a manner similar to Figure 20 [52]; (2) researchers reported [53] the "zeta-potential" control for coagulation of three clay minerals by alum; (3) electron photomicrographs led to the discovery that the efficiency of coagulation of a dickite clay by alum was affected more by adsorption of potential determining ions (Al^{3+}, OH^-) and by bridging of particles by $Al(OH)_{3(s)}$ than by a decrease of the zeta potential [54]; and (4) investigators released [55] a comprehensive study of the reactions of various aluminum species with montmorillonite, kaolinite, volcanic ash and feldspathic sand.

Naturally Occurring Organic Color

Considerable information is available about the chemical nature of organic color in natural waters [56]. Yet, the precise structure of color

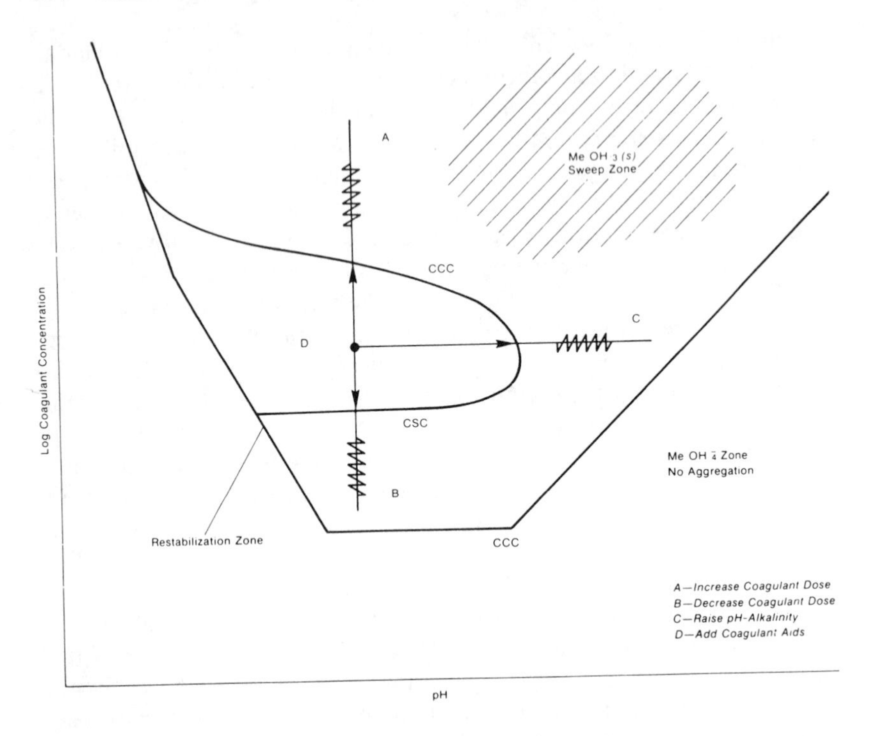

Figure 21. Overcoming the problem of restabilization. Reproduced from Rubin and Kovac [51], courtesy of Ann Arbor Science Publishers.

molecules remains unsolved. It is known, however, that organic color in water has the physical properties of a negatively charged colloid [57]. Also, the colloidal particles of the organic color have very small sizes, 3.5–10. mμ [58].

Two highly colored surface waters were coagulated successfully with both Al(III) and Fe(III) sulfates [59]. Figure 22a shows the electromobility and residual color values when a highly colored water was coagulated with 60 mg/l of alum. Figure 22b presents similar data for coagulation of the same water with 50 mg/l of ferric sulfate. These two figures, in an almost classic manner, show that charge reversal of the floc particles and optimum color removal occurs at or very near the same pH value.

A stoichiometric relation between a measurable raw water property and the coagulant dosage required for treatment of that water was developed [60]. Electrophoretic measurements were utilized to establish the optimum conditions for removing color from six different soft and

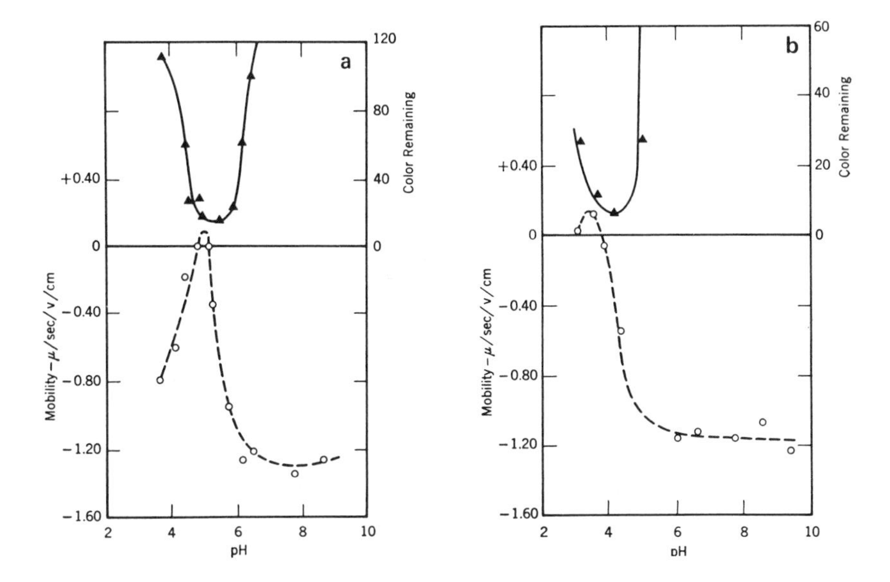

Figure 22. **(a)** Effect of pH and constant alum dosage on coagulation of water B. A constant alum dosage of 60 mg/l was used, and the pH was varied with HCl and $Ca(OH)_2$. Although the mobility curve is shown with a small positive loop, actually no definitely positive particles were found in any jar. **(b)** Effect of pH and constant ferric sulfate dosage on coagulation of water B. A constant $Fe_2(SO_4)_3$ dosage of 50 mg/l was used, and the pH was adjusted with HCl and $Ca(OH)_2$. Reproduced from Black and Williams [59], courtesy of the American Water Works Association.

highly colored natural waters. That ferric sulfate dosage is a function of raw water color is seen in Figure 23, and of pH value is seen in Figure 24. Also, the pH value for optimum color removal, 3.78, is related closely to the pH value at which a zero electromobility value is recorded (Figure 25). The best color removal is recorded at a pH value just below the isoelectric point where mobility of the floc particles is positive, and just above the pH of the maximum positive charge. In Figure 26, the optimum pH value of coagulation for alum is well above that for ferric sulfate.

A comprehensive study has described the coagulation of organic color with the "hydrolyzing coagulants," alum and ferric chloride, in the presence of dilute clay suspensions [61]. The characteristics of coagulation given in Table VII show that there is some agreement with the stoichiometry presented by Black et al. [60]. It is argued that organic

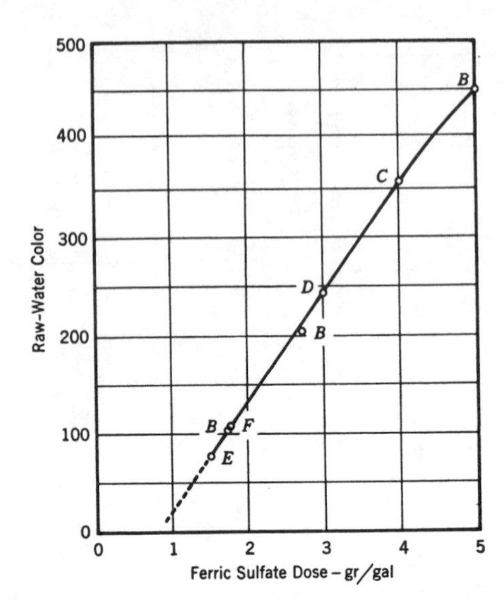

Figure 23. Minimum ferric sulfate dose as a function of raw water color. The data for the various waters are indicated by the point associated with the letter designation of the water. The data indicated that the minimum ferric sulfate dose necessary for good color removal is a simple function of the raw water color. Reproduced from Black et al. [60], courtesy of the American Water Works Association.

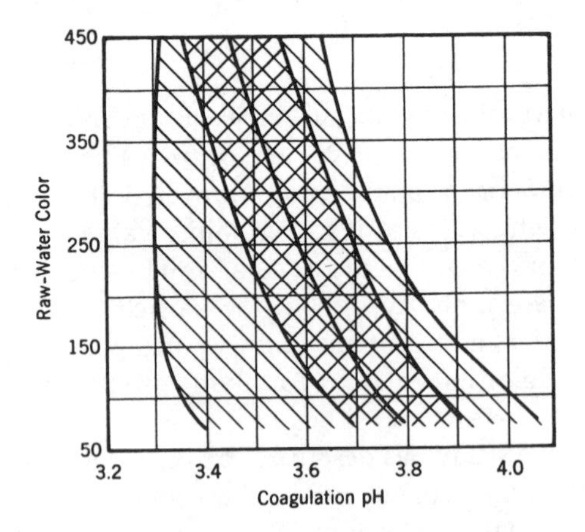

Figure 24. Range of optimum coagulation pH as a function of raw water color. The cross-hatched area is for best color removal, the diagonally lined area is for acceptable color removal. Reproduced from Black et al. [60], courtesy of the American Water Works Association.

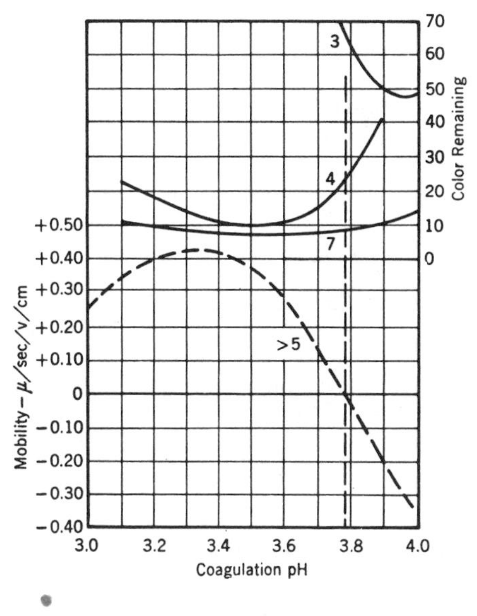

Figure 25. Relationship between zero mobility (---) and best color (—) removal for water C [60]. The number associated with each curve indicates the ferric sulfate dose at which the results were obtained. The best color removal at doses equal to or greater than the minimum occurs at a pH lower than pH_0 (vertical dashed line). Reproduced from Black et al. [60], courtesy of the American Water Works Association.

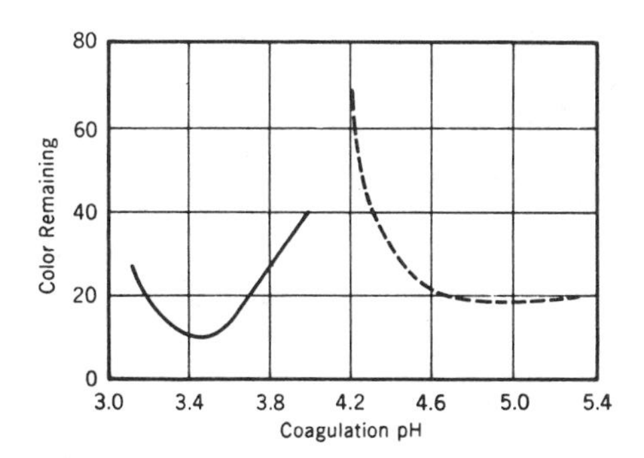

Figure 26.. Coagulation of water B with 5 gr/gal ferric sulfate (—) and 10 gr/gal alum (---). The results show two of the principal differences between ferric sulfate and alum as coagulants for organic color. The ferric sulfate is more effective and coagulates at a lower pH. Reproduced from Black et al. [60], courtesy of the American Water Works Association.

Table VII. Characteristics of Coagulation of Dilute Clay Suspensions and Humic Substances with Aluminum Sulfate[a]

Dilute Clay Suspensions	Humic Substances
Optimum pH 6.5–7.5	Optimum pH 5–6
Minimum residual turbidity independent of pH	Minimum residual color dependent on pH
Increased concentration slightly reduces coagulant dose	Increased concentration increases coagulant dose
Dose and optimum pH changed in presence of humic substances	Dose and optimum pH independent of presence of clay

[a]Reproduced from Hall and Packham [61], courtesy of the American Water Works Association.

color in the form of humic and fulvic acids is removed through a precipitation reaction with alum [61]. A precipitate apparently is formed "by the interaction of a partially hydrolyzed aluminum ion of empiric formula $Al(OH)_{2.5}$ and an ionic group, probably a carboxyl, on the organic molecule." This stoichiometric relation between coagulant dosage and color is explained by the formation of a soluble chelate with aluminum through secondary acidic groups on the organic molecule (probably, the phenolic-OH). These observations are slightly different than the traditional explanation of destabilization of a colloidal system by an oppositely charged particle.

Figure 27 shows the stability domains for a 5-mg/l humic acid-alum system [62]. These data reaffirm the roles that hydrolysis of aluminum ions and pH play in determining the stability of humic acid sols. Destabilization is fastest in region III (Figure 27) where $Al(OH)_{3(s)}$ forms. Between the pH values of 4 and 6 "adsorption of highly charged aluminum species can produce charge reversal and sol stability." The bivalent SO_4^{2-} coagulates positive sols at concentrations as low as 5.4×10^{-4} *M* [63]. This is believed to be the reason for existence of the upper boundary of region IV in Figure 27. Evidence was presented whereby region IV disappears upon addition of Ca^{2+} and SO_4^{2-} ions [62].

The stoichiometric relationship between humic and fulvic acids and alum and cationic polyelectrolytes was researched [64]. Table VIII shows the optimal doses of cationic polyelectrolyte required for "complete reaction" with humate and fulvate solutions as a function of their anionic groups. Reasonably good agreement is obtained between the cationic polyelectrolyte in μeq/l and the sum of carboxyl and hydroxy

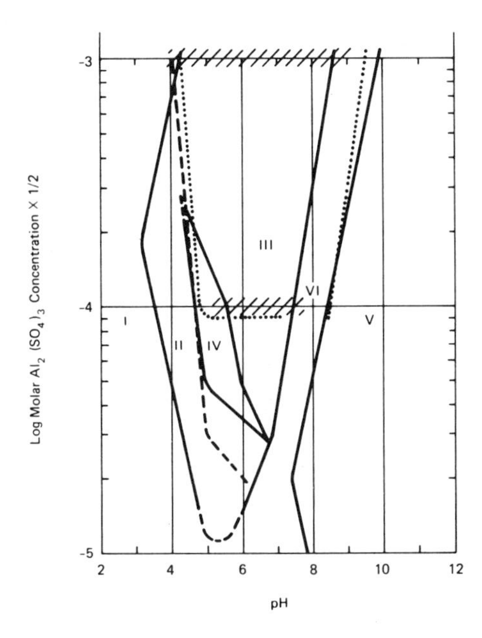

Figure 27. Aluminum sulfate-pH stability domain for the humic acid sol (5.0 mg/l) 24 hr after mixing the reacting components. I, V, VI = stable sols; II = sols coagulated by soluble aluminum species; III = sols destabilized, aluminum hydroxide present; IV = sols coagulated but stable. Dotted lines outline the region within which aluminum hydroxide precipitates in the absence of humic acid. Dashed line gives the pH of humic acid-$Al_2(SO_4)_3$ systems before pH adjustment. Systems with pH values greater than the dashed line had their pH adjusted with NaOH. Systems at pH values lower than the dashed line had H_2SO_4 added. Hatching gives pH range where 50% or greater humic acid removal occurred after 5 min of microflotation at 10^{-3} and 10^{-4} *M* aluminum [$Al_2(SO_4)_3$]. Reproduced from Mangravite et al. [62], courtesy of the American Water Works Association.

groups in μe/l for optimal coagulation at a pH value of 8.0, where both functional groups are in the anionic form. This is strong evidence for a charge-neutralization reaction between the polycation and the phenolate and carboxylate groups of humic and fulvic acids. A similar relationship is found for the optimal dose of alum for its coagulation of these two naturally occurring organic acids. Agreement of 45 μeq/l alum and 44.5 μeq/l of the carboxyl groups from 10 mg/l humic acid is good at pH 6.6. At this pH value, only the carboxyl groups are dissociated. At higher concentrations of humic acid, the stoichiometry between alum dosage and the two organic acids deviates somewhat. Figure 28 shows the effect

Table VIII. Cationic Polyelectrolyte and Alum Required for Reaction with Humic and Fulvic Acids Solutions as a Function of the Anionic Groups[a]

Acids	Concentration (mg/l)	Σ—COO— + —O— (μeq/l)	—COO— (μeq/l)	—O— (μeq/l)	Cationic Polyelectrolyte[b] Optimal Dose at pH 8.0 (μeq/l)	Alum Optimal Dose μeq/l	pH
Humic	5	43.5	22.2	21.3	43.1	27	6.7
	10	87.0	44.5	42.5	79.5	45	6.6
	25	218	111	107	190		
	50	435	222	213	342	97.7	6.0
Fulvic	25	118	41.0	77	101		
	50	236	82	154	228	90	5.8
	100	472	164	308	380		

[a]Reproduced from Narkis and Rebhun [64], courtesy of the American Water Works Association.
[b]Purifloc C-31. The Dow Chemical Co., Midland, Michigan.

of the cationic polyelectrolyte on removal of humic acid. The electrophoretic mobility becomes zero at the point where the humic acid content is minimal. This is additional evidence that negatively charged color molecules are destabilized by positively charged coagulants.

Recent concern about the role of humic and fulvic acids as precursors in the formation of trihalomethanes (see also Chapter 2) has prompted some reinvestigations into their removal from water by coagulation. For example, it was found that coagulation of humic and fulvic acids by alum: (1) reduces color and total organic compounds (TOC), which, in turn, reduces the formation of $CHCl_3$ on subsequent chlorination, and (2) selectively removes those portions of humic and fulvic acids most responsible for formation of $CHCl_3$ [65]. In another study, organic color is removed by alum in conjunction with high-molecular-weight polyelectrolytes [66]. When three cationic polyelectrolytes are used alone: Cat-Floc B, Nalcolyte 607 and Nalcolyte 8101, very little humic acid is removed. When these three coagulant aids are used with "high" doses of alum (50–150 mg/l), color removal is fair to good. Interestingly enough, when 5–50 mg/l alum is used with a nonionic polymer (Nalcolyte 8171) and an anionic polyelectrolyte (Betz 1120), color and turbidity removals are excellent (95–100%).

Two additional studies of note concern the removal of organic color

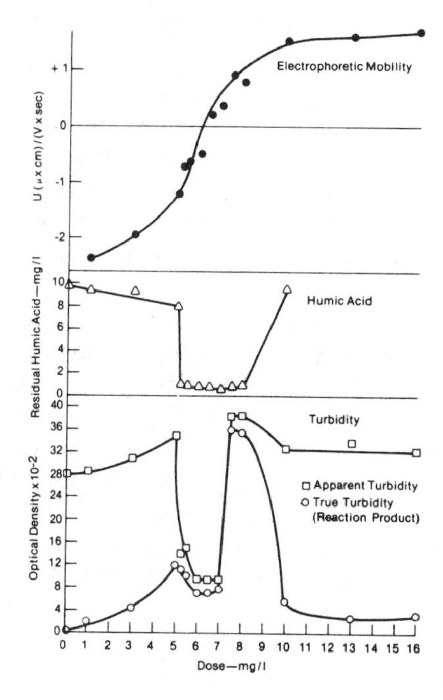

Figure 28. Effect of cationic polyelectrolyte on Na-humate solution; 10 mg/l humic acid; pH 8.0. Reproduced from Narkis and Rehburn [64], courtesy of the American Water Works Association.

with ferric sulfate [67] and the effects of organic solutes on chemical reactions of aluminum [68]. Additional information is given below concerning removal of organic color and THM precursors.

Role of Adsorption

The theory of electrical double-layer interactions is, perhaps, the simplest model that describes the reactions among colloidal particles in aqueous systems. The Gouy-Chapman double-layer model assumes that an equilibrium exists between the electrostatic forces attracting counterions from bulk solution to the charged surface, and diffusion that permits these ions to migrate back into bulk solution where the ionic concentration is lower. In this model, destabilization occurs when an indifferent electrolyte is added to a colloidal system. The content for destabilization depends on the valence of the indifferent electrolyte. Conformance to the Schulze-Hardy rule occurs. Also, restabilization of the colloids does not occur when the electrolyte concentration exceeds the CCC. There is considerable evidence that coagulation and restabilization of colloids occur also by noncoulombic forces. That is, coagulation by hydrolyzed metal ions (here, aluminum) and by "large" molecular species (polyelectrolytes, etc.) is greatly different than coagulation by simple ions (Na^+, Ca^{2+}, Al^{3+}, etc.). The hydrolyzed metal ions and the polyelectrolytes apparently are adsorbed on colloidal surfaces, which effects destabilization at contents considerably less than nonadsorbable coagulants.

Adsorption data have been compiled from a number of coagulation systems, some of which do not strictly apply to drinking water treatment [29]. These data are seen in Table IX and Figure 29. In short, coagulants that are adsorbed at "low" concentrations destabilize colloids, whereupon they are restabilized at higher concentrations. This occurs when the coagulant and colloid are oppositely charged. Restabilization is accompanied by charge reversal [for example, see $(PDADMA)^{n+}$ in Table IX]. According to the Gouy-Chapman double-layer model, charge reversal cannot occur because purely coulombic forces will not allow the attraction of counterions in excess of the original surface charge. Consequently, there must be additional chemical interactions to account for restabilization and charge reversal. This is attributed to adsorption.

The overall standard free energy of adsorption ($\Delta\bar{G}^0$) is the sum of the total chemical adsorption energy (ΔG^0), plus the electrochemical work involved [69]:

$$\Delta\bar{G}^0 = \Delta G^0 + zF\Delta\psi \tag{31}$$

where z = charge of the sorbate ion
$\Delta\psi$ = potential drop between the colloidal surface and the bulk solution
F = Faradays (23.06 kcal/eV)

When $\Delta\psi$ is negative, adsorption occurs spontaneously.

An indication of the relative contribution to $\Delta\bar{G}^0$ in Equation 31 by ΔG^0 and $zF\Delta\psi$ was provided by Stumm and O'Melia [29].

> For the adsorption of a monovalent organic ion to a surface of similar charge and against a potential drop ($\Delta\psi$) of 100 mV, the electrostatic term in Equation 31 is 2.3 kcal per mole. The standard chemical adsorption energy for typical sorbable monovalent organic ions is of the order of magnitude of -2 to -8 kcal per mole, thus indicating that the electrostatic contribution to adsorption easily can be smaller than the chemical contribution.

Table IX. Data For Figure 29[a]

Coagulant	Colloid	Charge Reversal	Restabilization	Remarks
Na^+	$(AgBr)^-$, 10^{-4} M	No	No	
Ca^{2+}	$(AgBr)^-$, 10^{-4} M	No	No	
Al^{3+}	$(AgBr)^-$, 10^{-4} M	No	No	
Hydrolyzed Al(III)$^{n+}$	$(AgI)^-$, 10^{-4} M	Yes	Yes	pH = 5
Hydrolyzed Fe(III)$^{n+}$	$(SiO_2)^-$, 0.2 g/l	Yes	Yes	pH = 5
Hydrolyzed Th(IV)$^{n+}$	$(AgBr)^-$, 10^{-4} M	Yes	Yes	pH = 4.5
$C_{12}H_{25}NH_3^+$	$(AgI)^-$, 8.1×10^{-3} M	Yes	Yes	
$C_{16}H_{33}NH_3^+$	$(AgI)^-$, 8.1×10^{-3} M	Yes	Yes	
Polydiallyldimethyl-ammonium^{n+}, (PDADMA)$^{n+}$	$(Kaolin)^-$, 73.2 mg/l	Yes	Yes	Coagulant dosage as moles $C_8H_{16}NCl$/l
NO_3^-	$(AgBr)^+$, 10^{-4} M	No	No	
SO_4^{2-}	$(AgBr)^+$, 10^{-4} M	No	No	
$CH_3-C_6H_4-SO_3^-$	$(FeO(OH))^+$, 2.6×10^{-3} M	No	No	
Polymeric or "Activated" Silica	$(AgBr)^+$, 5×10^{-4} M	Yes	Yes	
Polyacrylic Acid^{n-}, (PAA)$^{n-}$	$(AgBr)^+$, 2×10^{-3} M	Yes	Yes	
Polyethylene Oxide	$(Kaolin)^-$, 56.4 mg/l	No	Yes	
Dextran	*A. aerogenes* ca. 100 mg/l	No	Yes	pH = 5
Methyl Cellulose	$(AgBr)^-$, 2×10^{-3} M	No	Yes	
Polyacrylamide^{n-}, 30% Hydrolyzed (PAM-30)$^{n-}$	$(Kaolin)^-$, 33.3 mg/l	No	Yes	$Ca^{2+} = 2.25 \times 10^{-3}$ M

Table IX, continued

Coagulant	Colloid	Charge Reversal	Restabilization	Remarks
Polymeric or "Activated" Silica^{n-}	$(AgBr)^-$, 5×10^{-4} *M*	No	Yes	
Polystyrene Sulfonate^{n-}, (1)	$(AgBr)^-$, 5×10^{-4} *M*	No	Yes	$Ca^{2+} = 4.45 \times 10^{-4}$ *M*
Polystyrene Sulfonate^{n-}, (2)	*A. aerogenes*, ca. 100 mg/l	No	Yes	

[a]Reproduced from Stumm and O'Melia [29], courtesy of the American Water Works Association.

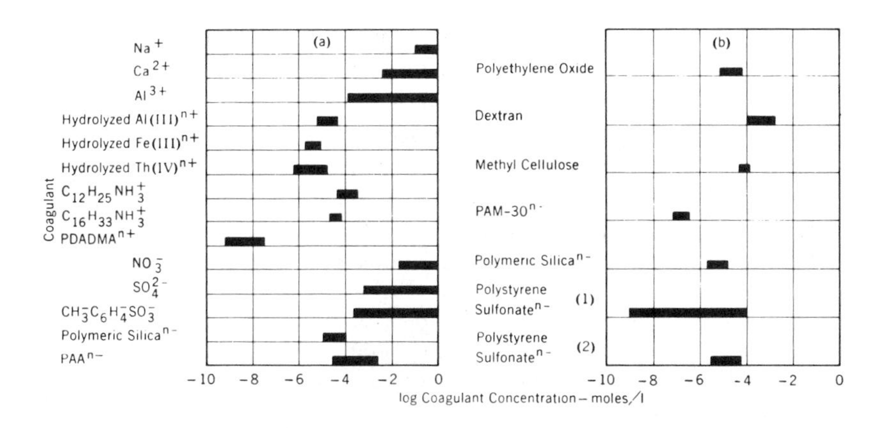

Figure 29. Typical destabilization regions for a few representative coagulants. **(a)** Contains coagulation regions observed when the colloid and the coagulant are of opposite charge. **(b)** Contains data obtained for the coagulation of colloids by uncharged molecules or by ions of like charge. These simplified figures are based on experimental results by various investigators who have used different experimental procedures and systems; the coagulation regions are therefore not always precisely comparable. The location of the coagulation regions are exemplifications, that is, they attempt to show essential features and are meaningful in a semiquantitative way only. With sorbable species, the exact location of the coagulation region on the concentration axis depends on the concentration, the surface potential of the colloids, pH, presence of other cations and anions, and temperature. Interactions with polymers depend on molecular weight and steric factors. Reproduced from Stumm and O'Melia [29], courtesy of the American Water Works Association.

Since the $\Delta\bar{G}^0$ term is negative, adsorption is spontaneous and opposite in direction to the electrochemical work. It should be apparent that destabilization and restabilization by noncoulombic forces are thermodynamically feasible. The thermodynamics of adsorption for organic compounds on inorganic surfaces is discussed thoroughly by Healy [70].

As mentioned above, much of the adsorption phenomena in coagulation comes from using natural and synthetic macromolecules as aggregating agents. Figure 10 depicts the "bridging" model that attempts to explain the destabilization of colloidal suspensions by polymers [26,71,72]. This model or theory explains such anomalies as the destabilization of negatively charged colloids with negative polymers. The optimum aggregation of colloids frequently occurs when the electromobility and/or zeta potential is not zero.

The effect of polymer adsorption on the electrokinetic stability of dilute clay suspensions was reported by Black et al. [73]. Experimental systems were kaolinite (12.2 meq/100 g) and bentonite (147.5 meq/100 g) and a cationic polymer with an empirical formula of $(C_8H_{16}NCl)_x$. Electrophoretic mobility measurements were performed by the technique of Black and Smith [19]. Adsorption measurements were made by radioisotope techniques, since the polymer was labeled with ^{14}C. Figures 30 and 31 are typical results. Several observations are made:

1. For each clay content, the electrophoretic mobility of the particles is reversed and approaches a limiting positive value.
2. Minimum residual turbidity is recorded when the particles are still negatively charged.
3. For each clay content, the amount of adsorbed polymer increases linearly at first, and then approaches a limiting value that is dependent on the initial clay content (see bottom graph in Figure 30).
4. The Langmuir adsorption isotherm equation (see Chapter 4) is fitted by the adsorption data (Figure 31).
5. There is an inverse relation between the quantity of polymer required for saturation of the surface and the clay content.
6. There appears to be a direct stoichiometric relation between the initial clay content and optimum polymer dose.
7. The zone of successful clay aggregation is broadened and the maximum degree of destabilization is increased as the initial clay content is increased.

This provides, perhaps, a strong argument for destabilization by adsorption when it is observed that there is an inverse relation between the quantity of polymer required for saturation of the surface and the clay content. This is explained by a greater probability of interparticle collisions in the more concentrated clay suspensions than in the less concentrated suspensions. This affords a greater probability of interparticle bridging (Reaction 2, Figure 10) where one polymer molecule or, at least, one site on the polymer must occupy at least one adsorption site on the clay's surface that is being bridged. Consequently, a larger saturation weight ratio of polymer to clay occurs in the lower clay suspensions. Another interpretation is that less bridging of particles occurs in dilute systems, whereas more bridging occurs in the concentrated systems.

Coagulation by Polyelectrolytes and Coagulant Aids

A number of methods and/or chemicals are used either as "aids" to coagulation or as the primary coagulant. Many difficulties with alum

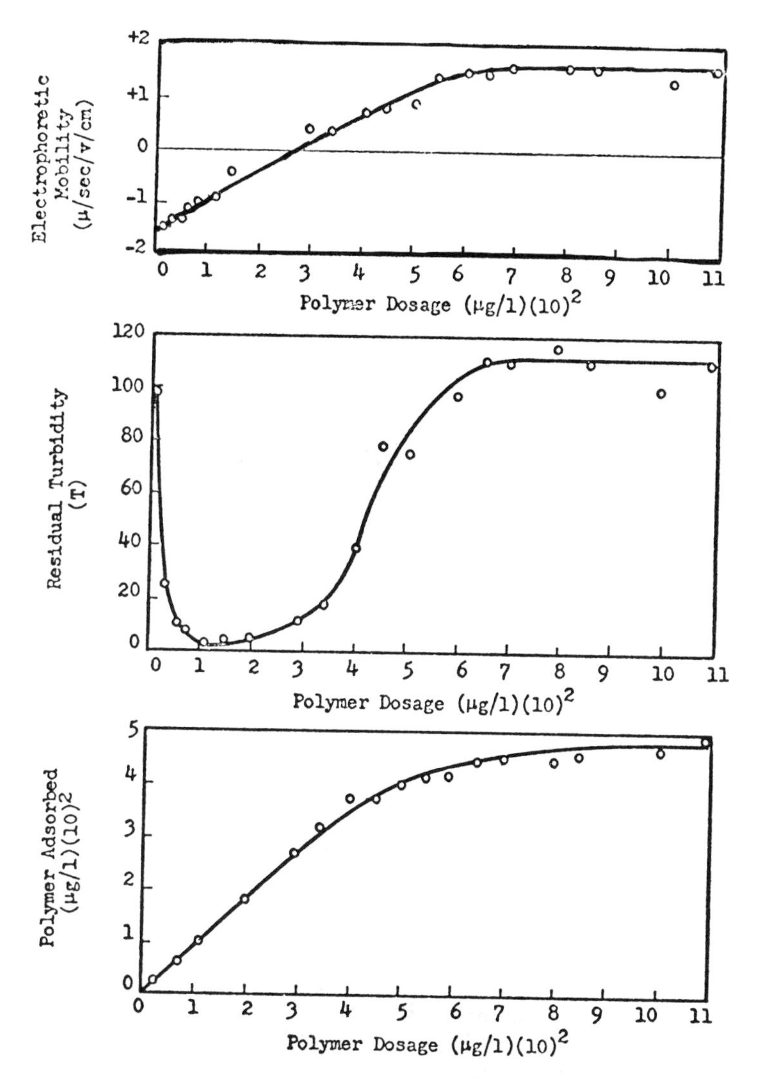

Figure 30. Destabilization of a kaolinite clay suspension with cationic polymer No. 4. Initial clay concentration = 73.2 mg/l. Reproduced from Black et al. [73], courtesy of the *Journal of Colloid Science.*

coagulation of natural waters have been cited above. An additional problem is posed by the small and slow-settling flocs from low-temperature coagulation and/or those in "soft," colored waters. Fragmentation of "fragile" flocs under hydraulic forces in basins and sand filters, and the inability to obtain clear water in the presence of interfering substances

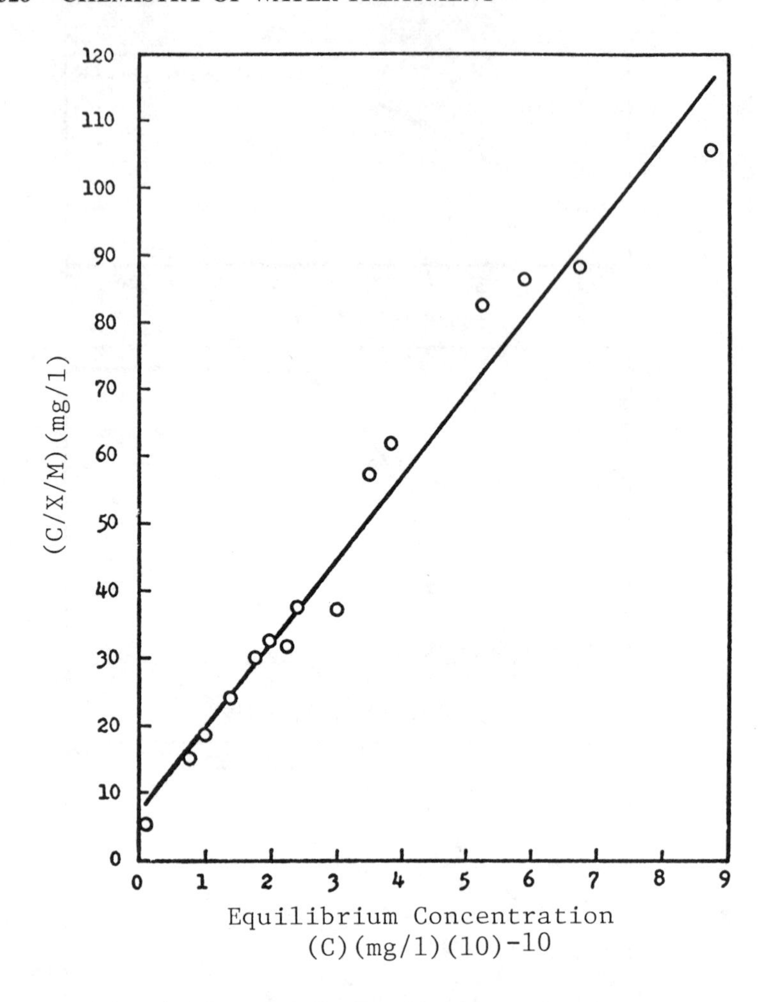

Figure 31. Langmuir adsorption isotherm. Cationic polymer No. 4, kaolinite concentration = 73.2 mg/l. Reproduced from Black et al. [73], courtesy of the *Journal of Colloid Science.*

also are problems [8]. It is necessary, therefore, to overcome these difficulties and to improve the settleability and "toughness" of the flocs. This will increase the production of potable water without requiring expansion of the plant's physical facility. The materials that find utilization as "coagulant aids," i.e., any substance that improves the floc's quality, are, for the most part, polyelectrolytes or activated silica.

Polyelectrolytes

These compounds are mostly synthetic organic compounds of high molecular weight. They are polymers composed of a chain of monomers. In turn, these monomers are varied frequently within a given polymer, which results in compounds with different molecular weights. These polymers are linear or branched. If the monomer contains an ionizable group, such as carboxyl, amino or sulfonic, then the polymer is called a polyelectrolyte. There are cationic, anionic or ampholytic (has both positive and negative groups) which, of course, depends on the nature of the functional groups within the monomer. Nonionic polymers are those compounds without any ionizable groups. Some examples of polymers and polyelectrolytes are [20]:

1. **Nonionic polymers:**
   polyethylene oxide:

$$[-CH_2-CH_2-O-]_n$$

polyacrylamide (PAM):

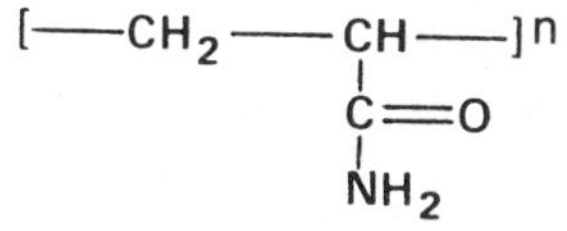

$$[-CH_2-\underset{\underset{\displaystyle NH_2}{|}}{\underset{\displaystyle C{=}O}{\underset{|}{CH}}}-]_n$$

2. **Anionic polyelectrolytes:**
   polyacrylic acid (PAA):

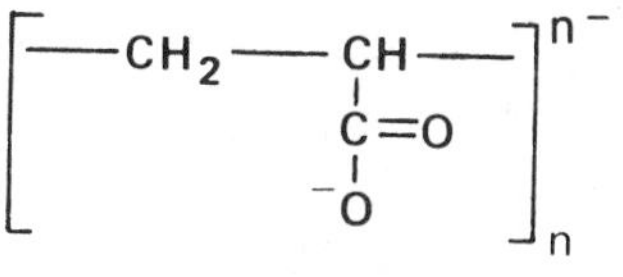

$$\left[-CH_2-\underset{\underset{\displaystyle ^{-}O}{|}}{\underset{\displaystyle C{=}O}{\underset{|}{CH}}}-\right]_n^{n-}$$

hydrolyzed polyacrylamide (HPAM):

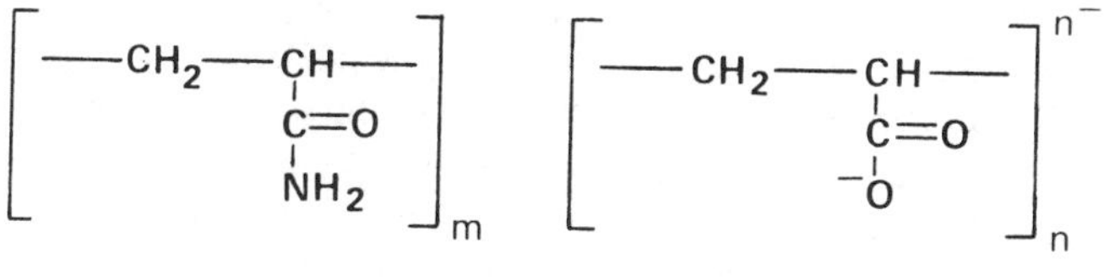

$$\left[-CH_2-\underset{\underset{\displaystyle NH_2}{|}}{\underset{\displaystyle C{=}O}{\underset{|}{CH}}}-\right]_m \left[-CH_2-\underset{\underset{\displaystyle ^{-}O}{|}}{\underset{\displaystyle C{=}O}{\underset{|}{CH}}}-\right]_n^{n-}$$

In the case of HPAM, the negative charge on the polymer is a function of the number of acrylamide groups that are hydrolyzed to acrylic acid. In turn, the degree or percentage of hydrolysis determines the total charge on the HPAM. That is, a 10% hydrolyzed compound means that only 10 of 100 monomer units have become negatively charged through hydrolysis.

polystyrene sulfonate (PSS):

$$\left[-CH_2-CH(C_6H_4SO_3^-)- \right]_n^{n-}$$

3. Cationic polyelectrolytes:
polydiallyldimethylammonium (PDADMA, Cat-Floc):

$$\left[-CH-CH_2-CH(-CH_2-)-CH_2-N^+(CH_3)_2-CH_2- \text{(ring)} \right]_n^{n+}$$

As mentioned above, these polymers and polyelectrolytes are able to flocculate colloidal particles due to adsorption. In most cases, the bonding mechanism between a functional group on the polymer and a site on the colloid's surface is quite specific. In addition, molecular weight and degree of branching of the polymer play a mechanistic role in their ability to flocculate. Some case histories follow.

The use of high-molecular-weight organic compounds as flocculants is not necessarily a recent development. Many natural organic materials, such as starch, starch derivatives, cellulose compounds, polysaccharide gums and proteinaceous materials, have proven useful as aids to flocculation [8,74].

An "anionic A" polymer (acrylamide), a "cationic B" polymer (organic cation), and a "nonionic C" (natural cellulose derivative) were used in the laboratory coagulation of five natural waters [74]. Representative results from the jar-test are seen in Figure 32 for Ohio River Water (initial turbidity = 1000 units) and for a pond water (initial turbidity = 300 units), respectively. It is apparent that greater turbidity removals are effected in less settling time with the cationic B polyelectrolyte than with alum alone or with the anionic A and the nonionic C. Similar results were obtained by Black et al. [75], who investigated 17 polyelectrolytes and polymers for their ability under laboratory conditions to flocculate suspensions of two clay minerals: kaolinite and illite.

A Georgia kaolinite suspension (CEC = 12.2 meq/100 g, surface area of 15.8 m^2/g) was flocculated by a cationic polymer (PDADMA) and two anionic polymers that were partially hydrolyzed polyacrylamides (4% HPAM 4 and 30% HPAM 30) [76]. It is noted in Figure 33 that

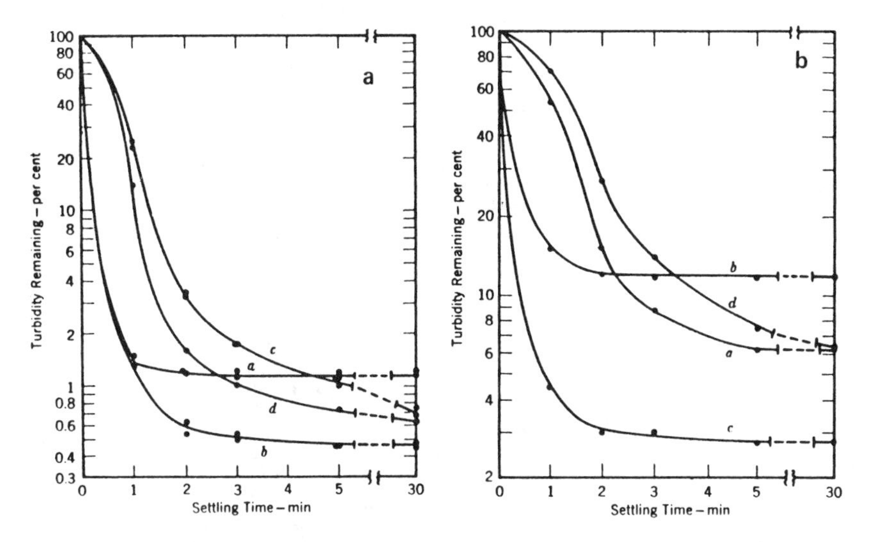

Figure 32. **(a)** Effect of coagulant aids on Ohio River water; 40 mg/l alum was added to each test. Curves, each one an average of duplicate tests, represent coagulations made with aids added in optimum amounts, as follows: a = 1 mg/l anionic a; b = 6 mg/l cationic b; c = 20 mg/l nonionic c; d = alum alone. **(b)** effect of coagulant aids on pond water. Each solution contained 20 mg/l alum. Curves represent coagulations obtained with: a, alum alone; b, 0.5 mg/l anionic a; c, 2 mg/l cationic b; and d, 2 mg/l nonionic C. Reproduced from Cohen et al. [74], courtesy of the American Water Works Association.

flocculation of the negatively charged kaolinite is effected by the two anionic polymers where electrophoretic mobility values remained negative. These studies also included the calcium content as a variable where 250 mg/l was a factor in the flocculation mechanism. According to Black et al. [76]:

> Divalent calcium ions may affect anionic polymer clay systems in the following ways: a. By compressing the thickness of the double layer of the clay particles, thereby reducing interparticle repulsive forces. b. By reducing the repulsive forces between the anionic polymer and clay particle. c. By reducing interactions between polymer molecules adsorbed on clay particle surfaces.

Adsorption data for both anionic polymers were fitted to a linear form of the Langmuir equation (similar to Figure 31). Flocculation by the cationic polymer, PDADMA, was similar to Figure 30 [73].

Case Studies

Eight cationic polyelectrolytes were evaluated as primary coagulants in Potomac River water and five other natural waters by means of the jar-test [77]. It was concluded that these compounds can serve effectively as

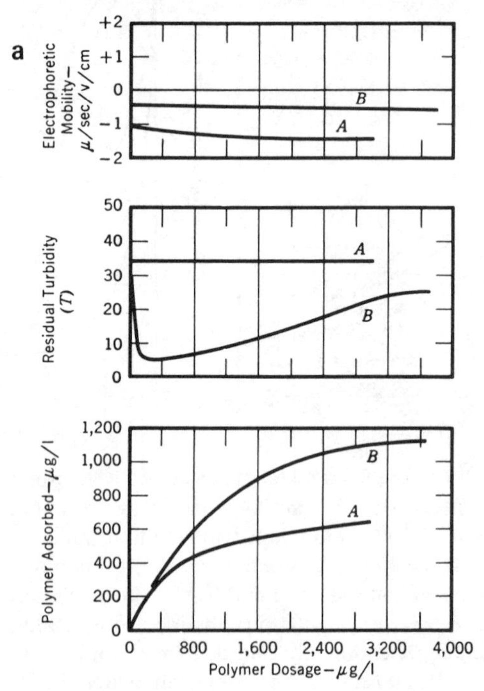

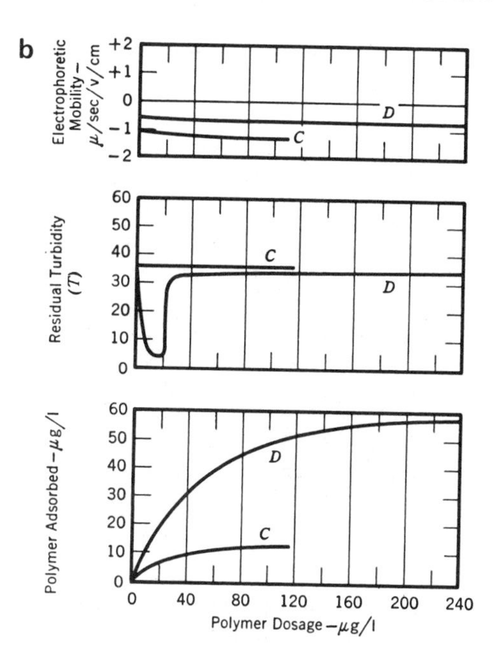

Figure 33. Effect of $CaCl_2$ on destabilization of kaolinite clay suspensions with anionic polymers **(a)** HPAM-4 [clay suspension concentration = 33.4 mg/l, $CaCl_2$ = 25.0 mg/l (curve A) and 250 mg/l (curve B)]; and **(b)** HPAM-30 [clay suspension concentration = 33.3 mg/l, $CaCl_2$ = 25.0 mg/l (curve C) and 250 mg/l (curve D)]. Reproduced from Black et al. [76], courtesy of the American Water Works Association.

prime coagulants for the treatment of natural water in a solids-contact clarifier and pressure diatomite filter system. Figure 34 shows that the optimum polymer dosage indicated by the minima of the turbidity curves are on the order of 2–4 ppm. These quantities are somewhat higher than those for similar synthetic suspensions with comparable turbidities [73,76].

The role of complex formation, principally with the Ca^{2+}, was investigated in the flocculation of negatively charged sols with anionic polyelectrolytes [78]. Flocculation occurs from adsorption of polymers on the colloidal surface and from bridging of polymer chains between solid particles if an appropriate content of an electrolyte is present. Complex formation occurs in the vicinity of the sol's surface between the counterion and functional groups of the polyelectrolyte. This mechanism apparently plays a major role in the attachment of anionic polyelectrolytes to negative hydrophobic sols.

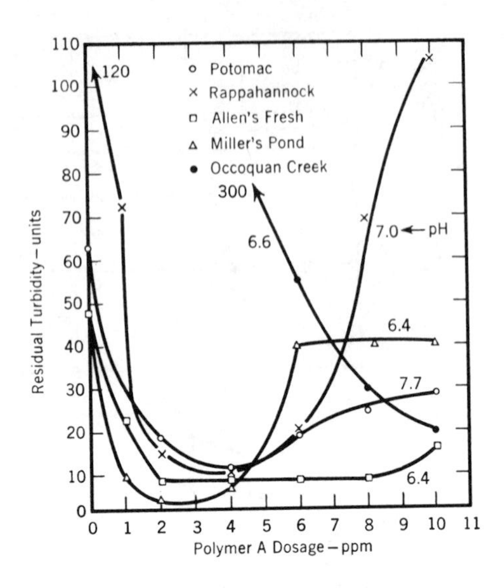

Figure 34. Flocculation of various waters treated with Polymer A. Reproduced from Pressman [77], courtesy of the American Water Works Association.

The cationic polyelectrolyte, poly-4-vinyl-N-methyl pyridonium iodide (PVMPI), flocculated various mixtures of clay minerals and fulvic and humic acids in laboratory studies [79]. Figure 35 shows the effectiveness of PVMPI on removal of humate, the clay mineral montmorillonite, and mixtures thereof. It appears that this cationic polyelectrolyte requires 1 mg/l or less for removal of the clay mineral, 2–6 mg/l for an organoclay complex (humate adsorbed on the clay), 12–14 mg/l for a suspension of the clay mineral in a 10-mg/l solution of humate, and 16–20 mg/l for the organoclay complex in a 7-mg/l solution of humate. All of these mixtures were designed to simulate natural water quality. In each system, a zero electrophoretic mobility is recorded at optimum removal.

An in situ study was conducted in which a combination of three organic polymers replaced the traditional inorganic coagulation chemicals [80]. Table X shows a summary of the quantities of all chemicals used at the Ithaca, NY water treatment plant over a 25-year period. Polymer A was Magnifloc 572-C, polymer B was Magnifloc 515-C and polymer C was Flocgel. After several years of experimentation with these three polymers and others, alum was completely eliminated in February 1975. The turbidity data in Table X show a slight decline, and there was an improvement in overall finished water quality with use of

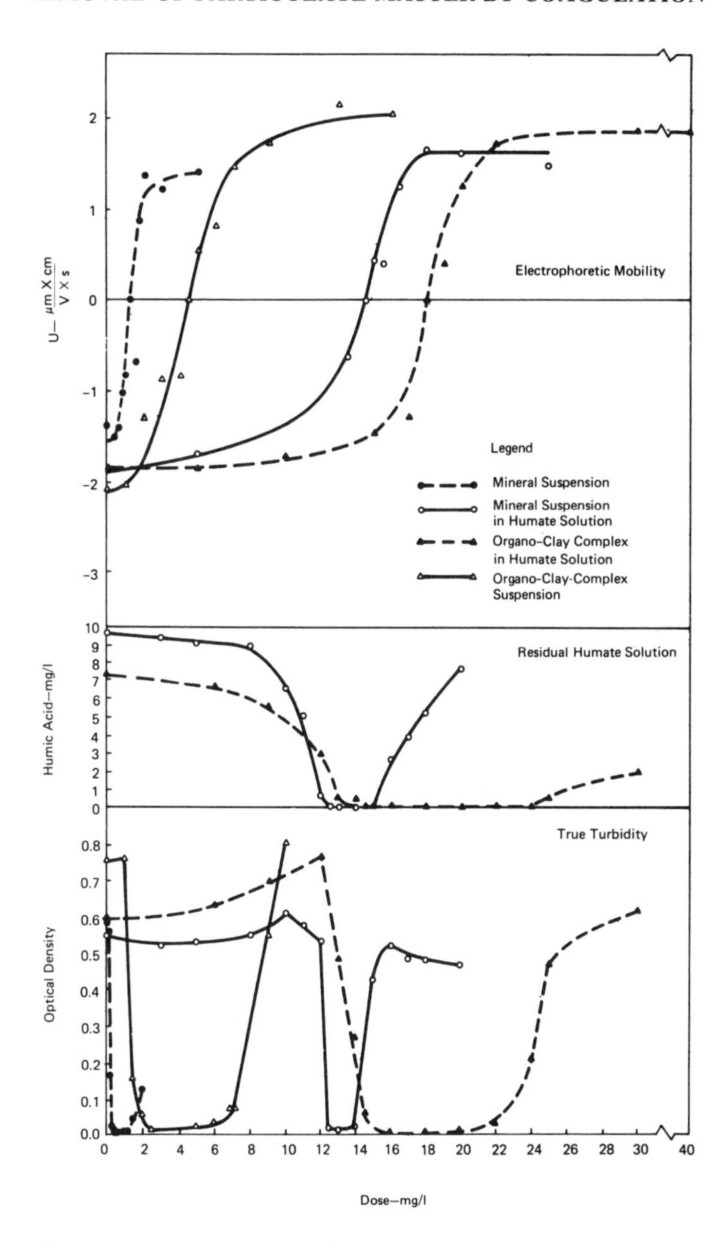

Figure 35. Flocculation curves of various systems of montmorillonite using PVPMI. Reproduced from Narkis and Rebhun [79], courtesy of the American Water Works Association.

Table X. Summary of Chemicals for Ithaca Water Plant[a]

	1951–1960	1961–1970	1971	1972	1973	1974	Feb 1975–Jan 1976	1976
Chlorine								
g/m^3	3.2	3.7	3.8	3.4	3.7	4.3	4.0	3.2
$lb/10^6$ gal	27	31	32	28	31	36	33	27
Sulfur Dioxide								
g/m^3	0.7	0.7	0.4					
$lb/10^6$ gal	6	6	3					
Alum								
g/m^3	22.5	23.4	33.0	31.8	19.7	6.6		
$lb/10^6$ gal	188	195	275	265	164	55		
Hydrated Lime								
g/m^3	7.8	5.8						
$lb/10^6$ gal	65	48						
Sodium Carbonate								
g/m^3	2.5	0.8						
$lb/10^6$ gal	21	7						
Sodium Hydroxide								
g/m^3		6.2	16.5	12	8.6			
$lb/10^6$ gal		52	138	97	72			
Polymer A								
g/m^3				0.5	1.8	3.6	1.9	1.9
$lb/10^6$ gal				4	15	30	16	16
Polymer B								
g/m^3							19.3	26.5
$lb/10^6$ gal							161	221
Polymer C								
g/m^3		0.2	0.6	1	0.5	0.4	0.2	0.1
$lb/10^6$ gal		2	5	9	4	3	2	1
Effluent Turbidity (NTU)	1.10	0.60	0.08	0.06	0.07	0.09	0.04	0.02

[a]Reproduced from Terrell [80], courtesy of the American Water Works Association.

the three polymers. Color, odor and iron were removed effectively by the chlorine, organic polymers and filters.

Summary of Mechanisms

Two mechanisms and/or chemical models are recognized for destabilization of colloidal systems in water. One is the so-called "double layer" model, which indicates the independence of the coagulant content to destabilize a given concentration of colloids. In this model, charge reversal and restabilization of the colloid does not occur. On the other hand, the "bridging" model indicates that there is a direct (stoichiometric) relation between the optimum flocculant content and the solid surface area of the colloids. The characteristics of these two models appear in Table XI [29]. Also, these characteristics are compared to those that describe the aggregation of colloids by hydrolyzed metal ions, e.g., Al or Fe (last column). Neither completely describes the coagulation process in drinking water treatment. "It is difficult, if not impossible to find a natural system that can be characterized by this model (the double-layer). Overemphasis on electrostatic phenomena in studies of coagulation in natural systems can produce results that are inefficient, uneconomical or both" [29]. Furthermore, it can be stated that it is not really necessary to utilize a particular mechanism for in-plant quality control of coagulation. It is the quality of the treated water and how it is achieved that is important.

Kinetics of Particle Aggregation

Flocculation and coagulation of colloids, by whatever mechanism, depend on the frequency of collisions and on the efficiency of particle contacts. Whenever suspended particles collide, there are, at least, three mechanisms of transport [81]:

1. Particles are in motion because of their thermal energy (Brownian motion). Any coagulation occurring by this means is called "perikinetic."
2. When the particles are large enough, or when the fluid shear rate is high enough, the relative motion from velocity gradients exceeds that by thermal effects. This is called "orthokinetic" coagulation.
3. In the sedimentation process, particles with different gravitational settling velocities may collide and aggregate.

In a monodisperse system, the decrease of the concentration of particles, N, with time due to collisions by Brownian motion is [20,82]:

$$\frac{dN_o}{dt} = -k_p N_o^2 \tag{32}$$

where N_o = total concentration of particles at time, t
k_p = the second-order rate constant

Or,

Table XI. Modes of Destabilization and Their Characteristics[a]

Phenomena	Physical Double-Layer Theory (Coagulation)	Chemical Bridging Model (Flocculation)	Aggregation by Hydrolyzed Metal Ions
Electrostatic interaction	Predominant	Subordinant	Important
Chemical interactions, and adsorption	Absent	Predominant	Important
Zeta potential for optimum aggregation	Near zero	Usually not zero	Not necessarily zero
Addition of an excess of destabilizing species	No effect	Restabilization due to complete surface coverage	Restabilization usually accompanied by charge reversal; may be blurred by hydroxide precipitation
Fraction of surface coverage (θ) by destabilization for optimum aggregation	Negligible	$\theta = 0.5$ (La Mer [26]) in general, $0 < \theta < 1$	$0 < \theta < 1, \theta$ not necessarily $= 0.5$
Relationship between optimum dosage of destabilizing species and the concentration of colloid (or concentration of colloidal surface)	CCC virtually independent of colloid concentration	Stoichiometry, a linear relationship between flocculant dose and surface area	Stoichiometry possible, but does not always occur
Physical properties of the aggregates which are produced	Dense, great shear strength, but poor filtrability in cake filtration	Flocs of three-dimensional structure; low shear strength, but excellent filtrability in cake filtration	Flocs of widely varying shear strength and density

[a]Reproduced from Stumm and O'Melia [29], courtesy of the American Water Works Association.

$$\frac{1}{N_o} - \frac{1}{N_o^o} = k_p t \tag{33}$$

where N_o^o = the total particle concentration at $t=0$.

For k_p, the Smoluchowski [83] equation is:

$$k_p = \alpha_p 4D\pi d \tag{34}$$

where α_p = the fraction of collisions leading to permanent aggregation
d = diameter of particle
D = Brownian diffusion coefficient, which can be expressed from the Einstein-Stokes equation:

$$D = \frac{kT}{3\pi\eta d} \tag{35}$$

where k = the Boltzmann constant
T = the absolute temperature
η = the absolute viscosity.

Substituting Equations 34 and 35 into Equation 33:

$$\frac{dN_o}{dt} = -\alpha_p \frac{4kTN_o^2}{3\eta} \tag{36}$$

k_p is the order of 2×10^{-12} cm^3/sec for water at 20°C when $\alpha_p = 1$ [81]. For a turbid water with 10^6 particles/cm^3, this concentration would be reduced by about one-half in approximately six days due to Brownian motion.

Integration of Equation 36 yields [20]:

$$N_o = \frac{N_o^o}{1 + (4\alpha_p kTN_o^o/3\eta)t} \tag{37}$$

where N_o^o = initial particle concentration when $t=0$.

When the quantity $(4\alpha_p kTN_o^o/3\eta) = 1/t_{1/2}$, Equation 37 becomes:

$$N_o = \frac{N_o^o}{1 + (t/t_{1/2})} \tag{38}$$

where $t_{1/2}$ = time necessary to reduce the concentration of particles by one-half.

For water at 25°C, the equation for $t_{1/2}$ is [20]:

$$t_{1/2} = \frac{1.6 \times 10^{11}}{\alpha_p N_o^o} \tag{39}$$

where $t_{1/2}$ has units of seconds, N_o^o, particles/ml and α_p is dimensionless. In Equation 39, $t_{1/2}$ is dependent on the initial particle content and the collision efficiency.

Agitation, of course, accelerates aggregation of particles in a moving fluid. Any spatial changes in the fluid's velocity causes a velocity gradient, and, since particles follow the fluid, there will be opportunities for interparticle contact. The rate of decrease in numbers of particles due to their aggregation under the effect of a mean velocity gradient G ($time^{-1}$) is [82]:

$$\frac{dNo}{dt} = -\tfrac{2}{3}\alpha_o \bar{G} d^3 N_o^2 \tag{40}$$

where α_o = the fraction of collisions resulting in permanent aggregation (α_o is similar to α_p)
d = the diameter of particles.

Equation 40 describes the aggregation of particles in motion as orthokinetic coagulation. The respective rates of orthokinetic and perikinetic coagulations may be ratioed:

$$\frac{(dN/dt)ok}{(dN/dt)pk} = \frac{\bar{G} d^3 \eta}{2kT} \tag{41}$$

This assumes $\alpha_o = \alpha_p$. This ratio is unity when the velocity gradient is 10 sec^{-1}, 25°C, and the colloidal particles have uniform diameters of 1 μm [20]. In a similar manner, when particle diameters are uniformly 10 μm, a velocity gradient of 0.01 sec^{-1} offers sufficient orthokinetic flocculation to equal the contacts resulting from diffusion. Mean velocity gradients of 10–100 sec^{-1} are common in water treatment plants [20].

The volume fraction of colloidal particles ϕ (volume per unit volume of suspension) is:

$$\phi = \frac{\pi}{6} d_o^3 N_o \tag{42}$$

where d_o = diameter of the particles at time = 0.

Substitution of Equation 42 into Equation 40 yields:

$$\frac{dN_o}{dt} = -\frac{4}{\pi}\alpha_o\bar{G}\phi N_o \tag{43}$$

Integration of Equation 43 for the boundary conditions $N_o = N_o^o$ at $t = 0$ and $N_o = N_o$ at time t yields:

$$ln\frac{N_o}{N_o^o} = -\frac{4}{\pi}\alpha_o\phi\bar{G}t \tag{44}$$

Thus, the rate of orthokinetic floccuation is apparently first-order with respect to N_o, $\bar{G}$ and ϕ, the floc-volume fraction.

In any system, the mean velocity gradient depends on the power dissipated within the water. Camp and Stein [84] gave this equation:

$$\bar{G} = \left(\frac{\bar{P}}{V\mu}\right)^{1/2} \tag{45}$$

where $\bar{P}$ = the power input to the fluid
V = the volume of the reactor or, in this case, flocculator
μ = the viscosity of the fluid

When flocculation chambers are baffled, interparticle collisions are accomplished by hydraulic mixing within the fluid as it flows through the tank. An equation from Camp [85] is:

$$\bar{P} = Q\rho_1 g h_f \tag{46}$$

where Q = flowrate
ρ_1 = fluid density
g = gravity acceleration constant
h_f = head loss in the tank.

Substitution of Equation 46 into 45 yields:

$$\bar{G} = \left(\frac{Q\rho_1 g h_f}{V\mu}\right)^{1/2} = \left(\frac{g h_f}{\nu\bar{t}}\right)^{1/2} \tag{47}$$

where ν = kinematic viscosity of the fluid
$\bar{t}$ = mean detention time of the fluid in the tank

The intent of the foregoing discussion is to apply the theories of particle transport to the design of flocculation units. O'Melia [20] provides an excellent discussion of this point. Design involves the selection of a velocity gradient $\bar{G}$, a reactor configuation and a contact time sufficient for aggregation of colloidal particles for subsequent removal by an appropriate treatment unit (e.g., settling tank, vacuum filter and sand filter). Orthokinetic flocculation is the mechanism underlying the coagulation process "in the field" for which Equations 25 and 26 are applicable. First-order removal of the particles is dependent on α_o, ϕ and $\bar{G}$ of Equation 26. These three factors are extremely difficult, if not impossible, to measure in the field. "Theoretical models which are not sufficiently well developed to be used directly can provide important information for the design of laboratory and pilot plant experiments" [20]. Among other things, the detention time of flocculators is inversely proportional to α_p or α_o, the collision efficiency factor. There are very few, if any, values of this factor available from field studies. Some are available from laboratory studies as seen in Table XII where, for example, latex spheres are hardly typical colloids in natural waters.

As seen in Table XII, large variations in the collision efficiency factor were reported that range from 0.01 to 0.448. That is, almost 50% of the collisions in one system were fruitful in producing aggregation, whereas only 1% was successful in another system. If this were occurring at a water treatment plant, there would be a tremendous effect on the deten-

Table XII. Observed Values of the Collision Efficiency Factor[a]

Colloid	Coagulant	Type of Flocculation	$\bar{G}$ (sec^{-1})	α_p	Reference
Polystyrene Latex Spheres	NaCl	Orthokinetic	11	0.448	[86]
Polystyrene Latex Spheres	NaCl	Orthokinetic	45	0.344	[86]
Polystyrene Latex Spheres	Polyethylenimine	Orthokinetic	11	0.217	[86]
Polystyrene Latex Spheres	Polyethylenimine	Orthokinetic	45	0.063	[86]
Silica	Al(III)	Perikinetic		0.010	[87]
Silica	Al(III)	Orthokinetic	10	0.011	[87]
Oil	$Ca(NO_3)_2$	Perikinetic		0.355	[82]
Polystyrene Latex Spheres	NaCl	Perikinetic		0.375	[82]
Polystyrene Latex Spheres	NaCl	Orthokinetic	1–80	0.364	[82]

[a]Reproduced from O'Melia [20], courtesy of John Wiley & Sons, inc.

tion time for a given reduction in turbidity. There would be a 45-fold difference in the detention time for the above two systems, all other factors remaining the same [20]. The research of Birkner and Morgan [86] suggests that the polyelectrolyte flocculation of latex spheres appears to be influenced by the velocity gradient. On the other hand, $\alpha_{p,o}$ is not affected by the type of flocculation when Al(III) salts are used as coagulants [87].

However, some data from pilot plant studies suggest that cationic polyelectrolytes can effectively destabilize colloidal and particulate matter without inorganic salts by the application of increased mixing energy [88]. Mechanical rapid-mixing units at water treatment plants are designed to provide 10–60 sec of detention time by intense mixing, and to yield velocity gradients in the order of 300 sec^{-1} or more. Flocculation basins that follow the rapid mix have a velocity gradient range of 5–100 sec^{-1}, wherein the destabilized particles coalesce and compact.

Pilot-plant studies were conducted at three sites in the United States [88]: (1) Beaver River, Pennsylvania, (2) Missouri River, Missouri, and (3) Mississippi River, Illinois. Figure 36 and Table XIII show the effect of velocity gradient on the cationic polyelectrolyte (Cat-Floc B, Calgon Corporation) flocculation of a low-turbidity water. The polyelectrolyte apparently can replace alum as the primary coagulant for low- and high-turbidity (data not shown) waters when $\bar{G}$ values of greater than 400 sec^{-1} are applied. There is an additional advantage to this process: sludge produced by polyelectrolyte treatment is heavier and denser than the alum sludge and has better compaction and sedimentation characteristics.

A bench-top batch reactor was used to evaluate rapid-mixing velocity gradients and time effects on flocculation characteristics [89,90]. A rapid mix of 650 sec^{-1} for 0.5–8 sec is optimum for various water turbidities coagulated wiht 3 mg/l alum, pH 6.4, and 0.15 mg/l anionic polyelectrolyte.

The kinetics of particle aggregation suggests that the total particle concentration rate of decrease can be represented by a first-order rate equation of the form [86]:

$$\frac{dN_t}{dt} = -K_E N_t \tag{48}$$

where K_E = experimental first-order rate constant comparable to the $K(4/\pi\alpha_o\bar{G}\phi)$ in Equation 43.

This first-order rate expression is applicable only at the optimum polymer dosage and only during the initial phase of the flocculation

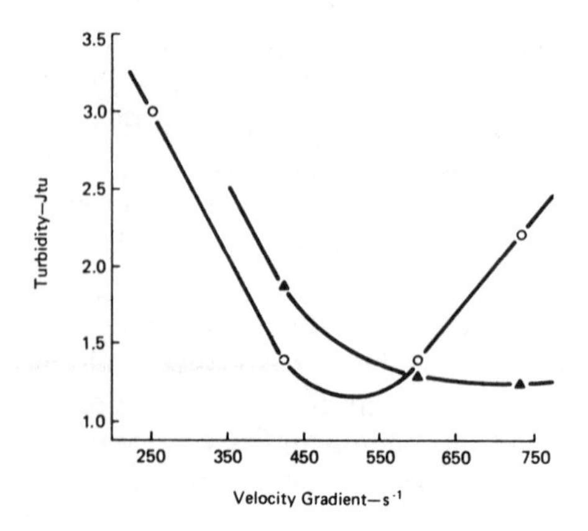

Figure 36. Results of application of velocity gradients greater than 400 sec^{-1}. ○ = 2.0 mg/l; ▲ = 1.0 mg/l. Reproduced from Morrow and Rausch [88], courtesy of the American Water Works Association.

Table XIII. Beaver River Study: Polyelectrolyte and Inorganic Coagulant Results[a]

Coagulant Treatment	Velocity Gradient (sec^{-1})	Coagulant Dosage (mg/l)	Raw Water Turbidity (JTU)	Average Settled Water Turbidity: Pilot Unit (JTU)	Average Settled Water Turbidity: Plant (JTU)
Alum	250	24.0	6–14	1.5	2.5
Alum/polymer[b] A	250	12.0/0.2	4–8	1.9	1.6
Polymer A	425	1.0	5–7	1.85	1.8
Polymer A	425	2.0	4–6	1.4	1.25
Polymer A	600	1.0	5–6	1.3	1.25
Polymer A	600	2.0	4–7	1.4	1.65
Polymer A	730	1.0	4–6	1.25	1.45
Polymer A	730	2.0	4–6	2.2	1.45
Polymer A	250	2.0	4–6	3.0	2.3

[a]Reproduced from Morrow and Rausch [88], courtesy of the American Water Works Association.
[b]Cat-Floc B, polyquaternary ammonium compound.

process (Figure 37). When the flocculant is insufficient or in excess, the reaction is definitely not first order. K_E values of 1.06×10^{-3} to 9.53×10^{-4} are observed for the optimum system when five $\bar{G}$ values ranged from 11 to 120 sec^{-1}. Furthermore, a 90% reduction in the initial particle concentration occurred after a reaction time of 424 min (7 hr).

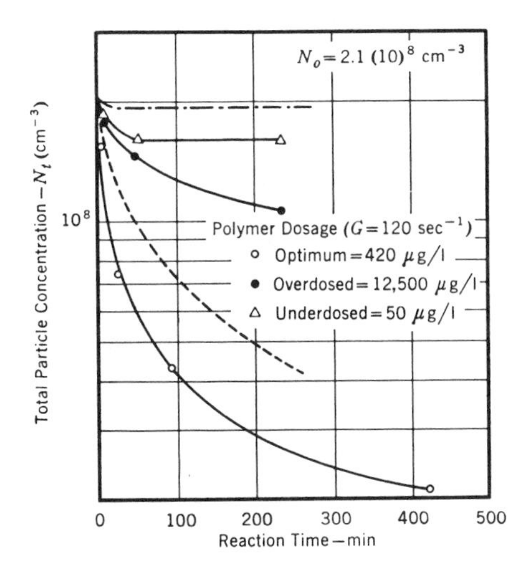

Figure 37. Effect of mixing intensity and polymer dosage on the flocculation kinetics of latex suspensions. --- = Brownian motion destabilization at the optimum polymer dosage; -·-·- = Brownian motion destabilization at under- and overdosed polymer conditions; —— = turbulent flocculation destabilization. Reproduced from Birkner and Morgan [86], courtesy of the American Water Works Association.

These observations apply, of course, to the experimental systems of Birkner and Morgan [86]. It is assumed that similar kinetics occur under optimum conditions at water treatment plants.

The jar-test was utilized to determine the effect of initial [HCO_3^-] on the efficiency of turbidity removal by flocculation with alum followed by settling [91]. This study indicated a major problem of the jar-test, namely, its inability to predict the performance of a continuous-flow flocculator. Nonetheless, kinetic experiments were designed that would yield an insight into the mechanism of the effect under observation. As the initial [HCO_3^-] was increased, the aggregation and erosion (floc breakup) rate constants became greater. The overall effect predicted by the kinetic model is that the performance of the continuous-flow flocculator improves as the [HCO_3^-] is increased. For example, when the [HCO_3^-] is increased from 20 to 90 mg/l at pH 7. and with 10 mg/l of alum, the predicted residual turbidity is decreased from 25 to 2 FTU.

Summary

Coagulation is a time-dependent process including several reaction steps: (1) hydrolysis of multivalent metal ions and subsequent polymerization to multinuclear hydrolysis species; (2) adsorption of hydrolysis species at the solid-solution interface to accomplish destabilization of the colloid; (3) aggregation of destabilized particles by interparticle bridging involving particle transport and chemical interactions; (4) aggregation of destabilized particles by particle transport and van der Waals forces; (5) "aging" of flocs, accompanied by chemical changes in the structure of Me—OH—Me linkages, concurrent change in floc sorbability and in the extent of floc hydration; and (6) precipitation of the metal hydroxide.

Some of these steps occur sequentially, some overlap (steps 1 and 2), and some may occur concurrently under certain conditions (steps 3 and 4, or step 6 with steps 1-5). It is reasonable to assume that different reaction steps may become rate controlling under different chemical conditions.

By careful choice of coagulant dosage, pH, and the possible addition of coagulation aids, the rates of pertinent chemical reactions can be adjusted so that particle transport controls the overall rate of aggregation. Chemical parameters also must be adjusted so that particle contacts induced by mass transport are efficient. The jar-test that has been used extensively in water treatment practice apparently has served this purpose. It has provided for rapid and efficient particle destabilization. In many natural systems particle transport is achieved by orthokinetic flocculation and therefore is influenced by the velocity gradient [86,87].

The coagulation process must be considered as one of several interdependent components of a water purification facility, and coagulation efficiency must be related to subsequent treatment units (settling, filtration, sludge disposal). More attention should be directed to the physical and chemical properties of the flocs: density, shear strength, compressibility, and filtrability. Trace quantities of impurities present in natural waters (color, silica, proteins) can have dramatic effects on the physical and chemical properties of the resulting flocs, and thereby alter significantly their settling and filtering characteristics. The recent development of synthetic polyelectrolytes suggests many potentially beneficial effects on both the kinetics of the formation and on the properties of the floc. The intelligent use of these materials with or without hydrolyzed metal ions may influence markedly the design of coagulation facilities.

It is important to reemphasize that coagulation phenomena in natural systems are quite specific. This specificity arises from the fact that colloid stability is affected by colloid-solvent, coagulant-solvent and colloid-coagulant interactions. The double-layer model, which considers only one type of colloid-coagulant interactions (coulombic forces) and neglects solvent interactions with the colloid and the coagulant, therefore can be used to describe only a very limited number of systems [29].

OPERATIONAL ASPECTS

Jar-Test

There is an extensive use of the jar-test for quality control of in-plant coagulation process. It is an extremely useful technique for determining parameters such as coagulant dosage, pH, alkalinity and floc time, but it is not particularly useful for "scaling up" information about flocculation rates from the jar-test to plant operation [20]. Nonetheless, some positive results have been obtained from these tests which, in turn, are translated to more efficient and effective plant operation.

Jar- and full-plant tests were conducted for an economical design and operation of a new 80-mgd water treatment facility at Phoenix, Arizona [92]. The raw water supplies for this city are the Salt and Verde Rivers. These parameters were investigated: (1) time required for good rapid mix; (2) need for rapid, uniform coagulant dispersion during rapid mix; (3) effects of diluting alum solutions upon coagulant effectiveness; (4) variances in flocculation characteristics of the two rivers supplied to Phoenix; and (5) effects of flocculation time on turbidity removal. Figure 38a shows that turbidity removal is affected, of course, by the basin overflow rate, but is independent of rapid-mix times for a given overflow rate. In this experiment, 18 mg/l alum were applied to each jar while the rapid mix occurred at a velocity gradient of 350 sec^{-1}. This was followed by a flocculation period of 30 min at a velocity gradient of 30 sec^{-1}.

Turbidity removals are not improved by a rapid mix greater than 5 sec. Figure 38b shows that the rapidity and uniformity with which alum is mixed with the entire volume of water are extremely important. The turbidity removals clearly show that alum should be added "instantaneously" to the entire volume of water.

That the jar-test results correlate reasonably well with plant tests is seen in Figure 39a [92]. The optimum flocculation velocity gradient is in the 20- to 30-sec^{-1} range. Another important operational variable is the detention time of the flocculation period. From plant operation, it was noted that basins with detention times of 55–60 min were producing water with lower turbidities than basins with 95–100 min of retention. Jar-tests (Figure 39b) confirmed this observation where optimum turbidity removals were obtained within 30–35 min. Results from this jar-test study of alum coagulation apparently were translated successfully into a more effective and efficient plant operation for the parameters reported, some of which have kinetic implications.

Jar-test results are significant for a number of quality control tech-

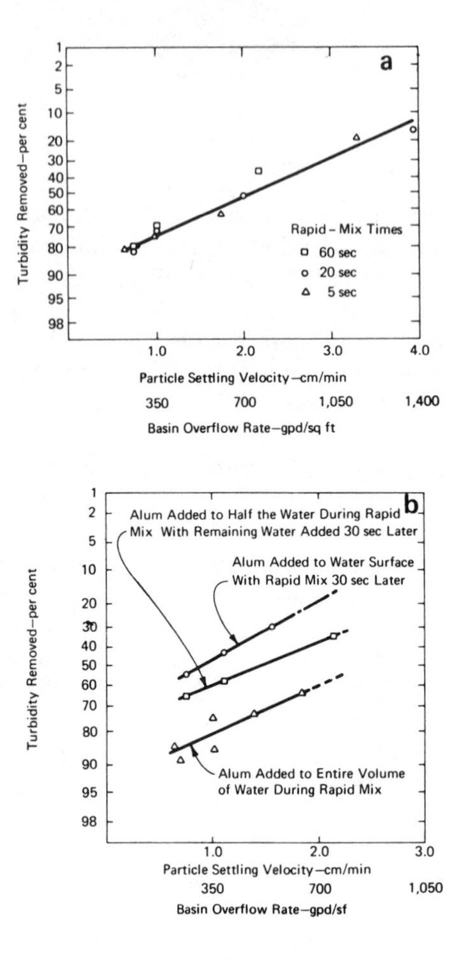

Figure 38. Turbidity removals vs settling for various rapid-mix **(a)** times and **(b)** conditions. Reproduced from Griffith and Williams [92], courtesy of the American Water Works Association.

niques for the coagulation-filtration process. They are: "time required for appearance of first floc, visual evaluation of floc size, rate of settling of floc, visual or photometric measurement of supernatant or filtrate clarity or color, and analytical determination of residual coagulant in supernatant or filtrate" [93,94]. The use of electrophoretic mobility measurements was suggested also for controlling the coagulation process. In-plant and continuous monitoring of turbidity is an extremely useful quality control for determining the optimum coagulant dosage.

The jar-test technique was used in conjunction with electromobility

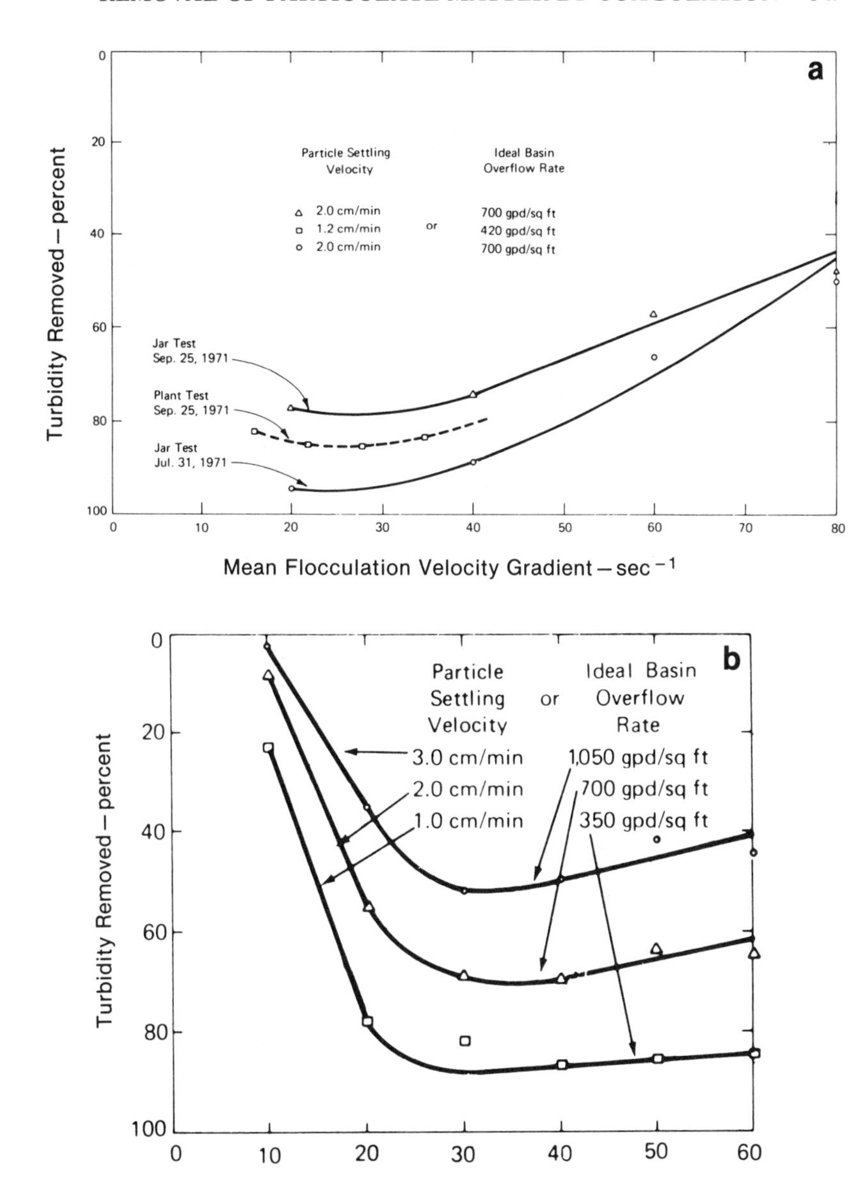

Figure 39. **(a)** Comparison of full-plant and jar-test turbidity removals after flocculation and sedimentation. **(b)** Effect of flocculation time on turbidity removal. Reproduced from Griffith and Williams [92], courtesy of the American Water Works Association.

and turbidity measurements in the coagulation of Little Miami River (Ohio) water whose turbidity is "low"—8.4 units [93]. Figure 40a shows the turbidity of coagulated water after sedimentation and sand filtration at two rates, 2 and 8 gpm/ft^2. Two turbidity curves are given, 15° and 90°. This refers to the angle of the light scattered by particulate matter in water when placed into a turbidimeter. "Large" particles scatter proportionately more light at 15° than at 90°. It is necessary to employ this technique when measusring the "low" turbidity of natural waters. In Figure 40a, note the numerically different ordinates for the turbidity plots. The turbidity minimum for settled water occurs at an alum dosage of 40 mg/l and at a negative zeta potential of −6 mV. Settled turbidity at zero ZP is three times the minimum value. Filtration rate is also a factor in the optimum alum dosage for removal of the "large" and "small" particles. Essentially all of the "large" particles are removed at 55 mg/l alum with

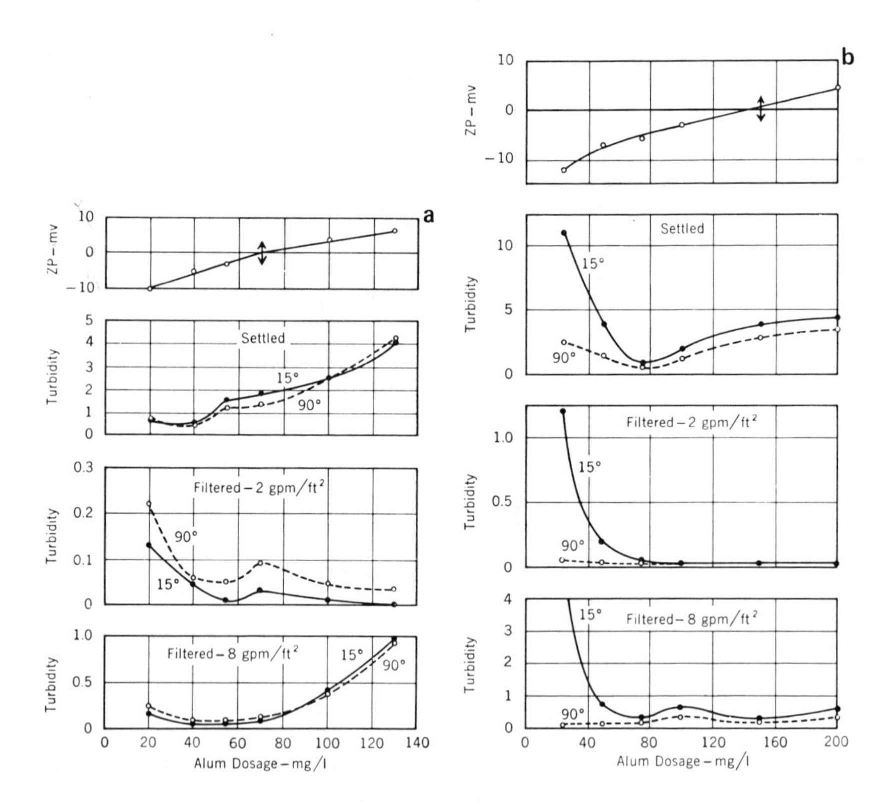

Figure 40. Coagulation with alum of **(a)** riverwater and **(b)** polluted riverwater. Reproduced from Hannah et al. [93], courtesy of the American Water Works Association.

a "few small" particles remaining. Zeta potential is still negative at this point. Considerably more alum (70 mg/l) is needed to remove and retain the floc particles at the higher filtration rate.

Figure 40b shows the response to coagulation of a mixture of six parts of Little Miami River water and one part of raw sewage in an effort to simulate a heavily polluted stream. The alum dosage for a minimum in the settled water's turbidity was increased to 75 mg/l with the particle's charge at −6 mV. Both filtration rates indicate that a dosage at least this size is needed to remove most of the "large" and "small" particles from the water. The effect of low temperatures (0.5–15°C) on turbidity removal by alum was investigated also [93]. Both sedimentation and filtration deteriorated with decreasing temperature.

Several observations were made on the operational aspects of coagulation [93]:

> No single monitoring or control procedure tested will serve all purposes.... The jar-test will continue to be an important aid in control of the coagulation-filtration processes.... Where raw water temperature is low, the jar-test must be run on samples held at the same temperature, if results are to be used in plant control. [This point is obvious, but frequently overlooked.] ...Measurements of electrophoretic mobility do not equivocally pinpoint the best dosage when alum alone is used.

Consequently, the jar-test does have a definite role in the operation of the coagulation process. Moffett came to the same conclustion: "The writer has been able to coordinate plant operation and jar-tests to ±1 ppm of alum and ±0.1 pH, where the chemical feeds in the plant are introduced in manner and locations that are dictated by the chemistry of coagulation" [95].

Additional discussions of the jar-test and operational aspects of coagulation are found in the literature [96–98].

Removal of Trihalomethane Precursors

Considerable information was given above for the removal of naturally occurring organic matter from wafer by chemical coagulation with Al(III) and Fe(III). Some of this organic matter takes the form of precursors to the formation of trihalomethanes via reactions with chlorine (see Chapter 2). This discovery has motivated several additional studies on the removal of these precursors by coagulation. While neither unique nor new, these studies provide useful operational data.

An "improved coagulation" for reducing THM precursors was reported for the waters of the Contra Costa County Water District, California [99]. Standard jar-test techniques determined the optimum coagulation conditions for reduction of total organic carbon (TOC). The coagulants, Al(III) and Fe(III), were tested over the pH range 3–8 and at various concentrations. Settled water samples were subjected to batch chlorination with 5 mg/l chlorine over 16–20 hr of contact. Figure 41 shows log-log plots of alum and iron dosages vs pH where the dashed lines are isoconcentration levels (isopleths) of TOC residuals after coagulation and sedimentation. The TOC isopleth in Figure 41a shows that the pH optimum lies between 5 and 6, which agrees with information cited above. For iron (Figure 41b) the pH optimum lies between 4 and 5. In turn, Figure 42 shows the THM isopleths as a function of pH value after coagulation with 20 mg/l alum. The TOC level is approximately 3 mg/l C over the entire pH range. Figure 43 shows clearly the dependence of THM formation on the precursor content (TOC). At pH values 5 and 6, a TOC of 2.5 mg/l and below yields THM contents of 100 μg/l and less. At pH 7, this minimum TOC level is reduced to 2 mg/l, whereas at pH 8, the yield of THM is increased greatly where the TOC minimum is 1 mg/l or less. It appears that coagulation at the acid pH values is preferable

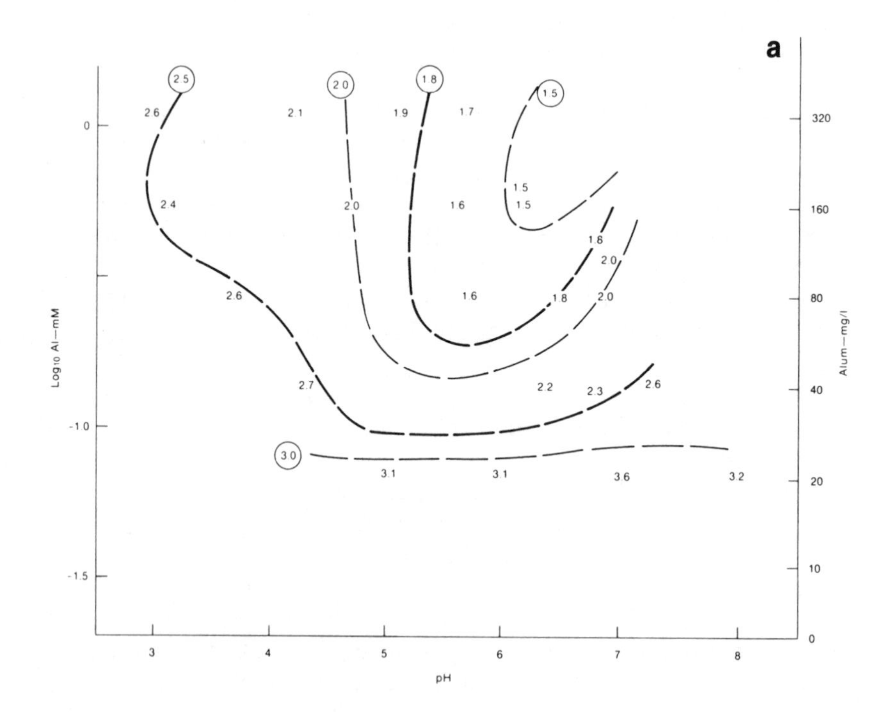

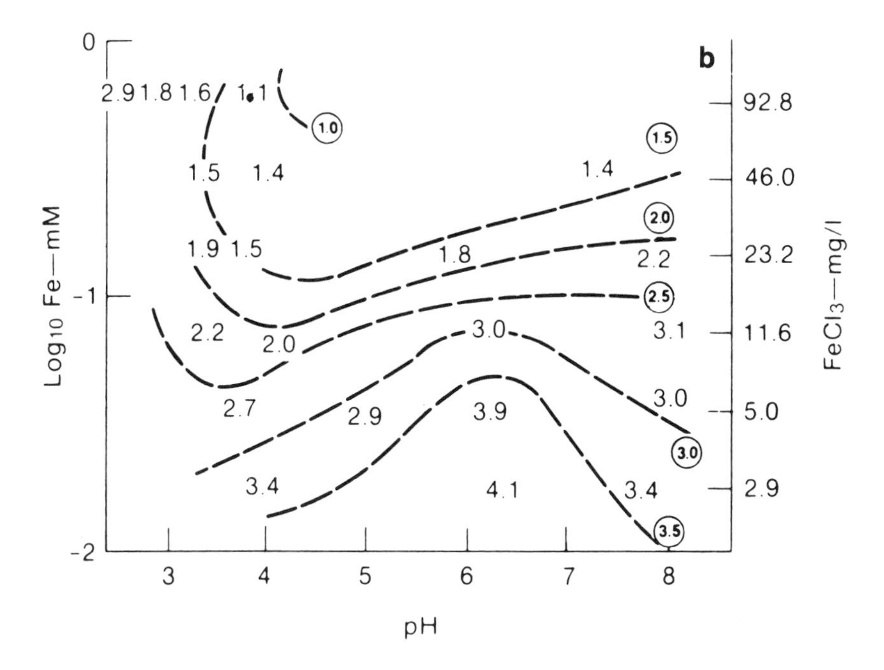

Figure 41. TOC isoconcentration lines following jar-test coagulation of Contra Costa canal water at 4.5 mg TOC/l (initial) and 18°C with **(a)** Al(III) and **(b)** Fe(III). Reproduced from Kavanaugh [99], courtesy of the American Water Works Association.

where THM formation is concerned. Also, this coagulation step should occur before chlorination. An excellent "up-to-date" review article was published on organics removal by coagulation [100].

A one-year study was reported for the removal by alum of organics from the waters of the Mississippi River at Minneapolis, Minnesota [101]. This laboratory study included the usual pH-alum dosage relationship and organic residuals measured by ultraviolet absorbance, TOC and potential formation of THM. There is a pronounced effect of turbidity on organics removal (Figure 44a). THM precursor removal is a function of pH and alum dose (Figures 44b and c). A useful, operational relation between alum dosage and residual TOC is seen in Figure 44d, at least for the Mississippi River at Minneapolis.

Removal of Specific Organic Compounds

There have been many unsuccessful attempts to achieve coagulation with Al(III) and Fe(III) salts for the removal of specific and dissolved

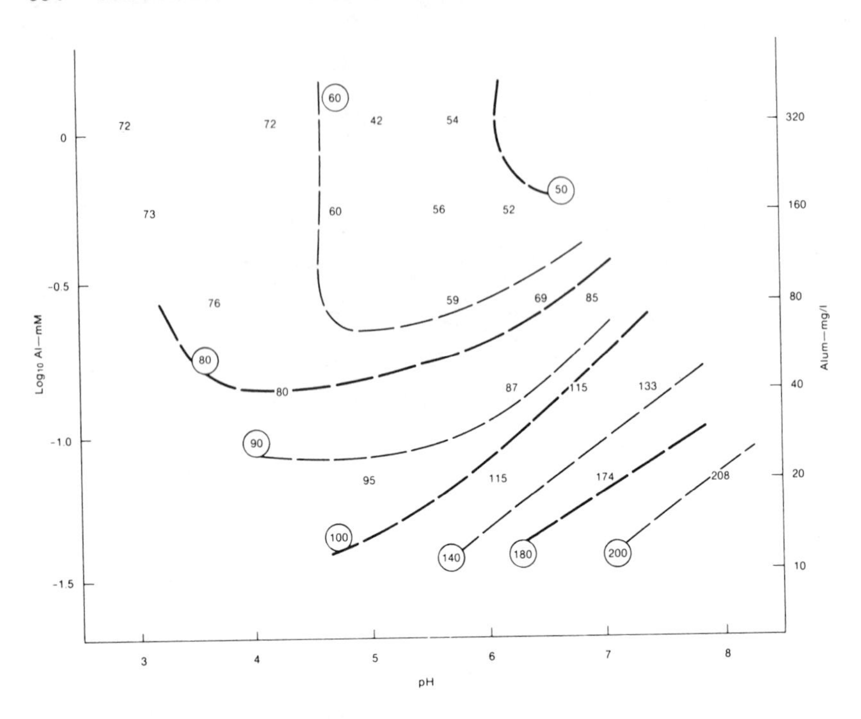

Figure 42. Total THM isopleths; coagulation with Al(III). Batch chlorination conditions: 5 mg/l Cl_2; temperature, 18°C; 20-hr contact time. Reproduced from Kavanaugh [99], courtesy of the American Water Works Association.

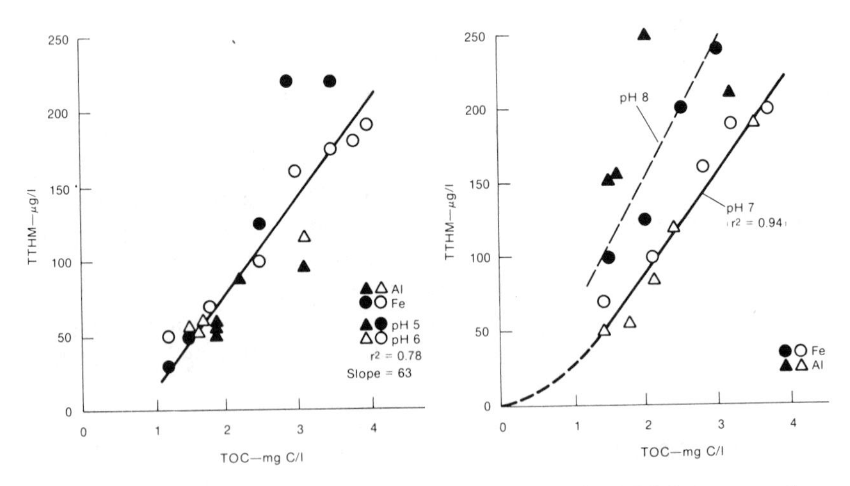

Figure 43. Total THM formed as a function of residual TOC concentration. Reproduced from Kavanaugh [99], courtesy of the American Water Works Association.

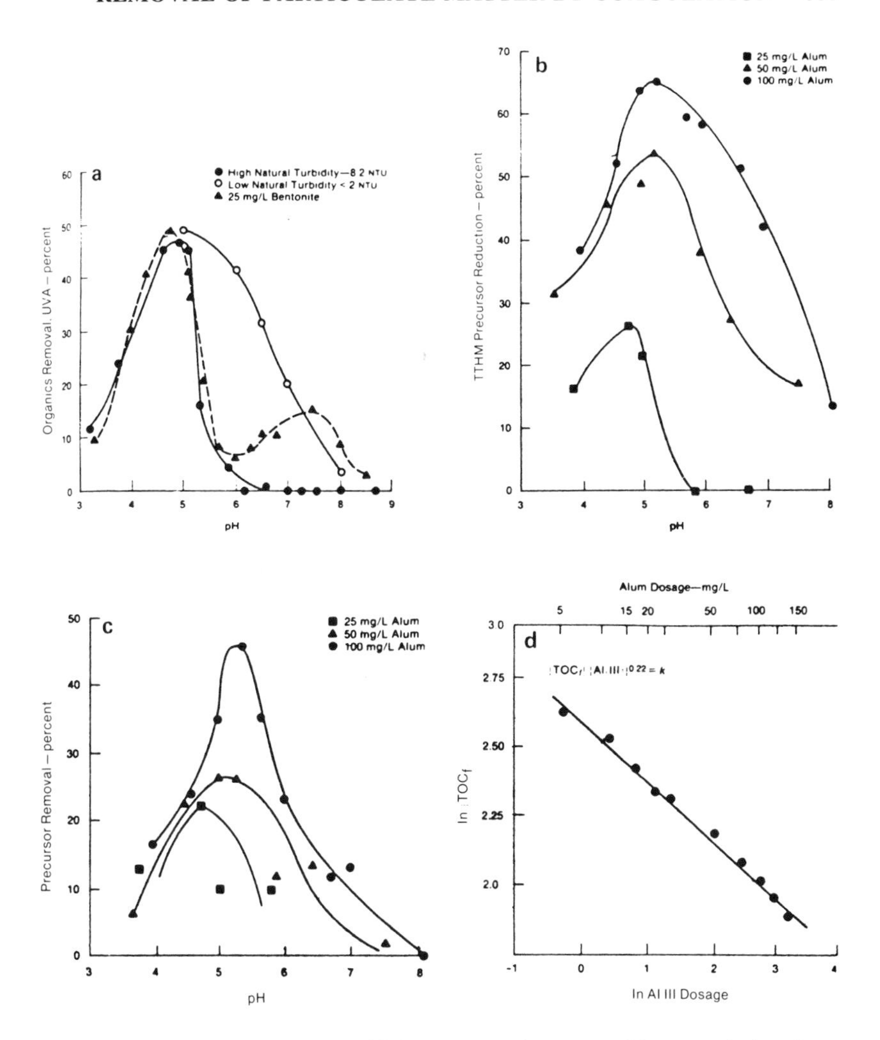

Figure 44. **(a)** Influence of turbidity on organics removal by coagulation for a 25-mg/l alum dose. **(b)** Total THM precursor removal as a function of pH and alum dose. **(c)** Percent removal of dichlorobromomethane precursors as a function of pH and alum dose. **(d)** Logarithms of TOC remaining vs Al(III) dosage at pH 5.0 for a water sample with an initial TOC of 16.0 mg/l. Reproduced from Semmens and Field [101], courtesy of the American Water Works Association.

organic compounds from water [100]. For example, the coprecipitation of phenols, citric acid and glycine from lakewaters spiked with iron was reported [102]. Excessive concentrations (287.–1437. mg/l) of iron are required to remove approximately 60–90% of the phenol, 20–25% of the glycine and 20–70% of citric acid in the spiked waters. Suspensions of DDT from water supplies were treated with various combinations of carbon and chemical coagulation with iron and aluminum hydrous oxides [103]. Removals of 80–98% of 0.1–10.0 mg/l of DDT were recorded. Complete removal of DDT was suggested when activated carbon was provided for a 15-min contact period after coagulation and sedimentation, but before sand filtration. In another study, alum and ferric sulfate with and without anionic and cationic polyelectrolytes were ineffective in removing malathion from water [104]. The removal from water of rotenone, toxaphene and various substances toxic to fish was investigated in a laboratory study [105]. Concentrations of alum as high as 100 mg/l produced no significant reduction in the quantities of these organics. Robeck et al. [106] obtained 55% removal of dieldrin and 35% removal of endrin via alum coagulation.

The effects of various conventional water treatment processes were reported on the removal of the sodium salt and several esters of 2,4-D and 2,4-dichlorophenol [107]. Coagulation with ferrous sulfate and alum up to 100 mg/l failed to remove these compounds. Activated carbon was the most effective treatment (see Chapter 4). Faust and Zarins [108] reported the removal of diquat and paraquat from water by: (1) activated carbon, (2) coagulation and (3) adsorption on clay minerals followed by coagulation. The last technique is reasonably effective. For example, a 96% removal of diquat is effected by 15.4 mg/l of bentonite followed by coagulation with 50. mg/l alum. Activated carbon is also very effective in removing diquat and paraquat. The inability of a conventional water treatment plant to remove toxaphene and BHC from the waters of Flint Creek, Alabama, was reported [109]. Ten phenylamide pesticides were not removed by 50 and 100 mg/l ferric sulfate and alum [110]. The greatest removal was 30% for linuron by 100 mg/l of ferric sulfate. Only a 62% reduction of benzopyrene, 5–400 μg/l, was effected with ferric chloride and basic aluminum polychloride [111]. Lesser removals of lindane, diethylphthalate and dibutylphthalate were reported.

Coagulation of Carbon Suspensions

Some circumstances require the employment of slurries of powdered activated carbon for taste and odor control, rapid removal of organics, etc. in the overall treatment process (see Chapter 4). It is frequently necessary to improve the sedimentation and filtration characteristics of this carbon by alum coagulation. Data summarizing laboratory study of this operational problem appear in Table XIV [112]. The initial residual

Table XIV. Coagulation of Carbon Suspensions (pH 7.0, Alkalinity 2 meq/l)[a]

		Alum Dose to Achieve			
Concentration of Carbon (mg/l)	Initial Residual Turbidity (units)	50% Turbidity Reduction	Residual Turbidity of One	Minimum Residual Turbidity (units)	Minimum Residual Turbidity (units)
2	0.36	7.2		25	0.04
5	1.3	7.0	5.8	30	0.06
25	5.6	6.8	8.0	25	0.14
50	5.5	3.9	7.8	30	0.27
100	5.9	3.9	7.6	25	0.22
150	6.0	2.8	9.0	25	0.21
300	6.2	2.9	10.2	35	0.45
500	5.9	5.2	12.2	25	0.75

[a]Reproduced from Letterman et al. [112], courtesy of the American Water Works Association.

turbidities were measured after pH adjustment and the subsequent rapid-mix, slow-mix, and sedimentation periods. The percentage turbidity reductions refer to reductions in the initial residual turbidities (~6 TU with carbon content >25 mg/l). Effective coagulation occurred over a broad range of alum doses.

Disposal of Coagulation Sludges

In recent years, disposal of sludges from the coagulation process is becoming more and more difficult because the most expedient method, direct discharge to surface waters, is forbidden by federal and state legislation. Two statistics indicate the magnitude of this sludge disposal problem [113,114]. First, there are 3600 or more water treatment plants in the United States. Second, there is an estimated quantity of solids from these plants in excess of 10^6 ton/yr [115].

Alum sludge is a non-Newtonian, bulky gelatinous material composed of hydrous aluminum oxide and such other inorganic particles as clay, sand or carbon, and such organics as color colloids, wastewater particulates and various types of microorganisms [113]. In addition, the sludge consists of other sediments from the clarifiers, filter wash water and sludge from wash water recovery. The total solids content is, of course, variable, but is in the range of 1000–20,000 mg/l of which 75–90% are suspended solids. Volatile solids are 20–35% of the total solids. Alum sludges tend to have neutral pH values. These sludges are readily settleable, but the resulting sludge volume and low solids content make them inconvenient to handle and subsequently to place in a landfill.

Few data have been published concerning the inorganic and organic contents of alum sludge and filter backwash water. These data undoubtedly exist, since permits are required for the discharge of these wastes to the environment. Table XV gives the elemental (primary and secondary drinking water constituents) contents of alum sludge and filter backwash water from the water treatment plant at Oak Ridge, Tennessee [116]. These contents are neither unusual nor exceedingly "high."

Considerable treatment of the alum sludge and the filter backwash water is required before ultimate disposal. To the extent that these two wastes are "low" in solids or "high" in water contents, it is necessary to dewater the sludge to achieve at least 20% solids (dry weight) prior to any subsequent handling [114]. The traditional methods of sludge dewatering are: sand drying beds, lagooning, centrifugation, vacuum filtration, pressure-filter presses and bed-filter presses. Other methods of disposal and/or treatment are alum recovery, discharge to sewerage system and "pellet" flocculation.

An innovative process that combines sludge dewatering and alum recovery was developed and patented by Fulton [117]. It can be utilized with or without recovery of the alum. A flow diagram of the Fulton

Table XV. Contents of Some Elements in Alum Sludge and Filter Backwash Water [116]

Constituent	Alum Sludge Average (ppm)	Backwash Water Average (ppm)
As	13.	15.
Ba	333.	450.
Cd	<1.	1.
Cr	200.	7,260.
Pb	47.	50.
Hg	<2.	<2.
Se	1.	1.
Ag		
Cu	7.	120.
Fe[a]		42,733.
Mn	983.	1,800.
Zn	167.	467.

[a]"Too high."

process is seen in Figure 45 [117]. Recovery of the alum should be close to 100%.

Figure 46 shows the flow diagram of a "pellet" flocculation process that was developed in Japan [117]. It involves multistage thickening of the sludge, treatment with sodium silicate and a polymer, and subsequent dewatering on a large horizontal rotating drum called the "dehydrum." Following addition of the sodium silicate, a polymer is added to the sludge just before it enters the dehydrum. This drum is compartmented into a settling chamber or pelletizing section at the inlet end. The decanting section is in the center of the drum. A sludge with 25–30% dry solids is discharged onto a conveyor. Additional drying to 65–70% solids may be accomplished in oil-fired dryers.

In 1976 the Passaic Valley Water Commission began operating an alum sludge and filter backwash water treatment and disposal process [118]. Figure 47 shows the complex flow diagram of these two wastes from the water treatment portion of the plant to the sludge facility. After the sludge is collected (storage tanks 1 and 2) from the coagulation process and from the filter backwash water settling tank, a polyelectrolyte is added to the sludge, which is thickened by centrifugation. At this point, lime and precoat are added to the thickened sludge, which

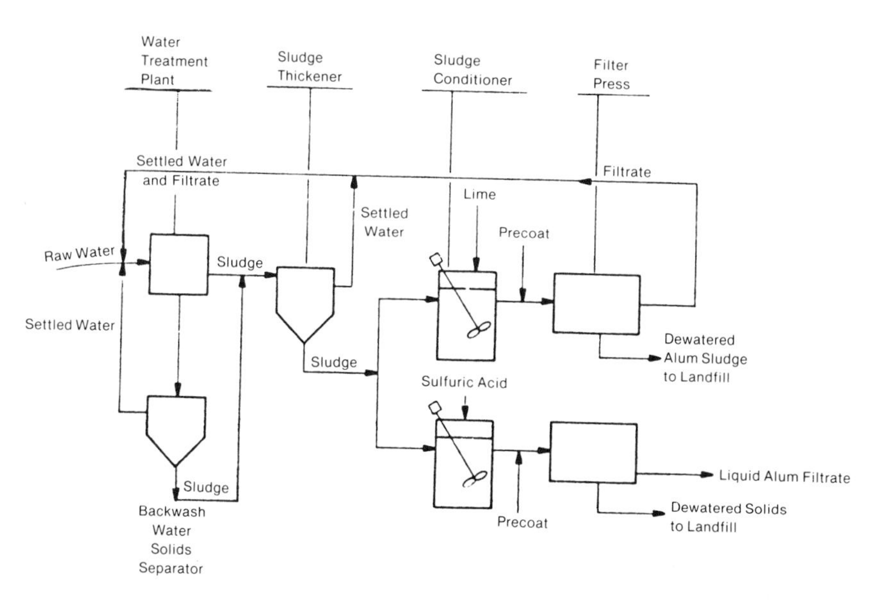

Figure 45. Sludge treatment system: Fulton process. Reproduced from Bishop [114], courtesy of the American Water Works Association.

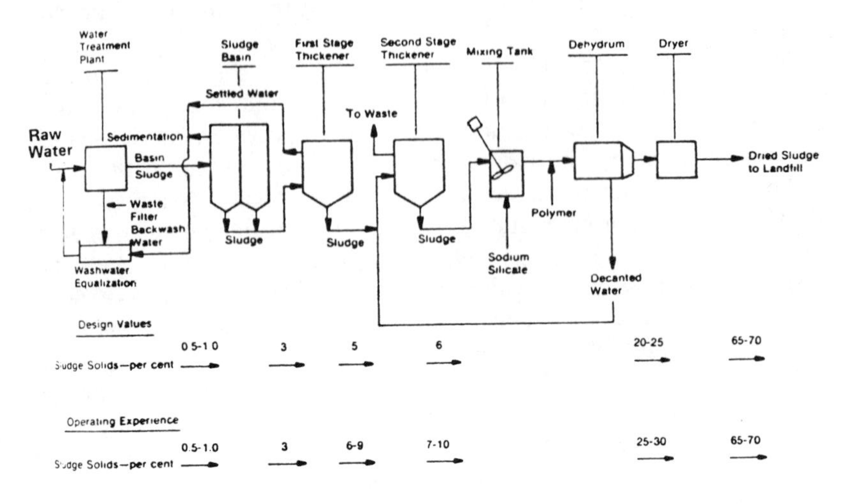

Figure 46. Sludge treatment system for Nishi-Nagasawa purification plant, Kanagawa Prefecture, Japan. Reproduced from Bishop [114], courtesy of the American Water Works Association.

then is concentrated further by an automatic filter press consisting of several septa. The conditioned sludge, 70% dry solids, is ready for ultimate disposal.

Coagulation Units

There are essentially two designs for coagulation units. The older conventional style is seen in Figure 48a [119] where the coagulating chemicals are added in the flash-mix portion of the tank. This is followed by flocculation in the baffled portion. Sedimentation follows in a rectangular tank, as seen in Figure 48b, after which the water is filtered. In cases where space is limited or a higher rate of treatment is required, the solids-contact clarifiers frequently are used. These are seen in Figure 48c along with the intimate mixing and recycling of the sludge with the incoming water. Automatic and selective recycling of the sludge occurs as well as its removal for additional treatment. These clarifiers can be operated at an upflow rate of 1 gpm/ft^2, which permits a higher rate of filtration to follow.

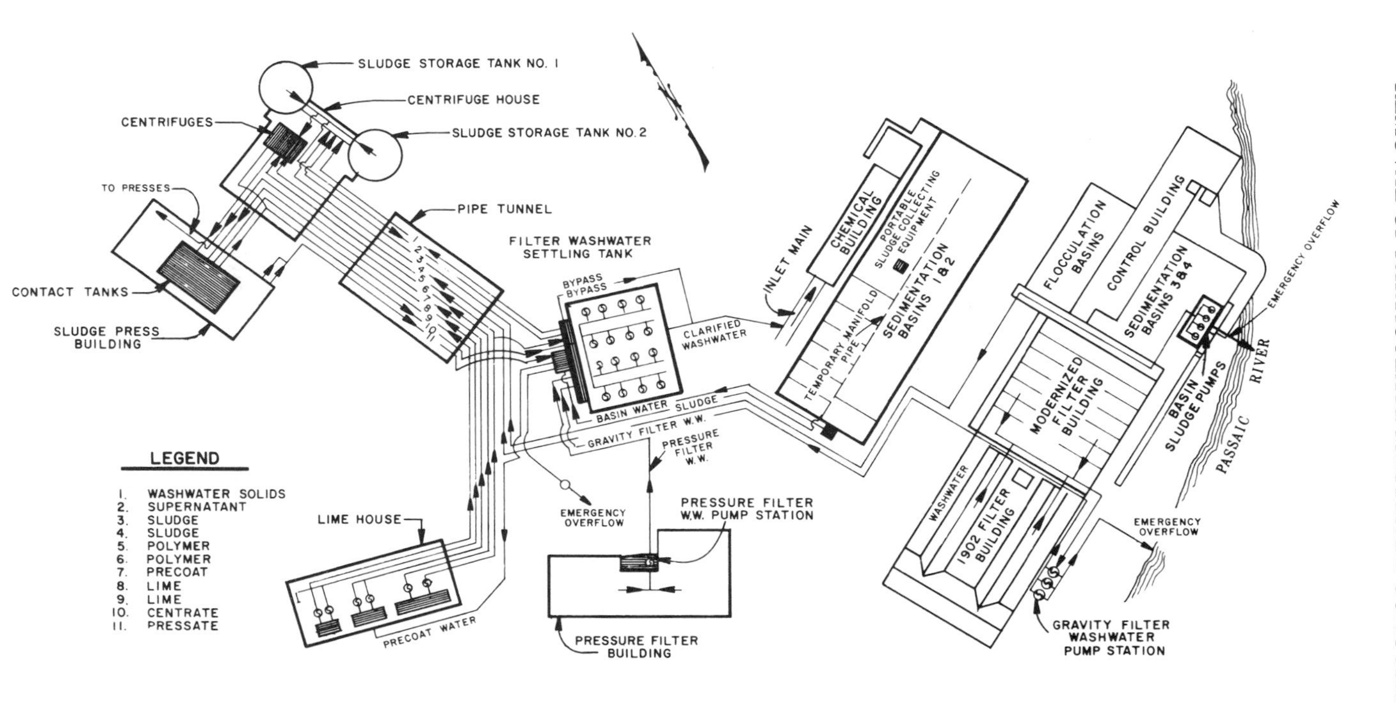

Figure 47. Plant schematic of sludge treatment at the Passaic Valley Water Commission, Little Falls, New Jersey.

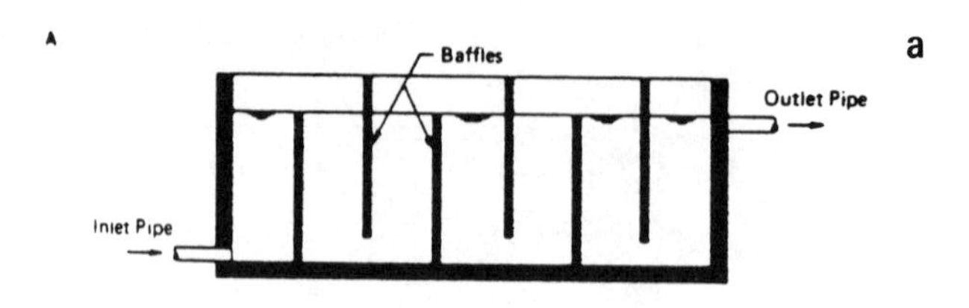
A
a
Baffles
Outlet Pipe
Inlet Pipe

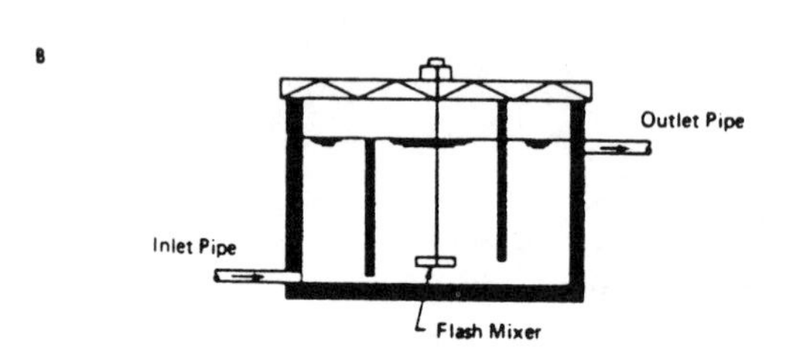
B
Outlet Pipe
Inlet Pipe
Flash Mixer

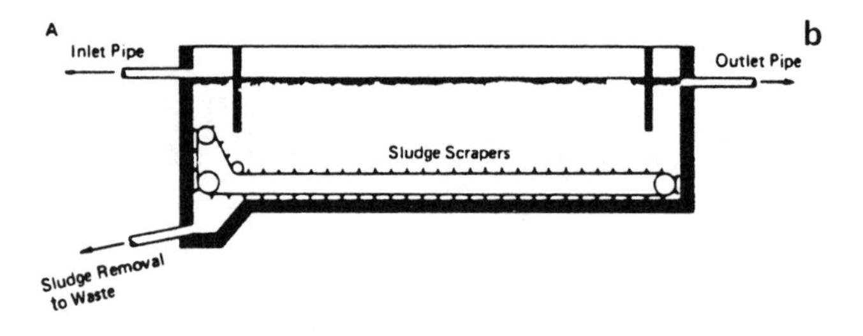
A
b
Inlet Pipe
Outlet Pipe
Sludge Scrapers
Sludge Removal
to Waste

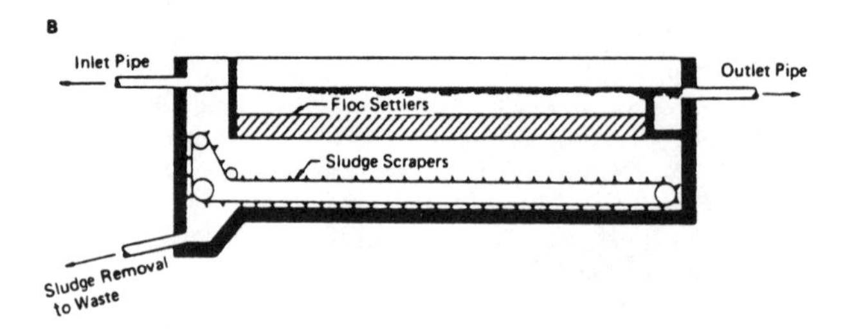
B
Inlet Pipe
Outlet Pipe
Floc Settlers
Sludge Scrapers
Sludge Removal
to Waste

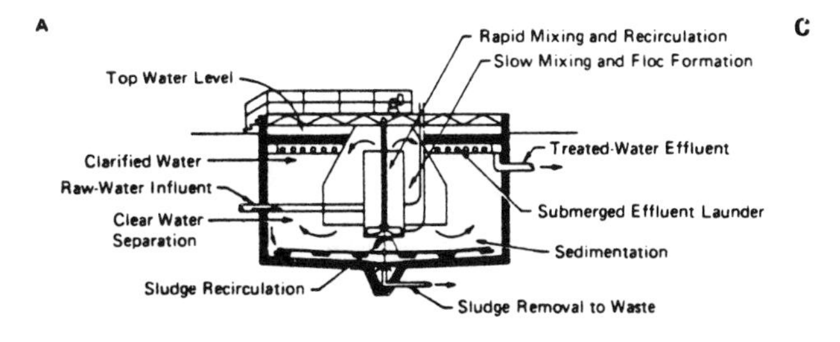

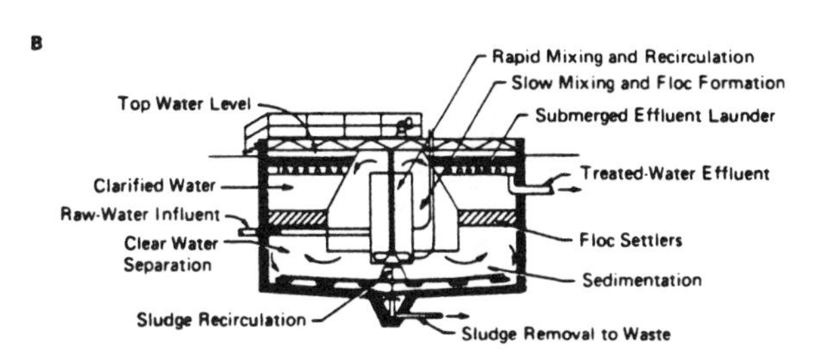

Figure 48. **(a)** Mixing tank; A = old style; B = modified. **(b)** Rectangular settling tank: A = standard; B = modified. **(c)** Clarifier: A = standard; B = modified. Reproduced from LaFontaine [119], courtesy of the American Water Works Association.

REFERENCES

1. LaMer, V. K., and T. W. Healey. *J. Phys. Chem.* 67:2417 (1963).
2. Baker, M. N. "The Quest for Pure Water," American Water Works Association, Denver, CO (1981).
3. Alpino, P. *Medicina Aegyptorum* (Venice, 1591).
4. "Alum: Medium of History," *Endeavour* 8(31):97 (1949).
5. Austen, P. T., and F. A. Wilbur. "Report of the State Geologist of New Jersey, 1884," Trenton, NJ (1885), p. 141.
6. Fuller, G. W. *Water Purification at Louisville* (New York: D. Van Nostrand Company, 1898).
7. Powell, S. T. *Water Conditioning for Industry* (New York: McGraw-Hill Book Company, 1954).
8. Cohen, J. M., and S. A. Hannah. In: *Water Quality and Treatment,* 3rd ed. (New York: McGraw-Hill Book Company, 1971).
9. Kemmer, F. N., Ed. *The Nalco Water Handbook* (New York: McGraw-Hill Book Company, 1979).
10. Stumm, W. *Environ. Sci. Technol.* 11:1066 (1977).

11. Priesing, C. P. *Ind. Eng. Chem.* 54:38 (1962).
12. Christman, C. H. *J. Am. Water Works Assoc.* 21:1076 (1929).
13. Mysels, K. J. *Introduction to Colloid Chemistry* (New York: John Wiley & Sons, Inc., 1959).
14. Black, A. P. *J. Am. Water Works Assoc.* 52:492 (1960).
15. Hückel, E. *Physik Z.* 25:204 (1924).
16. Briggs, D. R. *Ind. Eng. Chem.* (*Anal. Ed.*) 12:703 (1940).
17. Pilipovich, J. B. et al. *J. Am. Water Works Assoc.* 50:1467 (1958).
18. Black, A. P., and A. L. Smith. *J. Am. Water Works Assoc.* 57:485 (1965).
19. Black, A. P., and A. L. Smith. *J. Am. Water Works Assoc.* 58:445 (1966).
20. O'Melia, C. R. In: *Physicochemical Processes for Water Quality Control,* W. J. Weber, Jr., Ed. (New York: John Wiley & Sons, Inc., 1972).
21. Schulze, H. *J. Pr. Chem.* 25:431 (1882).
22. Hardy, W. B. *Proc. Roy. Soc.* Ser. A 66A:110 (1900).
23. Tamamuski, B., and K. Tamaki. *Kolloid. Zh.* 163:122 (1959).
24. Matijivic', E. et al. *J. Colloid. Sci.* 19:333 (1964).
25. Langelier, W. F., and H. F. Ludwig. *J. Am. Water Works Assoc.* 41:163 (1949).
26. La Mer, V. K., and T. W. Healy. *Rev. Pure Appl. Chem.* 13:112 (1963).
27. Black, A. P. et al. *J. Am. Water Works Assoc.* 57:1547 (1965).
28. Sillen, L. G., and A. E. Martell. "Stability Constants of Metal-ion Complexes," Special Publications 17 & 25, Chemical Society, London (1964, 1971).
29. Stumm, W., and C. R. O'Melia. *J. Am. Water Works Assoc.* 60:514 (1968).
30. Sullivan, J. H., Jr., and J. E. Singley. *J. Am. Water Works Assoc.* 60:1280 (1968).
31. Singley, J. C., and A. P. Black. *J. Am. Water Works Assoc.* 59:1549 (1967).
32. Singley, J. C., and J. H. Sullivan, Jr. *J. Am. Water Works Assoc.* 61:190 (1969).
33. Bartow, E., and B. H. Peterson. *Ind. Eng. Chem.* 20:51 (1928).
34. Black, A. P. et al. *Ind. Eng. Chem.* 25:811 (1933).
35. Packham, R. F. *J. Appl. Chem.* 12:564 (1962).
36. Larson, T. E., and A. M. Buswell. *Ind. Eng. Chem.* 32:132 (1940).
37. Modi, H. J., and D. W. Fuerstenau. *J. Phys. Chem.* 61:640 (1957).
38. Yopps, J. A., and D. W. Fuerstenau. *J. Colloid. Sci.* 19:61 (1964).
39. Mattson, S. *J. Phys. Chem.* 32:1532 (1928).
40. Black, A. P., and S. A. Hannah. *J. Am. Water Works Assoc.* 53:438 (1961).
41. Black, A. P. and C. L. Chen. *J. Am. Water Works Assoc.* 57:354 (1965).
42. Black, A. P. "Electrokinetic Characteristics of Hydrous Oxides of Aluminum and Iron," in *Principles and Applications of Water Chemistry,* S. D. Faust and J. V. Hunter, Eds. (New York: John Wiley & Sons, Inc., 1967).
43. Black, A. P., and J. V. Walters. *J. Am. Water Works Assoc.* 56:99 (1964).

44. Black, A. P., and C. L. Chen. *J. Am. Water Works Assoc.* 59:1173 (1967).
45. Kim, W. et al. *J. Am. Water Works Assoc.* 57:327 (1965).
46. Hall, E. S. *J. Appl. Chem.* 15:197 (1965).
47. Rubin, A. J., and H. Blocksidge. *J. Am. Water Works Assoc.* 71:102 (1979).
48. Cohen, J. M. *J. Am. Water Works Assoc.* 49:1425 (1957).
49. Packham, R. F. *J. Appl. Chem.* 12:564 (1962).
50. Hayden, P. L., and A. J. Rubin. In *Aqueous-Environmental Chemistry of Metals,* A. J. Rubin, Ed. (Ann Arbor, MI: Ann Arbor Science Publishers, Inc., 1974).
51. Rubin, A. J., and T. W. Kovac. In: *Chemistry of Water Supply, Treatment and Distribution,* A. J. Rubin, Ed. (Ann Arbor, MI: Ann Arbor Science Publishers, Inc., 1974).
52. McCooke, N. J., and J. R. West. *Water Res.* 12:793 (1978).
53. Gupta, V. S. et al. *J. Am. Water Works Assoc.* 67:21 (1975).
54. Ishibashi, T. *J. Am. Water Works Assoc.* 72:514 (1980).
55. Brown, D. W., and J. D. Hem. "Reactions of Aqueous Aluminum Species at Mineral Surfaces," U.S. Geological Survey Water Supply Paper 1827-F, Washington, DC (1975).
56. Faust, S. D., and O. M. Aly. *Chemistry of Natural Waters* (Ann Arbor, MI: Ann Arbor Science Publishers, Inc., 1981), Chapter 1.
57. Saville, T. *J. New England Water Works Assoc.* 31:79 (1917).
58. Black, A. P., and R. F. Christman. *J. Am. Water Works Assoc.* 55:753 (1963).
59. Black, A. P., and D. G. Willems. *J. Am. Water Works Assoc.* 53:589 (1961).
60. Black, A. P. et al. *J. Am. Water Works Assoc.* 55:1347 (1963).
61. Hall, E. S., and R. F. Packham. *J. Am. Water Works Assoc.* 57:1149 (1965).
62. Mangravite, Jr. F. J. et al. *J. Am. Water Works Assoc.* 67:88 (1975).
63. Herak, J., and B. Težak. *Ark. Kemi.* 26:1 (1954).
64. Narkis, N., and M. Rebhun. *J. Am. Water Works Assoc.* 69:325 (1977).
65. Babcock, D. B., and P. C. Singer. *J. Am. Water Works Assoc.* 71:149 (1979).
66. Haff, J. D. *J. Am. Water Works Assoc.* 70:520 (1978).
67. Maulding, J. S., and R. H. Harris. *J. Am. Water Works Assoc.* 60:460 (1968).
68. Lind, C. J., and J. D. Hem. "Effects of Organic Solutes on Chemical Reactions of Aluminum," U.S. Geological Survey Water Supply Paper 1827-G, Washington, DC (1975).
69. Stern, O. *Zeits. Electrochem.* 30:508 (1924).
70. Healy, T. W. In: *Organic Compounds in Aquatic Environments,* S. D. Faust and J. V. Hunter, Eds. (New York: Marcel Dekker, Inc., 1971).
71. Kane, J. C. et al. *J. Phys. Chem.* 67:1977 (1963).
72. Michaels, A. *Ind. Eng. Chem.* 46:1485 (1954).
73. Black, A. P. et al. *J. Colloid. Interface Sci.* 21:626 (1966).
74. Cohen, J. M. et al. *J. Am. Water Works Assoc.* 50:463 (1958).
75. Black, A. P. et al. *J. Am. Water Works Assoc.* 51:247 (1959).

76. Black, A. P. et al. *J. Am. Water Works Assoc.* 57:1547 (1965).
77. Pressman, M. *J. Am. Water Works Assoc.* 59:169 (1967).
78. Sommerauer, A. et al. *Kolloid-Z. Z. Polym.* 225:147 (1968).
79. Narkis, N., and M. Rebhun.*J. Am. Water Works Assoc.* 67:101 (1975).
80. Terrell, D. L. *J. Am. Water Works Assoc.* 69:263 (1977).
81. Stumm, W., and J. J. Morgan. *Aquatic Chemistry,* 2nd ed. (New York: John Wiley & Sons, Inc., 1981).
82. Swift, D. L., and S. K. Friedlander. *J. Colloid. Sci.* 19:621 (1964).
83. Smoluchowski, M. J. *Phys. Chem.* 92:129 (1917).
84. Camp, T. R., and P. C. Stein. *J. Boston Soc. Civ. Eng.* 30:219 (1943).
85. Camp, T. R. *Trans. Amer. Soc. Civ. Eng.* 120:1 (1955).
86. Birkner, F. B., and J. J. Morgan. *J. Am. Water Works Assoc.* 60:175 (1968).
87. Hahn, H. H., and W. Stumm. *J. Colloid. Interface Sci.* 28:134 (1968).
88. Morrow, J. J., and E. G. Rausch. *J. Am. Water Works Assoc.* 66:646 (1974).
89. Bratby, J. R. *J. Am. Water Works Assoc.* 73:312 (1981).
90. Bratby, J. R. *J. Am. Water Works Assoc.* 73:318 (1981).
91. Letterman, R. D. et al. *J. Am. Water Works Assoc.* 71:467 (1979).
92. Griffith, J. D., and R. G. Williams. *J. Am. Water Works Assoc.* 64:825 (1972).
93. Hannah, S. A. et al. *J. Am. Water Works Assoc.* 59:1149 (1967).
94. Black, A. P., and S. A. Hannah. *J. Am. Water Works Assoc.* 57:901 (1965).
95. Moffett, J. W. *J. Am. Water Works Assoc.* 60:1255 (1968).
96. Tekippe, R. J., and R. K. Ham. *J. Am. Water Works Assoc.* 62:594 (1970).
97. Tekippe, R. J., and R. K. Ham. *J. Am. Water Works Assoc.* 62:620 (1970).
98. Kawamura, S. *J. Am. Water Works Assoc.* 65:417 (1973).
99. Kavanaugh, M. C. *J. Am. Water Works Assoc.* 70:613 (1978).
100. Committee Report. *J. Am. Water Works Assoc.* 71:588 (1979).
101. Semmens, M. J., and T. K. Field. *J. Am. Water Works Assoc.* 72:476 (1980).
102. Sridharan, N., and G. F. Lee. *Environ. Sci. Technol.* 6:1031 (1972).
103. Carollo, J. A. *J. Am. Water Works Assoc.* 37:1310 (1945).
104. Whitehouse, J. D. University Kentucky—Water Resources Institute, Research Report No. 8, Lexington, KY (1967).
105. Cohen, J. M. et al. *J. Am. Water Works Assoc.* 52:1551 (1960).
106. Robeck, G. C. et al. *J. Am. Water Works Assoc.* 57:181 (1965).
107. Aly, O. M., and S. D. Faust. *J. Am. Water Works Assoc.* 57:221 (1965).
108. Faust, S. D., and A. Zarins. *Residue Rev.* 29:151 (1969).
109. Nicholson, H. P. et al. *Limnol. Oceano.* 9:310 (1964).
110. El-Dib, M. A., and O. A. Aly. *Water Res.* 11:611 (1977).
111. Thebault, P. et al. *Water Res.* 15:183 (1981).
112. Letterman, R. D. et al. *J. Am. Water Works Assoc.* 62:652 (1970).
113. Albrecht, A. E. *J. Am. Water Works Assoc.* 64:46 (1972).
114. Bishop, S. L. *J. Am. Water Works Assoc.* 70:503 (1978).
115. Hudson, H. E. Jr. *Proceedings of the 10th Sanitary Engineering Conference,* (Urbana, IL: University of Illinois, 1968).

116. Schmitt, C. R., and J. E. Hall. *J. Am. Water Works Assoc.* 67:40 (1975).
117. Fulton, G. P. *J. Am. Water Works Assoc.* 63:312 (1974).
118. Passaic Valley Water Commission, Little Falls, NJ (1976).
119. LaFontaine, E. O. *J. Am. Water Works Assoc.* 63:59 (1974).

JAWWA

64:825
59:1149
57:901
60:1225
62:594
62:620
65:417

CHAPTER 6

REMOVAL OF PARTICULATE MATTER BY FILTRATION

Christopher Uchrin
Department of Environmental Science
Cook College
Rutgers, The State University
New Brunswick, New Jersey

HISTORY OF THE FILTRATION PROCESS

Filtration is the process of removing solids from a fluid by passing it through a porous medium. It is widely used in water treatment to remove solids, including bacteria present in surface waters, precipitated hardness from lime-softened waters, and precipitated iron and manganese. Filtration is employed in wastewater treatment as a finishing process beyond secondary biological treatment or to remove solids remaining after precipitation or chemical coagulation in physicochemical processing. This chapter will focus primarily on deep granular filters (sand, dual-media and multimedia), although some attention will be given to precoat filters (diatomaceous earth).

The use of filtration in water treatment dates back to the slow sand filters used in London, England, as early as 1829. Filters of this type operated at a rate of 2–4 mgad (million gallons per acre per day) through a bed depth of 3 ft and served as the sole treatment device. The famous Broad Street (London) well epidemic in the early 1850s generated an international movement to require the filtration of all potable water. Rapid sand filters operating at rates of 1–4 gpm/ft^2 were subsequently designed in the United States in the late nineteenth century to minimize the required surface area and capital cost to achieve the desired capacity. Today's deep-bed (18–30 in.), granular filters utilize sand or sand in combination with other materials as the filter medium.

The use of precoat filters is quite limited. Diatomaceous earth filters are commonly used for swimming pool filtration and some industrial applications. Other applications have used various media, including perlite and powdered activated carbon.

FILTRATION MECHANISMS

Filtration involves a complex variety of mechanisms, which includes transport, attachment and detachment. Transport mechanisms bring a small particle from the bulk fluid to the surface of the filter medium. Important transport mechanisms include screening, interception, inertial forces, gravitational settling, diffusion and hydrodynamic conditions. These mechanisms and the efficiency of the removal of suspended solids depend on such physical characteristics as the size distribution of the filter medium, filtration rate, temperature and density and size of the suspended particles. Mathematical models for these transport mechanisms are presented below.

An attachment mechanism is required to retain the suspended particle once it has approached the surface of the filter medium or previously attached solids. Attachment mechanisms include straining, van der Waals forces, electrostatic interactions, chemical bridging and specific adsorption. The efficiency of the attachment mechanisms can be affected by coagulants applied in pretreatment and the chemical characteristics of the water and the filter medium.

The role of detachment as a filtration mechanism is also important. During deep granular filtration, sediments build in the filter, removed by the aforementioned mechanisms. The particles held in the filter are thus in equilibrium with hydraulic shearing forces that tend to tear them away and wash them deeper or through the filter. Since the presence of solids causes the interstitial volume to decrease, this causes an increase in the interstitial velocity and the clogged layers become less efficient at removing solids which then pass on to the lower, less clogged, portion of the media [1–3]. Several investigators including Tuepker and Buescher [4] and Cleasby et al. [5] examined detachment as a function of an increase in filtration rate. As this rate is increased, the hydraulic shearing forces also increase, which disturbs the existing equilibrium between the deposited solids and the water.

FILTRATION HYDRAULICS

Under conditions commonly employed in water treatment, the hydraulics of flow in a filter are the same as the hydraulics of groundwater flow. Thus, flow in a clean and even clogged, filter bed is laminar, and Darcy's law applies [6]. This law states that water velocity in a porous medium is proportional to the slope of the hydraulic gradient [7] or:

$$v = K_p S_l \tag{1}$$

where v = superficial flow velocity (L/t)
K_p = coefficient of permeability (L/t)
S_l = hydraulic gradient $= h_f/L$ (dimensionless)
h_f = head loss (L)
L = length (depth) of filter (L)
t = time

The consideration of head or energy lost during the filtration process is important. As solid particles are removed, the void spaces available for flow decrease. Since these void spaces can be considered analogous to small pipes, any conventional expression for head or energy loss can be applied, of which an example is the Darcy-Weisbach relationship [8]:

$$h_f = f \frac{L}{D} \frac{V^2}{2g} \tag{2}$$

where f = Darcy-Weisbach friction factor (dimensionless)
D = pipe diameter (L)
V = mean pipe flow velocity (L/t)
g = gravitational acceleration (L/t^2)

It can be noted that as the pipe diameter decreases, the head loss increases. It is likewise true in filtration: as filtration proceeds, the solids fill up the void spaces and the head loss increases. Application of Equation 2 to porous media yields the classical Carman-Kozney equation (7):

$$h_f = f_1 \left(\frac{L}{\phi d}\right)\left(\frac{1-\epsilon}{\epsilon^3}\right)\left(\frac{v^2}{g}\right) \tag{3}$$

where f_1 = friction factor $= 150(1-\epsilon)/N_R + 1.75$ (ft)
ϕ = particle shape factor (dimensionless); $\phi = 1$ for spherical particles
ϵ = bed porosity (dimensionless)
N_R = Reynolds number $= \phi(\rho v d)/\mu$ (dimensionless)

d = media particle diameter (L)
ρ = density of water, (M/L^3)
μ = absolute viscosity of water (M/Lt)

Equations have been suggested by other investigators, including Camp [9] and Rose [10].

Initial head losses in clean filters commonly range from less than 1 ft to 2.5 ft, depending on the particle size distribution of the media and the overflow rate (flow/filter surface area). The pressure distribution within the media of a typical gravity filter is displayed in Figure 1. Negative head (less than atmospheric pressure) occurs in a gravity filter when the summation of head loss from the media surface downward exceeds the available pressure. Air pockets that reduce the effective filtering area, thus increasing the filtration rate, and head loss through the remaining area can occur in the negative pressure zone. This results in a degradation of filter effluent quality. Negative head conditions can be eliminated by the use of pressure filters.

FILTRATION RATE PATTERNS

The efficiency of a filter is more sensitive to changes in rate rather than the actual rate itself. With pretreatment and proper design, Baumann

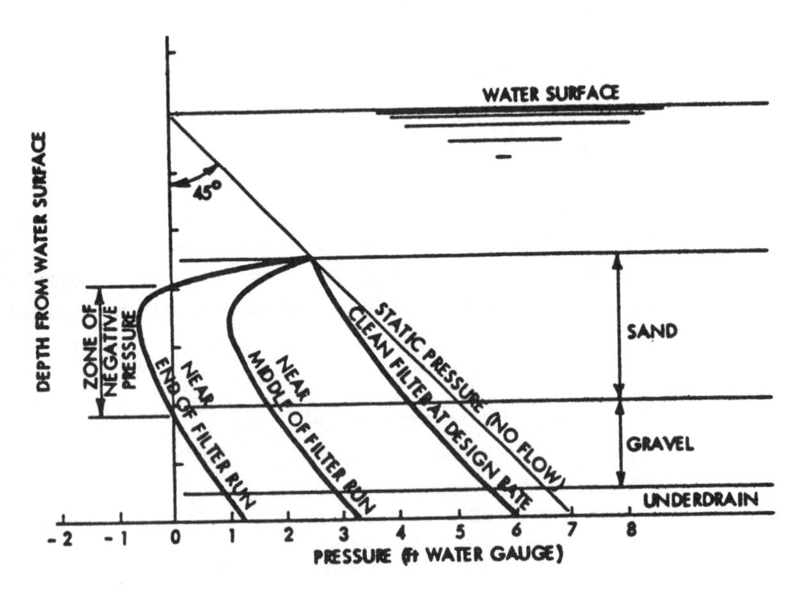

Figure 1. Pressure vs depth in a gravity filter at various times during a filter run. Reproduced from Weber [6], courtesy of John Wiley & Sons, Inc.

[11] showed that there is little difference in water quality from filters operated at rates between 2 and 6 gpm/ft^2 (0.08–0.24 m^3/m^2-min) as displayed in Figures 2 and 3. Cleasby et al. [5], however, investigated the effect of rate disturbances on quality. A representative run from their work is displayed in Figure 4. Tuepker and Buescher [4] evaluated the effect on performance by the severity of rate changes. An example of their work, displayed in Figures 5 and 6, shows that gradual rate changes did not affect performance as severely as rapid changes, and that the addition of a coagulant virtually damped out all effects. These characteristics are considered in the methods for filter operation. The three basic

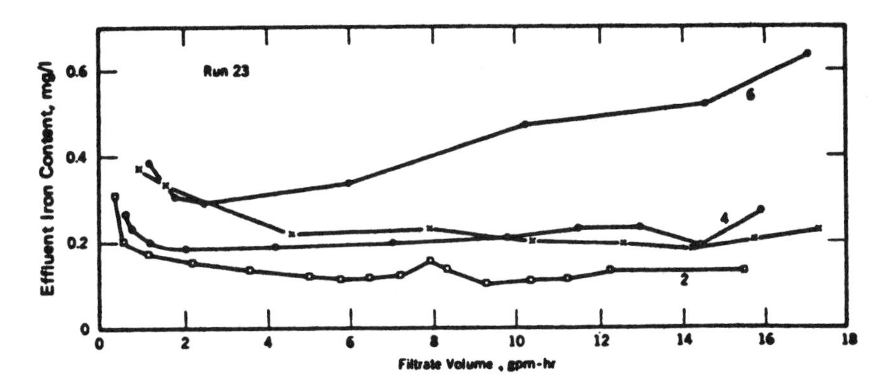

Figure 2. Effluent iron content and filtrate volume. Numbers on curves indicate filtration rate. Curve X indicates constant pressure. Initial rate was 6 gpm/ft^2, which decreased as head loss increased. Reproduced from Baumann [11], courtesy of Ann Arbor Science Publishers.

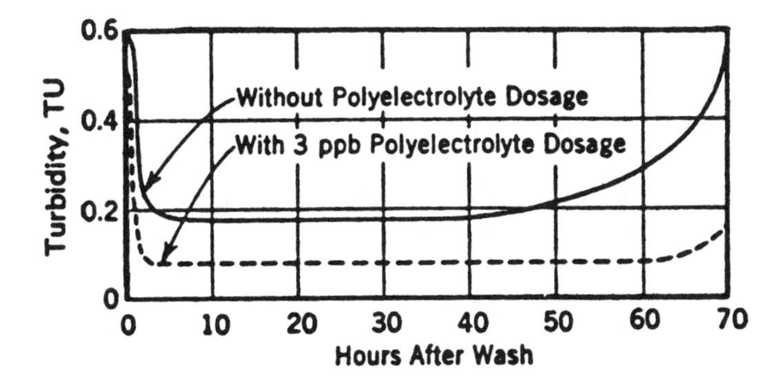

Figure 3. Effluent turbidity with constant filtration rate of 3.5 gpm/ft^2. Reproduced from Baumann [11], courtesy of Ann Arbor Science Publishers.

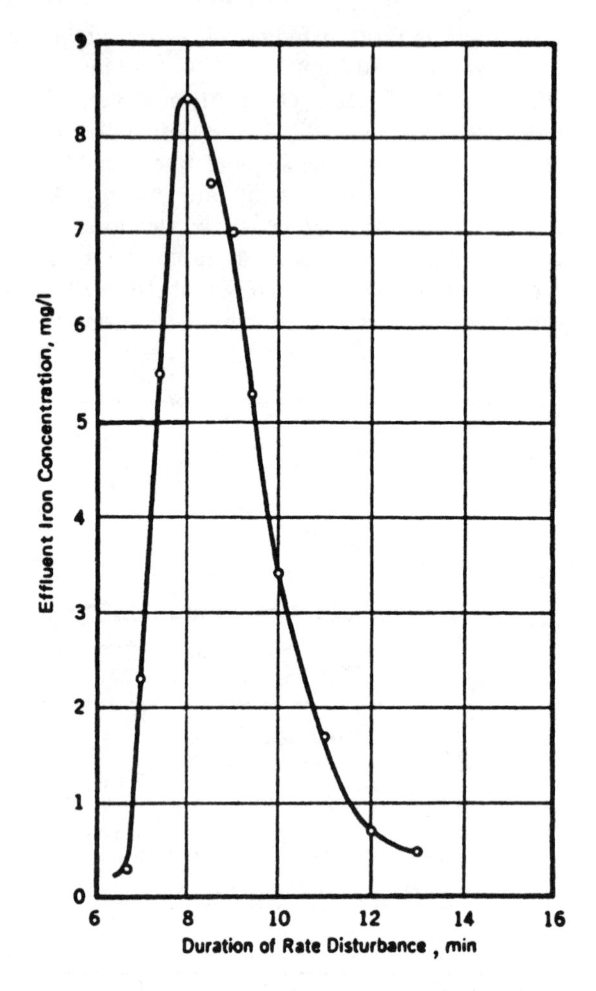

Figure 4. Effect of typical rate disturbance on effluent quality. This run had an instantaneous rate change from 2. to 2.5 gpm/ft^2. Reproduced from Cleasby et al. [5], courtesy of the American Water Works Association.

methods are constant-pressure filtration, constant-rate filtration and variable declining-rate filtration [12].

True constant-pressure filtration effectually provides declining-rate filtration. At the beginning of a filter run, the filter resistance is small. As the filter medium begins to clog, the flowrate decreases. Current practice does not employ such a method, since it requires the storage of a large volume of water [6]. This method is represented by trace A in Figure 7.

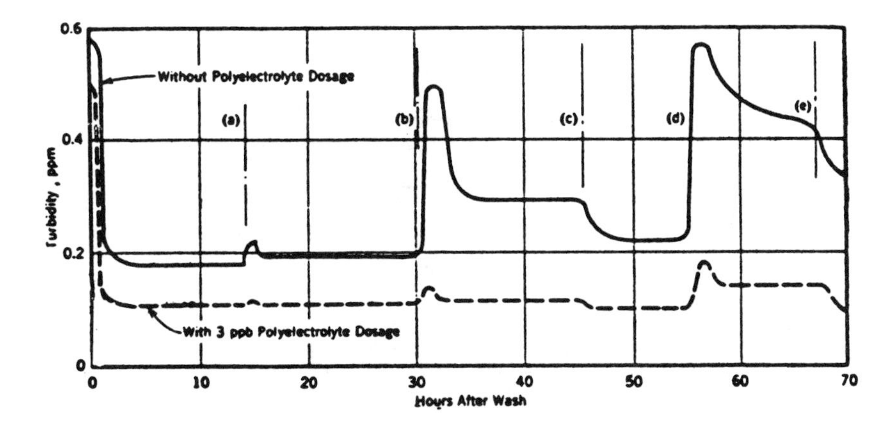

Figure 5. Effluent turbidity with rapid rate changes. **(a)** Rate increased to 2.5 gpm/ft^2 within 10 sec. **(b)** Rate increased to 3.5 gpm/ft^2 within 10 sec. **(c)** Rate decreased to 2.5 gpm/ft^2. **(d)** Rate increased to 3.5 gpm/ft^2 within 10 sec. **(e)** Rate decreased to 2.5 gpm/ft^2. Reproduced from Tuepker and Buescher [4], courtesy of the American Water Works Association.

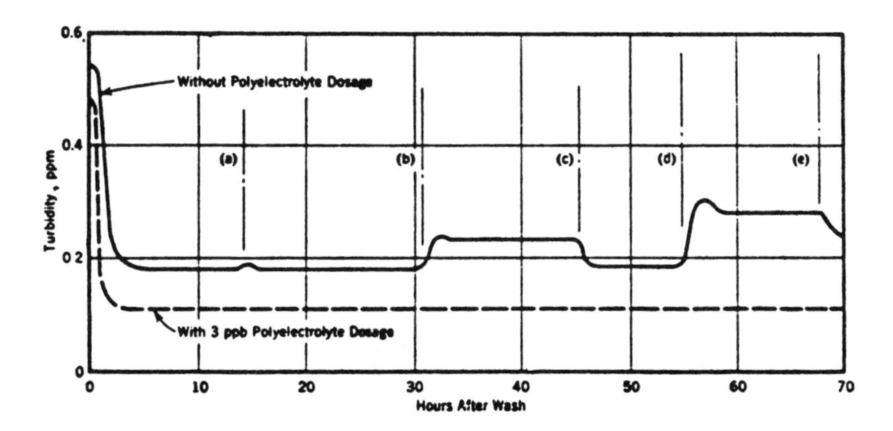

Figure 6. Effluent turbidity with gradual rate changes (over a 10-min period). Reproduced from Tuepker and Buescher [4], courtesy of the American Water Works Association.

Constant- and declining-rate filtration are more commonly employed in current practice. A constant pressure is maintained across the filter system, and the filtration rate is then held constant by means of a flow control valve. Trace B in Figure 7 represents this method. As the filter

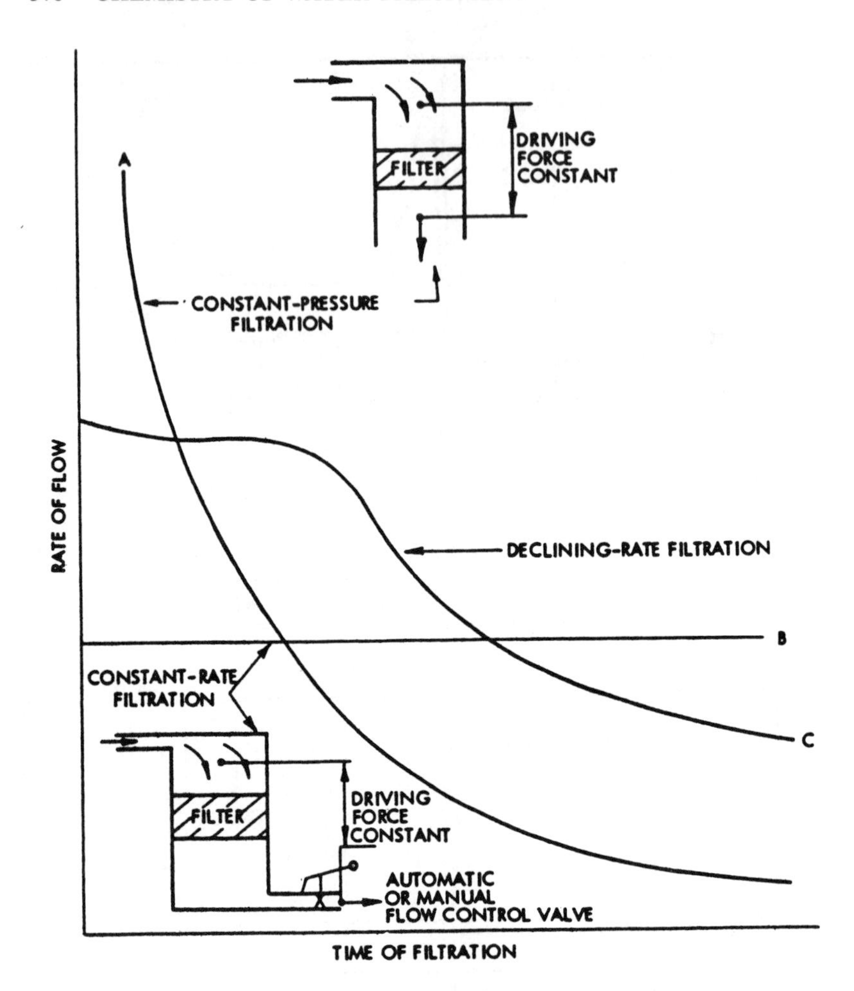

Figure 7. Typical rate of filtration patterns during a filter run. Reproduced from Baumann and Oulman [13], courtesy of the University of Texas Press.

resistance increases during operation (due to solids clogging), the flow control valve is gradually opened to maintain a constant flow. Several disadvantages have commonly been associated with this method, including high capital and maintenance costs, a poorer effluent quality than that obtained using declining-rate filtration [14,15] and frequency of equipment failure [11]. Alternative methods of flow control include flow splitting and declining-rate filtration.

Flow splitting usually involves the use of an influent weir box on each filter to split the flow nearly equally among the operating filters. Beside the obvious benefit of achieving constant-rate filtration without rate controllers, other benefits include gradual rate changes when bringing filters into or out of service [4,5] and head loss in the filter evidenced by the water level in the filter box. In addition, since the effluent control weir must be located above the sand to prevent dewatering of the bed, the possibility of a negative head in the filter is eliminated. The only disadvantage of this system is the additional depth required of the filter box to accommodate raising of the effluent outlet to a position above the filter media surface. A schematic of this method is displayed in Figure 8.

Variable declining-rate filtration is currently the most widely designed method of operation for gravity filters [11]. Its operation is similar to influent flow splitting as a comparison of its schematic (Figure 9) with Figure 8 will show. The principal differences involve the location and type of influent arrangement and the provision of less available head. The filter influent enters below the low water level (wash trough level) of the filters. The water level is essentially the same in all filters at the same time, except during backwashing. This is provided by a large influent header with large influent valves to each filter, thus resulting in small head losses. As a filter accumulates solids, consequently restricting its flow, the cleaner filters automatically pick up the lost capacity. The additional head required to accomplish this is provided by a rise in the water

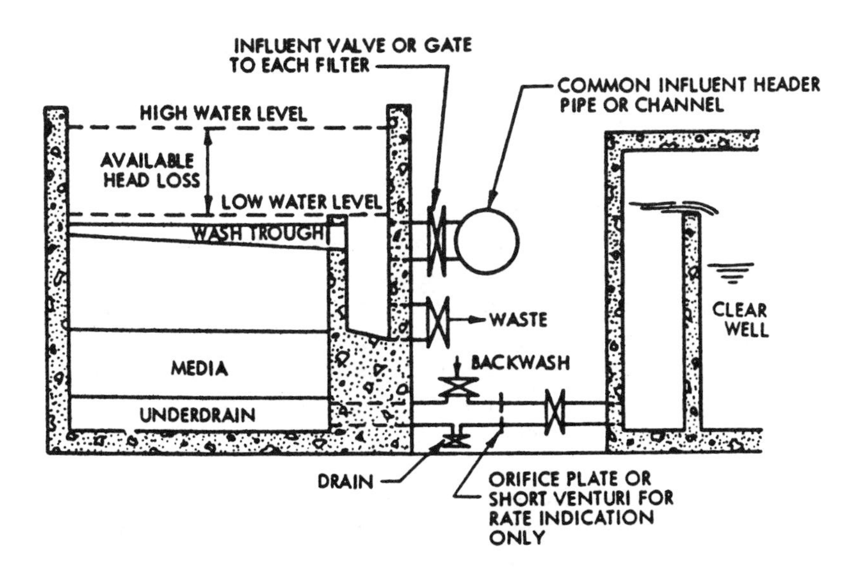

Figure 8. Gravity filter arrangement for rate control by influent flow splitting. Reproduced from Weber [6], courtesy of John Wiley & Sons, Inc.

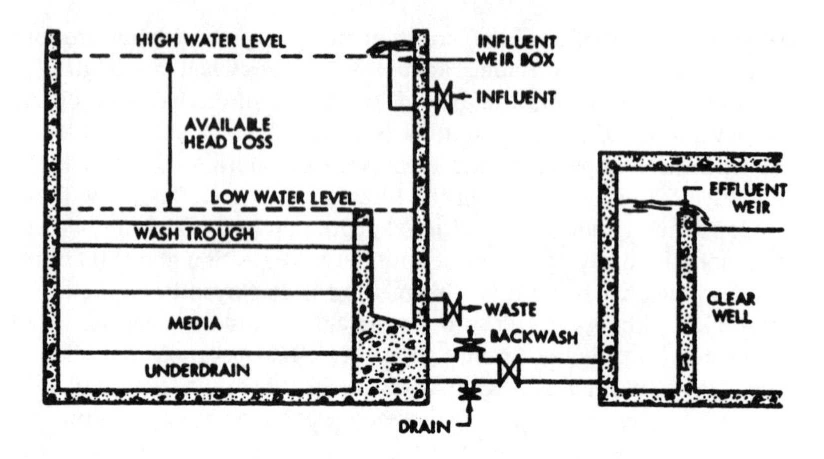

Figure 9. Gravity filter arranged for variable declining-rate of filtration. Reproduced from Weber [6], courtesy of John Wiley & Sons, Inc.

level of each filter. Typical water level variations in potable water plants range 1.5–2 ft [16]. Trace C in Figure 7 illustrates this method's rate pattern. Advantages of this method include a better effluent water quality toward the end of a run than that achieved by the constant-rate method, and a lesser head requirement due to the decrease in rate as the run progresses and head losses increase.

FILTER MEDIA

The quality of a filtered water is dependent on the media size and depth. Granular filter media commonly used in potable water filtration include silica sand [specific gravity (SG) = 2.65], anthracite coal (SG = 1.35–1.75) and garnet sand (SG = 4.0–4.4).

For many years, sand served as the sole medium for deep-bed filtration. Due to head loss considerations, it was found that a fine sand retained more material, but tended to shorter runs than those obtained using a coarse sand. Filter sand is classed by its effective size (10th percentile) and uniformity coefficient (60th percentile size/10th percentile size).

Because the media contains nonuniformly sized particles, size gradation will occur in a bed after the fluidization process of backwashing. Since the smaller particles will migrate to the top of the bed, greater removals and correspondingly higher head losses will be associated with this area. It would be more desirable to have the converse occur, with the finer particles serving for final polishing of the effluent. Upflow systems have been tried, but are not popular due to technical operation and design problems. To "reverse" this problem of size gradation, the practice of dual- or multimedia filtration was developed.

Conley and Pitman [17] were among the first to attempt dual-media filtration. They capped a sand filter with a layer of less dense but coarser anthracite coal. Thus, after backwashing, the coal will associate to a top layer, while the sand associates lower. Multimedia filters, with as many as five layers of hydraulically separable materials, have been used with success [11].

FILTER BACKWASHING

When the media become so clogged during filtration that head losses become excessive (generally about 8–9 ft) the unit must be removed from service and the media must be cleaned. This may be accomplished by scouring the clogged portion of the bed or by reversing the flow through the bed. This procedure of expanding the bed and washing out the trapped particles is called backwashing.

A schematic of a filter during backwashing is shown in Figure 10. Fluidization of a bed occurs when the upward or backwash velocity equals the critical velocity of the particles. Increasing the backwash velocity causes a concomitant expansion of the bed, which can allow open space for the trapped particles to be washed away. As the critical velocity, V_c, of the particles is reached, the frictional resistance of the particles will balance the head loss of the fluid expanding the bed, such that:

$$h_f \rho g = (\rho_s - \rho) g (1 - \epsilon_e) D_e \tag{4}$$

where h_f = head loss incurred during filter expansion (L)
ρ_s = density of media particles (M/L^3)
D_e = depth of expanded bed (L)
ϵ_e = porosity of expanded bed (dimensionless)

For a uniform bed of sand, Fair et al. [18] determined that:

$$\epsilon_e = (V_s/v_s)^{0.20} \tag{5}$$

where V_s = backwash velocity (L/t)
v_s = terminal fall velocity of media particles (L/t)

Therefore, a uniform bed of sand will expand when:

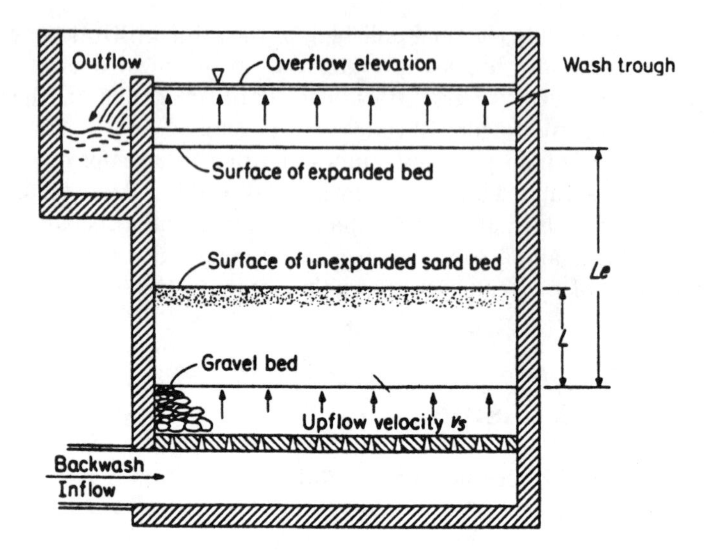

Figure 10. Expansion of a filter bed during backwashing. Reproduced from Clark and Viessman [7], courtesy of the International Textbook Co.

$$V_s = v_s \epsilon_e^{5.0} \tag{6}$$

Since the volume of sand particles of the expanded bed equals the volume of sand particles of the unexpanded bed:

$$D(1 - \epsilon)As = D_e(1 - \epsilon_e)A_s \tag{7}$$

where D = depth of unexpanded bed (L)
A_s = surface area of filter (L^2)

Therefore:

$$\frac{D_e}{D} = \frac{(1 - \epsilon)}{(1 - \epsilon_r)} \tag{8}$$

For multimedia beds, expansion occurs successively for each different layer. The upper layers will be expanded at lower backwash rates than the lower layers. Full expansion will be obtained when Equation 6 is satisfied for particles with the greatest size. The relative expansion of the bed is given by:

$$\frac{D_e}{D} = (1 - \epsilon) \sum_{i=1}^{n} \frac{P_i}{(1 - \epsilon_{ei})} \tag{9}$$

where i = layer number (dimensionless)
n = number of layers (dimensionless)
P_i = weight fraction of particles in layer (dimensionless)
ϵ_{ei} = expanded porosity of particles in layer (dimensionless)

FILTRATION MATHEMATIC MODELS

Several investigators have attempted mathematically to model deep-bed filtration processes. In general, two approaches have been tried, one of which employs measurable macroscopic variables of filtration, including filtration rate, grain size and water viscosity, to model head loss and filtrate quality as a function of time. The other approach uses a more deterministic, microscale approach, which, at the outset, employs submodels describing the mechanics of removal. Each is discussed briefly.

Iwasaki [19] originally noted that the rate of solids removal per unit depth of filter was proportional to the local concentration of solids, mathematically:

$$-\frac{\partial C}{\partial L} = \lambda C \tag{10}$$

where C = suspended solids concentration (M/L^3)
L = depth of filter (L)
λ = filter coefficient (1/L)

The filter coefficient λ is a function of both time and position in the filter.

Since filtration results in an accumulation of solids in the filter pores, the decrease of suspended solids in a filter lamina must balance the increase in solids depositing, or:

$$\upsilon \frac{\partial C}{\partial L} = \frac{\partial \Omega_d}{\partial t} + (\epsilon - \Omega_d)\frac{\partial C}{\partial t} \tag{11}$$

where Ω_d = volume of specific deposit per unit filter volume (dimensionless)

υ, C, L and ϵ are as before. The term $(\epsilon - \Omega_d)\partial C/\partial t$ is generally considered insignificant, since it represents the change in the amount of material in suspension within the pores with time.

Ives [2] proposed specific surface models for spherical particles and cylindrical capillaries, incorporating them into a general relationship between λ and Ω:

$$\lambda = \lambda_o \left[1 + \frac{\beta_p \Omega_d}{\epsilon}\right]^y \left[1 - \frac{\Omega_d}{\epsilon}\right]^z \left[1 - \frac{\Omega_d}{(\Omega)_\mu}\right]^x \tag{12}$$

where λ_o = filter coefficient for a clean filter (1/L)
β_p = packing constant of media (dimensionless)
$(\Omega_d)_\mu$ = ultimate volume of specific deposit per unit filter volume (dimensionless)
x,y,z = coefficients (dimensionless)

The coefficients, x, y and z enable this model to accommodate the work performed by other investigators. A value of unity for each leads to Ives' original [20] equation.

Substituting Equation 12 into Equation 10:

$$-\frac{\partial C}{\partial t} = \lambda_o \left[1 + \frac{\beta_p \Omega_d}{\epsilon}\right]^y \left[1 - \frac{\Omega_d}{\epsilon}\right]^z \left[1 - \frac{\Omega_d}{(\Omega_d)_\mu}\right]^x \tag{13}$$

yields an equation for filtrate quality as a function of time. Ives [2] also developed an expression for head loss:

$$\frac{\partial h_f}{\partial L} = \left(\frac{\partial h_f}{\partial L}\right)_o \left[1 + (2\beta_p + 1)\frac{\Omega_d}{\epsilon} + (\beta_p + 1)^2\left(\frac{\Omega_d}{\epsilon}\right)^2 + (\beta_p + 1)^3\left(\frac{\Omega_d}{\epsilon}\right)^3 + \cdots\right] \tag{14}$$

Several investigators, including O'Melia [21] and Yao et al. [22], have attempted more deterministic type models. Filtration is considered to be primarily dependent on particle transport and attachment. An analogy is drawn between filtration and coagulation/floccuation. The attachment mechanism involves particle destabilization as in coagulation. The transport mechanism is similar to flocculation where the particles must be moved from the bulk fluid to near the vicinity of the solid-liquid interface.

These models result in a single collector efficiency η_c which accounts for Brownian diffusion (for small particles about 1 μm in size), settling, and interception (both for larger particles). The equation for diffusional efficiency is given by:

$$(\eta_c)_D = 4.04\,Pe^{-2/3} \tag{15}$$

where Pe = Péclet number (dimensionless) given by:

$$Pe = \frac{\upsilon' d_c}{D_p} \tag{16}$$

where υ' = velocity at an infinite distance from the collector (L/T)
d_c = collector diameter (L)
D_p = particle diffusivity (L^2/t) given by:

$$D_p = \frac{\bar{k}T}{3\pi\mu d_p} \tag{17}$$

where $\bar{k}$ = Boltzmann's constant (ML^2/t^2)
T = absolute temperature (°K)
μ = absolute viscosity (M/Lt)
d_p = particle diameter (L)

The interception efficiency is:

$$(\eta_c)_I = -\frac{3}{2}\left(\frac{d_p}{d_c}\right)^2 \tag{18}$$

The gravitational efficiency is based on Stokes' law:

$$(\eta_c)_G = \frac{(\rho_s - \rho_l)gd_p^2}{18\mu\upsilon} \tag{19}$$

where ρ_s and ρ_l are the respective densities of the particle and the liquid.
The overall efficiency is then given by:

$$\eta_c = (\eta_c)_D + (\eta_c)_I + (\eta_c)_G \tag{20}$$

Iwasaki's equation (Equation 10) can be modified to:

$$\frac{\partial C}{\partial L} = -\frac{3}{2}\left[\frac{(1-\epsilon)}{d_c}\right]\eta\eta_c C \tag{21}$$

where η = collision efficiency factor (0 for no adhesion, 1 for total adhesion).

Integration of Equation 21 yields:

$$C = C_o \exp\left[-\frac{3}{2}(1-\epsilon)\eta\eta_c\frac{L}{d_c}\right] \tag{22}$$

Yao et al. [22] used this model to demonstrate the relative efficiencies with respect to particle size. Figure 11 shows some of their results.

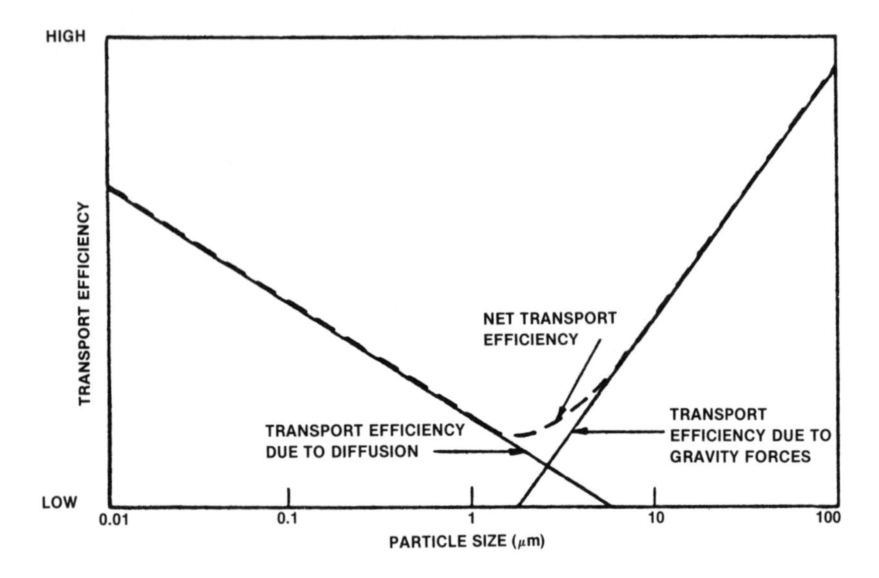

Figure 11. Relationship between transport efficiency and particle size. Reproduced from Yao [22], courtesy of the American Chemical Society.

TYPES OF FILTERS

Slow Sand Filters

A slow sand filter generally consists of a watertight basin containing a layer of sand over a layer of gravel. Common thicknesses for the sand are 3–5 ft, and 0.5–1 ft for the gravel. The filter is operated with a water depth of 3–5 ft above the sand surface. A *schmutzdecke* forms, in which most of the solids are removed. The *schmutzdecke* is a layer of microorganisms which facilitate removal of particles. When the head lost through the filter becomes excessive (commonly 3–5 ft), the filter is removed from service, drained and cleaned. Filter runs can range from 1 to 6 months.

Rapid Sand Filters

Rapid sand filters generally consist of a 2- to 2.5-ft layer of sand over a layer of gravel. Filter depths usually range from 8 to 9 ft. Multiple, rectangular units are generally used with a length/width ratio of 1.25. Individual units usually range in size from 450 to 4500 ft^2. Units are run until head losses become excessive (generally 8–9 ft) and are then backwashed.

Precoat Filters

Precoat filters generally consist of a septum (which supports the filter medium), the filter medium, a housing and a pump for operation in either pressure or vacuum mode. The filter medium is generally either diatomaceous earth or perlite. Diatomaceous earth (diatomite) is composed of fossil skeletons of diatoms which have been subjected to crushing, calcining, air classification and other processes. The resultant is a fine, porous, multishaped medium ranging in size from 5 to 50 μm.

Perlite is a siliceous rock, similar in outward appearance to diatomaceous earth. Perlite has a bulk dry density of 10–15 lb/ft as compared to 18–20 lb/ft for diatomite. Precoat filter applications include industrial and swimming pool filtration. They have been employed in some small-scale municipal applications.

RECENT DEVELOPMENTS IN FILTRATION

Direct Filtration

In the 1970s, direct filtration has received considerable attention in the treatment of drinking water [23–25]. This treatment is defined as filtration that is not preceded by sedimentation. This definition includes flow sheets that utilize either flocculation basins or contact basins without sludge collection equipment. Figure 12a shows a flow sheet typical of older rapid sand filter plants, whereas Figure 12b shows design trends in recent years for conventional filter plants.

Figures 12c and 12d show two options available for direct filtration. One flow sheet shows the addition of alum to a rapid-mix influent, which is followed by an application of a nonionic polymer or activated silica to the influent of dual or mixed media filters. The other flow sheet is a direct filtration scheme that uses a flocculation basin without sludge collection.

Process Design and Operation

In the direct filtration process, the rapid mix usually does not differ from conventional plants. A hydraulic jump in a Parshall flume may be

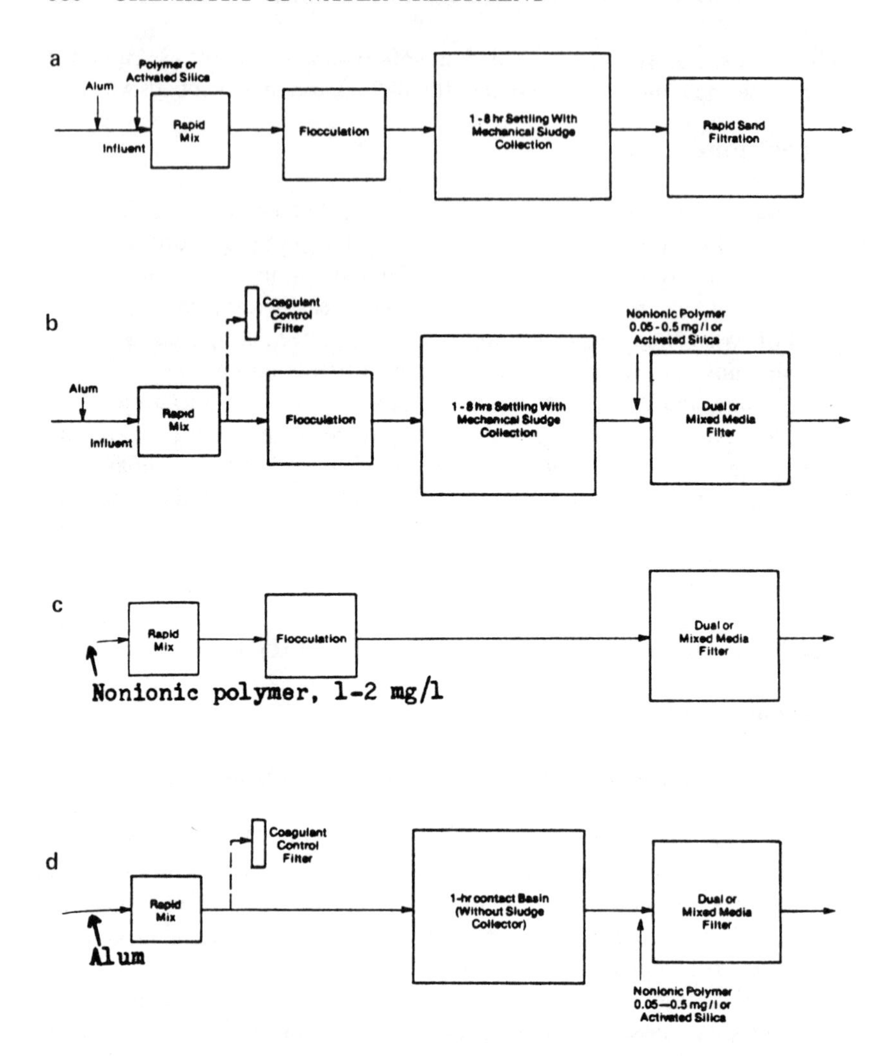

Figure 12. Flowsheets for **(a)** typical older conventional filter plants; **(b)** typical recent design trends for conventional filter plants; **(c)** typical direct filtration using a flocculation basin; **(d)** typical direct filtration using a contact basin. Reproduced from Culp [23], courtesy of the American Water Works Association.

used for the rapid mix [23]. Field experience has been good. Some designs extend the time for this type of mixing up to as much as five minutes, which is longer than conventional plants.

Filter media designs are basically the same for direct filtration as they are for conventional plants, with the important exception that sand should not be used as the medium. Rather, dual- and mixed-media beds are employed because they provide the space for storing solids removed from the water. Operating experiences have shown that a high quality of water is produced in direct filtration rates of 1–8 gpm/ft^2 ($0.7–5 \times 10^{-3}$ m/sec). The design rate usually is 5 gpm/ft^2 (3.3×10^{-3} m/sec).

Fine filter media are supported on a gravel bed, which is preferred over direct support on bottoms equipped with mechanical strainers or nozzles. The latter are not recommended [23]. Surface wash or an air scour is a necessity in direct filtration. Backwash rates of 15 gpm/ft^2 (.01 m/sec^2) are adequate.

Culp [23] argues for a mixed-media bed over a dual-media bed. The former has the advantage of providing more filter grains and more surface area of grains for a given head loss across the bed than dual-media beds. This is seen from these data:

Type of Filter Bed	Total Surface Area of Grains/mm^2 of Bed Surface Area	Total Number of Grains/mm^2 of Bed Surface Area
Dual-Media, 76 cm (30 in.)	2090	1200
Mixed-Media, 76 cm (30 in.)	2680	2770

It is essential, however, to use a polymer that increases the shearing strength of floc and its adherence to the surface of filter grains. An excellent 36-in. (91-cm)-deep mixed-media filter would consist of:

1. 16 in. (41 cm) of −4. to +14 mesh coal, SG of 1.48;
2. 8 in. (20 cm) of −9 to +18 mesh coal, SG of 1.62;
3. 9 in. (23 cm) of −30 to +40 mesh sand, SG of 2.4; and
4. 3 in. (8 cm) of 40 to 80 mesh garnet, SG of 4.2.

A typical 36-in. (91-cm)-deep dual-media bed would consist of:

1. 25 in. (64 cm) of −4 to +20 mesh coal, SG of 1.55, and
2. 11 in. (28 cm) of −20 to +50 mesh sand, SG of 2.4.

According to Culp [23], this dual-media bed produces the same quality of water as a mixed-media bed, but the chemical dosages will be higher and the filter runs shorter. The key for successful operation of direct filtration apparently lies in optimum coagulant and polymer dosages. This point is discussed below.

Flocculation Prior to Direct Filtration

A laboratory study was conducted to optimize the time of flocculation of the raw water supply from Deer Creek reservoir on the Provo River in Utah [26]. Optimization was measured by turbidity residual, particle size distribution and filterability. Flocculation time is extremely important prior to direct filtration for these reasons:

1. The aggregates may be too large for deep penetration into the filter.
2. Energy is wasted when continuous agitation is applied beyond the optimum floc sizes.
3. "Large" aggregates may be sheared, as has been observed by La Mer and Healy [27,28].

Table I shows the results from coagulation of Deer Creek reservoir water with 3 mg/l alum and 0.25 mg/l cationic polymer. It should be noted that the raw water turbidity was "low" and in the range of 0.77–2.9 turbidity units (TU). Frequently these waters with low turbidities are difficult to coagulate. Subsequent filtration through a 30-cm bed of sand was conducted at 4.8 gpm/ft^2 (3.2 mm/sec). For this system, a flocculation time of 7 min was optimum. Longer times did not improve the turbidity quality of the filter effluent. All residual turbidities were lower than the drinking water standard of 1. TU. Flocculation times shorter than 7 min were not sufficient to produce an aggregation of singlet flocs required for their removal from water. No data were given on the length of the filter runs.

Applications and Operational Aspects

Color was removed effectively from a raw water supply in Australia by coagulation with alum (10 mg/l) and cationic polymer (1 mg/l) and with a direct filtration rate of 6 gpm/ft^2 (4 mm/sec) [24]. A typical removal was from 40 to 5 ACU in a pilot plant operation. Also, turbidity was reduced from 2.5 TU to values of 0.2 TU, and iron (0.8 mg/l) and manganese (0.13 mg/l) were reduced to less than 0.05 mg/l by direct filtration. In a related study using Owens River water for the city of Los Angeles, pretreatment with ozone (1–2 mg/l) improved the turbidity quality of direct filtered water.

Table I. Effect of Flocculation Time on Filter Effluent Turbidity[a]

Raw Water Turbidity (TU)	Flocculation Velocity Gradient, $\bar{G}$ (sec^{-1})	Time of Flocculation, t (sec)	Gt	Flocculated Water Turbidity (TU)	Filter Effluent Turbidity (TU)		
					2-L Influent	4-L Influent	6-L Influent
0.77	100	0	0	0.77	0.76	0.76	0.76
0.88	100	120	12,000	1.20	0.22	0.17	0.11
2.40	100	420	42,000	2.30	0.12	0.12	0.12
2.60	100	900	90,000	2.20	0.13	0.12	0.12
1.80	100	1800	180,000	1.70	0.10	0.10	0.10
2.90	100	2700	270,000	2.10	0.13	0.13	0.13

[a]Reproduced from Treweek [26], courtesy of the American Water Works Association.

Direct filtration proved to be a viable alternative to conventional sedimentation-filtration system for removal of turbidity and diatoms from the waters of Lake Huron [29], that is, if the coagulant demand (alum) was less than 15 mg/l and the diatom concentration was less than 1000 areal standard units (asu)/ml. A pilot plant study was conducted to investigate the effects of raw water turbidity, alum dosage, flocculation time and velocity gradient on the effectiveness of subsequent direct filtration. Typical results are seen in Table II, where the intensity of the alum mixing stage is reported as the mean velocity gradient, $\bar{G}$ (sec^{-1}):

$$\bar{G} = \left(\frac{P}{\mu V}\right)^{0.5} \tag{23}$$

where P = power dissipated in the water (ft-lb/sec)
μ = absolute viscosity of the water (gram mass/cm-sec)
V = volume of the water (ft^3, liters)

In general, $\bar{G}$ is a measure of the contact times of floc particles. The larger the $\bar{G}$ value, the shorter is the contact time, which has the result of producing smaller floc particles. In this study, one of the experiments was to determine whether higher flocculation speeds would create a smaller floc, which, in turn, would lead subsequently to longer filter runs.

Table II. Flocculation Velocity Gradient Study[a,b]

Run Number	Filter Coal es (mm)	Turbidity Raw (FTU)	Turbidity Effluent (FTU)	Gradient (sec^{-1})	Final Head Loss (ft)	Run Time (hr)
53	0.9	16.0	0.14	20	8+	16.0
54	0.9	12.5	0.14	100	8+	15.0
55	0.9	14.0	0.14	300	7.3[c]	17.0
53	1.05	16.0	0.15	20	6.0[c]	14.5
54	1.05	12.5	0.13	100	6.0[c]	16.5
55	1.05	14.0	0.15	300	4.5[c]	13.0
53	1.55	16.0	0.13	20	6.0[c]	13.5
54	1.55	12.5	0.15	100	6.0[c]	13.5
55	1.55	14.0	0.15	300	4.5[c]	11.5

[a]Reproduced from Hutchison [29], courtesy of the American Water Works Association.
[b]Flocculation rate—288 m/day; alum—12 mg/l; flocculation time—14.5 min.
[c]Filter breakthrough.

Hutchinson [29] reported that, with velocity gradients greater than 100 sec^{-1} and with an effective size (es) of 0.9 mm for filter coal, there was a greater chance of "early" breakthrough of turbidity (Table II). The filtration rate for this study was held constant at 4.8 gpm/ft^2 (288 m/day). Also, there was some evidence of a smaller floc volume indicated by a longer time to reach a certain level of head loss at the higher flocculation mixing intensity (see 300 sec^{-1} G value, 0.9 mm es coal in Table II). As the effective size of coal was increased, however, final head losses were decreased and filter runs were shorter. This suggests deeper penetration of the floc into the bed. The turbidity quality of the filter's effluent was unaffected by either velocity gradient or the es of the filter coal.

The effect of flocculation time on head loss and filter run time was investigated also [29]. As seen in Table III, little advantage was obtained when the flocculation time was carried beyond 4.5 min. Filter breakthrough occurred sooner, and filter runs were shorter when flocculation times were greater than 4.5 min. These observations generally agree with those of Treweek [26] cited above. Turbidity quality, however, was unaffected by any one combination of the variables seen in Table III.

Some extremely important observations about direct filtration were reported in the above study [29]. The optimum effective size of the coal filtering medium was near 1.05 mm—that is, with regard to effluent quality, length of filter runs and floc distribution within the bed and in the absence of diatoms. Also, there was little or no change in the

Table III. Flocculation Time Study[a,b]

Run Number	Filter Coal es (mm)	Turbidity Raw (FTU)	Turbidity Effluent (FTU)	Floc Time (min)	Final Head Loss (ft)	Run Time (hr)
56	0.9	13	0.16	4.5	8+	13.4
53	0.9	16	0.14	14.5	8+	16.0
58	0.9	15	0.16	28.0	7.0[c]	14.0
56	1.05	13	0.16	4.5	8+	16.8
53	1.05	16	0.15	14.5	6.0[c]	14.5
58	1.05	15	0.17	28.0	4.7[c]	11.5
56	1.55	13	0.16	4.5	8+	15.5
53	1.55	16	0.13	14.5	6.0[c]	13.5
58	1.55	15	0.16	28.0	4.5[c]	10.0

[a]Reproduced from Hutchison [29], courtesy of the American Water Works Association.
[b]Filtration rate—288 m/day; alum—15 mg/l; flocculation velocity gradient—20 s^{-1}.
[c]Filter breakthrough.

effluent's turbidity for filtration rates of 2.4–7.2 gpm/ft^2 (144–432 m/day). Diatoms were a treatment problem in this study [29]. Filter coal with an effective size of 1.5 mm operated successfully when the diatom levels averaged 2500 asu/ml. Filter runs in excess of 12 hr at 4.8 gpm/ft^2 (288 m/day) were obtained. Nonionic polymers were added at the filter inlet and prevented breakthrough on all media sizes up to 1.55 mm under all conditions.

That naturally occurring color and other $CHCl_3$ precursors are removed by direction filtration was reported by Scheuch and Edzwald [30]. Synthetic and natural waters from the Raquette River, New York, were utilized in a bench-scale pilot-plant operation. For the natural water, turbidity values ranged from 0.68 to 1.4 TU, whereas color values ranged from 40 to 70 color units (CU). The experimental filter was 2.54 cm (id) × 14-cm deep. Sieved and washed Ottawa sand was the filter medium with a geometric mean grain size of 0.6 mm and a bed porosity of 0.4. The filtration rate was 2 gpm/ft^2 (1.3 mm/sec). A cationic polymer was used as the only coagulant.

Typical results from the Raquette River study are seen in Figure 13 [30]. The residual turbidity and color after filtration were within their drinking water standards. A 75–80% color removal was achieved by direct filtration. However, head losses reached 60–100 in. within a relatively short period of time—4 hr. Chloroform formation was monitored

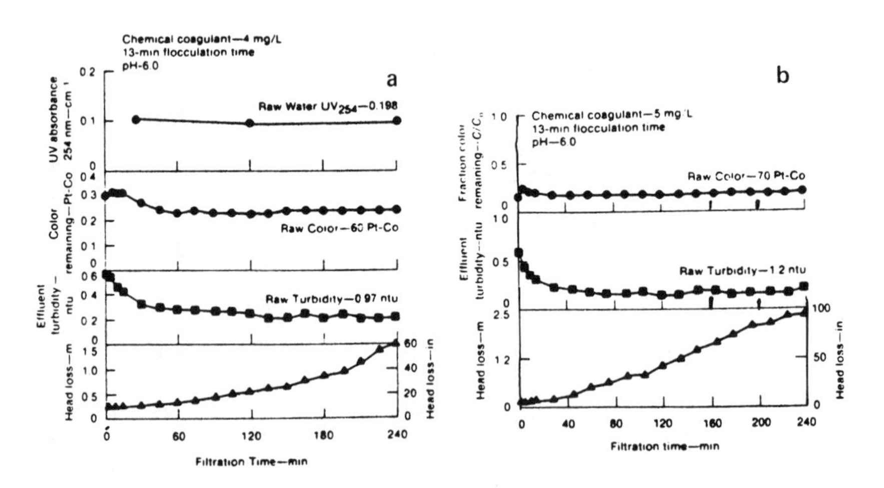

Figure 13. Filtration performance for Raquette River. Head loss data are in meters (in.) of water. Reproduced from Scheuch and Edzwald [30], courtesy of the American Water Works Association.

also in the direct filtration experiments. There was a concurrent reduction in [$CHCl_3$] with color removal.

An extensive comparative study was conducted on the performance of dual- and mixed-media filters at the treatment plant in Minneapolis, Minnesota [31]. Other sites throughout the United States were examined also as seen in Table IV. The filtration rate was held constant at 2 gpm/ft^2 (1.3 mm/sec). The raw water supply was obtained from the Mississippi River, which had a turbidity of 1.9–2.0 TU in the experimental time frame. However, the water was softened by the addition of lime that was followed by alum coagulation and direct filtration.

Table IV. Media Design Specifications[a]

Plant	Filter Design	Media	Depth mm	Depth in.	Effective Size (mm)	Uniformity Coefficient
Duluth Plant[b]	Mixed	Coal	41.25	16.5	1.0–1.1	1.7 max.
		Sand	22.5	9	0.42–0.45	1.35–1.7
		HDS[c]	11.25	4.5	0.21–0.32	1.8 max.
	Dual	Coal	52.5	21	1.0–1.1	1.7 max.
		Sand	22.5	9	0.42–0.45	1.35–1.7 max.
Minneapolis Plant[b]	Mixed	Coal	48.75	19.5	1.03–1.05	1.6
		Sand	22.5	9	0.50	1.6
		HDS	3.75	1.5	0.20	1.8
	Dual	Coal	50	20	1.00	1.5
		Sand	30	12	0.4	1.6
Clinton, Connecticut, Pilot Plant	Mixed	Coal	41.25	16.5	1.0–1.2	<1.7
		Sand	22.5	9	0.45–0.50	1.5
		HDS	11.25	4.5	0.2–0.3	<2.0
	Dual	Coal	65	26	0.95	<1.5
		Sand	25	10	0.45	<1.35
New York (Croton) Pilot Plant	Mixed	Coal	45	18	1.05–1.15	1.6–1.8
		Sand	22.5	9	0.5–0.6	1.4–1.6
		HDS	7.5	3		
	Dual	Coal	45	18	1.05–1.15	1.6–1.8
		Sand	30	12	0.5–0.6	1.4–1.6
Seattle, Washington, Pilot Plant	Mixed	Coal	45	18	1.0–1.1	1.7
		Sand	22.5	9	0.42–0.55	1.8
		HDS	7.5	3	0.18–0.32	2.2
	Dual	Coal	50	20	0.92	1.28
		Sand	25	10	0.40	1.3

[a]Reproduced from Burns et al. [31], courtesy of the American Water Works Association.
[b]Surface wash was provided for all filters.
[c]HDS = high-density sand.

Typical results for two runs at the Fridley Filter Plant (FFP) at Minneapolis are seen in Figure 14 [31]. In the initial stages, the mixed-media produced a filtered water turbidity equal to or slightly less than the dual-media. After 24 hr, a breakeven point was reached after which the dual-media produced a more consistent effluent quality than the mixed-media. The gradual increase in the water's turbidity from the dual-media beyond the breakpoint indicated a progressive saturation of the intersti-tial pore spaces and deep penetration into the bed. For the mixed-media, the rapid increase in turbidity after the breakpoint indicates a break-through condition. Head losses were about the same for both media. Since the shapes of the headloss curves are linear, this signifies filtration with depth. According to Qureshi, the performance of the dual-media

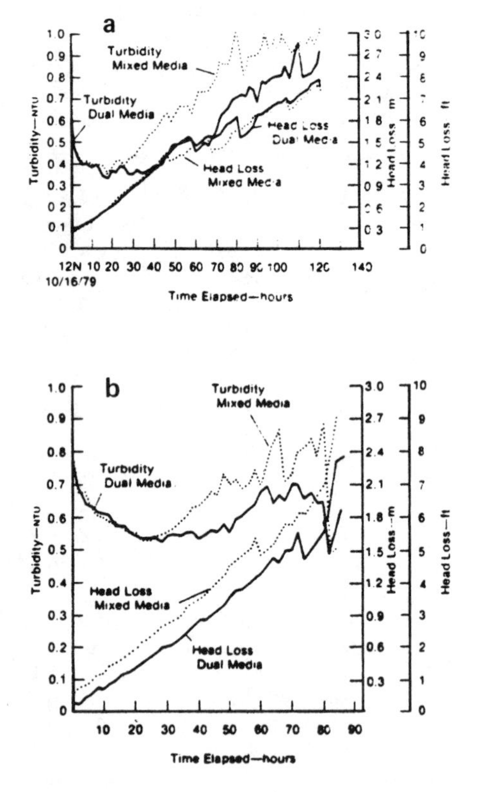

Figure 14. Average values for **(a)** first and **(b)** second sets of runs at FFP. Reproduced from Qureshi [31], courtesy of the American Water Works Association.

was superior to mixed-media in terms of head loss, filter effluent turbidity and unit filter run.

Miscellaneous

Coarse grades of diatomite material can be coated with such substances as alum or polyelectrolytes to remove fine particulate matter [32]. Filtration rates, ranging from 2 to 10 gpm/ft^2, were applied to mixed-media filters at Erie County, New York [33]. Asbestiform fiber removal by dual- and mixed-media was studied in a pilot-plant operation at Duluth, Minnesota [34]. The costs of operating several direct-filtration plants were reported [35]. Declining-rate filtration apparently produces a better quality water, and requires less available head loss than constant-rate filtration [36]. The design and operation of high-rate filters were presented from a mathematical basis in three papers [37]. Sand filters were treated with anionic, cationic and nonionic polymers in a research study to determine the effects of pH and various cations on the surface properties of the filtering medium [38].

REFERENCES

1. Ives, K. J. *J. San. Eng. Div.* (*ASCE*) 87(SH3):23 (1961).
2. Ives, K. J. "Proc. Int. Water Supply Cong.," International Water Supply Association, London, Vienna, Austria (1969).
3. Mackile, V., and S. Mackile. *J. San. Eng. Div.* (*ASCE*) 87(SA5):17 (1961).
4. Tuepker, J. L., and C. A. Buescher, Jr. *J. Am. Water Works Assoc.* 60:1377 (1968).
5. Cleasby, J. L. et al. *J. Am. Water Works Assoc.* 55:869 (1963).
6. Weber, W. J., Jr. *Physicochemical Processes for Water Quality Control* (New York: John Wiley & Sons, Inc., 1972).
7. Clark, J. W., and W. Viessman, Jr. *Water Supply and Pollution Control* (New York: International Textbook Co., 1966).
8. Streeter, V. L., and E. B. Wylie. *Fluid Mechanics* (New York: McGraw-Hill Book Company, 1979).
9. Camp, T. R. *J. San. Eng. Div.* (*ASCE*) 90(SA4):1 (1964).
10. Rose, H. E. *Proc. Instit. Mech. Eng.* (*London*) (1949), p. 154.
11. Baumann, E. R. In: *Water Treatment Plant Design,* R. L. Sanks, Ed. (Ann Arbor, MI: Ann Arbor Science Publishers, Inc., 1978).
12. Cleasby, J. L. *J. Am. Water Works Assoc.* 61:181 (1969).
13. Baumann, E. R., and C. S. Oulman. In: *Water Quality Improvement by Physical and Chemical Processes*, E. F. Gloyna and W. W. Eckenfelder, Eds. (Austin, TX: University of Texas Press, 1970).

14. Hudson, H. E. Jr. *J. Am. Water Works Assoc.* 51:1455 (1959).
15. Cleasby, J. L., and E. R. Baumann. *J. Am. Water Works Assoc.* 54:579 (1962).
16. Arboleda, J. R. *J. Am. Water Works Assoc.* 66:87 (1974).
17. Conley, W. R., and R. W. Pitman. *J. Am. Water Works Assoc.* 52:205 (1960).
18. Fair, G. M. et al. *Water and Wastewater Engineering, Vol. 2* (New York: John Wiley & Sons, Inc., 1968).
19. Iwasaki, T. *J. Am. Water Works Assoc.* 29:1591 (1937).
20. Ives, K. J. *Proc. Inst. Civil Eng.* 16:189 (1960).
21. O'Melia, C. R., and W. Stumm. *J. Am. Water Works Assoc.* 59:1393 (1967).
22. Yao, K. M. et al. *Environ. Sci. Technol.* 5:1105 (1971).
23. Culp, R. L. *J. Am. Water Works Assoc.* 69:375 (1977).
24. Tate, C. H., and R. R. Trussell. *J. Am. Water Works Assoc.* 72:165 (1980).
25. Bowers, A. E., and J. D. Beard, II. *J. Am. Water Works Assoc.* 73:457 (1981).
26. Treweek, G. P. *J. Am. Water Works Assoc.* 71:96 (1979).
27. Healy, T. W., and V. K. LaMer. *J. Phys. Chem.* 66:1835 (1962).
28. La Mer, V. K., and T. W. Healy. *Pure Appl. Chem.* 13:112 (1963).
29. Hutchinson, W. R. *J. Am. Water Works Assoc.* 68:292 (1976).
30. Scheuch, L. E., and J. K. Edzwald. *J. Am. Water Works Assoc.* 73:497 (1981).
31. Qureshi, N. *J. Am. Water Works Assoc.* 73:490 (1981).
32. Burns, D. E. et al. *J. Am. Water Works Assoc.* 62:121 (1970).
33. Westerhoff, G. P. *J. Am. Water Works Assoc.* 63:376 (1971).
34. Logsdon, G. S., and J. M. Symonds. *J. Am. Water Works Assoc.* 69:499 (1977).
35. Logsdon, G. S. et al. *J. Am. Water Works Assoc.* 72:134 (1980).
36. Cleasby, J. L. *J. Am. Water Works Assoc.* 73:484 (1981).
37. Kawamura, S. *J. Am. Water Works Assoc.* 67:535, 653, 705 (1975).
38. Loganatban, P., and W. J. Maier. *J. Am. Water Works Assoc.* 67:336 (1975).

CHAPTER 7

REMOVAL OF HARDNESS AND SCALE-FORMING SUBSTANCES

SCALES AND THEIR FORMATION

Types of Scale

Hardness of natural waters results from dissolution of geologic formations (minerals) containing calcium, magnesium and silica. In turn, these waters may be supersaturated with respect to one or more of the original minerals, whereupon "scale" is formed in a distribution system, hot water heater, boiler, etc. Hardness has been explained traditionally by the occurrence of calcium and magnesium compounds: bicarbonates, carbonates, sulfates, chlorides and nitrates in water. Silica originates in the dissolution or chemical weathering of amorphous or crystalline $SiO_{2(s)}$ and the major clay minerals: kaolinite, illite and montmorillonite. Some typical dissolution reactions at 25°C are:

$$\underset{\text{Calcite}}{CaCO_{3(s)}} + H^+ = Ca^{2+} + HCO_3^- \qquad K_1 = 10^{1.856} \tag{1}$$

$$\underset{\text{Magnesite}}{MgCO_{3(s)}} + H^+ = Mg^{2+} + HCO_3^- \qquad K_2 = 10^{2.235} \tag{2}$$

$$\underset{\text{Gypsum}}{CaSO_4 \cdot 2\,H_2O_{(s)}} = Ca^{2+} + SO_4^{2-} + 2\,H_2O \qquad K_3 = 10^{-5.3} \tag{3}$$

$$\underset{\text{Amorphous}}{SiO_{2(s)}} + 2\,H_2O = H_4SiO_4 \qquad K_4 = 10^{-2.77} \tag{4}$$

$$Al_2Si_2O_5(OH)_{4(s)} + 5\,H_2O = 2\,Al(OH)_{3(s)} + 2\,H_4SiO_4 \quad K_5 = 10^{-9.93} \tag{5}$$
Kaolinite Gibbsite

$$Mg(OH)_{2(s)} = Mg^{2+} + 2\,OH^- \quad K_6 = 10^{-10.74} \tag{6}$$
Brucite

A more comprehensive list of natural chemical weathering reactions has been compiled [1].

That there are regional variations in the hardness of ground and surface waters is seen in Figures 1a and 1b [2]. There may be locally hard waters within the generalized soft water areas. To date, no primary potable water standards exist at the federal level; however, some states, e.g., New Jersey (50–250 mg/l as $CaCO_3$) may have secondary standards for hard water.

The semantics of water hardness is confusing, to say the least. The terms "carbonate and noncarbonate" hardness attempt to distinguish sources of hardness:

Classification	Carbonate Hardness (CH)	Noncarbonate Hardness (NCH)
Calcium hardness	$Ca(HCO_3)_2$ $CaCO_3$	$CaSO_4$ $CaCl_2$
Magnesium hardness	$Mg(HCO_3)_2$ $MgCO_3$	$MgSO_4$ $MgCl_2$

It is somewhat difficult to assign anions exclusively to the Ca^{2+} and Mg^{2+} cations. The compounds listed above distinguish carbonate and noncarbonate hardnesses.

There are four essential types of scales: calcium carbonate, magnesium hydroxide, calcium sulfate and silica. Each is formed through precipitation reactions with their solubilities at ordinary surface and groundwater temperatures, regulated by dissolution Reactions 1–6. Whenever water is heated, e.g., in domestic hot water heaters or boilers, the tendency to form scales becomes more severe because of their inverse solubility relation with temperature (Table I). On the other hand, solubility of silica increases at higher temperatures [3]:

Temp. (°C)	Solubility (mg/l)	
	Quartz	Amorphous
0		60–80
25	6.0	115.
84	26.	
100		370.

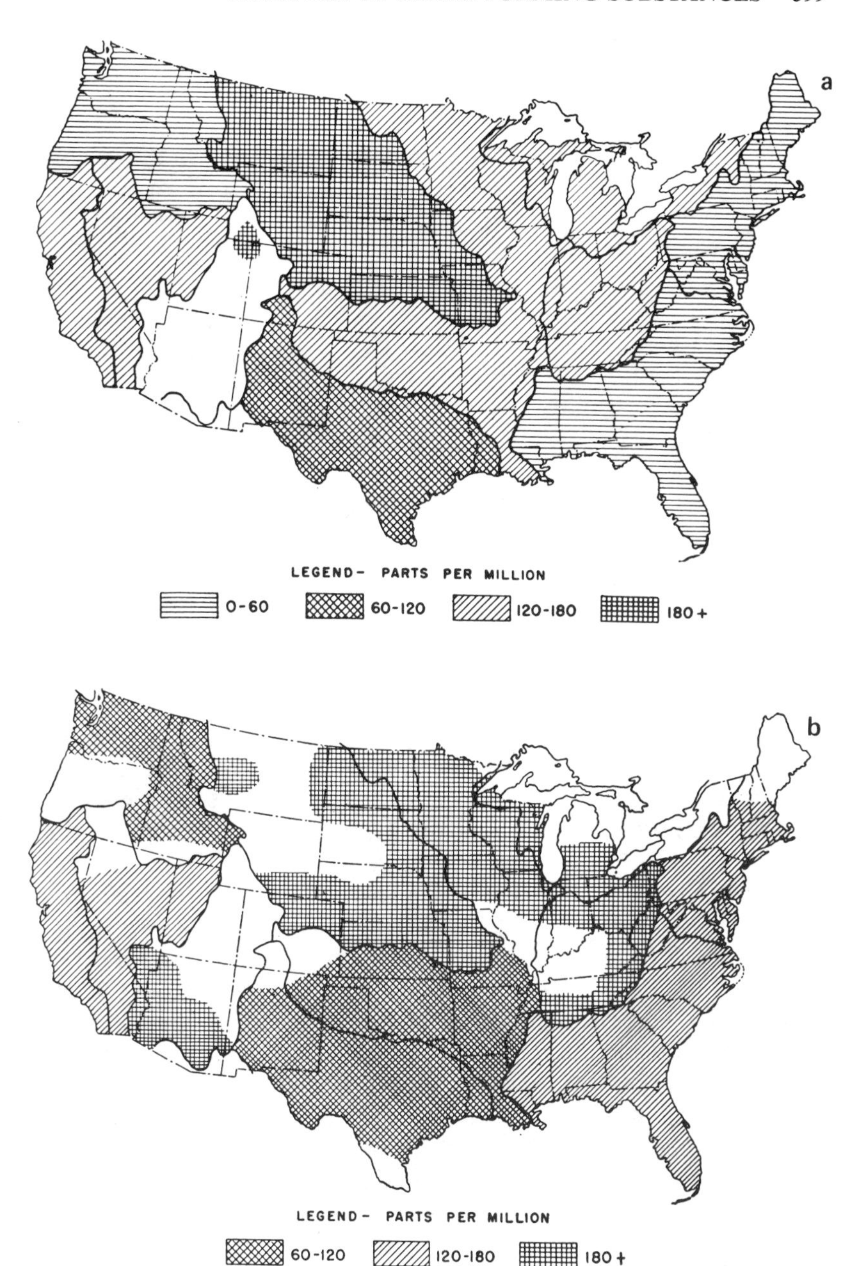

Figure 1. Map showing average hardness of **(a)** surface water and **(b)** groundwater weighted against population served in major drainage areas. Reproduced from Powell [2], courtesy of the McGraw-Hill Book Company.

Table I. Solubilities of Hard Water Scales

Temperature		$CaCO_{3(s)}$		$Mg(OH)_{2(s)}$		$MgCO_{3(s)}$		$CaSO_{4(s)}$	
°C	°F	$-\log Ks$[a]	$[CA^{2+}]$ (mg/l)	$-\log Ks$	$[Mg^{2+}]$ (mg/l)	$-\log Ks$	$[Mg^{2+}]$ (mg/l)	$-\log Ks$	$[Ca^{2+}]$ (mg/l)
0	32	8.023	3.90		1.21		24.5[b]		409.[c]
5	41	8.087							
10	50	8.150							
15	59	8.215							
20	68	8.280	2.90						
25	77	8.342		11.6[d]	2.67	7.46[d]	4.52	5.3	89.8
30	86	8.395							
40	104	8.515							
50	122	8.625	1.95						
80	176	8.975	1.30						
100	212				1.75		18.2		362.

[a]Harned and co-workers [4,5].
[b]Stumm and Morgan [6].
[c]Nordell [7]. Values may be in error.
[d]*Lange's Handbook of Chemistry* [8].

In waters where HCO_3^- and CO_3^{2-} anions predominate over Cl^-, SO_4^{2-} and NO_3^-, $CaCO_{3(s)}$ (calcite) is the principal scale-former in condensers and other water-jacketed equipment, water heaters, hot-water piping and fixtures, etc. Where there is a sufficient quantity of Mg^{2+} cations, $Mg(OH)_{2(s)}$ (brucite) will form a scale rather than $MgCO_{3(s)}$, because the former is more insoluble than the latter (See Table I). In waters where SO_4^{2-} predominates over Cl^-, NO_3^-, HCO_3^-, and CO_3^{2-}, $CaSO_{4(s)}$ (gypsum) will form a scale. It too has an inverse solubility-temperature relation. $CaSO_{4(s)}$ is somewhat more soluble than $CaCO_{3(s)}$, so that NCH waters have lesser scale-forming tendencies than CH waters. Silica is rather objectionable in high-pressure boilers, where it has pronounced scale-forming tendencies [7]. If calcium hardness is present, the scale may be a calcium silicate; if aluminum is present, an aluminosilicate scale may be formed (for example, Reaction 5), or the scale may consist almost entirely of silica (SiO_2). These silica scales are usually very hard, glassy, adherent and difficult to remove. Silica frequently is carried over with the steam and forms a scale in superheater tubes and on turbine blades. Since the solubility of silica increases with temperature, it may be soluble to a certain extent in high-pressure steam. Normally, silica scale is not a problem in a domestic hot water heater. Some waters with substantial quantities of iron and manganese will form scales. Removal of these two constituents is discussed in Chapter 8.

Calculation of Potential Scale Formation

The calcium and magnesium contents of natural waters do, of course, vary considerably [1]. Consequently, scale-forming tendencies also vary. A water is considered to be hard whenever one or more of the solubility product constants cited above is exceeded. Furthermore, Reactions 1 and 2 show the $[H^+]$ dependence of the solubility of calcium and magnesium carbonate scales. In this case, stability and/or solubility diagrams may be employed to ascertain the hardness of water [1].

The solubility product constant for $CaCO_{3(s)}$ is:

$$Ks = [Ca^{2+}][CO_3^{2-}] \qquad Ks = 10^{-8.342},\ 25°C \tag{7}$$

Substituting for $[CO_3^{2-}]$:

$$Ks = [Ca^{2+}]C_t\alpha_C^{2-} \tag{8}$$

where $C_t = [H_2CO_3] + [HCO_3^-] + [CO_3^{2-}]$

$\alpha_C^{2-} = [CO_3^{2-}]/C_t$

Since $[Ca^{2+}] = C_t$:

$$Ks = [Ca^{2+}]^2\alpha_C^{2-} \tag{9}$$

To construct a stability diagram for $CaCO_{3(s)}$, it is convenient to divide the pH scale into three regions: 0.00–6.35 (I); 6.35–10.33 (II) and 10.33–14.0. In region I, $\alpha_C^{2-} \simeq K_1K_2/[H^+]^2$ [see Faust and Aly [1] for full derivation of the alpha equations for the carbonate system], whereupon:

$$Ks = \frac{[Ca^{2+}]^2K_1K_2}{[H^+]^2} \tag{10}$$

where K_1 and K_2 are the primary and secondary dissociation constants for H_2CO_3, respectively. Taking logs of Equation 10 and rearranging:

$$\log[Ca^{2+}]eq = -pH + 0.5(pK_1 + pK_2 - pK_s) \tag{11}$$

This equation represents the equilibrium boundary line for $CaCO_{3(s)}$ solubility in region I (Figure 2). In region II, $\alpha_C^{2-} \simeq K_2/[H^+]$, whereupon:

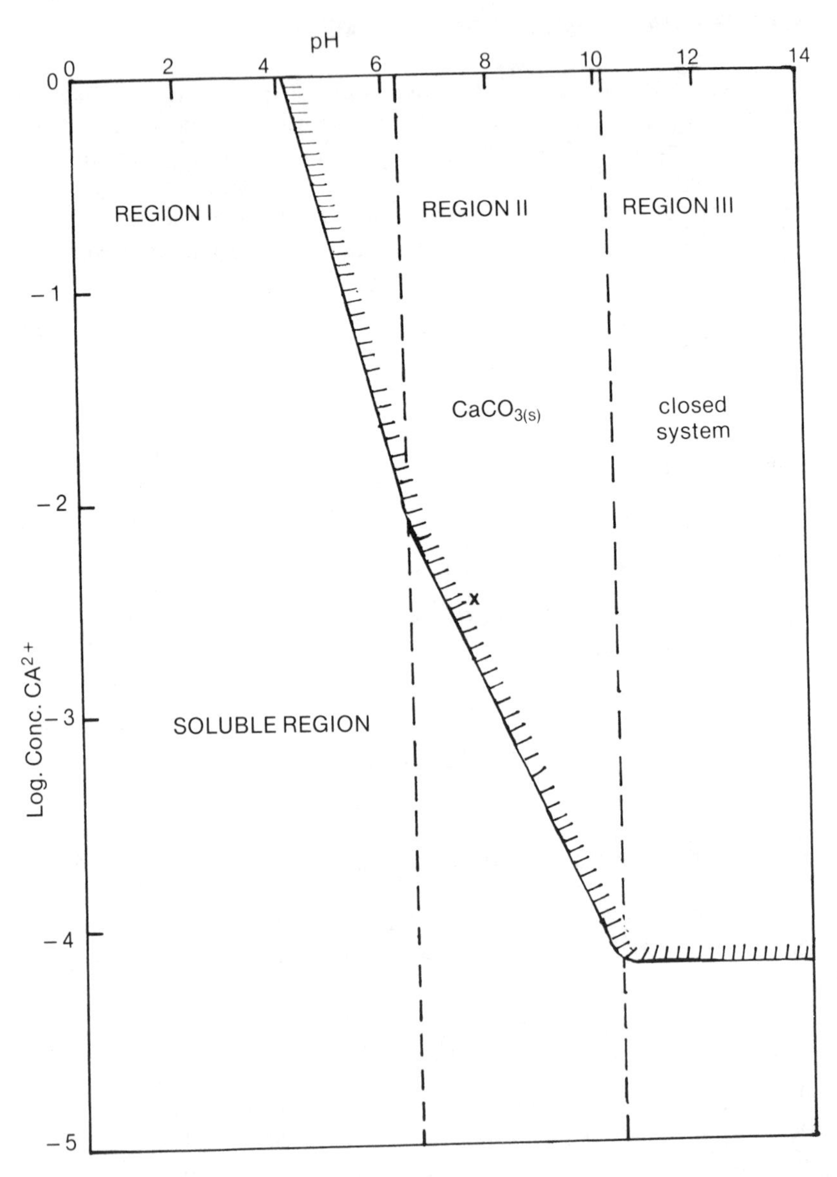

Figure 2. Solubility diagram for calcium carbonate in a closed system at 25°C.

$$K_s = \frac{[Ca^{2+}]^2 K_2}{[H^+]} \tag{12}$$

or

$$\log[Ca^{2+}]eq = -0.5\,pH + 0.5(pK_2 - pK_s) \tag{13}$$

In region III, $\alpha_C^{2-} \simeq K_1K_2/K_1K_2 \simeq 1.0$, whereupon:

$$K_s = [Ca^{2+}]^2 \tag{14}$$

or

$$\log[Ca^{2+}]eq = -0.5pK_s \tag{15}$$

Equations 11, 13 and 15 are the linear relations between the equilibrium concentrations of $[Ca^{2+}]$ and pH that represent the boundary conditions between soluble and insoluble $CaCO_{3(s)}$.

Three examples shall suffice to demonstrate the use of Figure 2 to predict the scale-forming tendencies of natural waters.

Example A.

A natural groundwater has a pH value of 7.8 and a $[Ca^{2+}]$ of 96.0 mg/l [1]. This is a region II water. A plot of pH 7.8 and log $[Ca^{2+}] = -2.62$ places this water (x on Figure 2) in the solid region. Therefore, this water should form a $CaCO_{3(s)}$ scale at 25°C.

Example B.

A natural groundwater has a pH value of 7.3 and a $[Ca^{2+}]$ of 79.0 mg/l [9]. This is a region II water. At 15°C the $[Ca^{2+}]eq$ is calculated from Equation 13:

$$\log[Ca^{2+}]eq = -0.5 \times 7.3 + 0.5(10.43 - 8.215)$$

$$[Ca^{2+}]eq = 2.87 \times 10^{-3}M, \ 115 \text{ mg/l}$$

Since the equilibrium content of Ca^{2+} is not exceeded, this water should not form a $CaCO_{3(s)}$ scale at 15°C.

Example C.

Raise the temperature of the water in Example B to 50°C (122°F). In this case, $[Ca^{2+}]eq$ is:

$$\log[Ca^{2+}]eq = -0.5 \times 7.3 \times 0.5(10.172 - 8.625)$$

$$[Ca^{2+}]eq = 1.33 \times 10^{-3}M, 53\,mg/l$$

The equilibrium content of $CaCO_{3(s)}$ is exceeded at 50°C (simulating a hot water heater), and the water is a potential scale-former.

Similar calculations may be performed for $Mg(OH)_{2(s)}$, $MgCO_{3(s)}$ and $CaSO_{4(s)}$. Equilibrium constants are given above and elsewhere [1] for these solid phases. The protolysis constants for carbonic acid are given in Table II for temperatures 0–80°C [4,5].

CHEMISTRY OF SOFTENING PROCESSES

Lime-Soda Ash

In the lime-soda ash process of softening hard waters, Ca^{2+} is precipitated as $CaCO_{3(s)}$ and Mg^{2+} as $Mg(OH)_{2(s)}$. Lime, $Ca(OH)_2$, and soda ash, $NaCO_3$, are employed for reduction of the $[Ca^{2+}]$ and $[Mg^{2+}]$. Consequently, the solubility product constants of $CaCO_{3(s)}$ and $Mg(OH)_{2(s)}$ control the softening reactions. That is, stoichiometric quantities of lime and soda ash raise the pH value of the water so that $CaCO_{3(s)}$ (see Figure 2) and $Mg(OH)_{2(s)}$ are at their minimum solubilities.

Table II. Protolysis Constants of Carbonic Acid [3,4]

Temperature			
°C	°F	pK_1	pK_2
0	32	6.577	10.625
5	41	6.517	10.557
10	50	6.465	10.490
15	59	6.420	10.430
20	68	6.382	10.377
25	77	6.351	10.329
30	86	6.327	10.290
40	104	6.296	10.220
50	122	6.287	10.172
80	176	(6.315)	(10.122)

Stoichiometric chemical reactions of the lime-soda process (Equations 16–21) are given in Table III. These reactions are essentially a pH adjustment whereupon:

$$[Ca^{2+}] = C_t = [H_2CO_3] + [HCO_3^-] + [CO_3^{2-}] \quad (22)$$

and

$$[Mg^{2+}] = 2[OH^-] \quad (23)$$

At a pH value of 10.3, Equation 22 becomes:

$$[Ca^{2+}] = C_t \simeq [CO_3^{2-}] = 10^{-4.171}\ M,\ 2.7\ mg/l \quad (24)$$

and

$$[Mg^{2+}] = 2 \times 10^{-3.7} \text{ or } 4 \times 10^{-4} M,\ 9.72\ mg/l.$$

Equations 16 –21 also show that the following must be achieved after addition of lime and soda ash:

$$[Ca^{2+}]_{original} + [Ca^{2+}]_{added} = C_{t_{original}} + [CO_3^{2-}]_{added} \quad (25)$$

Equation 16 shows the neutralization of carbonic acid (CO_2 acidity) in the natural water. If an analytical value of H_2CO_3 is not available, then it should be calculated because one equivalent of lime is required [1]. Equation 17 is the removal of the CaCH, whereas Equation 18 is removal of the CaNCH. Equation 20 is the removal of the MgCH, whereas Equation 21 is removal of the MgNCH. These reactions are assumed to become complete in plant practice.

Throughout the lime-soda softening reactions, electroneutrality conditions must be maintained:

Table III. Reactions of the Lime-Soda Softening Process

Equation	No.
$H_2CO_3 + Ca(OH)_2 = CaCO_{3(s)} + 2\ H_2O$	16
$Ca^{2+} + 2\ HCO_3^- + Ca(OH)_2 = 2\ CaCO_{3(s)} + H_2O$	17
$Ca^{2+} + Na_2CO_3 = CaCO_{3(s)} + 2\ Na^+$	18
$HCO_3^- + Ca(OH)_2 = CaCO_{3(s)} + OH^- + H_2O$	19
$Mg^{2+} + 2\ HCO_3^- + 2\ Ca(OH)_2 = 2\ CaCO_{3(s)} + Mg(OH)_{2(s)} + 2\ H_2O$	20
$Mg^{2+} + Ca(OH)_2 + Na_2CO_3 = CaCO_{3(s)} + Mg(OH)_{2(s)} + 2\ Na^+$	21

$$2[Ca^{2+}] + 2[Mg^{2+}] + [H^+] = [HCO_3^-] + 2[CO_3^{2-}] + 2[SO_4^{2-}] + [Cl^-] + [OH^-] \quad (26)$$

From Equations 25 and 26, it is convenient to calculate the quantities of lime and soda ash required for softening, as well as to sort out the CH from the NCH.

Table IV contains two examples of calculations for the lime-soda softening of natural groundwater cited above. First, it is necessary to calculate the milliequivalents and millimolar quantities of the major cations and anions. Second, the C_t for the carbonate species is calculated from the $[HCO_3^-]$ and the appropriate alpha value [1]. Third, the $[H_2CO_3]$ is taken as the difference between C_t and $[HCO_3^-]$. The $[CO_3^{2-}]$ is assumed to be negligible. Fourth, a meq scale is drawn above and below, on which the meq of cations and anions is drawn. It is convenient to represent the cations in this order: Ca^{2+}, Mg^{2+}, Na^+ and K^+ (if necessary), and the anions in this order: HCO_3^-, SO_4^{2-}, Cl^- and NO_3^- (if necessary). Consequently, CaCH and MgCH take precedence over CaNCH and MgNCH. If there is no MgCH, then the order is CaCH, CaNCH and MgNCH. If there is no CaNCH, then the order is CaCH, MgCH and MgNCH. There occasionally are waters with only noncarbonate hardness. For example A, there are 2.18 meq CaCH, 2.61 meq CaNCH and 1.56 meq MgNCH. The appropriate lime and soda-ash quantities are given. It is not necessary to draw a scale for H_2CO_3; just make certain that its lime requirements are included with the others. For example B, there are 3.94 meq CaCH, 0.44 meq MgCH and 1.86 meq MgNCH. Both examples satisfy the conditions of Equation 26.

The calculations in Table IV assume, of course, that Reactions 16–21 satisfy the stoichiometric requirements. This is frequently not the case because of an insufficient contact time allotted in plant reactors to allow the completion of these reactions. In order to increase the rates of precipitation, lime and soda ash are added in excess of the calculated values. Whatever this excess, it may be added to the examples in Table IV. Higher water temperatures increase the rates of precipitation and also provide lower residual contents of Ca and Mg. This is the "hot" lime-soda process, and is employed for boiler waters. There are also occasions when it is not necessary to utilize the full quantities of lime and soda ash. For example, municipal drinking water does not need softening to $[Ca^{2+}] = 2.7$ mg/l and $[Mg^{2+}] = 9.72$ mg/l. The quantities of lime and soda ash could be lowered, or the technique of split treatment could be instituted (see below).

Table IV. Calculation of Carbonate Hardness, Noncarbonate Hardness and Lime-Soda Quantities

From example A, a natural water contains:

Constituent	mg/l	meq	m*M*
pH	7.8		
Ca^{2+}	96.	4.79	2.395
Mg^{2+}	19.	1.56	0.78
Na^+	18.	0.78	0.78
K^+	1.5	0.04	0.04
HCO_3^-	133.	2.18	2.18
SO_4^{2-}	208.	4.33	2.165
Cl^-	25.	0.705	0.705
H_2CO_3	5.2		0.084

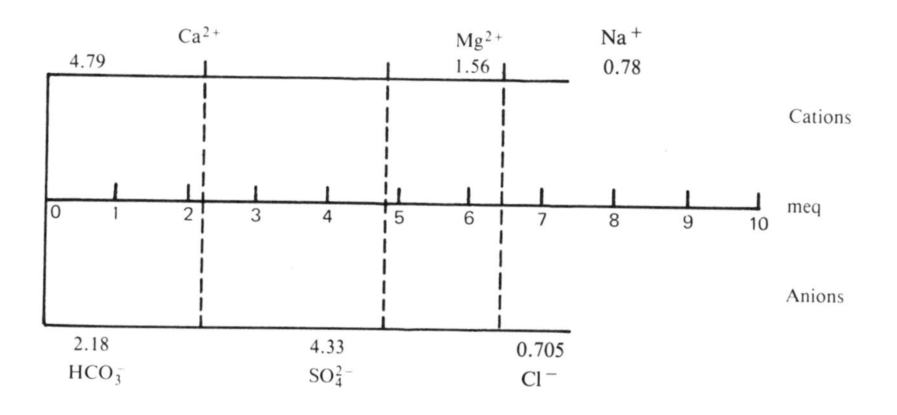

Type Hardness	meq	Equation	$Ca(OH)_2$ meq	$Ca(OH)_2$ m*M*	Na_2CO_3 meq	Na_2CO_3 m*M*
H_2CO_3		16		0.084		
CaCH	2.18	17	2.18	1.09		
CaNCH	2.61	18			2.61	1.305
MgCH						
MgNCH	1.56	21	1.56	0.78	1.56	0.78
	6.35		3.74	1.954	4.17	2.085

From Equation 25:

2.395 m*M* $Ca^{2+}_{original}$ + 1.954 m*M* lime = 2.264 m*M* C_t + 2.085 m*M* soda ash added

4.349 = 4.349

Table IV, continued

From example B, a natural water contains:

Constituent	**mg/l**	**meq**	**m*M***
pH	7.3		
Ca^{2+}	79.	3.94	1.97
Mg^{2+}	28.	2.30	1.15
Na^{+}	8.1	0.35	0.35
K^{+}	5.7	0.146	0.146
HCO_3^-	267.	4.38	4.38
SO_4^{2-}	51.	1.06	0.53
Cl^-	29.	0.82	0.82
H_2CO_3	30.8		0.50

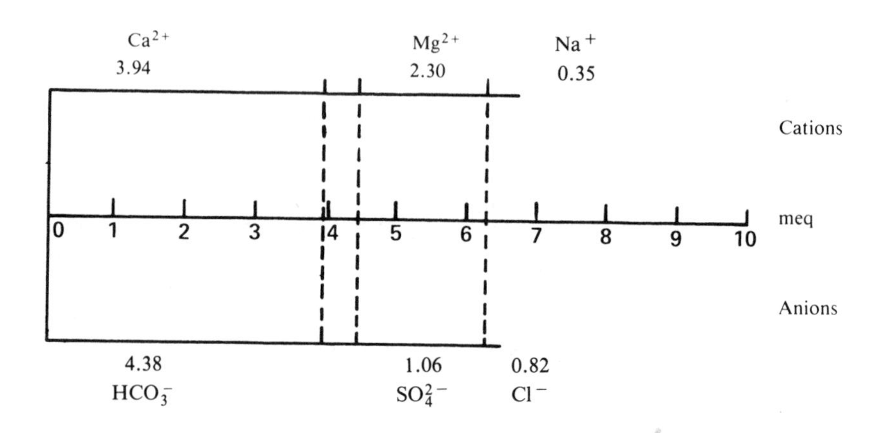

Type Hardness	**meq**	**Equation**	**$Ca(OH)_2$ meq**	**$Ca(OH)_2$ m*M***	**Na_2CO_3 meq**	**Na_2CO_3 m*M***
H_2CO_3		16		0.50		
CaCH	3.94	17	3.94	1.97		
CaNCH						
MgCH	0.44	20	0.88	0.44		
MgNCH	1.86	21	1.86	0.93	1.86	0.93
	6.24		5.80	3.84	1.86	0.93

From Equation 25:

1.97 m*M* $Ca^{2+}{}_{original}$ + 3.84 m*M* lime = 4.88 m*M* C_t + 0.93 m*M* soda ash added

5.81 = 5.81

Alkalinity of the Lime-Soda Process

In the operation of water-softening treatment units, guidelines for quality control are based upon the total (TALK) and phenolphthalein (PhPhALK) alkalinities. For the purpose of this discussion, TALK is defined as the quantity of 0.02 *N* H_2SO_4 required for titration to a methyl orange or fixed pH endpoint. It is expressed mathematically as:

$$\frac{\text{TALK}}{50{,}000} = [HCO_3^-] + 2[CO_3^{2-}] + [OH^-] - [H^+] \quad (27)$$

where the brackets indicate mol/l. The coefficient before each term converts the mol/l to eq/l. TALK has units of mg/l as $CaCO_3$, which are converted to eq/l by the factor 50,000. In turn, PhPhALK is defined as the quantity of 0.02 *N* H_2SO_4 required for titration to the phenolphthalein endpoint (pH 8.5–9.0). It is approximated by:

$$\frac{\text{PhPhALK}}{50{,}000} = [CO_3^{2-}] + [OH^-] - [H^+] \quad (28)$$

Since the pH values of the lime-soda process are in the range of 10–11 from addition of $Ca(OH)_2$ and Na_2CO_3, it is necessary to have reasonably accurate values of TALK and PhPhALK for quality control. Optimum removal of Ca^{2+} occurs, apparently, when TALK/PhPhALK (as $CaCO_3$) = 2.0 [10]. This has been a traditional "rule-of-thumb" in water softening [2,7].

Several methods are available for estimating the various kinds of alkalinity and the concomitant $[Ca^{2+}]$. For example, *Standard Methods* [11] still uses the alkalinity relationships shown in Table V. Also, the nomographs of Dye [12] are utilized frequently. There are, however, easier and less confusing mathematical methods for calculating the three major types of alkalinity and $[Ca^{2+}]$ in Equation 27 [1,10].

The appropriate equilibrium equations are:

$$K_s = [Ca^{2+}][CO_3^{2-}], 10^{-8.342} (25°C) \quad (7)$$

$$K_w = [H^+][OH^-], 10^{-14} (25°C) \quad (29)$$

$$K_{2H_2CO_3} = \frac{[H^+][CO_3^{2-}]}{[HCO_3^-]}, 10^{-10.329} (25°C) \quad (30)$$

In these and subsequent equations, calcium reduction is the only objective. This is frequently the situation because either the raw water $[Mg^{2+}]$

Table V. Alkalinity Relationships[a] [11]

Result of Titration	Hydroxide Alkalinity as $CaCO_3$	Carbonate Alkalinity as $CaCO_3$	Bicarbonate Concentration as $CaCO_3$
P=0	0	0	T
P<½T	0	2P	T-2P
P=½T	0	2P	0
P>½T	2P-T	2(T-P)	0
P=T	T	0	0

[a]P: phenolphthalein alkalinity; T: total alkalinity.

is acceptable, or chemical costs can be saved where Mg removal is not necessary. In any event, Schierholz et al. [10] derived, by combining Equations 7, 27, 29 and 30, expressions for calculating $[HCO_3^-]$, $[CO_3^{2-}]$ and $[Ca^{2+}]$ at appropriate pH and TALK values:

$$[HCO_3^-] = \frac{TALK/50{,}000 + [H^+] - Kw/[H^+]}{1 + 2K_2/[H^+]} \quad (31)$$

$$[CO_3^{2-}] = \frac{TALK/50{,}000 + [H^+] - Kw/[H^+]}{2(1 + [H^+]/2K_2)} \quad (32)$$

$$[Ca^{2+}] = \frac{K_s}{[CO_3^{2-}]} = \frac{2K_s(1 + [H^+]/2K_2)}{TALK/50{,}000 + [H^+] - Kw/[H^+]} \quad (33)$$

By measurement of pH, TALK, temperature and total dissolved solids (TDS), each ion concentration and the three types of alkalinity can be calculated. These values, for example, are given in Table VI for a temperature of 20°C and a TDS of 200. mg/l. (It is noted that "maximum carbonate" apparently means the $[HCO_3^-] + [CO_3^{2-}]$ for a given pH value and $[Ca^{2+}]$, i.e., maximum $CaCO_{3(s)}$ precipitation.) These concentrations are, of course, those that should occur for the specified conditions from the thermodynamic equations. There may be differences between the calculated values and those observed at water softening plants due to analytical errors, "slow" kinetics of precipitation, etc. Schierholz et al. [10] give TALK and ionic concentrations at 5°, 10° and 15°C.

Cadena et al. [13] argue that formation of the ion pair, $CaCO_3^0$, limits the reduction of $[Ca^{2+}]$ to 13.5 mg/l as $CaCO_3$ (5.41 mg/l as Ca^{2+}) in the lime-soda softening process. This $[Ca^{2+}]$ is twice the quantity calculated from the Ks value of $10^{-8.342}$. The ion pair is formed:

$$Ca^{2+} + CO_3^{2-} = CaCO_3^0, K_{34} = 10^{4.467}\ 25°C. \quad (34)$$

Table VI. pH and Concentration (as Calcium Carbonate) of Carbonate, Bicarbonate and Hydroxyl Alkalinity at the pH of Maximum Carbonate Concentration; Total Dissolved Solids = 200 mg/l. Temperature = 20°C[a]

Alkalinity (mg/l)	pH	Carbonate (mg/l)	Bicarbonate (mg/l)	Hydroxyl (mg/l)	Calcium (mg/l)
20	10.15	8.8	5.6	5.6	10.0
22	10.18	10.1	6.0	6.0	8.7
24	10.20	11.4	6.3	6.3	7.8
26	10.22	12.7	6.7	6.7	7.0
28	10.25	14.0	7.0	7.0	6.3
30	10.27	15.3	7.3	7.3	5.8
32	10.28	16.7	7.7	7.7	5.3
34	10.30	18.1	8.0	8.0	4.9
36	10.32	19.5	8.3	8.3	4.5
38	10.33	20.9	8.6	8.6	4.2
40	10.35	22.3	8.9	8.9	4.0
42	10.36	23.7	9.1	9.1	3.7
44	10.37	25.2	9.4	9.4	3.5
46	10.39	26.6	9.7	9.7	3.3
48	10.40	28.1	9.9	9.9	3.1
50	10.41	29.6	10.2	10.2	3.0
52	10.42	31.1	10.5	10.5	2.8
54	10.43	32.6	10.7	10.7	2.7
56	10.44	34.1	11.0	11.0	2.6
58	10.45	35.6	11.2	11.2	2.5
60	10.46	37.1	11.4	11.4	2.4
62	10.47	38.7	11.7	11.7	2.3
64	10.48	40.2	11.9	11.9	2.2
66	10.48	41.8	12.1	12.1	2.1
68	10.49	43.3	12.3	12.3	2.0
70	10.50	44.9	12.6	12.6	2.0
72	10.51	46.4	12.8	12.8	1.9
74	10.51	48.0	13.0	13.0	1.8
76	10.52	49.6	13.2	13.2	1.8
78	10.53	51.2	13.4	13.4	1.7
80	10.53	52.8	13.6	13.6	1.7
82	10.54	54.3	13.8	13.8	1.6
84	10.55	55.9	14.0	14.0	1.6
86	10.55	57.5	14.2	14.2	1.5
88	10.56	59.1	14.4	14.4	1.5
90	10.57	60.8	14.6	14.6	1.5
92	10.57	62.4	14.8	14.8	1.4
94	10.58	64.0	15.0	15.0	1.4
96	10.58	65.6	15.2	15.2	1.3
98	10.59	67.2	15.4	15.4	1.3
100	10.59	68.9	15.6	15.6	1.3

[a]Reproduced from Schierholz et al. [10], courtesy of the American Water Works Association.

Combination of Equation 34 with Equation 7 leads to [13]:

$$CaCO_3^0 = \frac{KsCaCO_3}{K_{34}} = \frac{4.58 \times 10^{-9}}{3.41 \times 10^{-5}} = 13.4 \text{ mg/l as } CaCO_3$$

This revelation may be important in the treatment of boiler-fed waters where $[Ca^{2+}]$ must be as low as possible.

Kinetics of $CaCO_{3(s)}$ Precipitation

Large (i.e., several MGD) water softening plants using the lime-soda process are designed around a chemical reaction time of 40–60 min [14]. That is, hydraulic design of the flocculating basins is of this order of magnitude of time. Consequently, the chemical kinetics of the $CaCO_{3(s)}$ and $Mg(OH)_{2(s)}$ precipitation must "fit" into this time frame. Kinetic studies have indicated that 25 min are required to reach equilibrium levels of the Ca^{2+} in the precipitation of $CaCO_{3(s)}$ [14]. In addition, the initial pH value in the range of 9–12 appeared to be a factor affecting the kinetics of precipitation. The higher the pH value, the quicker the attainment of the equilibrium content of Ca^{2+}. It was concluded that this 25-min reaction time is compatible with the traditional hydraulic design of 40–60 min [14].

A kinetic study was conducted [15] on the precipitation and crystallization of $CaCO_{3(s)}$ under typical water softening practice conditions, i.e., a high degree of oversaturation, a "moderately high" concentration of seed crystals and a pH range of 8–10. Data were obtained, first to establish a model for Ca^{2+} removal that had been proposed previously [16]:

$$-\frac{d[Ca^{2+}]}{dt} = K \cdot [CaCO_{3(s)}] \cdot ([Ca^{2+}][CO_3^{2-}]) - \frac{K_s}{\gamma_D^2} \tag{35}$$

where

K = rate constant for calcite crystallization (1/mol/min/mg/l seed)
$[CaCO_{3(s)}]$ = crystal concentration mg/l at any time, t, in the
K_s = solubility product constant for $CaCO_{3(s)}$
γ_D^2 = activity coefficient for divalent ions

Typical kinetic data are seen in Figure 3 where the reaction was followed by pH measurements with time and subsequent calculation of $[Ca^{2+}]$, $[CaCO_{3(s)}]$ and $[CaCO_3^0]$. Equation 35 was verified as the mathematical model for the crystallization and precipitation process. Data were ob-

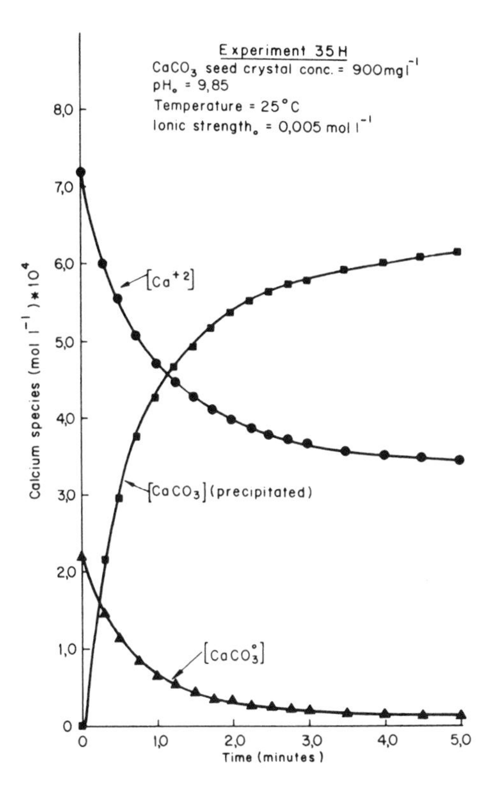

Figure 3. Plot of derived calcium species vs time, from experimentally measured pH values. Reproduced from Wiechers et al. [15], courtesy of the International Association on Water Pollution Research.

tained to evaluate factors influencing the value of K, of which $[H^+]$ is an important component (Figure 4). The value of K, the apparent rate constant, was dependent on, or, at least, varied with, the initial pH value. No explanation was offered for this observation. Temperature is also a variable influencing K for which a typical Arrhenius plot was obtained in the 10–40°C range. An activation energy of 10.3 ± 0.9 kcal/mol was computed. In summary, Equation 35 appears to be valid for Ca^{2+} and CO_3^{2-} concentrations ranging from 1.4 to 3.29×10^{-4} *M* and 1.83 to 5.29×10^{-4} *M*, respectively, and $[CaCO_{3(s)}]$ ranging from 100 to 1000 mg/l. An important consequence of this study is the close fit of experimentally derived and theoretically calculated kinetic plots. Thus, the K value is reliable and will be useful in designing chemical reactors for water softening.

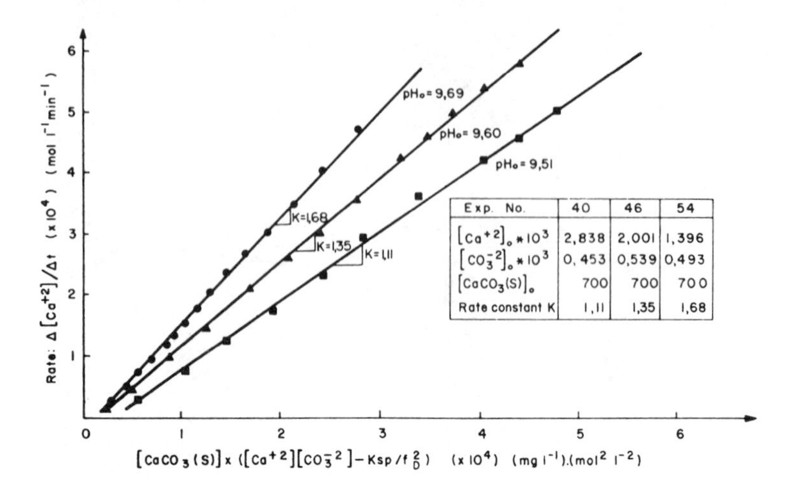

Exp. No.	40	46	54
$[Ca^{+2}]_o \times 10^3$	2,838	2,001	1,396
$[CO_3^{-2}]_o \times 10^3$	0,453	0,539	0,493
$[CaCO_3(S)]_o$	700	700	700
Rate constant K	1,11	1,35	1,68

Figure 4. Effect of initial pH (pH_0) on rate constant K. Reproduced from Wiechers et al. [15], courtesy of the International Association on Water Pollution Research.

pH Control

Lime-soda softened waters usually have pH values of approximately 10.4 and are supersaturated with $CaCO_{3(s)}$ and $Mg(OH)_{2(s)}$. Consequently, it is necessary to acidify these waters to prevent postprecipitation of these two substances in filters following the softening or in the water distribution system. Adjustment of the pH value toward neutrality is accomplished either by addition of $CO_{2(g)}$ (i.e., recarbonation) or by sulfuric acid. The former process may be preferred over the latter because it is less hazardous to use than H_2SO_4.

In the recarbonation process, $CO_{2(g)}$ is bubbled into the water in order to effect these reactions:

$$CaCO_{3(s)} + CO_{2(g)} + H_2O = Ca(HCO_3)_2 \quad (36)$$

$$Ca(OH)_2 + 2\,CO_{2(g)} = Ca(HCO_3)_2 \quad (37)$$

Liquid CO_2 is the most common source, with onsite generation as an alternative process. Economy ultimately dictates the choice. A detailed description of the recarbonation process is available in Haney and Hamann [17]. Figure 5 shows typical $CO_{2(g)}$ feed systems.

Recarbonation is especially important where substantial quantities of $Mg(OH)_{2(s)}$ have been precipitated [18]. This occurs because $Ca(OH)_2$

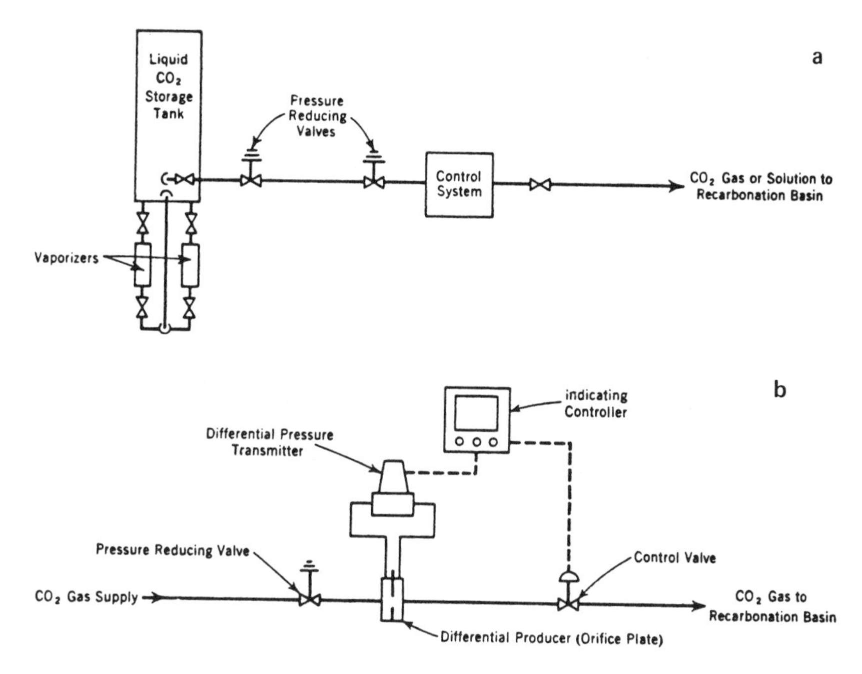

Figure 5. Schematics of **(a)** typical carbon dioxide feed system and **(b)** typical control system for carbon dioxide gas feed. Reproduced from Haney and Hamann [17], courtesy of the American Water Works Association.

must be employed to reach pH values in the range of 10. to 11. according to Equation 21 and Figure 6. Larson et al. [18] recommended a saturation pH value of 8.5 for 77°F water and 9.5 for 33°F water.

In the reduction of excessive pH values and alkalinities, sulfuric acid is also utilized:

$$H_2SO_4 + 2\,CaCO_{3(s)} = Ca(HCO_3)_2 + CaSO_4 \tag{38}$$

$$H_2SO_4 + Ca(OH)_2 = CaSO_4 + 2\,H_2O \tag{39}$$

Needless to say, extreme caution must be utilized in handling and adding concentrated sulfuric acid to water. Powell [2] and Nordell [7] describe many kinds of acid-feeding equipment.

Silica Removal

Briefly, the aqueous chemistry of silica is:

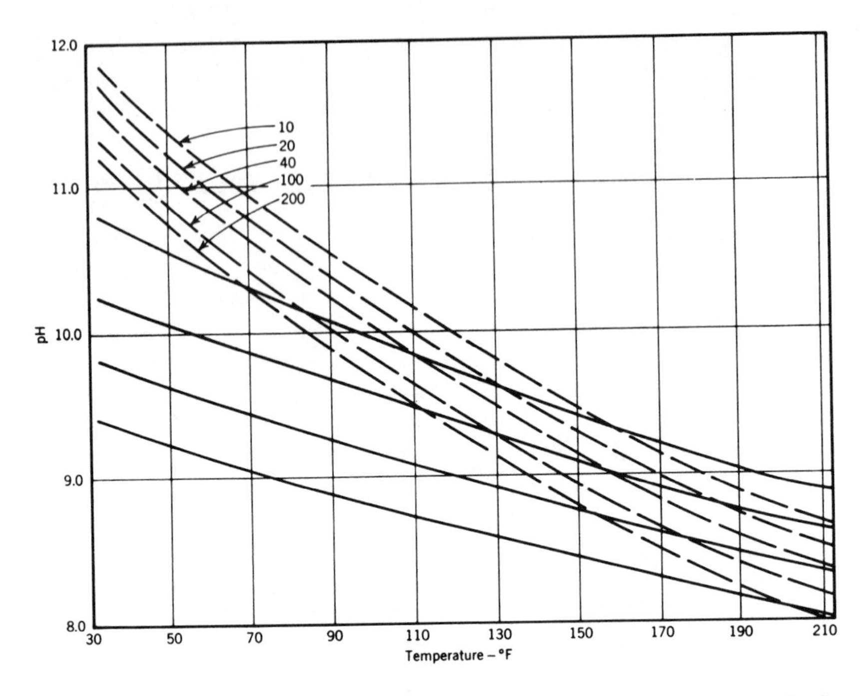

Figure 6. Influence of temperature on magnesium solubility at 200 ppm TDS. Dashed curves represent magnesium solubility (as $CaCO_3$), and the solid curves represent pH variation. Reproduced from Larson et al. [18], courtesy of the American Water Works Association.

$$SiO_{2(s)} + 2H_2O = H_4SiO_4 \qquad K_4 = 10^{-2.77} \tag{4}$$
Amorphous

$$SiO_{2(s)} + 2H_2O = H_4SiO_4 \qquad K_{40} = 10^{-3.76} \tag{40}$$
Quartz

$$H_4SiO_4 = H^+ + H_3SiO_4^- \qquad K_{41} = 10^{-9.46} \tag{41}$$
Silicic acid

$$H_3SiO_4^- = H^+ + H_2SiO_4^{2-} \qquad K_{42} = 10^{-12.56} \tag{42}$$

For the most part, dissolved silica is a neutral molecule (H_4SiO_4 or $Si(OH)_4$ [5]). Consequently, a removal method involves adsorption of silica onto a surface such as ferric hydroxide, rather than precipitation as a discrete compound. Also, silica can be removed by a highly basic anion-exchange unit [2].

Ferric hydroxide (or ferric oxides) is precipitated from ferric sulfate by sodium or calcium hydroxide for silica removal. In this case, a pH value of 9.0 or greater is required. Dosages of ferric hydroxide are somewhat empirical [2]: for waters with less than 10 ppm of silica, the quantity of ferric sulfate is 15–20 ppm/ppm of silica removed. At higher silica contents, this ratio may be lowered to 10:1. According to plant experience, a residual of about 2 ppm silica is obtainable, and the process is suitable mainly for normal water temperatures.

An alternative and, perhaps, the most efficient process for silica removal is adsorption by an insoluble magnesium compound, i.e., $Mg(OH)_{2(s)}$ or $MgCO_{3(s)}$. If a natural water contains a considerable portion of Mg hardness, then silica will be removed to a certain extent with the concurrent precipitation of $Mg(OH)_2$. The efficiency of silica removal is influenced to a large degree by the water temperature. At 50°F (10°C), the removal of 1 ppm silica requires the precipitation of 30 ppm of magnesium hardness. This empirical ratio applies when the initial silica content is greater than 15 ppm. At 70°F (21.1°C), the ratio is 7 ppm Mg hardness to 1 ppm silica. In the event that there is an insufficient quantity of magnesium, dolomitic lime $[CaMg(CO_3)_2]$ or MgO is added to the process.

The most efficient technique for silica removal is the "hot-process" softening [2]. In this method, the water is heated to 100°C (212°F) where the rate of hydration of MgO to $Mg(OH)_{2(s)}$ increases greatly and the solubility of $Mg(OH)_2$ is also lower. This is appropriate for boiler feed waters and for steam generators. Figure 7 shows a family of curves depicting silica removal for initial contents up to 50 ppm, and for various quantities of Mg hardness removed or MgO added. The chemistry of silica removal by $Mg(OH)_{2(s)}$ is unclear. Magnesium silicate [e.g., Mg_2SiO_4 (forsterite)], may be formed, but the most probable mechanism is simple adsorption of H_4SiO_4 onto the surface of $Mg(OH)_{2(s)}$.

DESCRIPTIVE WATER SOFTENING PROCESSES

Cold Lime-Soda Ash

Current water softening processes by the lime-soda ash method are traced to Thomas Clark, a Scottish professor of chemistry at Aberdeen University [2,7]. In 1841 Clark found that hard waters could be softened by the addition of lime. Later, John Henderson Porter used soda ash in addition to lime. Thus, water softening by this method becomes, essentially, a process of handling the considerable quantities of $CaCO_{3(s)}$ and $Mg(OH)_{2(s)}$ "sludge" from the precipitation reactions. The "cold"

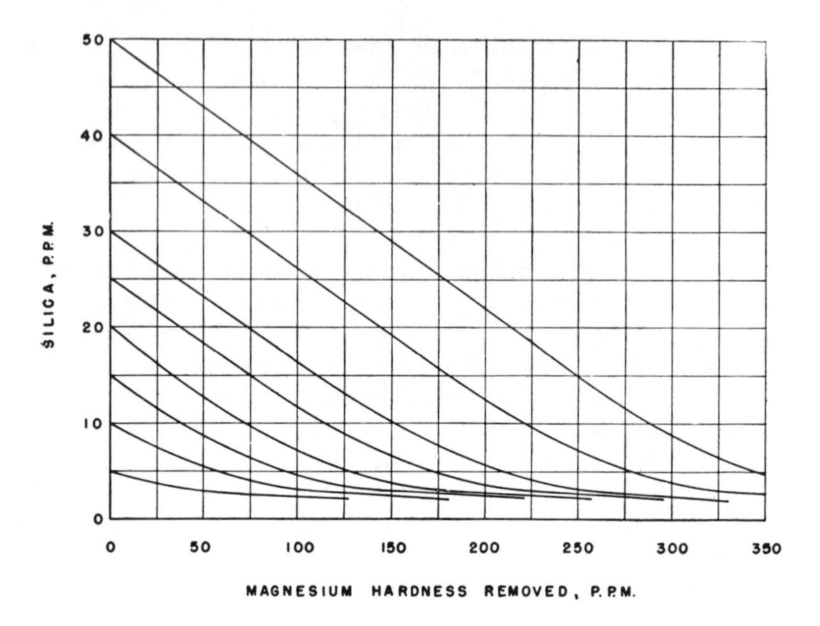

Figure 7. Silica removed by precipitation of magnesium in hot-process softening. Reproduced from Powell [2], courtesy of the McGraw-Hill Book Company.

process refers to the natural, ambient water temperatures. The "hot" process refers to situations where the water is heated to 100°C or so. Following softening, these waters are employed for boilers or steam generators.

There are four basic types of cold lime-soda ash processes: the sludge blanket, "conventional," "catalyst" and intermittent or batch process. The first three are continuous-flow, whereas the fourth is as the name indicates.

Sludge Blanket Type

There are a number of different designs, but the basic difference between this and the older "conventional" type of process is that this type filters the treated water upward through a suspended sludge blanket of previously formed precipitate [2,7,19]. One of the most common of this type of unit is the Spaulding precipitator (Figure 8). In size it may range from capacities of a few thousand to 10 million gal/day per unit. The raw water and treating chemicals are introduced into the top of the inner chamber where they are thoroughly mixed by means of a centrally

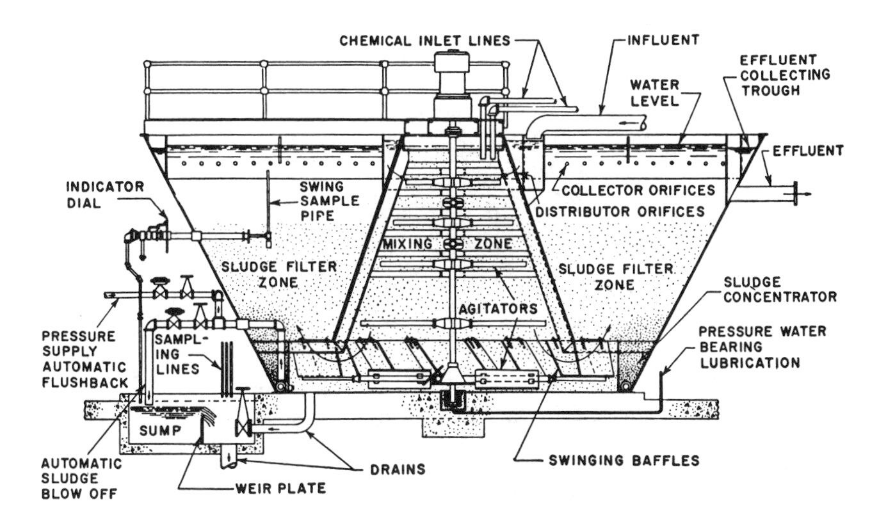

Figure 8. Spaulding upward-flow coagulating and clarification tank. Reproduced from Powell [2], courtesy of the McGraw-Hill Book Company.

disposed, motor-driven, mechanical agitator. Either wet or dry types of feeders may be used. If the wet type is used, the hydrated or slaked lime is fed as a 5% slurry. Where soda ash is also required with the wet type of feeders, it is usually dissolved in and fed with the lime suspension rather than being fed separately. The coagulant is dissolved in and fed from a separate chemical feeder. When dry feeders are employed, separate feeders are used for the lime, soda ash and coagulant.

The hardness constituents in water react with the added chemicals, and the precipitates are kept in suspension by the agitator. The treated water with the suspended precipitates flows to the bottom of the inner chamber and then rises in the outer chamber. At some point in the outer chamber, a given flowrate prevents the further expansion of the sludge blanket. A sludge blanket level is kept by bleeding off sludge at the same rate as that at which new sludge is formed. The bleeding-off operation is usually performed through a sludge concentrator. The density of the sludge in the blanket ranges about 1–2% solids; however, it will range about 10–15% solids in the concentrator.

Another type of tank based upon the rapid-flow principle is seen in Figure 9. It differs in design and operation from the Spaulding precipitator in that the chemicals are added directly to a relatively thick concentration of "sludge" near the bottom of the tank. The raw water is treated with this mixture and is stirred at a velocity higher than, for example, a coagulation tank (see Chapter 5). The purpose of this design is to

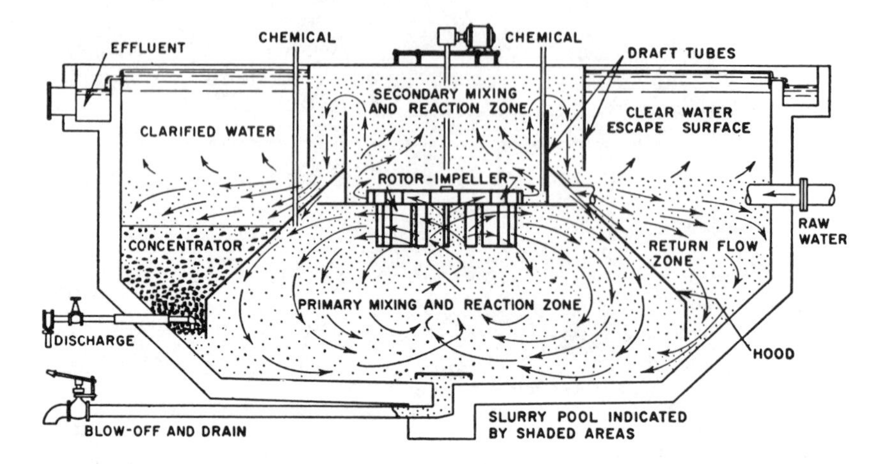

Figure 9. Rapid-flow slurry-pool tank. Reproduced from Powell [2], courtesy of the McGraw-Hill Book Company.

introduce precipitated material that acts as a "seed." Consequently, larger sized particles result with a higher dry solids content.

The advantages of the "sludge blanket type" over the "conventional" type include: (1) more intimate and thorough mixing of the lime suspension with the raw water, thereby increasing the rates of chemical reaction and decreasing the waste lime; (2) elimination of "after-precipitation" due to supersaturation and/or incomplete reactions due to the intimate contact of the treated water with a large mass of the solid phase; (3) absence of "after-precipitates," resulting in the production of an effluent clear enough (turbidity usually under 10 ppm) for many industrial applications, often eliminating the use of subsequent filtration equipment; and (4) a briefer detention period (1 hr) than that required by the "conventional" type of equipment (2–4 hr), resulting in a more compact plant occupying less space.

Conventional Type

The term "conventional type" is commonly used to designate a continuous cold lime-soda water softener that does not employ a sludge blanket, but allows sludge to settle to the bottom of a softener. In the design of the "conventional" type, the raw water and chemicals are mixed in a chamber usually provided with a mechanical type of agitator. From the mixing chamber, the treated water with its precipitates flows into a settling basin or basins by gravity. The sludge settles to a solids concentration of 3–7%.

In general, the detention period employed is 4 hr. Dry chemical feeders are usually employed. Filters are generally used to remove turbidity. Recarbonation with flue gas is generally used to reduce after-precipitation of the carbonates.

Catalyst Type—"Spiractor"

As seen in Figure 10, the Spiractor consists of a conical tank which is about two-thirds filled with a finely divided, granular "catalyst." This so-called "catalyst" may be almost any finely granular (0.3–0.6-mm diameter) insoluble mineral substance. Crushed and graded calcite was originally employed, but waste greensands are now often used in place of calcite. The tank may be closed to operate under pressure, or it may be of the open type for gravity operation. In either case, the raw water and chemicals, lime or lime and soda ash, enter near the bottom of the cone and spiral upwardly through the suspended catalyst bed.

Calcium carbonate formed by the reactions then deposits on the catalyst grains in an adherent form, so that these granules increase greatly in size. Magnesium hydroxide, however, does not form adherent deposits, nor are coagulants of any value in making magnesium deposits adhere. The Spiractor process is, therefore, largely limited to the removal of calcium hardness.

The detention period in a Spiractor is very much less than in any other

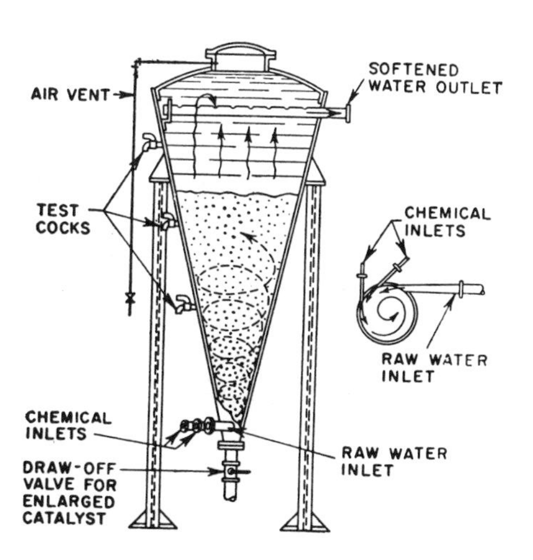

Figure 10. Cross-sectional view of the Permutit Spiractor, showing flow. Reproduced from Powell [2], courtesy of the McGraw-Hill Book Company.

type of equipment—8-12 min as compared with 2-4 hr for the conventional type and 1 hr for the sludge blanket type. Also, the by-product is a granular material that drains rapidly to less than 5% moisture. Consequently, its disposal is a much simpler matter than that of the usual lime-soda sludge, which is a pasty material composed of about 90% water.

Hot Lime-Soda Ash

This process differs from the cold lime-soda ash mainly in that the reactions are conducted at elevated temperatures, approximating the boiling point of water. This, of course, increases the rate of the precipitation reactions considerably. Also, the solids are larger and denser, and sedimentation (where appropriate) occurs faster. Finally, this process eliminates the use of lime for neutralization of the $CO_{2(g)}$. The reactors used in the hot lime-soda ash process are the same as those seen in Figures 8-10 [2,7,19].

Ion Exchange

Water softening by ion exchange, i.e., the exchange of Ca^{2+} and Mg^{2+} ions for Na^+, H^+, etc., evolved from the 1756 "discovery" of zeolites by Cronstedt, a Swedish geologist [7]. These naturally occurring minerals are hydrated double silicates consisting of an alkali or alkali-earth oxide, alumina, silica and water. An example is thomsonite: $(Na_2O, CaO) \cdot Al_2O_3 \cdot 2\ SiO_2 \cdot 2.5\ H_2O$. Almost 100 years later, J. Thomas Way, a chemist for the Royal Agricultural Society of England, published his extensive studies of the ion-exchange properties of zeolites. This work paved the way for today's ion-exchange chromatographic techniques. In 1905, Robert Gans, a German chemist, discovered that zeolites could be used to soften hard waters. He also invented processes for synthesizing zeolites and designed, perhaps, the first zeolite water softener. In 1934 and 1935 new types of cation exchanges became available: the so-called carbonaceous types, made by the sulfonation of coal and the sulfonated synthetic resins. Later work led to the development of sulfonated styrene types of resins. These are made by sulfonation of a resin produced in a bead form by the copolymerization of styrene and divinylbenzene (Figure 11).

Principles of Operation

There are four major classes of ion-exchange resins: strongly acidic and weakly acidic cation exchangers, and strongly basic and weakly basic

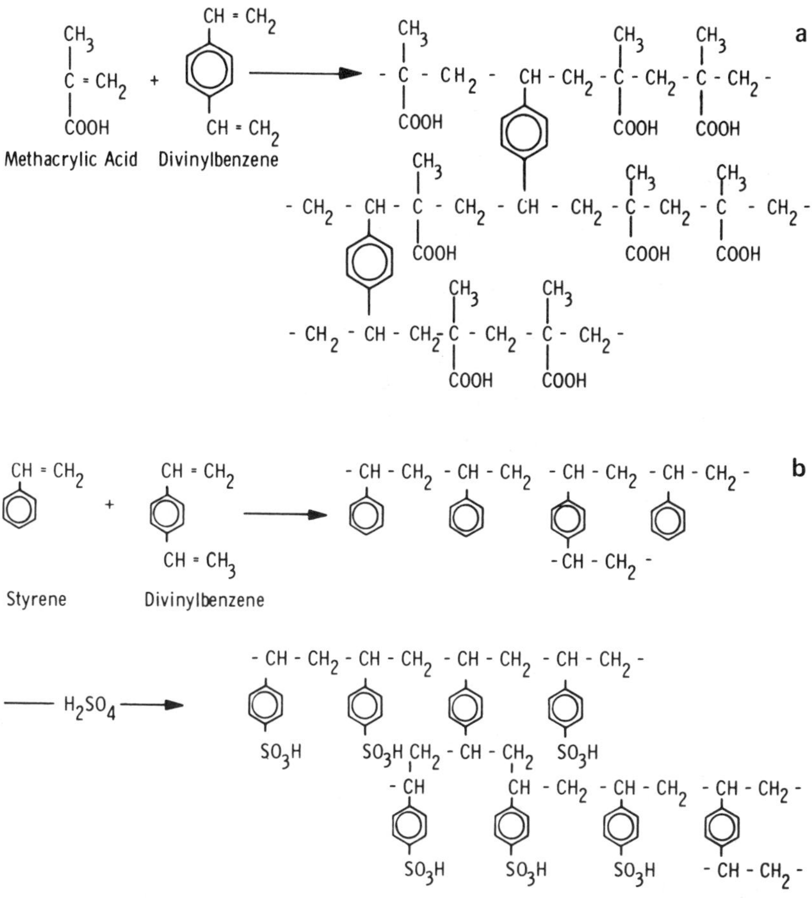

Figure 11. Preparation of a cation exchange resin by polymerization of **(a)** an organic electrolyte (polymerization of methacrylic acid, with divinylbenzene cross-linking); and **(b)** a neutral organic molecule followed by addition of functional groups (polymerization of styrene, with divinylbenzene cross-linking followed by sulfonation). Reproduced from Weber [25], courtesy of John Wiley & Sons, Inc.

anion exchangers. The "strength" of the resin refers to the type of functional group that holds the exchangeable cation or anion. The exchangeable counter ion of an acidic cation resin is either H^+ or a monovalent cation such as Na^+. For a basic anion resin, the exchangeable counterion is either OH^- or some other monovalent anion, e.g., Cl^-. These resins are regenerated, therefore, with an acid, a base or a salt. Typical exchange reactions are:

Strongly acidic cation exchangers:

(1) hydrogen form, regenerate with HCl or H_2SO_4

$$2R{-}SO_3H + Ca^{2+} = (R{-}SO_3)_2Ca + 2H^+ \quad (43)$$

(2) sodium form, regenerate with NaCl

$$2R{-}SO_3Na^+ + Ca^{2+} = (R{-}SO_3)_2Ca + 2Na^+ \quad (44)$$

Weakly acidic cation exchangers:

(1) hydrogen form, regenerate with HCl or H_2SO_4

$$2R{-}COOH + Ca^{2+} = (R{-}COO)_2Ca + 2H^+ \quad (45)$$

(2) sodium form, regenerate with NaOH

$$2R{-}COONa + Ca^{2+} = (R{-}COO)_2Ca + 2Na^+ \quad (46)$$

Strongly basic anion exchangers:

(1) hydroxide form, regenerate with NaOH

$$2R{-}R_3'NOH + SO_4^{2-} = (R{-}R_3'N)_2SO_4 + 2OH^- \quad (47)$$

(2) chloride form, regenerate with NaCl or HCl

$$2R{-}R_3'NCl + SO_4^{2-} = (R{-}R_3'N)_2SO_4 + 2Cl^- \quad (48)$$

Weakly basic anion exchangers:

(1) free base or hydroxide form, regenerate with NaOH, NH_4OH or Na_2CO_3:

$$2R{-}NH_3OH + SO_4^{2-} = (R{-}NH_3)_2SO_4 + 2OH^- \quad (49)$$

(2) chloride form, regenerate with HCl

$$2R{-}NH_3Cl + SO_4^{2-} = (R{-}NH_3)_2SO_4 + 2Cl^- \quad (50)$$

The R refers to one of the polymers in Figure 11 that hold the functional groups.

The general order of selectivity of ion exchange is given in Table VII for waters less than 1000 mg/l TDS [19]. In general, selectivity is affected

Table VII. General Order of Ion Selectivity in Waters with Less Than 1000 mg/l TDS [19]

Cations		Anions
Fe^{3+}	Exchanged First	CrO_4^{2-} [a]
Al^{3+}		SO_4^{2-} [a]
Pb^{2+}		SO_3^{2-} [a]
Ba^{2+}		HPO_4^{2-} [a]
Sr^{2+}		CNS^-
Cd^{2+}		CNO^-
Zn^{2+}		NO_3^-
Cu^{2+}		NO_2^-
Fe^{2+}		Br^-
Mn^{2+}		Cl^-
Ca^{2+}		CN^-
Mg^{2+}		HCO_3^-
K^+		$HSiO_3^-$
NH_4^+		OH^-
Na^+		F^-
H^+		
Li^+	Exchanged Last	

[a]These may be displaced as they are protonated at low pH to $HCrO_4^-$, HSO_3^- and $H_2PO_4^-$.

by: (1) ionic valence: 3 > 2 > 1, (2) atomic numbers: Ba > Sr > Ca > Mg in Group IIA and (3) hydrated ionic radius: the larger the radius, the lower is the selectivity and exchange capacity. A simple example is:

$$\left.\begin{matrix} Ca^{2+} \\ Mg^{2+} \\ Fe^{2+} \\ 2\,NH_4^+ \\ 2\,Na^+ \end{matrix}\right\} + H_2R = \left\{\begin{matrix} CaR \\ MgR \\ FeR \\ (NH_4)_2R \\ Na_2R \end{matrix}\right. + 2\,H^+ \tag{51}$$

If this ion-exchange process were continued to exhaustion, the first ions to appear in the effluent would be NH_4^+ and Na^+. The general order of preference of the cation or anion exchanger is given in Table VII. The TDS concentration is important wherein the exchanger has preference for one ion over another. For example, in the Na^+ cycle, an exchanger has a preference for Ca^{2+} over Na^+ at 1000 mg/l; but at 100,000 mg/l, the preference is Na^+ over Ca^{2+}. This is the underlying principle for regenerating exchangers with concentrated NaCl solutions (i.e., brine).

Equilibria

Equations 43–51 are written as equilibrium reactions. That is, the reactions are reversible for the exchange of ions in solution and ions attached to a resin matrix. There are several approaches for the quantitative description of ion-exchange equilibria. In short, these equilibria are treated either as a chemical reaction, i.e., the law of mass action is applied to describe the distribution of ions between the resin phase and solution phase [20], or they are treated as membrane exclusion phenomena, i.e., the Donnan equilibrium model for membranes is applied [21].

A generalized reaction suffices for a cation exchange:

$$\underset{\text{solution}}{C^{x+}} + \underset{\text{resin}}{x(R^-)D^+} = \underset{\text{solution}}{xD^+} + \underset{\text{resin}}{(R^-)_xC^{x+}} \tag{52}$$

The law of mass action accords the relation:

$$K_D^C = \frac{(D^+)^x((R^-)_xC^{x+})}{(C^{x+})((R^-)D^+)^x} \tag{53}$$

where $(D^+)^x, (C^{x+})$ = the activities of D and C, respectively, in solution
$((R^-)_xC^{x+}), ((R^-)D^+)$ = the corresponding activities of the ions in the resins

The K_D^C in Equation 53 is not strictly an equilibrium constant, but a ratio of ions in solution and in the resin phase. An "apparent equilibrium constant" may be derived from Equation 53 by inserting the appropriate activity coefficients, γ_i. This may be simplified by assuming that the γ_i values for ions in natural freshwaters are close to unity. Thus, the activities of C and D may be replaced by their respective molar concentrations $[C^{x+}]$ and $[D^+]^x$. It is somewhat more difficult to assume unity for the γ values for ions in the resin phase because ion contents may be several molal. This is circumvented by an experimentally determined equivalent ionic fraction, y_i. The equivalent ionic fraction of a counterion species in a given phase (solution or resin) is defined as the ratio of equivalents of that species to all counterion species in the given phase. Consequently, Equation 53 may be written:

$$K_D^C = \frac{[D^+]^x(y_{RxC})(\gamma_{RxC})}{[C^{x+}](y_{RD})^x(\gamma_{RD})^x} \tag{54}$$

Rearranging,

$$\frac{[D^+]^x(y_{RxC})}{[C^{x+}](y_{RD})^x} = K_D^C \frac{(\gamma_{RD})^x}{(\gamma_{RxC})} \tag{55}$$

The left side of Equation 55 is obtained experimentally and is the "apparent equilibrium constant," K_{app}. The literature contains values of K_{app} for specific systems which, in turn, may have limited use for water softening. Nevertheless, the equilibrium aspect of ion-exchange phenomena is emphasized here [22,23]. The reader is referred to the literature [22–24] for details of the Donnan membrane theory.

Types and Capacities of Resins

Typical ion-exchange resins have spherical shapes, i.e., beads, and have an approximate size of 20–50 mesh (0.8–0.4 mm diameter). Some of the current types of resins are seen in Table VIII. The capacity for these resins to exchange ions is expressed in meq/ml (same as normality), meq/g (dry), kgr/ft^3 and lb/ft^3. Furthermore, the meq and kgr are

Table VIIIa. Commercial Designations of Typical Ion-Exchange Resins[a]

	Manufacturer			
Resin Type	**Diamond Shamrock Corp.**	**Dow Chemical Co.**	**Ionac Chemical Corp.**	**Rohm and Haas Co.**
Cation				
Standard Gel	C-20	HCR-S	C-249	IR-120
High Crosslinked Gel	C-20 × 10	HGR	C-250	IR-122
Macropore	C-26	MSC-1	CFP-110	IR-200
Weak Acid	C-433	CCR-2	CC	IRC-84
Anion				
Type I Gel		SBR	ASB1	IRA-400
Type I Gel (less crosslinker)	A-101D	SBR-P	ASB1-P	IRA-402
Type I Macropore	A-161	MSA-1	A-641	IRA-900
Type II Gel	A-102D	SAR	ASB2	IRA-410
Type II Macropore	A-162	MSA-2	A-651	IRA-910
Weak Base	A-340	WGR-2	A-305	IRA-68
Weak Base Macropore	A-378	MWA-1	AFP-329	IRA-93

Table VIIIb. Characteristics of Cation Exchangers[b]

Physical and Chemical Data					Recommended Operating Limits			
Group	Description	Exchanger	Trade Name[c]	Polar Exchange Groups	Flowrate gpm/ft² normal and max[d]	pH[e]	Temp., °F	Chlorine
1	Natural greensand zeolite (sodium aluminosilicate)	1 2	Zeo-Dur Inversand		3–4	6.8–8.0	110	No limit
2	Synthetic gel zeolite (sodium aluminosilicate)	3 4 5	Decalso Super Nalcolite Aridzone		3–4	6.0–8.3	140	No limit
3	Carbonaceous zeolite (sulfonated coal)	6 6 7 8	Zeo-Karb Cochranex CCA Catex-55 (Na) Catex-12 (H)	Methylene sulfonic, carboxylic, phenolic OH	4–6	No limit	120	Zero
4	Phenolic resin	9	Duolite C-3	Nuclear sulfonic	5–8	No limit	120	Limited
5	Styrene, medium capacity (styrene polyvinyl resin)	10 11 12	Amberlite IR-112 Chempro C-25 Nalcite MCR	Nuclear sulfonic	5–8	No limit	[f]	No limit
6	Styrene, high capacity (styrene polyvinyl resin)	13 14 15 16	Amberlite IR-120 Chempro C-20 Nalcite HCR Permutit Q	Nuclear sulfonic	5–8	No limit	250	No limit

			Capacity Data—Sodium Cycle			Capacity Data—Hydrogen Cycle[g,h]			
Group	Description	Exchanger	Regenerant (lb/ft^3)	Capacity (kgr/ft^3)	Efficiency (lb/kgr)	Acid	Regenerant (lb/ft^3)	Capacity (kgr/ft^3)	Efficiency (lb/kgr)
1	Natural greensand zeolite	1,2	1.25	2.8	0.45				
2	Synthetic gel zeolite	3	3.2	8	0.40				
		3	4	9	0.45				
		4,5	4	12	0.33				
		3	5	10	0.50				
		4	5	14	0.36				
		4,5	6	15	0.40				
		5	9	17	0.53				
3	Sulfonated coal	6,7,8	3.15	7	0.45	H_2SO_4	2	8	0.25
4	Phenolic resin	9				H_2SO_4	4	8.2	0.49
							6	10.0	0.60
							8	11.1	0.75
							10	12.0	0.80
						HCl	4.5	10.5	0.43
							8.0	12.5	0.64
							12.5	14.5	0.86
							17.0	16.0	1.06
							21.0	17.0	1.23
5	Styrene, medium capacity	10,11,12	3	12	0.25	H_2SO_4	2.5	9.5	0.26
			4	13.8	0.29		4.0	11.8	0.34
			6	16.4	0.37		5.0	13.5	0.37
			8	18.2	0.44		6.0	14.2	0.42
			10	21.5	0.47		8.0	16.2	0.49

Table VIIIb. continued

		Capacity Data—Sodium Cycle				Capacity Data—Hydrogen Cycle[g,h]			
Group	Description	Exchanger	Regenerant (lb/ft^3)	Capacity (kgr/ft^3)	Efficiency (lb/kgr)	Acid	Regenerant (lb/ft^3)	Capacity (kgr/ft^3)	Efficiency (lb/kgr)
6	Styrene, high capacity	13,14	5	18	0.28		2.5	9.0	0.28
		15,16	6	24	0.25		4.0	10.5	0.38
			6.6	22	0.30		5.0	11.0	0.45
			7.5	22	0.34		6.0	12.0	0.50
			8.4	24	0.35		7.5	13.5	0.55
			10.0	25	0.40		8.0	15.0	0.53
			13.5	27	0.50		10.0	25.0	0.40
			15	30	0.50				
						HCl	10.0	19.0	0.53
							15.0	21.6	0.70
							20.0	25.0	0.80

[a]Courtesy of the Permutit Company.
[b]Reproduced from Powell [2], courtesy of the McGraw-Hill Book Company.
[c]Trade names are identified with the following manufacturers: Arizona Mineral Corp. (5); Chemical Process Co. (9,11); Dow Chemical Co., Nalco Inc., distributors (12,15); Nalco Inc. (4); Infilco Inc. (7,8); the Permutit Co. (1,3,6,16); Rohm & Haas Inc. (10,13).
[d]Based on minimum bed depth of 30 in. Deeper beds may permit higher rates.
[e]Where no upper limit is given for pH it is assumed that the water is nonscaling.
[f]Group 6 exchangers are preferred for hot water service.
[g]Capacity values for cation exchangers in the hydrogen cycle in the higher regeneration levels are median values selected from manufacturers' data, based on treating a water in which alkalinity is 50% of the anions and sodium is 50% of the cations. It is assumed that leakage must not exceed 2%. Better capacity and efficiency are attainable when the sodium is lower or the specifications are less rigorous.
[h]Weights of acid based on 66°Bé H_2SO_4 and 30% HCl.

expressed usually in terms of $CaCO_3$. In the early history of ion-exchange softening, it was convenient to express water hardness in gr/gal, where 1 gr/gal = 17.1 mg/l. Consequently, the capacity of ion exchangers was expressed in kgr/ft^3. The factor for conversion of resin normality to kgr/ft^3 is approximately 22. For example, a cation resin with a capacity of 2 meq/ml has an exchange capacity of approximately 44 kgr/ft^3.

Applications

For water softening, the columnar flow-through system is utilized throughout municipal, industrial and domestic situations. A typical unit is seen in Figure 12. For most situations, the sodium form of the resins is employed. In accord with Equation 44, Ca^{2+} and Mg^{2+} ions in the raw water are exchanged for the Na^+ ion. Such other multivalent ions as Fe^{2+} are exchanged also for Na^+. Most of the Na^+ ions in the resin eventually are exchanged by the counter ions, whereupon the capacity for

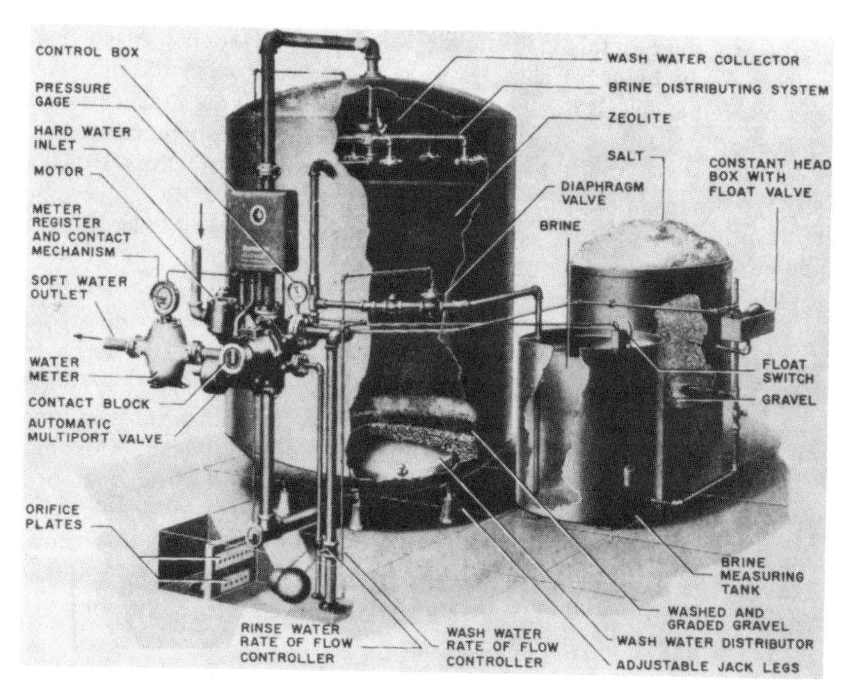

Figure 12. Automatic-pressure zeolite softener, showing brine distributor, mineral bed and underdraining system. Reproduced from Powell [2], courtesy of the McGraw-Hill Book Company.

softening becomes "exhausted." The unit then is removed from service, backwashed to remove any particulate matter and regenerated to the Na cycle by passing a concentrated solution of NaCl through the bed. A rinse with water is needed to remove excess NaCl. Downflow rates of the hard water are in the range of 2–8 gpm/ft^2 of resin.

It is necessary to know the capacity of the resin bed and the process efficiency. The "theoretical" capacity of a resin is the equivalent number of exchangeable sites per unit volume or unit weight of resin. However, the practice is to operate the unit at a level less than theoretical. The exchange reactions, Ca^{2+} and Mg^{2+} for Na^+, and the regeneration reactions, Na^+ for Ca^{2+} and Mg^{2+}, are equilibrium reactions. It would be time consuming either to operate the bed to complete exhaustion or to regenerate to 100% capacity. The latter would require large quantities of salt for the Na cycle, or acid for the H cycle.

The degree of theoretical capacity is called the "degree of column utilization," which is the ratio of the practical operation to the theoretical exchanger capacity [25]. "Efficiency" designates the degree of utilization of the regenerant. Column efficiency, therefore, is the ratio of the operating exchange capacity of a unit to the exchange that, theoretically, could be obtained from a given weight of applied regenerate. Consequently, higher efficiencies are realized with lower levels of regeneration, but this is at the loss of bed capacity. In actual practice, some efficiency is sacrificed in order to obtain a reasonable column utilization. According to Nordell [7], levels of regeneration are usually 3–10 lb of NaCl/ft^3 of resin, which represent efficiencies of 70–45% and column utilizations of 30–65%. Table IX shows some typical relationships among regeneration level, efficiency and column utilization. They were calculated [25] from some data from Kunin [24] for Amberlite IR-120, a raw water hardness of 500 mg/l as $CaCO_3$, a hydraulic loading of 2 gpm/ft^3 and a regenerate solution of 10% NaCl.

A typical calculation for the sizing of a sodium ion-exchange system is given in Table X [19]. These calculations take into consideration "leakage" of the hardness cations into the softened water. This may occur either from incomplete regeneration of the resin or from operation of the bed to or near complete exhaustion. The average hydraulic flowrate is 833 gpm/196 ft^3 = 4.25 gpm/ft^3, or 6.12 gpm/ft^3 at peak flow. The reader is referred to the literature [2,7,19] for complete details on the design of many kinds of available ion-exchange systems for domestic, municipal and industrial applications.

Typical water softening data are seen in Table XI for the Mississippi River [2]. This is an example of a combined treatment, i.e., the raw water is softened first by the cold lime-soda ash process, which is followed by

Table IX. Efficiency and Column Utilization as a Function of Regeneration Level[a]

Regeneration Level (lb NaCl/ft³ resin)	Hardness Removed (lb $CaCO_3$/ft³ resin)		Regeneration Efficiency (%)	Column Utilization[c] (%)
	Theoretical[b]	Actual		
1	0.85	0.83	98	14
2	1.70	1.32	78	22
3	2.55	1.83	72	30
5	4.25	2.85	67	47
10	8.50	3.90	46	64
15	12.75	4.65	36	76
20	17.00	5.00	29	82

[a]Reproduced from Weber [25], courtesy of John Wiley & Sons, Inc.
[b]Based on quantity of NaCl used.
[c]Based on total exchange capacity of resin of 6.1 lb/ft³.

ion-exchange. It should be noted that the sodium content of the softened water has been increased and 2.2 mg/l of silica has been removed. Here, again, operational results will vary in accord with the choice of the ion-exchange system.

Selection of the softening process(es) is dictated largely by economic considerations. Many combinations of chemical precipitation and ion-exchange systems can be devised. The cold lime-soda ash method can reduce concentrations of Ca^{2+} and Mg^{2} to approximately 50 ± 20 mg/l as $CaCO_3$. Ion exchange can reduce this to near the zero level. In general, the 50-mg/l level is acceptable for municipal drinking waters, whereas many industrial applications require zero hardness. Almost all domestic waters are softened by ion exchange where it is unnecessary to achieve a zero hardness level. If the latter prevails, the water frequently becomes corrosive to copper pipes due to $CO_{2(g)}$ acidity (see Chapter 9). Domestic softeners should be designed to reduce hardness levels to the 50- to 75-mg/l as $CaCO_{3(s)}$ range.

According to Weber [25]:

> Raw water conditions and operating factors favoring the use of ion exchange include: (a) low color and turbidity levels, (b) hardness largely not associated with carbonate alkalinity, (c) raw water hardness variable, (d) equipment is relatively simple to operate, and (e) no sludge disposal problems. Factors favoring use of chemical precipitation include: (a) raw water requires clarification in any event, and (b) hardness largely balanced by carbonate alkalinity.

Table X. Sample Calculation of Sodium Exchange System[a]

Water Analysis—Total hardness = 200 mg/l, allowable leakage = 3 mg/l, TE[b] = 300 mg/l.

Plant Requirements: 1,200,000 gpd = average flow of 1,200,000/1440 = 833 gpm; must accommodate extended peak flow of 1200 gpm.

1. Calculation of required size to handle 1200 gpm. Use three 10-ft-diameter units.
2. Calculation of required resin volume.
   (a) Required capacity in kilograins:
   Hardness = 200 mg/l; 200/17.1 = 11.7 gr/gal,
   1,200,000 × 11.7/1000 = 14,035 kgr/day,
   Schedule all 3 units to be regenerated each shift.
   ∴ Required capacity = 14,035/3 = 4678 kgr total
   4678/3 units = 1560 kgr/unit
   (b) Allowable leakage = 3 mg/l
   This leakage can be achieved @ 6 lb NaCl/ft^3.
   (Actual leakage @ 6 lb/ft^3 will be only 1.5 mg/l H)[c]
   Capacity at 6 lb/ft^3 = 20 kgr/ft^3
   ∴ Required resin = 1560/20 = 78 ft^3 minimum
   (c) Bed depth: Area of 10-ft-diameter unit = 78.5 ft^2. Therefore, bed would be approximately 1 ft deep. To achieve the desired leakage, minimum bed depth is 30 in. Therefore, each unit requires:
   2.5 ft deep × 78.5 ft^2 area = 196 ft^3
   (d) The actual regeneration period will be:
   $$\frac{196\ \text{ft}^3 \times 3\ \text{units} \times 20\ \text{kgr/ft}^3}{14{,}035\ \text{kgr/day}} = 0.84\ \text{day}$$
   0.84 × 24 = 20 hr,
   or one unit will be regenerated every 6.66 hr
3. Salt consumption @ 6 lb salt—20 kgr/ft^3
   6/20 = 0.3 lb/kgr (2.1 mg/l salt/mg/l hardness),
   14,035 kgr/day × 0.3 lb NaCl/kgr = 4200 lb NaCl,
   this requires bulk deliveries for practical handling by plant labor.

[a]Reproduced from Kemmer [19], courtesy of the McGraw-Hill Book Company.
[b]TE = total electrolyte.
[c]H = hardness.

Ion-exchange processes do have the brine disposal problem, which may be solved by dilution with surface waters, discharge into a sanitary sewer, discharge into evaporation ponds or injection into brine disposal wells.

Variations of Exchanger Systems

Figure 13 shows six cycles or modes of ion-exchange systems for municipal and industrial operations. Cycle 1 is used for boiler feed

Table XI. Cold-Process Lime Softening, Silica Removal, and Zeolite Softening of Mississippi River Water[a]

Constituents	Raw (mg/l)	Lime Softened (mg/l)	Zeolite Softened (mg/l)
As $CaCO_3$:			
Calcium	118	91	1
Magnesium	40	18	1
Total hardness	158	109	2
Sodium	84	84	191
Total cations	242	193	193
Sulfate	44	73	73
Chloride	96	96	96
Mineral-acid ions	140	169	169
Bicarbonate	102	0	0
Carbonate		24	24
Total anions	242	193	193
Silica	5.7	3.5	3.5
pH	8.0	10.0	10.0

[a]Reproduced from Powell [2], courtesy of the McGraw-Hill Book Company.

waters, whereas cycle 2 is applied for the same purpose when a "high" quality water is required. Cycles 3 and 4 are used for $CO_{2(g)}$ and silica removal. Here the water passes through a H-cation exchanger that converts all carbonates to H_2CO_3 which, in turn, is removed by neutralization on the strongly basic anion exchanger or by degasification. Silica is reduced to the range of 0.05–0.2 mg/l by the strongly basic anion exchanger. Cycle 5 is a highly unique and speciality system. Cycle 6 is a mixed bed of cation and anion exchangers. The resins frequently are in the H and OH cycles, respectively. This yields demineralized water. Many of these cycles carry trade names from the various manufacturers of ion-exchange resins and the concomitant equipment.

Case Studies—Split Treatment

One of the treatment options for water softening by lime-soda ash (and ion exchange for that matter) is split treatment. Only a portion of the water is softened and then is combined with unsoftened water before being discharged into the distribution system. The case study below points to the several advantages of this method over the conventional lime-soda ash treatment.

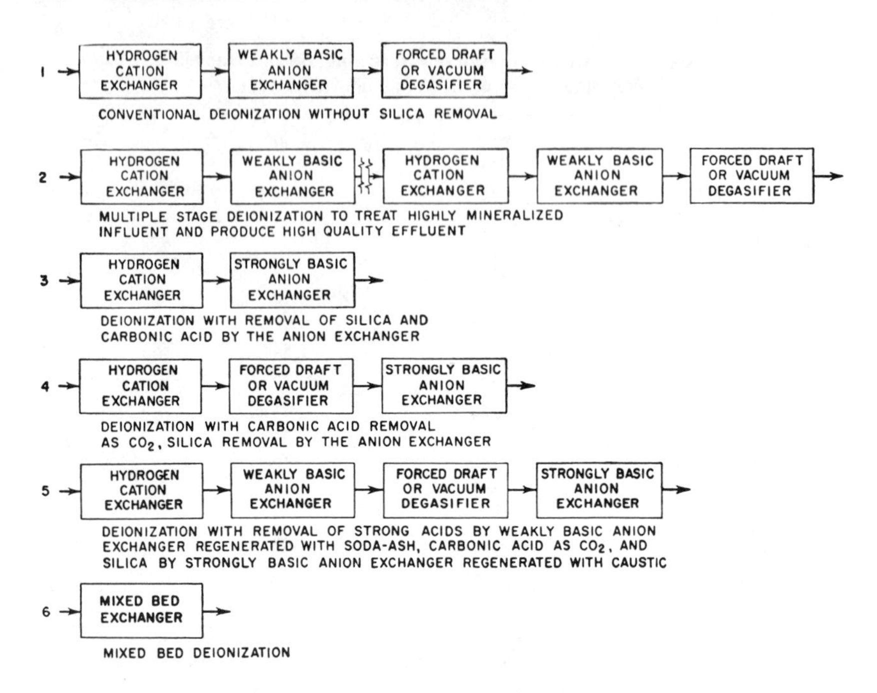

Figure 13. Flowsheet showing arrangements of deionization cycles used for water purification. Reproduced from Powell [2], courtesy of the McGraw-Hill Book Company.

Owosso, Michigan [26] draws its water supply from several deep wells wherein the Ca^{2+} hardness is approximately 290 mg/l as $CaCO_3$ and the Mg^{2+} hardness is about 130 mg/l as $CaCO_3$. Most is carbonate hardness. In 1972 their facilities were upgraded for conventional and split (75%) treatments with lime-soda ash (Figure 14). Prior to this, there were operating problems caused by sludge deposits in the recarbonation basins and filter-influent piping which were cleaned every six months or so. Filter runs were as short as 12 hr. In-plant tests were conducted to account for these operational problems. First, it was discovered that the Langelier Index (LI) was in the +0.2–0.8 range, which was consistent with the sludge deposits (mainly $CaCO_3$) (see Chapter 9). Second, an attempt was made to redissolve this sludge through a negative LI by recarbonation. A poor $CO_{2(g)}$ transfer efficiency caused the failure of this test, whereupon it was decided to soften about 75% of the water.

Split treatment requires a $[OH^-]$ of sufficient magnitude to precipitate all of the magnesium to prevent additional formation of $Mg(OH)_2$ after the waters are recombined. Appropriate calculations were made for the

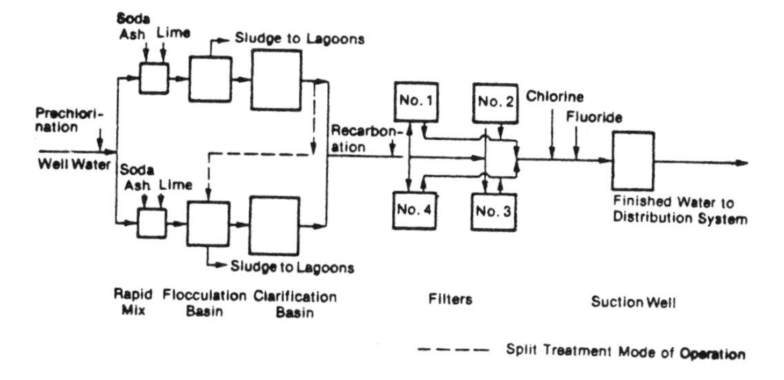

Figure 14. Schematic process flow diagram, Owosso, Michigan, water treatment plant. Reproduced from Singhal [26], courtesy of the American Water Works Association.

split-treatment dosages of lime, soda ash and $CO_{2(g)}$. After several months of experimentation, these conclusions were drawn:

1. Optimum $[OH^-]$ should yield a pH value of 12.1 or 12.2 in the first-stage effluent. $Mg(OH)_2$ apparently was precipitated first.
2. HCO_3^- alkalinity of 50–70 mg/l as $CaCO_3$ is optimum for the secondary effluent. Any content of HCO_3^- above this quantity will result in a greater effluent hardness caused by dissolution of $CaCO_{3(s)}$.
3. The optimum pH value of the secondary effluent should be 9.2–9.5.

Table XII shows the raw and treated water hardnesses before and after institution of split treatment. In addition to the improved water quality, there was a greater efficiency of filter performance.

Split treatment resulted in considerable savings in chemical costs: 25% less lime was used, 13.7% more soda ash was used and $CO_{2(g)}$ was completely eliminated. All of this yielded a net reduction in chemical costs.

A split-treatment softening has been reported for waters with a "high" magnesium content that removes only a portion of the hardness [27,28]. Groundwaters are appropriate because these waters generally do not require a separate stage of treatment for taste, odor, etc. As depicted in Figure 15, the water is divided into two streams: one is treated with an excess dose of $Ca(OH)_2$ for "complete" precipitation of the magnesium and some calcium, and one is untreated. These two streams are recombined, whereupon Na_2CO_3 is added to remove additional CaNCH, if desired.

In order to effect split-treatment softening, Zipf and Luthy [27] derived nine basic equations to calculate the lime dosage for Mg removal

Table XII. Summary of Raw- and Treated-Water Quality at Owosso Michigan, Water Treatment Plant (April 1973–July 1974)[a]

Year and Month	Raw Water Quality (mg/l $CaCO_3$)			Treated Water Quality (mg/l $CaCO_3$)			Remarks
	Ca^{2+}	Mg^{2+}	Alk.	Ca^{2+}	Mg^{2+}	Alk.	
1973							
April	339	164	368	137	18		Conventional treatment from Apr. 1973 to Nov. 1973
May	302	147	370	135	20		
June	289	152	356	121	44	102	Average soda ash 1006 lb/mil gal 1722×10^3 kg/m^3
July	278	143	340	111	56	109	Average lime 4146 lb/mil gal, 7100×10^3 kg/m^3
August	288	161	361	112	24	83	Average final hardness 148 mg/l
September	284	137	339	108	20		
October	289	139	341	121	19		
November	289	141	337	105	49	84	
December	290	123	340	85	34	93	Split treatment from Dec. 1973 to Jul. 1974
1974							
January	292	121	343	98	37	105	Average soda ash 1144 lb/mil gal 1900×10^3 kg/m^3
February	286	126	341	89	44	113	Average lime 3093 lb/mil gal 5280×10^3 kg/m^3
March	284	132	343	63	52	98	Average final hardness 128 mg/l
April	279	115	331	71	55	118	
May	280	134	340	64	52	101	
June	299	133	350	72	58		
July	291	124	345	77	78	81	

[a]Reproduced from Singhal [26], courtesy of the American Water Works Association.

(ten basic equations are needed if Na_2CO_3 is included). Figure 16 shows the acidity mass balance for this treatment as well as the symbolism for the three waters. All concentrations and equilibrium constants (H^+ excepted) are expressed in equivalents of $CaCO_3$ at 25°C and 0.01 *M* ionic strength. The equilibrium equations are:

$$HCO_3^- = H^+ + CO_3^{2-}, \quad K_{56} = 10^{-9.85} \text{ (as } CaCO_3\text{)}, 25°C \tag{56}$$

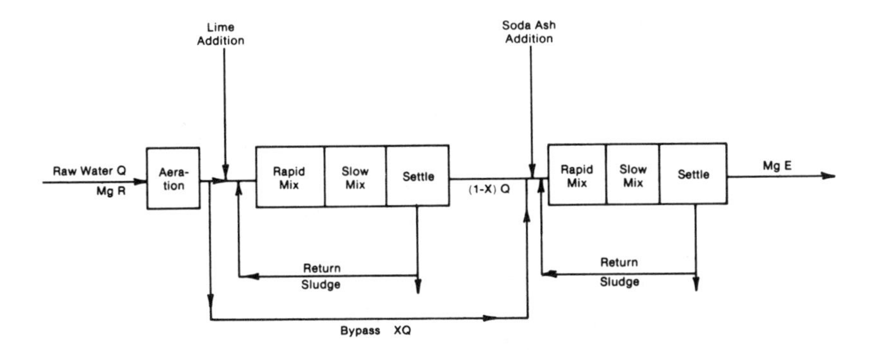

Figure 15. Flow diagram for typical split-treatment lime-soda ash softening process. Reproduced from Zipf and Luthy [27], courtesy of the American Water Works Association.

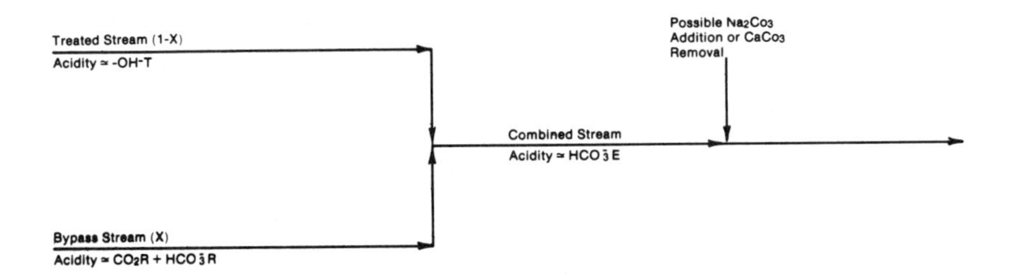

Figure 16. Acidity mass balance relationships during split treatment. Reproduced from Zipf and Luthy [27], courtesy of the American Water Works Association.

$$CaCO_{3(s)} = Ca^{2+} + CO_3^{2-}, \quad K_{57} = 10^2 \,(\text{as } CaCO_3) \tag{57}$$

Combining Equations 56 and 57 gives:

$$\frac{K_{57}}{K_{56}} = \frac{(Ca^{2+})(HCO_3^-)}{(H^+)} \tag{58}$$

From this, a parameter F_1 is obtained:

$$F_1 = (Ca^{2+})(HCO_3^-) \tag{59}$$

and

$$F_1 = \frac{K_{57}}{K_{56}}(H^+) \tag{60}$$

Neglecting (H^+) and (OH^-), alkalinity is expressed:

$$ALK = HCO_3^- + CO_3^{2-} \qquad (61)$$

and

$$ALK = (HCO_3^-)\left(1 + \frac{K_2}{(H^+)}\right) \qquad (62)$$

From this, a second parameter, F_2, is obtained:

$$(HCO_3^-) = ALKF_2 \qquad (63)$$

where

$$F_2 = \frac{1}{1 + K_2/(H^+)} \qquad (64)$$

Equations 59, 60, 63 and 64 are four of the basic equations. A fifth basic equation is derived from the assumption that all Mg^{2+} in the treated stream is precipitated as $Mg(OH)_{2(s)}$. Some Mg will remain in solution due to the solubility product constant of $Mg(OH)_{2(s)}$. For purposes of calculation, it can be neglected. Consequently, the effluent $[Mg^{2+}]$ becomes equal to the fraction of Mg in the bypass stream:

$$Mg^{2+}E = X(Mg^{2+}R) \qquad (65)$$

where X is the bypass fraction. The sixth basic equation is:

$$\text{Hardness } E = Ca^{2+}E + Mg^{2+}E \qquad (66)$$

The seventh basic equation is derived from a mass balance for total acidity (ACY) at the point where the treated and untreated streams are combined:

$$ACY = CO_2 + HCO_3^- + H^+ - OH^- \qquad (67)$$

As seen in Figure 16, ACY in the treated stream is $-OH^-T$, whereas the ACY in the untreated stream is $CO_2R + HCO_3^-R$. When the pH values are ≥ 8.2, CO_2R is small compared to HCO_3^-R, whereupon:

$$HCO_3^-E = X(CO_2R + HCO_3^-R) - (1 - X)(OH^-T) \qquad (68)$$

This equation is valid for lime or lime-soda ash treatment because the addition or removal of CO_3^{2-} does not affect the effluent's total acidity statement.

The eighth and ninth basic equations are modified conventional lime and soda ash expressions. The lime dosage is modified to account for some Mg^{2+} in the treated water and for HCO_3^- in the effluent. Neglecting CO_2E:

$$Ca(OH)_2 = CO_2R + HCO_3^-R - HCO_3^-E + (1 - X)Mg^{2+}R \tag{69}$$

Soda ash dosage is:

$$Na_2CO_3 = \text{Hardness R} - \text{Hardness E} - \text{ALKR} + \text{ALKE} \tag{70}$$

Equations 69 and 70 apply either to lime or lime-soda ash treatment. A tenth basic equation is needed for lime-soda ash treatment to provide a relation between the effluent hardness and alkalinity. Combining Equations 59, 64 and 65 yields:

$$\text{Hardness E} = \frac{F_1}{HCO_3^-E} + X(Mg^{2+}R) \tag{71}$$

From Equation 68:

$$X = \frac{HCO_3^-E + OH^-T}{HCO_3^-R + CO_2R + OH^-T} \tag{72}$$

Combining Equation 71 and 72 through the X term:

$$\text{Hardness E} = \frac{F_1}{HCO_3^-E} + (Mg^{2+}R)\left[\frac{HCO_3^-E + OH^-T}{HCO_3^-R + CO_2R + OH^-T}\right] \tag{73}$$

Minimum hardness is obtained by taking the derivative of Equation 73 and equating it to zero:

$$\frac{d\text{Hardness E}}{dHCO_3^-E} = \frac{-F_1}{(HCO_3^-E)^2} + \frac{Mg^{2+}R}{HCO_3^-R + CO_2R + OH^-T} = 0 \tag{74}$$

Rearranging gives the tenth basic equation:

$$\frac{F_1}{Mg^{2+}R}[HCO_3^-R + CO_2R + OH^-T] - (HCO_3^-E)^2 = 0 \tag{75}$$

All of this leads to the 10 basic equations and 12 operational variables. Since there are 2 variables more than the 10 equations, they must be selected from the 12, and their values must be set arbitrarily. For example, these "decision variables" may be: hardness E and pH, ALK E and pH, and lime dosage and bypass fraction X.

A solution set for lime treatment only is shown by selection of hardness E and pHE as "decision variables" with no soda ash. This results in a matrix of variables in the basic equations as seen in Table XIIIa. It also gives the partitioning of each variable and the appropriate equation as well as the precedence order of solving these equations. This precedence is achieved by inspection where functions that have only one unknown are computed and eliminated from the matrix. This process continues until all variables and equations are accounted for. The order in which the variables and equations are eliminated becomes the order of the solution set. The matrix is rearranged to reflect this consideration, which gives the correct partitioning and precedence order for the solution set (Table XIIIb).

A sample calculation is seen in Table XIV for a lime-soda ash treatment with hardness E and pHE as "decision variables." Other examples, in the form of exponential curves, are given in Zipf and Luthy [27]. A computer program is available to solve a wide variety of "decision variables." In addition, some general observations about split-treatment were: (1) the bypass fraction may be as high as 0.9; (2) effluent pH values range from 8.0 to 9.4, with 7.8–8.0 as minimum values; and (3) $[OH^-]$ in the treated water is usually a limiting factor with 50 mg/l as $CaCO_3$ as a practical lower limit. F_1 and F_2 values are given in Table XV [29].

Table XIIIa. Matrix Solution to Lime-Only Split Treatment with Effluent Hardness and pH As Decision Variables[a]

	Presence of Indicated Process Variables in Basic Equations								
Equations	F_1	F_2	OH^-T	$Ca^{2+}E$	$Mg^{2+}E$	HCO_3^-E	Alkalinity E	X	Lime
59	x			x		x			
60	x								
63		x				x	x		
64		x							
65					x			x	
66				x	x				
68			x			x		x	
69						x		x	x
70							x		

Table XIIIb. Partitioning and Precedence Order for Solution Set to Lime-Only Split Treatment with Effluent Hardness and pH As Decision Variables[a]

	Process Variables								
Equations	F_1	F_2	Alkalinity E	HCO_3^-E	$Ca^{2+}E$	$Mg^{2+}E$	X	Lime	OH^-T
60	[x][b]								
64		[x]							
70			[x]						
63		x	x	[x]					
59	x			x	[x]				
66					x	[x]			
65						x	[x]		
69				x			x	[x]	
68				x			x		[x]

[a]Reproduced from Zipf and Luthy [27], courtesy of the American Water Works Association.
[b]Brackets indicate the precedence order of equations for the solution set.

Table XIV. Sample Calculation for Lime and Soda Treatment with Effluent Hardness and pH as Decision Variables[a]

Raw Water Characteristics (mg/l as $CaCO_3$ except pH)		Equation	Result
Hardness	350		
Ca^{2+}	250	60	1774.2
Mg^{2+}	100	64	0.9466
Alkalinity	220.6	[b]	86.2
HCO_3^-	220	59	20.6
CO_2	78.5	66	49.4
pH	7.28	65	0.494
Equilibrium Constants			
log K_2	−9.849	68	121.5
log K_5	2.00 (LSI = 0.0)	63	91.
Select Decision Variables			
Effluent Hardness	70.0	69	263.
Effluent pH, pH E	8.60	70	150.4

[a]Reproduced from Zipf and Luthy [27], courtesy of the American Water Works Association.
[b]Equation not given.

Sludge Handling and Disposal

The problem posed by sludge disposal from water treatment plants was discussed in two committee reports for the American Water Works Association [30,31]. PL 92–500, the Water Pollution Control Act

Table XV. Values of Constants F_1 and F_2 for Various Values of pH[a]

pH	$F_1 = K_{57}(H^+)/K_{56}$	$F_2 = 1/[1 + K_2/(H^+)]$	F_1/F_2
8.0	7080	0.9861	7180
8.2	4467	0.9781	4567
8.4	2818	0.9657	2918
8.6	1778	0.9468	1878
8.8	1122	0.9106	1232
9.0	708	0.8762	808
9.2	447	0.8171	547

[a]Reproduced from Rossum [29], courtesy of the American Chemical Society.

Amendments of 1972, included these sludges as an industrial waste requiring a discharge permit under the National Pollutant Discharge Elimination System (NPDES). Thus, the previous practice of disposal by direct discharge into a waterway was forbidden.

Sludge Characteristics

Sludges from the lime-soda ash process have a volume ranging from 0.3 to 5.% of the raw water treated. The solids content of settled sludge varies widely—from 2. to 30.% [30]. Table XVI shows the calculated composition of lime-soda ash sludge, which is more or less typical [32]. For this example, $CaCO_{3(s)}$ was 84.4% of the sludge's total weight. The composition, sludge volumes and sludge solids, of course, will vary from plant to plant according to raw water hardness, degree of treatment, use of split treatment, etc. Any means of treatment and disposal becomes a problem of handling a by-product which, in most cases, is an added cost to treatment of the water. In general, handling and disposal methods are lagooning, drying beds, mechanical concentration by vacuum filters, centrifuges, belt filters, pressure filters, etc., and recalcining.

Sludge Dewatering

Reduction of the moisture content of lime-soda ash sludges is a prerequisite for ultimate disposal. That is, the lesser the sludge volume, the lower is the cost. Lime sludges can be dewatered to about 50% within a reasonable time period on drying beds [31]. This practice, like lagooning, has the disadvantage of requiring substantial surface area, a climate conducive to drying and a location for ultimate disposal. An advantage may

Table XVI. Calculated Composition of Primary Sludge[a]

Sludge Components	ppm	lb/mil gal	% by weight
$CaCO_3$			
From Free CO_2	32	266	
From CH	530	4,420	
From NCH	20	167	
Subtotal	582	4,853	84.4
$Mg(OH)_2$	81	675	11.7
SiO_2	4	33	0.6
Insolubles from Lime		190	3.3
Total lb/mil gal		5,750	100.0
Total primary and secondary sludge per mil gal = 6,900 lb			

[a]Reproduced from Black et al. [32], courtesy of the American Water Works Association.
[b]Calculations based on free CO_2 in raw water—14 ppm; carbonate hardness in raw water—265 ppm; noncarbonate hardness removed—20 ppm; insolubles in lime, SiO_2, R_2O_3, and MgO—8%.

be the relatively low cost. The Louisville Water Company [33] has had experience with sludge disposal lagoons. An extensive field study revealed the life expectancy of three existing lagoons to be about 21.2 yr instead of the 30–40 yr projected prior to construction. Difficulties with dewatering of the sludge was the apparent reason. This study also provided operational data for the lagoons at temperatures less than 0°C.

Dewatering lime-soda ash sludges by vacuum filtration is reasonably successful, and yields a crumbly cake with a solids content up to 65% [31]. Vacuum filtration of softening sludges was evaluated by the specific resistance method [34]. There are two general considerations: (1) sludges high in magnesium content are more difficult to dewater than sludges low in magnesium, and (2) calcium hardness has no significant effect on the filterability of lime-softening sludges.

In other processes, Schwayer and Luttinger [35] describe a dual-cell gravity (DCG) concentrator for mechanical gravity dewatering of softening sludges, especially metal hydroxides. Reh [36] reported the state-of-the-art of water treatment sludge disposal and handling.

Recalcination

A process for recovery of lime and magnesium carbonate from softening sludges was developed at Dayton, Ohio [32]. Briefly described, a mixed sludge of $CaCO_{3(s)}$ and $Mg(OH)_{2(s)}$ is treated with stack gas [20%

$CO_{2(g)}$] from a lime kiln. This selectively dissolves the gelatinous $Mg(OH)_{2(s)}$ from the crystalline $CaCO_{3(s)}$, which, in turn, is calcined to $CaO_{(s)}$ (92–93%). The flow diagram of the $MgCO_3$ recovery portion of the process is seen in Figure 17 [32]. Several benefits were noted from the recovery treatment: (1) cost of softening the water was reduced substantially (by 1970 dollar values and energy costs); (2) for each ton of lime used for softening, 1.2–1.3 tons of lime were recovered; the excess was sold; (3) it eliminated disposal of the $CaCO_{3(s)}$ sludge; the recovered $MgCO_{3(s)}$ was discharged into a sanitary sewer; (4) it eliminated also the need for extra-plant supplies of $CO_{2(g)}$ for recarbonation. There was a sufficient supply of this gas from the lime kiln. Thus, recalcination is largely an economic decision.

A pilot-plant operation has been designed to recover and dispose of alum as well as softening sludges [37]. The flow diagram is seen in Figure 18. At a 70-mgd rate of flow, the softening plant produced 350,000 gpd of sludge with 4% solids from the removal of 55 mg/l hardness. This sludge had been lagooned and dried to 30% solids, whereupon it was disposed to landfill. This handling and disposal method eventually became too costly, and landfill sites disappeared. In the pilot-plant operation, the softening sludge was recarbonated, concentrated and centrifuged. The centrifuge cake was recalcined in an electric furnace. This resulted in a quantity of $CaO_{(s)}$ content the equivalent of the original lime used in the softening process.

Recarbonation is accomplished with a 25/75 $CO_{2(g)}$–air mixture. This yields a $Mg(HCO_3)_2$ solution of 16,200 mg/l as $CaCO_3$ in accord with:

$$Mg(OH)_{2(s)} + 2\,CO_{2(g)} = Mg(HCO_3)_2 \tag{76}$$

The recarbonation can be controlled with the 25% $CO_{2(g)}$ in the air mixture to avoid formation of solid $MgCO_3 \cdot 3\,H_2O_{(s)}$ and $Ca(HCO_3)_2$, and to keep the pH of the sludge at 7.3–7.5.

After the recarbonation step, the residual sludge, mainly $CaCO_{3(s)}$, is centrifuged to a cake with 50–60% solids. This cake subsequently is burned at temperatures in the 871–1204°C range. A typical analysis of the recalcined lime is: CaO, 90.8%; MgO, 4.1%; SiO_2, 2.1%; and R_2O_3, 2.5%. The slakable CaO is 82.9%. Silica content of the sludge cake must be carefully controlled to avoid formation of a CaO-Si compound that will not slake.

This recarbonation-recalcination process produces two waste streams that require treatment: the overflow streams from the sludge thickener and the centrifuge centrate stream. These streams have a soluble color concentration of 300 units and a [$Mg(HCO_3)_2$] of 4380. mg/l. Discharge

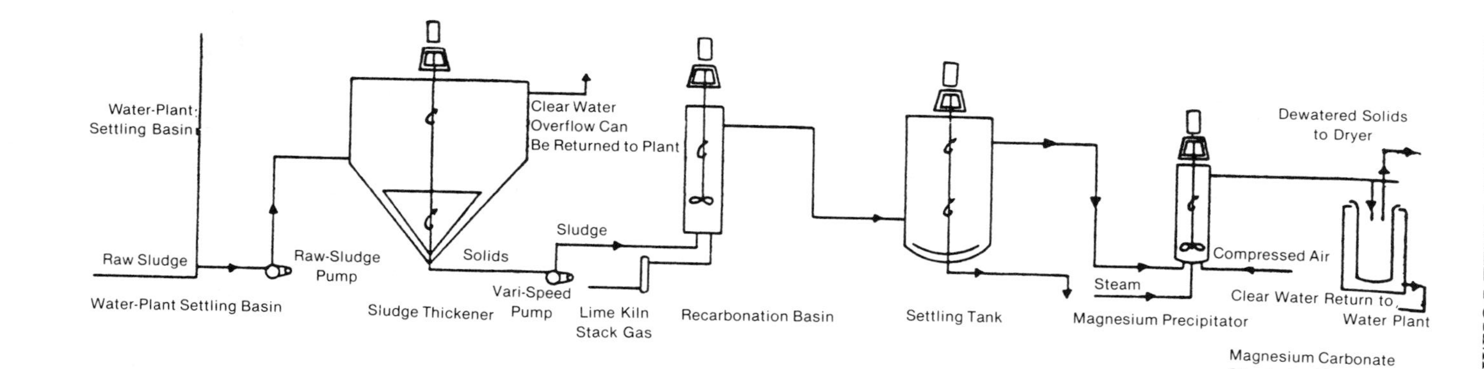

Figure 17. Total recovery of lime-softening sludge: magnesium recovery pilot plant. Reproduced from Black et al. [32], courtesy of the American Water Works Association.

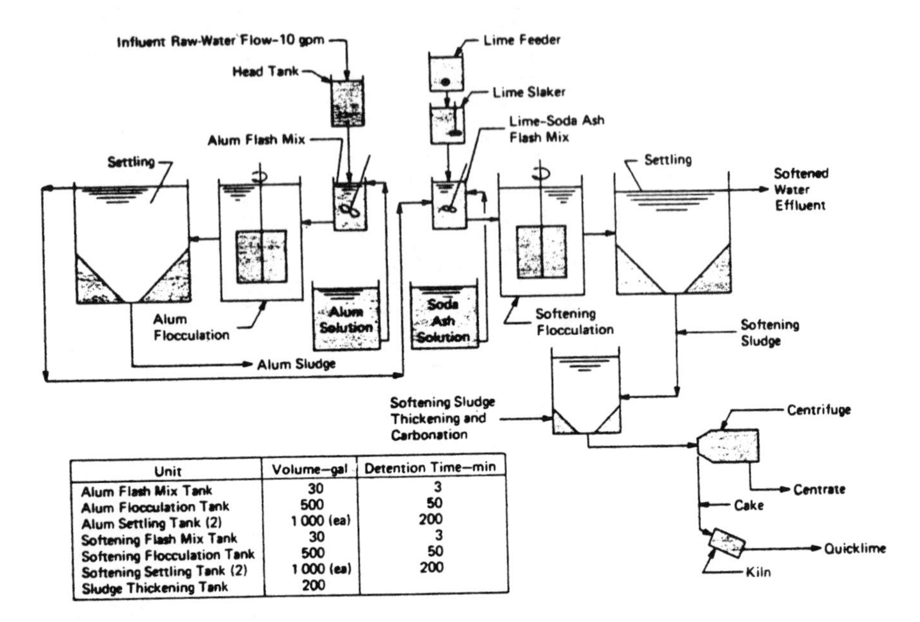

Unit	Volume–gal	Detention Time–min
Alum Flash Mix Tank	30	3
Alum Flocculation Tank	500	50
Alum Settling Tank (2)	1 000 (ea)	200
Softening Flash Mix Tank	30	3
Softening Flocculation Tank	500	50
Softening Settling Tank (2)	1 000 (ea)	200
Sludge Thickening Tank	200	

Figure 18. Pilot-plant schematic for Columbus, Ohio. Reproduced from Burris et al. [37], courtesy of the American Water Works Association.

of these two waste streams into a natural waterway is prohibited, and recylcing to the raw water influent is undesirable because of Mg buildup. Consequently, these streams are resoftened by the addition of $Ca(OH)_2$ to a pH value of 11. This removes "almost all of the color" and produces a sludge of $CaCO_{3(s)}$ and $Mg(OH)_{2(s)}$. This sludge is ultimately combined with the lime-treated alum sludge and dewatered by a filter press. This filter cake, 40–50% solids, is disposed by landfill or whatever.

Coagulation of Sludges

The electrophoretic characteristics of $CaCO_{3(s)}$ and $Mg(OH)_{2(s)}$ have been researched [38] and the electrophoretic mobility values given in Figure 19. These values are related to the alkaline pH values at which they are formed. For $CaCO_{3(s)}$, electromobility values are in the range of -0.9 to -1.0 μm/sec/V/cm, whereas $Mg(OH)_{2(s)}$ has a slightly positive charge. This positive value is attributed to the formation of $Mg(OH)_{2(s)}$ in the presence of Ca^{2+} ions. Subsequent experiments confirmed this point.

Coagulation of $CaCO_{3(s)}$ with alum and four coagulant aids has been

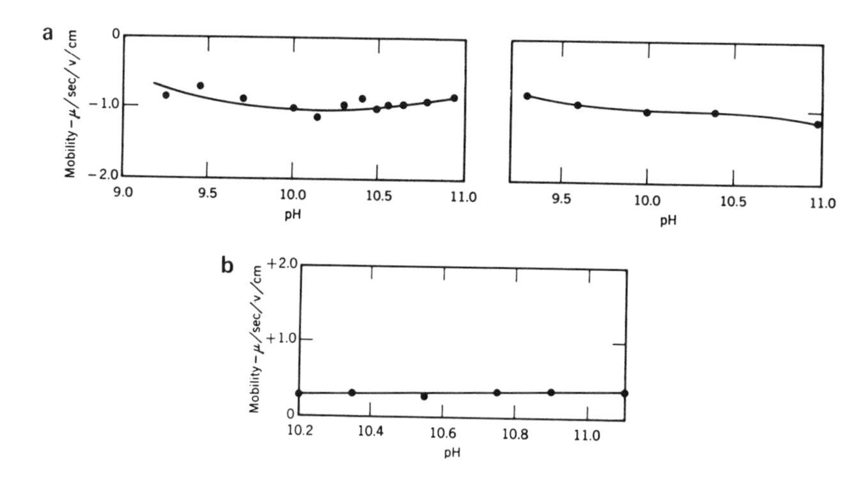

Figure 19. **(a)** Effect of pH on mobility of $CaCO_3$. The curve at the left pertains to $CaCO_3$ sludge produced by softening water A with varying dosages of lime in the pH range 9.0–11.0. The curve at the right shows the effect of pH on the same sludge suspended in deionized water. **(b)** Effect of pH on mobility of $Mg(OH)_2$. The relatively constant mobility throughout the pH range shown is probably the result of the protective action of the water molecules surrounding the strongly hydrophilic $Mg(OH)_2$ particles. Reproduced from Black and Christman [38], courtesy of the American Water Works Association.

studied [38]. Typical results are seen in Figure 20, where aid B is an anionic acrylamide (Separan NP10), aid C is a nonionic potato starch and aid D is a high-molecular-weight cationic polymer (Nalco 600). Coagulation occurs in all systems, and is accompanied with and without charge reversal. Researchers have drawn the following conclusions from 20a and b: "the mobility value in itself is not a reliable indication of the degree of coagulation to be expected with the use of a coagulant aid and softening sludges.... The best coagulation was usually obtained at substantially negative mobility values" [38].

REFERENCES

1. Faust, S. D., and O. M. Aly. *Chemistry of Natural Waters* (Ann Arbor, MI: Ann Arbor Science Publishers, Inc., 1981).
2. Powell, S. T. *Water Conditioning for Industry* (New York: McGraw-Hill Book Company, 1954).
3. Hem, J. D. "Study and Interpretation of the Chemical Characteristics of

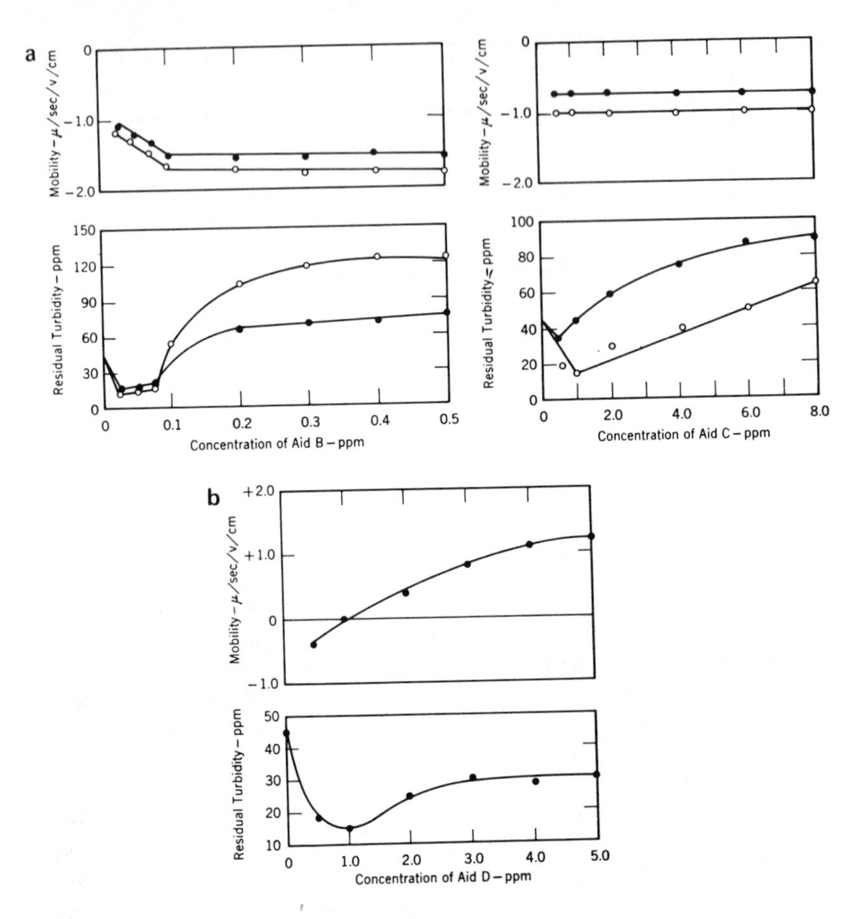

Figure 20. Effects of **(a)** aids B and C, and **(b)** aid D on mobility and coagulation of $CaCO_3$ at pH 9.8. Aid D was added after the lime. Coagulation was accompanied by charge reversal. Reproduced from Black and Christman [38], courtesy of the American Water Works Association.

Natural Water," Geological Survey Water-Supply Paper 1473, U.S. Government Printing Office (1970).

4. Harned, H. S., and R. Davies. *J. Am. Chem. Soc.* 65:2030 (1943).
5. Harned, H. S., and S. R. Scholes. *J. Am. Chem. Soc.* 63:1706 (1941).
6. Stumm, W., and J. J. Morgan. *Aquatic Chemistry*, 2nd ed. (New York: John Wiley & Sons, Inc., 1981).
7. Nordell, E. *Water Treatment* (New York: Van Nostrand Reinhold Company, 1961).
8. *Lange's Handbook of Chemistry*, 11th ed., (New York: McGraw-Hill Book Company, 1973).

9. White, D. E. et al. "Chemical Composition of Subsurface Waters," Geological Survey Professional Paper 440-F, U.S. Government Printing Office (1963).
10. Schierholz, P. M. et al. *J. Am. Water Works Assoc.* 68:112 (1976).
11. *Standard Methods for the Examination of Water and Wastewater,* 15th ed. (Washington, DC: American Public Health Association, 1981).
12. Dye, J. F. *J. Am. Water Works Assoc.* 50:800 (1958).
13. Cadena, C. F. et al. *J. Am. Water Works Assoc.* 66:524 (1974).
14. Alexander, H. J., and M. A. McClanahan. *J. Am. Water Works Assoc.* 67:618 (1975).
15. Wiechers, H. N. S. et al. *Water Res.* 9:835 (1975).
16. Reddy, M. M., and G. H. Nancollas. *J. Colloid Interface Sci.* 36:166 (1971).
17. Haney, P. D., and C. L. Hamann. *J. Am. Water Works Assoc.* 71:512 (1969).
18. Larson, T. E. et al. *J. Am. Water Works Assoc.* 51:1551 (1959).
19. Kemmer, F. N., Ed. *The Nalco Water Handbook* (New York: McGraw-Hill Book Company, 1979).
20. Boyd, G. E. et al. *J. Am. Chem. Soc.* 69:2818 (1947).
21. Bauman, W. C., and A. Eichhorn. *J. Am. Chem. Soc.* 69:2830 (1947).
22. Helfferich, F. *Ion Exchange* (New York: McGraw-Hill Book Company, 1962).
23. Rieman, W., and H. F. Walton. *Ion Exchange in Analytical Chemistry* (New York: Pergamon Press, 1970).
24. Kunin, R. *Ion Exchange Resins*, 2nd ed. (New York: John Wiley & Sons, Inc., 1958).
25. Weber, W. J., Jr. *Physicochemical Processes for Water Quality Control* (New York: John Wiley & Sons, Inc., 1972).
26. Singhal, A. K. *J. Am. Water Works Assoc.* 69:158 (1977).
27. Zipf, K. A., Jr., and R. G. Luthy. *J. Am. Water Works Assoc.* 73:304 (1981).
28. Cleasby, J. L., and J. H. Dillingham. *J. San. Eng.* 92:942, 1 (1966).
29. Rossum, J. R. *Ind. Eng. Chem.* 47:2313 (1955).
30. Committee Report, *J. Am. Water Works Assoc.* 70:498 (1978).
31. Committee Report, *J. Am. Water Works Assoc.* 64:814 (1972).
32. Black, A. P. et al. *J. Am. Water Works Assoc.* 63:616 (1971).
33. Hubbs, S. A., and J. L. Pavoni. *J. Am. Water Works Assoc.* 66:658 (1974).
34. Hawkins, F. C. et al. *J. Am. Water Works Assoc.* 66:653 (1974).
35. Schwayer, W. L., and L. B. Luttinger. *J. Am. Water Works Assoc.* 65:399 (1973).
36. Reh, C. W. *J. Am. Water Works Assoc.* 72:115 (1980).
37. Burris, M. A. et al. *J. Am. Water Works Assoc.* 68:247 (1976).
38. Black, A. P., and R. F. Christman. *J. Am. Water Works Assoc.* 53:737 (1961).

CHAPTER 8

REMOVAL OF INORGANIC CONTAMINANTS

PROBLEMS WITH INORGANIC CONTAMINANTS

The nearly ubiquitous distribution in natural waters of the inorganic constituents Ag, As, Ba, Cd, Cr, Cu, F, Fe, Hg, H_2S, Mn, Na, Pb, Se and Zn has created concern about their physiological and other effects on humans [1] (see also Chapter 1). Three of these ions—Cu, Fe and Mn—clearly are nuisance types of constituents. Cu affects the taste quality of drinking water, and the oxidative products of Fe and Mn impart turbidity to the water, stain clothing in laundering processes and interfere with the brewing of tea and coffee and the flavor of other beverages [2]. Ferrous (II) and cupric (II) ions are corrosion products of distribution pipes (see Chapter 9). Because of their effects on drinking water quality, it is necessary to lower the contents of these constituents to the current maximum contaminant level (MCL) standards set forth by the U.S. Environmental Protection Agency (EPA) (see Chapter 1). For a description of the geologic sources and typical contents of these metals and semimetals in water, see Faust and Aly [2].

IRON

Aqueous Chemistry

The aqueous chemistry of iron is rather complex, since this metal enters into several protolysis and oxidation-reduction reactions [2–4]. Some of the more pertinent reactions are given in Table I from which the pE-pH stability diagram was drawn for Figure 1. Construction details for this figure are given in Faust and Aly [2].

Table I. Some Aqueous Reactions for Iron at 25°C

Reaction No.	Reaction	log K[a]	V^0
1	$Fe^{2+} + 2\ e = Fe^0_{(s)}$	−14.9	−0.440
2	$Fe^{3+} + e = Fe^{2+}$	+13.17	+0.771
3	$FeOH^{2+} + H^+ + e = Fe^{2+} + H_2O$	+15.5	+0.914
4	$Fe(OH)_{3(s)} + 3\ H^+ + e = Fe^{2+} + 3\ H_2O$	+17.8	+1.052
5	$Fe^{3+} + H_2O = FeOH^{2+} + H^+$	−2.2	
6	$FeOH^{2+} + H_2O = Fe(OH)_{2(s)} + H^+$	−4.6	
7	$FeOH^{2+} + 2\ H_2O = Fe(OH)_{3(s)} + 2\ H^+$	−2.41	
8	$O_{2(g)} + 4\ H^+ + 4\ e = 2\ H_2O$	+83.3	+1.229
9	$2\ H_2O + 2\ e = H_{2(g)} + 2\ OH^-$	−28.1	−0.828
10	$Fe_{(s)} + 0.5\ O_{2(g)} + 2\ H^+ = Fe^{2+} + H_2O$	+56.4	+1.67
11	$O_{2(g)} + 4\ Fe^{2+} + 10\ H_2O = 4\ Fe(OH)_{3(s)} + 8\ H^+$	+11.6	+0.177

[a]$K = K_{RED}$ for reduction reaction and = Keq for protolysis reaction.

Briefly explained, the pE-pH diagram gives the boundaries in which a given iron species is stable or predominant. The diagram also is useful for determining the pE-pH conditions under which Fe(II) is oxidized to Fe(III) with $O_{2(g)}$ as the electron acceptor, and is subsequently precipitated as $Fe(OH)_{3(s)}$. Most natural waters have pH values ranging from 5.0 to 8.5, and pE values ranging from +2 to +12. Thus, Fe(II) would be the predominant iron species in the absence of an electron acceptor such as $O_{2(g)}$. In order to oxidize Fe(II) to Fe(III), it is necessary to raise both the pE and pH values. The former may be raised by adding such an electron acceptor as $O_{2(g)}$, Cl_2 or $KMnO_4$, whereas the pH value is increased through addition of OH^- from $Ca(OH)_2$ or NaOH. This diagram also shows that the pE value required for oxidation is lowered as the pH value is increased. This interpretation has very pragmatic operational applications in iron removal from natural waters. This is also a factor in the oxygenation kinetics of Fe(II) discussed below. The thermodynamic instability of iron pipes is seen in Equation 1 (see also Chapter 9).

Removal

Current water treatment practice employs three general methods for reducing iron contents to less than the MCL. The primary method uses oxidation of the Fe(II) and subsequent precipitation to $Fe(OH)_{3(s)}$ followed by sedimentation and filtration. Ion exchange is frequently employed where iron contents are less than 10 mg/l and where "low"

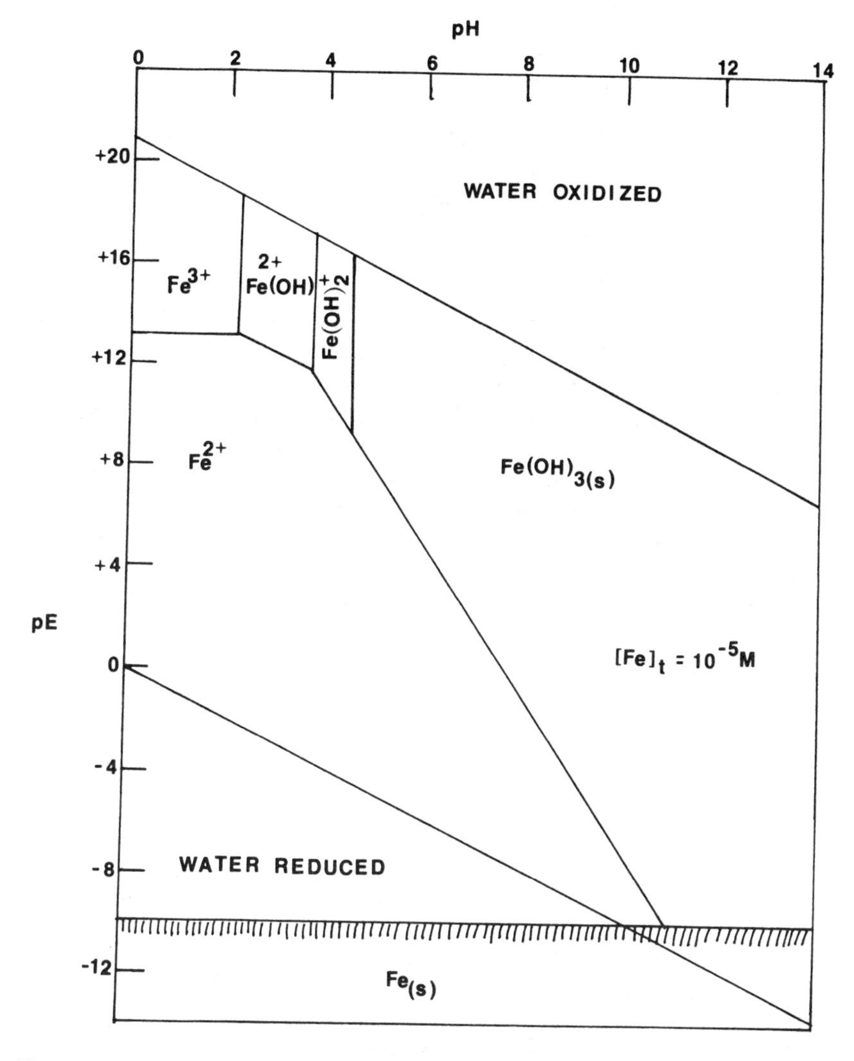

Figure 1. pE-pH diagram for the iron system. Reproduced from Faust and Aly [2], courtesy of Ann Arbor Science Publishers.

volumes exist at municipal plants or for domestic purposes. A third method utilizes the stabilization of iron in a suspended state by a dispersing agent, usually a polyphosphate or a silicate, to prevent deposition. This is not a removal method per se, and is limited to iron contents of 1.0 mg/l and less. All of this may be summarized as follows:

1. oxidation and precipitation methods: aeration ($O_{2(g)}$), sedimentation and filtration; oxidation by $KMnO_4$, Cl_2, O_3 or ClO_2, followed by sedimentation and filtration; and calcined magnesite-diatomaceous earth filtration;
2. ion-exchange methods: "zeolite" softening (also see Chapter 7), and "manganese-zeolite" process; and
3. stabilization with polyphosphates and silicates.

Kinetics of Oxygenation

Reactions 10 and 11 from Table I indicate that the oxidations of Fe(0) to Fe(II) and Fe(II) to Fe(III) are thermodynamically feasible with oxygen as an electron acceptor. In addition to the use of oxygenation of iron for its removal at water treatment plants, these reactions are also encountered in waters polluted by acid mine drainage or iron pickling wastes, and, of course, in the corrosion of iron water-distribution pipes (see Chapter 9). Consequently, the oxygenation of the lower valence states of iron has been studied extensively. For example, Weiss [5] suggested that the oxygenation of Fe(II) follows this sequence:

$$Fe(II) + O_{2(g)} = Fe(III) + HO_2^{\cdot} \tag{12}$$

$$Fe(II) + HO_2^{\cdot} = Fe(III) + H_2O_2 \tag{13}$$

$$Fe(II) + H_2O_2 = Fe(III) + HO^{\cdot} + H_2O \tag{14}$$

$$Fe(II) + HO^{\cdot} = Fe(III) + H_2O \tag{15}$$

These reactions are not balanced with respect to H^+, and do not specify any dependence of the reaction rate on this variable. It is believed that the rate-limiting step is Equation 12, which indicates that the rate of oxidation is first order with respect both to [Fe(II)] and $[O_{2(g)}]$:

$$-\frac{d[Fe(II)]}{dt} = k[Fe(II)][O_{2(g)}] \tag{16}$$

This reaction is considerably more complex than Equations 12–15 indicate. Among other things, the rate of oxidation is dependent on the nature of the anion present in water and increases as the complexing affinity of the anion to ferric iron increases [6].

Manometric techniques were employed to investigate the kinetics of oxygenation of Fe(II) [6]. The stoichiometric reaction is:

$$Fe^{2+} + 0.25\,O_{2(g)} + 2\,OH^- + 0.5\,H_2O = Fe(OH)_{3(s)} \tag{17}$$

In bicarbonate solutions (alkalinities, $2.9\text{–}3.9 \times 10^{-2}$ eq/l), the general rate law is:

$$-\frac{d[Fe(II)]}{dt} = k[OH^-]^2 P_{O_2}[Fe(II)] \tag{18}$$

where k = rate constant (M^{-2}-atm^{-1}-min^{-1})
$[OH^-]$ = hydroxyl ion concentration
$[Fe(II)]$ = concentration of total ferrous iron

Equation 18 can be reduced to first order:

$$-\frac{d[Fe(II)]}{dt} = k_1[Fe(II)] \tag{19}$$

where $k_1 = k[OH^-]^2 P_{O_2}$ (holding OH^- and O_2 constant) and has units of time^{-1}. Integration of Equation 19 yields

$$[Fe(II)] = [Fe(II)]_o e^{-k_1 t} \tag{20}$$

Figure 2a shows typical rate data from Stumm and Lee [6], which suggested a first-order reaction with respect to Fe(II) and independent of [Fe(III)]. However, Figure 2b shows the effect of pH value or, more precisely, $[OH^-]$. Here, the log of the rate constant k_1 divided by the P_{O_2} is plotted as a function of the solution's pH. Slopes of the straight lines indicate a 100-fold increase in oxygenation rate with an increase of one pH unit. This relationship is seen in another manner by calculating the $t_{1/2}$ value from:

$$t_{1/2} = \frac{0.693}{k_1} \tag{21}$$

where $t_{1/2}$ = the time required for ½ of the remaining reaction to occur
0.693 = a constant (2.303 log 2)
k_1 = a first-order rate constant that must be based on natural logarithms

Using data from Stumm and Lee [6]:

when $P_{O_2} = 0.107$ atm, $[OH^-] = 1.86 \times 10^{-8}$ (pH 6.27)

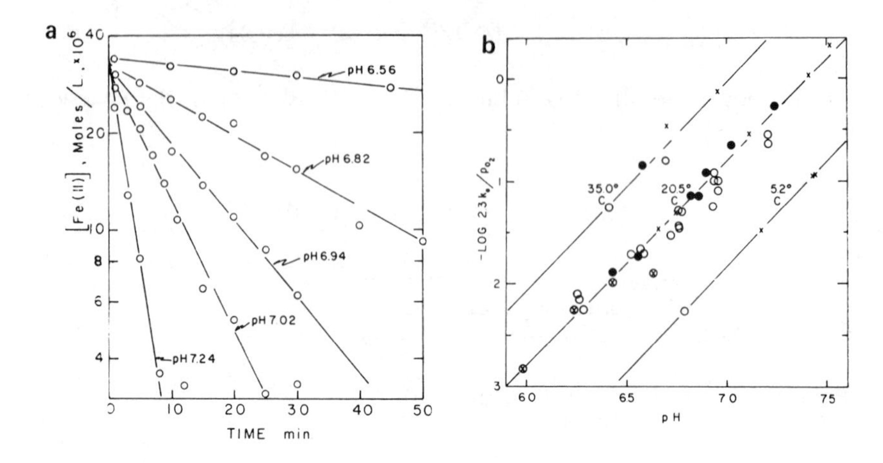

Figure 2. **(a)** Oxygenation rate of ferrous iron is proportional to Fe(II) and is strongly influenced by pH (20.5°C, P_{O_2} = constant). **(b)** An increase by one pH unit must increase oxygenation rate 100-fold. Reproduced from Stumm and Lee [6], courtesy of the American Chemical Society.

$$t_{1/2} = \frac{0.693}{3.33 \times 10^{-3}\ \text{min}^{-1}} = 208.1\ \text{min}$$

when $P_{O_2} = 0.195$ atm, $[OH^-] = 2.24 \times 10^{-7}$ (pH 7.35)

$$t_{1/2} = \frac{0.693}{0.942\ \text{min}^{-1}} = 0.73\ \text{min}$$

These calculations, of course, demonstrate the effect of $[OH^-]$ on the oxidation of Fe(II) in accord with the data of Stumm and Lee [6]. There are some water treatment plant design implications in these chemical kinetic data. Catalytic effects of Cu^{2+}, $MnO_{2(s)}$ and $H_2PO_4^-$ were observed also. The presence of these constituents increased the rate of oxygenation of Fe(II).

This research was extended to include the effects of ionic media, alkalinity and temperature on the kinetics of ferrous iron oxygenation and to identify oxidative product(s) [7]. In this study, the general rate law, Equation 18, was confirmed. Table II summarizes previous efforts on these kinetics, and Table III shows that an increase of ionic strength decreases the rate constant and increases the $t_{1/2}$ value. This effect may be due to complexation of Fe^{2+} by ClO_4^-. The effect of temperature on the rate of oxygenation was expected to follow the normal course, as can be seen in Figure 3. However, when the data were normalized with

Table II. Summary of Previous Results on Oxygenation Kinetics of Ferrous Iron[a]

Investigators	Solution Composition	Reported Rate Information	Remarks
Stumm and Lee	29–39 meq/l of alk as $NaHCO_3$, P_{O_2} varies, $T = 20.5\,°C$	$k = 8.0 \pm 2.5 \times 10^{13}\ M^{-2}\text{-atm}^{-1}\text{-min}^{-1}$	Assume ionic strength $\approx 34 \times 10^{-3} M$, $\gamma_{OH} \approx 0.84$, K_H oxygen at $20.5\,°C = 0.00138\ M\text{-atm}^{-1}$
Morgan and Birkner	$P_{O_2} = 0.6$ atm, $T = 25\ °C$, alk = 32 meq/l	$\tau_{1/2} \approx 16$ min at pH 6.70→ $k = 2.0 \times 10^{13} M^{-2}\text{-atm}^{-1}\text{-min}^{-1}$ $\tau_{1/2} \approx 47$ min at pH 6.52→ $k = 1.7 \times 10^{13} M^{-2}\text{-atm}^{-1}\text{-min}^{-1}$	
Schenk and Weber	$P_{O_2} = 0.21$ atm, $T = 25\ °C$, alk = 30–50 meq/l	$k = 2.1 \pm 0.5 \times 10^{13} M^{-2}\text{-atm}^{-1}\text{-min}^{-1}$	Assume ionic strength $\approx 40 \times 10^{-3} M$
Theis	$P_{O_2} = 0.5$ atm, $T = 25\,°C$, alk = 0.0158 M as NaOH initially	$\tau_{1/2} = 25.4$ min at pH 6.3→ $k = 1.36 \times 10^{14}\ M^{-2}\text{-atm}^{-1}\text{-min}^{-1}$	
Kester et al.	$T = ?$, $P_{O_2} = 0.21$ atm, Narragansett seawater Sargasso seawater	$\tau_{1/2} = 5.5$ min at pH 8.0→ $k = 6 \times 10^{11}\ M^{-2}\text{-atm}^{-1}\text{-min}^{-1}$ $\tau_{1/2} = 3.3$ min at pH 8.0→ $k = 1 \times 10^{12}\ M^{-2}\text{-atm}^{-1}\text{-min}^{-1}$	Kinetics were followed by ferric iron absorbance in the UV region
Tamura et al.	$T = 25\ °C$, P_{O_2} varies, alk $= 10^{-2} M$ $NaHCO_3$, total ionic strength = 0.11 M	0.1 M ClO_4^-: $k = 2.38 \times 10^{14}\ M^{-3}\text{-sec}^{-1} = 1.8 \times 10^{13}\ M^{-2}\text{-atm}^{-1}\text{-min}^{-1}$ 0.1 M NO_3^-: $k = 2.04 \times 10^{14}\ M^{-3}\text{-sec}^{-1} = 1.6 \times 10^{13}\ M^{-2}\text{-atm}^{-1}\text{-min}^{-1}$ 0.1 M Cl^-: $k = 1.63 \times 10^{14}\ M^{-3}\text{-sec}^{-1} = 2 \times 10^{13}\ M^{-2}\text{-atm}^{-1}\text{-min}^{-1}$	

Table II. continued

Investigators	Solution Composition	Reported Rate Information	Remarks
		0.1 M Br^-, 0.1 M I^-, 0.033 M SO_4^{2-}: $k = 1.36 \times 10^{14}$ M^{-3}-sec^{-1} $= 1.0 \times 10^{13}$ M^{-2}-atm^{-1}-min^{-1}	
Murray and Gill	$T = ?$, Puget Sound seawater	$\tau_{1/2} = 3.9$ min at pH 8.0→ $k = 8.9 \times 10^{11}$ M^{-2}-atm^{-1}-min^{-1}	First 5 min linear on first-order plot

[a]Reproduced from Sung and Morgan [7], courtesy of the American Chemical Society.

Table III. Variation of k with Ionic Strength (Adjusted by $NaClO_4$)[a,b]

Ionic Strength (M)	$\tau_{1/2}$ (min), Least Squares	k (M^{-2}-atm^{-1}-min^{-1})
0.009	18.0	$4.0 \pm 0.6 \times 10^{13}$
0.012	18.2	$3.1 \pm 0.7 \times 10^{13}$
0.020	18.5	$2.9 \pm 0.6 \times 10^{13}$
0.040	21.5	$2.2 \pm 0.5 \times 10^{13}$
0.060	26.8	$1.8 \pm 0.3 \times 10^{13}$
0.110	37.9	$1.2 \pm 0.2 \times 10^{13}$

[a]Reproduced from Sung and Morgan [7], courtesy of the American Chemical Society.
[b]$T = 25$ °C; alkalinity $= 9 \times 10^{-3}$ M $NaHCO_3$; $[Fe(II)]_0 = 34.7$ μM; $P_{O_2} = 0.20$ atm; pH $\approx$ 6.84.

respect to changes in Kw and O_2 solubility, the rate constant varied slightly with increasing temperature (data not given here). Sung and Morgan [7] were uncertain about the role of bicarbonate alkalinity in the oxygenation kinetics. Their data showed that effects of alkalinity variation from 9 to 50 mM were explained reasonably well by ionic strength.

When the dependency of the oxygenation rate on $[H^+]$ was reinvestigated, the researchers noted that "one of the major uncertainties in the numerical value for the rate constant arises from the variation of pH within an experiment" [7]. Figure 4 shows typical variation by which pH drops initially 0.1 unit following the addition of the stock Fe(II) solution which introduces 1 mM protons. This pH decrease effect was not mentioned in the work of Stumm and Lee [6].

A semikinetic study of the aeration of iron-bearing groundwaters in Illinois was reported [8]. Table IV shows $t_{1/2}$ values, in minutes, obtained for eight natural waters. It is rare for researchers to use natural waters. Most prefer to utilize synthetic waters. In any event, a comparison of the $t_{1/2}$ values of 4.3–54 min with a theoretical detention time of 60 min for aerators shows the inadequate design of most iron removal units.

The study of Ghosh et al. [8] provoked an interesting discussion by Stumm and Singer [9], Winklehaus et al. [10] and Morgan and Birkner [11]. The major objections of Stumm and Singer [9] were: (1) the uncertainty of the analytical method for determining Fe^{2+} in the presence of solid iron-bearing phases and organic matter, and (2) the incorrectness of interpreting the effect of bicarbonate alkalinity on the kinetics of Fe(II) removal as the result of the precipitation of $FeCO_{3(s)}$. Winklehaus et al. [10] also were concerned about the assertion that the formation of

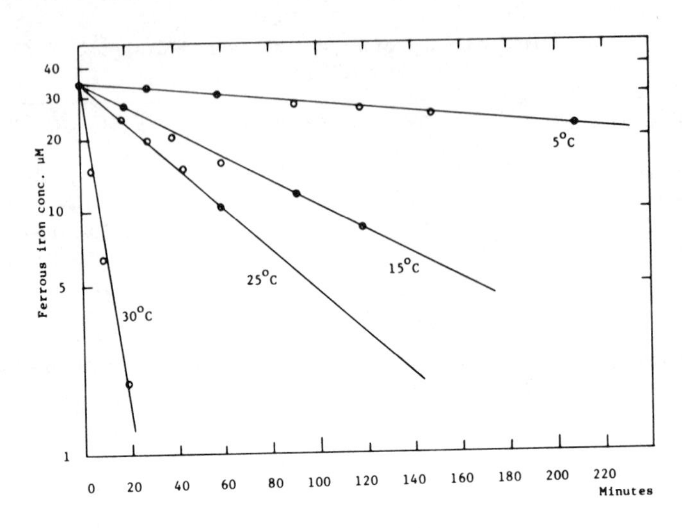

Figure 3. Effect of temperature on the oxygenation of kinetics of ferrous iron. All experiments were conducted in 0.11 *M* ionic strength adjusted with $NaClO_4$; alkalinity = 9 m*M* as $NaHCO_3$; $P_{O_2} = 0.2$ atm; pH ≃ 6.82; $[Fe(II)]_o =$ 34.7 μM. Reproduced from Sung and Morgan [7], courtesy of the American Chemical Society.

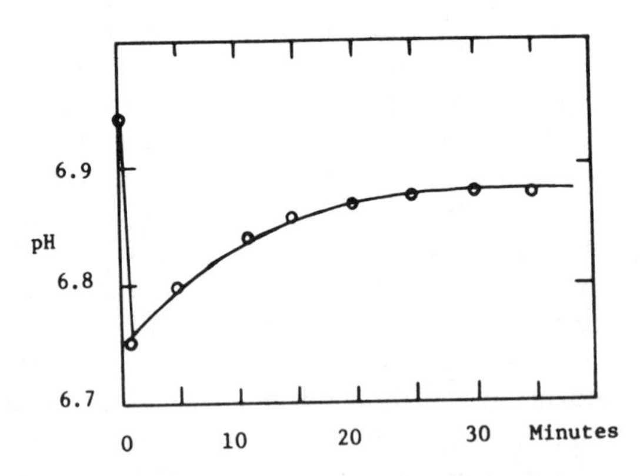

Figure 4. Variation in pH during an oxygenation experiment. Initial alkalinity = 10^{-2} *M*; $P_{O_2} = 0.2$ atm; $P_{CO_2} = 0.071$ atm; T = 25°C. The initial drop in pH coincides with the addition of stock ferrous solution, which introduces 1 m*M* protons. Reproduced from Sung and Morgan [7], courtesy of the American Chemical Society.

Table IV. Characteristics of Illinois Groundwaters[a]

	Raw Water				Aerated Water				
Plant	pH	Fe(II), (10^4 mol/l)	CO_3^{2-} (10^6 mol/l)	Molar conc.[b] product $[Fe(II)][CO_3^{2-}]$	Half-life,[c] $t_{1/2}$ (min)	Equil. pH[d]	Alk. (mg/l as $CaCO_3$)	Dissolved Oxygen (mg/l)[e]	Temp. (°C)
Clinton	7.58	0.33	15.3	5.06×10^{-10}	4.3	7.78	610	7.35	13.0
Danvers	7.47	0.36	16.3	5.88×10^{-10}	6.5	7.68	512	6.40	13.5
Windsor	7.10	0.90	6.91	6.22×10^{-10}	13.2	7.48	520	7.20	14.5
Forrest	7.40	0.46	11.2	5.15×10^{-10}	16.0	7.72	475	7.72	10.5
Deland	7.10	0.72	5.3	3.82×10^{-10}	22.5	7.67	456	7.67	13.0
Cisco	7.45	0.63	12.5	7.90×10^{-10}	25.6	7.71	458	6.70	13.5
Wapella	7.32	0.85	9.0	7.65×10^{-10}	36.0	7.67	410	6.40	12.0
Arcola	7.30	0.75	6.7	5.01×10^{-10}	54.0	7.49	355	6.90	12.5

[a]Reproduced from Ghosh et al. [8], courtesy of the American Society of Civil Engineers.
[b]Theoretical K_{sp} for $FeCO_3 = 2.11 \times 10^{-11}$ at 25°C.
[c]The half-life, $t_{1/2}$, is the time required to reduce the Fe (II) to one-half its original value.
[d]Equilibrium pH refers to the pH obtained following aeration of the raw water.
[e]Dissolved oxygen following aeration of the raw water.

$FeCO_{3(s)}$ affected the oxygenation kinetics, since the reported values for pH and alkalinity (Table IV) do not show any correlation of $t_{1/2}$ to $[CO_3^{2-}]$. Difficulties arise from attempts to compare directly Ghosh's [8] rate data to those of Stumm and Lee [6]. First, the two experimental systems were different, and no rate constants were presented by Ghosh et al. [8]. Also, there was a mathematical error in the rate constants presented by Stumm and Lee [6]. Another objection to the $FeCO_{3(s)}$ explanation [11] concerned the inadequacy of the thermodynamic data as a basis for computing the supersaturation of the eight groundwaters. In addition, Morgan and Birkner [11] presented and compared their own kinetic data with those of Stumm and Lee [6] and Ghosh et al. [8]. These data are given in Table V, where there are 4- to 18-fold differences in the observed and computed $t_{1/2}$ values. No rational explanation was given for these differences [11].

Buffer value β (eq/pH unit) of natural waters has an effect on the oxygenation kinetics of Fe(II) [12,13]. Buffer value is [2]:

$$\beta = 2.3([H^+] + [OH^-] + C_t[\alpha_1(\alpha_o + \alpha_2) + 4\alpha_2\alpha_o]) \tag{22}$$

where
$$\alpha_o = [H_2CO_3]/C_t$$
$$\alpha_1 = [HCO_3^-]/C_t$$
$$\alpha_2 = [CO_3^{2-}]/C_t$$
$$C_t = [H_2CO_3] + [HCO_3^-] + [CO_3^{2-}]$$

Kinetic data were obtained to indicate that, for values of β between 10^{-4} and 4×10^{-3} eq/pH, there was no effect on the pseudo-first-order rate constant. However, at higher values of β, there was a definite correlation between these two values. A regression analysis gave a correlation coefficient of 1.00. Therefore, a β term should be included in the Fe(II) oxygenation rate equation:

$$-\frac{d[Fe(II)]}{dt} = k[OH^-]^2 P_{O_2}[Fe(II)][\beta]^n \tag{23}$$

from which a first-order rate constant is obtained:

$$-\frac{d[Fe(II)]}{dt} = k_1[Fe(II)] \tag{24}$$

where $k_1 = k[OH^-]^2 P_{O_2}[\beta]^n$

The empirical value of n was estimated to be 0.51. Table VI shows the variation of k and k_1 with alkalinity and β. There appears to be an

Table V. Data Related to Rate of Fe(II) Removal from Aerated Illinois Groundwater[a]

Plant Number	Temperature (°C)	$[O_2]$ (10^4 mol/l)	$[OH^-]$ (10^4 mol/l)	k_m[b] (10^{-16} sec)	k_o[c] (10 min^{-1})	$t_{1/2}$ (min)		Apparent Supersaturation
						Computed	Observed[d]	
1	13.0	2.30	2.14	5.4	5.7	1.2	4.3	9.2
2	13.5	2.00	1.90	5.5	4.0	1.7	6.5	6.4
3	14.5	2.25	1.26	5.6	2.5	2.7	13.2	11.5
4	10.5	2.41	1.66	5.0	3.3	2.1	16.0	7.8
5	13.0	2.40	1.66	5.4	3.6	1.9	22.5	10.4
6	13.5	2.09	2.04	5.5	4.8	1.4	25.6	10.5
7	12.0	2.00	1.55	5.2	2.5	2.8	36.0	11.9
8	12.5	2.16	1.07	5.4	1.3	5.2	54.0	5.9

[a]Reproduced from Morgan and Birkner [11], courtesy of the American Society of Civil Engineers.
[b]Estimated from rate data of Stumm and Lee [6].
[c]Computed first-order rate constant for oxidation: $k_o = k_m[O_2]\,(OH^-)^2$.
[d]From Table 3 of Ghosh et al. [8].

Table VI. Variation of k and k_1 with Alkalinity and Buffer Value[a]

Alkalinity (eq/l)	Buffer Value (eq/pH)	k 10^{-13} (l^3/mol^2-atm-min)	k_1[b] (min^{-1})
3.0×10^{-4}	1.32×10^{-4}	2.80	0.02231
1.12×10^{-3}	0.48×10^{-3}	3.07	0.02438
4.4×10^{-3}	1.84×10^{-3}	2.73	0.02166
6.0×10^{-3}	2.48×10^{-3}	3.45	0.02659
1.2×10^{-2}	0.48×10^{-2}	3.21	0.02549
1.8×10^{-2}	0.72×10^{-2}	3.96	0.03150
2.8×10^{-2}	1.07×10^{-2}	4.73	0.03759
4.07×10^{-2}	1.52×10^{-2}	5.29	0.04200

[a]Reproduced from Jobin and Ghosh [12], courtesy of the American Water Works Association.
[b]$k_1 = k[P_{O_2}][OH^-]^2$.

increase in the oxygenation rate of Fe(II) when β values are greater than 4×10^{-3} eq/pH. Note that the k and k_1 values in Table VI do not include the $[\beta]^n$ term.

The oxidation of iron pyrite (FeS_2) with subsequent release of acidity into waters draining through coal mines has been investigated [14,15]. Figure 5 shows that the rate of oxidation of Fe^{2+} by oxygen in abiotic systems is a function of $[H^+]$ when the pH value is greater than 4.5. Equation 18 is valid for these conditions. At pH values less than 3.5, the reaction proceeds at a rate independent of $[H^+]$:

$$-\frac{d[Fe^{2+}]}{dt} = k'[Fe^{2+}][O_{2(g)}] \quad (25)$$

where $k' = 1.0\times10^{-7}$ atm^{-1}-min^{-1} at 25°C

Another study was conducted on the oxygenation kinetics of Fe^{2+} and the effect of dissolved silica upon this oxidation [16]. The overall rate constant, $k = 2.1 \pm 0.5\times10^{13} M^{-2}$-atm^{-1}-min^{-1} agreed, of course, with previous work [6]. The addition of sodium metasilicate, in concentrations up to 120 mg/l, to the Fe^{2+} systems increased the rate of oxygenation within a pH range of 6.6–7.1. The dissolved silica apparently produced a catalytic effect that was expressed in terms of the rate of oxidation:

$$-\frac{d[Fe^{2+}]}{dt} = \left(kP_{O_2}[OH^-]^2 + k_{si}[H_4SiO_4]^{0.5}[OH^-]^{0.5}\right)[Fe^{2+}] \quad (26)$$

in which k = the value given above, and $k_{si} = 1.8 \pm 0.3\times10^3 M^{-1}$.

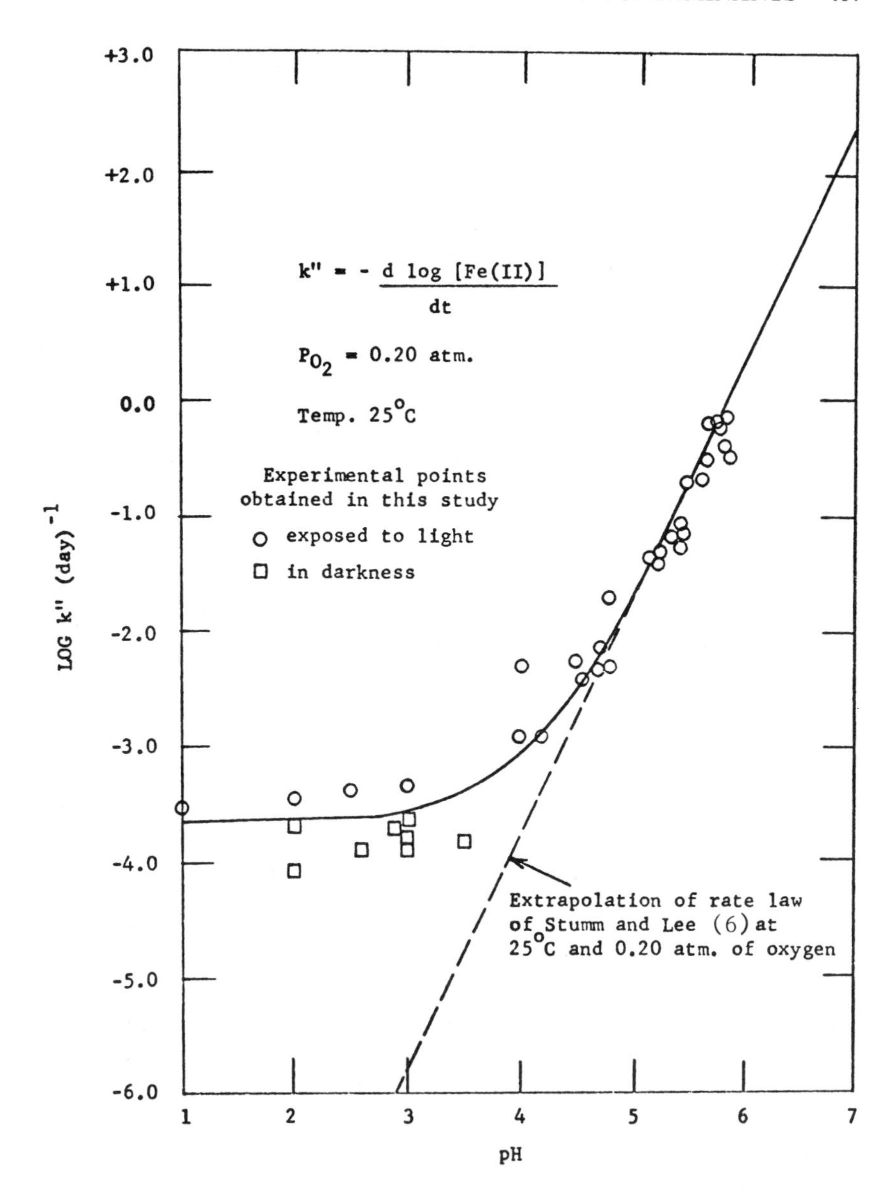

Figure 5. Oxygenation rate of ferrous iron as a function of pH [15].

Alternative Oxidants

Ozone, chlorine dioxide, chlorine and potassium permanganate are feasible alternative oxidants for iron removal. The half-cell reduction reactions for these four oxidants are given in Table VII. Each of these

Table VII. Reactions of Fe(II) with Alternative Oxidants at 25°C

Reaction Number	Reaction	V^0
4	$Fe^{2+} + 3\ H_2O = Fe(OH)_{3(s)} + 3\ H^+ + e$	+1.052[a]
27	$O_{3(g)} + 2\ H^+ + 2\ e = O_{2(g)} + H_2O$	+2.07
4 and 27	$2\ Fe^{2+} + 5\ H_2O + O_{3(g)} = 2\ Fe(OH)_{3(s)} + 4\ H^+ + O_2$	+1.06
28	$ClO_{2(g)} + 4\ H^+ + 5\ e = Cl^- + 2\ H_2O$	+1.511
4 and 28	$ClO_{2(g)} + 5\ Fe^{2+} + 13\ H_2O = 5\ Fe(OH)_{3(s)} + 11\ H^+ + Cl^-$	+0.501
29	$Cl_{2(g)} + 2\ e = 2\ Cl^-$	+1.3595
4 and 29	$2\ Fe^{2+} + 6\ H_2O + Cl_{2(g)} = 2\ Fe(OH)_{3(s)} + 2\ Cl^- + 6\ H^+$	+0.385
30	$MnO_4^- + 4\ H^+ + 3\ e = MnO_{2(s)} + 2\ H_2O$	+1.695
4 and 30	$3\ Fe^{2+} + 7\ H_2O + MnO_4^- = 3\ Fe(OH)_{3(s)} + MnO_{2(s)} + 5\ H^+$	+0.685

[a] As a reduction reaction.

four reactions are combined with Equation 4 to indicate the thermodynamic feasibility of Fe(II) oxidation to $Fe(OH)_{3(s)}$. Kinetic data for these reactions are rather sparse. Nordell [17] reported that "with chlorine it is possible to oxidize iron rapidly at a lower pH than with dissolved oxygen." A rather qualitative laboratory study indicated that with chlorine, 10 ppm of Fe^{2+} "was completely oxidized in less than 15 minutes at a pH of 5.0, whereas with air a pH of 7.0 was required." In a pilot plant test, Matthews [18] reported that a 9.5 ppm dosage of chlorine "completely removed" 4.5 ppm iron and 1.8 ppm manganese as measured after the reaction time, sedimentation and filtration. Willey and Jennings [19] concluded that "dissolved iron and manganese can be effectively removed from water by the continuous feeding of $KMnO_4$ to a water before it is passed through a manganese greensand filter." A permanganate "demand curve" (not shown here) was drawn in which the implied stoichiometry of the reaction was 6 ppm $KMnO_4$ for 10 ppm of ferrous iron. Cromley and O'Connor [20] investigated the use of $O_{3(g)}$ as an alternative to aeration for iron removal where organic compounds (see below) interfered. Some relevant kinetic information is seen in Figure 6 where there is a comparison of ferrous iron removal by continuous aeration and ozone.

Filtration

Rapid sand filtration usually follows the oxidation of Fe(II) for removal of $Fe(OH)_{3(s)}$. This filtration, of course, removes such particu-

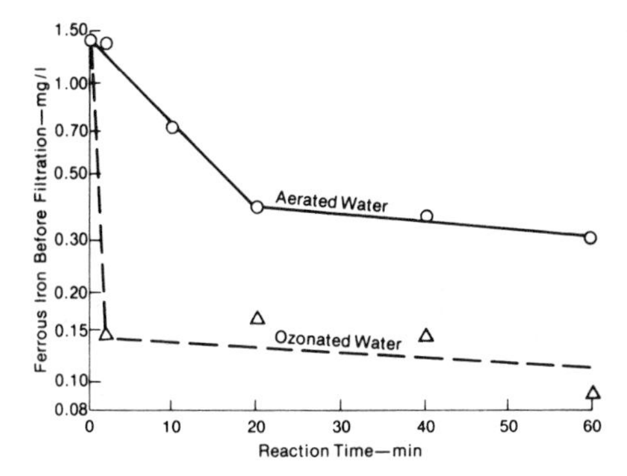

Figure 6. Comparison of the ferrous iron concentrations in an aerated and ozonated groundwater. Reproduced from Cromley and O'Connor [20], courtesy of the American Water Works Association.

late matter as flocculents from chemical coagulation, softening sludges, etc. (see Chapter 6). There are occasions, however, when filtration is unique for iron removal, especially groundwaters. A diatomite filtration process has been developed for iron and manganese removal from the groundwaters of Massachusetts [21]. The flow diagram of a pilot plant is seen in Figure 7 where aeration and $KMnO_4$ are used to oxidize the Fe(II) before diatomite filtration. In view of the kinetic data cited above, the detention times may be inadequate. Iron and manganese contents of the raw water typically were 6–11 ppm and 0.25–0.45 ppm, respectively. The pressure filter was precoated with 0.15 lb/ft^2 of diatomaceous earth (body feed in Figure 7) and was operated at a flow of 1 gpm/ft^2. Typical results are given in Table VIII where the 0.02 ppm of iron in the filtrate was obtained at least 50% of the time. One of the operational difficulties arose from the use of lime for alkalinity control. Filter runs, as measured by head loss, were short with lime, but were lengthened considerably when soda ash was employed.

Bell [22] also reported upon the diatomite filtration removal of iron and manganese from groundwaters. Pilot-plant experiments with several "filter aids" (the body feeds) supplemented with MgO produced substantial reductions in iron contents (<0.01 ppm) and in manganese contents (<0.05 ppm). Filter runs up to 400 hr were obtained. No preaeration is needed for this process.

A pilot plant composed of (1) a constant-flow head regulator, (2) aerator, (3) reaction-sedimentation basin, and (4) a bank of four rapid sand filters with different depths of filtering medium (downflow rate = 2 gpm/ft^2) was employed for iron removal [23]. This study demonstrated

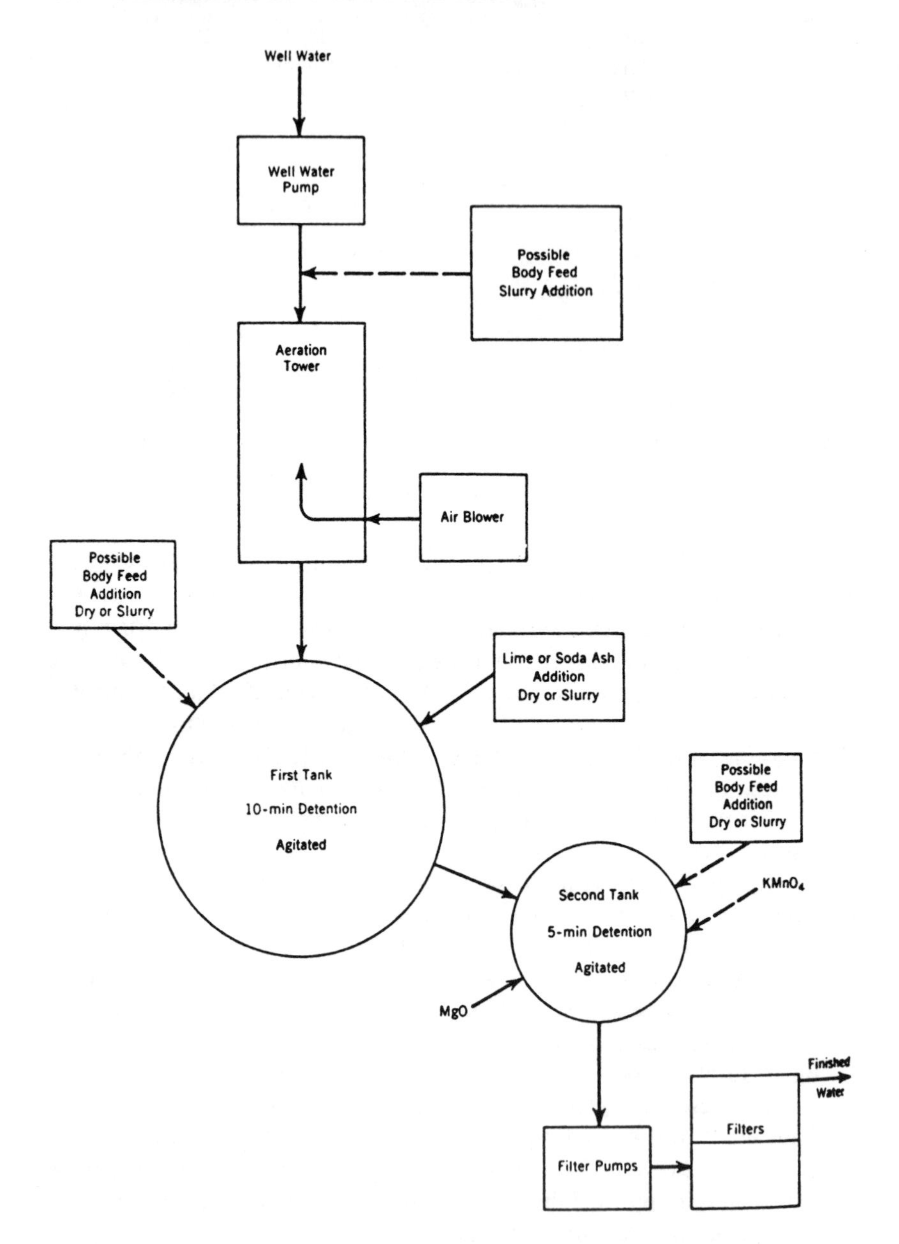

Figure 7. Flow diagram of diatomite pilot plant. Reproduced from Coogan [21], courtesy of the American Water Works Association.

Table VIII. Iron Removal at Amesbury[a]

Body Feed					Results	
Dosage[b] (ppm)	Grade	Feed	pH	Rate of Increase in Head Loss (lb/hr)	Iron in Raw Water (ppm)	Iron in Filtrate (ppm)
60	1[c]	As slurry in pump line	7.7–8.1	1.3	6–10	0.02
60	1[c]	As slurry in pump line	8.0–8.4	0.65	8.	0.02
60	1[c]	Dry in first tank	8.0–8.4	1.7	6.5	0.02
60	2[d]	Dry in second tank	8.0–8.3	1.4	6.	0.02

[a]Reproduced from Coogan [21], courtesy of the American Water Works Association.
[b]On pressure filters.
[c]Celite 503.
[d]Celite 545.

a difficulty frequently associated with the operation of sand filters for iron removal, that is, the growth and accumulation of bacterial slimes in the filters. Typical operational results are seen in Figure 8 for unripened (i.e., no bacterial slime) sand filters [23]. Only the 30-in. filter reduced total iron to acceptable contents. In Figure 9, iron removal after 16–18 hr of filtration was extremely erratic [23]. For the 18- and 30-in. filters, the iron content in the effluent was higher than in the influent. These "ripened" filters contain heavy bacterial slimes that create anaerobic or reducing conditions, leading to the reduction of previously removed Fe(III) to Fe(II). Only an excessively high dosage of chlorine, 50 mg/l, removed the bacterial slime and maintained oxidizing conditions in the sand filters.

Bituminous coal, 1.0-mm geometric size, has been proposed for removing iron from groundwater without preaeration [24]. Figure 10 shows a comparison of coal and sand filters for iron removal over a 48-hr period. Better removal by coal is apparent. Furthermore, the coal filter can be operated at four times (8 gpm/ft^2) the rate of a sand filter (2 gpm/ft^2) for comparable head loss and acceptable effluent quality up to about 12 hr. Also, the coal filter was more effective than the sand filter in removing soluble Fe(II), apparently by adsorption. No data were given on backwashing or cleaning of the coal filters.

Ion Exchange

Ferrous iron is removed by ion-exchange processes, provided the water is reasonably free from particulate matter. The cation exchangers are usually the high-capacity resin types in the sodium cycle:

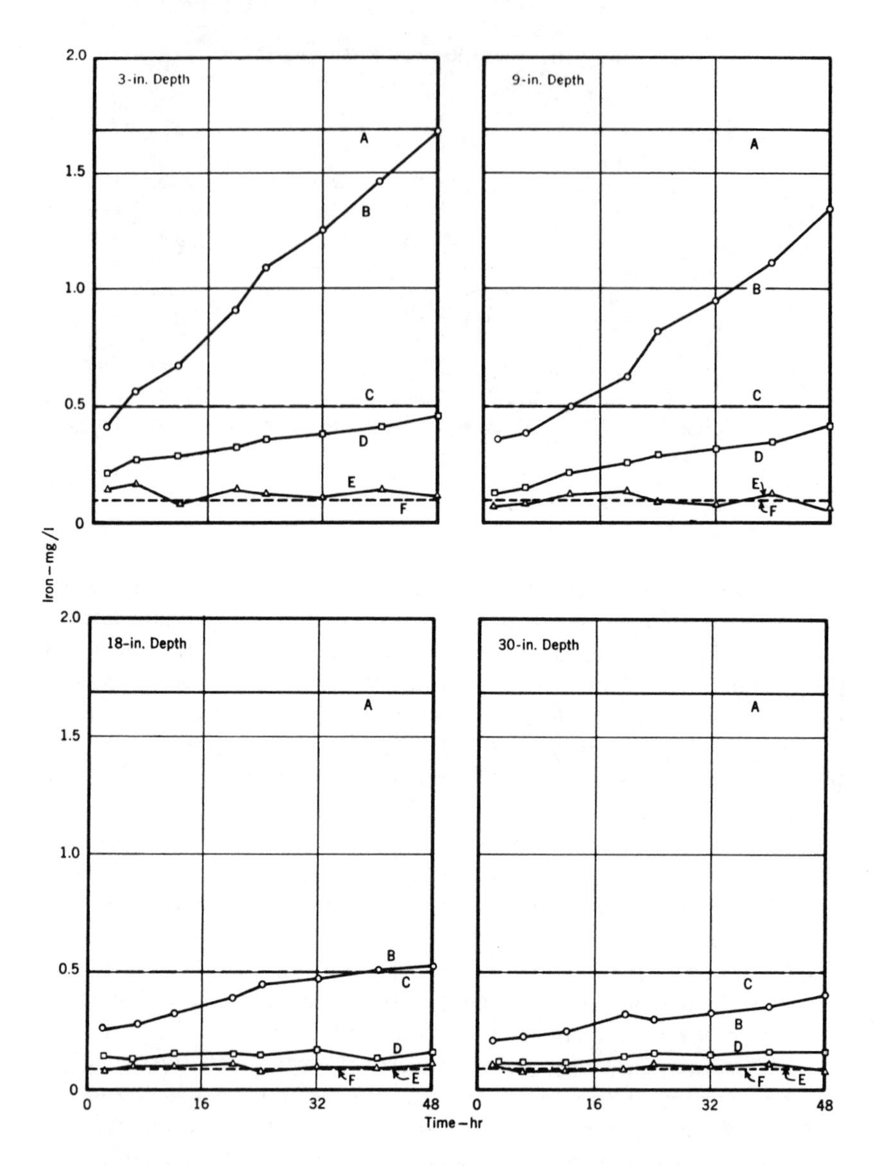

Figure 8. Iron concentration in filtered water from "unripe" filters. A = average influent total iron; B = total iron; C = average influent ferrous iron; D = total ferrous iron; E = filterable ferrous iron; F = average influent filterable ferrous iron. Only the 30-in. filter reduced the total iron to an acceptable level. Reproduced from Ghosh, et al. [23], courtesy of the American Water Works Association.

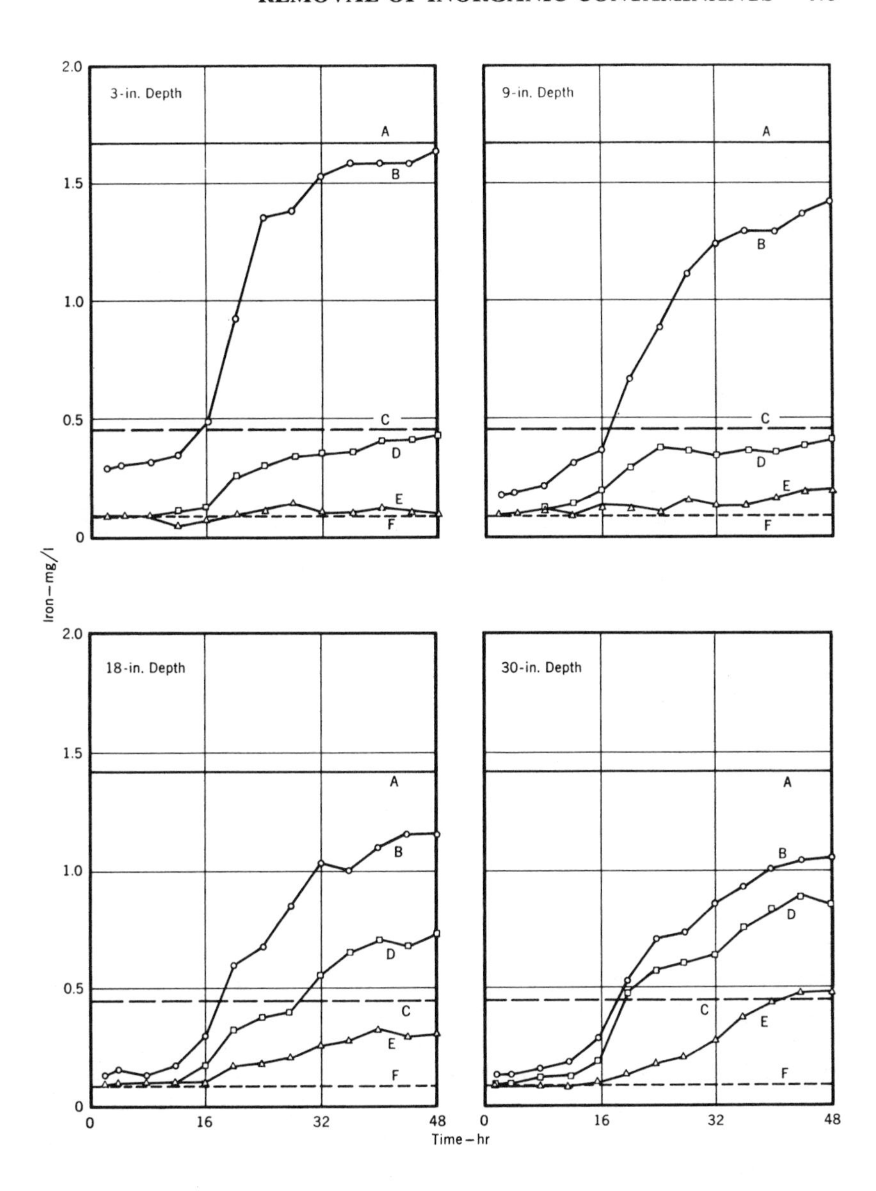

Figure 9. Iron concentration in filtered water from "ripe" filters. Lines A to F as in Figure 8. After 16–18 hr of filtration, the filterable ferrous iron in the effluent from the 18- and 30-in. filters was higher than that present in the influent. Reproduced from Ghosh et al. [23], courtesy of the American Water Works Association.

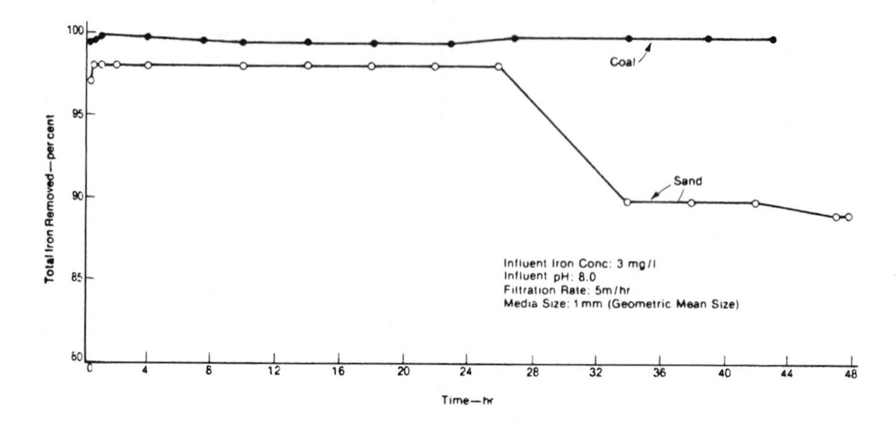

Figure 10. Performance of coal and sand filters of equal media size during 48-hr filtration run. Reproduced from George and Chanduri [24], courtesy of the American Water Works Association.

$$Fe^{2+} + 2\,NaR = FeR_2 + 2\,Na^+ \tag{31}$$

Regeneration is accomplished with NaCl. It is necessary to consider the hardness content when sizing and designing the capacity of the ion exchanger for Fe^{2+} removal. A "rule of thumb" is 0.5 ppm Fe/ppm of hardness up to a maximum of 50 ppm [17]. The raw water should not be aerated before the ion-exchange unit, because $Fe(OH)_{3(s)}$ would precipitate and "foul" the bed. All of this limits the ion exchange to groundwaters.

Ferrous iron may also be removed by the so-called manganese zeolite process. This is not an ion-exchange process per se, but the filtering medium is a mixture of anthracite and zeolite or greensand material. According to Nordell [17], the manganese zeolite is prepared by alternate treatments of a greensand zeolite with manganous sulfate and potassium permanganate. This results in the precipitation of $MnO_{2(s)}$ in the zeolite bed where it presumably catalyzes the oxidation of Fe(II) by continuous addition of $KMnO_4$ with subsequent filtration of the $Fe(OH)_{3(s)}$. The zeolite is regenerated first by backwashing and then by treating with $KMnO_4$. This process is limited to groundwaters with iron contents of 5 mg/l or less. This process may be utilized to remove manganese and hydrogen sulfide from water. Figure 11 shows a typical municipal unit operated at a flowrate of 3 gpm/ft^2. Water treatment plants range from 30 to 2800 gpm.

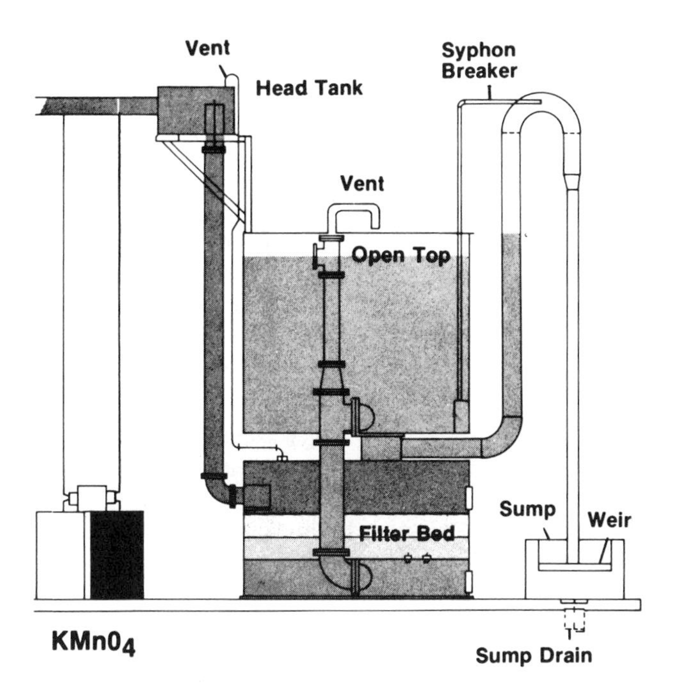

Figure 11. Schematic drawing of Zeo-Rex iron removal system showing chemical feed system. Reproduced courtesy of the Permutit Co., Inc.,

Effect of Organic Matter

That iron is complexed by naturally occurring organic acids has been reported in the limnological [25] and water treatment [17] literature for decades. This is the so-called filterable iron associated with waters "high" in organic color. This type of iron is difficult to remove by the conventional treatment methods cited above. Considerable research [26,27] has been conducted to explain the chemical nature of this complexed iron. For example, Shapiro [26] reported the Fe^{3+} holding capacity of "yellow" organic acids (Figure 12). The $[H^+]$ apparently is a factor, since the acids showed a maximum at a pH of 10. This is not necessarily the maximum capacity of yellow organic acids to hold iron, rather it was the maximum response to $[H^+]$ in the experimental system. From 5 to 50 ppm of tannic (digallic) acid reduces dissolved Fe(III) to Fe(II) when pH values are less than 4 [27]. Above this pH value, a "black material" containing Fe(III) and tannic acid was precipitated. In a 500-ppm solution of tannic acid, a fairly stable complex with Fe(II) is

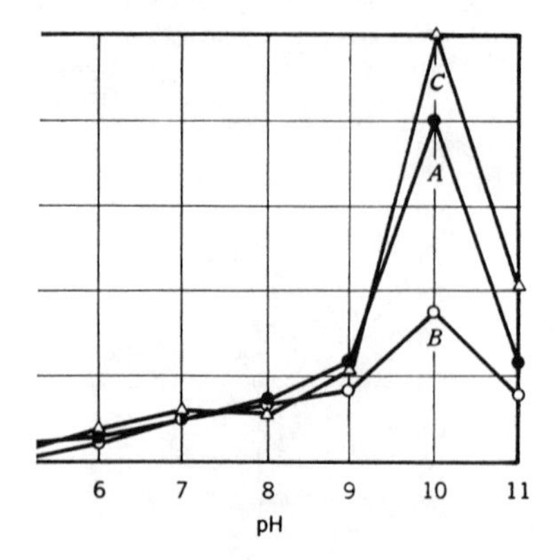

Figure 12. Iron-holding capacity of yellow acids as a function of pH. The pH during filtration and the pH at which the iron was added were the same. The three curves are for the following experimental conditions: A = iron added as $FeCl_3$ to 1 mg acids in 50 ml water; B = iron added as $FeCl_3$ to 2 mg acids in 50 ml water; C = iron added as $Fe(NO_3)_3$ to 1 mg acids in 50 ml water. Reproduced from Shapiro [26], courtesy of the American Water Works Association.

formed—stable, that is, toward air oxidation. Tannic acid is present in many types of plant material.

The reductive and complexation properties of humic acids (HA) for Fe^{3+} and Fe^{2+} have been investigated [28]. These so-called humic acids [2] were prepared from well decomposed leaves. A typical result was that 50 mg/l HA, pH 5.0, did not complex any Fe^{2+}, but with an increase to pH 8.0, 2.1 mg/l iron was complexed. This increase in complexing ability was attributed to greater dissociation of carboxyl groups of the HA. Also, it was proposed that humic acids could chemically reduce Fe^{3+} to Fe^{2+} with subsequent complexation [28].

It is difficult to oxygenate Fe(II) in the presence of humic and tannic acids [12]. In Figure 13, tannic acid appears to retard the oxidation of Fe(II) more than the humic acids. The mechanism is probably:

$$\text{Fe(II)} + 0.25\,\text{O}_2 + \text{organics} = \text{Fe(III)} - \text{organic complex} \quad (32)$$

$$\text{Fe(III)} - \text{organic complex} = \text{Fe(II)} + \text{oxidized organic} \quad (33)$$

$$\text{Fe(II)} + 0.25\,\text{O}_2 + \text{organics} = \text{Fe(III)} - \text{organic complex} \quad (34)$$

Such organic compounds as humic acid and tannic acid with —OH and

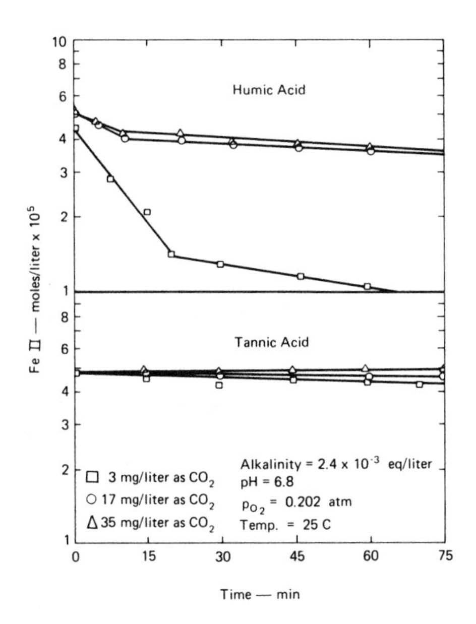

Figure 13. Oxygenation of Fe(II) in the presence of organic acids. Reproduced from Jobin and Ghosh [12], courtesy of the American Water Works Association.

—COOH groups apparently can reduce Fe(III) reasonably fast in synthetic solutions [9,15]. Reactions 32 and 33 suggest a catalytic effect of the Fe(II)-Fe(III) couple in the aerobic oxidation of organic matter with subsequent oxidation of Fe(II) in Reaction 34. The relative rates of Reactions 32 and 33 dictate the eventual oxidation of Fe(II) in the presence of organics. This explanation is offered for the apparent two rates of Fe(II) oxygenation in the presence of humic acid as seen in Figure 13. Or, it may be that humic acid, at a pH of 6.8, is simply a "weaker" complexer of ferrous iron than tannic acid.

Products of natural vegetative decay can retard the rate of oxygenation of Fe(II) [29,30]. Typical results are seen in Figure 14. Manometric studies showed that the rate of oxygen consumption by the Fe(II)-tannic acid system at a pH value of 7.0 was appreciably slower than the rate in the absence of organic material. By the same token, the presence of Fe(II) retarded the oxidation of tannic acid at pH values of 7.0, 9.5 and 10.8.

A schematic diagram was offered by Theis and Singer [29] to explain the behavior of iron in the presence of organic matter and oxygen:

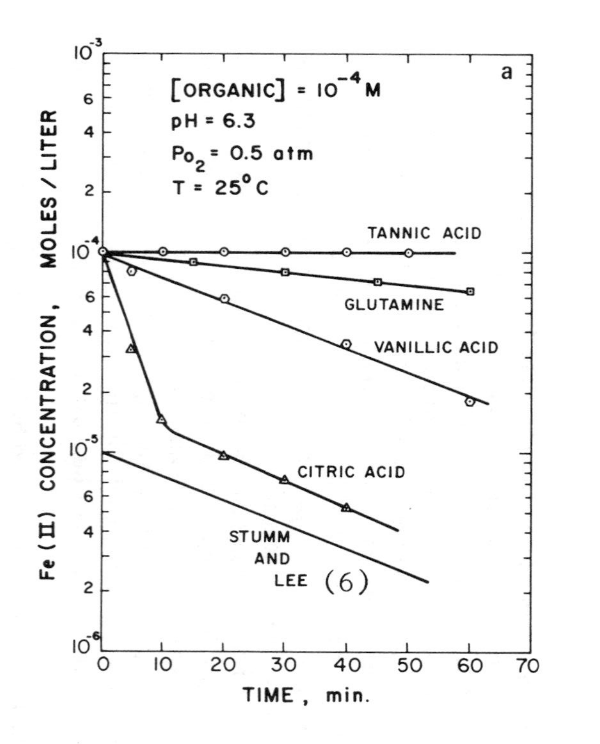

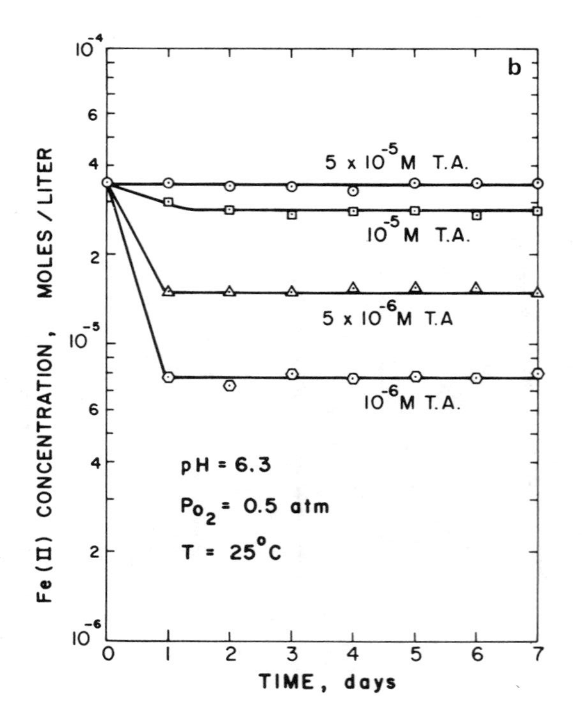

Figure 14. **(a)** Effects of representative organic compounds on rate of oxidation of ferrous iron at pH 6.3, $P_{O_2} = 0.5$ atm and T = 25°C as compared with the rate in simple aqueous media. **(b)** Inhibition of ferrous iron oxidation in the presence of various concentrations of tannic acid at pH 6.3, $P_{O_2} = 0.5$ atm and T = 25°C. Reproduced from Theis and Singer [29,30], courtesy of the American Chemical Society and Ann Arbor Science Publishers.

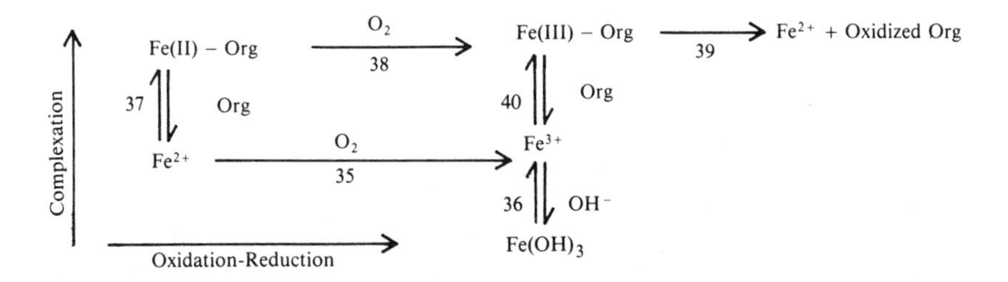

Reactions 35 and 36 represent the rapid oxidation of ferrous iron to ferric iron in the absence of dissolved organic matter. When significant quantities of organic matter are present, Reactions 35 and 37 compete for the Fe(II). Whatever fraction of Fe(II) is complexed or oxidized depends on $[H^+]$, and the quantity and type of organic matter present. Reaction 38 proceeds slowly, on the order of several days. The Fe(III)-organic complex is unstable, whereupon the iron is reduced to Fe(II) (Reaction 39), and the cycle begins again until all of the organic matter is oxidized. Reaction 40 is competitive to Reaction 36. This model (Reactions 36–40) differs slightly from Reactions 32–34 by the addition of Reactions 37 and 38. That is, the complexation of Fe(II) and its slow oxidation to the corresponding Fe(III) complex apparently provides an explanation for the slow rate of oxygen consumption by the iron-organic system.

What does all of this mean for the removal of iron in highly colored waters? Obviously, oxygenation is not effective since the kinetics are too slow for normal treatment plant design. The obvious solution to this is to remove the organic color by carbon adsorption and/or chemical coagulation (see Chapters 4 and 5), hence the iron is removed also.

Ferrous Carbonate, Ferrous and Ferric Hydroxide

In the aqueous chemistry of Fe(II) and Fe(III) in surface and groundwaters or in a water treatment plant, three solids phases may form: $FeCO_{3(s)}$ (siderite), $Fe(OH)_{2(s)}$ (ferrous hydroxide) and $Fe(OH)_{3(s)}$ (ferric hydroxide). In natural waters under reducing conditions, the stable form of iron is, of course, the +II oxidation state. Here, the solubility of Fe(II) is controlled by $Fe(OH)_{2(s)}$ in waters with "low" $CO_{2(g)}$ contents (Figure 15a) [31]. On the other hand, in natural waters with alkalinities greater than 5×10^{-3} eq/l and 5×10^{-3} *M* of total carbonate species, the solubility of Fe(II) may be controlled by $FeCO_{3(s)}$ (Figure 15b) [31]. The appropriate reactions are given in Table IX. At pH values below 10.5, $FeCO_{3(s)}$ obviously controls the $[Fe^{2+}]$ in solution, as seen in Figure 15b. The solubility product constant in Table IX, Reaction 44, was computed

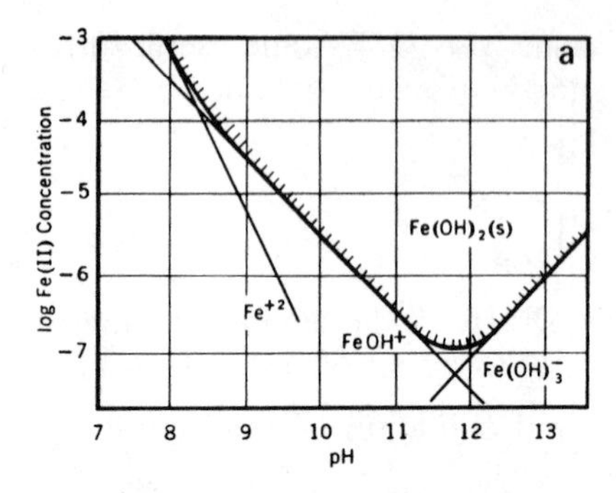

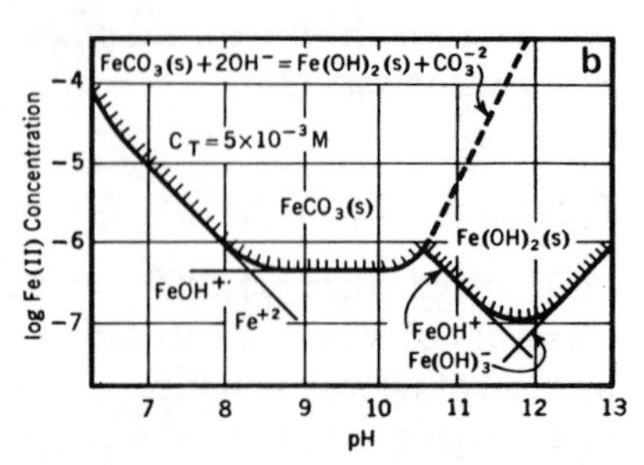

Figure 15. Solubility of ferrous iron in **(a)** waters free from appreciable alkalinity and **(b)** carbonate-bearing waters. Reproduced from Singer and Stumm [31], courtesy of the American Water Works Association.

Table IX. Equilibria Defining Fe(II) Solubility [31]

Reaction Number	Reaction	log Keq 25°C	Reference
41	$Fe(OH)_{2(s)} = Fe^{2+} + 2\ OH^-$	−15.1	32
42	$Fe(OH)_{2(s)} = Fe^+OH + OH^-$	−9.4	32
43	$Fe(OH)_{2(s)} + OH^- = Fe(OH)_3^-$	−5.08	33
44	$FeCO_{3(s)} = Fe^{2+} + CO_3^{2-}$	−10.68	34

from the free energy data tabulated by Latimer [34]. This value of 2.1×10^{-11} was suspect, since it could not account for the high concentrations of Fe(II) in natural groundwaters [31], whereupon a value of 5.7×10^{-11} at 25°C and an ionic strength of 0.0 were obtained [31]. It is argued, then, that groundwaters previously oversaturated by Latimer's constant are now "in or near saturation with siderite ($FeCO_{3(s)}$)" [31]. Evidence (not given here) was gathered to support this statement.

The question of the formation of $FeCO_{3(s)}$ in carbonate hard waters and its "interference" with oxygenation of Fe(II) (Figure 15a) was reexamined [35] with the possibility that $FeCO_{3(s)}$ may be solid phase rather than $Fe(OH)_{3(s)}$ when iron removal is by aeration. Siderite apparently is stable against oxidation to Fe(III) [35]. The precise nature of the solid phase is an important consideration because $FeCO_{3(s)}$ is a crystalline substance, whereas $Fe(OH)_{2(s)}$ and $Fe(OH)_{3(s)}$ are amorphous and gelatinous in form. In short, the original question was not answered because both the Fe^{2+} and Fe^{3+} yielded large crystalline precipitates in the presence of high concentrations of carbonates [35]. Some useful filtration data for the three solid phases cited above were obtained by Olson and Twardowski [35].

Effects of Dissolved Silica

Dissolved silica [H_4SiO_4 or $Si(OH)_4$] affects the aqueous treatment chemistry of iron by catalyzing the rate of oxygenation of Fe(II) [16] and by preventing iron deposition in distribution systems [36]. The solubility of amorphous silica in water at 0°C is approximately 1.–1.3 m*M* (60.–80. mg/l as SiO_2), and at 25°C it is 1.7–2.3 m*M* (100–140 mg/l) [37]. The following reaction demonstrates that dissolved silica reacts with iron [38]:

$$Fe^{3+} + Si(OH)_4 + H_2O = FeSiO(OH)_3^{2+} + H_3^+O \qquad (45)$$

Using a stability constant of 1.8×10^9 for Reaction 45, it may be calculated that an acid mine water with a $[Si]_t$ of 60 mg/l and a $[Fe]_t$ of 5.6 mg/l, at a pH value of 3. would have approximately one-third of the iron complexed with silica.

Figures 16a and 16b show typical rates of Fe(II) oxidation in the presence of silica at pH values of 6.7 and 6.9, respectively. Equation 26, cited above, describes the linear relationship noted in the rate curves. Figure 16c gives the effect of dissolved silica concentration on the Fe(II) oxidation. Figure 16d shows that the catalytic effect is a function of $[H^+]$. No explanation was offered for the greater rates of oxidation of Fe(II) in the presence of silica. A complex cation undoubtedly is formed, but its precise role is uncertain.

Dart and Foley [36] present some actual operational experiences that appear to be opposite to the conclusions drawn by Schenk and Weber [16]. Reports had noted that "iron removal problems" were experienced in "waters with 30 or 40 mg/l SiO_2 [that] often released very little of their iron content on either aeration or chlorination, followed by filtration." The silica apparently reacts with the $Fe(OH)_{3(s)}$ and holds it in suspension. This supposition was tested at four water treatment plants in the province of Ontario, Canada. Wellwaters served four communities whose iron contents ranged from 0.2 to 1.45 mg/l. Natural silica contents were 15–20 mg/l. These iron contents are insufficient to install removal equipment, yet they are high enough to precipitate in the presence of chlorine added for disinfection. A trial-and-error procedure was employed in one of the communities to establish the optimum silica concentration for prevention of iron deposition in the distribution system. This content

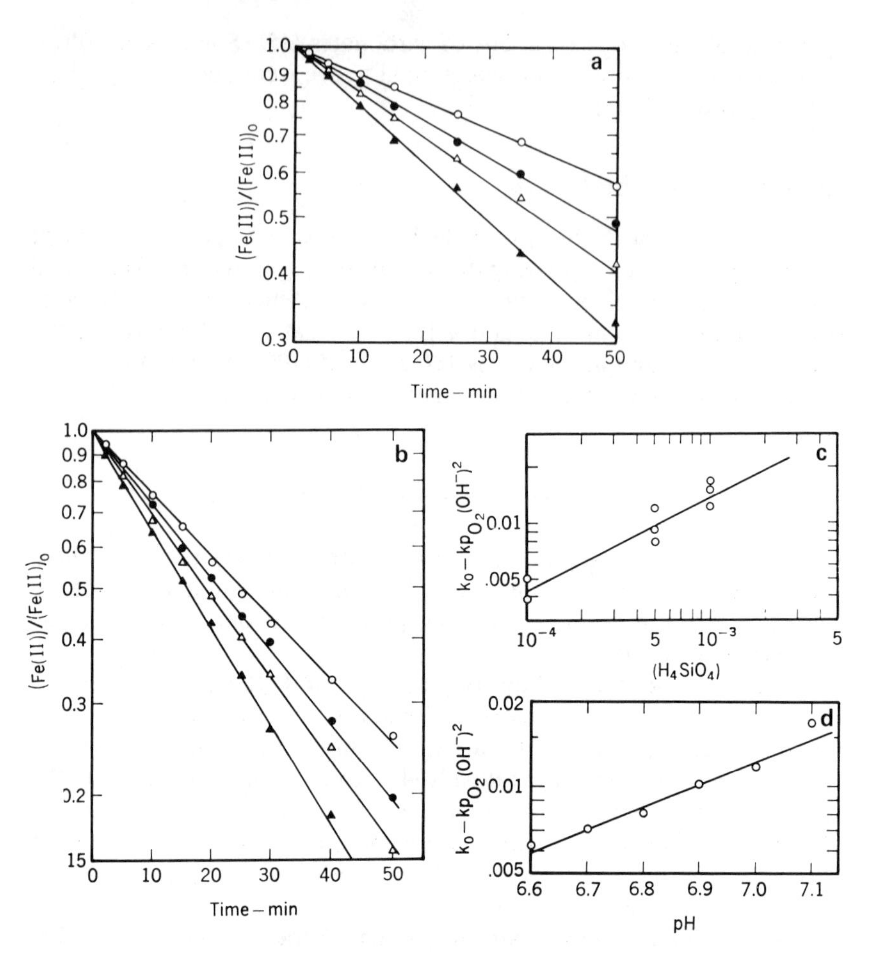

Figure 16. **(a)** Rates of iron(II) oxidation at pH 6.7, as a function of silica concentration, and **(b)** rates of iron(II) oxidation at pH 6.9 as a function of silica concentration. $T = 25°C$, $P_{O_2} = 0.2095$ atm. H_4SiO_4 concentrations are: ○ = zero; ● = 0.1, △ = 0.5; ▲ = 1 mmol/l. **(c)** Catalytic effect of silicic acid on iron(II) oxidation. A plot of the logarithm of the apparent increase in oxidation rate vs the logarithm of H_4SiO_4 for various pH values results in a family of lines with a common slope of 0.5, indicating a 0.5-order dependence of oxidation rate on the silicic acid concentration. One such line, for pH 6.9, is plotted here to illustrate the observed dependence. **(d)** Catalytic effect of silicic acid as a function of pH. A plot of the logarithm of the increase in oxidation rate vs pH results in a family of parallel lines for various H_4SiO_4 concentrations with a slope of 0.5. The line obtained for $[H_4SiO_4] = 5 \times 10^{-5}$ *M* is plotted in this figure as an example to illustrate the observed relationship. Reproduced from Schenk and Weber [16], courtesy of the American Water Works Association.

was 2.7 mg/l of sodium silicate to 0.5 mg/l of iron, and was fed continuously into the system. No staining of water fixtures nor complaints of "red water" were received. For another community, the optimum $[H_4SiO_4]$ was 4.6 mg/l for iron contents varying between 0.2 and 0.7 mg/l. It was noted also that the pH value should be above 7.5 for "effective and efficient" treatment. Application of sodium silicate should be simultaneous with chlorination or other oxidative treatments. Manganese was stabilized with greater effectiveness than iron. The mechanism for prevention of iron deposition apparently is a complex formation with Fe^{3+} similar to the one proposed by Weber and Stumm [38]. Attempts to complex Fe^{2+} were unsuccessful [36].

Polyphosphate Treatment

There are occasions where the addition of sodium hexametaphosphate stabilizes or disperses iron in water distribution systems. The pragmatic dosage is 5 mg/l of phosphate per 1 mg/l of iron. Also, this treatment is limited to iron concentrations less than 1 mg/l. When the water is heated, polyphosphate reverts to orthophosphate and the dispersive property is lost. Application of the polyphosphate must be ahead of any oxidative step because $Fe(OH)_{3(s)}$ cannot be stabilized.

MANGANESE

Aqueous Chemistry

Manganese has nine oxidation states: (0):Mn, (II):Mn^{2+}, +2.67 $-Mn_3O_{4(s)}$, (III):$Mn_2O_{3(s)}$, (V):MnO_4^{3-}, (VI):MnO_4^{2-} and (VII):MnO_4^-. Not all of these are relevant to the aqeuous chemistry of manganese. Some of the more pertinent reactions are given in Table X, from which the pE-pH diagram in Figure 17 was drawn [41]. It is apparent that $Mn^{2+}{}_{(aq)}$ occupies a large portion of the normal pE-pH values in natural waters. It is also obvious that oxidation of Mn^{2+} to one or more of the three oxides of manganese is thermodynamically feasible under the appropriate pE-pH conditions.

Removal

Kinetics of Oxygenation

That the oxygenation kinetics of Mn^{2+} to $MnO_{2(s)}$ or to $MnO_{x(s)}$ does not follow the same rate law as Fe(II) is seen by comparing Figure 18a to

Table X. Some Aqueous Reactions for Manganese at 25°C

Reaction Number	Reaction	log K[a]	V^{0c}
46	$Mn^{2+} + 2\ e = Mn_{(s)}$	−40.	−1.179
47	$Mn^{2+} + 2\ OH^- = Mn(OH)_{2(s)}$	+12.96[b]	
48	$MnO_{2(s)} + 4\ H^+ + 2\ e = Mn^{2+} + 2\ H_2O$	+41.6	+1.228
49	$MnO_4^- + 4\ H^+ + 3\ e = MnO_{2(s)} + 2\ H_2O$	+86.	+1.692
50	$MnO_4^{2-} + 4\ H^+ + 2\ e = MnO_{2(s)} + 2\ H_2O$	+76.5	+2.257
51	$Mn_2O_{3(s)} + 6\ H^+ + 2\ e = 2\ Mn^{2+} + 3\ H_2O$	+48.9	+1.443
52	$Mn_3O_{4(s)} + 8\ H^+ + 2\ e = 3\ Mn^{2+} + 4\ H_2O$	+61.8	+1.824
53	$2\ MnO_{2(s)} + 2\ H^+ + 2\ e = Mn_2O_{3(s)} + H_2O$	+34.4[c]	+1.014
54	$O_2 + 2\ Mn^{2+} + 2\ H_2O = 2\ MnO_{2(s)} + 4\ H^+$		0.0

[a]$K = K_{RED}$ for a reduction reaction and = Keq for a protolysis reaction.
[b]From Morgan [39].
[c]From Pourbaix [41].

18b [4,39,42]. Rather, the decrease of $[Mn^{2+}]$ with time suggests an autocatalytic reaction. The rate equation is:

$$-\frac{d[Mn(II)]}{dt} = k_1[Mn(II)] + k_2[Mn(II)][MnO_{2(s)}] \tag{55}$$

The integrated form of Equation 55 was fitted to the experimental data extremely well (Figure 18c). According to Morgan [39], the reaction sequence is:

$$Mn(II) + 0.5\,O_2 \xrightarrow{\text{slow}} MnO_{2(s)} \tag{56}$$

$$Mn(II) + MnO_{2(s)} \xrightarrow{\text{fast}} Mn(II) \cdot MnO_{2(s)} \tag{57}$$

$$Mn(II) \cdot MnO_{2(s)} + 0.5\,O_2 \xrightarrow{\text{slow}} 2\,MnO_{2(s)} \tag{58}$$

When these three equations are summed, a stoichiometry similar to Reaction 54 is observed. However, Stumm and Morgan [4] indicate that the experimental data are consistent with an autocatalytic model, because the extent of Mn(II) removal is not in accord with the stoichiometry. Some Mn(II) is adsorbed onto higher oxides of Mn in slightly alkaline solutions. Also, oxidation to MnO_2 is not complete, since the solid phase from the oxygenation has been found to range from $MnO_{1.3}$ to $MnO_{1.9}$. In any event, the rate dependence on the $P_{O_{2(g)}}$ and $[OH^-]$ is the same as for iron. Therefore, k_2 in Equation 55 is:

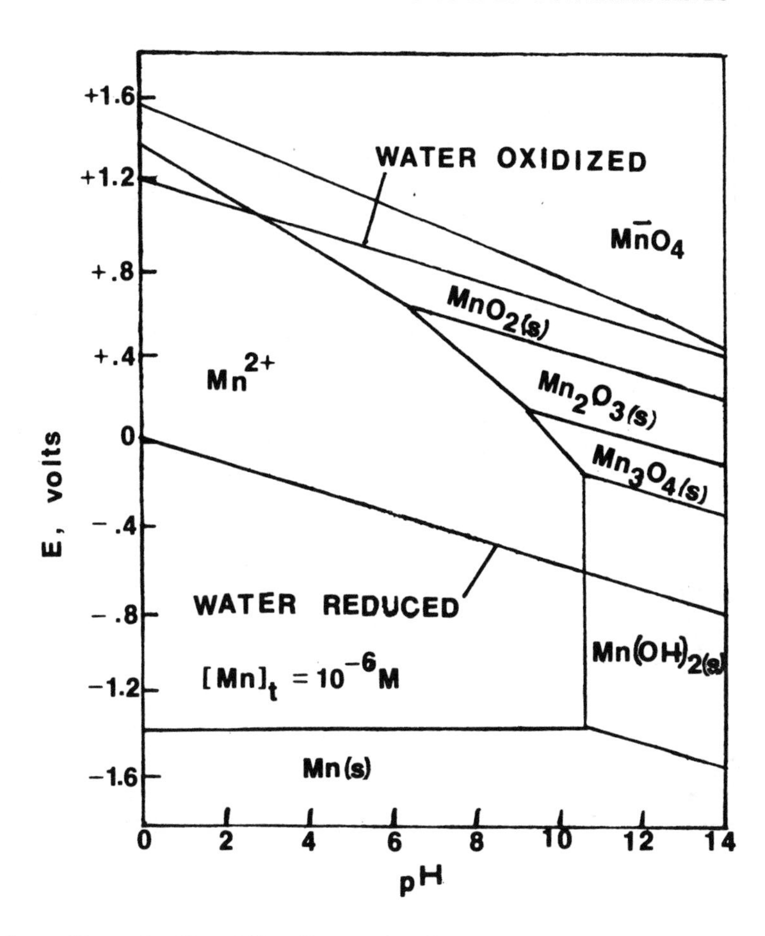

Figure 17. pE-pH stability diagram for the system manganese-water at 25°C. Reproduced from Pourbaix [41], courtesy of Pergamon Press, Inc.

$$k_2 = k_3[OH^-]^2P_{O_2} \tag{59}$$

This is seen in Figure 18d. It should be noted that the maximum rates of oxygenation of Fe(II) and Mn(II) are separated by 3 pH units. This may cause some operational difficulties at treatment plants where both of these constituents occur. According to Stumm and Morgan [4], such metal ions as Cu^{2+} and such metal complexes as SO_4^{2+} do not have an appreciable effect on the oxygenation rate of Mn(II). This is somewhat at variance with the experimental data from Hem [40], who observed a considerable decrease in the rate of oxygenation at a pH value of 9.0 in the presence of 2000 mg/l SO_4^{2+}.

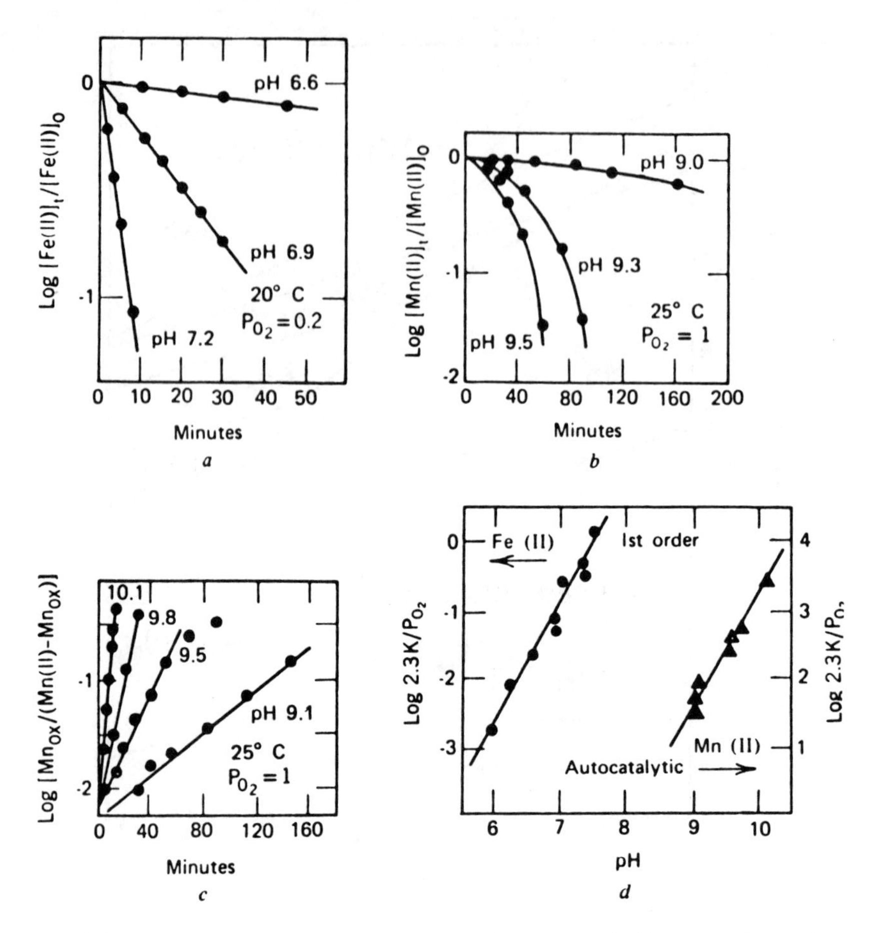

Figure 18. Oxidation of Fe(II) and Mn(II) by oxygen. All experiments were conducted with dissolved Fe(II) or Mn(II) concentrations of less than 5×10^{-4} *M*. In each series of experiments, the pH was controlled by continuously bubbling CO_2- and O_2-containing gas mixtures through HCO_3^- solutions of known alkalinity. **(a)** Oxygenation of Fe(II) in bicarbonate solutions. **(b)** Removal of Mn(II) by oxygenation in bicarbonate solutions. **(c)** Oxidation of Mn(II) in HCO_3^- solutions (autocatalytic plot). **(d)** Effect of pH on oxygenation rates. Reproduced from Stumm and Morgan [4], courtesy of John Wiley & Sons, Inc.

Alternative Oxidants

Chlorine, chlorine dioxide and potassium permanganate are employed also for oxidation of Mn(II). Table XI shows the thermodynamic feasibility for each oxidative reaction. Kinetic data are sparse and many are empirical and pragmatic in nature. For example, Griffin [43] published a review paper on the removal of Mn(II) from water which stated:

> Chlorine oxidizes Mn(II) to Mn(IV) over a wide range of pH values. At pH values of 8.0 or more and at alkalinities of 50 ppm or more, the oxidation is relatively rapid. At pH 8.0, the time requirements may be approximately 2–3 hours. As the pH increases, the time requirements diminish to the pH values in the softening zone, where the oxidation appears to be complete within minutes. At pH values lower than 8.0, the rate of oxidation appears to be progressively slower as the pH is depressed, until at pH 6.0 the time requirement can be 12 hours or longer.

These statements undoubtedly were made from operational experiences. Some qualitative kinetic data are seen in Table XII [17]. Also, it is necessary to have a minimum of 0.5 ppm free available chlorine residual throughout the treatment plant for effective removal [43]. This point had been observed, years earlier, by Mathews [44] in an experimental water

Table XI. Reactions of Mn(II) with Alternative Oxidants at 25°C

Reaction Number	Reaction	V^0
48	$Mn^{2+} + 2\ H_2O = MnO_{2(s)} + 4\ H^+ + 2\ e$	+1.228
28 and 48	$5\ Mn^{2+} + 6\ H_2O + 2\ ClO_{2(g)} = 5\ MnO_{2(s)} + 12\ H^+ + 2\ Cl^-$	+0.283
29 and 48	$Mn^{2+} + 2\ H_2O + Cl_{2(g)} = MnO_{2(s)} + 4\ H^+ + 2\ Cl^-$	+0.131
30 and 48	$3\ Mn^{2+} + 2\ H_2O + 2\ MnO_4^- = 5\ MnO_{2(s)} + 4\ H^+$	+0.467

Table XII. Manganese: Effect of pH on the Oxidation of Manganous Manganese by Chlorine, Without Stirring[a]

Raw Water		Manganese in Treated and Filtered Water (ppm)		
pH	Manganese (ppm)	15 min	30 min	60 min
8.05	10.0			9.0
9.0	10.0	8.0	6.5	5.0
9.45	10.0	4.5	2.5	1.3
10.0	10.0	0.3	0.1	<0.02

[a]Reproduced from Nordell [17], courtesy of Van Nostrand Reinhold Company.

treatment plant. Combined chlorine, i.e., chloramines (see Chapter 10), has very little oxidative effect on Mn(II).

"Chlorine dioxide, generally speaking, oxidizes Mn(II) much more rapidly than chlorine." Griffin's pragmatic statement [43] has no data to support it. "Best results are obtained when the pH is more than 7.0."

"Potassium permanganate oxidizes Mn(II) to Mn(IV) very rapidly" [43]. The optimum pH range apparently is 7.2–8.3 based on some operational experiences [45]. Here again, some qualitative experimental kinetic data suggest a "rapid" reaction, less than 5 min, occurs when 1.0 ppm of Mn^{2+} is treated with 0.5 ppm total residual chlorine and 1.0–1.8 ppm $KMnO_4$ over the pH range of 5.0–9.0. An interesting observation was reported in 1960 [46] that is related to the autocatalytical model cited above [39,42]. Samples of manganese deposits were taken from a distribution system and, when added to solutions of soluble manganese, completely removed this constituent from water.

Ion Exchange

In a manner similar to iron, manganese is removed from water by the manganese zeolite process [17,43]. The capacity of this zeolite is 0.09 lb/ft^3, and 0.18 lb/ft^3 of $KMnO_4$ is required to regenerate the bed [43]. The usual rate of flow is 3 gpm/ft^2. Here again, the zeolite acts as an oxidizing contact medium, an adsorber of soluble manganese and a filter medium. For small industrial and household uses, manganese zeolite filters remove Mn or Fe up to 10 mg/l. For large municipal and industrial uses, this process is limited to 1.0 mg/l of these two constituents [43]. The pH value of the water should be 7.5 or greater for the most effective removal.

Manganous Carbonate and Manganous Hydroxide

In the absence of dissolved oxygen and other oxidants, Mn(II) is the stable form in aquatic systems. The principal equilibria that control $[Mn(II)]_t$ in groundwaters, for example, are the solid phases: $Mn(OH)_{2(s)}$ (pyrochroite), $MnCO_{3(s)}$ (rhodochrosite) and $MnS_{(s)}$ (alabandite). Only the first two are considered here. In systems where carbonates and sulfides are either absent or in low contents, the solubility of Mn(II) is limited by $Mn(OH)_{2(s)}$. Equations 47, 60 and 61, Table XIII, were used by Morgan [39] to construct the solubility diagram in Figure 19. The maximum insolubility appears to be approximately 10^{-7} *M* or 54. μg/l Mn at a pH value of approximately 12. However, $Mn(OH)_2$ is extremely soluble in water at pH values less than 12. In waters containing a C_t for

Table XIII. Equilibria Defining Mn(II) Solubility

Reaction Number	Reaction	log Keq 25°C
47	$Mn^{2+} + 2\ OH^- = Mn(OH)_{2(s)}$	+12.96[a]
60	$Mn(OH)_{2(s)} = MnOH^+ + OH^-$	−9.95[a]
61	$Mn(OH)_{2(s)} + OH^- = Mn(OH)_3^-$	−5.35[b]
62	$MnCO_{3(s)} = Mn^{2+} + CO_3^{2-}$	−10.41[a]
63	$MnCO_{3(s)} + OH^- = MnOH^+ + CO_3^{2-}$	−7.00[a]
64	$MnCO_{3(s)} + 3\ OH^- = Mn(OH)_3^- + CO_3^{2-}$	−2.80[a]
65	$MnCO_{3(s)} + H^+ = MnHCO_3^+$	+1.91[a]

[a]From Morgan [39].
[b]From Fox et al. [46].

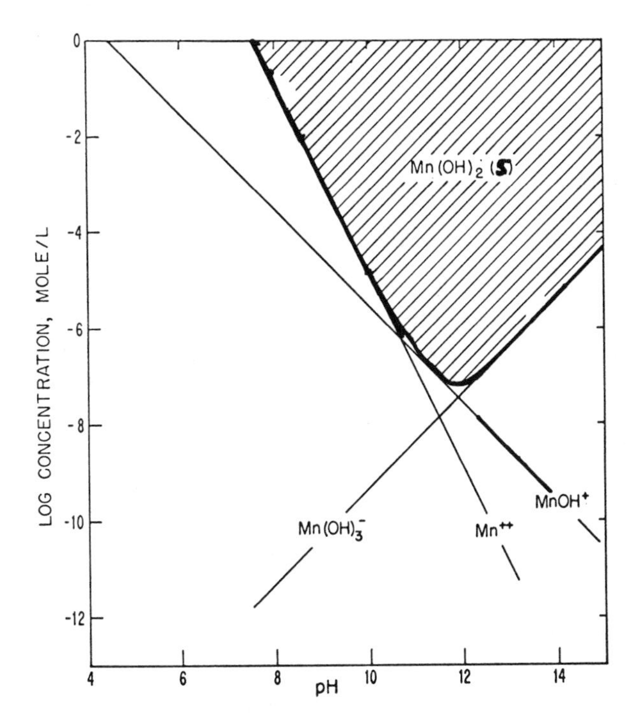

Figure 19. Solubility of manganous hydroxide. Reproduced from Morgan [39], courtesy of John Wiley & Sons, Inc.

carbonate species of 2×10^{-3} M, the solubility of $[Mn]_t$ is decreased somewhat through the formation of $MnCO_{3(s)}$. For example, at pH 8., more than 1000 times as much Mn^{2+} can be retained in a carbonate-free water than in one containing the 1×10^{-3} $M\,C_t$ of carbonates (Figure 20).

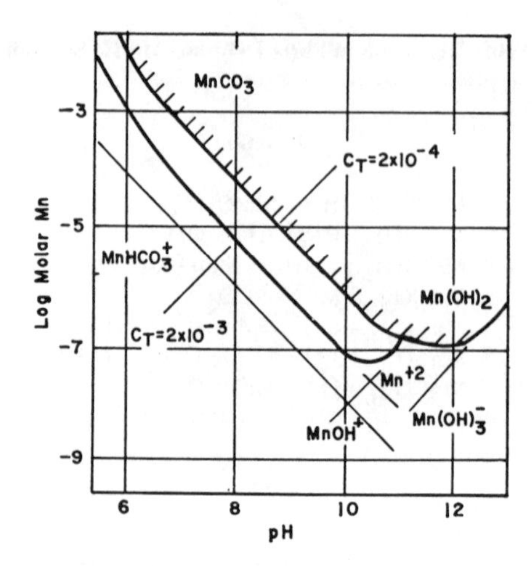

Figure 20. Solubility of Mn(II) in water containing 2×10^{-3} *M* total carbonic species. Reproduced from Morgan [39], courtesy of John Wiley & Sons, Inc.

Settelability and Filterability Difficulties of $MnO_{2(s)}$

Under the proper conditions of pH, the oxidation of Mn(II) by $O_{2(g)}$, Cl_2 and $KMnO_4$ is fairly rapid kinetically. However, the formation and precipitation of $MnO_{2(s)}$ and/or $MnO_{x(s)}$ are slow. Furthermore, these oxides have colloidal properties. Some of the operational problems with manganese at a water treatment plant employing chemical coagulation, sedimentation, sand filtration and postchlorination are indicated here: "Filter runs averaged 24 hours. The fine sand grains were coated with $MnO_{2(s)}$ deposits; the coarse sand grains were stained dark brown with $MnO_{2(s)}$, and the gravel beds were heavily coated and cemented with it. The distribution mains, service line, and meters carried heavy deposits of $MnO_{2(s)}$. Consumer complaints about the water, with regard to the laundering of clothes and the staining of porcelain fixtures, were numerous" [43].

$MnO_{2(s)}$ suspensions can be colloidally dispersed for "several months" in neutral and alkaline pH ranges and in the absence of multivalent cations [48]. Some flocculation occurs slowly when the pH value is lowered to 3.0 and below. Mn(II) ions and other multivalent cations destabilize $MnO_{2(s)}$ dispersions. Mn^{2+} concentrations of 10^{-4}–10^{-5} *M* cause rapid flocculation of $MnO_{2(s)}$ colloids within the pH range of normal stability.

Chemical coagulation, sedimentation and sand filtration are imperative treatments following the oxidative step for Mn^{2+}. Briefly stated, alum $[Al_2(SO_4)_3]$ and a coagulant aid (activated silica, for example) are used for flocculation of colloidal $MnO_{2(s)}$. This $MnO_{2(s)}$ occurs as a negatively charged colloid over a broad range of pH values (5.–11.). A rather comprehensive study of the electrophoretic characteristics of and the coagulation of $MnO_{2(s)}$ was reported by Posselt et al. [47]. For the former, an electromobility value of -2.5 μm/sec/V/cm at a pH value of 5. was measured. This negative value was reduced to -1.0 with 10 mg/l and higher of Ca^{2+} with concurrent destabilization of colloidal $MnO_{2(s)}$. This clearly demonstrates the role of surface charge in the coagulation of this colloid. Subsequent experiments were conducted at pH values of 5., 7. and 9. with the polycation PDADMA, ferric sulfate, alum and various combinations of these coagulants and aids. Figure 21 shows typical results with PDADMA and alum. The V-shaped curves in Figure 21a show that destabilization of the colloid occurs from sorption of the PDADMA, and restabilization occurs from an excessive quantity of the polycation. This was confirmed by separate electromobility measurements where complete restabilization corresponded to a more or less constant value of $+1.5$ to 2. μm/sec/V/cm. Figure 21b shows the coagulation of $MnO_{2(s)}$ by alum, for which the optimum dosage was somewhat dependent on the pH value. Similar results were obtained for ferric sulfate. Parallel experiments were conducted for each of the coagulants in the presence of Ca^{2+} ions. In each case, the coagulant dose was lowered considerably due to neutralization of negative surface charges by the Ca^{2+}, especially at pH values of 7. and 9.

Sodium hexametaphosphate has been used to stabilize Mn^{2+} where concentrations are 2.0 mg/l and below [48]. A 2:1 ratio of phosphate usually is effective for stabilization, and pH values are in the 7.–8. range. This treatment is confined mostly to groundwaters.

Case Studies of Iron and Manganese Removal

An experimental treatment plant for the wellwaters of Sioux Falls, South Dakota consisted of: chlorine, coke-tray aeration, contact filter, sedimentation and rapid sand filtration (2 gpm/ft^2) [44]. The most effective scheme was obtained with free residual chlorination, a 9.5 mg/l dosage for iron, 4.5 mg/l and manganese, 1.8 mg/l, in the raw water. Neither constituent could be detected in the water after the treatment sequence. No difficulties with colloidal $MnO_{2(s)}$ were reported.

Adams [45] reported treating the Allegheny River at Wilkinsburg,

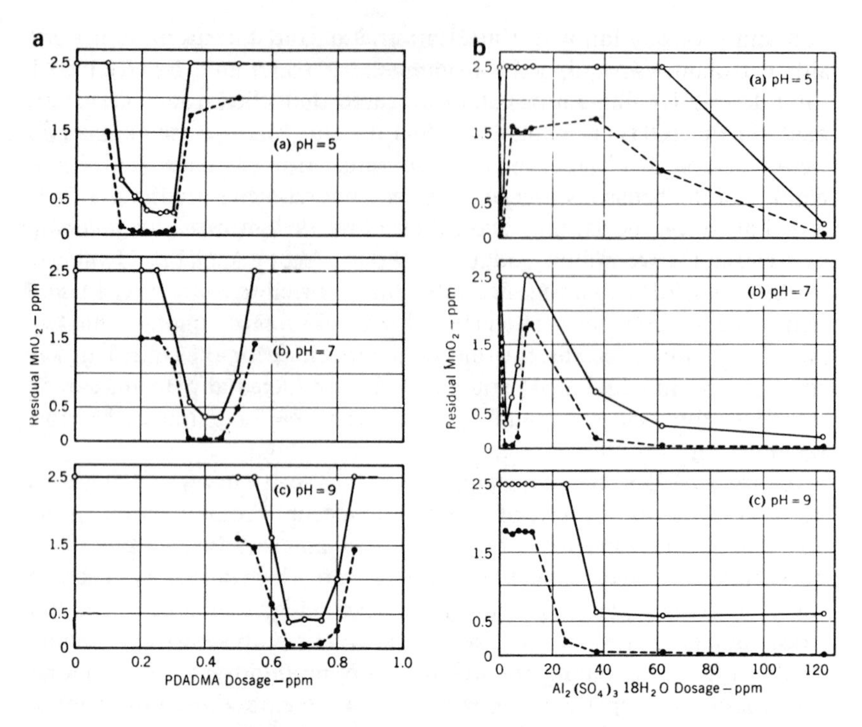

Figure 21. **(a)** Effect of a polycation on the stability of colloidal hydrous $MnO_{2(s)}$. The polyelectrolyte used in these experiments was PDADMA polycation. The solid lines represent unfiltered samples, and filtered samples are represented by dashed lines. HCl was used for adjustment and maintenance of pH 5; Na_2CO_3 was used for pH 7 and 9. **(b)** Coagulation of hydrous $MnO_{2(s)}$ with aluminum sulfate. Solid lines represent unfiltered samples; dashed lines represent filtered samples. Reproduced from Posselt et al. [47], courtesy of the American Water Works Association.

Pennsylvania for removal of 0.5–4.0 mg/l Mn at an annual average of 1.2 mg/l. Experimentation on a pilot-plant scale led to this order of chemical addition: chlorine, lime, $KMnO_4$, alum and polyelectrolyte (coagulation aid). pH values were maintained in the 7.2–8.3 range. Considerable experimentation with the quantities and order of the addition of these chemicals produced a 10-yr average of 0.05 mg/l Mn in the treated water.

A split-flow treatment was used for Fe and Mn removal from groundwater supplies (12 wells) of Eau Claire, Wisconsin. Iron contents ranged from 0 to 2.4 mg/l, and manganese contents ranged from 0 to 5.66 mg/l. The treatment process consisted of pre- and postchlorination, coagulation and precipitation with $Ca(OH)_2$, and activated silica for coagulant aid. The plant was equipped with two solids contact units (see Chapter 5 for a descriptive illustration) and four rapid sand filters, each rated at 6 mgd. The $Ca(OH)_2$ was the essence of the treatment where Fe and Mn were removed by coagulation and precipitation in the suspended sludge blanket in the solids contact unit. It was necessary to raise the pH value to 9.5 or so for 100% removal of Mn. Split-flow treatment was achieved by treating the wellwaters containing the Fe, Mn and "high" alkalinity. With the aid of activated silica, a controlled sludge blanket was maintained for these waters. The effluent from the solids contact unit was mixed with other wellwaters with low alkalinities and no Fe and Mn. The combined waters had a pH value of 8.5, and required very little or no recarbonation. The combined waters were 45% treated and 55% raw water.

A survey of 29 iron and manganese removal plants in eastern Nebraska was conducted for conformance to the drinking water standards of 0.3 and 0.05 mg/l, respectively [51]. This study also provided an excellent opportunity for a concurrent evaluation of various treatment schemes. A compilation and summary of the data are given in Table XIV. From this study, it was concluded that the percentage of plants meeting the standards were: 45% where only aeration, contact time and filtration were used (data not shown), 58% where aeration, chemical addition, contact time and filtration were used (Table XIVa) and 100% where aeration and partial softening were used (Table XIVb).

Cleasby [52] presents an excellent case study of the significance of feed position of the chemical oxidant as it affects the nature and filtrability of the precipitates. The essential treatment was diffused aeration followed by a baffled detention tank, a detention tank and rapid sand filters. $KMnO_4$ was added at various points along the baffled and unbaffled detention tanks until a location just before the sand filters. Whatever oxides of iron and manganese were formed were filtrable, and remained on the filter rather than appearing in the distribution system.

ARSENIC

Aqueous Chemistry

Arsenical compounds are among the most widely distributed elements in the earth's crust and in the biosphere. The concern about arsenic in the environment is given in Chapter 1 and in Faust and Aly [2]. For the most

Table XIVa. Iron and Manganese Data on Nebraska Water Supplies (Treatment Provided: Aeration, Chemical Addition, Contact Time and Filtration)[a]

Municipality		Treatment Provided[c]	Iron (mg/l)		Manganese (mg/l)		Chlorine Residual (mg/l) Eff.	pH Eff.
No.	Population[b]		Inf.	Eff.	Inf.	Eff.		
13	3,650	ApCclDFgs	0.50	0.03	0.15	<0.02	0.35	7.4
14	642	ApCcsDFps	2.40	0.85	0.50	0.45	0.30	7.8
15	323	ApCcDFps	1.15	0.10	0.55	0.20	0.30	7.6
16	149,518	CcAcCcDFgsCn	0.10	0.05	0.10	<0.02	1.00	7.8
17	2,669	ApCcDFgs	0.70	0.65	0.20	0.08	0.10	8.0
18	7,441	CclApDFgsCcf	(3.80)[d]	0.05	(1.50)	<0.02	0.40	8.4
19	1,355	ApCcDFp	(0.90)	0.20	(0.20)	<0.02	0.20	6.9
20	6,371	ACcDFgs	(6.50)	0.20	(2.00)	0.02	0.25	7.9
21	291	ApCcFp	10.00	0.10	2.50	0.02		7.5
9[e]	7,920	AtCcDFgs	(3.00)	0.10	(0.60)	<0.02	0.20	
22	1,215	ACcDFgs	0.06	0.06	0.25	0.12		
23	285	ApCcDFps	3.55	0.60	0.95	<0.02	1.00	7.0

[a]Reproduced from Andersen et al. [51], courtesy of the American Water Works Association.
[b]Preliminary U.S. census figures for 1970.
[c]Explanation of treatment symbols:
A—Aeration or oxidation
c—contact beds or trap filled with coke or other material
p—patented aerator
t—overflow trays, cascade, or other splash aerator

C—Chemicals added
- c—chlorine
- f—fluoride
- l—lime
- n—ammonia

D—Detention

F—Filter
- g—gravity
- p—pressure
- s—sand

[d]Concentrations in parentheses are average values obtained for wells pumping at the time of plant visitation.

[e]Municipality has two iron- and manganese-removal plants.

Table XIVb. Iron and Manganese Data on Nebraska Water Supplies (Treatment Provided: Aeration and Partial Lime Softening)[a]

Municipality		Treatment	Iron (mg/l)		Manganese (mg/l)		Hardness (mg/l as $CaCO_3$)		Chlorine Residual (mg/l)	pH
No.	Population[b]	Provided[c]	Inf.	Eff.	Inf.	Eff.	Inf.	Eff.	Eff.	Eff.
24	19,449	AtSlFgsCcFp	7.50	0.05	0.40	<0.02	472	160	0.60	9.4
25	6,106	ACcSlFgs	(4.40)[d]	0.03	(0.75)	<0.02			0.20	8.5
26	5,444	ApCcDSalFgs	4.00	0.03	1.10	<0.02	330	196		8.4
27	1,177	ADSalFgs	0.60	0.03	1.05	<0.02	300	152	0.00	8.5
28	2,779	ApCcSalFgs	(0.75)	0.05	(0.75)	<0.02	310	106	0.70	8.5

[a]Reproduced from Andersen et al. [51], courtesy of the American Water Works Association.
[b]Preliminary U.S. census figures for 1970.
[c]Explanation of treatment symbols:

- A—Aeration or oxidation
  - p—patented aerator
  - t—overflow trays, cascade, or other splash aerator
- C—Chemicals added
  - c—chlorine
- D—Detention
- F—Filter
  - g—gravity
  - p—pressure
  - s—sand
- S—Softening unit
  - a—alum
  - l—lime

[d]Concentrations in parentheses are average values obtained for wells pumping at the time of plant visitation.

part, arsenic occurs in an inorganic form in aquatic environments resulting from the dissolution of such solid phases as $As_2O_{3(s)}$ (arsenolite), $As_2O_{5(s)}$ (arsenic anhydride) and $AsS_{2(s)}$ (realgar). Two weak acids may occur in the water phase: $HAsO_2$ (arsenious acid, +III oxidation state) and H_3AsO_4 (arsenic acid, +V oxidation state). The acid-base equilibria are given in Table XV. Since the solubilities of the oxides are "high," arsenic occurs in the anionic form, which is dependent on the pH value of the water. There are environments of pH values and redox values where reduced oxidation states and methylated forms of arsenic occur: $CH_3H_2AsO_3$ monomethylarsonic acid), $(CH_3)_2HAsO_2$ (dimethylarsinic acid), $(CH_3)_2HAs$ (dimethylarsine), $(CH_3)_3As$ (trimethylarsine) and AsH_3 (arsine). These arsenical compounds, however, are not stable in the oxidizing conditions of water treatment plants. Faust and Aly [2] give more information about the environmental chemistry of arsenic.

Removal

A laboratory evaluation of conventional water treatment processes for removal of As contents ranging from 0.6 to 2.0 mg/l in Taiwanian groundwaters has been reported [53]. Coagulation by alum and lime (separately) effected removals of 32 and 20%, respectively, of 1.0 mg/l As_t at a pH of 6.8. A 50-mg/l dosage of $FeCl_3$ reduced 1.0 mg/l As_t by 90% at a pH of 6.8. The addition of an oxidant, $KMnO_4$ or Cl_2, to coagulation by $FeCl_3$ produced As removals on the order of 95–98%. Table XVI shows that a combined process of chlorine and ferric chloride can effect As removal up to 100%. Slow and rapid filtration through anthracite or sand were ineffective for As removal. Based on these results from laboratory studies, a pilot plant was constructed in which a combination

Table XV. Acid-Base Equilibria of Arsenious and Arsenic Acids at 25°C [41]

Reaction No.	Reaction	log K_a
	+3	
66	$AsO^+ + H_2O = HAsO_2 + H^+$	0.34
67	$HAsO_2 = AsO_2^- + H^+$	−9.21
	+5	
68	$H_3AsO_4 = H_2AsO_4^- + H^+$	−3.60
69	$H_2AsO_4^- = HAsO_4^{2-} + H^+$	−7.26
70	$HAsO_4^{2-} = AsO_4^{3-} + H^+$	−12.47

Table XVI. Results of Oxidation and Coagulation Test for As Removal[a]

Property	Chlorine Added (mg/l)					Remarks
	5	10	20	30	40	
Ferric chlorine added (mg/l)	50	50	50	50	50	Fe in raw water: 1.3 mg/l
Free residual Cl_2 (mg/l)	0	0.3	1.0	2.0	3.0	As in raw water: 0.8 mg/l+
Fe in supernatant (mg/l)	0.5	0.2	0.2	0.2	0.2	
As in supernatant (mg/l)	0.12	0.08	0.01	Trace	Trace	
As removal (%)	82.5	90.0	98.7	~100	~100	

[a]Reproduced from Shen [53], courtesy of the American Water Works Association.

of conventional schemes was tried. The most effective treatment (100% removal) was 20 mg/l Cl_2 followed by 50 mg/l $FeCl_3$, sedimentation and sand filtration. The sand filter eventually had to be washed with NaOH to remove the As. A permanent water treatment plant eventually was constructed consisting of: aeration, mixer, sedimentation tank, two slow sand filters, storage tank, elevation tank and a sand-washing basin.

Adsorption of As(V) onto the hydrous oxides of Al and Fe was examined [54]. The results, in summary, are seen in Figure 22 where $FeCl_3$ is the more effective treatment.

Ferguson and Anderson [55] researched the adsorption of As(III) and As(V) on $Al(OH)_{3(s)}$ and $FeOOH_{(s)}$. These experiments were conducted in a manner similar to carbon adsorption. That is, the arsenics were presented to preformed quantities of the two oxides. A summation of these studies is given in Table XVII. The adsorption data for As(V) were fitted to the Langmuir equation for graphical purposes. These data in Table XVII can be extrapolated to lower As contents relevant to conditions of water treatment. "It can be calculated that for a dose of 100 mg/l of alum or 50 mg/l of ferric chloride, removal of an initial arsenic concentration of 0.1 mg/l would be 34% for As(V) on $Al(OH)_{3(s)}$, 3.8% for As(III) onto $Al(OH)_{3(s)}$, and 11% for As(III) or (V) onto $FeOOH_{(s)}$." These results are somewhat at variance with those cited above.

An alumina (Al_2O_3) column was reasonably effective for As removal (the oxidation state was not cited) from a natural groundwater with [As] = 0.106 mg/l [56]. No quantitative data were given.

An alumina column was used for As and F removal from a groundwater at Why, Arizona [57]. The As content was 0.10–0.17 mg/l and the F content was 2.4–3.3 mg/l in the raw water. The laboratory water treatment plant consisted of a test column of 28–48 mesh alumina and a pretreatment of H_2SO_4 for adjustment of the pH to 5.4. Dilute NaOH (4%)

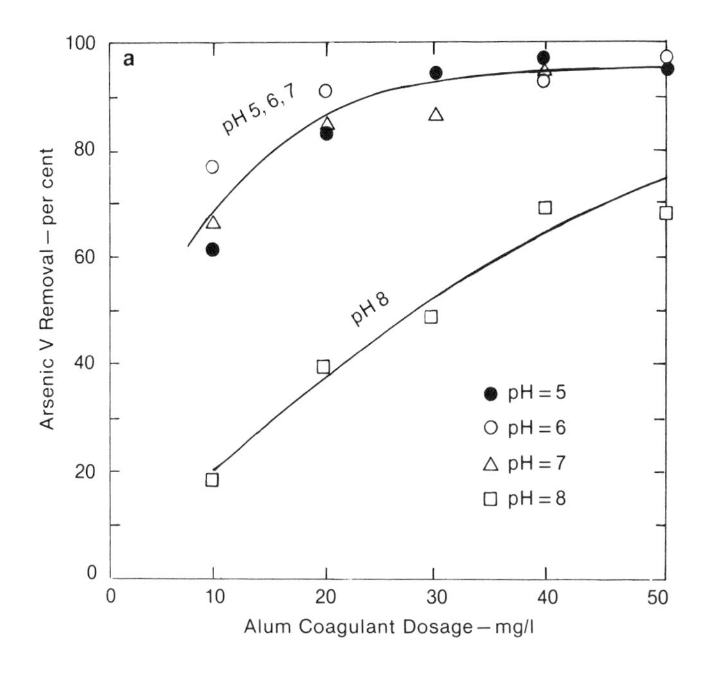

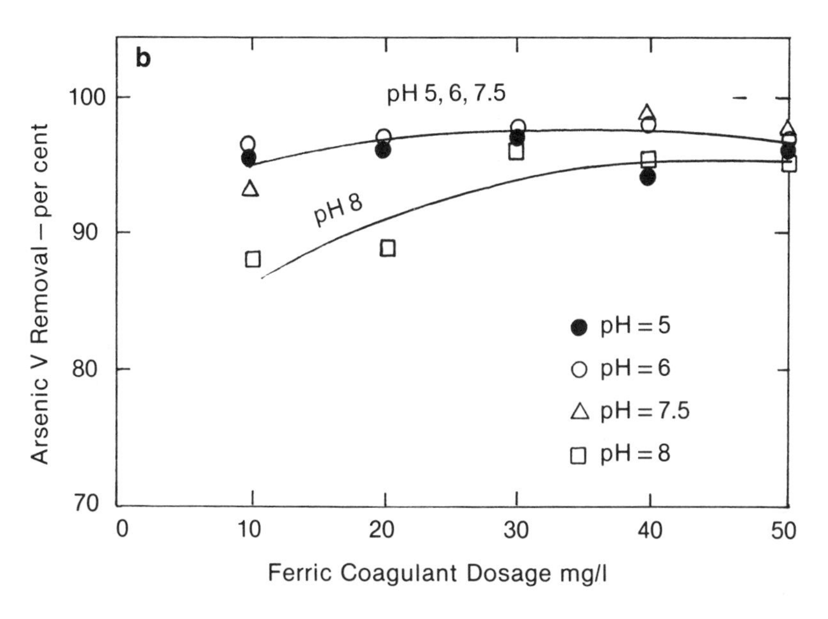

Figure 22. Removal of As(V) by adsorption on **(a)** hydrous aluminum oxide and **(b)** hydrous ferric oxide. Reproduced from Gulledge and O'Connor [54], courtesy of the American Water Works Association.

Table XVII. Adsorption Isotherm Experimental Conditions[a]

Adsorption	Precipitate Concentration (g/l)	pH Range		Final Concentration Range (mol/l)
		Initial	Final	
As(V) on $Al(OH)_3$	0.29–1.40	6.9–8.0	7.2–8.3	5×10^{-6}–5×10^{-3}
As(V) on FeOOH	0.29–2.90	5.4–6.4	5.6–6.6	5×10^{-6}–5×10^{-3}
As(III) on $Al(OH)_3$	0.20–1.30	6.8–7.8	6.3–7.3	2×10^{-4}–6×10^{-4}
As(III) on FeOOH	0.12–0.78	7.0–8.0	6.5–7.5	1×10^{-4}–6×10^{-4}

[a]Reproduced from Ferguson and Anderson [55], courtesy of Ann Arbor Science Publishers.

was used to regenerate the alumina. Downflow rate was not given. Figure 23 shows typical breakthrough curves for As and F. It is not clear why the pH value of the raw water was adjusted to an acid value where the weak acid of As^{3+} would be associated and the monovalent ion of $H_2AsO_4^-$ of As^{5+} would occur. In any event, this process was reasonably effective for removing the two constituents that may be limited to small supplies and groundwaters. Also, there is the added cost of adjusting the pH after passage through the column.

As(III) and As(V) have been removed from fresh- and seawater by adsorption on alumina (28 × 48 mesh), bauxite (30 × 60 mesh) and GAC (8 × 30 mesh) [58]. Batch systems, 25°C, were shaken continuously for 36 hr for alumina and bauxite and 48 hr for activated carbon. Table XVIII shows the efficiencies of As(III) and As(V) removal by the three adsorbents. Activated alumina removed the highest quantity of As(V), whereas As(III) was adsorbed equally by the bauxite and alumina. As(III) removal on activated carbon apparently was negligible. Table XVIII also shows the arsenic removal necessary to attain the 0.05-mg/l drinking water standard. All of this led to the conclusion that these adsorbents were technically feasible. Additional data of As removal is given below in the Se section, since their chemistries are similar. (Figure 35 shows the effect of pH on As removal by lime softening.)

BARIUM

Aqueous Chemistry

This element is one of the inorganics whose content in drinking waters is limited to 1.0 mg/l (see Chapter 1). Trace concentrations, 9.0–152

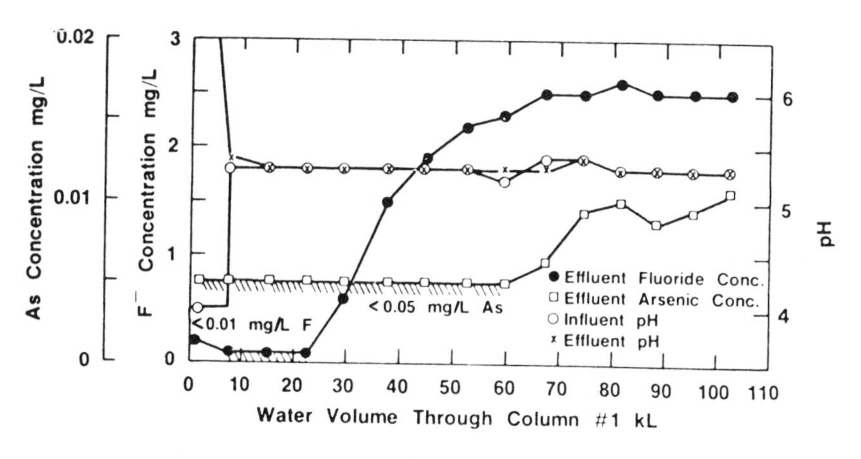

Figure 23. Plots of phase II, run 3 water concentrations. Raw water contained 2.8 mg/l fluoride, 0.14 mg/l As and pH 7.5. The column held 0.028 m^3 of activated alumina [57].

μg/l, were found in several surface waters in North America [2]. Barium contents in the groundwaters of northern Illinois have been as high as 19. mg/l with 2.-7. mg/l being a more common range [59]. Since barium is one of the alkaline earth elements, its chemical properties are similar to Ca and Mg. Consequently, the treatments of lime-soda softening and ion exchange are appropriate also for Ba. The softening reactions at 25°C are:

$$Ba^{2+} + CO_3^{2-} = BaCO_{3(s)} \quad \text{(witherite)} \tag{71}$$

$$\Delta G^\circ_{REX} = -18.189 \text{ kcal/mol}$$

$$\log K_s = 13.33$$

$$Ba^{2+} + SO_4^{2-} = BaSO_{4(s)} \quad \text{(barite)} \tag{72}$$

$$\Delta G^\circ_{REX} = -13.96 \text{ kcal/mol}$$

$$\log K_s = 10.23$$

$$Ba^{2+} + 2\,OH^- = Ba(OH)_{2(s)} \tag{73}$$

$$\Delta G^\circ_{REX} = 4.488 \text{ kcal/mol}$$

$$\log K_s = -3.29$$

Equation 71 represents the formation of $BaCO_{3(s)}$, whose solubility is about 30. μg/l at pH values greater than 10. Reaction 72 is the formation

Table XVIII. Efficiency of As(III) and As(V) Removal by Different Adsorbents under Different Solution Conditions[a,b]

	q_{100}[c]	q[d]	%	pH	Arsenic Removed[e] (mg/g of Solids)
Activated Bauxite, 2 g/l					
As(III)					0.12
6.54 μM	3.27	2.6	79.6	7.99	
13.4 μM	6.65	4.86	73.0	7.69	
26.7 μM	13.3	9.31	70.0	7.80	
As(V)					2.0
24.7 μM	12.3	12.3	100	6.5	
60.1 μM	30.0	29.76	99	6.7	
104 μM	52.0	50.2	97	6.8	
Activated Alumina, 2 g/l					
As(III)					0.2
6.65 μM	3.26	2.9	88.8	8.47	
13.4 μM	6.66	5.53	83.0	8.04	
26.7 μM	13.3	10.5	79.3	8.20	
As(V)					4.1
32 μM	16.0	16.0	100	6.9	
66.7 μM	33.4	33.3	99.7	6.8	
133 μM	66.5	65.9	99.0	6.9	
Activated Carbon, 3 mg/l					
As(V)					0.34
12.8 μM	4.26	4.11	96.5	3.1	
25.4 μM	8.46	7.82	92.4	3.1	
34.7 μM	11.6	9.69	85.5	3.2	

[a] Reproduced from Gupta and Chen [58], courtesy of the Water Pollution Control Federation.
[b] Solvent matrix: water.
[c] q_{100} = adsorption density at 100% removal.
[d] q = adsorption density (μM/g of solids).
[e] At optimum pH, 50-μg/l arsenic concentration.

of $BaSO_{4(s)}$, whose solubility is about 1 mg/l. Barium hydroxide is too soluble for precipitation at Ba contents of natural waters. It appears that lime-soda softening is the preferred treatment by precipitation.

Removal

Laboratory jar-tests were conducted for an evaluation of the softening treatment [59]. It was found that maximum Ba removal was effected in the pH range of 9.5–11.5. The equilibrium content of 30 μg/l apparently was not obtained because of the coprecipitation of Ca and Mg in the synthetic water. However, attainment of the MCL of 1.0 mg/l is feasible in the softening process (Table XIX). There is, of course, a sludge disposal problem (see Chapter 7).

Ion exchange is also a feasible treatment for Ba removal [59]. A short field study was conducted on two full-scale ion-exchange softening plants in northern Illinois. The flow diagram for plant C is seen in Figure 24a, and barium and hardness removals are given in Figure 24b. The raw water for plant C contained a hardness of 218 mg/l as $CaCO_3$, and a Ba content of 10. mg/l. The data in Figure 24b show that the Ba content was reduced to less than 1 mg/l through 153 bed volumes (188,000 gal). Breakthrough for either Ba or hardness was not reached, because the ion-exchange units were operated on a flow basis (i.e., the treatment cycle was completed before breakthrough).

Other studies have shown that powdered (PAC) and granular activated carbon (GAC) are ineffective methods for Ba removal [59]. Less than 7% was removed by PAC, and none was removed by GAC in laboratory studies. Reverse osmosis (RO) and electrodialysis (ED) were effective methods, giving 95–98% removal of 9.15 mg/l Ba in tapwater [59].

CADMIUM

Aqueous Chemistry

The MCL for cadmium in drinking water is 0.01 mg/l (see Chapter 1). Surface water contents of Cd in North America [2] ranged from 21. to 130. μg/l. It is not especially widespread from natural geologic sources, but may occur in substantial concentrations due to the discharge of wastewaters, especially plating wastes. The aqueous chemistry of cadmium is, for the most part, dominated by Cd^{2+}, $CdCO_{3(s)}$ (otavite) and $Cd(OH)_{2(s)}$ [2,60]. The appropriate reactions are:

$$Cd^{2+} + CO_3^{2-} = CdCO_{3(s)} \quad \text{(otavite)} \tag{74}$$

$$\Delta G^\circ_{REX} = -15.25 \text{ kcal/mol}$$

$$\log K_s = 11.18$$

$$Cd^{2+} + 2OH^- = Cd(OH)_{2(s)} \tag{75}$$

$$\Delta G^\circ_{REX} = -19.62 \text{ kcal/mol}$$

Table XIX. Barium Removal from Well Water by Full-Scale Lime-Softening Water Treatment Plants[a]

Treatment Plant	Treatment pH	Water Sample	Barium Concentration (mg/l)				Hardness Concentration (as $CaCO_3$) (mg/l)	
			Raw Water	Settled Water	Recarbonated Water	Filtered Water	Raw Water	Filtered Water
A	10.5	1	7.5	0.91	1.08	0.63	276	72
		2	7.5	0.95	1.18	1.00	268	68
		3	7.5	0.95	1.00	0.73		
Average concentration			7.5	0.94	1.09	0.88	272	70
(Average percent removed)				88	86	88		
B	10.3	1	17.3	1.60		0.90	240	72
		2	18.0	1.90		0.84	252	80
		3	17.0	2.05		0.69		
Average concentration			17.4	1.85		0.81	246	76
(Average percent removed)				89.4		95.3		69

[a]Reproduced from Sorg and Logsdon [59], courtesy of the American Water Works Association.

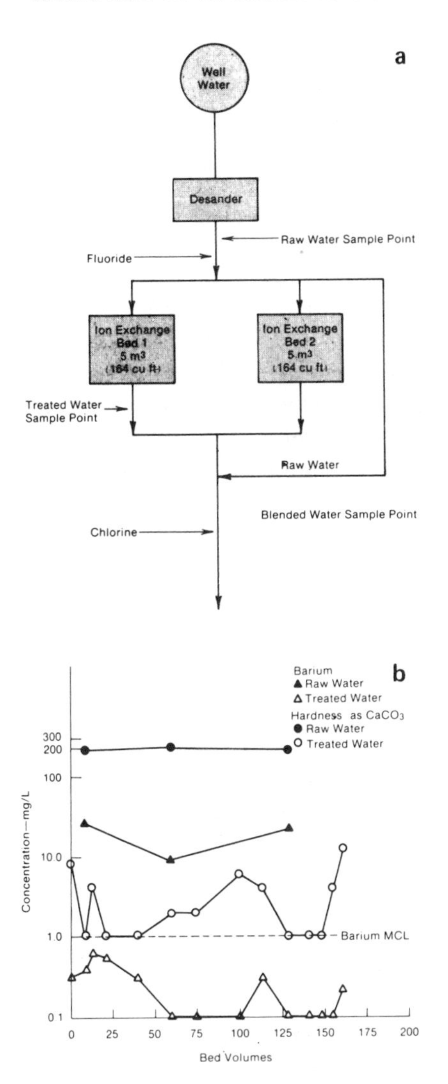

Figure 24. **(a)** Flow diagram of ion exchange softening plant C. **(b)** Barium and hardness removal from wellwater by full-scale ion exchange treatment plant C. Reproduced from Sorg and Logsdon [59], courtesy of the American Water Works Association.

$$\log K_s = 14.4 \text{ (aged)}$$

$$\log K_s = 13.7 \text{ (fresh) [61]}$$

Reaction 74 represents the formation of $CdCO_{3(s)}$ whose solubility is about 300. μg/l for pH values greater than 10. Reaction 75 is the formation of cadmium hydroxide whose solubility, at pH 10., is 44 μg/l for the "aged" form and 225. μg/l for the "fresh" precipitate. In the Cd-S-CO_2-H_2O system, Cd solubility is below 10 μg/l in the pH value range 8.9–10.7, according to Hem's thermodynamic calculations [62].

Removal

Few studies have reported the specific removal of Cd from water because it is an infrequent problem. Consequently, it is necessary to evaluate the effectiveness of conventional water treatments before seeking processes specifically for Cd. Chemical coagulation (jar-tests) was investigated for Cd-spiked (0.03 mg/l) Ohio River water and a wellwater in Ohio [60]. Alum and ferric sulfate removals with "high alkalinity" [200 mg/l as $CaCO_{3(s)}$] wellwater were about 10% at pH 7.; this was increased to about 100% at pH 9. The dosages were 20 mg/l for ferric sulfate and 30 mg/l for alum. Removals from spiked Ohio River water, for ferric sulfate (30 mg/l), were 20–30% at pH 7. That was increased to nearly 100% at pH 8.5–9.0. Removals by alum were not reproducible for pH values greater than 8.0. Low turbidities (1–20 TU) and the alkalinity (50–60 mg/l) of the riverwater were the apparent reason for poor removals of Cd by alum. Pilot-plant studies, in general, confirmed the coagulation results from the jar-tests. Also, jar-tests conducted for the effect of coagulant dosage showed that an increase in the quantity of alum effected greater removal. However, a slight increase in Cd removal was observed with ferric sulfate (Figure 25a). When the coagulant doses were held constant and [Cd^{2+}] was increased (Figure 25b), the percentage removal decreased, even at the optimum pH. All of this limits chemical coagulation as a viable treatment for soluble Cd con-

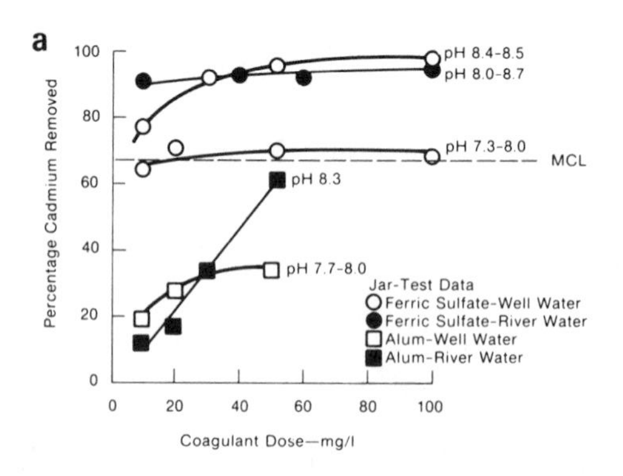

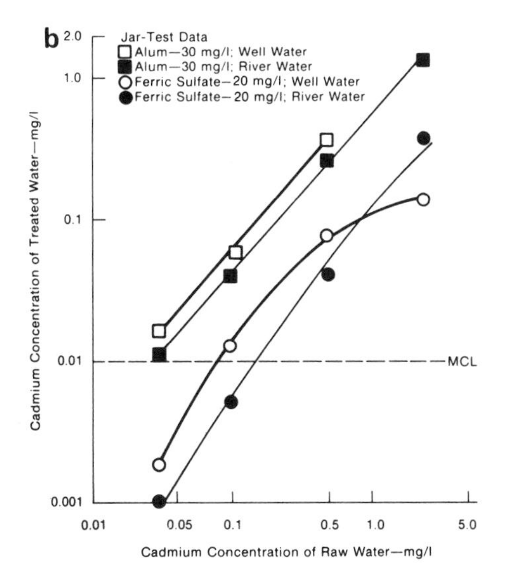

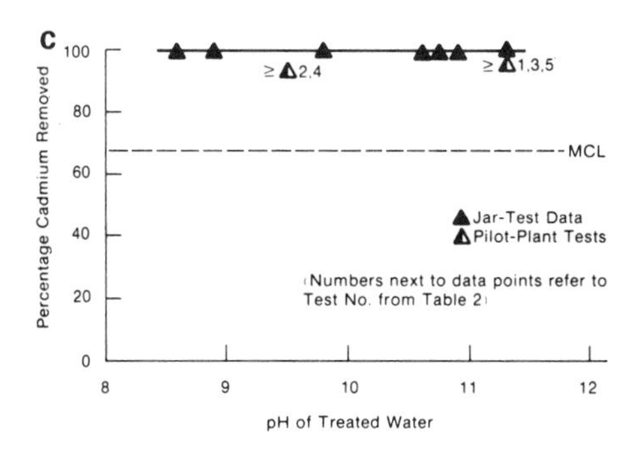

Figure 25. **(a)** Effect of coagulant dose on cadmium removal by coagulation treatment. Cadmium concentration = 0.03 mg/l. **(b)** Effect of initial concentration on cadmium removal by coagulation treatment. **(c)** Effect of pH on cadmium removal by lime softening. Cadmium concentration = 0.03 mg/l. Reproduced from Sorg et al. [60], courtesy of the American Water Works Association.

tents to 0.1 mg/l and below, according to Sorg et al. [60]. Furthermore, it is not feasible to treat clear groundwaters for specific removal of soluble constituents by coagulation. On the other hand, some Cd removal is, perhaps, incidental to the coagulation of turbid surface waters. Here again, specific coagulation of surface waters for Cd is not warranted.

Reaction 74 suggests that the lime-soda softening treatment of hard waters should also reduce [Cd^{2+}]. Here again, laboratory jar-tests demonstrated that the spiked Ohio wellwater was treated for complete Cd removal by the softening process (Figure 25c) [60]. No chemical dosages were given, so it is presumed that they were added in excess of stoichiometric quantities. Pilot-plant tests confirmed the results seen in Figure 25c. Furthermore, complete removal was observed for Cd contents up to 10. mg/l. Thus, lime-soda softening is a feasible process for Cd removal, provided, of course, it is applied to hard waters (see Chapter 7).

Ion-exchange resins should be able to remove Cd from ground- and surface waters if it is not overwhelmed by such other cations as Ca^{2+}, Mg^{2+}, Hg_2^{2+}, Pb^{2+} and Ba^{2+} [60]. Domestic and plating wastewaters have been treated successfully for Cd where its concentration is considerably higher than in slightly contaminated waters.

Some Cd removal is experienced with PAC. For example, two types of PAC gave 29 and 26% removal of 0.05 mg/l Cd at pH 7. at 100 mg/l doses. When the pH value was increased to 9., removals were increased to 53 and 56% [60]. This is a very unlikely pH value for PAC treatment of drinking water. As seen in Table XX, some Cd is removed by GAC, either virgin or spent. Here again, the percentage removals indicate that any Cd removal in a conventional water treatment plant using GAC would be incidental to this process.

Two PAC (Nuchar S-N and Nuchar S-A) and two GAC (Darco HD 3000 and Filtrasorb 400) were examined in the laboratory for CD removal from plating wastes [63]. Figure 26 shows the adsorptive capacity of the four carbons, 1. g/l, over the pH range of 3.–10. These data were obtained from batch experiments with a reaction time of 2 hr. It is obvious that PAC is more effective than GAC. That the [H^+] affects the adsorption of Cd is seen also. The convergence of the four curves to the maximum adsorption at pH 10. is explained by "substantial precipitation of $Cd(OH)_{2(s)}$ and/or $CdCO_{3(s)}$." Typical adsorption isotherms are seen in Figure 27 for two carbons and for Cd-BF_4 and $Cd(CN)_2$—components of plating solutions. Here again, the PAC was more effective. In this study, Langmuir adsorption isotherms and pH effects were also presented [63]. This study tends to confirm the data presented by Sorg et al. [60]. Cadmium adsorption by 10. g/l of three GAC (Nuchar C-190N, Nuchar 722 and Filtrasorb 400) were reported [64,65]. These studies were oriented toward treatment of plating wastes. This strongly suggests that GAC is not a practical drinking water treatment for Cd removal.

Cadmium sorption by the oxides of Mn(IV), Fe(III) and Al(III) was

Table XX. Results of Pilot Tests on Removal of Cadmium by Granular Activated Carbon[a]

Test No.	Length of Test Run (hr)	Average Raw Water Concentration (mg/l)[b]	Average Percentage Removed by Granular Activated Carbon Filters	
			GAC[c]	Exhausted GAC[d]
1	103	0.028	22	54
2	104	0.027	18	48
3	104	0.032	19	43
4	81	0.031	8	26
5	101	0.027	30	40
6	102	0.029	29	34
7	102	0.029	10	34
8	101	0.029	7	18

[a]Reproduced from Sorg et al. [60], courtesy of the American Water Works Association.
[b]Spiked Cincinnati tapwater.
[c]Filtrasorb 100, Calgon Corp., Pittsburgh, PA.
[d]Filtrasorb 200, Calgon, Corp., Pittsburgh, PA.

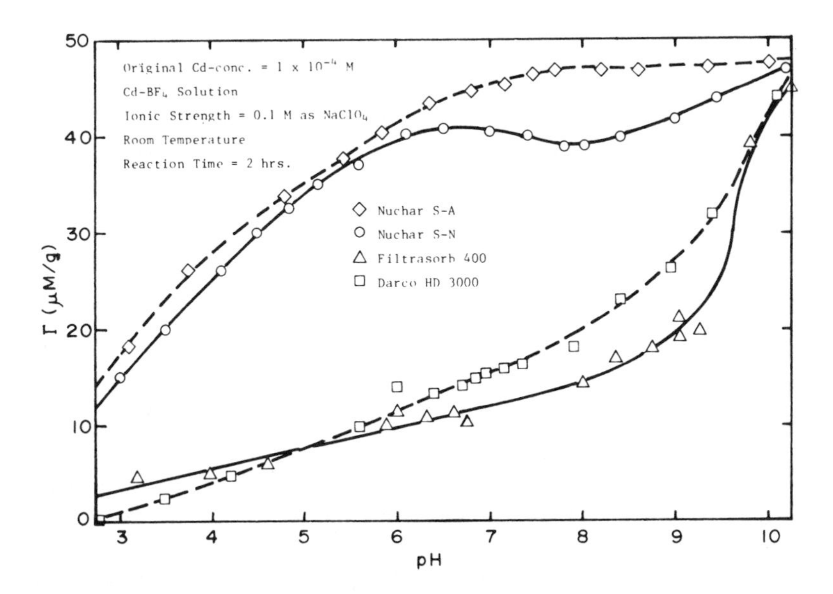

Figure 26. Comparison of Cd(II) adsorption capacity by GAC and PAC, as affected by pH. Reproduced from Huang and Smith [63], courtesy of Ann Arbor Science Publishers.

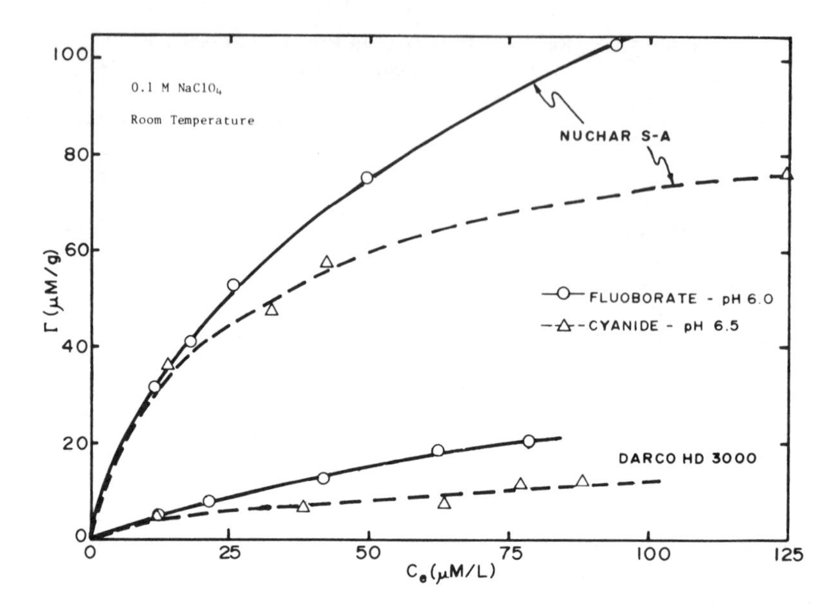

Figure 27. Typical adsorption isotherm—comparison of various systems. Removal efficiency is generally better with Cd-BF_4 than Cd-CN solution. Reproduced from Huang and Smith [63], courtesy of Ann Arbor Science Publishers.

reported [66]. It was indicated that sorption on hydrous oxides of Mn and Fe is feasible for removal of Cd from water and wastewater. Manganese dioxide was the most effective of the three, and sorption was influenced by $[H^+]$, with greater removal of Cd at pH 8.3 than at 5.

CHROMIUM

Aqueous Chemistry

The MCL for total chromium in drinking water is 0.05 mg/l [1]. Chromium concentrations in natural waters are limited by the low solubility of Cr(III) oxides. Chromium contents in 15 rivers of North America ranged from 0.7 μg/l (Sacramento River) to 84 μg/l (Mississippi River) [2]. There is the possibility of contamination from discharge of industrial wastewaters. The essential aqueous chemistry of Cr is its occurrence in either the +3 or +6 oxidation state. Cr(III) occurs as a cation, with $Cr(OH)_3$ being very insoluble in water:

$$Cr^{3+} + 3\,OH^- = Cr(OH)_{3(s)} \tag{76}$$

$$\Delta G^\circ_{REX} = -51.018 \text{ kcal/mol}$$

$$\log K_s = 37.4$$

At pH 7., the solubility of chromic hydroxide has been reported to be $10^{-14.3}$ *M* as Cr[41]. Cr(VI) occurs as anions in water:

$$H_2CrO_4 = \underset{\text{acid chromate ion}}{HCrO_4^-} + H^+ \tag{77}$$

$$\log Keq = -0.75$$

$$H_2CrO_4 = \underset{\text{chromate ion}}{CrO_4^{2-}} + 2\,H^+ \tag{78}$$

$$\log Keq = -7.20$$

$$2\,H_2CrO_4 = \underset{\text{dichromate ion}}{Cr_2O_7^{2-}} + H_2O + H^+ \tag{79}$$

$$\log Keq = -0.18$$

Removal

The effects of conventional water treatments have been evaluated for Cr removal, although Cr is not expected to be a serious drinking water quality problem [67]. It is not commonly found in ground- or surface waters. Any removal of Cr in a conventional water treatment plant, therefore, would be incidental. Figures 28a and 28b give the results of jar and pilot plant tests for chemical coagulation of Cr-spiked(0.15 mg/l) well- and riverwaters. Cr(III) undoubtedly was precipitated as its hydroxide rather than removed by coagulation per se. Coagulation by ferric sulfate and alum was ineffective for Cr(VI), as seen in Figure 28b. However, when ferrous sulfate was employed, nearly 100% Cr removal was achieved. Cr(VI) was reduced to Cr(III) by the Fe(II) ion, with subsequent precipitation as $Cr(OH)_{3(s)}$. However, $FeSO_4$ is not a commonly used coagulant or reductant for drinking water treatment. There is a disparity between the pH of optimum coagulation by $FeSO_4$ (8.–10.) and the pH for Cr(VI) reduction by Fe(II)(acid medium). It would not be very pragmatic to require two pH adjustments in a water treatment plant

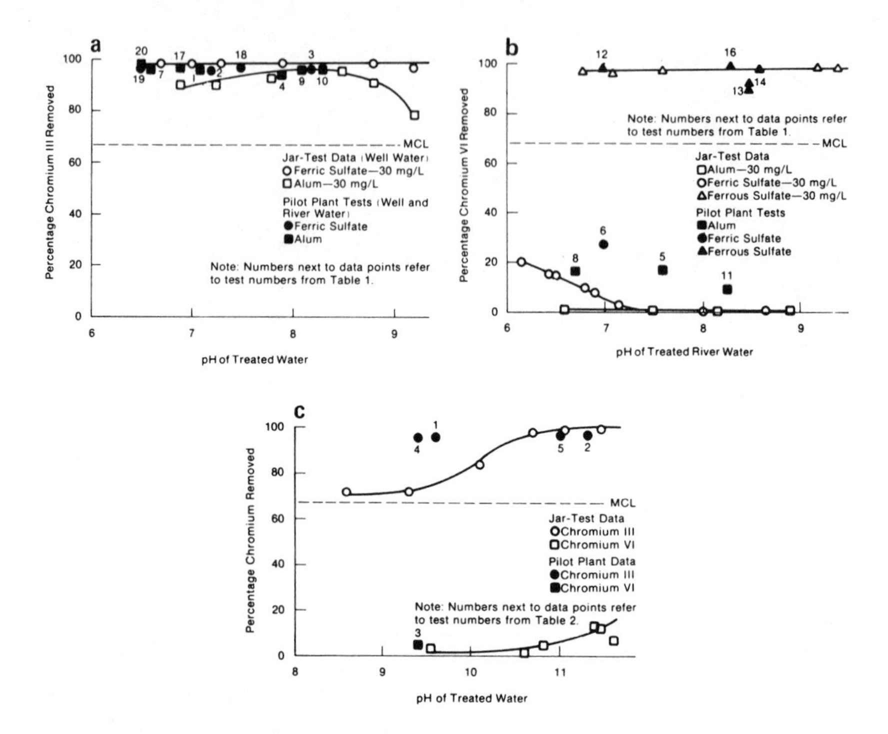

Figure 28. Effect of pH on **(a)** 0.15-mg/l Cr(III) removal by coagulation treatment; **(b)** 0.15-mg/l Cr(VI) removal by coagulation treatment; and **(c)** 0.15-mg/l Cr removal from wellwater by lime softening. Reproduced from Sorg [67], courtesy of the American Water Works Association.

specifically for Cr removal. That the lime-soda softening process is effective for Cr(III) removal, but ineffective for Cr(VI), is seen in Figure 28c.

Of the several studies of the removal of Cr by PAC and GAC [64, 67–70], some are oriented to the treatment of industrial wastewaters. For example, a 10-g/l batchwise quantity of a GAC (a calcined coke) was used for Cr(VI) removal over the pH range of 2.–10. [68]. Greatest removals occurred at pH 2., whereupon there was an exponential decrease of effectiveness to a pH value of about 6., where there was virtually no reduction in [Cr]. In another batch study, a 10,000-mg/l quantity of Filtrasorb 400 was utilized for Cr(VI) and Cr(III) removal [69]. This GAC responded somewhat differently to pH effects. Virtually no adsorption of Cr (either valence) occurred below pH 2. or above pH 8., with the maximum occurring at pH values 5.–6. Loosely packed columns of various quantities of Filtrasorb 400 were operated continuously at 2 gpm/ft^2 at pH 2.5 and 10^{-3} M Cr(VI) [70]. No Cr was removed within the first 50 or so bed volumes because of the number of H^+ ions required to hydrolyze the carbon surface. All of this work was directed to treating industrial wastewaters and has limited application to drinking waters. Sorg [67] reported that several jar-test studies with PAC (Filtrasorb 400) dosages of 10–200 mg/l showed very low removals, about 10% of 0.15 mg/l Cr(VI), from pH 6.8 to 8.3. It was concluded [67] "that GAC treatment probably is not a very effective method for Cr(VI) [removal] from drinking water with low organic concentrations." The same conclusion may be drawn for PAC. Sorg [67] cited some laboratory data which suggested that reverse osmosis was reasonably effective for removing both oxidation states of Cr.

COPPER

Aqueous Chemistry

The MCL for copper is 1.0 mg/l in drinking water, which is a secondary standard. Copper is not distributed widely in natural waters. In the famous study of 15 North American rivers [2], copper contents ranged from 0.83 to 105. μg/l. Discharge of industrial wastewaters may increase copper contents to greater than 1.0 mg/l. As cited in Chapter 9, copper ions may originate from the acidic corrosion of copper pipes. As high as 3.0 mg/l has been measured in a household wellwater [72]. In these cases, neutralization of the water with calcite beds will eliminate the corrosive condition and the copper ions. The aqueous chemistry of copper is covered in Chapter 9. Here, only the treatments of waters with copper contents arising from sources other than corrosion are considered.

Removal

Since the concern about copper in drinking water is secondary, there are few reports on its specific removal. There are studies of Cu(II) removal by six activated carbons from synthetic seawater, 1.65 *M* NaCl [64]. This is not applicable to treatment of freshwaters. Some research has been conducted on adsorption of Cu(II) on hydrous oxides of Mn and Fe. Copper(II) adsorption on $MnO_{2(s)}$ was observed over the pH range of 0.5–3.0, which is not very practical for water treatment [73]. The hydrous oxide of Fe(III) sorbed Cu(II) over the pH range 4.3–6.5 [74]. The removal of Cu(II) and other metals from water by silicon alloys

Mg_2Si, CaSiBa, MgFeSi and SiF was reported [75]. These alloys and metal exchangers were placed on a shaker for 30 min, presumably at room temperature. Table XXI shows Cu removal by ion exchange with the alloys at a pH value of 4.5. Parallel studies were conducted with GAC and a cation and an anion exchanger. It should be noted that the heavy metals were contained in a single solution, hence, competitive effects should be noted. The alum coagulation of Cu^{2+}-, Cd^{2+}- and Zn^{2+}-fulvic acid complexes was investigated [76]. The conditions were 0.5–1.0 mg/l Me^{2+}, 30–50 mg/l alum and a pH value of 7. A 96% removal of the Cu^{2+}- complex, 59% for Cd^{2+}- and 82% for Zn^{2+}- were effected with 30-mg/l alum.

LEAD

Aqueous Chemistry

The primary drinking water standard for lead is 0.05 mg/l [1]. Natural sources of lead include: $PbCO_{3(s)}$ (cerussite), $PbS_{(s)}$ (galena) and $PbSO_{4(s)}$ (anglesite). Lead contents in natural waters may range from <1.0 to 890. μg/l [2]. The principal problem with lead in drinking water comes from corrosion in water distribution systems [60] (see Chapter 9). Other sources of lead would come from weathering of lead paints, metallic products, industrial wastewaters, etc. [2]. The appropriate chemistry is (25°C):

$$Pb^{2+} + 2OH^- = \underset{\text{plumbous hydroxide}}{Pb(OH)_{2(s)}} \qquad (80)$$

$$\Delta G^\circ_{REX} = -20.182 \text{ kcal/mol}$$

$$\log K_s = 14.8$$

$$Pb^{2+} + CO_3^{2-} = \underset{\text{(cerussite)}}{PbCO_{3(s)}} \qquad (81)$$

$$\Delta G^\circ_{REX} = -18.83 \text{ kcal/mol}$$

$$\log K_s = 13.8$$

At a pH value of 10., Pb contents would be 33 μg/l from Reaction 80 and 26 μg/l from Reaction 81. These values are below the MCL value of 50 μg/l.

Table XXI. Heavy Metal Extraction by Exchange Materials[a]

Material[b]	Metals Remaining after Contact (all at 20 ppm to start) (ppm)									
	Cd	Cu	Hg	Zn	Na	Fe	Mg	Ca	Si	Ba
Activated Carbon (Darco-G 60)	19.7	7.4	0.4	19.9						
Cation Exchange Resin (Dowex HCR-W)	0.4	0.4	19.3	0.5	pH = 2.0					
Anion Exchange Resin (R-H IRA-93)	19.4	14.5	0.8	18.5	77.1					
Mg_2Si	0.3	0.3	0.1	0.4			38.0		1.6	
CaSiBa	<0.1	<0.1	0.1	<0.05				10.0	45.0	129.0
MgFeSi	3.1	0.5	1.3	0.6		0.1	27.0		0.3	
SiFe	16.8	2.3	0.4	6.9		36.2			0.5	

[a]Reproduced from McKaveney et al. [75], courtesy of the American Chemical Society.
[b]0.50 g of material in contact with 100-ml solution for 30 min.

Removal

Since lead is not a widespread problem in drinking water, very few papers have been published for the specific treatment of Pb removal. Conventional water treatments were tested for Pb removal through laboratory jar-tests and by a pilot plant [60]. That alum and ferric sulfate coagulation were effective over the pH range 6.–10. is seen in Figure 29a. Turbidity affects the removal wherein Pb is adsorbed by particulate matter that is eventually coagulated. An 80% removal of Pb, 0.15 mg/l, was observed with a 48-TU water, whereas a 20–60% reduction was observed with a 9.5-TU water. Since no coagulant was added in these two experiments Pb removals were obtained by adsorption on the suspended particles with subsequent precipitation. The lime softening process was extremely effective for Pb removal via Reaction 80 (Figure 29b). Sorg

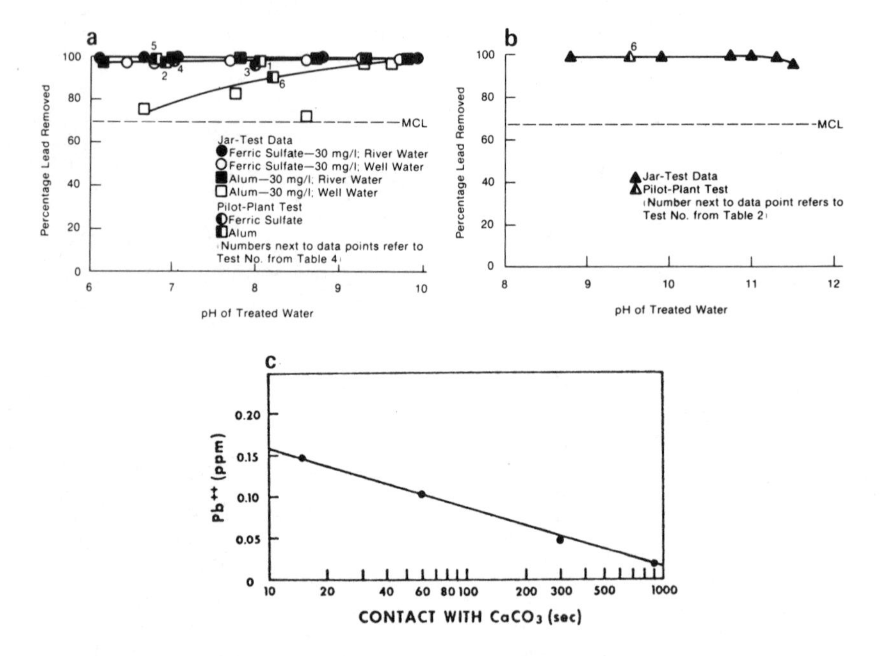

Figure 29. **(a)** Effect of pH on lead removal by coagulation treatment. Lead concentration = 0.15 mg/l. **(b)** Effect of pH on lead removal by lime softening. Lead concentration = 0.15 mg/l. Reproduced from Sorg et al. [60], courtesy of the American Water Works Association. **(c)** Residual lead in solution after treatment with $CaCO_3$ as a function of $CaCO_3$ contact time before filtration. Reproduced from Hautala et al. [78], courtesy of the International Association on Water Pollution Research.

[60] speculated on the use of a strong acid cation exchange process for Pb removal from groundwaters. The selectivity series for cations indicates that this process should be effective [77]: $Ba^{2+} > Pb^{2+} > Sr^{2+} > Ca^{2+} > Ni^{2+} > Cd^{2+} > Cu^{2+} > Co^{2+} > Zn^{2+} > Mg^{2+}$ and $Ag^{+} > Cs^{+} > Rb^{+} > K^{+} > NH_4^{+} > Na^{+} > H^{+} > Li^{+}$. Few studies have been conducted with the removal of Pb by PAC or GAC. In laboratory jar-tests, it was reported that 10 mg/l PAC effected an 89% removal of 0.15 mg/l Pb, and 200 mg/l PAC effected a 98% reduction at pH 7.3–7.4 [60]. In a parallel study using GAC (Filtrasorb 200) in a pilot plant, Pb removals were 95% and greater from 0.11 to 0.20 mg/l Pb in the raw water. The use of powdered $CaCO_{3(s)}$ (0.1% w/w solids to solution) for Pb removal from dilute wastewaters was investigated [78]. Figure 29c shows the residual $[Pb^{2+}]$ as a function of contact time in a batch process. A 10-mg/l Pb^{2+} solution was made basic to pH 8., and was filtered to yield 2.0 mg/l Pb in the filtrates. In turn, $CaCO_{3(s)}$ was added to the filtrate from which the data for Figure 29c was obtained. Swallow et al. [74] reported that the hydrous oxide of Fe(III) sorbed Pb(II) over the pH range of 4.3–7.0.

MERCURY

Aqueous Chemistry

The MCL for mercury in drinking water is 0.002 mg/l [1]. There are very little historical data on the occurrence of Hg in ground-, surface and finished waters in the United States. This is due mainly to the cumbersome analytical techniques for Hg in water. Mercury contents of <0.5–$6.8\ \mu g/l$ were reported for several surface waters in the United States [2]. Of the several mercury-bearing minerals in nature, only a few exist abundantly. The most common are the sulfides [cinnabar and metacinnabar (HgS)] and native mercury. In aqueous systems, mercury can exist in one of three oxidation states: as the free metal, Hg^0, as the mercurous ion, Hg_2^{2+} or as the mercuric ion, Hg^{2+}. Mercury readily forms complexes with many inorganic anions and organic constituents for which the methylation reactions have been documented [2]. An important characeristic of Hg is its tendency to adsorb and adhere to various types of surfaces. This, of course, has immediate application to water treatment.

Removal

Conventional water treatment processes have been evaluated for Hg removal [67,79,80]. The coagulation work [67] is seen in Figure 30. The

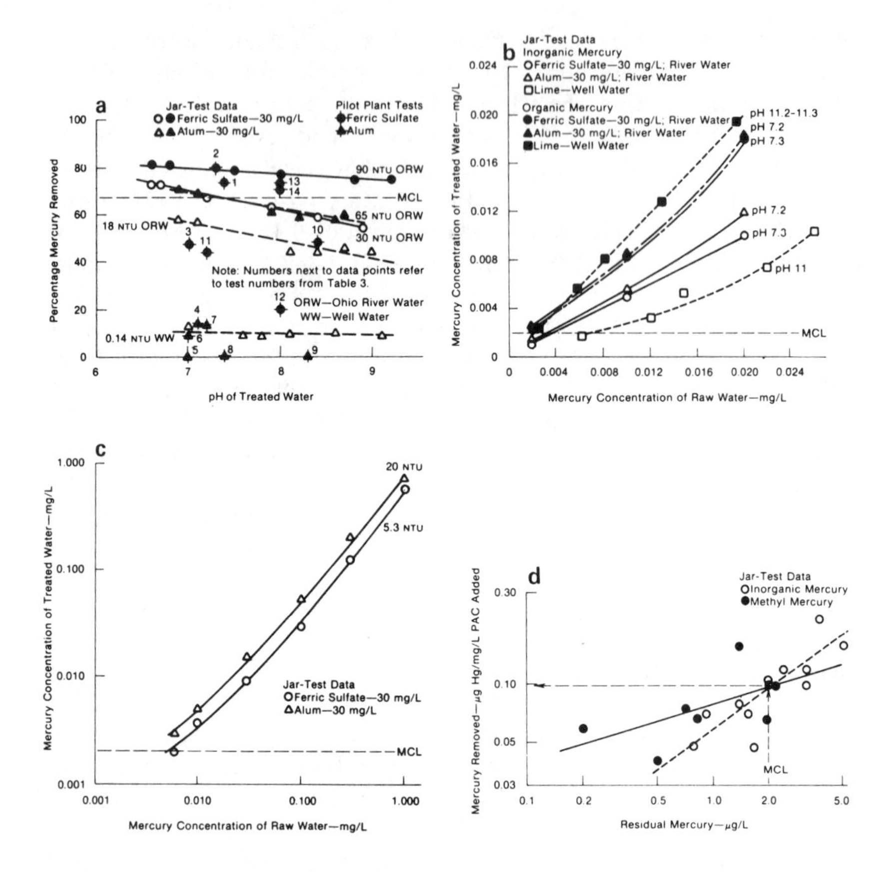

Figure 30. (a) Effect of pH and turbidity on inorganic mercury (0.006 mg/l) removal by coagulation treatment. (b) Effect of initial concentration on mercury removal by coagulation and lime softening treatment. (c) Effect of initial concentration on inorganic mercury removal from riverwater by conventional coagulation. (d) Freundlich isotherms for mercury adsorption with PAC. Reproduced from Sorg [67] and Logsdon and Symons [79], courtesy of the American Water Works Association.

$[H^+]$ apparently exerts little, if any, influence on Hg removal over the pH range of 6.5 to 9. (Figure 30a). That is, either there was not very much adsorption of Hg on the oxides of Al and Fe or there was no coprecipitation of mercury. Only when turbidity was present at the 90-NTU (nephelometric turbidity units) level was there an 80% removal of Hg. This observation was at variance with the 100% removal at pH 8. reported in another study [80]. It can be seen also that, for a given level of turbidity, ferric sulfate was more effective than alum. This is also at variance with the results of Logsdon and Symons [79], who observed a 60% Hg reduction in a coagulation system of turbidity, 100 NTU and alum. To summarize, the MCL for Hg is reached only if the initial concentration does not exceed 0.003 mg/l for alum and about 0.006 mg/l for iron (Figure 30) [67]. That organic mercury is not removed at all by coagulation is seen in Figure 30b. Lime softening is more effective than coagulation for a spiked wellwater treated to a pH of 11, as seen in Figure 30b. However, the $[Hg]_t$ should not exceed 0.006–0.007 mg/l in order to achieve the MCL for Hg by lime softening. No organic Hg is removed in this treatment.

Several investigations have been conducted on the utilization of PAC and GAC for Hg removal. Figure 30d shows the log-log plot of the adsorption data for inorganic Hg and CH_3Hg^+ [79]. These data indicate that each mg/l of PAC (Nuchar Aqua A) removes 0.0001 mg of inorganic Hg and CH_3Hg^+. Pilot-plant studies with PAC (Darco-B and Darco-M) and GAC (Hydrodarco 1030 and Filtrasorb 100) were conducted [67]. When PAC was used in conjunction with ferric sulfate, Hg removals were greater than by either treatment alone. Cumulative removals for inorganic mercury by GAC filtration ranged from 49 to 97%, whereas organic Hg removals ranged from 80 to nearly 100% [67]. Exhausted GAC was as effective as fresh GAC. Figure 31a shows the effect of pH on the removal of inorganic mercury by PAC (Darco HDB) [81]. Figure 31b shows the adsorption of CH_3Hg^+ by three different types of PAC: PAC-1 lignite base (Darco HDB), PAC-2, petroleum base (Amoco PX-21) and PAC-3, bituminous base (Nuchar Aqua) [81]. With the exception of PAC-3, methyl mercury removals were decreased as the pH value was increased from 7 to 9. PAC-1 apparently was the most effective of the three powdered carbons. At pH 7., the MCL for Hg was reached by 80 and 100 mg/l PAC-1 [81]. Batch and column tests were conducted for Hg removal by PAC (Filtrasorb 300, 270 mesh) and GAC (Filtrasorb 300, 8×30 mesh) [82]. Typical log-log plots of the adsorption data from batch tests for pH values of 4. and 10. are seen in Figure 32. Experimental conditions were: 2.0 g carbon, 200 ml of 10.0-mg/l Hg(II), and 6 hr of agitation at 20°C. Greater adsorption occurred at the lower pH value of 4. For example, at 1 mg/l Hg, the quantity adsorbed was 0.14 mg Hg/g carbon at pH 10., whereas the quantity at pH 4. was 1 mg Hg/g. Breakthrough curves for the carbon columns are seen in Figure 33 for pH values 4. and 10. For the latter system, Hg broke through within the first day, whereas no breakthrough occurred within 5 days at pH 4.

The coprecipitation of Hg(II) with iron(III) hydroxide over the pH range of 4.–12. was observed [83]. At pH 8., a 95% removal of 100-mg/l Hg was effected by 388 mg/l Fe. Formation and coprecipitation of $Hg(OH)_{2(s)}$ was the apparent mechanism for the removal.

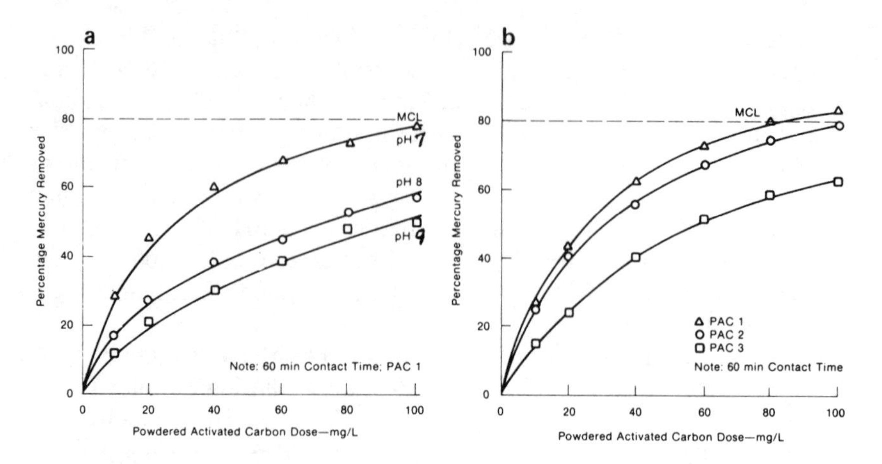

Figure 31. **(a)** Effect of pH on removal of inorganic mercury (0.010 mg/l) with PAC treatment. **(b)** Removal of methyl mercury (0.010 mg/l as Hg) by PAC at pH 7. Reproduced from Thiem et al. [81] and Sorg [67], courtesy of the American Water Works Association.

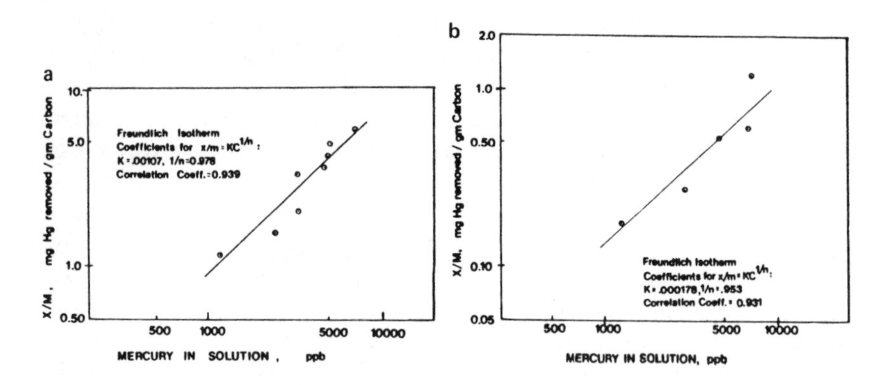

Figure 32. Adsorption isotherm for Hg(III) and PAC at 20°C and **(a)** pH 4 and **(b)** pH 10. Reproduced from Humenick and Schnoor [82], courtesy of the American Society of Civil Engineers.

SELENIUM

Aqueous Chemistry

The MCL for selenium in drinking water is 0.01 mg/l. Sulfides or native sulfur deposits often contain Se in significant amounts because of the chemical similarity between S and Se. Since Se is not evenly distrib-

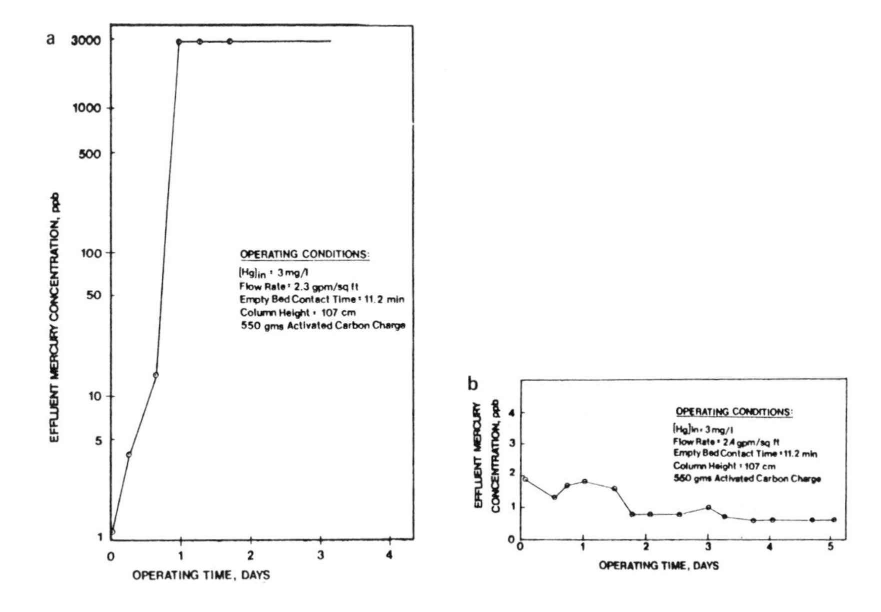

Figure 33. Carbon-only system: column run at 25°C and **(a)** pH 10 and **(b)** pH 4. Reproduced from Humenick and Schnoor [82], courtesy of the American Society of Civil Engineers.

uted in geologic materials, its occurrence in natural waters is usually less than 1. μg/l. Sorg and Logsdon [84] noted a survey of 139 groundwaters in Nebraska where 40.3% exceeded the MCL, and the two highest contents were 0.48 and 0.103 mg/l. Se exists in four oxidation states in aqueous systems: −2, 0, +4 and +6. The reduced state, −2, is represented by the selenide species, and the elemental state, Se(0), exists in several allotropic forms, whereas the two higher oxidation states are the selenites, +4, and selenates, +6.

Removal

Very little work has been reported on the specific removal of Se in potable water supplies. Figure 34a shows the attempts to remove Se(IV), 0.03 mg/l, from wellwater and the Ohio River by chemical coagulation [84]. Ferric sulfate appears to be the better coagulant, with greatest removals occurring in the 6.–7. range of pH values. However, opposite results were obtained when the effect of coagulant dose on Se(IV), 0.1 mg/l, was studied (Figure 34b) [84]. Insignificant removals of Se(VI)

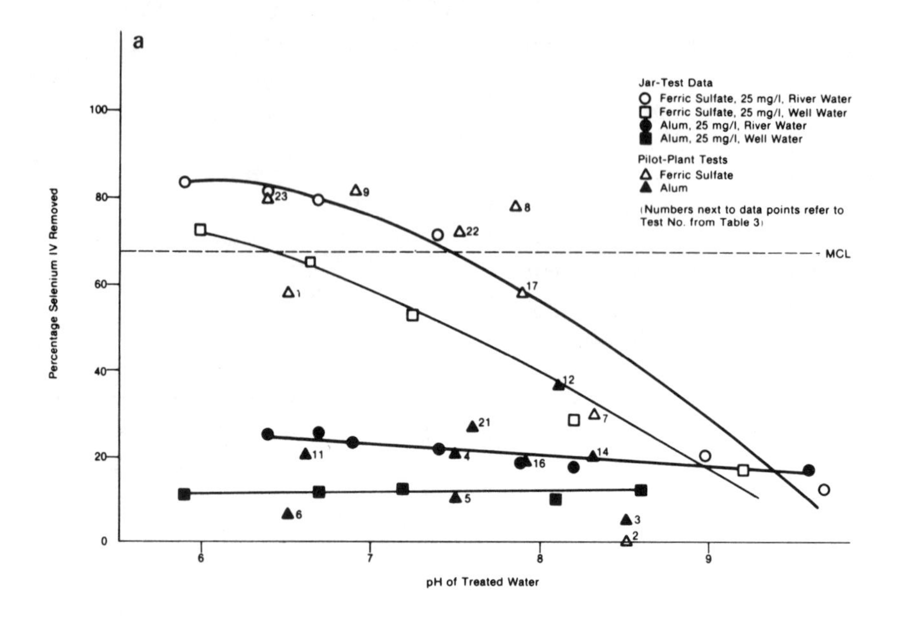

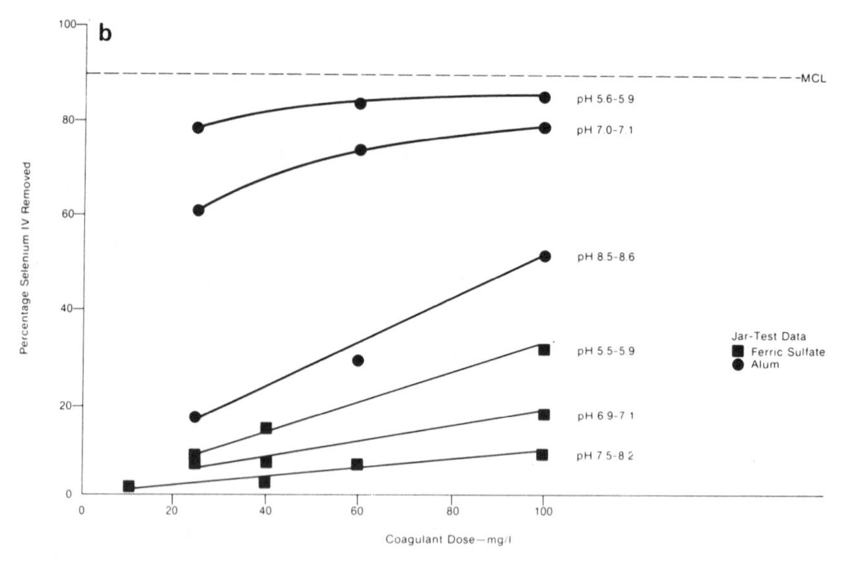

Figure 34. **(a)** Effect of pH on Se(IV) (0.03 mg/l) removal by coagulation treatment. **(b)** Effect of coagulant dose on Se(IV) (0.1 mg/l) removal from wellwater. Reproduced from Sorg and Logsdon [84], courtesy of the American Water Works Association.

were effected by the two coagulants in jar and pilot-plant tests. Lime softening was not a very effective treatment for either Se(IV) or Se(VI) [84]. The maximum removal (50% at pH 11.5) was obtained with a hard wellwater spiked with 0.1 mg/l Se(IV).

Laboratory tests with alumina columns as an adsorber or ion exchanger have been reasonably efficient with a removal of 95% of 0.3 mg/l of Se(IV) and Se(VI) [84,85]. There are, however, interferences by other anions in water. For example, the order of preference of alumina for anions is: $OH^- > H_2PO_4^- > F^- > H_2AsO_4^- > HSeO_3^-$.

Several experiments have been conducted with PAC removal of Se(IV) and Se(VI) [84]. The ineffectiveness of this treatment is observed with removals less than 4% with PAC dosages up to 100 mg/l. Nor did GAC work. Some limited experiments with reverse osmosis (RO) indicated 97% removals of Se(IV) and Se(VI) in tapwater spiked with 0.1 mg/l. It would appear, therefore, that only a strong anion exchanger and RO are effective for the specific removal of the two Se species from water [84].

That arsenic can be removed from hard water by lime softening is seen in Figure 35 [84]. The $[H^+]$ obviously is a factor with 100% removal of

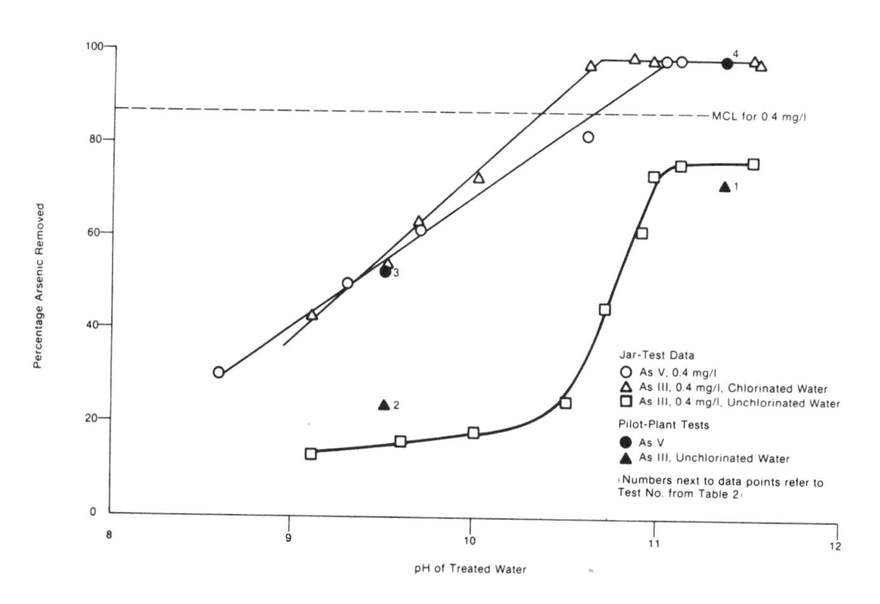

Figure 35. Effect of pH on arsenic removal by lime softening. Reproduced from Sorg and Logsdon [84], courtesy of the American Water Works Association.

As(V) occurring at pH values greater than 10.5. The formation of calcium arsenate may be responsible for this removal:

$$3\,Ca^{2+} + 2\,AsO_4^{3-} = Ca_3(AsO_4)_{2(s)} \tag{82}$$

$$\Delta G^{\circ}_{REX} = -24.78 \text{ kcal/mol}$$

$$\log K_s = 18.17$$

As(III) removal was not as effective as As(V) with the lime treatment. Figure 35 also shows that As(III) and As(V) removals are nearly the same in pilot-plant tests as with the laboratory jar-tests.

Table XXII summarizes pilot-plant tests designed to evaluate the effectiveness of coagulation and filtration through dual media (anthracite and sand) and two types of GAC (Filtrasorb 200 and Hydrodarco). Very little As(III) and As(V) were removed. Removals in excess of 90% were obtained with ferric sulfate coagulation and subsequent filtration. There was little difference between the dual-filter media and the GAC for As removal. These tests also indicate that GAC, virgin and exhausted, has an adsorptive capability for As which has been confirmed [58].

SODIUM

Concern

Sodium is, of course, a constituent in all natural freshwaters. Its content varies with geology and possible contamination by wastewaters [2]. Sodium compounds and exchange resins are utilized extensively in drinking water treatment (Table XXIII). Concern has been expressed frequently about the health effects of sodium [1]. There is some indirect evidence linking excessive sodium intake to hypertension. For most people, the contribution of drinking water to sodium intake is small relative to total dietary intake. According to the report from the National Academy of Sciences [86], "it appears that at least 40% of the total population would benefit if total sodium ion intake were maintained [at] not greater than 2,000 mg/day.... [W]ith sodium ion concentration in the water supply of not more than 100 mg/l, the contribution of water to the desired total intake of sodium would be 10% or less for a daily consumption of two liters."

Table XXII. Results of Conventional Coagulation Pilot-Plant Tests for Arsenic Removal: Ohio River Water[a]

Coagulant			Arsenic Contaminant		Percentage Arsenic Removal				
						Filtered Water			
							Granular Activated Carbon Filters		
Type	Dose (mg/l)	pH of Treated Water	Form	Raw Water Concentration (mg/l)	Settled Water	Dual-Media Filter	Virgin GAC	Exhausted GAC	Virgin GAC
Alum	24	8.9	As(V)	0.37	14	14	11		
Ferric sulfate	25	6.7	As(V)	0.39	81	96	98		
Ferric sulfate	26	8.8	As(V)	0.33	46	62	61		
Ferric sulfate	29	7.0	As(III)	0.12	36	64	90	91	
Ferric sulfate	28	8.3	As(III)	0.12	61	82	80	84	
Ferric sulfate	30	8.0	As(V)	0.26	80	95	97	95	
Alum	29	7.0	As(III)	0.16	14	21		35	44

[a]Reproduced from Sorg and Logsdon [84], courtesy of the American Water Works Association.

Table XXIII. Sodium Addition During Water Treatment [86]

Constituent			Typical Range of Resulting Sodium Concentration Addition
Name	Formula	Common Usage	(mg/l)
Sodium Exchange Resins	NaR	Softening	70–340
Sodium Carbonate	Na_2CO_3	Softening	30–200
Sodium Hypochlorite	NaClO	Disinfection	2–10
Sodium Carbonate	Na_2CO_3	pH adjustment	5–18
Sodium Hydroxide	NaOH	pH adjustment	2–9
Sodium Fluoride	NaF	Fluoridation	0.8–1.5
Sodium Fluosilicate	Na_2SiF_6	Fluoridation	0.3–0.5
Sodium Aluminate	$NaAlO_2$	Coagulant aid	1–15
Sodium Silicate	Na_4SiO_4	Coagulant aid/ corrosion control	0.54–8.4
Sodium Hexametaphosphate	$(NaPO_3)_6$	Sequestering agent	0.2–0.7
Bimetallic Glassy Phosphates	Na—Zn—PO_4	Corrosion control	0.05–1.4
Sodium Glassy Phosphates	Na—PO_4	Corrosion control	0.6–4.0

Removal

The sodium content of drinking water can be controlled through reducing or eliminating its use in the various types of municipal and domestic water treatments. A study of this problem led to these conclusions [87]:

> Utilities that use ion exchange softening could switch from sodium exchange to hydrogen exchange followed by pH adjustment, if necessary, with a nonsodium compound.
>
> Private home owners who have ion exchange softeners could change the plumbing in their homes so drinking water is not softened. Changing home softeners to an acid cycle is not advisable.
>
> Utilities that use lime-soda ash softening could reduce or eliminate the addition of soda ash (Na_2CO_3) used to removed NCH. A residual hardness of 125 mg/l is usually not objectionable to water customers, and most of this could be left as NCH.
>
> The use of sodium-containing compounds for pH adjustment, coagulation, and disinfection can easily be minimized. Many compounds that do not contain sodium are readily available to perform these functions.
>
> The addition of sodium to drinking water from compounds added for fluoridation and corrosion control is usually insignificant (<2

mg/l) because only a small quantity of the compound is required to perform these functions.

Specific treatment methods for the removal of sodium from drinking water do not exist. Existing methods for desalination of seawater and treatment of brackish waters are used for the reduction of total dissolved solids (TDS), including sodium; for example, distillation is being used for the desalination of seawater, and reverse osmosis is being used to treat brackish and seawaters with TDS concentrations ranging from 500 to 36,000 mg/l. Ion exchange is being used to deionize water for laboratory and industrial use. A thermally regenerable ion exchange process is producing favorable results on water with a TDS concentration of 500 mg/l. One manufacturer has an ion exchange process for treating brackish water with TDS concentrations from 500 to 5000 mg/l. Another process, electrodialysis, has a TDS removal application range for concentrations from 1000 to 5000 mg/l. Blending is also a feasible approach for the reduction of TDS and sodium in drinking water if an additional source of water lower in sodium is available.

SILVER

Aqueous Chemistry

The interim drinking water standard for silver is 0.05 mg/l [1]. Silver is an infrequent constituent of natural waters with a content in 15 major North American rivers ranging from 0 to 1.0 μg/l [2]. It has been reported that less than 7% of 1577 samples from their rivers and lakes survey had detectable amounts of silver, with the maximum value at 0.038 mg/l and a mean value of 0.0026 mg/l [88]. Consequently, silver should not be a problem in either ground- or surface waters unless there is a wastewater source.

Since silver is a very noble metal, its domain of stability covers a large portion of a pE–pH diagram [41]. The reduction reaction for metallic silver is:

$$Ag^+ + e^- = Ag_{(s)}; \quad V^0 = +0.799 \text{ V} \qquad (83)$$

Consequently, the redox value in water must exceed +0.799 V before Ag ions form from the corrosion of silver. Under these conditions, argentous oxide (Ag_2O) can form:

$$2\,Ag^+ + H_2O = Ag_2O + 2\,H^+, \text{ log Keq} = -6.33 \qquad (84)$$

The solubility and pH value of a saturated solution of Ag_2O are 16.2 mg/l as Ag_2O and 10.15, respectively, at 25°C [41].

Removal

Data are scarce for the specific removal of silver from water. Figures 36a and b illustrate the removal of silver from water by coagulation, while Figure 36b shows the use of PAC and 36c shows the results of lime softening [60]. Natural turbidity was a factor in the coagulation removal of silver from Ohio River water. Ferric sulfate was slightly more effective than alum. While pH is a factor with alum, it is not for ferric sulfate. Very little additional silver is removed when either coagulant dosage exceeds 30 mg/l. In order to reach the MCL for silver from 0.15 mg/l, 100 mg/l PAC was required (Figure 36b). Lime softening was moderately effective requiring pH values in excess of 9.0. Since no lime dosages were given, it is presumed that treatment to a specified pH value was the operational mode.

ZINC

Aqueous Chemistry

The secondary interim drinking water standard for zinc is 5.0 mg/l. The concern over zinc in potable waters lies with the milky turbidity imparted by the relatively insoluble $Zn(OH)_{2(s)}$. In the citation for the secondary regulations [89], it is stated: "Zinc, like copper, is an essential and beneficial element in human metabolism. Zinc can also impart an undesirable taste to water. At higher concentrations, zinc salts impart a milky appearance to water." The precipitation reaction is [90]:

$$Zn^{2+} + 2OH^- = Zn(OH)_{2(s)} \tag{85}$$

$$\Delta G^\circ_{REX} = -23.08 \text{ kcal/mol}$$

$$\log Keq = 16.92$$

At a pH value of 8.0, dissolution of $Zn(OH)_{2(s)}$ would yield 0.78 mg/l Zn^{2+} ions. Additional chemistry of zinc is found in Chapter 9 where the occurrence of this metal results from corrosion of water distribution pipes. In the survey of 1577 surface waters, Zn contents ranged 2–1200 μg/l, and was detected in 77% of these samples [88]. In drinking water samples, the content range was 3–2000 μg/l.

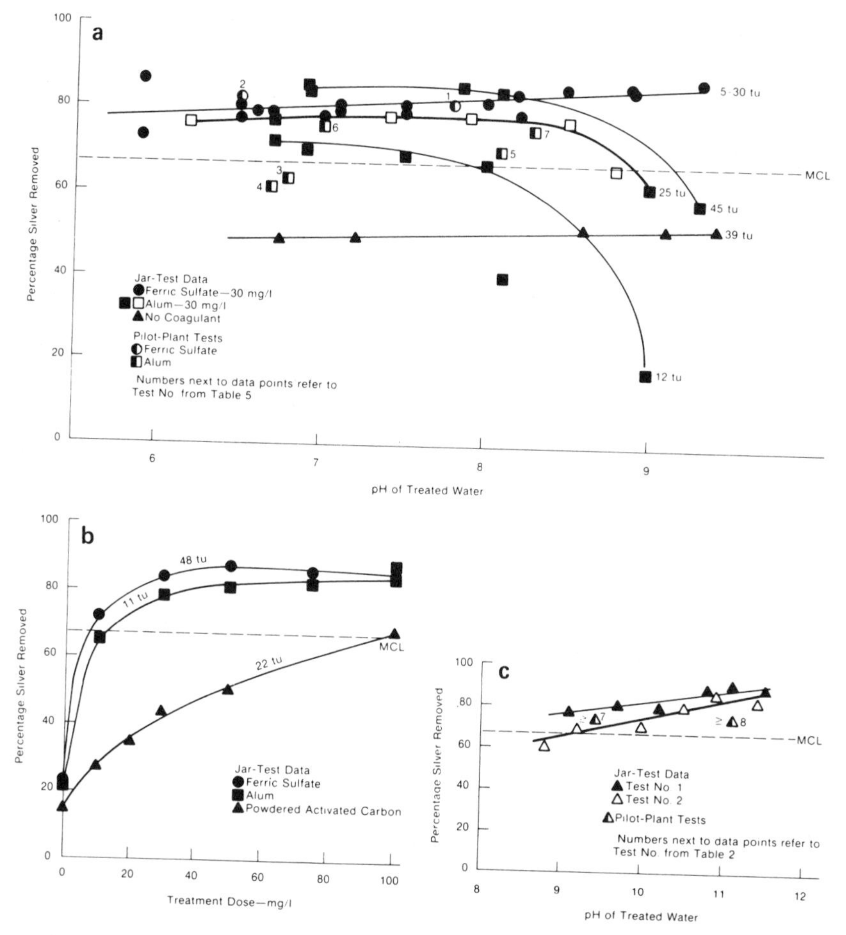

Figure 36. **(a)** Effect of pH and turbidity on silver removal from riverwater by coagulation treatment. Silver concentration = 0.15 mg/l **(b)** Effect of dose on silver removal from riverwater with ferric sulfate, alum and PAC. Silver concentration = 0.15 mg/l. **(c)** Effect of pH on silver removal by lime softening. Silver concentration = 0.15 mg/l. Reproduced from Sorg et al. [60], courtesy of the American Water Works Association.

Removal

Since zinc is a secondary drinking water standard and its content in natural waters rarely exceeds the MCL of 5.0 mg/l, there have been few studies of its specific removal from water. Zinc undoubtedly is effectively removed by conventional treatments that employ pH adjustment into the alkaline range where $Zn(OH)_{2(s)}$ would precipitate. In situations where it is necessary to remove more than one metal, a "collective" treatment may be employed. Some of these treatments are cited below. It should be noted also that since the chemistries of Zn and Cd are similar, it would be expected that Zn would react toward treatments with carbon, for example, in a similar manner.

FLUORIDE

Concern

The MCL for F^- in drinking water are given in Table I, Chapter 1. It is considered an essential nutrient in human metabolism [91]. Fluoride is added to many drinking waters in small quantities to prevent dental caries. Above the optimum value for this prevention, the mottling of teeth, i.e., dental fluorosis, occurs. The intake of excessively large amounts of fluoride over prolonged periods of time may produce other health effects: (1) bone changes, (2) crippling fluorosis and (3) death from a single dose of 2250–4500 mg [91]. In most natural waters, fluoride contents range from μg/l to 1–2 mg/l. It was discovered that natural groundwaters with about 1.0 mg/l prevented dental caries. There are occasional natural waters with unusually high $[F^-]$, for example, 32 mg/l in southeastern Arizona and 67 mg/l in the Union of South Africa [2].

Removal

The conventional water treatments of coagulation and lime softening require such excessive quantities of chemicals that it is impractical to use them [91]. The most frequently used treatment for F^- removal is an ion-exchange process with media such as activated alumina (see Figure 23 and Rubel and Williams [57]), bone char or granular tricalcium phosphate [91]. The latter two were used in the 1930–1950 period, but were abandoned for one reason or another. Activated alumina (Al_2O_3) has been used successfully as a defluoridation medium in Bartlett, Texas, from 1952 to 1977. There are several advantages of using alumina in downflow columns: (1) it is somewhat specific for F^- and has a relatively high exchange capacity for this ion; (2) this capacity is not affected by the SO_4^{2-} or Cl^- contents of water (the order of decreasing preference for anions exchange on alumina is: OH^-, PO_4^{3-}, $Cr_2O_7^{2-}$, F^-, SO_3^{2-}, CrO_4^{2-}, NO_2^-, Cl^-, NO_3^-, MnO_4^-, and SO_4^{2-} [92]); and (3) it has a relatively low cost—about 10% of the cost of a synthetic anion resin (1978 prices). After the alumina bed is exhausted, it is regenerated with 1% NaOH, rinsed with a dilute acid, usually 0.05 *N* H_2SO_4, and a water rinse. The operational aspects of a 1.5-mgd defluoridation plant at Desert Center, California, were reported [91] where fluoride contents are lowered to less than 1.0 mg/l from 8.0 mg/l.

Strong anion-exchange resins are not usually considered for fluoride removal because of their low capacity (88 g/m^3 for Amberlite XE-75) and relatively high cost. There is also an extensive competition by other anions for the exchange sites. Fluoride is usually listed last in the selectivity series of preference of anions.

Four activated carbons, three activated bauxites and two activated aluminas were investigated for defluoridation of distilled water, seawater and simulated geothermal water [93]. All systems were continuously mixed batches with 25 g/l of the adsorbents at 25°C. Rate studies indicated that adsorption was completed within 12 hr. The researchers, however, selected a 48-hr contact time. Table XXIV shows a summary of pH studies in which percentage removals are reported. The initial [F^-] was 20 mg/l. Adsorbent 1 was Hydrodarco 3000, adsorbent 2: Filtrasorb 300; adsorbent 6: Alcoa F-1, and adsorbent 9: Milwhite. Optimum pH values in the acid range are expected for the alumina and bauxite because of the competition of OH^- and F^- for the positively charged sites. It appears that these two adsorbents can, at least, remove F^- to the MCL from a 20-mg/l content and are, obviously, superior to the two GAC.

COLLECTIVE TREATMENTS FOR TRACE METALS

Many of the studies cited above were specific treatments for an individual constituent. On occasion, however, several contaminants may coexist in the raw water supply, in which case a "collective" treatment is needed—a single treatment applied for the collective removal of two or more elements whose chemistries are similar. Several studies cited below have been oriented to wastewater treatment, but can be applied to potable water treatment.

Carbonate and Hydroxide Precipitation

An extensive theoretical and experimental study of the precipitation and solubilities of the carbonates and hydroxides of Zn, Cd and Pb has

Table XXIV. Optimum pH_{48}[a] for the Maximum Removal of Fluoride[b]

Background Solution	Activated Carbon (Adsorbent 1) Optimum pH_{48}	Activated Carbon (Adsorbent 1) Percentage Maximum Removal Efficiency	Activated Carbon (Adsorbent 2) Optimum pH_{48}	Activated Carbon (Adsorbent 2) Percentage Maximum Removal Efficiency	Activated Alumina (Adsorbent 6) Optimum pH_{48}	Activated Alumina (Adsorbent 6) Percentage Maximum Removal Efficiency	Activated Bauxite (Adsorbent 9) Optimum pH_{48}	Activated Bauxite (Adsorbent 9) Percentage Maximum Removal Efficiency
DDW[c]	6.2	84	6.0	45	4.9–8.0	95	5.5–6.8	95
Dil. (1:3) SW[d]	6.2	72	6.0	24	4.9–8.0	95	5.5–6.8	95
Dil. (1:3) SGW[e]	6.2	72	5.9	24	4.9–8.0	95	5.5–6.8	95

[a]pH_{48} is the pH after 48 hr of contact between adsorbent and solution.
[b]Reproduced from Choi and Chen [93], courtesy of the American Water Works Association.
[c]Deionized-distilled water.
[d]Fourfold dilution of seawater.
[e]Fourfold dilution of simulated geothermal water.

been reported [94]. Nickel was included in the study, but is deleted here because it has neither a primary nor a secondary drinking water standard. The generalized solubility reactions for the metals in Table XXV are:

$$K_{s_{OH}} = [Me^{2+}][OH^-]^2 \tag{86}$$

$$K_{s_C} = [Me^{2+}][CO_3^{2-}] \tag{87}$$

The $[H^+]$ affects the solubilities of the metal hydroxides and carbonates, the details of which are not given here. See Faust and Aly [2] and Patterson [94] for solubility diagrams. This study was designed to prepare fresh precipitates (after 4 hr) of the hydroxides and carbonates in individual systems and compare a metal's solubility against the theoretical values calculated from thermodynamic data. There was an experiment in which Zn, Cd and Pb were combined into one solution and precipitated. Tables XXVI and XXVII summarize the appropriate data for drinking water treatment of Zn, Cd and Pb. It appears that the optimum treatment for zinc is precipitation as its hydroxide to pH 9.5, where the 0.25-mg/l residual is considerably less than the MCL of 5.0 mg/l. Cadmium and lead can effectively be removed either by the hydroxide or the carbonate at pH values greater than 10.0. The MCL of 0.01 mg/l for Cd and of 0.05 mg/l for Pb are not reached. On the other hand, Sorg et al. [60] have demonstrated through jar-tests and pilot-plant studies that lime softening can remove nearly 100% of Cd and Pb, and meet their MCL. Naylor and Daque [95] found that an effective system for lead removal was the lime-soda, solids-contact softening process. "Virtually all lead

Table XXV. Metal Carbonate and Hydroxide Solubility Product Constants at 25°C [94]

Metal	log K_{sOH}[a]	pH[b]	log K_{sC}[c]
Zinc			
Aged	−17.0	9.8	−10.7
Fresh	−16.0		
Cadmium			
Aged	−14.4	10.5	−11.2
Fresh	−13.7		
Lead			
Aged	−15.3	10.0	−13.0
Fresh	−14.9		

[a]Hydroxide.
[b]Minimum hydroxide solubility.
[c]Carbonate, aged or fresh not specified.

Table XXVI. Soluble Metal Concentration for a Combined Zinc-Cadmium-Lead System[a]

Test System	Filtered pH	Soluble Zinc Concentration			Soluble Cadmium Concentration			Soluble Lead Concentration		
		M	Coll.[b] (mg/l)	Ind.[c] (mg/l)	*M*	Coll. (mg/l)	Ind. (mg/l)	*M*	Coll. (mg/l)	Ind. (mg/l)
Hydroxide	7.3	$10^{-2.6}$	150		$10^{-2.2}$	800		$10^{-4.3}$	11.4	25.6
	8.8	$10^{-5.1}$	0.5		$10^{-3.2}$	65		$10^{-4.3}$	10.8	6.0
	9.1	$10^{-5.4}$	0.25		$10^{-3.7}$	22.5		$10^{-5.0}$	2.0	
	11.0	$10^{-5.1}$	0.55	0.68	$10^{-6.1}$	0.1		$10^{-4.7}$	4.0	
Carbonate	8.4	$10^{-4.7}$	1.2	0.95	$10^{-4.8}$	1.8	1.2	$10^{-4.6}$	5.0	2.0
	8.6	$10^{-4.6}$	1.6		$10^{-5.2}$	0.8	1.7	$10^{-4.7}$	3.8	
	9.1	$10^{-4.9}$	0.8	0.75	$10^{-5.7}$	0.25		$10^{-4.8}$	3.2	3.6
	10.1	$10^{-5.1}$	0.55	0.60	$10^{-5.6}$	0.3	0.25	$10^{-4.5}$	6.0	
	10.7	$10^{-5.1}$	0.55	0.85	$10^{-5.9}$	0.15	0.25	$10^{-4.6}$	5.2	

[a]Reproduced from Patterson et al. [94], courtesy of the Water Pollution Control Federation.
[b]Coll. = collective.
[c]Ind. = individual.

Table XXVII. Summary of Treatment Results[a]

Metal	Treatment System	Optimum Treatment pH	Observed Concentration (mg/l)
Zinc	Hydroxide	9.5	0.25
Cadmium	Hydroxide	10.4	0.20
	Carbonate ($10^{-2.7}$)	10.7	0.35
	Carbonate ($10^{-1.2}$)	10.0	0.25
Lead	"Hydroxide"	10.5	0.60
	Carbonate ($10^{-2.7}$)	10.1	0.60
	Carbonate ($10^{-1.1}$)	7.5	1.00

[a]Reproduced from Patterson et al. [94], courtesy of the Water Pollution Control Federation.

was removed from solutions containing 2 mg/l of lead and with a contact solids concentration of 1% and greater." The lowest levels of Pb, <0.05 mg/l, were obtained in the pH range of 9.2–10.4.

Adsorption by Hydrous Iron and Manganese Oxides

The adsorption of lead(II), zinc(II) and cadmium (II) (thallium was included, but is not relevant to drinking water at this time) on hydrous iron (HFO) and manganese (HMO) oxides was investigated [96]. That there was significant adsorption of these three metals on HMO is seen in Figure 37a. As usual, $[H^+]$ affects adsorption and, in general, it increases with increasing pH value. There was a 100% removal of 0.1 m*M* Pb over the pH range of 4.0–8.0. Zn and Cd were completely removed by the HMO within the pH values of 6.0–8.0. Considerably lesser quantities of these metals were adsorbed by HFO than by HMO (Figure 37b). Reaction time in these experiments was 3 hr. Adsorption isotherms are seen in Figure 37c for HMO. These results suggest that HMO certainly could be employed as a specific treatment for Pb, Zn and Cd, or that these metals would be incidentally removed, if present, during Fe and Mn treatment.

SUMMARY OF WATER TREATMENT EFFECTIVENESS—PRIMARY CONSTITUENTS

A summary of the effectiveness of various water treatments for removing primary drinking water contaminants is seen in Tables XXVIII,

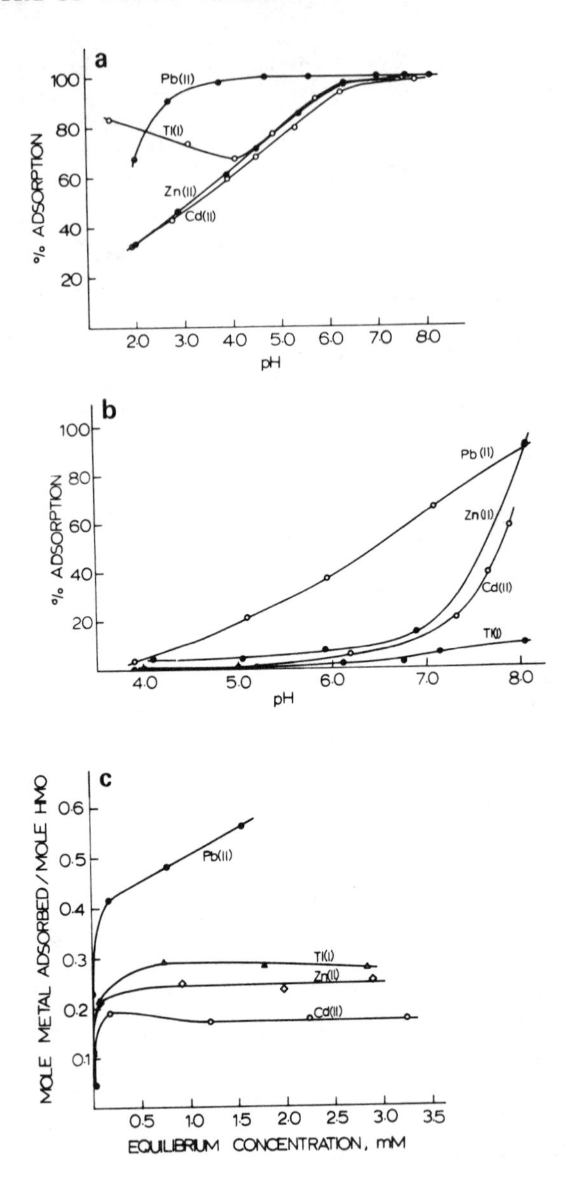

Figure 37. **(a)** Effect of pH on adsorption of heavy metal ions on hydrous manganese oxide (HMO). HMO = 0.436 mmol; heavy metal ion = 0.1 mmol; solution volume = 100 ml. **(b)** Effect of pH on adsorption of heavy metal ions on hydrous ferric oxide (HFO). HFO = 0.625 mmol; heavy metal ion = 0.1 mmol; solution volume = 100 ml. **(c)** Effect of heavy metal ion concentration at equilibrium on its adsorption on HMO (0.436 mmol). Reproduced from Gadde and Laitinen [96], courtesy of the American Chemical Society.

Table XXVIII. Summary of Effectiveness of Water Treatment Processes for the Removal of Inorganic Contaminants[a,b]

Contaminant	Conventional Coagulation: Alum	Conventional Coagulation: Iron[c]	Lime Softening	Ion Exchange: Cation	Ion Exchange: Anion	Activated Alumina	Activated Carbon: PAC	Activated Carbon: GAC	Reverse Osmosis	Electrodialysis
Arsenic[d]										
+III	P	F	P pH<10.5 L-G pH>10.5	P	L	F	P	F (bone char)[e]	G-F	G-F[e]
+V	G pH<7.5 F pH>7.5	E	F pH<10 G pH 10–10.8 E pH>10.8	P	E	E	P	E (bone char)	E	E[e]
Barium	P	P	E pH 9.5–10.8	E	P	P[e]	P	P	E	E
Cadmium	G>pH 8.5 F-P<pH 8.5	E pH>8 F-L pH<8	E	E	P	P[e]	L	L	E	E[e]
Chromium										
+III	E	E	E pH>10.5 G pH<10.5	E	P	P[e]	P	P	E	E[e]
+VI	P	P E (ferrous)	P	P	E	P[e]	P	P	E	E[e]
Fluoride	P	P	P	P	L	E	P	E (bone char)	E	E
Lead	E	E	E	[f]	P	P[e]	P	P	E	E[e]
Mercury										
Organic	F-P[g]	F-P[g]	P	ID	ID	P[e]	G-P	E	G-F[e]	G-L[e]
Inorganic	F-P[g]	F-G[g]	F	[f]	[f]	P[e]	G-P	E-G	G-F	G-L[e]
Nitrate	P	P	P	P	E	P	P	P	G	G
Selenium										
+IV	P	F pH<7.5 L pH>7.5	L-P	P	E	E-G	P	P	E	E[e]

Table XXVIII. continued

Contaminant	Conventional Coagulation		Lime Softening	Ion Exchange		Activated	Activated Carbon		Reverse	Electrodialysis
	Alum	Iron[c]		Cation	Anion	Alumina	PAC	GAC	Osmosis	
+VI	P	P	P	P	E	L-G	P	P	E	E[e]
Silver	G pH<8	G	G	E	P	P[e]	L	L	E	E[e]
Radium	P	P	E pH 9.5–10.8	E	P	P[e]	P	P	E	E[e]

[a]Reproduced from Sorg and Logsdon [59], courtesy of the American Water Works Association.
[b]E—excellent, 90–100%; G—good, 70–90%; F—fair, 40–70%; L—low, 20–40%; P—poor, 0–20%; ID—insufficient data.
[c]Results based on ferric iron coagulant except as noted.
[d]Oxidation of As(III) to As(V) will result in As(III) removals similar to As(V).
[e]Estimated.
[f]Possible under controlled conditions, but not practical to water treatment.
[g]Removal dependent on turbidity.

XXIX and XXX [59,60,67,84,91]. Additional information has been provided from sources other than those given in these references. Some general comments are appropriate here. No single treatment is effective for all contaminants, although osmosis and electrodialysis have some good removals for all contaminants. These two techniques, however, are used mainly for desalting seawater and brackish groundwaters. Any water that is not free from turbidity requires pretreatment. The economics of treatment for these primary and other contaminants must be considered, that is, the capital and operating costs of an existing facility or the construction of a single treatment or even a new plant will influence choice of treatment. EPA commissioned a study of the costs of water treatments for the primary contaminants that resulted in four publications [98–101].

Table XXIX. Most Effective Treatment Methods for Removal of Inorganic Contaminants[a]

Contaminant	Most Effective Treatment Methods
Arsenic	As (V)—iron coagulation, pH 6–8; alum coagulation, pH 6–7; excess lime softening; activated alumina, pH 5–6
	As(III)—oxidation treatment of As(III) to As(V); use same treatment list for As(V)
Barium	Lime softening, pH 11; ion exchange softening
Cadmium	Iron coagulation, above pH 8; lime softening; excess lime softening
Chromium	Cr(III)—iron coagulation, pH 6–9; alum coagulation, pH 7–9; excess lime softening
	Cr—ferrous sulfate coagulation, pH 7–9.5
Fluoride	Ion exchange with activated alumina or bone char
Lead	Iron coagulation, pH 6–9; alum coagulation, pH 6–9; lime or excess lime softening
Mercury	Inorganic—ferric sulfate coagulation, pH 7–8; granular activated carbon
	Organic—granular activated carbon
Nitrate	Ion exchange with anion resin
Radium	Lime softening; ion exchange with cation resin
Selenium	Se(IV)—ferric sulfate coagulation, pH 6–7; ion exchange with anion resin or activated alumina; reverse osmosis
	Se(VI)—ion exchange with anion resin or activated alumina; reverse osmosis
Silver	Ferric sulfate coagulation, pH 7–9; alum coagulation, pH 6–8; lime or excess lime softening

[a]Reproduced from Sorg and Logsdon [59], courtesy of the American Water Works Association.

Table XXX. Most Probable Applications of Water Treatment Processes for Inorganic Contaminant Removal[a]

Treatment Process	Principal Application for Water Treatment	Inorganic Contaminant Treatment Capability Effectiveness[b]			Most Probable Application for Inorganic Removal
		High	Moderate	Low	
Convention Coagulation	Clarification of surface waters	Cd Cr(III) Cr(VI) As(V) Ag Pb	As(III) Se(IV) Hg(0) Hg(I)	Ba F NO_3 Ra Se(VI)	Removal of Cd, Cr, As, Ag, or Pb from surface waters
Lime Softening	Removal of hardness from ground and surface water	Ba Ra Cd Cr(III) As(V) Pb	Se(IV) As(III) Hg(I) F	Cr(VI) NO_3 Se(VI) Hg(0)	Removal of Ba or Ra from groundwaters; removal of Cd, Cr(III), F, As(V), or Pb from hard surface waters requiring softening
Cation Exchange	Removal of hardness from groundwaters	Ba Ra Cd Pb Cr(III)		As Se NO_3 F Cr(VI)	Removal of Ba or Ra from groundwaters
Anion Exchange	Removal of nitrate from groundwaters	NO_3 Cr(VI) Se	Ba Ra Cd Pb Cr(III)		Removal of NO_3 from groundwaters
Activated alumina	Removal of fluoride from groundwaters	F As Se		Ba Ra Cd	Removal of F, As, or Se from groundwaters

Powdered Activated Carbon	Removal of taste and odors from surface waters		Hg(I) Hg(0) Cd	Ba Ra Cr(III) F NO_3 Ag	Removal of Hg from surface waters during emergency spills
Granular Activated Carbon	Removal of taste, odors and organics	Hg(I) Hg(0)	Cd	Ba Ra Cr(III) F NO_3	Removal of Hg from surface or groundwaters
Reverse Osmosis and Electrodialysis	Desalting of seawater or brackish groundwaters	As(V) Ba Cr Pb Cd Se Ag F Ra Hg	NO_3 As(III)		Removal of all inorganics from groundwaters

[a]Reproduced from Sorg and Logsdon [59], courtesy of the American Water Works Association.
[b]High—greater than 80%; moderate—20–80%; low—less than 20%

HYDROGEN SULFIDE

This constituent is listed in the National Secondary Drinking Water Regulations as a contributor to odorous water, i.e., it is responsible for a rotten egg smell [89] (see also Chapter 3). Also, H_2S reacts with many of the metals cited above to produce, for example, black stains or black deposits of iron sulfide. H_2S cannot be considered a normal constituent of natural waters, but is associated with waters "high" in sulfate and organic matter [2]. The microbially mediated oxidation of organic matter utilizes SO_4^{2-} as an electron acceptor, which results in reduced sulfur as H_2S. Also, this odorous substance is associated frequently with Fe and Mn from groundwaters in or near former swampy areas. In either case, H_2S can be removed easily from water by aeration, chemical oxidation or adsorption onto carbon.

Aqueous Chemistry

$$H_2S = H^+ + HS^-; \quad \log Ka_1 = 10^{-6.88} \tag{88}$$

$$HS^- = H^+ + S^{2-}; \quad \log Ka_2 = 10^{-12.2} \text{ to } 10^{-17.1} \tag{89}$$

Thus, H_2S is a "weak" acid, and the secondary dissociation is not realized in a practical sense. This gaseous substance can be quite soluble in water: at 760 mm Hg and 25°C, the solubility of H_2S is 3380 ppm [102]. There are 30 or more ionic and molecular sulfur species in existence, of which 5 are thermodynamically stable in water at 25°C and 1 atm: HSO_4^-, SO_4^{2-}, S^0, H_2S and HS^- [42]. Such other species as thiosulfate, polysulfides and polythionates are unstable in water. A pE-pH diagram is seen in Figure 38 where H_2S and S^0 are stable, generally, under reducing conditions and below pH 7. Sulfate is stable under oxidizing conditions and over the entire pH range. Additional sulfur chemistry is given in the literature [2,41,102].

Oxygenation and/or Aeration

Aeration has been the traditional treatment for volatile substances causing tastes and odors in water. Aerators range from simple mechanical devices to more complex diffusers [103]. However, aeration is not an efficient method for removing tastes and odors because many are not volatile enough. Aeration does, of course, provide dissolved oxygen

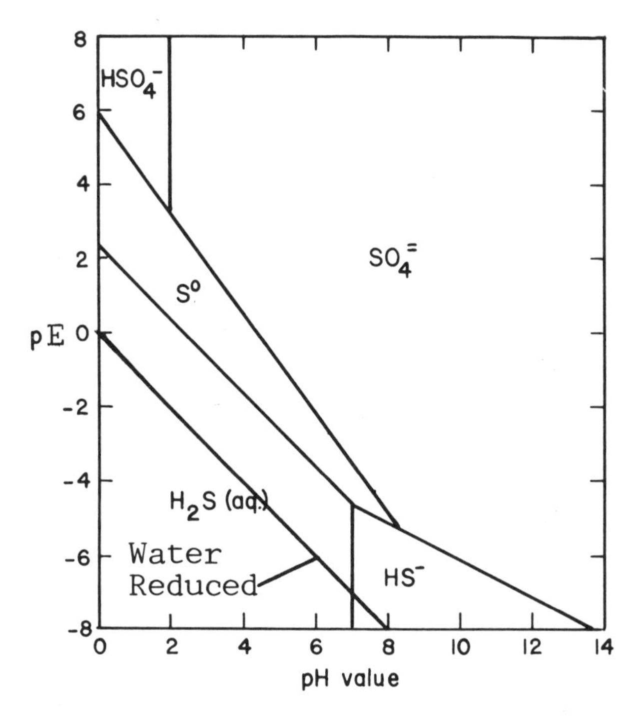

Figure 38. pH-pE predominance diagram of sulfur species. Reproduced from Chen [102], courtesy of Ann Arbor Science Publishers.

which can act as an electron acceptor for reduced sulfur species. Chen [102] has summarized several studies of the kinetics and mechanism of the oxygenation of sulfide [104–107]. The reaction of reduced sulfur with oxygen apparently is complex, is slow in the absence of a catalyst and is dependent on pH. A reaction pathway for slightly acid, alkaline and neutral solutions is proposed [104]:

$$HS^- + O_2 = HS^\cdot + O_2^- \qquad (90)$$

$$HS^\cdot + O_2 = HO_2 + S \qquad (91)$$

$$HS^\cdot + O_2^- = S + HO_2^- \qquad (92)$$

$$HS^- + (x - 1)S = H^+ + S_x^{2-}, \text{ x can be 2–5} \qquad (93)$$

Equation 93 indicates that the reaction proceeds through polysulfide formation which, in turn, is pH-dependent [105]. Oxidation of sulfide may produce or consume H^+ depending on the products or other items. The following reactions may represent the stoichiometry of the oxidation:

$$2HS^- + O_2 + 2H^+ = 2H_2O + 2SO \tag{94}$$

$$2HS^- + 2O_2 = H_2O + S_2O_3^{2-} \tag{95}$$

$$2HS^- + 4O_2 = 2SO_4^{2-} + 2H^+ \tag{96}$$

Figure 39 shows typical rate curves for oxidation of HS^- by O_2 [106]. The pH dependence of this oxidation is seen in Figure 40. At pH values less than 6.0, the oxidation reaction is extremely slow where H_2S is the predominant species. As the $[H^+]$ is decreased, the observed specific rate is increased, and reaches a maximum between the pH values of 8. and 8.5. Beyond these pH values, the reaction rate is lowered to a minimum near pH 9., whereupon it is increased again to a second maximum near pH 11., after which it is decreased again.

The rate of sulfide oxidation was described by this expression:

$$-\left[\frac{d[\Sigma S^{2-}]}{dt}\right]_{t=0} = Ri = k[\Sigma S^{2-}]_o^m[O_2]_o^n \tag{97}$$

where R_i is the initial rate and the brackets represent molar concentrations. The exponents m and n were evaluated from which the empirical equation evolved [104]:

$$-\left[\frac{d[\Sigma S^{2-}]}{dt}\right] = k[\Sigma S^{2-}]_0^{1.34}[O_2]_0^{0.56} \tag{98}$$

The catalytic or inhibitive effects of various metallic cations and organic compounds on sulfide oxidation were reported [107]. The decreasing order of reaction rate acceleration by cations is: $Ni^{2+} > Co^{2+} > Mn^{2+} > Cu^{2+} > Mg^{2+} > Ca^{2+}$. A 5×10^{-5} *M* quantity of Ni^{2+} effected a 50-fold increase in the reaction rate at pH 7.29. Such organic compounds as anisole, citrate, EDTA and nitrilotriacetic acid inhibited the oxygenation of sulfide.

The oxygenation of reduced sulfur species in aqueous solutions at pH values of 4., 7.55 and 10. was researched [108]. The reaction rate was, of course, first-order with respect to reduced sulfur at all pH values, and was nearly first-order with respect to O_2 at pH 7.55. The same rate equation as in 97 was used [108], whereby the empirical values of the exponents, m and n, were 1.02 and 0.80, respectively. The stoichiometric reaction was:

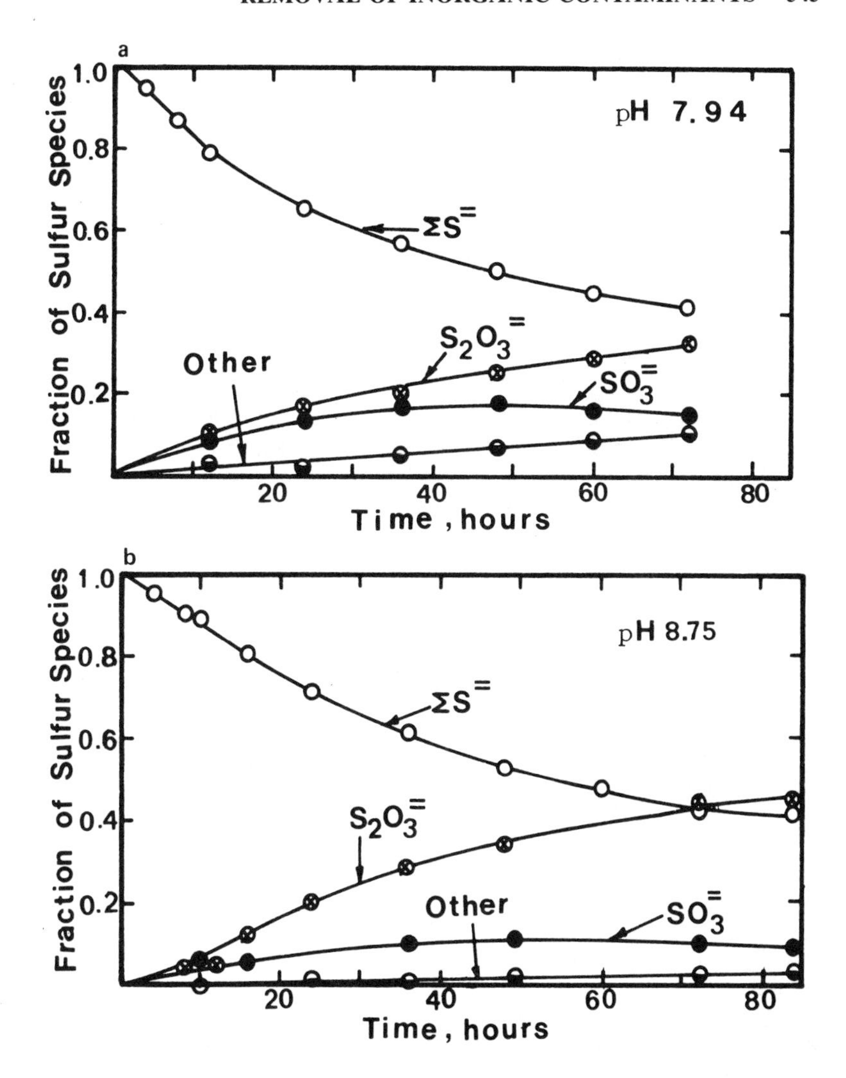

Figure 39. Kinetics of sulfide oxidation by O_2 at **(a)** pH 7.94 and **(b)** pH 8.75. $[\Sigma S^{2-}] = 1 \times 10^{-4}$ *M*, $[O_2] = 8 \times 10^{-4}$ *M*. Reproduced from Chen and Morris [106], courtesy of Pergamon Press, Ltd.

$$4\,HS^- + 5.5\,O_2 = S_2O_3^{2-} + SO_3^{2-} + SO_4^{2-} + 2\,H^+ + 2\,H_2O \qquad (99)$$

Data were obtained to confirm the molar ratio of 1.375, $[O_2]/[HS^-]$. Table XXXI is a summary of the reaction products from several investigations. It is, indeed, a complex reaction.

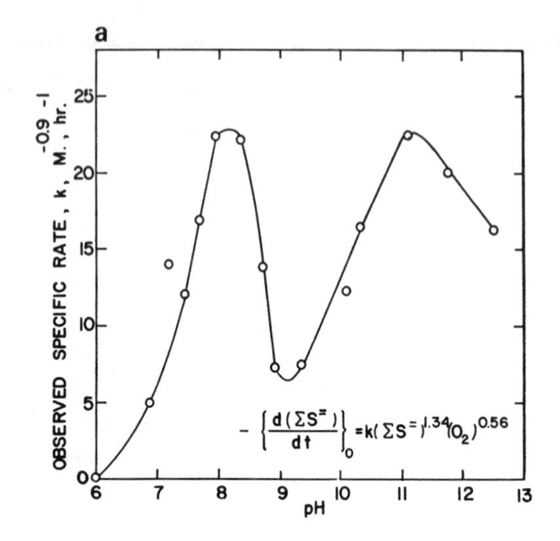

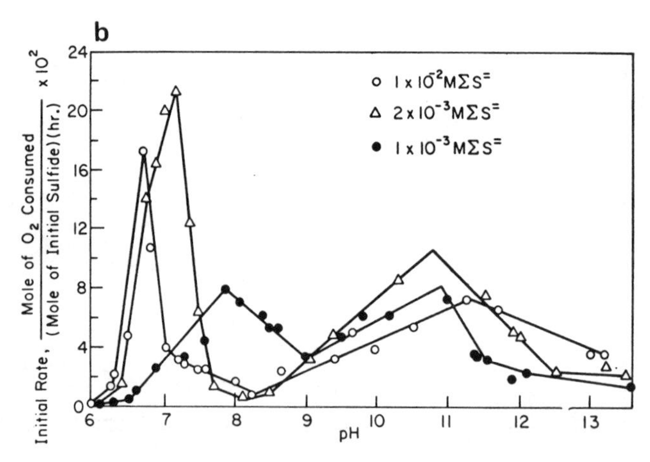

Figure 40. pH dependence of **(a)** observed specific rate of sulfide oxidation and **(b)** initial oxygen uptake by sulfide ($O_2 = 0.21$ atm, $T = 25°C$). Reproduced from Chen and Morris [104], courtesy of the American Chemical Society.

Chlorination

Reduced species of sulfur effectively are oxidized by Cl_2 in aqueous solutions. The stoichiometry is [109]:

$$Cl_2 + S^{2-} = 2\,Cl^- + S^0 \tag{100}$$

$$4\,Cl_2 + S^{2-} + 4\,H_2O = 8\,HCl + SO_4^{2-} \tag{101}$$

These reactions occur very rapidly, and 8.5 parts of Cl_2 are required per part of H_2S for complete oxidation. If the predominant reaction is 101, then the acidity produced would have to be neutralized either by natural or added alkalinity. A pH range of 6.5–7.3 appears to be optimum.

Ozonation

$O_{3(g)}$ can be used to oxidize reduced sulfur species:

$$S^{2-} + O_{3(g)} + H_2O = S^0 + 2OH^- + O_2 \quad (102)$$

$$S^{2-} + 4O_{3(g)} = SO_4^{2-} + 4O_2 \quad (103)$$

The theoretical molar ratios of $O_{3(g)}/S^{2-}$ of 1:1 for S^0 and 4:1 for SO_4^{2-} have been confirmed [102]. The kinetics of the reaction has been reported to be "instantaneous" [110].

Permanganate

Potassium permanganate has been used successfully to remove H_2S from water:

$$8MnO_4^- + 3S^{2-} + 4H_2O = 8MnO_{2(s)} + 3SO_4^{2-} + 8OH^- \quad (104)$$

Another possible reaction is [111]:

$$4MnO_4^- + 3H_2S = 2SO_4^{2-} + S^0 + 3MnO_{(s)} + MnO_{2(s)} + 3H_2O \quad (105)$$

The molar ratio of 4:3 was observed in laboratory titrations of several wellwaters at neutral pH values. Attempts to document the kinetics of $KMnO_4$ reaction with Na_2S were unsuccessful because it is claimed to go to completion within five seconds.

Hydrogen Peroxide

A kinetic study [112] reveals that H_2O_2 has found application for treatment of industrial and municipal wastewaters for H_2S removal, and that it has potential for drinking water. The oxidation reaction is:

$$H_2S + H_2O_2 = 1/x\ S_x + H_2O \quad (106)$$

Frequently, $x = 8$. Another reaction may be:

$$HS^- + 4H_2O_2 = SO_4^{2-} + 4H_2O + H^+ \tag{107}$$

A two-term rate law, consistent with the observed kinetic data, is:

$$-\frac{d[H_2S]}{dt} = k_1[H_2S][H_2O_2] + k_2Ka_1\left(\frac{[H_2S][H_2O_2]}{[H^+]}\right) \tag{108}$$

where Ka_1 is from Equation 88. The values for k_1 and k_2 were determined to be 29.0 M^{-1}-min^{-1} and 0.48 M^{-1}-min^{-1}, respectively, at 25°C. The reactions, 106 and 107, and the rate law, 108, were manipulated into pseudo-first-order kinetics. Figure 41 shows the pH dependence of the reaction and the first-order reaction rate constant. Table XXXII summarizes the kinetic data. A stoichiometric excess of H_2O_2 operationally should be added to the system in order to effect reaction 107 within 15 min or so.

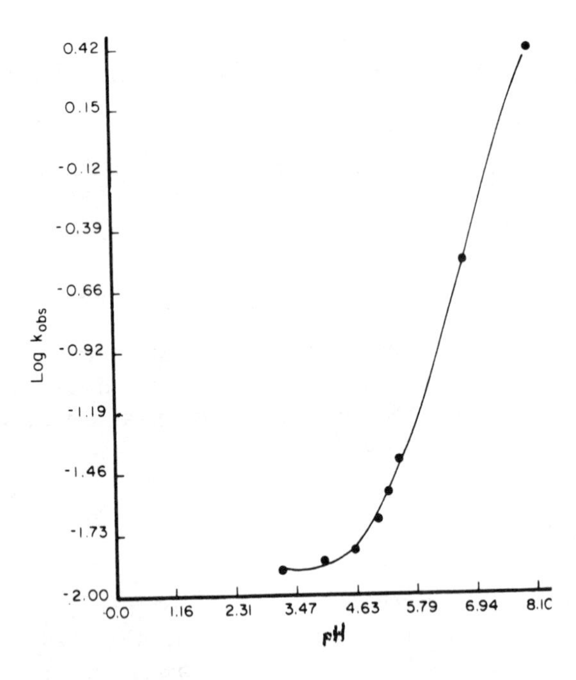

Figure 41. Plot of log k_{obs} vs pH. Reproduced from Hoffman [112], courtesy of the American Chemical Society.

Table XXXI. Summary of Reaction Products Observed in Investigations of Oxygenation of Reduced Sulfur Species[a]

Investigator	pH	Reaction Solution	$[S(-II)t]_0/[O_2]_0$	Products Obsd
Chen and Morris	6–12	Controlled	0.06–1.25	S_x^{2-}, SO_3^{2-}, S^0, $S_2O_3^{2-}$, SO_4^{2-}
Avrahami and Golding	11–14	Controlled	0.08–0.67	S^0 (occasionally), $S_2O_3^{2-}$, SO_4^{2-}
Cline and Richards	7.8	Seawater	0.125–0.5	SO_3^{2-}, $S_2O_3^{2-}$, SO_4^{2-}
Skopintsev et al.	8.2	Seawater	0.2–8.0	SO_3^{2-}, $S_2O_3^{2-}$
Demirjian	7, 8.6	Controlled	0.03–5.0	S^0, SO_3^{2-}, $S_2O_3^{2-}$, SO_4^{2-}
Alferova and Titova	9–13	Controlled	20	SO_3^{2-}, $S_2O_3^{2-}$, SO_4^{2-}
O'Brien and Birkner	4–10.7	Controlled	1.0–1.37	SO_3^{2-}, $S_2O_3^{2-}$, SO_4^{2-}

[a]Reproduced from O'Brien and Birkner [108], courtesy of the American Chemical Society.

Table XXXII. Summary of Kinetic Data at 25°C and $\mu = 0.4$[a]

$[S^{2-}]_0 \times 10^3$ *M*	$[H_2O_2]_0 \times 10^2$ *M*	pH	T(°C)	k_{obs} (min^{-1})	σ_x
1.5	1.5	6.81	25.0	0.149	0.012
1.5	3.0	6.81	25.0	0.293	0.034
1.5	4.5	6.78	25.0	0.453	0.013
1.5	6.0	6.76	25.0	0.620	
1.5	8.0	6.75	25.0	0.813	
1.5	10.0	6.80	25.0	0.990	
1.5	3.0	8.10	25.0	2.609	
1.5	3.0	6.81	25.0	0.293	0.034
1.5	3.0	5.47	25.0	0.03	
1.5	3.0	5.05	25.0	0.021	
1.5	3.0	4.58	25.0	0.016	
1.5	3.0	4.01	25.0	0.014	
1.5	3.0	3.20	25.0	0.013	

[a]Reproduced from Hoffman [112], courtesy of the American Chemical Society.

Adsorption

Powdered and granular activated carbon are capable of adsorbing H_2S from water. Consequently, any H_2S that occurs in surface and groundwaters will be removed when other constituents are treated by carbon. There are occasions where domestic wellwaters are contaminated with H_2S. In this situation, a granular activated carbon filter will suffice.

REFERENCES

1. "National Interim Primary Drinking Water Regulations," *Federal Register* 40:(248):59566 (1975).
2. Faust, S. D., and O. M. Aly. *Chemistry of Natural Waters* (Ann Arbor, MI: Ann Arbor Science Publishers, Inc., 1981).
3. Hem, J. D., and W. H. Cropper. "Chemistry of Iron in Natural Water," U.S. Geological Survey Water Supply Paper 1459, Washington, DC (1962).
4. Stumm, W., and J. J. Morgan. *Aquatic Chemistry,* 2nd ed. (New York: John Wiley & Sons, Inc., 1981).
5. Weiss, J. *Naturwissenschaften* 23:64 (1935).
6. Stumm, W., and G. F. Lee. *Ind. Eng. Chem.* 53:143 (1961).
7. Sung, W., and J. J. Morgan. *Environ. Sci. Technol.* 14:561 (1980).
8. Ghosh, M. M. et al. *J. San. Eng. Div., ASCE* 92(SA1):199 (1966).

9. Stumm, W., and P. C. Singer. *J. San. Eng. Div., ASCE* 92(SA5):120 (1966).
10. Winklehaus, C. et al. *J. San. Eng. Div., ASCE* 92(SA6):129 (1966).
11. Morgan, J. J., and F. B. Birkner. *J. San. Eng. Div., ASCE* 92(SA6):137 (1966).
12. Jobin, R., and M. M. Ghosh. *J. Am. Water Works Assoc.* 64:590 (1972).
13. Ghosh, M. M. In: *Aqueous Environmental Chemistry of Metals,* A. J. Rubin, Ed. (Ann Arbor, MI: Ann Arbor Science Publishers, Inc., 1974).
14. Singer, P. C., and W. Stumm. *Science* 167:1121 (1970).
15. Singer, P. C., and W. Stumm. "Oxygenation of Ferrous Iron," Report 14010-06/69, FWQA, U.S. Department of the Interior, Washington, DC (1970).
16. Schenk, J. E., and W. Weber, Jr. *J. Am. Water Works Assoc.* 60:199 (1968).
17. Nordell, E. *Water Treatment,* 2nd ed., (New York: Van Nostrand Reinhold Company, 1961).
18. Matthews, E. R. *J. Am. Water Works Assoc.* 39:680 (1947).
19. Willey, B. F., and H. Jennings. *J. Am. Water Works Assoc.* 55:729 (1963).
20. Cromley, J. T., and J. T. O'Connor. *J. Am. Water Works Assoc.* 68:315 (1976).
21. Coogan, G. J. *J. Am. Water Works Assoc.* 54:1507 (1962).
22. Bell, G. R. *J. Am. Water Works Assoc.* 57:458, 655 (1965).
23. Ghosh, M. M. et al. *J. Am. Water Works Assoc.* 59:878 (1967).
24. George, A. D., and M. Chanduri. *J. Am. Water Works Assoc.* 69:385 (1977).
25. Hutchinson, G. E. *A Treatise on Limnology, Vol. I* (New York: John Wiley & Sons, Inc., 1957).
26. Shapiro, J. *J. Am. Water Works Assoc.* 56:1062 (1964).
27. Hem, J. D. "Complexes of Ferrous Iron with Tannic Acid," U.S. Geological Water Supply Paper 1459 D, Washington, DC (1960).
28. Oldham, W. K., and E. F. Gloyna. *J. Am. Water Works Assoc.* 61:610 (1969).
29. Theis, T. L., and P. C. Singer. *Environ. Sci. Technol.* 8:569 (1974).
30. Theis, T. L., and P. C. Singer. In: *Trace Metals and Metal-organic Interactions in Natural Waters,* P. C. Singer, Ed. (Ann Arbor, MI: Ann Arbor Science Publishers, Inc., 1973).
31. Singer, P. C., and W. Stumm. *J. Am. Water Works Assoc.* 62:198 (1970).
32. Leussing, D. L., and I. M. Kolthoff. *J. Am. Chem. Soc.* 75:2476 (1953).
33. Gayer, K. H., and L. Woontner. *J. Phys. Chem.* 60:1509 (1956).
34. Latimer, W. E. *The Oxidation States of the Elements and Their Potentials in Aqueous Solutions,* 2nd ed. (Englewood Cliffs, NJ: Prentice-Hall Inc., 1952).
35. Olson, L. L., and C. J. Twardowski, Jr. *J. Am. Water Works Assoc.* 67:150 (1975).
36. Dart, F. J., and P. D. Foley. *J. Am. Water Works Assoc.* 62:663 (1970).
37. Krauskopf, K. B. *Geochim. Cosmochim. Acta* 10:1 (1956).
38. Weber, W. J., Jr., and W. Stumm. *J. Inorg. Nucl. Chem.* 27:237 (1965).
39. Morgan, J. J. In: *Principles and Applications of Water Chemistry,* S. D.

Faust and J. V. Hunter, Eds. (New York: John Wiley & Sons, Inc., 1967).
40. Hem, J. D. "Chemical Equilibria and Rates of Manganese Oxidation," U.S. Geological Survey Water Supply Paper 1667–A, Washington, DC (1963).
41. Pourbaix, M. *Atlas of Electrochemical Equilibria in Aqueous Solutions* (New York: Pergamon Press, Inc., 1966).
42. Morgan, J. J., and W. Stumm. *Proceedings 2nd International Water Pollution Research Conference* (Tokyo, 1964) (New York: Pergamon Press, 1965).
43. Griffin, A. E. *J. Am. Water Works Assoc.* 52:1326 (1960).
44. Mathews, E. R. *J. Am. Water Works Assoc.* 39:680 (1947).
45. Adams, R. B. *J. Am. Water Works Assoc.* 52:219 (1960).
46. Fox, R. K. et al. *J. Am. Chem. Soc.* 63:1779 (1941).
47. Posselt, H. S. et al. *J. Am. Water Works Assoc.* 60:48 (1968).
48. Morgan, J. J., and W. Stumm. *J. Colloid. Sci.* 19:347 (1964).
49. Illig, G. L., Jr., *J. Am. Water Works Assoc.* 52:867 (1960).
50. Owens, L. V. *J. Am. Water Works Assoc.* 55:721 (1963).
51. Andersen, D. R. et al. *J. Am. Water Works Assoc.* 65:635 (1973).
52. Cleasby, J. L. *J. Am. Water Works Assoc.* 67:147 (1975).
53. Shen, Y. S. *J. Am. Water Works Assoc.* 65:543 (1973).
54. Gulledge, J. H., and J. T. O'Connor. *J. Am. Water Works Assoc.* 65:548 (1973).
55. Ferguson, J. F., and M. A. Anderson. In: *Chemistry of Water Supply Treatment and Distribution,* A. J. Rubin, Ed. (Ann Arbor, MI: Ann Arbor Science Publishers, Inc., 1974).
56. Bellack, E. *J. Am. Water Works Assoc.* 63:454 (1971).
57. Rubel, F., Jr., and F. S. Williams. "Pilot Study of Fluoride and Arsenic Removal from Potable Water," U.S. EPA Report–600/2–80–100, Cincinnati, Ohio (1980).
58. Gupta, S. K., and K. Y. Chen. *J. Water Poll. Control Fed.* 50:493 (1978).
59. Sorg, T. J., and G. S. Logsdon. *J. Am. Water Works Assoc.* 72:411 (1980).
60. Sorg, T. J. et al. *J. Am. Water Works Assoc.* 70:680 (1978).
61. Patterson, J. W. et al. *J. Water Poll. Control Fed.* 49:2397 (1977).
62. Hem, J. D. *Water Res.* 8:661 (1972).
63. Huang, C. P., and E. H. Smith. In: *Chemistry in Water Reuse,* W. J. Cooper, Ed. (Ann Arbor, MI: Ann Arbor Science Publishers, Inc., 1981).
64. Huang, C. P. In: *Carbon Adsorption Handbook,* P. N. Cheremisinoff and F. Ellerbusch, Eds. (Ann Arbor, MI: Ann Arbor Science Publishers, Inc., 1978).
65. Huang, C. P., and F. Ostovic. *J. Environ. Eng. Div. (ASCE)* 104(EE5):863 (1978).
66. Posselt, H. S., and W. J. Weber, Jr. In: *Chemistry of Water Supply Treatment and Distribution,* A. Rubin, Ed. (Ann Arbor, MI: Ann Arbor Science Publishers, Inc., 1974).
67. Sorg, T. J. *J. Am. Water Works Assoc.* 71:454 (1979).
68. Huang, C. P., and M. H. Wu. *J. Water Poll. Control Fed.* 47:2437 (1975).

69. Huang, C. P., and M. H. Wu. *Water Res.* 11:673 (1977).
70. Bowers, A. R., and C. P. Huang. *Prog. Water Tech.* 12:629 (1980).
71. "National Secondary Drinking Water Regulations," *Federal Register* 42(62):17143 (March 31, 1977).
72. Faust, S. D. Personal observation in Lebanon Township, NJ (1980).
73. Murray, D. J. et al. In: *Adsorption from Aqueous Solution,* Advances in Chemistry Series 79 (Washington, DC: American Chemical Society, 1968).
74. Swallow, K. C. et al. *Environ. Sci. Tech.* 14:1326 (1980).
75. McKaveney, J. P. et al. *Environ. Sci. Tech.* 6:1109 (1972).
76. Truitt, R. E., and J. H. Weber. *Water Res.* 13:1171 (1979).
77. Rothbart, H. L. In: *An Introduction to Separation Science,* B. L. Karger, Ed. (New York: John Wiley & Sons, Inc., 1973).
78. Hautala, E. et al. *Water Res.* 11:243 (1977).
79. Logsdon, G. S., and J. M. Symons. *J. Am. Water Works Assoc.* 65:554 (1973).
80. Ebersole, G., and J. T. O'COnnor. "Removal of Mercury from Water by Conventional Treatment Processes," American Water Works Association, Chicago, IL (1972).
81. Thiem, L. et al. *J. Am. Water Works Assoc.* 68:447 (1976).
82. Humenick, M. J., Jr., and J. L. Schnoor. *J. Environ. Eng. Div., ASCE* 100EE6:1249 (1974).
83. Inoue, Y., and M. Munemori. *Environ. Sci. Technol.* 13:443 (1979).
84. Sorg, T. J., and G. S. Logsdon. *J. Am. Water Works Assoc.* 70:379 (1978).
85. Trussell, R. R. et al. "Selenium Removal from Ground Water Using Activated Alumina," U.S. EPA Report 600/2-80-153, Cincinnati, Ohio (1980).
86. "Drinking Water and Health," Recommendations of the National Academy of Sciences, *Federal Register* 42(132):35764 (July 11, 1977).
87. Lauch, R. P., and T. J. Sorg. *J. Am. Water Works Assoc.* 74:256 (1981).
88. Kopp, J. F., and R. T. Kroner. "Trace Metals in the Waters of the United States," Federal Water Quality Administration, Cincinnati, Ohio (1970).
89. "National Secondary Drinking Water Regulations," *Federal Register* 42(62):17143 (March 31, 1977).
90. Dean, J. H., Ed. *Lange's Handbook of Chemistry,* 11th ed. (New York: McGraw-Hill Book Company, 1973).
91. Sorg, T. J. *J. Am. Water Works Assoc.* 70:105 (1978).
92. Kubli, H. A. *Helv. Chim. Acta* 3:453 (1947).
93. Choi, W. W., and K. Y. Chen. *J. Am. Water Works Assoc.* 71:562 (1979).
94. Patterson, J. W. et al. *J. Water Poll. Control Fed.* 49:2397 (1977).
95. Naylor, L. M., and R. R. Dague. *J. Am. Water Works Assoc.* 67:560 (1975).
96. Gadde, R. R., and H. A. Laitinen. *Anal. Chem.* 46:2022 (1974).
97. "Manual of Water Treatment Techniques for Meeting the Interim Primary Drinking Water Regulations," U.S. EPA Report-600/8-77-005, Cincinnati, Ohio (1977).
98. Gumerman, R. C. et al. "Estimating Water Treatment Costs" (Vol.

Summary), U.S. EPA Report-600/2-79-162a, ORD, Cincinnati, Ohio (August 1979).

99. Gumerman, R. C. et al. "Estimating Water Treatment Costs, Vol. 2: Cost Curves Applicable to 1 to 200 mgd Treatment Plants," EPA Report-600/2-79-162b, ORD, Cincinnati, Ohio (August 1979).
100. Hansen, S. P. et al. "Estimating Water Treatment Costs, Vol. 3: Cost Curves Applicable to 2500-gpd to 1-mgd Treatment Plants," U.S. EPA Report-600/2-79-162c, ORD, Cincinnati, Ohio (August 1979).
101. Lineck, T. S. et al. "Estimating Water Treatment Costs, Vol. 4: Computer User's Manual For Retrieving and Updating Cost Data," U.S. EPA Report-600/2-79-162d, ORD, Cincinnati, Ohio (August 1979).
102. Chen, K. Y. In: *Chemistry of Water Supply Treatment and Distribution,* A. J. Rubin, Ed. (Ann Arbor, MI: Ann Arbor Science Publishers, Inc., 1974).
103. *Water Quality and Treatment,* 3rd ed. (New York: McGraw-Hill Book Company, 1971).
104. Chen, K. Y., and J. C. Morris. *Environ. Sci. Technol.* 6:529 (1972).
105. Chen, K. Y., and S. K. Gupta. *Environ. Lett.* 4:187 (1973).
106. Chen, K. Y., and J. C. Morris. *Adv. Water Poll. Res.* 3:32/1 (1972).
107. Chen, K. Y., and J. C. Morris. *J. San. Eng. Div., ASCE* SA1:215 (1972).
108. O'Brien, D. J., and F. B. Birkner. *Environ. Sci. Technol.* 11:1114 (1977).
109. Black, A. P., and J. B. Goodson, Jr. *J. Am. Water Works Assoc.* 44:309 (1952).
110. Kirk-Othmer. *Encyclopedia of Chemical Technology, Vol. 19* (New York: John Wiley & Sons, 1969).
111. Wiley, B. F. et al. *J. Am. Water Works Assoc.* 56:475 (1964).
112. Hoffmann, M. R. *Environ. Sci. Technol.* 11:61 (1977).

CHAPTER 9

REMOVAL OF CORROSIVE SUBSTANCES

CORROSION AS AN ECONOMIC AND ESTHETIC CONCERN

Concern about the corrosive effect of drinking water on distribution and plumbing systems traditionally has been economic and esthetic. It has been estimated that the annual economic loss from water corrosiveness is approximately $700 million [1]. In recent times, however, the corrosion of metallic and asbestos-cement distribution systems has posed a significant threat to health. The presence of contaminants such as lead, cadmium and asbestos in drinking water has prompted the U.S. Environmental Protection Agency (EPA) to require identification of these corrosion by-products and to report such corrosive characteristics as pH, alkalinity, hardness, temperature, total dissolved solids and the Langelier index (LI) (see Chapter 1) [2].

In addition to chemical characteristics of the water, the following construction materials must be identified and reported if present in distribution systems [2]:

- lead from piping, solder, caulking, interior lining of distribution mains, alloys and home plumbing;
- copper from piping and alloys, service lines and home plumbing;
- zinc from galvanized piping, service lines and home plumbing;
- ferrous piping materials such as cast iron and steel; and
- asbestos-cement pipe.

Also, it may be necessary to identify and report vinyl-lined asbestos pipe and coal tar-lined pipes and tanks.

Consequently, it is imperative to establish a monitoring system for the many corrosion by-products. Reference [2] gives the details of the requirements for the surveillance of corrosive drinking waters.

CHEMISTRY OF CORROSION

The corrosion of metals is an extremely complex chemical and electrochemical phenomenon. In these oxidative reactions, countless local galvanic couples form on the surface of the corroding metal in which the metal is spontaneously oxidized and its oxidant is reduced. Each couple is actually a microbattery where the corrosion reaction proceeds with a flow of electric current between anodic and cathodic sites on the metal. The basic requirements of metallic corrosion are: (1) differences of electrical potential between adjacent areas on an exposed metallic surface to provide anodes and cathodes; (2) moisture to provide an electrolyte; (3) an oxidizing agent to be reduced at the cathode; and (4) an electrical path in the metal for electron flow from anodes to cathodes.

The conditions under which an electrochemical reaction can occur at the interface of a metal and an electrolyte depend on the relative values of the electrode potential, E, of the metal (experimentally determined) and the thermodynamic equilibrium potential, E^0 (or V^0, IUPAC convention [3]), of the reaction (calculated). That is, oxidation occurs only if the electrode potential of the metal is more positive (i.e., greater) than the equilibrium potential of the reaction. Likewise, a reduction reaction occurs only if the electrode potential of the metal is more negative (i.e., less) than the equilibrium potential. References [3] and [4] give a more thorough explanation of the electrochemistry of corrosion.

Table I shows the "classical" electromotive force series of pure metals under standard state conditions (25°C, 1 atm, 1 *M* solution) [6]. Metals lying below hydrogen spontaneously corrode in an aqueous environment, whereas those that lie above do not corrode. This generalization applies when the metal's half-cell reaction is combined with the hydrogen half-cell reaction. There are conditions, of course, where the metals lying above hydrogen, especially copper, spontaneously corrode. Table I also presents a practical galvanic series of metals and alloys [6] based on many laboratory and field tests using several corrosive solutions under a variety of service conditions.

Iron

The corrosion chemistry of iron is exceedingly complex since Fe(0), Fe(II) and Fe(III) may enter into many oxidation-reduction and precipitation reactions [5,7]. A few of the more relevant reduction reactions are given in Table II at standard-state conditions. In turn, these reactions can be used to construct a pH–pE diagram as seen in Figure 1 [8]. The corrosion reaction (6 in Table II) Fe(0)-Fe(II) is set at a pE value of

Table I. Electromotive Force and Practical Galvanic Series[a]

Electromotive Force Series	Volts	Practical Galvanic Series of Metals and Alloys
		Protected End (Cathodic or most noble)
1 Au^{3+}	+1.36	1 Platinum
		2 Gold
		3 Graphite
2 Ag^{2+}	+0.8	4 Silver
		5 Hastelloy C (passive)
		6 18-8-3 Chromium-nickel-molybdenum-iron (passive)
3 Cu^{2+}	+0.34	7 18-8 Chromium-nickel-iron (passive)
4 Sb^{3+}	+0.1	8 Chromium-iron (passive)
		9 Inconel (passive)
5 $H_{2(g)}$	0.00	10 Nickel (passive)
		11 Silver solder
		12 Monel
6 Pb^{2+}	−0.12	13 Copper-nickel alloys
		14 Bronzes
7 Sn^{2+}	−0.14	15 Copper
		16 Brasses
8 Ni^{2+}	−0.23	17 Hastelloy C (active)
		18 Inconel (active)
9 Co^{2+}	−0.29	19 Nickel (active)
		20 Tin
10 Cd^{2+}	−0.40	21 Lead
		22 Lead-tin solders
11 Fe^{2+}	−0.44	23 18-8-3 Chromium-nickel-molybdenum-iron (active)
12 Cr^{2+}	−0.56	24 18-8 Chromium-nickel-iron (active)
13 Zn^{2+}	−0.76	25 Ni-Resist
		26 Chromium-iron (active)
14 Al^{3+}	−1.33	27 Cast-iron
		28 Steel or Iron
15 Mg^{2+}	−1.55	29 Aluminum 17ST
		30 Cadmium
		31 Aluminum 2S
		32 Zinc
		33 Magnesium alloys
		34 Magnesium

[a]Reproduced from Obrecht and Pourbaix [6], courtesy of the American Water Works Association.

−9.95 when an assumed concentration of 10^{-5} *M* is used for the soluble Fe(II). Since this reaction is the oxidative portion of the corrosion reaction for metallic iron, it is combined with Equation 9 (Table II) as the reductive reaction, and yields an E^0 value of +0.44. This shows that metallic iron pipes are thermodynamically unstable in water, i.e., they

Table II. Some Corrosion Reactions for Fe(0)–Fe(II)–Fe(III) at 25°C

Reaction Number	Reaction	V^0	pE^0
1	$O_{2(g)} + 4\ H^+ + 4\ e = 2\ H_2O$	+1.229	+83.3
2	$Fe^{3+} + e = Fe^{2+}$	+0.771	+13.07
3	$FeOH^{2+} + H^+ + e = Fe^{2+} + H_2O$	+0.914	+15.51
4	$Fe(OH)_2^+ + 2\ H^+ + e = Fe^{2+} + 2\ H_2O$	+1.191	+20.2
5	$Fe(OH)_{3(s)} + 3\ H^+ + e = Fe^{2+} + 3\ H_2O$	+1.052	+17.8
6	$Fe^{2+} + 2\ e = Fe_{(s)}$	−0.440	−14.9
7	$FeCO_{3(s)} + H^+ + 2\ e = Fe_{(s)} + HCO_3^-$	−0.413	−14.0
8	$2\ H_2O + 2\ e = H_{2(g)} + 2\ OH^-$	−0.828	−28.1
9	$2\ H^+ + 2\ e = H_{2(g)}$	0.0	0.0
10	$Fe_{(s)} + 2\ H^+ = Fe^{2+} + H_{2(g)}$	+0.440	+14.9
11	$Fe_{(s)} + 0.5\ O_{2(g)} + 2\ H^+ = Fe^{2+} + H_2O$	+1.67	+56.4
12	$O_{2(g)} + 4\ Fe^{2+} + 10\ H_2O = 4\ Fe(OH)_{3(s)} + 8\ H^+$	+0.177	+11.6

have an inherent tendency to corrode. The remaining reactions yield the pH–pE conditions at which Fe(II) is oxidized to one or more of the Fe(III) species. The Fe(II)-Fe(III) boundary is computed from Equation 2 (Table II) where the pE^0 value is +13.07. At pE values > +13.07, Fe(III) is the predominant species, whereas at pE values < +13.07, Fe(II) is the predominant species. This reaction and the $Fe_{(s)}$–Fe^{2+} reaction are independent of $[H^+]$ over the pH range of 0.0–2.2. Above the pH value of 2.2, Fe(III) enters into a series of acid-base reactions. For example:

$$Fe^{3+} + H_2O = FeOH^{2+} + H^+; \quad \log K_{13} = -2.2 \qquad (13)$$

Consequently, $FeOH^{2+}$ is the predominant species over the pH range of 2.2–3.7 and above pE values of +12. to +13. $Fe(OH)_2^+$ is the predominant species in the pH range of 3.7–4.8. Above the latter pH value, $Fe(OH)_{3(s)}$ is the major ferric species.

That the $[H^+]$ influences the pE value at which $Fe^{2+}(aq)$ is oxidized to any one of the three Fe(III) species is seen also in Figure 1. For example, the linear form of Equation 5 is:

$$pE = 22.8 - 3pH \qquad (14)$$

This equation indicates that there is a three-unit decrease in pE for a one-unit increase in pH. If an appropriate electron acceptor is present, then

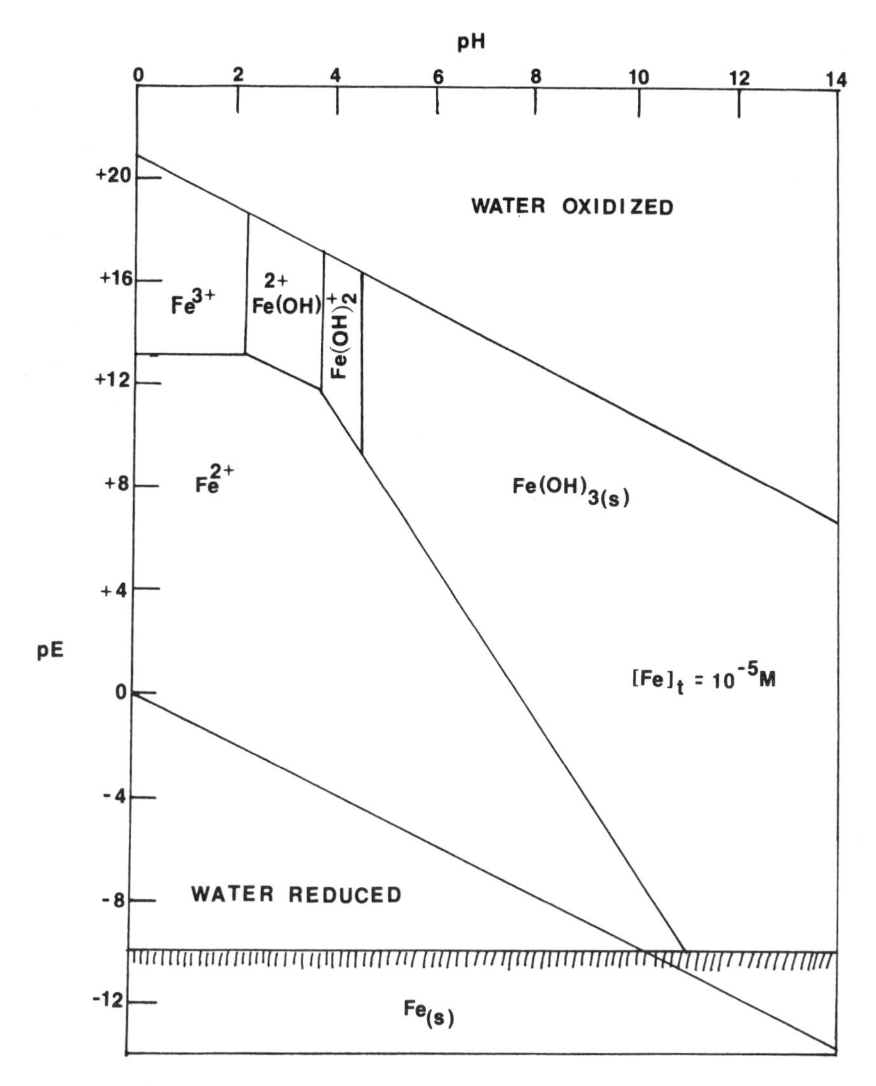

Figure 1. pH–pE stability diagram for the iron system. Reproduced from Faust and Aly [8], courtesy of Ann Arbor Science Publishers.

the pE value of the corrosion reaction to $Fe(OH)_{3(s)}$ is lowered as the pH value is increased. Equation 12 (Table II) is an example of a corrosion reaction at higher pH values. pH–pE diagrams do not give any information on the kinetics, i.e., rate of corrosion of metals. They convey only the pH–pE conditions under which a given oxidation state of iron should be stable, and a given oxidation or reduction reaction should occur.

pE–pH diagrams have been used extensively to relate laboratory data to the various theoretical conditions of corrosion, immunity and passivation [6]. Immunity is defined as the thermodynamic stability of the metal, whereas passivation is formation of a stable solid oxide or any other insoluble stable product. Corrosion is formation of stable soluble metallic products.

Figure 2a shows the pH–pE (E) diagram of Obrecht and Pourbaix [6]. It varies slightly from Figure 1 because $Fe_2O_{3(s)}$ and $Fe_3O_{4(s)}$ were used for the solid corrosion products rather than $Fe(OH)_{3(s)}$. Figure 2b shows the thermodynamically defined areas of corrosion, passivation and immunity of iron, whereas Figure 2c relates these quantities to the experimentally determined rates of iron corrosion. Obrecht and Pourbaix [6] interpret Figure 2c as acceptable agreement between the experimental results and theoretical predictions. For example, in the area of immunity, metallic iron is stable theoretically and should not corrode. In the area of passivation, iron is covered by an oxide film, which offers some protection against additional corrosion. This is seen in Figure 2c, where the rates of corrosion are extremely slow or nonexistent in the immunity and passivation areas.

The triangular shape of the corrosion area in Figure 2b results from the solubility of ferrous oxide greater than ferric oxide. Consequently, it is responsible for the corrosion of iron in water which, in turn, leads to the formation of craters covered with tubercles of rust. These tubercles result from localized attack at the discontinuity of the mill scales. This is seen in Figure 3 which is the "classic" sketch of pit corrosion. Since an electrode potential difference exists between the inside and outside of the crater, a current is produced that leads to highly localized corrosion. Oxygen is the electron acceptor in this representation of iron corrosion.

Some natural waters may contain a significant amount of bicarbonate ions to form $FeCO_{3(s)}$ (siderite):

$$FeCO_{3(s)} + H^+ = Fe^{2+} + HCO_3^-; \quad \log K_{15} = -0.34 \qquad (15)$$

This precipitation reaction and the reduction reaction 7 in Table II would yield an additional corrosion product. Figure 4 [9] shows the pH–pE (E) diagram with the presence of 100 ppm (solid lines) and 1000 ppm (dashed lines) of bicarbonate. Formation of $FeCO_{3(s)}$ occurs under alkaline conditions ($pH > 8.3$) and negative pE values. According to Hem [9], ferrous carbonate is readily oxidized to ferric hydroxide at pE values of an aerated water. This reaction may be [10]:

$$Fe(OH)_{3(s)} + CO_{2(g)} + H^+ + e = FeCO_{3(s)} + 2\,H_2O; \quad \log K_{16} = 10.4 \qquad (16)$$

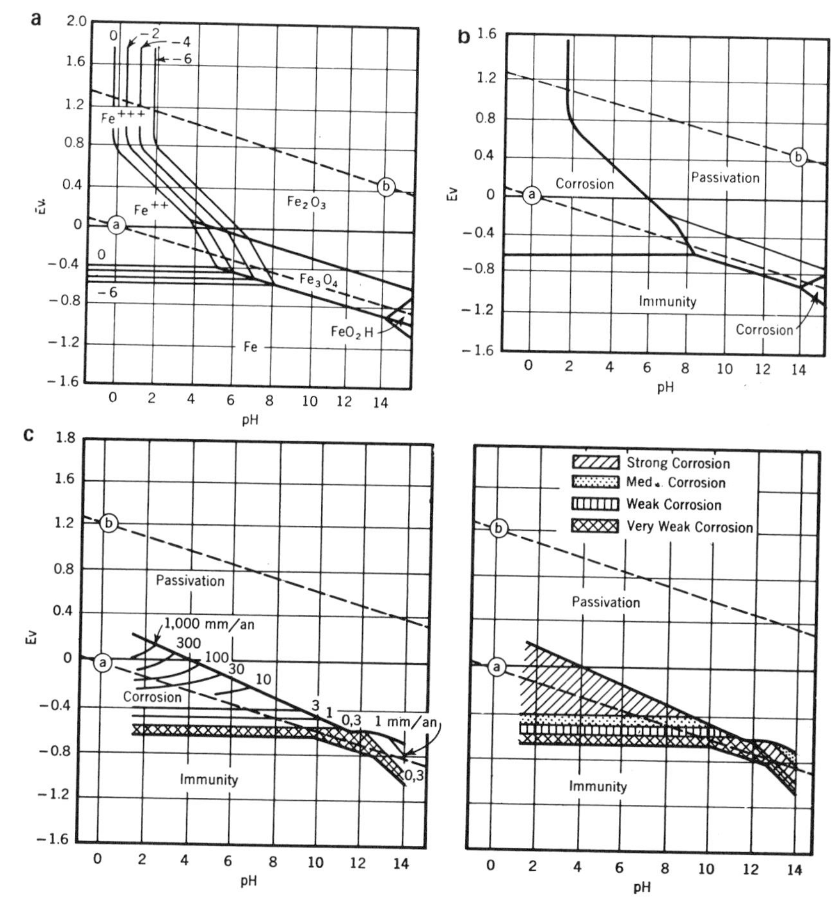

Figure 2. **(a)** Equilibrium potential-pH diagram. **(b)** Theoretical conditions for iron. **(c)** Experimental conditions of corrosion, immunity and passivation for iron. Reproduced from Obrecht and Pourbaix [6], courtesy of the American Water Works Association.

Copper

Since copper is a noble metal, it shows excellent resistance to corrosion. That is, the corrosion of metallic copper to cupric ions via the hydrogen ion reaction (Equation 23, Table III) is not thermodynamically spontaneous. On the other hand, Equation 24, Table III, shows that copper should be corrosive in an oxygenated environment. The theoretical corrosion diagram is seen in Figure 5a for a "pure" water system. At a pH of 6–7, it is completely uncorroded if the E value does not exceed

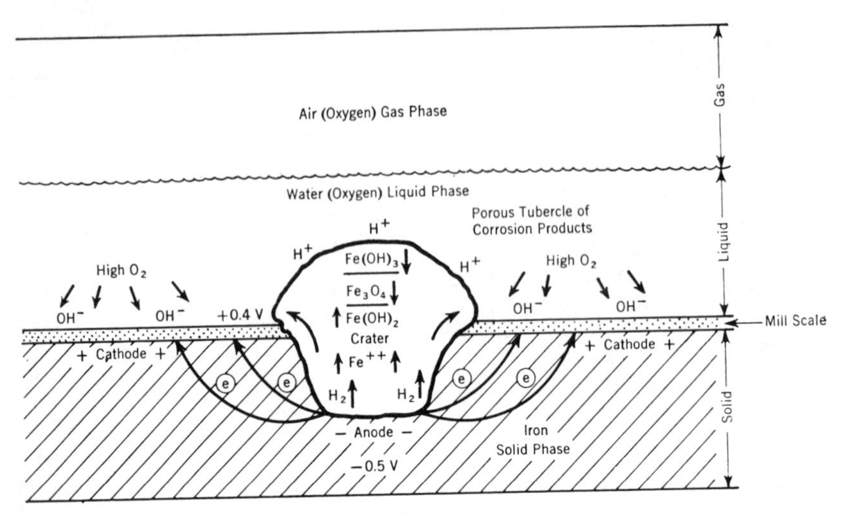

Figure 3. Pitting of iron by tuberculation and oxygen-concentration cells. Reproduced from Obrecht and Pourbaix [6], courtesy of the American Water Works Association.

+0.2. Above this E value and for a given pH value, the corrosion products are one or more of the following: Cu^+ (not shown) and Cu^{2+} ions, $Cu_2O_{(s)}$ and $CuO_{(s)}$. It should be noted also from Figure 5a that the immunity area for metallic copper extends well into the stability area of water, which is an additional reason for its thermodynamic stability. Figure 5b was drawn from carefully controlled laboratory experiments by Obrecht and Pourbaix [6] where a "close correlation" was obtained between total corrosion and oxygen consumption. These data demonstrate that copper does corrode in the presence of oxygen, as predicted by Equation 24, Table III.

There is evidence that copper pipes are corroded under acid conditions ($pH < 6.5$) and in the presence of carbon dioxide acidity [13]. Two corrosion products—Cu^{2+} ions and $Cu_2(OH)_2CO_{3(s)}$ (malachite)—have been observed. Equation 24, Table III accounts for the cupric ions that impart a bitter, metallic taste to the water, whereas Equation 25 accounts for formation of the solid, blue-green corrosion product. The chemistry of this type of copper corrosion apparently is rather complex. "Carbon dioxide added to a pure copper-water system in the absence of oxygen did not cause copper corrosion" [6]. However, when both oxygen and carbon dioxide were present in an experimental system (Figure 6) rather than oxygen alone (Figure 5b), an increase in corrosion rate was observed. For example, the total corrosion after six days was 0.65 mg

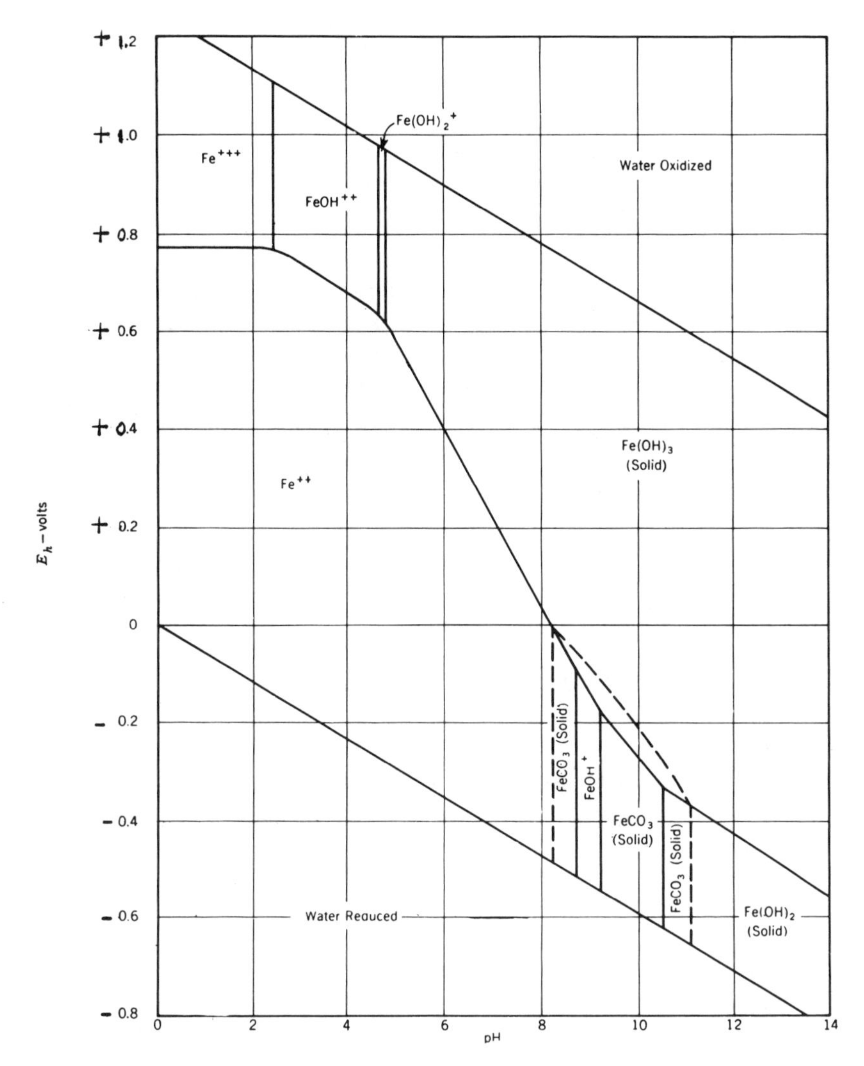

Figure 4. Stability fields of ferrous and ferric species for activity of 0.01 ppm dissolved iron. $[HCO_3^-] = 100$ ppm (solid boundaries); or $[HCO_3^-] = 1000$ ppm (dashed boundaries). Reproduced from Hem [9], courtesy of the American Water Works Association.

Cu/cm^2 (Figure 6) for the Cu-O_2-CO_2 system, whereas it was 0.27 mg Cu/cm^2 (Figure 5b) for the Cu-O_2 system. In Figure 6, total corrosion was obtained from the summation of Cu in the precipitate (perhaps malachite), in the surface film (copper oxides) and the copper ions in solution.

Table III. Some Corrosion and Precipitation Reactions for Copper at 25°C

Reaction Number	Reaction	V⁰[a]	pE⁰
17	$Cu^+ + e = Cu_{(s)}$	+0.520	+8.81
18	$Cu^{2+} + 2\ e = Cu_{(s)}$	+0.337	+11.4
19	$Cu^{2+} + e = Cu^+$	+0.153	+2.59
20	$Cu_2O_{(s)} + 2\ H^+ + 2\ e = 2\ Cu_{(s)} + H_2O$	+0.471	+15.97
21	$2\ Cu^{2+} + H_2O + 2\ e = Cu_2O_{(s)} + 2\ H^+$	+0.203	+6.88
22	$2\ CuO_{(s)} + 2\ H^+ + 2\ e = Cu_2O_{(s)} + H_2O$	+0.669	+22.7
23	$Cu_{(s)} + H^+ = Cu^{2+} + H_{2(g)}$	−0.337	−11.4
24	$O_{2(g)} + 4\ H^+ + 2\ Cu_{(s)} = 2\ H_2O + 2\ Cu^{2+}$	+0.892	+60.5
25	$2\ Cu^{2+} + HCO_3^- + 2\ H_2O = Cu_2(OH)_2CO_{3(s)}$ (malachite)		−4.50[b,c]
26	$Cu^{2+} + CO_3^{2-} = CuCO_{3(s)}$		6.77[b,c]

[a]From Pourbaix [11].
[b]Log Ks.
[c]From Stiff [12].

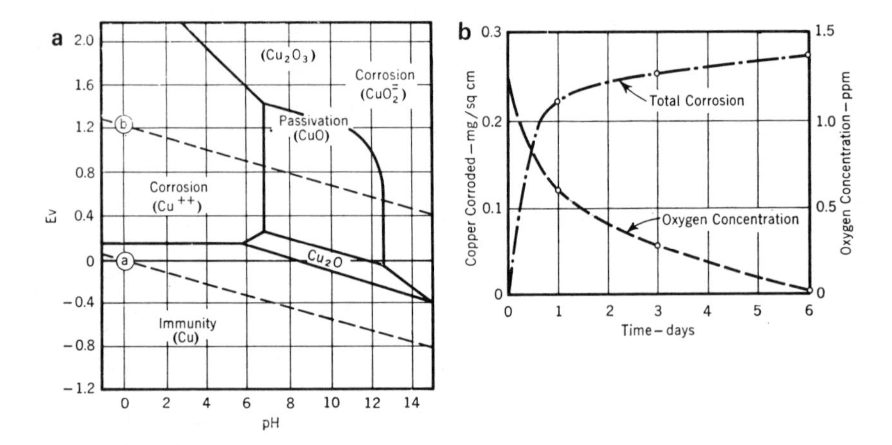

Figure 5. **(a)** Theoretical conditions for copper, 25°C. **(b)** Corrosion of high-purity copper in a closed system at 140°F (60°C) containing oxygen. Reproduced from Obrecht and Pourbaix [6], courtesy of the American Water Works Association.

Lead

Lead pipe and lead-based solders have been employed for many years in water distribution systems. Where its use is predominant—in many

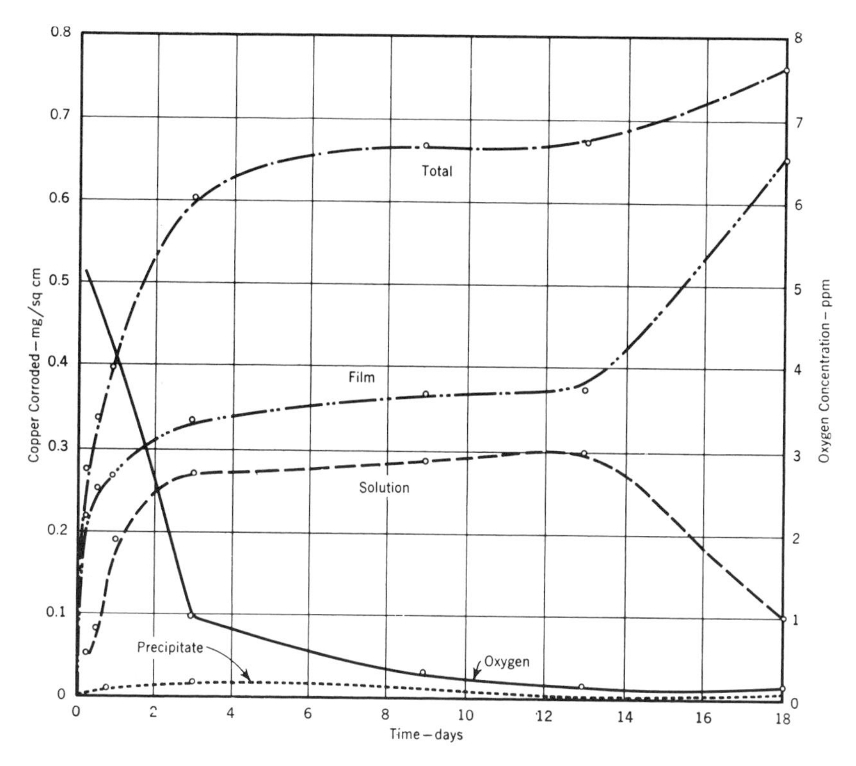

Figure 6. Effect of oxygen and $CO_{2(g)}$ on corrosion of pure copper. Closed system, 150°F (60°C), $[CO_{2(g)}] = 70.$ ppm. Reproduced from Obrecht and Pourbaix [6], courtesy of the American Water Works Association.

parts of Europe and North America—lead pipe is used primarily for service and residential lines. Lead service lines are still being installed today (e.g., in Chicago and New England areas) [14,15]. This is surprising in view of the EPA interim primary drinking water standard of 0.05 mg/l for lead [2]. Where the water is corrosive, concentrations of lead can greatly exceed this drinking water standard (e.g., Table IV).

That lead pipe is readily corroded is seen in the theoretical corrosion diagram in Figure 7 [11] and Reaction 27, Table V. The oxidation of elemental lead is thermodynamically spontaneous in water to Pb^{2+} ions which, in turn, form the oxides of lead: $PbO_{(s)}$, $PbO_{2(s)}$ and $Pb_3O_{4(s)}$. In the presence of carbon dioxide acidity, these compounds are fairly soluble with subsequent formation of the more insoluble lead carbonate (reactions 33 and 34, Table V) or lead hydroxycarbonate (reaction 35). For these reasons, lead must not be used for soft drinking waters, but may have some use in hard, scale-forming waters. The lead carbonates provide protection against additional corrosion by forming a "hard"

Table IV. Reported Lead Levels in Drinking Water [14]

Location	Maximum Lead Concentration (mg/l)
Boston, MA	1.51
Worcester, MA	1.90
Malborough, MA	0.25
Chatham, MA	0.10
New Bedford, MA	0.26
Bennington, VT	0.86
Seattle, WA	0.17
Victoria, BC	3.00
Oslo, Norway	2.00
Glasgow, Scotland	0.36
Sutherland, Scotland	3.14

scale on the interior surface of the pipe. Also, these carbonates form a "bridge" between the immunity area and the passivation area created by the stability of $PbO_{2(s)}$ (Figure 8).

Zinc

The corrosion of galvanized pipe by aggressive potable waters continues to be a problem [2,6]. That zinc is corroded spontaneously is seen in Equation 36, Table VI. The theoretical graph is seen in Figure 9. Since the $Zn_{(s)}$-Zn^{2+} lies outside of the water stability field, there is a rather "large" area for corrosion. It should be noted that the zinc corrosion reaction lies below the iron corrosion reaction (see Table I); hence, it is "more" corrosive than iron. For galvanized pipes, this provides protection for the underlying iron. Passivation protection for zinc is provided by $ZnO_{(s)}$ Equation 40, Table VI, or $Zn(OH)_{2(s)}$ as seen in Figure 9. This area of passivation may be expanded through formation of the insoluble $ZnCO_{3(s)}$ (not shown; see Equation 39, Table VI).

Asbestos-Cement Pipe

Much concern has been expressed by EPA [2] about release of asbestos fibers from corroding asbestos-cement (A/C) pipes. Occupational studies have indicated that inhaled asbestos is a human carcinogen. Also, an increased incidence of gastrointestinal cancer has been found in

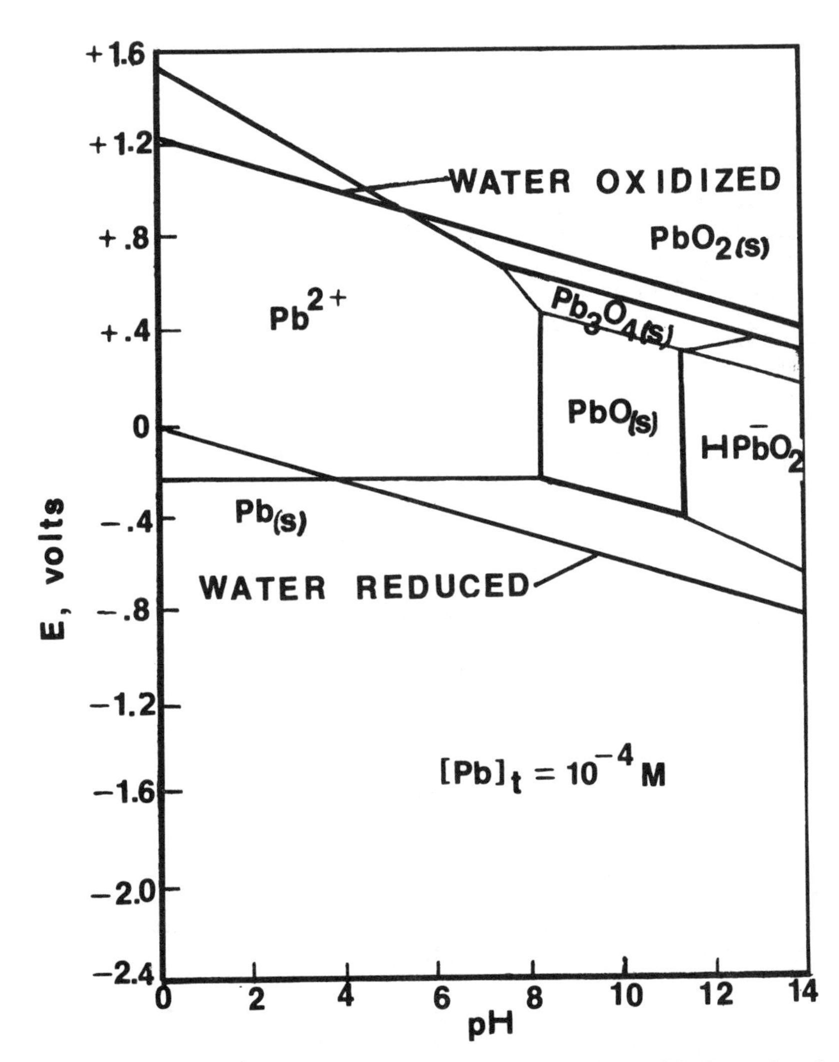

Figure 7. Potential-pH diagram for the Pb-H_2O system, 25°C. Reproduced from Pourbaix [11], courtesy of Pergamon Press Ltd.

occupational groups who have been exposed heavily to asbestos. Evidence for a relationship between asbestos in drinking water and increased cancer risk is inconclusive. However, EPA feels that there is justification to prevent the distribution of asbestos in drinking water.

The American Water Works Association [17,18] has established criteria for determining the quality of water that can be transported through A/C pipe without any adverse effects. An empirical indicator formula

Table V. Some Corrosion and Precipitation Reactions for Lead at 25°C [11]

Reaction Number	Reaction	V^0	pE^0
27	$Pb^{2+} + 2\ e = Pb_{(s)}$	−0.126	−4.27
28	$PbO_{(s)} + 2\ H^+ + 2\ e = Pb_{(s)} + H_2O$	+0.248	+8.41
29	$Pb_3O_{4(s)} + 2\ H^+ + 2\ e = 3\ PbO_{(s)} + H_2O$	+0.972	+32.95
30	$3\ PbO_{2(s)} + 4\ H^+ + 4\ e = Pb_3O_{4(s)} + 2\ H_2O$	+1.127	+76.4

Reaction Number	Reaction	log Ks
31	$Pb^{2+} + H_2O = PbO_{(s)} + 2\ H^+$	−12.65
32	$PbO_{(s)} + H_2O = HPbO_2^- + H^+$	+15.36
33	$Pb^{2+} + H_2O + CO_{2(g)} = PbCO_{3(s)} + 2\ H^+$	−5.31
34	$Pb^{2+} + H_2CO_3 = PbCO_{3(s)} + 2\ H^+$	−3.87
35	$3\ Pb^{2+} + CO_3^{2-} + 2\ H_2O = Pb_3(CO_3)_3(OH)_{2(s)} + 2\ H^+$	+18.8

was developed for a water's aggressiveness towards the corrosion of A/C pipes. The so-called aggressiveness index (AI) is:

Formula	Comment	Type of A/C Pipe[b]
$AI = pH + \log(AH)^a \geq 12.0$	Nonaggressive	I or II
$AI = pH + \log(AH) = 10.0$ to 11.9	Moderately aggressive	II
$AI = pH + \log(AH) \leq 10.0$	Highly aggressive	Neither

[a]A = alkalinity as mg/l as $CaCO_3$; H = Ca hardness as mg/l $CaCO_3$.
[b]Type I A/C pipe = not autoclaved, no limit on uncombined $Ca(OH)_2$. Type II A/C pipe = autoclaved, uncombined $Ca(OH)_2$ limited, resistant to all levels of soluble sulfates.

A study concerning the behavior of A/C pipe under various water quality conditions has been reported [17]. A summary of the data is seen in Table VII where it is somewhat obvious that the "low" values of pH, alkalinity and calcium hardness have a corrosive effect on A/C pipes. Asbestos fibers counts are given also for systems A–J cited in Table VII. Control of A/C pipe corrosion is given below.

Control of Corrosion

Corrosion control generally is accomplished by chemical treatment of the water and/or electrochemical treatment of the metallic or asbestos-

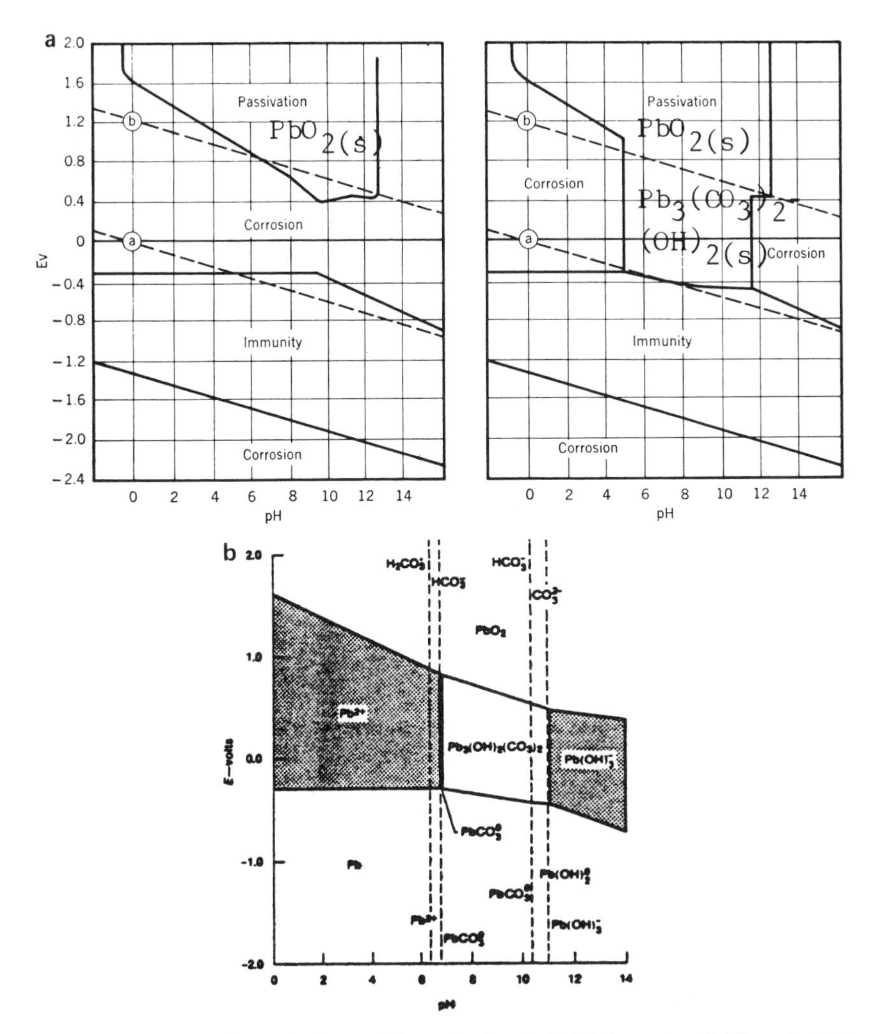

Figure 8. **(a)** Theoretical conditions for lead, 25°C in water (left) and 1 *M* $CO_{2(g)}$ (right). Reproduced from Obrecht and Pourbaix [6], courtesy of the American Water Works Association. **(b)** Potential-pH diagram for $C_t = 10^{-3.7}$ *M* (20. mg/l as $CaCO_3$) and dissolved Pb(II) species at 10^{-6} *M* (0.21 mg/l). Ionic strength = 0.0, T = 25°C, areas of passivation and immunity are unshaded. Reproduced from Schock [15], courtesy of the American Water Works Association.

cement pipe. It is virtually impossible to eliminate the thermodynamic corrosion of iron and copper pipes. Rather, a treatment or combination of treatments are employed to reduce the rate of corrosion to barely

Table VI. Some Corrosion and Precipitation Reactions for Zinc at 25°C [11]

Reaction Number	Reaction	V^0	pE^0
36	$Zn^{2+} + 2\ e = Zn_{(s)}$	−0.763	−25.9
37	$ZnO_{(s)} + 2\ H^+ + 2\ e = Zn_{(s)} + H_2O$	−0.439	−14.9
38	$HZnO_2^- + 3\ H^+ + 2\ e = Zn_{(s)} + 2\ H_2O$	+0.054	+1.83

Reaction Number	Reaction	log Ks
39	$Zn^{2+} + HCO_3^- = ZnCO_{3(s)} + H^+$	0.45[a]
40	$Zn^{2+} + H_2O = ZnO_{(s)} + 2\ H^+$	−10.96
41	$ZnO_{(s)} + H_2O = HZnO_2^- + H^+$	16.68

[a]From Kelley and Anderson [16].

detectable levels (see Figure 2c). Chemical control is usually an adjustment of the water's pH value (i.e., neutralization of acidity), removal of dissolved oxygen and an adjustment of the calcium content to provide a protective coating of $CaCO_3$. Electrochemical control is accomplished by avoiding the unfavorable range of the equilibrium potential pH for the iron-water-ion system. The theoretical protection diagram for the iron-water system is seen in Figure 10. Two principles of electrochemical corrosion control are: (1) to lower the potential into the area of immunity by galvanizing the iron with zinc or magnesium alloys (applied, of course, to the exterior of water pipes) or to apply an electric current to lower the potential below −0.62 V (Figure 10); and (2) to raise the potential into an area of passivation in order to provide a protective oxide film. Obrecht and Pourbaix [6] suggest aeration of the water to form an iron oxide film. This appears theoretically to be an attractive approach to corrosion control. However, it is not practical to do because the iron oxide films are not bound tightly and can be removed easily. Alternative anodic protective films are $CaCO_{3(s)}$ and phosphates. These alternatives are discussed below.

STABILITY INDICES

Several chemical indices indicate the corrosiveness of raw and finished water toward metallic and A/C pipes. For the most part, they are derived from chemical equilibrium equations for the $CaCO_{3(s)}$ system. In turn,

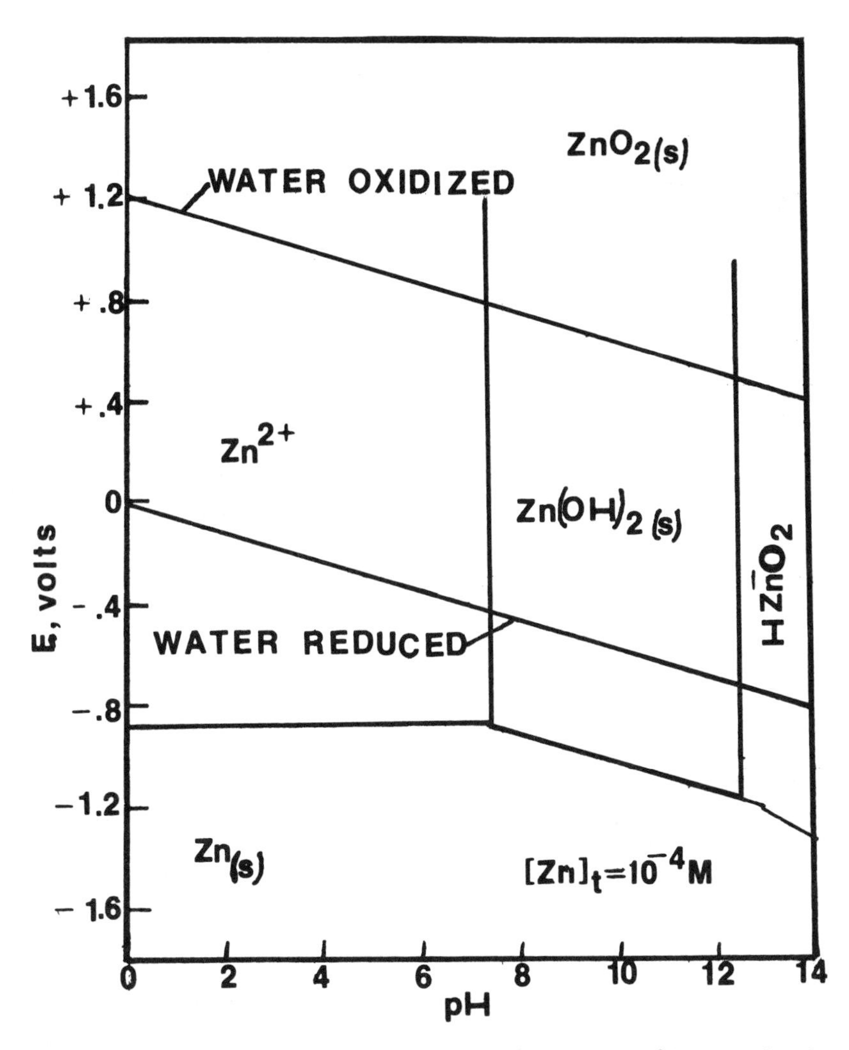

Figure 9. Potential-pH diagram for the system Zn-H_2O at 25°C. Reproduced from Pourbaix [11], courtesy of Pergamon Press Ltd.

these indices are related to some symptom of corrosion, i.e., "red water" complaints due to the presence of iron by-products. Langelier [19,20] developed a saturation index (SI) for corrosion protection by a thin film of $CaCO_{3(s)}$ on the interior walls of pipes. This index is calculated from readily obtainable analytical values and indicates the tendency of a natural or finished water either to deposit or dissolve calcium carbonate.

Table VII. Corrosion of Asbestos Cement Pipes in Various Water Qualities[a]

System	Initial Aggres-siveness Index	pH	Alkalinity as $CaCO_3$ (mg/l)	Calcium Hardness as $CaCO_3$ (mg/l)	Consistently Quantifiable Fibers	Pipe Wall Deteriorated as Deter-mined by Inspection
A	5.34	5.2	1.0	1.4	yes	yes
J	5.67	4.8	3.0	2.5	yes	yes
I	7.46	6.0	4.0	7.5	no	no
H	8.74	7.1	89	0.5	yes	yes
B	9.51	7.2	14	14.5	yes	yes
G	10.48	8.3	20	7.5	yes	[b]
C	11.56	7.5	88	82	no	[b]
D	12.54	7.8	220	250	no	[b]
E	12.74	9.4	50	44	no	[b]

[a] Reproduced from Buelow et al. [17], courtesy of the American Water Works Association.
[b] Not inspected.

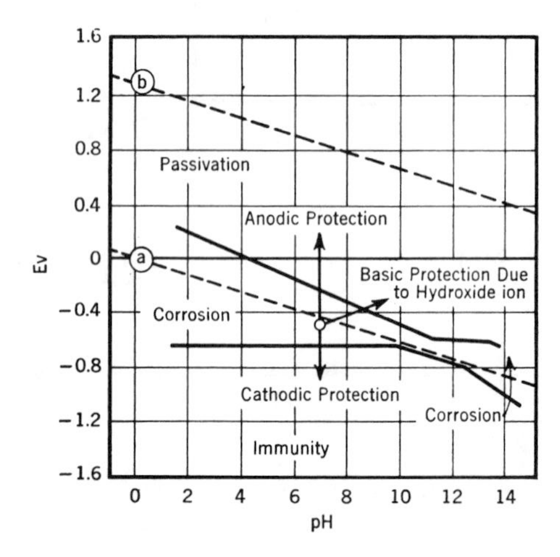

Figure 10. Theoretical protection diagram for the Fe-H_2O system. Reproduced from Obrecht and Pourbaix [6], courtesy of the American Water Works Association.

Langelier's Saturation Index

The fundamental reaction in the SI of Langelier is:

$$CaCO_{3(s)} + H^+ = Ca^{2+} + HCO_3^- \tag{42}$$

When equilibrium is reached:

$$K_{43} = \frac{[Ca^{2+}][HCO_3^-]}{[H^+]} \tag{43}$$

Rearranging, taking logs, etc.:

$$pHs = pCa^{2+} + pHCO_3^- - pK_{43} \tag{44}$$

where pHs = equilibrium pH value for Equation 42
pCa^{2+} = equilibrium calcium content
$pHCO_3^-$ = total alkalinity when the pH value is less than 9.5
pK_{43} = arithmetic difference between pK_2 (second protolysis constant for H_2CO_3) and pK_s (solubility product constant for $CaCO_{3(s)}$); i.e., $pK_{43} = pK_s - pK_2$.

Thus, the pHs is a calculated value from the three terms on the right-hand side of Equation 43.

The saturation index is calculated from:

$$SI = pHac - pHs \tag{45}$$

where pHac = actual pH value of the water.

When pHac is less than pHs, negative SI values are obtained, and the water tends to be corrosive. When the pHac is greater than the pHs, positive SI values are obtained, and the water tends to be $CaCO_{3(s)}$ scale forming. When the SI value is 0, the water is in equilibrium with respect to Equation 42.

Occasionally, there is a need to "correct" Equation 43 for alkalinity due to carbonate above a pHac of 9.5 and hydroxyl ions above a pHac of 10.5. In this case, an equation can be derived:

$$pHs = -pK_{43} + pCa^{2+} + p\left[alk - [H^+] - \frac{Kw}{[H^+]}\right] + \log\left[1 + \frac{2\,K_2}{[H_s^+]}\right] \tag{46}$$

Also, corrections should be made in Equations 43 and 46 for the ionic strength μ of the water due to dissolved and charged ions. The concentration terms should read activities (). For example:

$$K_s' = [Ca^{2+}]\gamma_{Ca^{2+}}[CO_3^{2-}]\gamma_{CO_3^{2-}} = (Ca^{2+})(CO_3^{2-}) \tag{47}$$

where K_s' = thermodynamic infinite dilution solubility product constant for $CaCO_{3(s)}$

The activity coefficients $\gamma_{Ca^{2+}}$ and $\gamma_{CO_3^{2-}}$ are calculated from the ionic strength of the water. In turn, it is necessary to know the ionic composition of the water in order to calculate the ionic strength. Since the ionic composition is infrequently known for raw or finished waters, there are several pragmatic and generalized methods for calculating μ and γ [5,8,21]. In Table VIII, values of pK_2' and pK_s' are given for ionic strengths ranging from 0.0000 to 0.020 (total dissolved solids, 0 to 800 ppm) and for temperatures ranging from 0 to 90°C. The 25°C K_s' value of 4.8×10^{-9} comes from Frear and Johnston [22] and the K_2' values come from MacInnes and Belcher [23].

The SI concept can be extended to magnesium compounds, although they are infrequently found in pipe scales. There are, however, hard waters containing magnesium salts that require "softening" via the lime-soda process (see Chapter 7). This treatment precipitates magnesium hydroxide in accord with:

$$K_{sMg} = [Mg^{2+}][OH^-]^2 \tag{48}$$

from which this equation is derived:

$$pHsMg = \tfrac{1}{2}[pMg^{2+} - pK_{sMg}] + pKw \tag{49}$$

Equation 49 gives the equilibrium pH value for the $Mg(OH)_2$ system.

Applications of Langelier's Saturation Index

There are several techniques for the application and use of the SI. Langelier [19] constructed rather complex stability diagrams for calculating pHs from the water quality variables in the right-hand side of Equation 44. Figure 11 shows a log-log plot of $[Ca^{2+}]$ and $[Mg^{2+}]$ vs pH (25°C) for a series of total alkalinity curves. Two example uses of the diagram are given. The example of $[Ca^{2+}] = 30.$ ppm, total alkalinity = 60. ppm, and pHac = 7.9 yields a pHs of 8.6 and a SI value of −0.7. This water would be undersaturated with respect to $CaCO_{3(s)}$, and would dissolve any previously precipitated film. At a pHac of 7.9, there is speculation about the corrosiveness of the water.

Another use of the diagram is to determine the equilibrium concentration of calcium at the specified pHac and total alkalinity. From the example above, proceed upward on the pHac 7.9 line to the total alkalinity curve of 60 ppm. Under these conditions, $[Ca^{2+}]$ should be 150 ppm. Consequently, it would be necessary to add 120 ppm of calcium.

Table VIII. Values of pK_2' and pK_s' at 25°C for Various Ionic Strengths and of the Difference ($pK_2' - pK_s'$) for Various Temperatures[a]

Ionic Strength	Total Dissolved Solids	25°C pK_2'	25°C pK_s'	25°C $pK_2'-pK_s'$	$pK_2'-pK_s'$[b] 0°C	10°C	20°C	50°C	60°C	70°C	80°C	90°C
0.0000	0	10.26	8.32	1.94	2.20	2.09	1.99	1.73	1.65	1.58	1.51	1.44
0.0005	20	10.26	8.23	2.03	2.29	2.18	2.08	1.82	1.74	1.67	1.60	1.53
0.001	40	10.26	8.19	2.07	2.33	2.22	2.12	1.86	1.78	1.71	1.64	1.57
0.002	80	10.25	8.14	2.11	2.37	2.26	2.16	1.90	1.82	1.75	1.68	1.61
0.003	120	10.25	8.10	2.15	2.41	2.30	2.20	1.94	1.86	1.79	1.72	1.65
0.004	160	10.24	8.07	2.17	2.43	2.32	2.22	1.96	1.88	1.81	1.74	1.67
0.005	200	10.24	8.04	2.20	2.46	2.35	2.25	1.99	1.91	1.84	1.77	1.70
0.006	240	10.24	8.01	2.23	2.49	2.38	2.28	2.03	1.94	1.87	1.80	1.73
0.007	280	10.23	7.98	2.25	2.51	2.40	2.30	2.05	1.96	1.89	1.82	1.75
0.008	320	10.23	7.96	2.27	2.53	2.42	2.32	2.07	1.98	1.91	1.84	1.77
0.009	360	10.22	7.94	2.28	2.54	2.43	2.33	2.08	1.99	1.92	1.85	1.78
0.010	400	10.22	7.92	2.30	2.56	2.45	2.35	2.10	2.01	1.94	1.87	1.80
0.011	440	10.22	7.90	2.32	2.58	2.47	2.37	2.12	2.03	1.96	1.89	1.82
0.012	480	10.21	7.88	2.33	2.59	2.49	2.39	2.13	2.04	1.97	1.90	1.83
0.013	520	10.21	7.86	2.35	2.61	2.50	2.40	2.15	2.06	1.99	1.92	1.85
0.014	560	10.20	7.85	2.36	2.62	2.51	2.41	2.16	2.07	2.00	1.93	1.86
0.015	600	10.20	7.83	2.37	2.63	2.52	2.42	2.17	2.08	2.01	1.94	1.87
0.016	640	10.20	7.81	2.39	2.65	2.54	2.44	2.19	2.10	2.03	1.96	1.89
0.017	680	10.19	7.80	2.40	2.66	2.55	2.45	2.20	2.11	2.04	1.97	1.90
0.018	720	10.19	7.78	2.41	2.67	2.56	2.46	2.21	2.12	2.05	1.98	1.91
0.019	760	10.18	7.77	2.41	2.67	2.57	2.47	2.21	2.12	2.05	1.98	1.91
0.020	800	10.18	7.76	2.42	2.68	2.58	2.48	2.22	2.13	2.06	1.99	1.92

[a]Reproduced from Langelier [20], courtesy of the American Water Works Association.
[b]See *J. Am. Water Works Assoc.* 30:1806 (1938) for correction to pK_2'-pK_s'.

This is not quite accurate, because it is difficult to add calcium to water without affecting the pH and total alkalinity values. There are additional and more complex stability diagrams of plots of CO_3^{2-}, OH^-, and H_2CO_3 vs pH [19,20] and at temperatures other than 25°C. pHs may also be calculated from [24]:

$$pHs = A + B - \log[Ca^{2+}] - \log[alk] \quad (50)$$

A and B are constants related to water temperature and total dissolved residue, respectively. Table IX gives values for A and B, and logs of $[Ca^{2+}]$ and total alkalinity. Calculation of pHs is, of course, a summation of appropriate values from the tables for the terms in Equation 50.

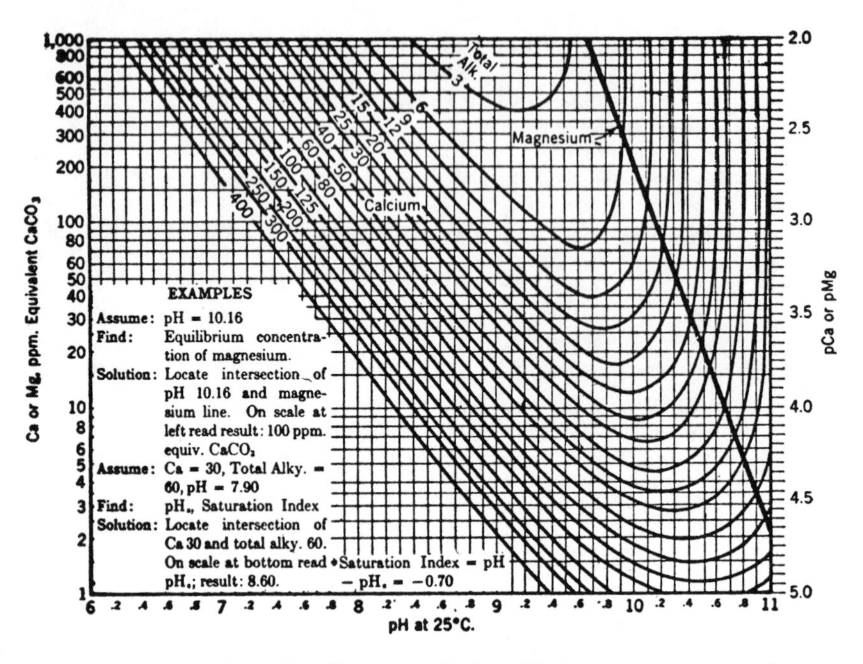

Figure 11. Use of stability diagram to find equilibrium concentration of magnesium or calcium. Reproduced from Langelier [19], courtesy of the American Water Works Association.

Calculation of pH from $CaCO_{3(s)}$ Solubility Diagram

The $CaCO_{3(s)}$ solubility diagram may be used to calculate pHs values (see Faust and Aly [8] for construction of diagram). Figure 12 shows this solubility diagram and an example calculation using water quality data from the Delaware River at Trenton, New Jersey, and at 15°C. The pHs value is obtained by plotting the actual pH and calcium content from

Table IXa. Constant A as a Function of Water Temperature[a]

Water Temperature (°C)	A[b]
0	2.60
4	2.50
8	2.40
12	2.30
16	2.20
20	2.10

Table IXb. Constant B as a Function of Total Dissolved Residue[a]

Total Dissolved Residue (mg/l)	B
0	9.70
100	9.77
200	9.83
400	8.86
800	9.89
1000	9.90

Table IXc. Logarithms of Calcium Ion and Alkalinity Concentrations[a]

Ca^{2+} or Alkalinity (mg/l as $CaCO_3$) equiv	log
10	1.00
20	1.30
30	1.48
40	1.60
50	1.70
60	1.78
70	1.84
80	1.90
100	2.00
200	2.30
300	2.48
400	2.60
500	2.70
600	2.78
700	2.84
800	2.90
900	2.95
1000	3.00

[a]Reproduced from *Standard Methods* [24], courtesy of the American Public Health Association.

[b]Calculated from $K_2(H_2CO_3)$ values of Harned and Scholes, and $Ks(CaCO_3)$ values of Larson and Buswell reported by Faust and Aly [8].

which a line is drawn to the equilibrium solubility line. This intersection yields, for the example, a pHs of 9.0 an a SI value of −1.5. This technique also gives the calcium deficiency at the actual pH value by drawing a line vertically until it intersects the equilibrium boundary line. For the example, 89.6 mg/l of calcium would be required to saturate the water at

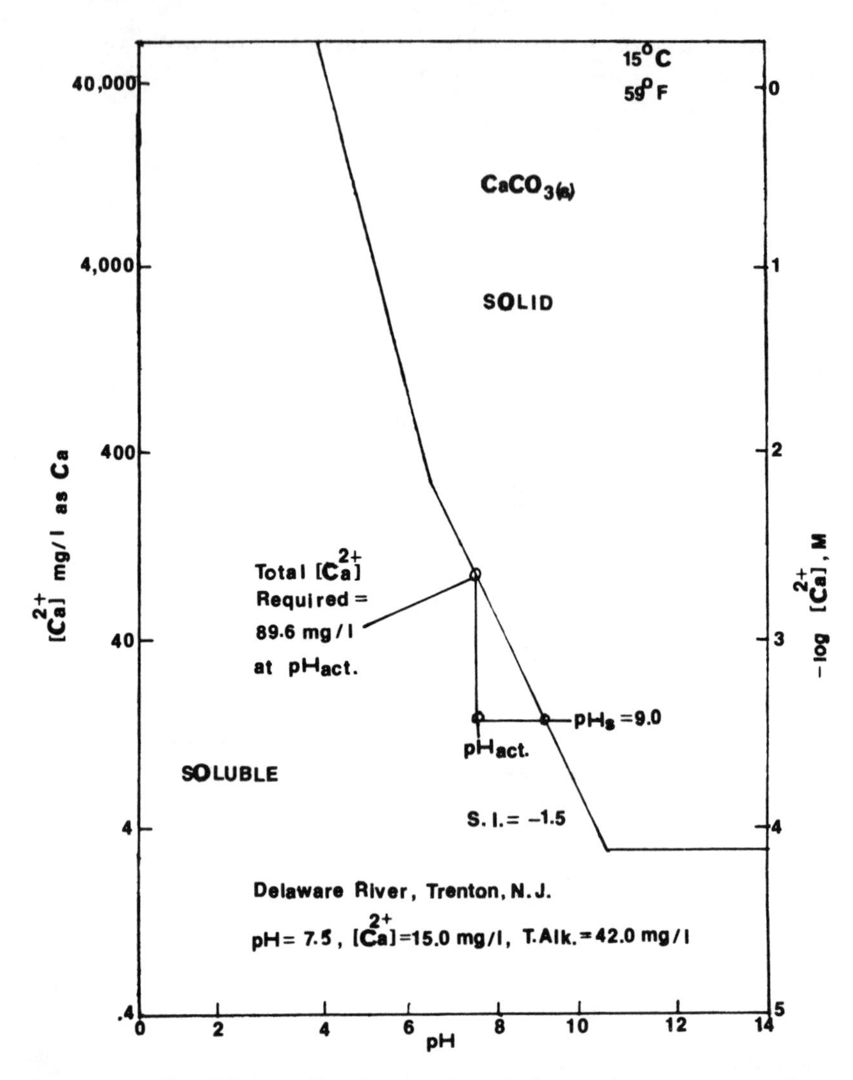

Figure 12. Simplified stability diagram for calculation of pHs, SI and $[Ca^{2+}]_t$ for saturation.

the actual pH of 7.5. An approximation of the concomitant alkalinity is calculated from the relation [8]:

$$2\,[Ca^{2+}] = [alk] \tag{51}$$

This method yields essentially the same pHs value as the Langelier diagram and is less confusing. The empirical method of Larson [24] is at variance with the Langelier and $CaCO_{3(s)}$ solubility diagrams. Also, there will be differences in calculated SI values due to variations in the thermodynamic constants used in Equations 44 and 46.

There are, of course, severe limitations of the thermodynamic methods described above for predicting the scale-forming or scale-dissolving tendencies of raw and finished waters via Equations 42 and 43. They provide no information about the kinetics of the precipitation of $CaCO_{3(s)}$ in water distribution systems. It is argued, perhaps, that kinetics is a factor, since rapid precipitation would lead to scale formation within the first foot or so of the water distribution system, whereas slow precipitation may protect only the extremities of the system. The rate of scale formation ideally should provide a "thin" film throughout the distribution system. Support for this argument comes from Kemp [25]:

> Unfortunately there is a tendency for water engineers to suppose that the pHs has a bearing on the corrosion of metals as well as that of cement and concrete. This is based on the presumption, not necessarily true, that a scale-forming water is likely to deposit a film of calcium carbonate over metal pipes and fittings in contact with it and so protect them from corrosion, and this is often given as a further (if not the prime reason) for giving the water a small positive saturation index.

A modified Langelier pHs equation has been developed for waters whose sulfate contents are greater than 1000 mg/l [26]. This kind of water may exist in open, recirculating cooling water systems where pH is controlled by sulfuric acid addition. Formation of $CaSO_4^0$ ion pairs can significantly reduce free $[Ca^{2+}]$, which, in turn, would cause the calculated pHs value to be low. The relevant equations are:

$$K_{52} = \frac{[CaSO_4^0]}{[Ca^{2+}][SO_4^{2-}]\gamma_1^8} = 10^{2.31} \tag{52}$$

$$[Ca^{2+}]_t = [Ca^{2+}] + [CaSO_4^0] \tag{53}$$

$$[SO_4^{2-}]_t = [SO_4^{2-}] \tag{54}$$

Combining Equations 52, 53 and 54 yields the free calcium content:

$$[Ca^{2+}] = \frac{[Ca^{2+}]_t}{1 + K_{52}\gamma_1^8[SO_4^{2-}]_t} \tag{55}$$

Taking the logarithm of Equation 55 and substituting into:

$$pHs = -pK_{43} - \log[Ca^{2+}] - \log[alk] - 5\log\gamma_1 \quad (56)$$

(Equation 56 is an ionic strength correction of Equation 44.) The corrected form of the pHs equation is:

$$pHs = -pK_{43} - \log[Ca^{2+}]_t - \log[alk] - \log\left[\frac{\gamma_1^5}{1 + K_{52}\gamma_1^8[SO_4^{2-}]_t}\right] \quad (57)$$

where γ_1 is the appropriate activity coefficient. The last term in Equation 57 is the correction for $CaSO_4^0$ formation. Data are given in Table X to demonstrate the significant difference between pHs values calculated from Equations 56 and 57.

Other Indices

Ryznar [27] stated that "the saturation index, however, is not always reliable in predicting this (scale-forming or corrosive) because some waters with a positive index actually may be quite corrosive." It was stated also that "one reason for this is the fact that the Langelier index does not indicate how much calcium carbonate will deposit: whether a state of supersaturation will be present which will be great enough to produce a precipitate, and whether it is great enough to give a protective film." Whereupon, Ryznar proposed an empirical expression that was labeled the stability index (StI for convenience):

$$StI = 2pHs - pH \quad (58)$$

In this system, the stability index will be positive for all waters. Two examples are:

Water A, 75°C	Water B, 75°C
pHac 6.5	pHac 10.5
pHs 6.0	pHs 10.0
SI +0.5	SI +0.5
StI +5.5	StI +9.5

According to Ryznar, water A will give an appreciable amount of $CaCO_{3(s)}$ scale, whereas water B will form a limited amount of scale and may be severely corrosive, especially at higher temperatures. Experi-

Table X. Comparison of pH_s Predictions for Literature Cooling Water Data[a,b]

Species	I	II	III
Na + K	2,648	23,690	1,342
Ca	141	300	116
Mg	31	61	12
HCO_3	56	61	299
Cl	750	12,136	1,498
SO_4	4,500	33,800	1,239
SiO_2	50		20
A (ppm $CaCO_3$)	46	50	245
TDS	8,000	70,000	4,300
pH	6.5	6.0	7.1
pH_s (Equation 56)	7.84	7.35	7.10
pH_s (Equation 57)	8.26	8.74	7.25

[a]Reproduced from McGaughey and Matson [26], courtesy of the International Association on Water Pollution Research.
[b]All concentrations are mg/l^{-1} unless otherwise noted.

mental data are given to support this interpretation (Figure 13). That waters with high pH values will form less scale and have positive StI values, is consistent with the fact that $CaCO_{3(s)}$ solubility decreases with an increase in pH value (see Figure 12) and an increase of temperature [8]. These waters simply do not have the Ca^{2+} and CO_3^{2-} to form $CaCO_{3(s)}$.

Caldwell and Lawrence [28,29] published a series of diagrams that are similar to Langelier's, one of which is seen in Figure 14. These diagrams can be used for both corrosion control and water softening by the lime-soda process (see Chapter 7). Two examples are cited for use of the C-L diagrams:

	Water C	Water D
Temperature (°C)	15.0	15.0
Total Dissolved Solids (mg/l)	96.0	115.0
pH	8.6	9.5
Ca^{2+} (mg/l as $CaCO_3$)	70.0	28.0
alk (mg/l as $CaCO_3$)	37.0	70.0

When the values for pH, $[Ca^{2+}]$ and [alk] from water C are plotted on the C-L diagram (Figure 14), they form a triple point A. This is inter-

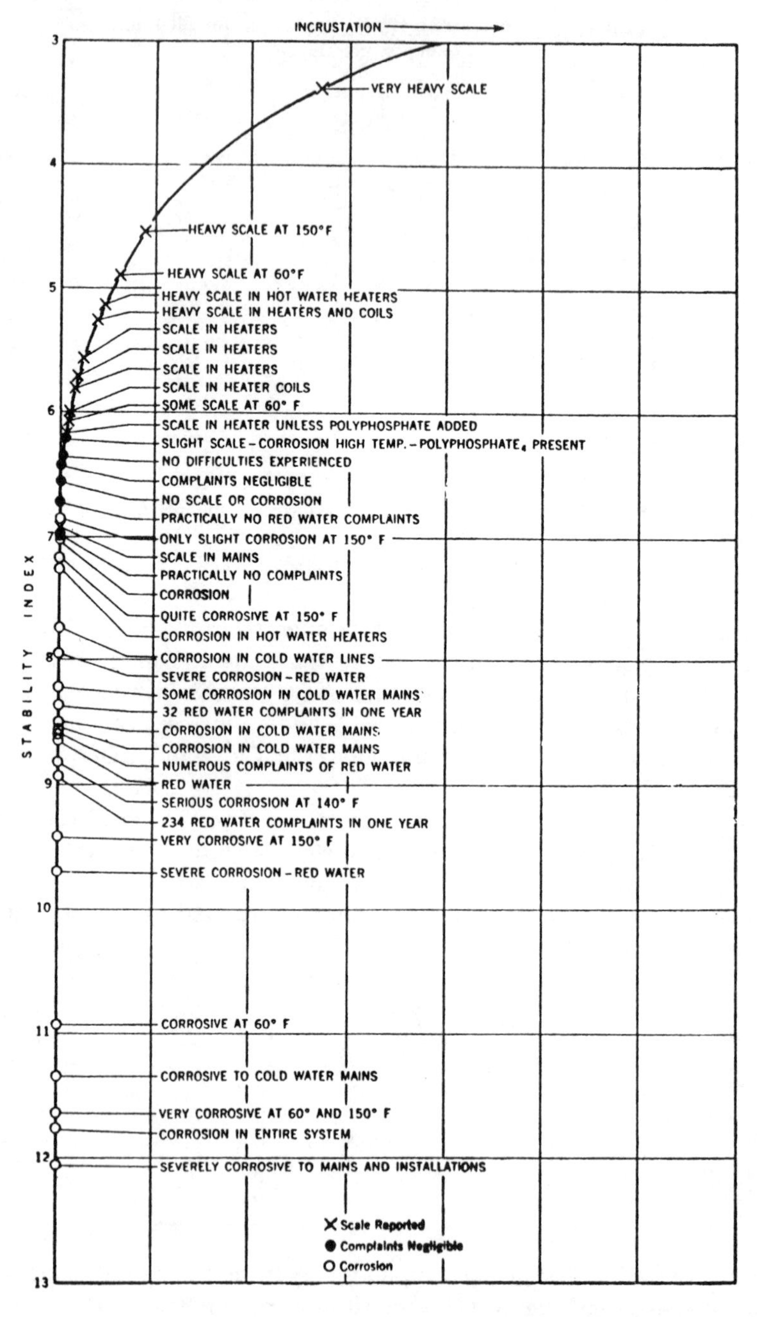

Figure 13. Field observations of Ryznar's stability index. Reproduced from Ryznar [27], courtesy of the American Water Works Association.

preted to be a water whose Ca^{2+} and CO_3^{2-} are in equilibrium. When these three values from water D are plotted, they do not form a triple point. Rather, three double points B, C and D result. In order to determine the over- and undersaturation, the $[Ca^{2+}]$ at the intersection of the pH and alkalinity values, point C, is compared with the actual $[Ca^{2+}]$. If the measured $[Ca^{2+}]$ is greater than the graphical value, the water is oversaturated and $CaCO_{3(s)}$ should form. For this example, the actual $[Ca^{2+}]$ is 28 mg/l, whereas the graphical value is 6 mg/l. Undersaturation is denoted where the $[Ca^{2+}]_{ac} < [Ca^{2+}]_{graph}$. C-L diagrams are available at 2, 5, 15 and 25°C and at 40, 400 and 1200 mg/l TDS.

Another statement was critical of the SI concept: "Langelier's saturation index is not a proper measure of the forces tending to develop calcite coatings in a water line" [30]. Instead, a more useful and exact measure of calcium carbonate excess would be the total excess of Ca^{2+} and CO_3^{2-} that must be precipitated from a water for complete stability. For this purpose, a driving-force index (DFI) has been devised. It is:

$$\mathrm{DFI} = \frac{[Ca^{2+}][CO_3^{2-}]}{KsCaCO_{3(s)}} \tag{59}$$

This equation uses multiplication of actual concentrations of calcium and carbonate divided by the solubility product constant, Ks, for $CaCO_{3(s)}$. Corrections are made for temperature and dissolved solids. When the DFI is 1.0, the water is in equilibrium with $CaCO_{3(s)}$ or it is stable. When the DFI is >1.0, $CaCO_{3(s)}$ should precipitate, and when it is <1.0, the scale should dissolve and the water may be corrosive. McCauley expresses the terms in Equation 59 as equivalents of $CaCO_3$. This is simply incorrect from a thermodynamic point of view. The DFI values should be equal, less than or greater than 1.0, and molar concentrations should be used in Equation 59. In any event, the concept of DFI is valid, although McCauley's use is empirical. In principle, the DFI is the same as the saturation index employed by geochemists to determine the equilibrium conditions of various dissolved constituents within geologic formations [8].

CASE STUDIES OF CORROSION CONTROL

Application of Stability Indices and $CaCO_{3(s)}$ Films

There is a considerable amount of pragmatic experience and data from application of various stability indices. Flentje [31] reported an empirical relation between Langelier's SI and Ryznar's index (RI) and corrosion at

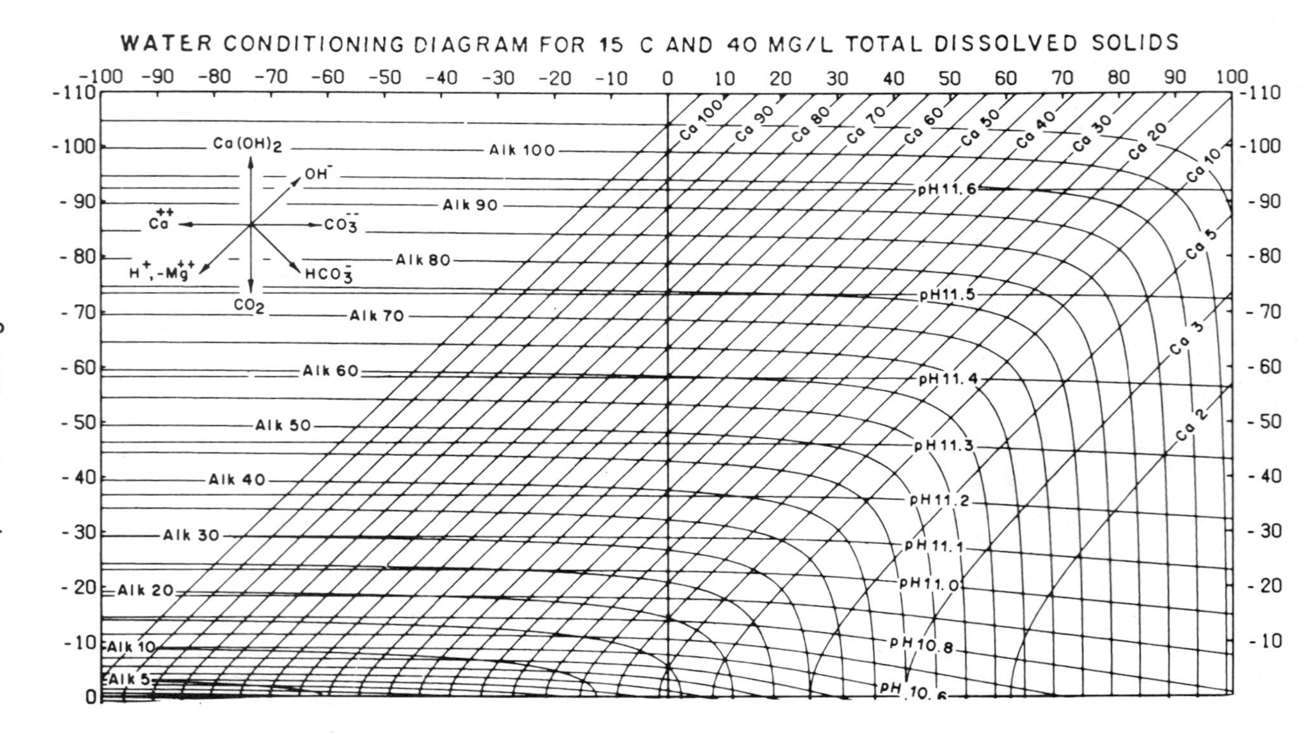

WATER CONDITIONING DIAGRAM FOR 15 C AND 40 MG/L TOTAL DISSOLVED SOLIDS
ACIDITY, MG/L AS CACO3
Ca(OH)2
OH-
CO3--
HCO3-
CO2
Ca++
H+, -Mg++
Alk 100
Alk 90
Alk 80
Alk 70
Alk 60
Alk 50
Alk 40
Alk 30
Alk 20
Alk 10
Alk 5
pH11.6
pH11.5
pH11.4
pH11.3
pH11.2
pH11.1
pH11.0
pH10.8
pH 10.6
Ca 100
Ca 90
Ca 80
Ca 70
Ca 60
Ca 50
Ca 40
Ca 30
Ca 20
Ca 10
Ca 5
Ca 3
Ca 2

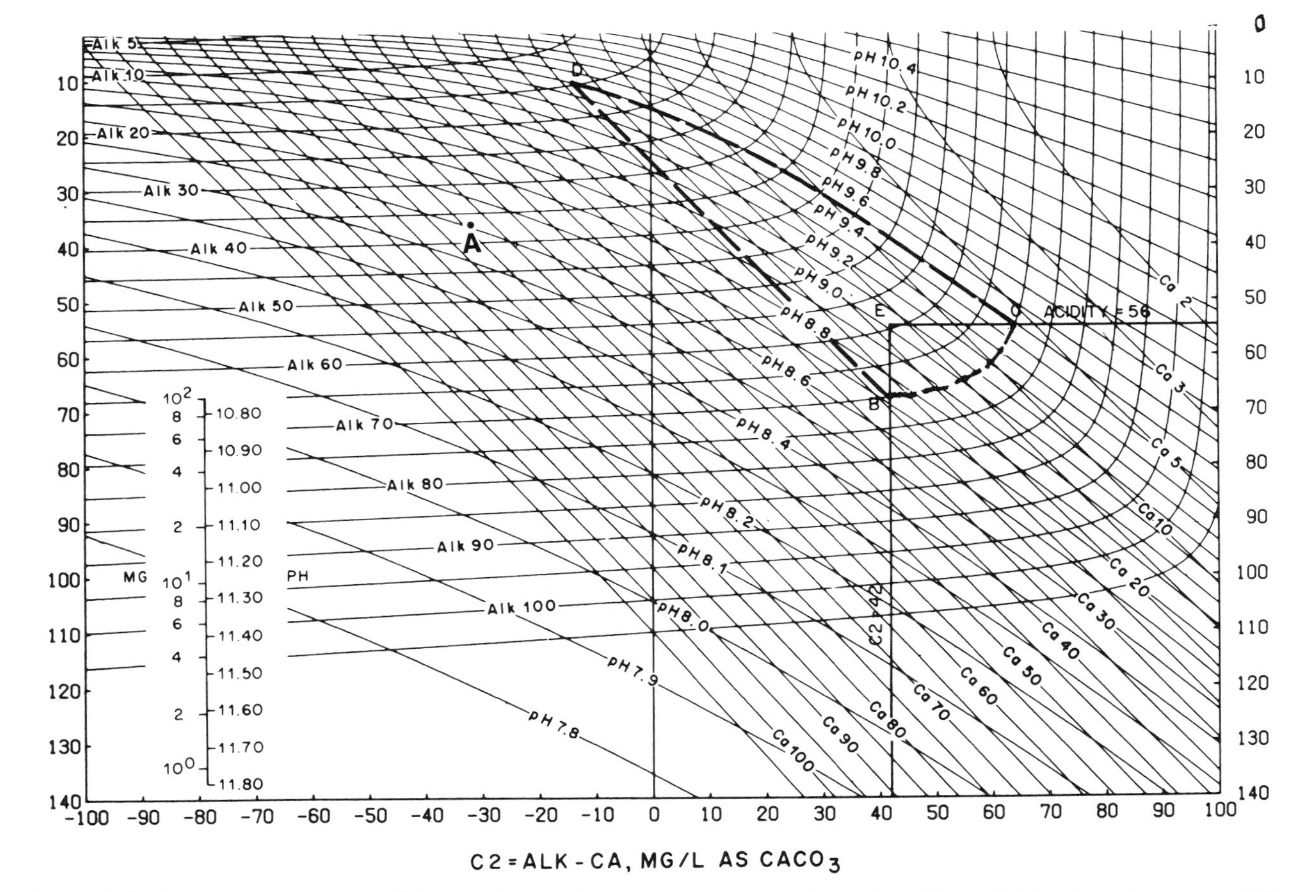

Figure 14. Caldwell-Lawrence water conditioning diagram for corrosion control. Reproduced from Caldwell and Lawrence [28], courtesy of the American Chemical Society.

various locations in the U.S. In this case, corrosion was measured by the "iron pickup," i.e., an increase in the iron content as the water flows through the distribution system from a water treatment plant. Table XI shows these iron concentrations and the concomitant SI and RI values of the various waters. Negative SI values correlate reasonably well with the occurrence of iron. There is lesser correlation with the RI value. See, for example, the data from Uniontown, Pennsylvania. The SI values are −3.60 and −2.09, whereas the RI values are 3.60 and 11.09, respectively. The iron contents were 0.24 and 0.22 ppm. The SI appears to be more consistent in indicating corrosive water as seen in Figure 15 [31]. Similar experiences were reported by Dye [32] for several Midwestern cities. For the best anticorrosion treatment of softened surface waters, Dye recommends the maintenance of a DFI of 15–20 and a SI of 1.0 or more in the pH range of 9.5–10.0. Since these waters may be too scale-forming, 0.5 ppm polyphosphate should be applied.

Experimental data have been reported on the relation of $CaCO_{3(s)}$ deposition on iron surfaces, dissolved oxygen content and the SI [33]. This study was catalyzed by reports from European literature that a heterogeneous layer of rust and $CaCO_{3(s)}$ gives a better protection for corrosion than does a layer of either rust or $CaCO_{3(s)}$ alone. In turn, this raises another question concerning the role of dissolved oxygen content of water. Since DO is an important factor in promoting corrosion (Equations 11 and 12), it would seem desirable to remove it from water. Consequently, there are divergent opinions about the permissible level of DO in the water.

In Figure 16, the influence of pH on the deposition of $CaCO_{3(s)}$ (mg/10 cm^2) on iron surfaces is seen [33]. The experimental water had an initial hardness and alkalinity of 205 mg/l as $CaCO_3$, pHs of 7.35, and the DO content was held at 9.0 mg/l. The plot shows that as the pH value is increased greater amounts of $CaCO_{3(s)}$ are deposited (curves 1–5). In curve 6, a stainless steel coupon was used instead of iron. Practically no $CaCO_{3(s)}$ was deposited on this coupon in five days, despite a SI value of +1.05. The relation between $CaCO_{3(s)}$ deposition and SI values is seen in Figure 17a. There is no discontinuity or change in the trend as the curves pass through the SI value of zero. That there is an increase of corrosion as the SI values become more negative is seen in Figure 17b. Figure 17c shows the $CaCO_{3(s)}$ deposition from hard water (205 mg/l) at a pH value of 8.4 and at four DO levels. These curves clearly show that DO plays an important role in the deposition of $CaCO_{3(s)}$. Curve 1, Figure 17d, indicates an increase in corrosion with an increase in DO content for a soft water, whereas curve 2 shows lesser corrosion for a hard water. Where positive SI values are maintained, the effect of DO content is relatively

Table XI. Langelier Index and Iron Pickup in Distribution Systems[a]

Location	Langelier Index ($pH-pH_s$) at 60°F	Ryznar Index ($2pH_s-pH$)	Average Iron Pickup in System (ppm)	Maximum Iron in System Test (ppm)
Alexandria, VA	−2.05	11.10	0.17	0.60
Ashtabula, OH	−0.71	8.80	0.15	0.75
Ashtabula, OH	−0.59	8.48	0.08	0.30
Bernardsville, NJ	−0.60	9.50	0	0.09
Chattanooga, TN	−1.24	10.18	0	0.04
Clinton, IA	0.08	7.34	0.08	0.15
Davenport, IA	−1.01	9.12	0.06	0.10
East St. Louis, IL	−0.31	8.02	0.01	0.08
East St. Louis, IL	−0.17	7.74	0.03	0.25
Ellwood City, PA	−0.13	8.20		
Huntington, W VA	0.32	8.26	0	0.10
Monmouth, NJ	−1.18	10.62	0.03	0.14
Monongahela, PA	−0.14	8.70	0	0.12
Muncie, IN	0.20	7.30	0.05	0.18
Newark, NJ	−2.40	11.70		
New Castle, PA	−0.57	8.94	0.04	0.07
Noroton, CT	−2.38	11.46		
Old Dominion, VA	−1.86	11.82	0.10	0.20
St. Joseph, MO	0	7.70		
St. Joseph, MO	0.29	7.22	0	Trace
South Pittsburgh, PA	−1.00	9.70	0.04	0.05
Uniontown, PA	−3.60	3.60	0.24	0.35
Uniontown, PA	−2.09	11.09	0.22	0.65
White Deer, PA	−2.97	12.94	0.72	1.68

[a]Reproduced from Flentje [31], courtesy of the American Water Works Association.

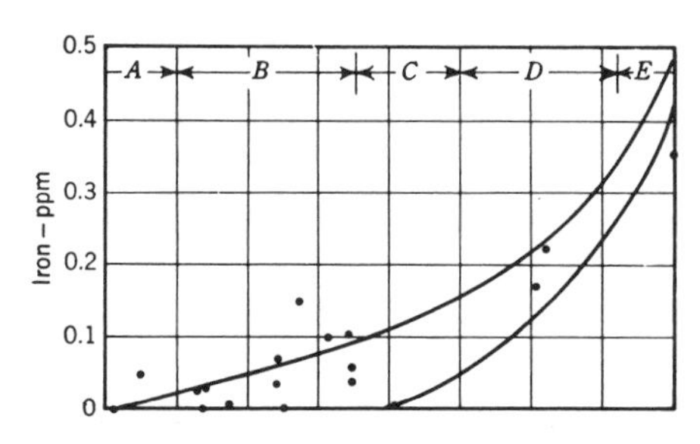

Figure 15. Langelier index and iron pickup in distribution systems. A is zone of $CaCO_3$ deposition; B, of little pickup; C, of red and dirty water; D, of red water unless polyphosphates are used; E, of red and dirty water even if polyphosphates are used. Reproduced from Flentje [31], courtesy of the American Water Works Association.

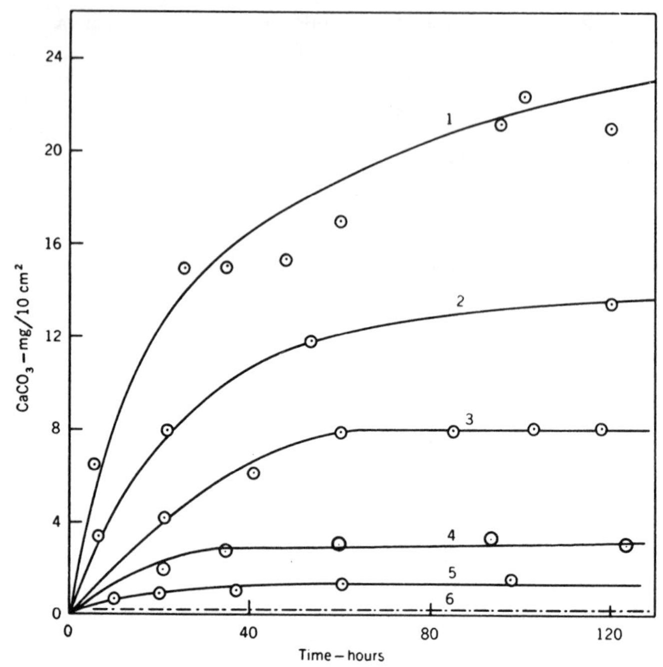

Figure 16. Influence of pH on $CaCO_{3(s)}$ deposition at iron surfaces. Curve 1: pH 8.40, SI = +1.05; curve 2: pH 7.65, SI = +0.30; curve 3: pH 7.35, SI = ±0.00; curve 4: pH 6.95, SI = −0.40; curve 5: pH 6.55, SI = −0.80; curve 6: pH 8.40, SI = +1.05. Curve 6 represents data from tests in which stainless steel specimens were used instead of cast iron. Ordinate values for curves 5 and 6 are multiplied by a factor of 3 to make them visible on the graph. Reproduced from Stumm [33], courtesy of the American Water Works Association.

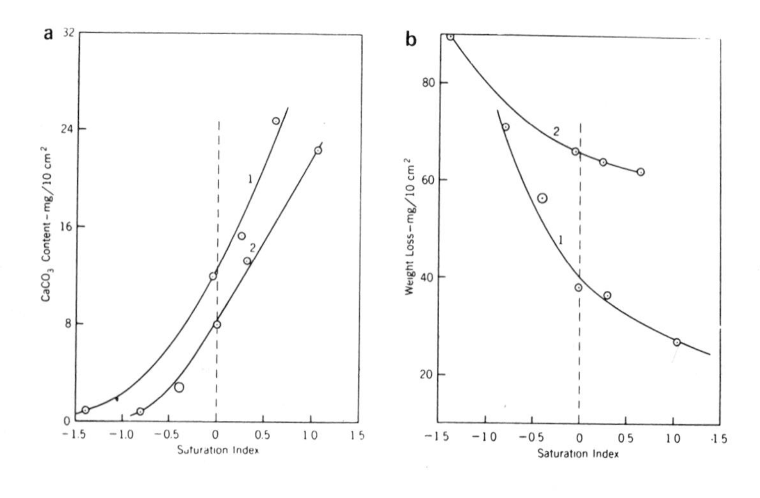

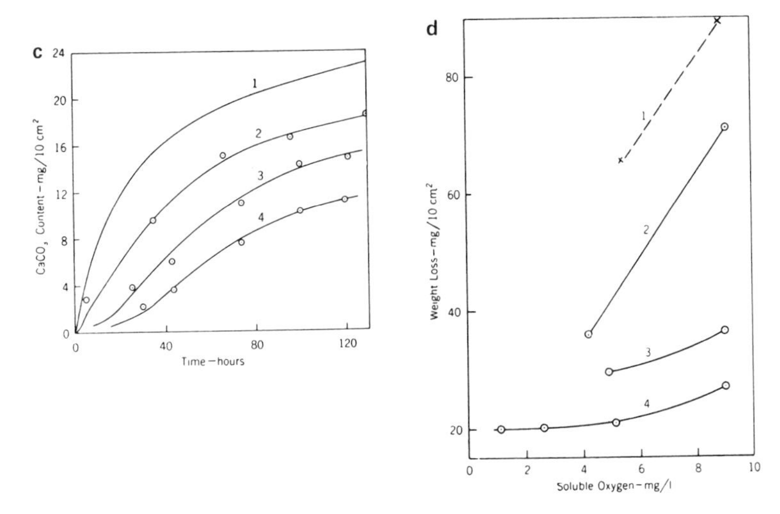

Figure 17. **(a)** $CaCO_{3(s)}$ deposition as a function of SI. Hardness and alkalinity were: curve 1, 117; curve 2, 205 mg/l (as $CaCO_3$). DO was 9.0 mg/l for both curves. The $CaCO_3$ content for both curves is after five days of immersion. **(b)** Corrosion as a function of SI. Corrosion is weight loss after 5 days of immersion. Hardness and alkalinity were: curve 1, 205; curve 2, 117 mg/l (as $CaCO_3$). DO was 9.0 mg/l for both curves. **(c)** Influence of DO on $CaCO_{3(s)}$ deposition. Hardness and alkalinity = 205 mg/l (as $CaCO_3$), pHs = 7.35, SI = +1.05. DO was: curve 1, 9.0; curve 2, 5.2; curve 3, 2.6; curve 4, 1.4 mg/l. **(d)** Corrosion as a function of $[O_{2(g)}]$. Corrosion is weight loss after five days of immersion. Hardness and alkalinity (as $CaCO_3$) was 117 mg/l for curve 1 and 205 mg/l for curves 2–4. pH was: curve 1, 6.40; curve 2, 6.55; curve 3, 7.65; curve 4, 8.40. SI was: curve 1, −1.40; curve 2, −0.80; curve 3, +0.30; curve 4, +1.05. Reproduced from Stumm [33], courtesy of the American Water Works Association.

insignificant (curves 3 and 4, Figure 17d). The essence of this study [33] is that, in natural waters saturated by $CaCO_3$, DO may not have an appreciable accelerating influence on the corrosion rate because it promotes film formation. Also, waters with "high" DO levels may produce better protective coatings than waters with "low" quantities of DO.

McCauley [34,35] confirmed and extended Stumm's work with protective $CaCO_{3(s)}$ films. For example, a protective coating of calcite ($CaCO_{3(s)}$), limonite (hydrous iron oxide) and siderite ($FeCO_{3(s)}$) occurred best when cast iron coupons were corroded, or when an outside current was impressed on the coupons. Water flowrate seemed to be a factor, with "high" flowrates producing a hard, durable film, whereas static tests produced a soft film [34]. In the second study [35], 0.5 ppm polyphosphate was added to waters with the "proper" level of $CaCO_{3(s)}$ supersaturation to form a thin, dense film on cast iron pipes.

Polyphosphates

One of the first reports of control of $CaCO_{3(s)}$ deposits by metaphosphates [$(NaPO_3)n$] was made in 1939 [36]. It was discovered that phosphates were useful for preventing cementation of filter sand in lime-soda softening plants. Subsequent work by Hoover [36] and other investigators [37,38] led to these results:

1. The presence of 1.0–1.5 ppm of hexametaphosphate ($Na_6P_6O_{18}$) prevented precipitation of $CaCO_{3(s)}$ from a water containing about 200 ppm Ca^{2+} (as $CaCO_3$) and 160 ppm carbonate alkalinity at pH 10.16–10.62. Lower concentrations of 0.3–0.6 ppm of hexametaphosphates did not prevent precipitation but lengthened the time for it to occur.
2. It was reported that 90 ppm $CaCO_3$ was held in solution by 2 ppm sodium hexametaphosphate and also prevented additional precipitation from a water with 400 ppm calcium hardness at 60°C [37].
3. Approximately 600 ppm $CaCO_3$ was prevented from precipitating at 40°C by 2.0 ppm metaphosphate at 100°C [38]. This same concentration held 200 ppm $CaCO_3$ in solution and prevented its precipitation.
4. Continuous treatment with metaphosphate was able to reduce the thickness of a previously deposited $CaCO_3$ scale in a lime softening plant line [36].

It appears, therefore, that hexametaphosphates can prevent or control the precipitation of $CaCO_{3(s)}$. The critical point is the proper phosphate dosage.

A rather comprehensive laboratory study of the chemical and hydraulic conditions underlying the use of polyphosphate for developing protective calcite coatings was reported [39]. The relevant data are seen in Table XII where the optimum conditions are: a flow velocity of 3.0 ft/sec, pH values of 9.4–9.7, a metaphosphate content of about 0.5 ppm, and a DFI in the "neighborhood of 150 ppm." In the 11 tests reported in Table XII, good-to-excellent coatings were produced on sandblasted black iron pipe walls and on cast iron specimens. These are, of course, laboratory results obtained under ideal and carefully controlled conditions.

A combined laboratory and field study was conducted on the control of corrosion with 2.5 mg/l zinc orthophosphate (0.5 mg/l Zn) and/or 2 mg/l sodium-zinc glassy phosphate in the distribution system of the

Table XII. Coating Development on Iron Pipe[a,b]

Run Number	Metaphosphate (ppm)	pH	Time (hr)	Added Ca^{2+} (ppm)	Total Excess (ppm)	DFI	Flow Velocity	Coating
50	0.45	9.4	3	0		137	1.5	Good, rust on seam
51	0.42	9.4	3	0	112	140	1.5	Good, rust on seam
52	0.30	9.4	2	0	145	147	3.0	Good, rust on seam
53	0.31	9.4	2	0	143	141	3.0	Good, rust on seam
54[c]	0.02	9.5	2	0	116	168	3.0	Coarse, fine rust
55[d]	0.10	9.4	2	0	140	148	3.0	Excellent, some rust
56[d]	0.48	9.6	2	0	153	199	3.0	Excellent, some rust
57	0.38	9.7	2	0	165	244	3.0	Excellent, no rust
58[e]	0.52	9.4	2	110	220	220	3.0	Very good, no rust
59	0.87	9.4	2	120	170	229	3.0	Very good, no rust
60[c]		9.8	2				3.0	Coarse, fine rust

[a]Reproduced from McCauley [39], courtesy of the American Water Works Association.
[b]Rusty water during flushing of distribution mains.
[c]Started metaphosphate 3.80 ppm, calcium added 0.00 over 2 hr of run, metaphosphate decreased to 0.35 ppm; calcium increased to add 244 ppm.
[d]Cast iron sample only; no pipe nipple in line.
[e]Stainless steel specimen also placed in line.

Middlesex Water Co. (New Jersey) [40,41]. The water treatment plant was designed to adjust the pH of the finished water with NaOH for corrosion control. Initial side-by-side bench tests with the bimetallic phosphate offered very little additional protection over that provided by caustic soda pH adjustment (Table XIII). Additional tests were made with mild steel certified test coupons inserted into the plant's effluent line (Table XIV) and with zinc phosphate only. There was an obvious reduction in the corrosion rate when the zinc concentration was held at approximately 0.5 mg/l. Optimum results were obtained when the pH value was adjusted to an average of 7.8. Lower pH values, ~7.2, were not as effective in reduction of corrosion.

Shull [42] reported somewhat different results with use of the bimetallic (Na-Zn) phosphate for corrosion control than those reported above. In an experimental situation under various conditions, corrosion control was measured by head loss and "iron pickup" in the water. Figure 18 shows the essence of a 35-day study. The pH curves of 9.52 and 8.18 showed the greatest head loss that "was probably due to the deposition of calcium carbonate scale." The bimetallic phosphate-treated water showed a loss of head 63% less than the control. No appreciable quantity of iron was found in the effluent waters of the experimental pipes.

Corrosion of A/C pipes, apparently, can be controlled by pH control and/or the addition of zinc orthophosphate [17,43]. Table XV summarizes a laboratory study on the corrosion of A/C pipe coupons [17]. Experiments 1 and 2 are typical examples wherein corrosion was controlled by 0.3–0.5 mg/l (as Zn) of zinc orthophosphate. In experiments 3 and 4, the pH value of 7.0 and the AI of 9.3 were lower than those values in experiments 1 and 2. These data indicate that better control is obtained at a higher pH value. Experiments 6 and 7 were designed to demonstrate the effect of a $CaCO_{3(s)}$ film wherein no corrosion was observed in the supersaturated system (#7). Experiment 8 was designed to show only the effect of $[H^+]$ which it did.

Summary

There are several types of corrosion and corrosion control in water distribution systems, related water-using equipment and boiler and condensate systems that are not considered here. The reader is given several references for obtaining this information. The *Nalco Water Handbook* [44] carries descriptions of various types of corrosion and control methods. Corrosion of copper pipes is described in Shull [42], Cruse [45], Task Group [46] and Bennet et al. [47]. Geld and McCaul [48] describe the

Table XIII. Bench Studies for Monitoring Corrosion of pH-Adjusted Water[a]

Date (1971)	Number of Days	Average pH	Temperature		Corrosion Rate						Corrosion Reduction (%)	
					NaOH		$Zn_3(PO_4)_2$		$NaZnPO_4$			
			°C	°F	10^{-6} m/year	mil/year	10^{-6} m/year	mil/year	10^{-6} m/year	mil/year	$Zn_3(PO_4)_2$	$NaZnPO_4$
7/26–8/26	30	8.1	26	78	3.4	13.80	1.37	5.49	2.7	11.0	60	20
9/10–10/6	26	7.8	22	71	2.6	10.60	1.38	5.54	2.52	10.1	48	5

[a]Reproduced from Mullen and Ritter [41], courtesy of the American Water Works Association.

Table XIV. Peak Corrosion Rate Data[a]

Month and Year	Corrosion Rate 10^{-6} m/year	Corrosion Rate mil/year	Zn Dosages Average (mg/l)	Zn Dosages Maximum (mg/l)	Reduction of Corrosion (%)
July					
1973	1.4	5.7	0.46	0.50	58[b]
1974	0.9	3.6	0.49	0.56	86
July–September					
1975	0.8	3.5	0.43	0.53	83
1976	1.35	5.4	0.49	0.52	72
1977	0.9	3.6	0.49	0.52	81
1978	1.32	5.3	0.47	0.54	75

[a]Reproduced from Mullen and Ritter [41], courtesy of the American Water Works Association.
[b]Five months' data.

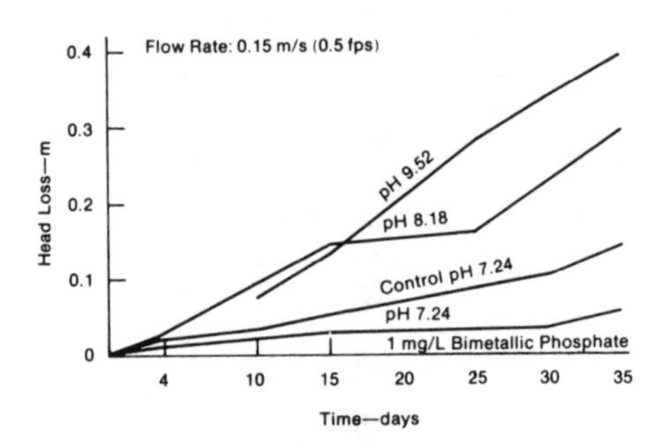

Figure 18. Corrosion studies of Crum Creek plant effluent at varying pH values and with bimetallic phosphate addition. Reproduced from Shull [42], courtesy of the American Water Works Association.

corrosion of bronzes, nickel-copper, stainless steel, and wrought and galvanized iron. Home plumbing corrosion is described in Hoyt et al. [49].

REFERENCES

1. Ryder, R. A. *J. Am. Water Works Assoc.* 72:267 (1980).
2. U.S. EPA. "Interim Primary Drinking Water Regulations," *Federal Register* 45:57332 (August 27, 1980).
3. International Union of Pure and Applied Chemistry. *Nomenclature of Inorganic Chemistry,* 2nd ed. (London: Butterworths, 1970).

Table XVa. Water Quality Conditions for Bench-Scale Experiments[a]

Experiment Number	pH	Calcium as $CaCO_3$ (mg/l)	Total Alkalinity as $CaCO_3$ (mg/l)	Aggressiveness Index	Corrosion Control Method
1	8.2	6	20	10.28	None
2	8.2	6	20	10.28	Zinc orthophosphate
3	7.0	10	20	9.30	None
4	7.0	10	20	9.30	Zinc orthophosphate
5	8.2	6	20	10.28	Zinc chloride
6	7.5	145	125	11.76	None
7	7.9	145	125	12.16	Slightly positive LI
8	9.0	25	40	12.00	$CaCO_3$ saturation

Table XVb. Summary of A/C Pipe Coupon Tests[a]

Experiment Number	pH	Aggressiveness Index	Additive	Inner Wall of Coupon Softened
3	7.0	9.3	None	Yes
4	7.0	9.3	Zinc orthophosphate	Slightly
1	8.2	10.28	None	Yes
2	8.2	10.28	Zinc orthophosphate	No
5	8.2	10.28	Zinc chloride	No
6	7.5	11.76	None (slightly under pHs)	Slightly
8	9.0	12.00	None (at pHs)	Very slightly
7	7.9	12.16	None (slightly over pHs)	No

[a]Reproduced from Buelow et al. [17], courtesy of the American Water Works Association.

4. Morris, J., and W. Stumm. In: *Equilibrium Concepts in Natural Water Systems,* Advances in Chemistry Series 67 (Washington, DC: American Chemical Society, 1967), p. 270.
5. Stumm, W., and J. J. Morgan. *Aquatic Chemistry,* 2nd ed. (New York: John Wiley & Sons, 1981).
6. Obrecht, M. F., and M. Pourbaix. *J. Am. Water Works Assoc.* 59:977 (1967).
7. Hem, J. D., and W. H. Cropper. U.S. Geological Water Supply Papers 1459A (1962), 1459B (1960), and 1459C (1960), Washington, DC.
8. Faust, S. D., and O. M. Aly. *Chemistry of Natural Waters* (Ann Arbor, MI: Ann Arbor Science Publishers, Inc., 1981), p. 177.
9. Hem, J. D., *J. Am. Water Works Assoc.* 53:211 (1961).
10. Sato, M. *Econ. Geol.* 55:298 (1960).
11. Pourbaix, M. *Atlas of Electrochemical Equilibria in Aqueous Solutions* (Oxford, England: Pergamon Press, 1966).
12. Stiff, M. J. *Water Res.* 5:171 (1971).

13. Senior Author. Personal observations in groundwaters of Lebanon Township, Hunterdon County, NJ.
14. Patterson, J. W., and J. E. O'Brien. *J. Am. Water Works Assoc.* 71:264 (1979).
15. Schock, M. R. *J. Am. Water Works Assoc.* 72:695 (1980).
16. Kelley, K. K., and C. T. Anderson. *U.S. Bur. Mines Bull.* 384:1–73 (1935).
17. Buelow, R. W. et al. *J. Am. Water Works Assoc.* 72:91 (1980).
18. "AWWA Standard for Asbestos-Cement Distribution Pipe," AWWA Report C–400–80, American Water Works Association, Denver, CO (1980).
19. Langelier, W. F. *J. Am. Water Works Assoc.* 38:169, 179 (1946).
20. Langelier, W. F. *J. Am. Water Works Assoc.* 28:1500 (1936).
21. Hem, J. D. "Calculation and Use of Ion Activity," U.S. Geological Survey Water Supply Paper 1535-C, Washington, DC (1961).
22. Frear, G. L., and J. Johnston. *J. Am. Chem. Soc.* 51:2082 (1929).
23. MacInnes, D. A., and D. Belcher. *J. Am. Chem. Soc.* 55:2630 (1933).
24. *Standard Methods for the Examination of Water and Wastewater,* 15th ed. (Washington, DC: American Public Health Association, 1980).
25. Kemp, P. H. *Water Res.* 5:735 (1971).
26. McGaughey, L. M., and J. V. Matson. *Water Res.* 14:1729 (1980).
27. Ryznar, J. W. *J. Am. Water Works Assoc.* 36:472 (1944).
28. Caldwell, D. H., and W. B. Lawrence. *Ind. Eng. Chem.* 45:535 (1953).
29. Merrill, D. T., and R. I. Sanks. "Corrosion Control by Deposition of $CaCO_3$ Films," American Water Works Association, Denver, CO (1978).
30. McCauley, R. F. *J. Am. Water Works Assoc.* 52:721 (1960).
31. Flentje, M. E. *J. Am. Water Works Assoc.* 53:1461 (1961).
32. Dye, J. F. *J. Am. Water Works Assoc.* 56:457 (1964).
33. Stumm, W. *J. Am. Water Works Assoc.* 48:300 (1956).
34. McCauley, R. F., and M. O. Abdullah. *J. Am. Water Works Assoc.* 50:1419 (1958).
35. McCauley, R. F. *J. Am. Water Works Assoc.* 52:1386 (1960).
36. Hoover, C., and O. Rice. *Water Sew. Works* 86:10 (1939).
37. Rice, O., and G. Hatch. *J. Am. Water Works Assoc.* 31:1171 (1939).
38. Illig, G. L. *J. Am. Water Works Assoc.* 49:805 (1957).
39. McCauley, R. F. *J. Am. Water Works Assoc.* 52:721 (1960).
40. Mullen, E. D., and J. A. Ritter. *J. Am. Water Works Assoc.* 66:473 (1974).
41. Mullen, E. D., and J. A. Ritter. *J. Am. Water Works Assoc.* 72.286 (1980).
42. Shull, K. E. *J. Am. Water Works Assoc.* 72:280 (1980).
43. Schock, M. R. et al. "Evaluation and Control of Asbestos-Cement Pipe Corrosion," EPA Report 600/D–81–067, U.S. EPA, Cincinnati, OH (1981).
44. *The Nalco Water Handbook,* F. N. Kemmer, Editor-in-Chief (New York: McGraw-Hill Book Company, 1979).
45. Cruse, H. C., and R. D. Pomeroy. *J. Am. Water Works Assoc.* 66:479 (1974).
46. Task Group Report. *J. Am. Water Works Assoc.* 52:1033 (1960).
47. Bennett, W. F. et al. *J. Am. Water Works Assoc.* 69:26 (1977).
48. Geld, I., and C. McCaul. *J. Am. Water Works Assoc.* 67:549 (1975).
49. Hoyt, B. P. et al. *J. Am. Water Works Assoc.* 71:720 (1979).

CHAPTER 10

REMOVAL OF PATHOGENIC BACTERIA AND VIRUSES

PURPOSES

Waterborne Diseases

The primary purpose for disinfection of drinking and bathing waters is, of course, prevention of waterborne diseases. Modern practice of disinfection began, perhaps, with chlorination of the waters from Boonton Reservoir of the Jersey City Water Works in 1908 [1]. Despite these 75 years of disinfection of public and other water supplies, Lippy [2] states: "The disturbing trend toward more frequent occurrence of outbreaks of waterborne diseases should serve as a warning to all who share in the responsibility for delivery of a safe and potable water." Bacterial waterborne diseases are seen in Table I [3,4], enteric viruses in Table II [4,5], and parasites in Table III [4].

During 1971–1978, 43 states and Puerto Rico reported 224 outbreaks of waterborne diseases that affected 48,193 individuals and included 2 deaths [6]. Since 1970 there has been a pronounced increase in outbreaks resulting in an average of 34/year. There are two reasons for this increase: (1) more active and accurate surveillance by regulatory agencies, and (2) modification of the definitions of municipal or community, semipublic or noncommunity and individual water systems by the U.S. EPA in 1976–1978 to correspond with those in PL 93-523, the Safe Drinking Water Act. Thus, more systems and people are under surveillance. Figure 1 shows the etiology of waterborne diseases for 1971–1978. There are several points about the data in this figure that should cause concern. First, the etiologic agent in 55% of the outbreaks could not be identified, but were categorized as acute gastrointestinal illnesses. Second, 10% of the outbreaks were chemical poisonings from arsenic,

Table I. Waterborne Bacterial Diseases [3,4]

Causative Agent	Disease	Symptoms	Reservoir
Salmonella typhosa	Typhoid fever	Incubation period 7–14 days. Headache, nausea, loss of appetite, constipation or diarrhea, insomnia, sore throat, bronchitis, abdominal pain, nose bleeding, shivering and increasing fever. Rose spots on trunk.	Feces and urine of typhoid carrier or patient.
S. paratyphii *S. schottmulleri* *S. hirschfeldii C.*	Paratyphoid fever	General infection characterized by continued fever, diarrheal disturbances, sometimes rose spots on trunk. Incubation period 1–10 days.	Feces and urine of carrier or patient.
Shigella flexneri *Sh. dysenteriae* *Sh. sonnei* *Sh. paradisinteriae*	Bacillary dysentery	Acute onset with diarrhea, fever, tenesmus and stool frequently containing mucus and blood. Incubation period 1–7 days.	Bowel discharges of carriers and infected persons.
Vibrio comma	Cholera	Diarrhea, vomiting, ricewater stools, thirst, pain, coma. Incubation period a few hours to 5 days.	Bowel discharges, vomitus, carriers.
Pasteurella tularensis	Tularemia	Sudden onset with pains and fever; prostration. Incubation period 1–10 days.	Rodent, rabbit, horsefly, woodtick, dog, fox, hog.
Brucella melitensis	Brucellosis (undulant fever)	Irregular fever, sweating, chills, pain in muscles.	Tissues, blood, milk, urine, infected animal.
Pseudomonas pseudomallei	Melioidosis	Acute diarrhea, vomiting, high fever, delirium, mania.	Rats, guinea pigs, cats, rabbits, dogs, horses.

Microorganisms	Gastroenteritis	Diarrhea, nausea, vomiting, cramps, possibly fever. Incubation period 8–12 hr, average.	Probably man and animals.
Leptospira icterohaemorrhagiae (spirochete)	Leptospirosis (Weil's disease)	Fevers, rigors, headaches, nausea, muscular pains, vomiting, thirst, prostration and jaundice may occur.	Urine and feces of rats, swine, dogs, cats, mice, foxes, sheep.
Enteropathogenic *E. coli*	Gastroenteritis	Watery diarrhea, nausea, prostration and dehydration.	Feces of carrier.

Table II. Human Enteric Viruses That Can Be Waterborne and Known Diseases Associated with These Viruses[a]

Group	Subgroup	No. of Types or Subtypes	Disease Entities Associated With These Viruses	Pathological Changes in Patients	Organs Where Virus Multiples
Enterovirus	Poliovirus	3	Muscular paralysis	Destruction of motor neurons	Intestinal mucosa, spinal cord, brain stem
			Aseptic meningitis	Inflammation of meninges from virus	Meninges
			Febrile episode	Viremia and viral multiplication	Intestinal mucosa and lymph
	Echovirus	34	Aseptic meningitis	Same as above	Same as above
			Muscular paralysis	Same as above	Same as above
			Guillain-Barre's syndrome[b]	Destruction of motor neurons	Spinal cord
			Exanthem	Dilation and rupture of blood vessels	Skin
			Respiratory diseases	Viral invasion of parenchymiatous of respiratory tracts and secondary inflammatory responses	Respiratory tracts and lungs
			Diarrhea		
			Epidemic myalgia	Not well known	
			Pericarditis and myocarditis	Viral invasion of cells with secondary responses	Pericardial and myocardial tissue
			Hepatitis	Same as above	Liver parenchyma
	Coxsackie virus	>24	Herpangina[c]	Viral invasion of mucosa with secondary inflammatory responses	Mouth
			Acute lymphatic pharyngitis	Same as above	Lymph nodes and pharynx
	A		Aseptic meningitis	Same as above	Same as above

			Muscular paralysis	Same as above	Same as above
			Hand-foot-mouth disease[d]	Viral invasion of cells of skin of hands and feet and mucosa of mouth	Skin of hands and feet and much of mouth
			Respiratory disease	Same as above	Same as above
			Infantile diarrhea	Viral invasion of cells of mucosa	Intestinal mucosa
			Hepatitis	Viral invasion of liver cells	Parenchyma cells of liver
			Pericarditis and myocarditis	Same as above	Same as above
	B	6	Pleurodynia[e]	Viral invasion of muscle cells	Intercostal muscles
			Aseptic meningitis	Same as above	Same as above
			Muscular paralysis	Same as above	Same as above
			Meningoencephalitis	Viral invasional invasion of cells	Meninges and brains
			Pericarditis, endocarditis, myocarditis	Same as above	Same as above
			Respiratory diseases	Same as above	Same as above
			Hepatitis or rash	Same as above	Same as above
			Spontaneous abortion	Viral invasion of vascular cells (?)	Placenta
			Insulin-dependent diabetes	Viral invasion of insulin-producing cells	Langerhans' cells of pancreases
			Congenital heart anomalies	Viral invasion of muscle cells	Developing heart
Reovirus		6	Not well known	Not well known	
Adenovirus		31	Respiratory diseases	Same as above	Same as above
			Acute conjunctivitis	Viral invasion of cells and secondary inflammatory responses	Conjunctival cells and blood vessels
			Acute appendicitis	Viral invasion of mucosa cells	Appendia and lymph nodes
			Intussusception	Viral invasion of lymph nodes (?)	Intestinal lymph nodes (?)
			Subacute thyroiditis	Viral invasion of parenchyma cells	Thyroid

Table II. continued

Group	Subgroup	No. of Types or Subtypes	Disease Entities Associated With These Viruses	Pathological Changes in Patients	Organs Where Virus Multiples
			Sarcoma in hamsters	Transformation of cells	Muscle cells
Hepatitis		>2	Infectious hepatitis	Invasion of parenchyma cells	Liver
			Serum hepatitis	Invasion of parenchyma cells	Liver
			Down's syndrome	Invasion of cells	Frontal lobe of brain, muscle, bones

[a]Reproduced from Taylor [5], courtesy of the American Water Works Association.
[b]Ascending type of muscular paralysis.
[c]Febrile episode with sores in mouth.
[d]Rash and blisters on hand-foot-mouth with fever.
[e]Pleurius type of pain with fever.

Table III. Waterborne Diseases from Parasites [4]

Causative Agent	Disease	Symptoms
Endamoeba histolytica	Amebiasis	Diarrhea alternating with constipation, chronic dysentery with mucus and blood
Giardia lamblia	Giardiasis	Intermittent diarrhea
Naegleria gruberi	Amoebic meningocephalitis	Death
Taenia saginata (beef tapeworm)		Abdominal pain, digestive disturbance, loss of weight
Ascario lumbricoides (round worm)	Ascariasis	Vomiting, live worms in feces
Schistosoma mansoni (blood fluke)	Schistosomiasis	Liver and bladder infection

chlordane, chromate, copper cutting oil, film developer fluid, ethyl acrylate, fluoride, fuel oil, fluradan, lead, kerosene, phenol, PCB, selenium and an unidentified herbicide. Third, there were 15 outbreaks of waterborne viral hepatitis during 1971–1978, and hepatitis A has been implicated in 68 outbreaks since 1946. Various causes can be cited for these outbreaks: crossconnections or backsiphonages, treatment deficiencies, and use of untreated, contaminated wellwaters. Consequently, there still is ample cause for concern about waterborne diseases and the continuation of disinfection.

Additional surveys of waterborne disease outbreaks include: 1946–60 [7]; 1971–1974 [8]; 1975–1976 [9]; waterborne giardiasis in the United States [10]; viruses in water supplies [11–13]; and the necessity of controlling bacterial populations in potable waters [14].

Indicator Organisms and Pathogenic Bacteria

Perhaps the classic paper of Kehr and Butterfield [15] established the relation between the indicator coliform organisms and enteric pathogens. Much of their data were collected from domestic sewage surveys conducted by the London Metropolitan Water Board during the years 1927–1938. The occurrence of *Salmonella* organisms was sought in these surveys. Kehr and Butterfield plotted the available data in order to establish some sort of a relationship between reported isolations and the normal prevalence of typhoid fever in various communities. Figure 2 shows

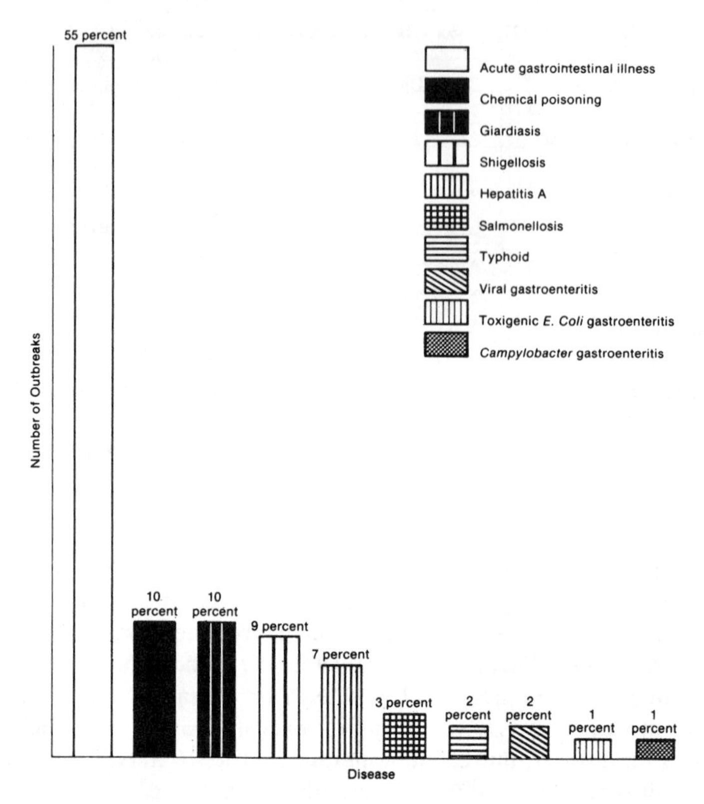

Figure 1. Etiology of waterborne disease in the United States, 1971–1978. Reproduced from Craun [6], courtesy of the American Water Works Association.

a log-log plot (for convenience) of the ratio of *Eberthella typhosa* (now *S. typhosa*) per 10^6 coliforms per ml vs the typhoid fever morbidity rate. The upper line is the suggested ratio (indicated by the data available), while the two lower lines indicate the ratios that might be expected (uncorrected for recovery losses) in polluted river waters and sewage. These ratios are the approximate minimum ratios to be expected in domestic sewage and surface waters polluted by sewage from reasonably large populations. This assumes, of course, that cases of typhoid fever and typhoid carriers exist as providers of *Salmonella* organisms in the sewage.

Additional evidence may be cited from fecal coliform correlations with *Salmonella* occurrence. This is seen in Table IV where data collected from numerous stream and estuarine surveys are grouped into fecal coliform density ranges that bracket recommended limits (for 1970) for

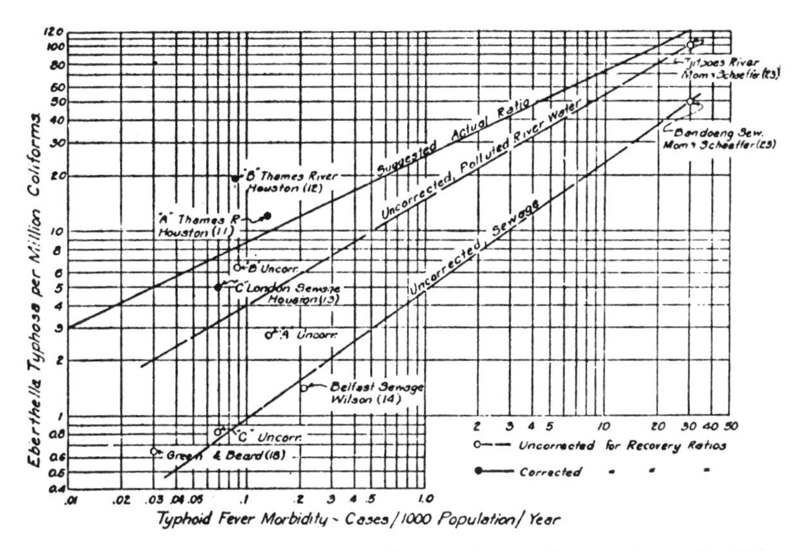

Figure 2. *Eberthella typhosa* per million coliforms for varying typhoid fever morbidity rates [15].

recreational waters and public water supply intakes [16]. These data suggest a sharp increase in the frequency of *Salmonella* detection when fecal coliform densities are greater than 200/100 ml of freshwater. Also, *Salmonella* isolations (positive occurrences) should occur with nearly a 100% frequency when the fecal coliform density exceeds 2000/100 ml. For estuarine waters, the 1–70 range of fecal coliforms has application to shellfish waters. In this range, only 6.7% of 184 *Salmonella* examinations were positive. In polluted estuarine waters with fecal coliforms ranging from 201 to >2,000,000/100 ml, recovery of *Salmonella* did not reach the 100% proportions observed in freshwater. These observations are illustrated further by a bacterial quality study at several water plant intakes along the Missouri River. When fecal coliforms exceeded 2000/100 ml, *Salmonella,* poliovirus types 2 and 3, and echovirus types 7 and 33 were detected [17].

Another major relation between indicator organisms and *Salmonella* is survival in freshwaters. These enteric organisms are not expected to survive for a considerable period of time in an unfavorable environment. An important consideration would be that the length of survival of the indicator organism and the pathogen should be approximately the same. Studies were conducted on the relative persistence of selected indicator organisms and a *Salmonella* strain in 52 urban stormwater samples [18]. Survival curves are seen in Figure 3 for selected enteric bacteria. It appears that fecal coliforms and *S. typhimurium* have approximately the

Table IV. Fecal Coliform Correlations with *Salmonella* Occurrence[a]

Source	Fecal Coliform Density (per 100 ml)	*Salmonella* Detection		
		Total Examinations	Positive Occurrence	
			Number	%
Freshwater	1-200	29	8	27.6
	201-2000	27	19	85.2
	>2000	54	53	98.1
Estuarine Water	1-70	184	12	6.5
	71-200	74	21	28.4
	201-2000	91	40	44.0
	>2000	75	45	60.0

[a]Reproduced from Geldreich [16], courtesy of the American Water Works Association.

same length of survival or rate of death in stormwater stored at 20°C. Similar survivals are expected in lakes, streams, rivers, etc., receiving these organisms. Kehr and Butterfield [15] and Heukelekian and Schulhoff [19] reported similar relationships for the survival of pathogens and indicator organisms in natural waters.

Statements above cite the two major arguments for the use of nonpathogenic enteric organisms to indicate the possible occurrence of pathogenic enteric bacteria. These are (1) the statistical correlations among coliform density, *Salmonella* density and typhoid death rate, and (2) the survival factor in natural waters. There are, however, several limitations to this practice, thoroughly discussed in Hoadley and Dutka [20].

CHEMISTRY OF DISINFECTION

Disinfection of water can be traced back to about 2000 BC, where two medical maxims in Sanskrit advised that water should be exposed to sunlight and filtered through charcoal and that "foul water" be treated by boiling and "by dipping seven times into it a piece of hot copper and then filtering it" [1]. There are also many references to boiling water and storage in silver flagons in ancient civilizations. Other early efforts of water disinfection were the use of copper, silver and electrolysis. The first American patent on chlorination of water was given to Albert R. Leeds, on May 22, 1888. This led eventually to the first large-scale chlorination of a public water supply in the United States at the Boonton Reservoir of the Jersey City Water Works in 1908 [1].

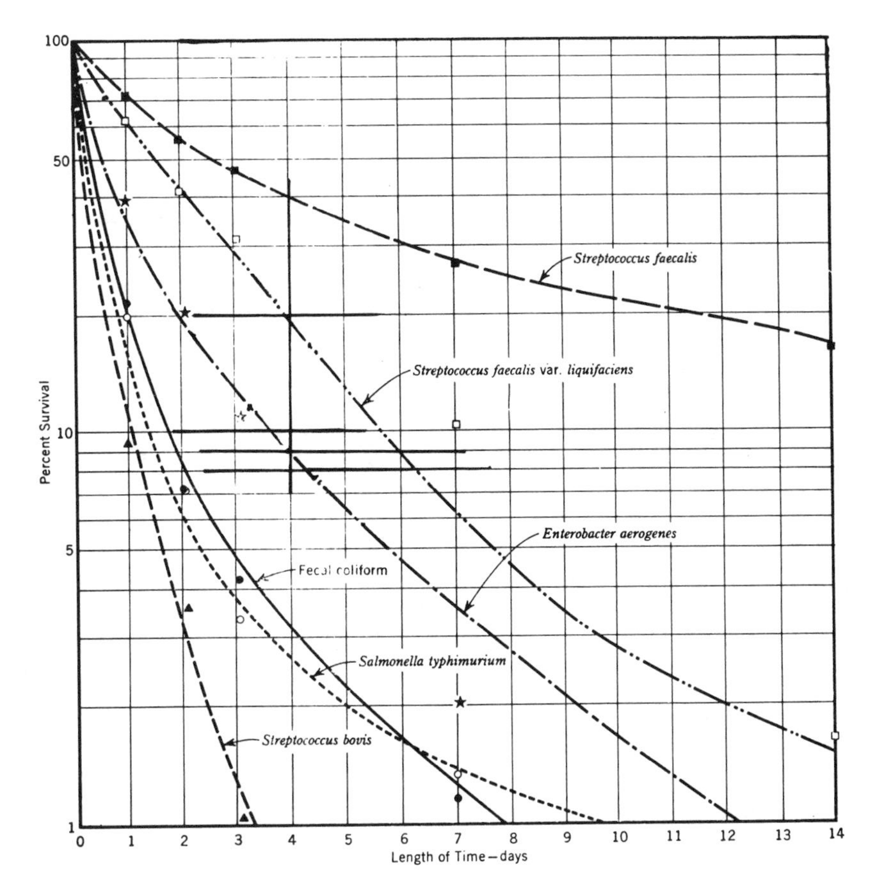

Figure 3. Persistence of selected enteric bacteria in sterilized stormwater stored at 20°C. Reproduced from Geldreich [16], courtesy of the American Water Works Association.

Types of Disinfectants

There are several means and types of disinfectants available for municipal and domestic water and swimming pool waters. It should be emphasized that these waters are disinfected for the destruction of pathogenic and indicator organisms. It is not necessary for the waters to be sterilized to achieve the complete absence of microorganisms. Some of the more specific disinfection processes for water are: (1) physical treatment by means of storage or the application of heat; (2) irradiation by ultraviolet light; (3) metals ions, for example, copper and silver; (4) strong acids and bases, i.e., pH adjustment; (5) surface active agents, for example, the quaternary ammonium compounds; and (6) chemical oxidants such as Cl_2, Br_2, I_2, O_3, ClO_2, $KMnO_4$, FeO_4^{2-} and organohalogen compounds [21]. Many of these are, of course, not suitable for the large-scale disinfection of public drinking water. For this water, Fair et al. [22] have established criteria for a potential disinfectant:

1. Ability of the disinfectant to destroy the kinds and numbers of organisms present within the contact time available, the range of water temperatures encountered, and the anticipated fluctuations in composition, concentration, and condition of the water being treated.
2. Ready and dependable availability of the disinfectant at reasonable cost and in a form conveniently, safely, and accurately applied.
3. Ability of the disinfectant, in concentrations employed, to accomplish the desired objectives without rendering the water toxic or objectionable, aesthetically or otherwise, for the purposes it is intended.
4. Ability of the disinfectant to persist in residual concentrations as a safeguard against recontamination where this might be important (such as potable-water distribution systems).
5. Adaptability of practical, duplicable, quick, and accurate assay techniques for determining disinfection concentration, for operating control of the treatment process, and as a measure of disinfecting efficiency.

Chlorine meets these criteria and the widespread practice of chlorination is, essentially, synonymous with disinfection.

Chemical Oxidants

Bromine. Br_2 is a dark, reddish-brown halogen that exists as a liquid at atmospheric pressure. It is about 3.2 times heavier than water, has a melting point of 7.3°C, and has a boiling point of 58.78°C. Its solubility in water at 20°C is 2.1×10^{-1} M (33.56 g/l). According to Laubusch [21], bromine is a good germicidal agent and can be detected in water rather easily. Bromine and monobromamine (NH_2Br) have been reported [23] to be nearly equal in germicidal properties and essentially equal to chlorine at comparable pH values. It is used primarily in disinfecting swimming pool and industrial waters [21]. It could be used, perhaps, in emergency situations for drinking water.

Iodine. This halogen, I_2, is a volatile, purplish-black, crystalline solid, which sublimates slowly at normal atmospheric conditions. It has a melting point at 114°C and a boiling point at 183°C. Its solubility in water is 1.33×10^{-3} *M* (.337 g/l). Here again, I_2 is an excellent disinfectant for drinking water, but is limited to emergency situations because of its obvious "iodine" taste. In the 1940s, research was conducted with such halogen compounds as Halazone for use as a field disinfectant in military operations. By 1945 Globaline (tetraglycine hydroperiodide) had been developed and demonstrated to be effective for the disinfection of enteric bacteria, amoebic cysts and cercariae of schistosoma with one or two tablets per quart and a 10-min contact period [21]. In turn, this led to a number of studies on the efficacy of I_2 as a disinfectant for public drinking waters. Some of these studies are presented below. It is used also for disinfecting swimming pool waters.

Ozone. This oxidant is a faintly blue, pungent smelling, unstable (in water) gas. It is an allotrophic form of oxygen in which three atoms are combined to form the molecule O_3. Because of the instability it usually is generated onsite. Some of the properties of ozone are: boiling point, $-112.4°C$ and melting point, $-193 \pm 0.5°C$. Its solubility in water is 0.494 cm^3/cm^3 of water at 0°C and 1 atm [24]. Ozone was discovered in 1783 by the Dutch scientist Van Marum. It is interesting to note that research was conducted late in the 19th century for municipal water treatment in Germany, Holland and France [21]. Ohmüller [25] reported that O_3 was effective in the destruction of typhoid and cholera bacteria at a semicommercial plant at Martinikenfeld in 1891. Several other investigations led to the installation of ozone treatment at water plants in Paris, Lille and Nice, France (1898–1904). By 1933 about 90 mgd of water was treated by O_3 in Paris. France still uses this treatment more than any other country. In the United States, O_3 was employed by the city of Philadelphia for taste and odor control as well as disinfection [21]. Today, ozone disinfection has been given serious consideration in order to avoid formation of the halogenated hydrocarbons (see Chapter 2). Some studies on the effectiveness of ozone as a disinfectant are cited below. The two major disadvantages of ozone treatment are: electric energy costs of generation are extremely high and its instability in water does not leave a residual for protecting the distribution system. Its use, therefore, is limited to emergencies and small-scale water systems.

Other. Other chemical oxidants for disinfection are: ClO_2, $KMnO_4$, FeO_4^{2-} and H_2O_2. Here again, their use is limited to specific treatment for destroying microorganisms. If these oxidants are used for such other purposes in water treatment as taste and odor control, and iron and manganese removal, there will be an incidental effect of disinfection. Chlorine is described below.

Alternative Disinfectants

Heat Treatment. It is common knowledge that boiling water for 15–20 min is sufficient for destruction of the pertinent microorganisms. This treatment is limited to emergency situations and is frequently recommended by utilities in such cases. There are continuous-flow water pasteurizers for small supplies, i.e., domestic applications [26].

Irradiation. The ultraviolet method of disinfection requires the exposure of a thin film of water, ~120-mm thick, to one or more quartz mercury vapor arc lamps that emit radiation in the range between 200 and 295 mμ. The operational aspects are: a wavelength of 2537 Å, flowrate of 750 gal/hr, a 4.2-sec contact time, and a flow depth of ⅜ in. [27]. Some of the advantages of this process of sterilization are: chemical treatment is avoided; contact times are short; it is extremely effective for the destruction of coliform organisms; and no tastes and odors are produced. Its disadvantages are: it is not very effective against spores, cysts and viruses; it requires pretreatment for turbidity removal; it has no residual disinfection capability; and the accompanying electrical energy costs may be high. Here again, ultraviolet irradiation is limited to small-scale water supplies.

Metal Ions. The bactericidal properties of such metal ions as Cu^{2+}, Ag^+ and Hg_2^{2+} have been known for many centuries [21]. Their effectiveness is noted at "low" concentrations. There have been numerous publications since 1869 or so on the disinfection properties of silver, etc. The algicidal properties of copper are well known also. Most of the research and applications of metallic disinfection have been with silver. There are reports of European and Russian uses of silver in village water supplies [21]. Some of the advantages are: "low" disinfectant concentrations and long residual bactericidal activity. Disadvantages are: pretreatment of the water is required for removal of turbidity, organic color, etc.; cysts and spores are resistant; cold temperatures lower germicidal activity; long contact times are required; and the treatment is costly. There may be some emergency and small-scale applications for disinfection by silver.

Chlorination

Semantics and Practices

Plain or simple chlorination is the application of chlorine to an untreated water supply as it enters into a distribution system either municipal or domestic. The intent is disinfection and no other treatment is required. Pre- and postchlorination are addition of chlorine before and after any other water treatment. Rechlorination is the addition of chlorine to the finished water at one or more points in the distribution system. The intent is to assure disinfection. Combined residual chlorination is the application of chlorine to water with a subsequent reaction with ammonia (NH_3) to yield and to maintain the chloramines NH_2Cl and $NHCl_2$ throughout the distribution system. Thus, the chlorine is combined with ammonia to produce two chloramines that also have disinfectant properties. Assurance of disinfection is the intent.

Free residual chlorination is the application of chlorine to water to produce a residual of hypochlorous acid HOCl and/or hypochlorite ion, OCl^-. No ammonia or chloramines are present. Thus, the residual chlorine in the water is "free" from any combined residual. It is necessary to satisfy the "chlorine demand" of the water in order for chlorine to be "free" for disinfection. This is frequently labelled "free available chlorine."

Breakpoint chlorination is the continuous addition of chlorine to water to a point or dosage whereby all of the chlorine demand has been "satisfied" and all of the ammonia has been oxidized, which leaves a "free" residual of chlorine. The quantitative aspects are given below. The primary intent is, of course, disinfection, but there are some other benefits. Control of tastes and odors may be one. To achieve the breakpoint, it is frequently necessary to practice "superchlorination."

Superchlorination is the addition of an extraordinary quantity of chlorine to water to achieve disinfection, taste and odor control, iron and manganese removal, etc. Chlorine dosages are several orders of magnitude greater than the 1.0-mg/l quantity for disinfection.

In accord with *Standard Methods* [28], chlorine demand is the quantity of chlorine reduced or converted "to inert or less active forms of chlorine by substances in the water." Since chlorine is a nonselective oxidant, almost any substance in water in a reduced valence state will react and consume chlorine. These substances may be NH_3, CN^-, organics, Fe^{2+}, Mn^{2+}, S^{2-}, SO_3^{2-}, etc. The chlorine demand presumably must be satisfied before disinfection is achieved. Such variables affecting the chlorine demand are pH, temperature, contact time and desired free residual. These must be stated. Knowledge of the chlorine demand from laboratory tests yields data that are readily converted to operational practice.

In accord with *Standard Methods* [28], chlorine requirement is the quantity of chlorine that must be added to a unit volume of water under specified conditions of pH, contact time and temperature for a specific result. The latter may be an iron or manganese residual, taste and odor level, or a bacterial count. It is, however, usually associated with a desired disinfection result requiring a specific chlorine residual.

Dechlorination is the partial reduction of chlorine residuals usually by sulfur dioxide or activated carbon prior to entry into the distribution system.

Physical and Chemical Properties

In its elemental form, chlorine is a greenish-yellow gas that is readily compressed into a clear, amber-colored liquid which, in turn, solidifies at an atmospheric pressure of about −102°C [29]. Gaseous chlorine forms a soft ice following contact with water at 9.5°C and at atmospheric pressure. This is chlorine hydrate, $Cl_2 \cdot 8H_2O$, so-called "chlorine ice." Chlorine has an atomic number of 17 and an atomic weight of 35.457. Molecular chlorine (Cl_2) has a weight of 70.914. Some of the physical properties of gaseous and liquid chlorine are given in Table V. A solubility curve for Cl_2 in water is seen in Figure 4.

Table V. Physical Properties of Gaseous and Liquid Chlorine [29]

Gas	
Density at 34°F and 1 atm (lb/ft³)	0.2006
Specific Gravity at 32°F and 1 atm	2.482[a]
Liquefying Point at 1 atm [°F (°C)]	−30.1 (−34.5)
Viscosity at 68°F (cP)	0.01325[b]
Specific Heat (Btu/lb-°F)	
At Constant (1 atm) Pressure and 59°F	0.115
At Constant Volume, 1 atm Pressure and 59°F	0.085
Thermal Conductivity at 32°F (Btu/hr-ft²-°F-ft)	0.0042
Solubility in Water (Figure 4) at 58°F [lb/1000] gal (g/l)]	60.84 (7.297)
Liquid	
Critical Temperature [°C (°F)]	144 (291.2)
Critical Pressure (psia)	1118.36
Critical Density [g/l (lb/ft³)]	573 (35.77)
Density at 32°F (lb/ft³)	91.67
Specific Gravity at 68°F	1.41[c]
Boiling (Liquefaction) Point at 1 atm [°C (°F)]	−34.5 (−30.1)
Freezing Point [°C (°F)]	−100.98 (−149.76)
Viscosity at 68°F (cP)	0.342[d]

[a]Specific gravity of air = 1.
[b]Approximately the same as saturated steam between 1 and 10 atm.
[c]Specific gravity of water = 1.
[d]Approximately 0.35 times that of water at 68°F.

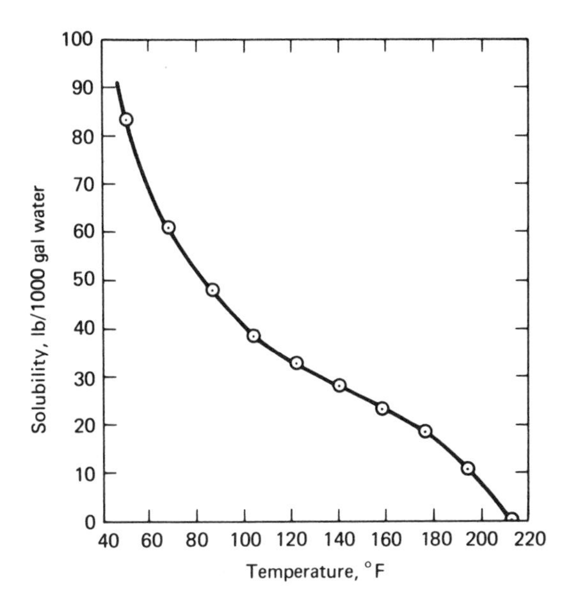

Figure 4. Solubility of chlorine in water. Reproduced from White [29], using the data of Winkler [30], courtesy of the Van Nostrand-Reinhold Company.

Chlorine Compounds

Several chlorine-containing compounds are available for disinfection of water (Table VI). Selection of the appropriate compound depends on the type of water facility. Gaseous chlorine is, today, employed exclusively for the large volumes of municipal and industrial waters. Small-scale facilities, domestic and small communities, frequently apply calcium or sodium hypochlorite. Large public swimming pools use gaseous chlorine, whereas smaller public and private pools apply either the hypochlorites or the chlorine derivatives of isocyanuric acid.

The term "available chlorine content" was employed in the early days of chlorination as a method of comparing the disinfection power of the various chlorinated compounds. Gaseous Cl_2 was thought to be entirely "available" for the disinfection reactions. It is, simply stated, a measurement of the oxidizing power expressed in terms of an equivalent quantity of chlorine. It is determined analytically by titration of the I_2 released by the chlorine compound from an acidic solution of KI. It is expressed as the percentage by weight of the original chlorine, 70.914. For example:

$$Cl_2 + 2I^- = 2Cl^- + I_2 \qquad (1)$$

Table VI. Percent Available Chlorine of Various Chlorine Materials[a]

Material	Available chlorine, %
Cl_2, Chlorine	100 (by definition)
Bleaching Powder (chloride of lime, etc.)	35–37
$Ca(OCl)_2$, Calcium Hypochlorite	99.2
Commercial Preparations	70–74
NaOCl, Sodium Hypochlorite (unstable)	95.2
Commerical Bleach (industrial)	12–15
Commercial Bleach (household)	3–5
ClO_2, Chlorine Dioxide	263.0
NH_2Cl, Monochloramine	137.9
$NHCl_2$, Dichloramine	165.0
NCl_3, Nitrogen Trichloride	176.7
$HOOCC_6H_4SO_2NCl_2$ (Halazone)	52.4
NClCONClCONCl CO, Trichloroisocyanuric Acid	91.5
CONClCONClCONH, Dichloroisocyanuric Acid	71.7
CONClCONClCON Na, Sodium Dichloroisocyanurate	64.5

[a]Reproduced from Laubusch [21], courtesy of the McGraw-Hill Book Company.

Since one mole of chlorine releases one mole of iodine:

$$\%AC = \frac{nMWCl_2}{MW\ compound} \times 100 = \frac{1 \times 70.914}{70.914} = 100\% \quad (2)$$

where AC = available chlorine
n = number of moles of released I_2

Another example:

$$OCl^- + H_2O + 2I^- = Cl^- + 2OH^- + I_2 \quad (3)$$

$$\%AC\ of\ NaOCl = \frac{70.914}{74.5} = 95.2\%$$

$$\%AC\ of\ HOCl = \frac{70.914}{52.} = 136.4\%$$

Hydrolysis and Protolysis of Chlorine

When gaseous chlorine is dissolved in water, the hydrolysis reaction occurs rapidly [31,32]:

$$Cl_2 + H_2O = H^+ + Cl^- + HOCl \quad (4)$$

The hydrolysis constant for this reaction is:

$$K_H = \frac{[H^+][Cl^-][HOCl]}{[Cl_2]} \tag{5}$$

The equilibrium constant values are:

	0°C	15°C	25°C	35°C	45°C	Ref.
$K_H \times 10^4$ (M^2/l^2)	1.46	2.81	3.94	5.10	6.05	32
$K_H \times 10^4$ (M^2/l^2)	1.56	3.16	4.48			31
$K_H \times 10^4$ (M^2/l^2)	1.45		4.47	5.68		32

The extent of chlorine hydrolysis is controlled by the $[H^+]$ in Equation 4. At pH values >3 and with $[Cl_2]_t < 100$ mg/l, very little molecular Cl_2 is present.

That the rate of hydrolysis is extremely rapid was determined by Morris [33] and Shilov and Solodushenkov [34]. The latter investigators found the hydrolysis essentially to be complete in less than 1 sec at 1°C. Morris [33], by means of kinetic considerations, found the reaction mechanism to be:

$$Cl_2 + OH^- = HOCl + Cl^- \tag{6}$$

and the rate of hydrolysis conforms to:

$$-\frac{d[Cl_2]}{dt} = \left[\frac{k_1[Cl_2]Kw}{[H^+]}\right] - k_2[HOCl][Cl^-] \tag{7}$$

The rate constant, k_1, is the order of 3.4×10^{14} at 1.2°C.

Equation 4 shows that, for each mole of Cl_2, there is formed one mole of HCl, a "strong" acid, and one mole of hypochlorous acid (HOCl), a "weak" acid. It should be noted that superchlorination of water will reduce the natural alkalinity in substantial quantities. For example, 1 mg/l Cl_2 produces the acidic equivalent of 1.41 mg/l of alkalinity as $CaCO_3$. Both HCl and HOCl were considered in this calculation. Since HOCl is a "weak" acid, it protolyzes in the Brønsted sense:

$$HOCl + H_2O = H_3^+O + Cl^- \tag{8}$$

The protolysis constant for HOCl has been reported by Morris [35] and several other investigators. Table VII shows the pKa values for HOCl for

Table VII. pK_a for Hypochlorous Acid at Even Temperature Intervals[a]

Temp (°C)	pK_a	pK_a (Caramazza)	pK_a (others)
0	7.825	7.82	7.72 (S and K)
5	7.754		7.63 (K and H)
10	7.690	7.72	
15	7.633	7.65	7.58 (B and D)
20	7.582		7.55 (A and M)
25	7.537	7.53	7.53 (H)
30	7.497		
35	7.463	7.50	

[a]Reproduced from Morris [35], courtesy of the American Chemical Society.

the temperature range from 0–35°C. It is, of course, the $[H^+]$ that determines the extent of Equation 8 and the predominance of [HOCl] or $[OCl^-]$ at a given pH value. For disinfection purposes, it is extremely important to know the distribution of chlorine between HOCl and OCl^- because they have considerably different capabilities for destruction of microorganisms. Figure 5 shows the percentage dissociation of HOCl over the pH range of 4.–11. at 0° and 20°C. The Ka values cited in the graph are earlier ones from Morris [35].

Oxidation States of Chlorine

Table VIII shows the oxidation states of chlorine and the free energies of formation for the dissolved and gaseous species. Some of the relevant reduction reactions (Equations 9–16) involving these species are given in Table IX at 25°C. The E-pH stability diagram for the $Cl_{2(g)}$–$H_2O_{(l)}$ system at 25°C is seen in Figure 6. It is obvious that, as an oxidant in water, Cl_2 is not stable.

Reactions with Ammonia

In aqueous systems, chlorine reacts with ammonia (NH_3) to form the chloramines:

$$NH_3 + HOCl = \underset{\text{monochloramine}}{NH_2Cl} + H_2O \quad (17)$$

$$NH_2Cl + HOCl = \underset{\text{dichloramine}}{NHCl_2} + H_2O \quad (18)$$

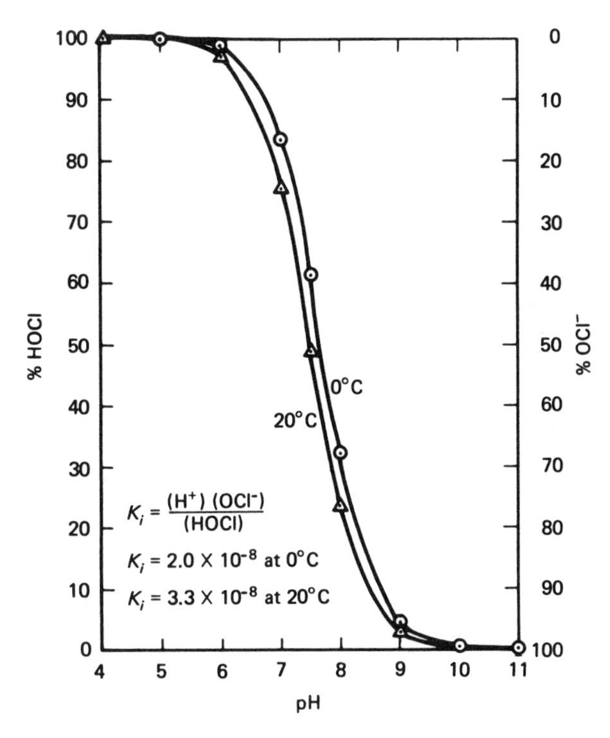

Figure 5. Protolysis of hypochlorous acid at 0 and 20°C. Reproduced from White [29], courtesy of the Van Nostrand-Reinhold Co.

$$NHCl_2 + HOCl = \underset{\substack{\text{trichloramine} \\ \text{(nitrogen trichloride)}}}{NCl_3} + H_2O \tag{19}$$

The ammonia occurs from natural decomposition of organic N compounds or denitrification reactions catalyzed by various microorganisms. Also, NH_3 may be added to water before (preammoniation) to prevent tastes and odors by phenols (see Chapter 3) or after (post ammoniation) addition of chlorine. The latter practice is the combined residual chlorination described above. In any event, NH_3 contributes to the chlorine demand of the water. Figure 7 shows a graphic illustration of the "breakpoint" reaction where only NH_3 is involved. The breakpoint curve results from reactions 17–19 whose products, the chloramines, are dependent on: pH value, relative concentrations of HOCl and NH_3 reaction time and temperature. Monochloramine usually is the only chloramine observed when pH values are greater than 8. and when the molar ratio of HOCl to NH_3 is 1:1 or less. At pH values less than 3., only trichloramine ordinarily is detected.

Table VIII. Oxidation States of Chlorine[a]

Oxidation Number (Z)	Considered	Not Considered	ΔG_f^o (kcal/mol)	Name, Color, Crystalline System
−1	Cl^-		−31.350	Chloride ion, colorless
0	Cl_2		1.650	Dissolved chlorine
+1	$HClO$		−19.110	Hypochlorous acid, colorless
"	ClO^-		− 8.900	Hypochlorite ion, colorless
+3	$HClO_2$		.070	Chlorous acid, colorless
"	ClO_2^-		2.740	Chlorite ion, colorless
+5	ClO_3^-		− .620	Chlorate ion, colorless
+7	ClO_4^-		− 2.470	Perchlorate ion, colorless
−1	*HCl*		−22.769	Hydrogen chloride, colorless
0	*Cl_2*		0	Chlorine, greenish-yellow
+1	*Cl_2O*		22.400	Chlorine monoxide, hypochlorous anhydride, brownish
+2		*ClO*		Chlorine oxide
+4	ClO_2		29.500	Chlorine dioxide, orange-yellow
+6		*ClO_3*		Chlorine trioxide, brown
+7		*Cl_2O_7*		Chloric heptoxide, perchloric anhydride, colorless
+8		*ClO_4*		Chlorine tetroxide, colorless

[a]Reproduced from Pourbaix [36], courtesy of the Pergamon Press, Ltd.

Table IX. Some Reduction Reactions of Chlorine at 25°C [36]

Reaction Number	Reaction	V°, (V)
9	$Cl_2 + 2\ e = 2\ Cl^-$	+1.395
10	$HOCl + H^+ + 2\ e = Cl^- + H_2O$	+1.494
11	$OCl^- + 2\ H^+ + 2\ e = Cl^- + H_2O$	+1.715
12	$ClO_3^- + 6\ H^+ + 6\ e = Cl^- + 3\ H_2O$	+1.451
13	$ClO_4^- + 8\ H^+ + 8\ e = Cl^- + 4\ H_2O$	+1.389
14	$ClO_{2(g)} + 4\ H^+ + 5\ e = Cl^- + 2\ H_2O$	+1.511
15	$Cl_{2(g)} + 2\ e = 2Cl^-$	+1.359
16	$ClO_{2(g)} + 2\ H^+ + 3\ e = ClO^- + H_2O$	+1.374

The rate of monochloramine formation was reported by Weil and Morris [38]. It is an extremely rapid reaction and extremely sensitive to changes in the $[H^+]$. Within the pH range of most water supplies, the reaction is usually 90% complete in one minute or less. As seen in Figure 8, the reaction rate is at its maximum at pH 8.5. The production of

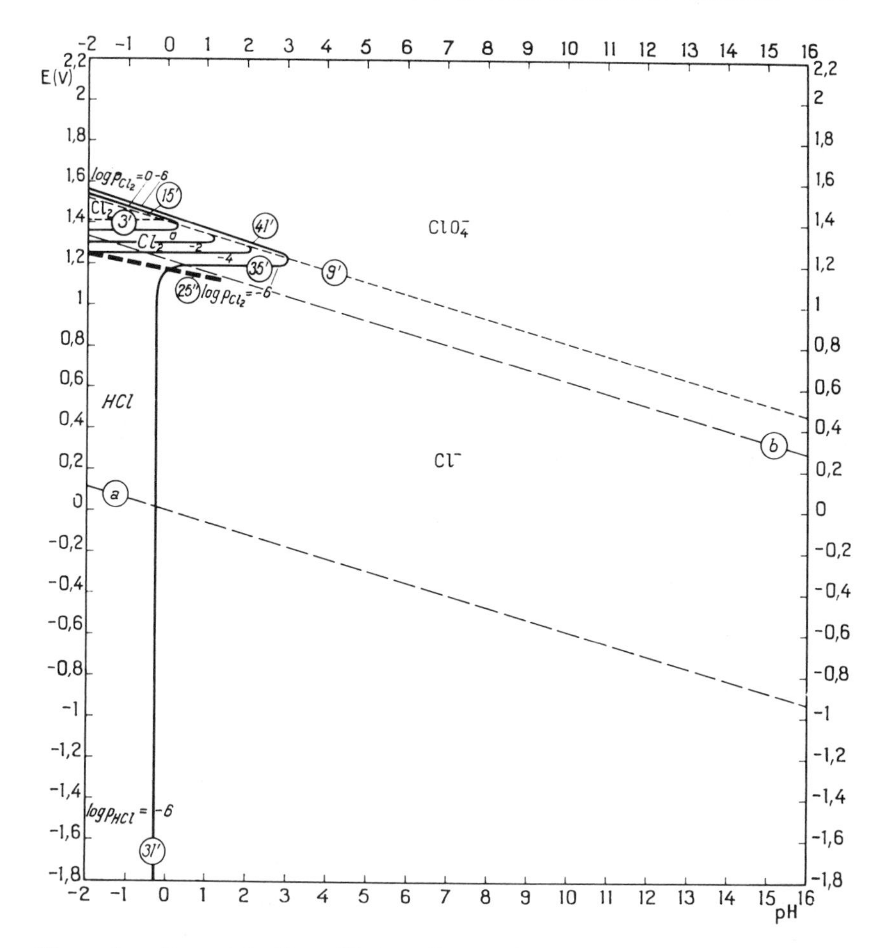

Figure 6. Potential-pH equilibrium diagram for the system Cl_2-H_2O at 25°C, for solutions containing 1 g-atom Cl/l. Equation numbers refer to original text. Reproduced from Pourbaix [36], courtesy of Pergamon Press Ltd.

$NHCl_2$ is optimized in the pH range of 7.–9. and at a HOCl:NH_3 ratio of 3:1 to 5:1 (by weight). At higher HOCl:NH_3 ratios or at lower pH values, di- and trichloramines are formed. Formation of monochloramine accounts for the chlorine residual in Figure 7 up to the "peak" of the curve.

It should be noted that reaction 17 could occur also between ionic molecules: $NH_4^+ + OCl^-$, between ionic and neutral molecules: $NH_4^+ +$ HOCl or $NH_3 + OCl^-$. Weil and Morris [38] favor the neutral reaction mechanism because it fits their kinetic data reasonably well. An increase in rate

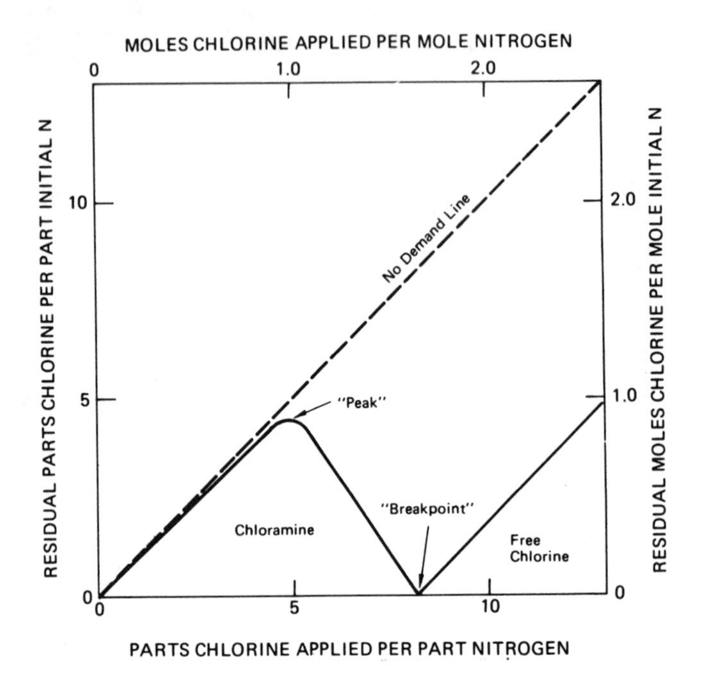

Figure 7. Diagrammatic representations of completed breakpoint reaction [37].

of the reaction as the pH value is increased in acid range is explained by a greater fraction of the analytical ammonia present as the free base (pKa = 9.2 for NH_4^+). Similarly, a decrease in the rate in the alkaline pH range can be attributed to ionization of the HOCl (pKa = 7.537). It is interesting to note that, at the pH of maximum rate (Figure 8), approximately 80% of the analytical chlorine is OCl^-. In any event, the rate of the ammonia reaction is:

$$v = k_1 [NH_3][HOCl] \gamma_{NH_3} \gamma_{HOCl} / \gamma_x \quad (20)$$

where v = velocity of reaction
k_1 = rate constant
γ = activity coefficients (γ_x is for the activated complex)

At 25°C, the second-order rate constant k_1 between the uncharged molecules is [39] 5.1×10^6 liter/mol-sec.

The specific formation rate of dichloramine is considerably slower than the formation of monochloramine, except at pH values less than 5.5. Specific rates are given in Table X for the formation of the three chloramines [40]. Since dichloramine forms more slowly than mono-

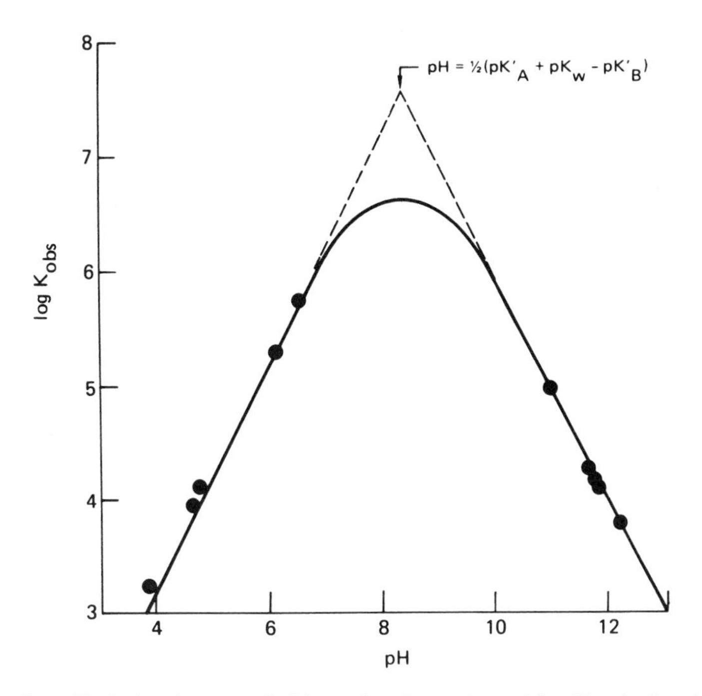

Figure 8. Variation in rate of chloramine formation with pH calculated with $\log K_n = 8.48$ [37].

chloramine at nearly neutral pH values, $NHCl_2$ does not constitute a large percentage of the available chlorine unless the water is quite acid or where the molar ratio of Cl_2:NH_3 is greater than 1.0. The relative percentages of di- and monochloramine for equimolar Cl_2 and NH_3 and for 25% excess NH_3 are seen in Figure 9 [41]. Dichloramine is somewhat unstable in water, more unstable than mono- and trichloramine. A possible decomposition reaction is [41]:

$$2\,NHCl_2 + H_2O = N_2 + HOCl + 3\,H^+ + 3\,Cl^- \tag{21}$$

Other possible decomposition reactions for $NHCl_2$ are:

$$NHCl_2 + H_2O = NOH + 2\,H^+ + 2\,Cl^- \tag{22}$$

Weil and Morris [42] reported that the decomposition of $NHCl_2$ in aqueous solutions is in accord with first-order kinetics wherein the specific rate is more than proportional to the $[OH^-]$ (also, see below). NOH is, perhaps, an intermediate in the decomposition of $NHCl_2$. Other competing reactions for NOH are [43]:

Table X. Comparison of Specific Rates for Chloramine Formation at 20°C[a]

	$k(M^{-1}\text{-sec}^{-1})$	General Expression for k (liter/mol-sec)
NH_2Cl	5.6×10^6 [b]	$k = 9.7 \times 10^8 \exp(-3000/RT)$
pH 3.2	4.9[c]	
pH 2.3	0.6[c]	
$NHCl_2$	2.7×10^2 [d]	$k = 7.6 \times 10^7 \exp(-7300/RT)$
NCl_3		
pH 2.3	14.9	$k = 1.15 \times 10^3 \exp(-2500/RT)$
pH 3.2	3.4	$k = 2.62 \times 10^4 \exp(-5200/RT)$
pH 3.97	1.6	$k = 7.83 \times 10^4 \exp(-6270/RT)$
pH 4.53	2.0	$k = 3.43 \times 10^5 \exp(-7000/RT)$

[a]Reproduced from Saguinsin and Morris [40], courtesy of Ann Arbor Science Publishers, Inc.
[b]Calculated assuming reaction between neutral molecules.
[c]Extrapolated K_{obs}.
[d]Rate for nonacid-catalyzed reaction.

$$NOH + NH_2Cl = N_2 + H_2O + H^+ + Cl^- \qquad (23)$$

$$NOH + NHCl_2 = N_2 + HOCl + H^+ + Cl^- \qquad (24)$$

$$NOH + 2\,HOCl = NO_3^- + 3\,H^+ + 3\,Cl^- \qquad (25)$$

In waters with pH <4, or where Cl_2 concentrations are considerably in excess of NH_3, trichloramine is formed. Since NCl_3 is formed from $NHCl_2$, (Reaction 19) the reaction occurs only under conditions in which the dichloramine is reasonably stable. Therefore, NCl_3 is the only chloramine at pH <3. Also, it may be formed when Cl_2:NH_3 molar ratios are >2. However, NCl_3 occurs in diminishing proportions as the pH value is increased from 3. to 7.5. Above the latter value, no NCl_3 is observed, regardless of the Cl_2:NH_3 ratio [43].

In Figure 7, the "breakpoint" becomes significant in the pH range of 6.–9. When the Cl_2:NH_3 molar ratios are 0 to 1, NH_2Cl is formed. This creates the "peak" seen in Figure 7. At molar ratios >1, $NHCl_2$ is formed, but it decomposes rapidly in accord with, perhaps, Reaction 22. Therefore, when the molar ratio exceeds 1. and up to about 1.65, the chlorine residual decreases until the "breakpoint" occurs after the NH_3 has been converted mainly to N_2 and some NO_3^- [42]. After the breakpoint, the chlorine residual exists as "free available" chlorine, that is, HOCl and OCl^-.

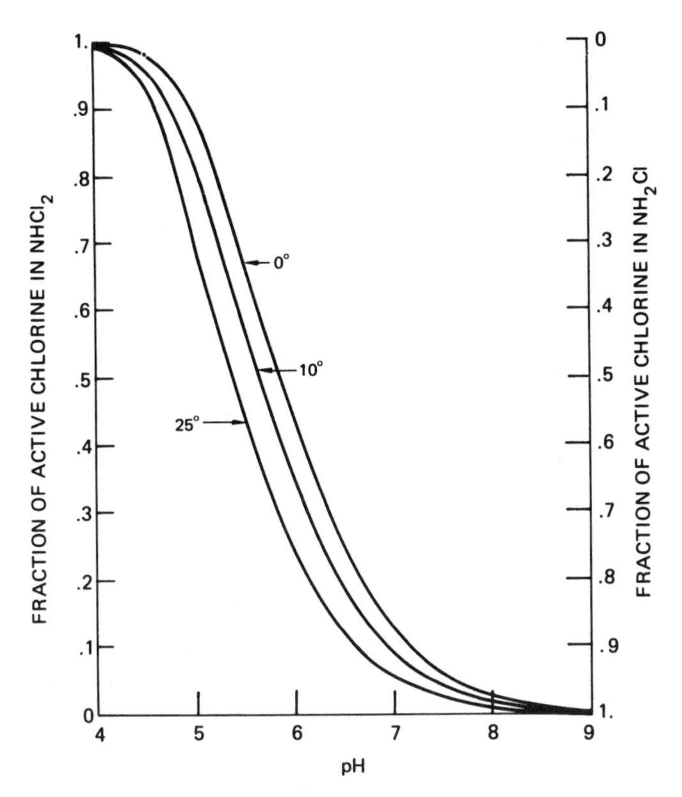

Figure 9. Proportions of monochloramine (NH_2Cl) and dichloramine ($NHCl_2$) formed in water chlorination with equimolar concentrations of chlorine and ammonia [37,41].

It is appropriate to examine the kinetics and stoichiometry of the "breakpoint" chlorination phenomenon. Figure 10 shows another schematic diagram of breakpoint chlorination [42]. It is based on the classical studies of Griffin and Chamberlain [44], Griffin [45] and Palin [46]. In Figure 10a, the residual oxidizing chlorine increases proportionately with chlorine dosage until the molar ratio of Cl_2:NH_3 reaches 1. This is up to point A. As seen in Figure 10b, very little, if any, combined nitrogen was oxidized in this range. Once the ratio exceeds 1., there is a decrease in residual oxidizing chlorine with the concurrent increase in chlorine dosage (points A to B, Figure 10a). At point B—the "breakpoint"—essentially all of the oxidizing chlorine is reduced, and all of the NH_3-N is oxidized (Figure 10b). An appropriate reaction is understood. Beyond point B, free chlorine residual exists. According to Wei and Morris [42], "the development of the breakpoint results from a redox

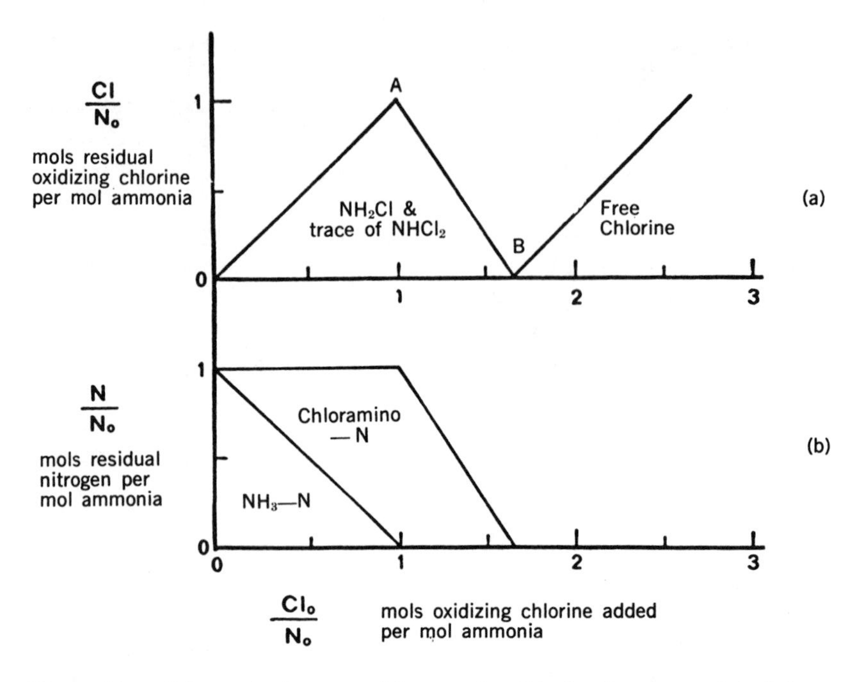

Figure 10. Schematic diagram of breakpoint chlorination. Reproduced from Wei and Morris [42], courtesy of Ann Arbor Science Publishers.

reaction between aqueous chlorine and ammonia with location of the breakpoint indicating the stoichiometry of the reaction." The molar ratio of reduced chlorine to oxidized nitrogen, equal to about 1.65 at the breakpoint, shows that the major oxidation of the ammonia is to gaseous nitrogen, which requires a ratio equal to 1.5. The excess over 1.5 is accounted for by a small side formation of nitrate as shown by Palin [46]. Except for the formation of NO_3^-, the stoichiometry is explained by two reactions:

$$HOCl + NH_3 = NH_2Cl + H_2O \tag{17}$$

and

$$2\,NH_2Cl + HOCl = N_2 + H_2O + 3\,H^+ + 3\,Cl^- \tag{26}$$

Although the stoichiometry of the breakpoint comes from the addition of reactions 17 and 26, the precise mechanism may not be [42]. This, however, is not a major consideration in the dynamics of the breakpoint process.

According to Wei and Morris [42], "it is the first hour of reaction that is most significant in terms of practical application of the breakpoint process, for disinfection of water supplies must almost always be accomplished within that period." This first hour of the Cl_2-NH_3 reaction was studied in great detail by Wei and Morris [42]. The experimental conditions were: (1) three pH values: 6.7, 7.0 and 7.2 (phosphate buffer); (2) four temperatures: 5, 10, 15 and 20°C; and (3) a varying of the reactant concentrations of NH_3-N from 0.25 to 1.5 mg/l; however, the molar ratio of Cl_2:NH_3 was held constant at 1.8 in order to leave some free chlorine after completion of the breakpoint.

Typical kinetic curves are seen in Figure 11. It is seen that formation of NH_2Cl is "rapid" (<1 min). Also, formation and decomposition of $NHCl_2$ are rapid, the latter of which is a first-order process. In each curve in Figure 11, a maximum [$NHCl_2$] is observed for which temperature affects the magnitude and rate of decomposition. $NHCl_2$ is more persistent at the lower temperatures of 10 and 5°C. The overall effect of pH on the breakpoint process as it is decreased from 7.2 to 7.0 to 6.7 is: (1) to increase the "hump" (magnitude) of the $NHCl_2$ curve; (2) to delay appearance of the hump on the time ordinate, i.e., it takes longer for the hump to form; and (3) to increase the persistence of $NHCl_2$. These observations are consistent with previous works of Morris et al. (see Figure 8 and Weil and Morris [38]).

Figure 12 summarizes the effect of initial [NH_3] on $NHCl_2$ formation at pH 7.0. Since the formation of $NHCl_2$ is a second-order reaction, its maximum concentration should occur sooner on the time ordinate and with a larger magnitude. This is seen in Figure 12. Also, the rate of decomposition of $NHCl_2$ is faster as the $[NH_3]_0$ is increased.

Table XI summarizes the experimental data of the stoichiometry of the Cl_2-NH_3 reaction in column 7. The R values are those obtained after 60 min of reaction and at the breakpoint. There was no significant effect of [H^+] on the R value within the limited pH range, 6.7–7.2. Wei and Morris [42] seem to think that there is an effect of temperature with an R value of 1.72 at 20°C and 1.66 at 5°C. This may or may not be significant.

The point in time at which $NHCl_2$ reaches its maximum concentration, i.e., the "hump," apparently governs the overall kinetics of the breakpoint. The kinetic pattern of this reaction indicates that $NHCl_2$ is a true intermediate in the breakpoint reaction. Since the formation of $NHCl_2$ by Reaction 18 is a second-order reaction [47] and its decomposition by Reaction 22 is a first-order reaction, the former is favored over the latter. Evidence for this statement is seen in Figure 11.

Another factor in the breakpoint reaction, especially in the last few minutes, is the resurgence of free chlorine. This may occur from:

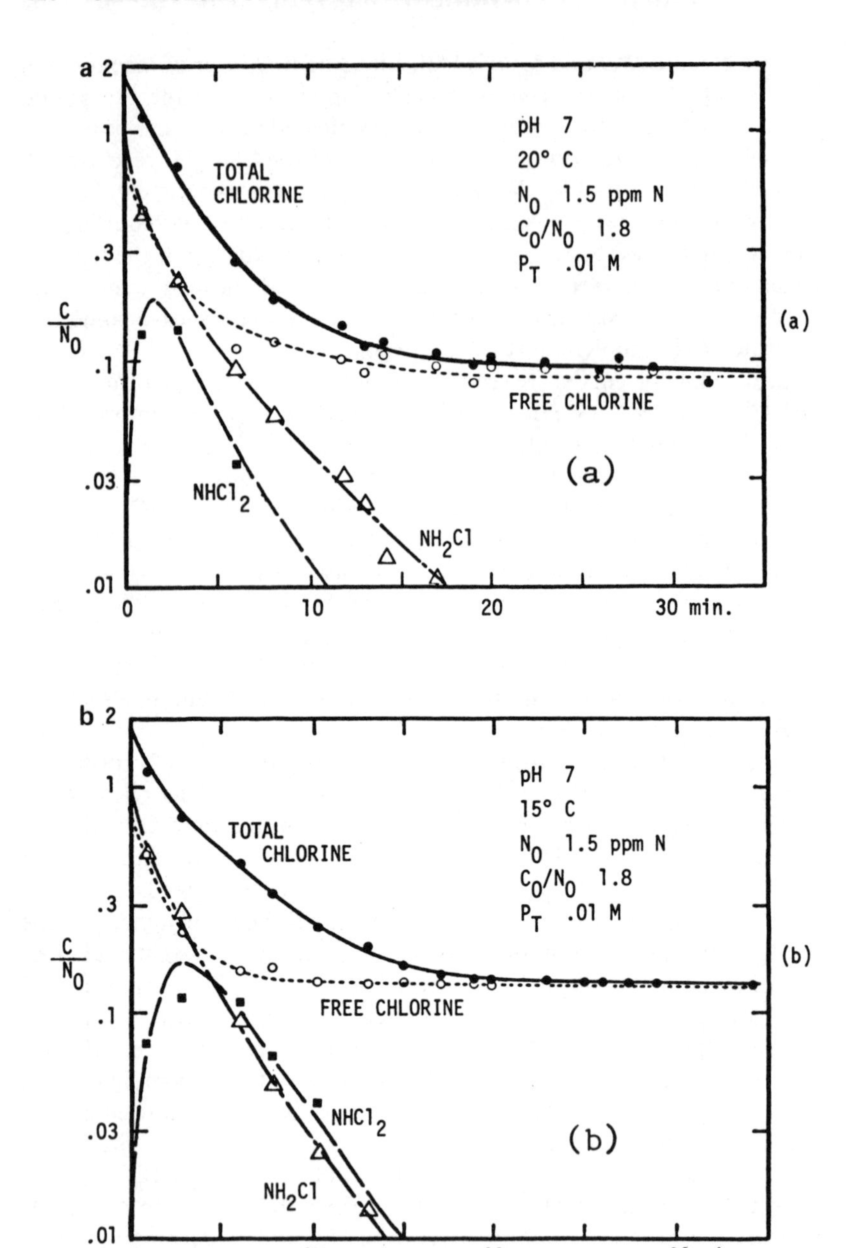

a
pH 7
20° C
N_0 1.5 ppm N
C_0/N_0 1.8
P_T .01 M
TOTAL CHLORINE
FREE CHLORINE
$NHCl_2$
NH_2Cl
(a)
$\frac{C}{N_0}$
30 min.
b
pH 7
15° C
N_0 1.5 ppm N
C_0/N_0 1.8
P_T .01 M
TOTAL CHLORINE
FREE CHLORINE
$NHCl_2$
NH_2Cl
(b)
30 min.

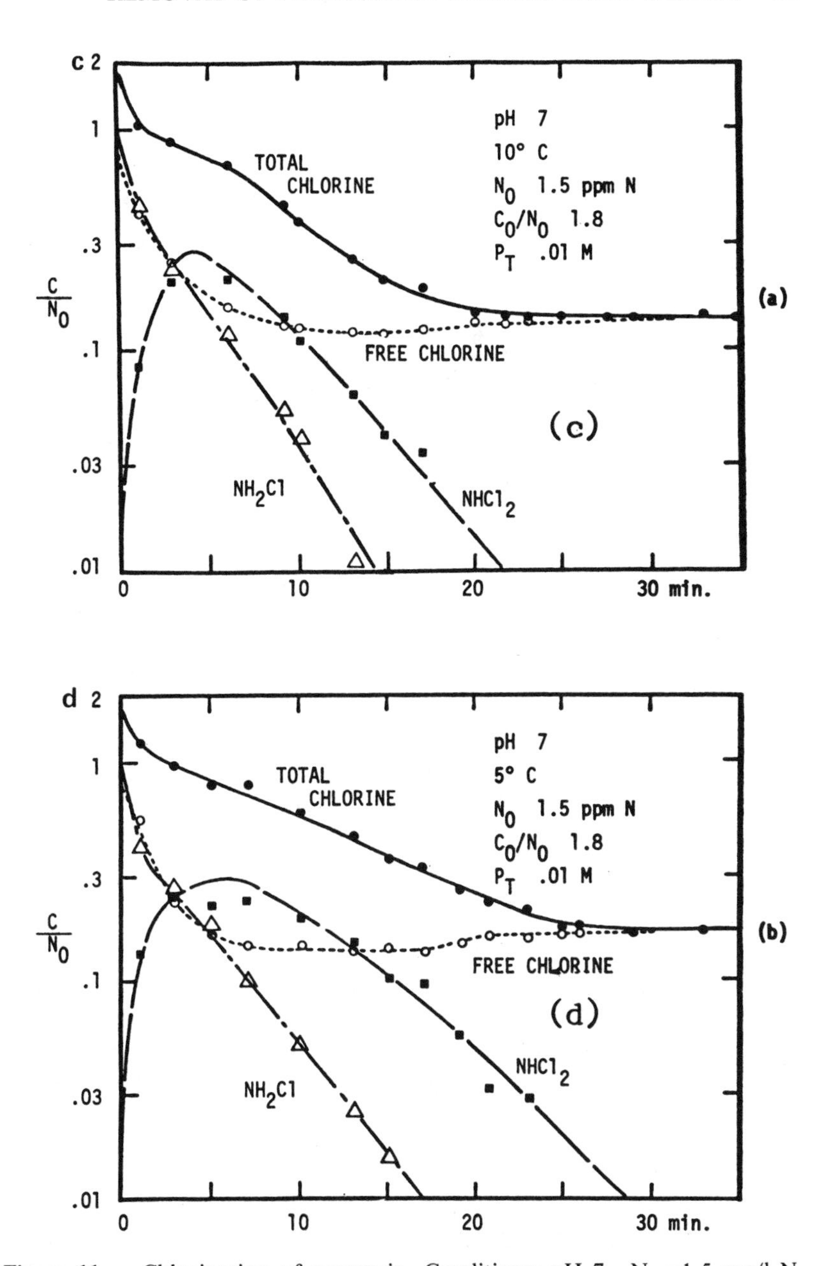

Figure 11. Chlorination of ammonia. Conditions: pH 7., $N_0 = 1.5$ mg/l N, $C_0/N_0 = 1.8$, $P_t = 0.01$ *M*, **(a)** 20°C, **(b)** 15°C, **(c)** 10°C, and **(d)** 5°C. Reproduced from Wei and Morris [42], courtesy of Ann Arbor Science Publishers.

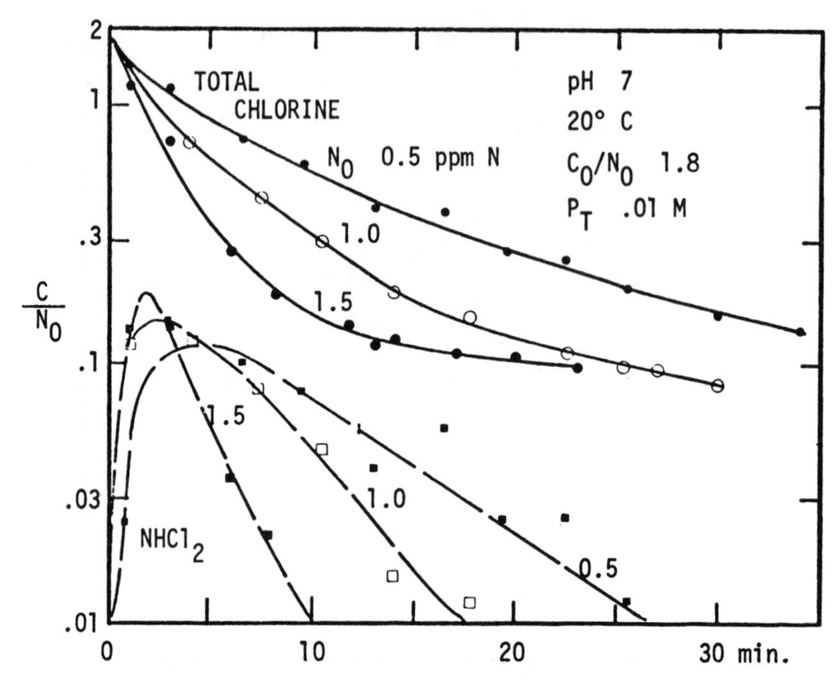

Figure 12. Effect of initial ammonia concentration on chlorination of ammonia at pH 7., 20°C, $C_0/N_0 = 1.8$, $P_t = 0.01$ *M*. Reproduced from Wei and Morris [42], courtesy of Ann Arbor Science Publishers.

$$2\,NHCl_2 + H_2O = N_2 + 3\,H^+ + 3\,Cl^- + HOCl \quad (27)$$

This resurgence of HOCl has been observed also by Palin [46]. It affects the kinetic process as well as the stoichiometry. Also it fits the experimental data whereby a molar ratio of 1.5 of Cl_2:NH_3 yields N_2 as a decomposition product in the breakpoint.

Saguinsin and Morris [40] reported the formation and decomposition of NCl_3 under a variety of conditions. That the kinetics of the formation reaction is much slower than formation of NH_2Cl or $NHCl_2$ was given in Table X. Values of the specific rate are seen in Figure 13 where it is a function of the $[H^+]$ at a given temperature. The formation reaction is not strongly dependent on the pH until a value of 3.2 is reached. When the ratio of formation of the three chloramines is compared (Table X) at 20°C, the specific rate for NH_2Cl is approximately 10^4 times that for $NHCl_2$, and about 10^6 times that for NCl_3, based on a direct reaction between HOCl and the uncharged amine. At the lower two pH values, 2.3 and 3.2, the effect of Cl_2 on the kinetics of the formation reaction

Table XI. Comparison of Stoichiometric Ratios[a]

pH	Temp. (°C)	N_0 (mg/l N)	C_0/N_0[b]	C_f/N_0[c]	R[d]
6.7	20	1	1.8	0.09	1.71
	15	1	1.8	0.10	1.70
	10	1	1.8	0.11	1.69
	5	1	1.8	[e]	[e]
7.0	20	1	1.8	0.07	1.73
	15	1	1.8	0.10	1.70
	10	1	1.8	0.11	1.69
	5	1	1.8	0.13	1.67
7.2	20	1	1.8	0.07	1.73
	15	1	1.8	0.08	1.72
	10	1	1.8	0.12	1.68
	5	1	1.8	0.12	1.68
7.0	20	1.5	1.8	0.09	1.71
	15	1.5	1.8	0.13	1.67
	10	1.5	1.8	0.14	1.66
	5	1.5	1.8	0.16	1.64
6.7	20	0.25	1.8	[e]	[e]
	20	0.5	1.8	[e]	[e]
7.0	20	0.5	1.8	[e]	[e]
7.2	20	0.5	1.8	[e]	[e]
	20	1.5	1.8	0.07(−)	1.73(+)
7.04–7.24	20	0.25	1.7	[e]	[e]
7.05–7.25	20	0.5	1.7	0.17	1.53
7.16–7.22	20	1	1.7	0.17	1.53

[a]Reproduced from Wei and Morris [42], courtesy of Ann Arbor Science Publishers.
[b]C_0/N_0 = molar ratio of initial chlorine to initial ammonia.
[c]C_f/N_0 = molar ratio of final free chlorine to initial ammonia.
[d]R = stoichiometric ratio as a molar ratio of chlorine reduced to ammonia oxidized. $R = C_0/N_0 - C_f/N_0$.
[e]Data are not available because the reaction had not reached completion at the time the experiment was terminated.

must be considered also. There was a tenfold increase in the reaction rate in the presence of 5×10^{-4} M Cl^-.

That the decomposition of NCl_3 in neutral and slightly alkaline solutions is a first-order process is seen in Figure 14a. Here again, the pH value, or, in this reaction, the $[OH^-]$, is a factor accelerating the decomposition. This is seen in Figure 14b. The decomposition was dependent directly on the $[OH^-]$ since a linear relation was observed between the specific rate constant and $[OH^-]$. Either one of these two reactions may be operative:

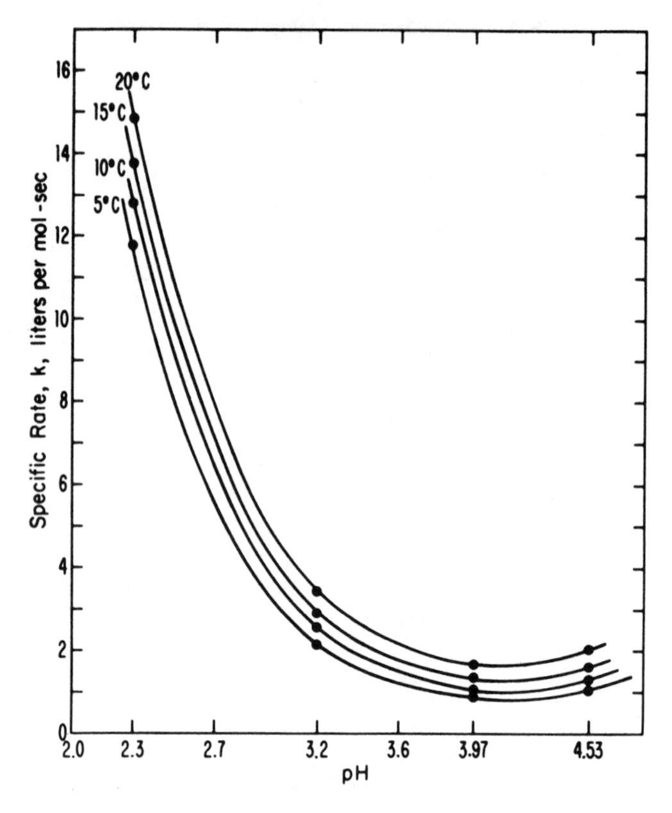

Figure 13. Specific rates of NCl_3 formation as a function of pH and temperature. Reproduced from Saguinsin and Morris [40], courtesy of Ann Arbor Science Publishers.

$$NCl_3 + H_2O \overset{OH^-}{=} NHCl_2 + HOCl \tag{28}$$

$$NCl_3 + OH^- = NHCl_2 + OCl^- \tag{29}$$

Reaction 28 is OH^--catalyzed, and Reaction 29 is a direct reaction with OH^-.

Sounier and Selleck [48] reported a rather comprehensive investigation into the kinetics of breakpoint chlorination in continuous flow systems via a pilot plant. Three types of chlorine contact chambers were employed: (1) a pipe reactor whose flow characteristics approached those of a plug flow reactor (PFR), (2) a tubular reactor, and (3) a tank reactor whose mixing characteristics approached those of an ideal continuous-flow, stirred-tank reactor (CSTR). Tapwater was employed in which the

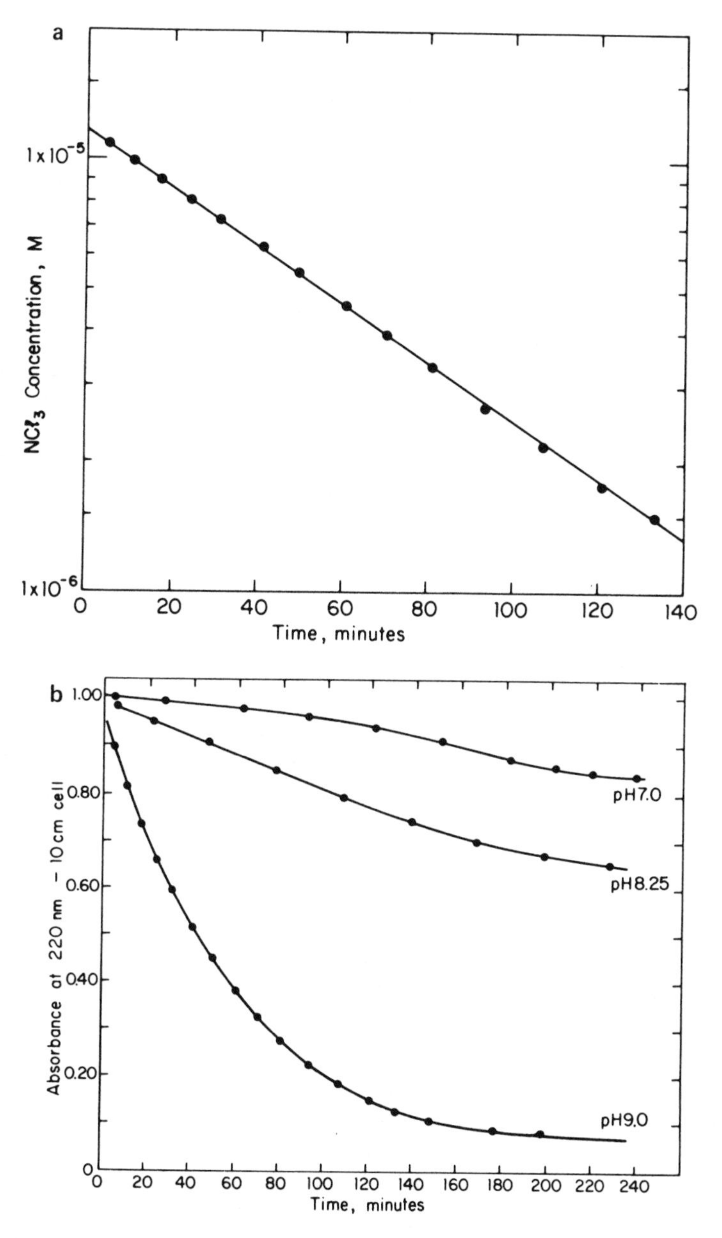

Figure 14. **(a)** Rate of decomposition of 1.15×10^{-5} M NCl_3 at 20°C and pH 9. **(b)** Effect of pH on the rate of decomposition of 1.2×10^{-5} M NCl_3 at 20°C. Reproduced from Saguinsin and Morris [40], courtesy of Ann Arbor Science Publishers.

molar ratio of chlorine to ammonia-N was held in the 1.6–2.78 range. The major variables in this were the $[H^+]$ and the $Cl_2:NH_3$-N molar ratio. Reaction times were <1 hr in which the formation and disappearance of the three chloramines were followed. The essential results were:

1. The major oxidation products of ammonia were N_2 and NO_3^-. The proportion of NH_3-N converted to NO_3^- "appeared to be independent of reaction pH" (6.65–9.2).
2. The maximum speed of the oxidations of mono- and dichloramine occurred at a pH of almost 7.5. These reactions were slower at pH values <7 and >8.
3. NCl_3 appeared very early in the reaction and reached its greatest content at a pH of about 7. in the PFR experiments. This observation is at variance with other researchers.
4. The oxidation of mono- and dichloramine occurred within a matter of minutes to unknown intermediate nitrogen compounds. According to Sounier and Selleck, the eventual appearance of N_2 and NO_3^- required substantially longer reaction times, i.e., several hours. No explanations were given for this observation.

Bromination

Bromine was applied for disinfection of water by Henderson in 1935 [49] and by Hildesheim for swimming pool water in 1936 [50]. Since these times, bromine has been used mainly for swimming pool disinfection.

Physical Properties

Bromine, Br_2, is a dark brownish-red, heavy liquid [29]. A heavy, brownish-red vapor with a sharp, penetrating and suffocating odor is yielded by this liquid at room temperature. Liquid bromine is extremely corrosive and destructive to organic tissues. It has an atomic number of 35, a mol wt of 159.832, a specific gravity of 3.12 and a boiling point of 58.78°C. Bromine is unique in that it is the only nonmetallic element that is a liquid at room temperature. It is produced by the oxidation of bromine-rich brines (0.05–0.6% Br^-) with chlorine. Bromine is then stripped with air or steam and is collected as liquid Br_2.

An interhalogen compound, bromine chloride is produced by mixing equal molar quantities of pure bromine and chlorine [51]. It condenses to a liquid below 5°C at 1 atm pressure or above 30 psig at 25°C. In the liquid phase, 80% or more of the liquid is bromine chloride, and the rest is Br_2 and Cl_2. In the gaseous phase, about 40% of the BrCl dissociates to Br_2 and Cl_2 at 25°C. BrCl is used occasionally for disinfection of water. The physical properties of bromine and bromine chloride are given in Table XII [52].

Hydrolysis and Protolysis Reactions

Bromine and bromine chloride hydrolyze rapidly:

$$Br_2 + H_2O = HOBr + H^+ + Br^- \quad (30)$$

$$BrCl + H_2O = HOBr + H^+ + Cl^- \quad (31)$$

In both cases, hypobromous acid is formed. The equilibrium constant for Reaction 30 is 5.8×10^9 at 25°C [52] and for Reaction 31, 7.21×10^{-5}

Table XII. Physical Properties of Bromine Chloride Compared to Bromine[a]

	BrCl	Br_2
Mol wt	115.37	159.83
% bromine	69.27	100
Fp, (°C)	−66	−7.27
Bp (°C)		
760 mm	5	58.78
226 mm	−30	25
100 mm	−56	6
10 mm		−32
5 atm	48	
10 atm	70	139.8
Density (g/cm^3)		
15°	2.352	3.1396
20°	2.339	3.1226
25°	2.324	3.1055
30°	2.310	3.0879
lb/gal, 25°C	19.3	25.8
Vapor density, (g/l), std conditions (0°C, 1 atm)	5.153	7.139
Latent heat of fusion, (cal/g)	17.6	15.8
Latent heat of vaporizn, (cal/g),	53.2	44.9
bp 25°	30.0	46.2
Heat capacity, (cal/°K-mol), 298°K	8.38	18.09
Entropy (cal/°K-mol), 298°K	57.34	36.38
Dipole moment, D	0.56	0

[a]Reproduced from Kanyaer and Shilan [52], courtesy Tekhvol. Institute of USSR.

at 25°C (calculated from free energy data in Latimer [53]). The extent of hydrolysis to HOBr is controlled by [H^+]. For a solution containing ≥10 mg/l Br and a pH of 6.3, 1% of the analytical bromine is Br_2. At lower bromine concentrations or higher pH values, HOBr is the predominant form. This acid is "weak" in the Brønsted sense:

$$HOBr + H_2O = H_3^+O + OBr^- \tag{32}$$

The protolysis constant for HOBr, at 25°C, is 2×10^{-9} [54]. Consequently, the pH value must exceed 8.7 before OBr^- predominates.

Oxidation States of Bromine

The relevant oxidation states of bromine in aqueous solutions are: 0, Br, −I, Br^- and +I, OBr^- [37]. Some reduction reactions (Equations 33–35) of bromine are given in Table XIII. An E-pH stability diagram is seen in Figure 15, where Br^- is the preferred oxidation state in water. Consequently, Br_2, HOBr and OBr^- are easily reduced.

Reactions with Ammonia

Bromine and bromine chloride react with NH_3 to form bromamines [55,56]:

$$NH_3 + HOBr = NH_2Br + H_2O \tag{36}$$

$$NH_2Br + HOBr = NHBr_2 + H_2O \tag{37}$$

$$NHBr_2 + HOBr = NHBr_3 + H_2O \tag{38}$$

According to Galal-Gorchev and Morris [55], NH_2Br predominates in alkaline solutions at "high" N:Br ratios; $NHBr_2$ is the major form in the

Table XIII. Some Reduction Reactions of Bromine at 25°C [36]

Reaction Number	Reaction	V^0 (V)
33	$Br_2 + 2\,e = 2\,Br^-$	+1.087
34	$HOBr + H^+ + 2\,e = Br^- + H_2O$	+1.331
35	$BrO^- + 2\,H^+ + 2\,e = Br^- + H_2O$	+1.589

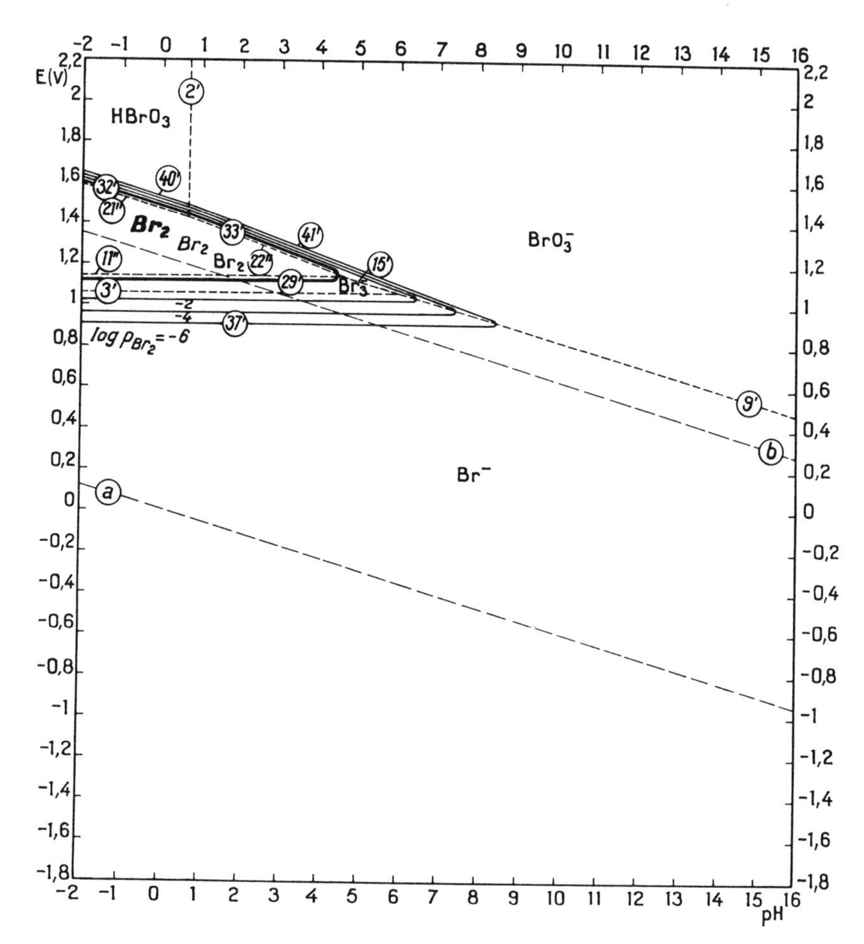

Figure 15. Potential-pH equilibrium diagram for the system Br_2-H_2O at 25°C, for solutions containing 1 g-atom Br/l. Equation numbers refer to the original text. Reproduced from Pourbaix [36], courtesy of Pergamon Press Ltd.

pH range 6–9 with N:Br ratios about 5–20, and NBr_3 predominates in more acid solutions. Nitrogen tribromide can occur in mixtures up to pH 8. when two to three moles of bromine per mole of ammonia are permitted to react.

The formation and subsequent decomposition of the bromamines apparently are extremely rapid [56]. When a mixture of NH_3-N and Br_2, a 2:3 molar ratio, was allowed to react at pH of 6., the following was observed:

Constituent	1 min (10^{-4} M Br_2)	12 min (10^{-4} M Br_2)
Dibromamine	0.75	0.25
Tribromamine	0.69	0.27

There is concurrent formation and decomposition of the two bromamines. When the pH value was held at 7.0 and at 20°C, the second-order rate constant for the decomposition of $NHBr_2$ was observed to be 17.5 liter/M-sec. This gives a half-life of 4.2 min for a 32-ppm solution. The temperature dependence of this decomposition was: $K = 1.38 \times 10^7$ $e^{-7510/RT}$ liter/M-sec. The kinetics of decomposition of NBr_3 are dependent on the $[H^+]$ and the molar ratio of NH_3-N:Br_2. When the pH is 6. and the molar ratio is 0.333:1, the decomposition rate is apparently third order with an observed rate constant of 9.5×10^3 liter^2/m^2-sec (20°C). When the pH is increased to 7. and the molar ratio to .5:1, the rate constant is increased to 1.2×10^5 liter^2/m^2-sec (20°C).

According to Johnson and Overby [56], kinetic data have shown that, at pH values of 6., 7., and 8., a breakpoint occurs for bromine at a N:Br mole ratio of 2:3. In the presence of excess Br_2, NBr_3 is the most stable species, whereas $NHBr_2$ is most stable in the presence of excess NH_3 (Figure 16). High pH values decrease the stability of both bromamines.

In addition, a breakpoint occurs when there is 17 mg/l Br_2 and 1 mg/l NH_3-N, which is the point of minimum bromamine stability. The apparent breakpoint reaction is [57]:

$$3\,HOBr + 2\,NH_3 = N_2 + 3\,HBr + 3\,H_2O \tag{39}$$

At or near neutral pH values, NH_2Br may form ammonia bromide which, in turn, dissociates into ammonia and the Br^+ ion [58]:

$$NH_2Br + H^+ = NH_3Br^+ \tag{40}$$

$$NH_3Br^+ = NH_3 + Br^+ \tag{41}$$

According to Johanneson [58], the NH_2Br and NH_3Br^+ are present in equal quantities at pH 6.5. These two reactions may account for the excellent germicidal efficiency of NH_2Br.

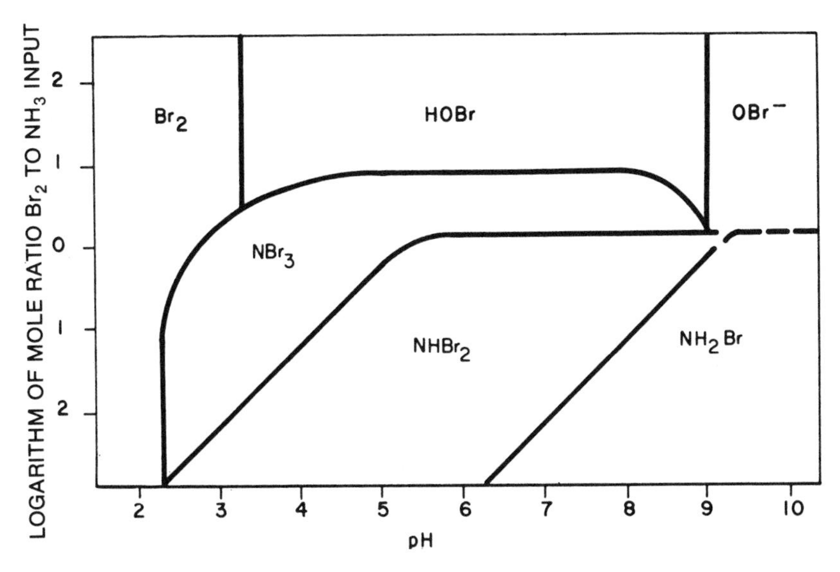

Figure 16. Principal species of bromine and bromamine predominating after 1–2 min at various pH and ammonia-to-bromine ratios. Lines represent equal equivalent concentrations. Hypobromous acid separation from bromine given for 10^{-3} *M* bromide. Reproduced from Johnson and Overby [56], courtesy of the American Society of Civil Engineers.

Iodination

Iodine, I_2, has a long history and record as a disinfectant for skin wounds and mucous surfaces of the human body. It is used, also, as a strong and universal sanitizing compound in hospitals and microbiology laboratories. The use of iodine for disinfection of drinking water, swimming pools and wastewaters has not been extensive. There are, however, some occasions, especially small-scale treatments and emergency situations, when it is utilized [59].

Physical Properites

Iodine has an atomic weight of 126.92 and a formula weight of 253.84. It is the only halogen that is solid at room temperature, and it sublimes. It is a blackish-gray solid with a specific gravity of 4.93, melts at 113.6°C, and boils at 184°C.

Iodine Compounds

One of the major uses of iodine-containing compounds is field sterilization of water, especially in military situations. During World War II, a series of studies led to the development of Globaline and other tablets for disinfection of small or individual supplies for the U.S. Army [60]. Table XIV gives several of these tablets and their iodine contents. The disinfective superiority of these iodine tablets over such chlorine-containing tablets as Halazone is discussed below.

Hydrolysis and Protolysis Reactions

Iodine hydrolyzes to form hypoiodous acid and hydroiodic acid:

$$I_2 + H_2O = HOI + H^+ + I^- \tag{42}$$

whereupon

$$K_H = \frac{[HOI][H^+][I^-]}{[I_2]} \tag{43}$$

K_H at 25°C is 3×10^{-13} M^2/l^2 [61]. Consequently, the extent of hydrolysis of I_2 is less than Cl_2 and Br_2 and is controlled, of course, by $[H^+]$.

Hypoiodous acid is "weak" in the Brønsted sense:

$$HOI + H_2O = H_3^+O + OI^- \tag{44}$$

with the protolysis constant $= 4.5 \times 10^{-13}$ at 25°C [61]. It is only slightly stronger than pure water as an acid. It is interesting to compare the effect of $[H^+]$ on the hydrolysis and protolysis of I_2 and HOI with Cl_2 and HOCl. This comparison is seen in Table XV. It is noted, for example, that at a pH value of 8., 78.5% of Cl_2 is OCl^-, whereas 88% of I_2 is in the HOI form and the remaining 12% is I_2. An iodate ion can be formed:

$$3\,HOI + 2\,OH^- = HIO_3 + 2\,H_2O + 2\,I^- \tag{45}$$

Formation of iodic acid is catalyzed by the OH^- ion and the occurrence of IO_3^- at alkaline pH values is significant since this anion has no disinfection properties [63].

Oxidation States of Iodine

Iodine has 12 oxidation states of which 2 are relevant to aqueous systems (−I and +I). Some reduction reactions (Equations 46–49) are

Table XIV. Characteristics of Some Polyiodides[a]

Commercial Name	Chemical Name	Formula	Active Iodine (%)	Solubility (g/l) at 25°C	Iodine Vapor Pressure[b] at 25°C (10^3 mm)	Relative[c] Stability
	Tetramethylammonium triiodide	$(CH_3)_4NI_3$	55.8	0.25	0.9	
Globaline	Tetraglycine hydroperiodide	$(NH_2CH_2COOH)_4HI \cdot 1.25I_2$	42.3	380	1.3	50
Potadine	Potassium tetraglycine triiodide	$(NH_2CH_2COOH)_4KI_3$	35.3	Large	0.5 (est)	95
Hexadines	Aluminum hexaurea sulfate triiodide	$Al[CO(NH_2)_2]_6SO_4I_3$	29.4	590	0.4	100
Hexadinen	Aluminum hexaurea dinitrate triiodide	$Al[CO(NH_2)_2]_6(NO_3)_2I_2$	28.5	390	0.12	

[a]Reproduced from Morris et al. [60], courtesy of the American Chemical Society.

[b]Determined by equilibration of solid with CCl_4 and titration of dissolved iodine, assuming validity of Henry's law.

[c]Percentages of initial active iodine remaining after 50 days when thin layers of powdered materials (<100-mesh) were maintained at 60°C for this period.

Table XVa. Effect of pH on the Hydrolysis of Iodine[a,b]

pH	Content of Residual (%)		
	I_2	HIO	IO^-
5	99	1	0
6	90	10	0
7	52	48	0
8	12	88	0.005

Table XVb. Effect of pH on the Hydrolysis of Chlorine

pH	Content of Residual (%)		
	Cl_2	HOCl	OCl^-
4	0.5	99.5	0
5	0	99.5	0.5
6	0	96.5	3.5
7	0	72.5	27.5
8	0	21.5	78.5
9	0	1.0	99.0

[a]Reproduced from Black et al. [62], courtesy of the American Water Works Association.
[b]Total iodine residual, 0.5 ppm.

Table XVI. Some Reduction Reactions of Iodine at 25°C [36]

Reaction Number	Reaction	V^0 (V)
46	$I_2 + 2\,e = 2\,I^-$	+0.621
47	$I^+ + 2\,e = I^-$	+0.946
48	$HOI + H^+ + 2\,e = I^- + H_2O$	+0.987
49	$OI^- + 2\,H^+ + 2\,e = I^- + H_2O$	+1.313

seen in Table XVI. The E-pH stability diagram is seen in Figure 17, where the iodide ion, I^-, is the stable species in most of the graph. I_2 predominates, for the most part, under acid conditions and above an E value of +0.6.

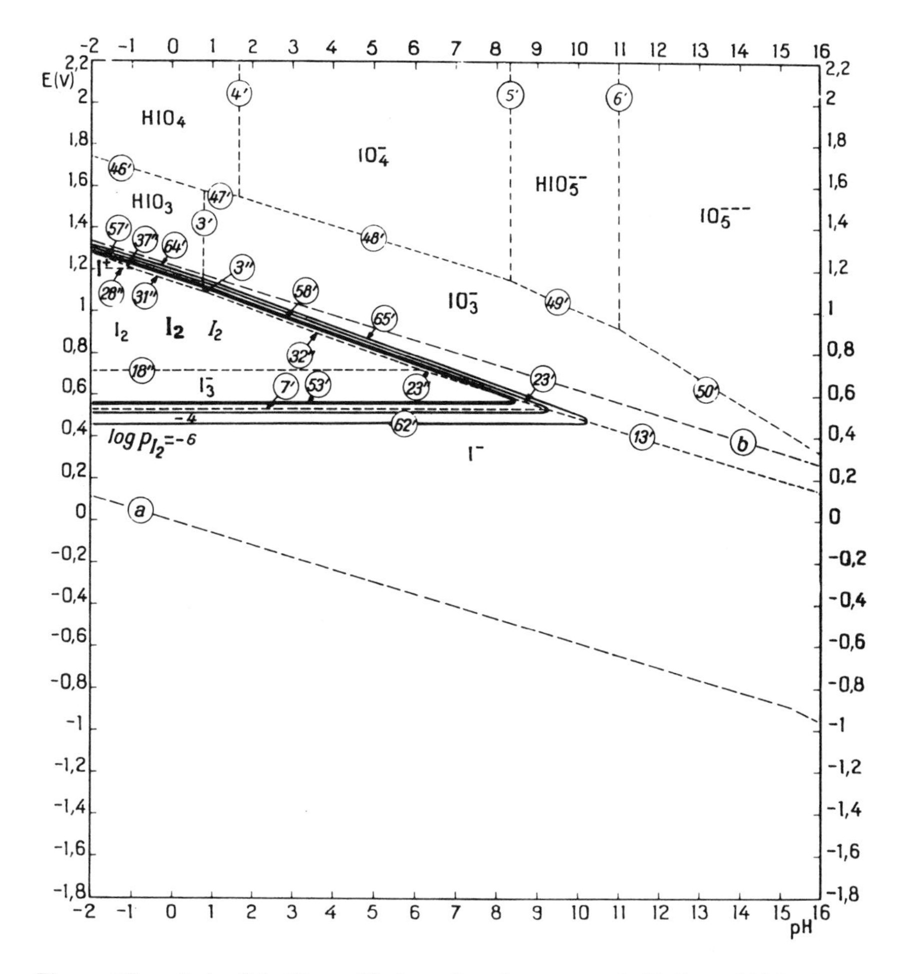

Figure 17. Potential-pH equilibrium for the system I_2-H_2O at 25°C, for solutions containing 1 g-atom I/l. Equation numbers refer to original text. Reproduced from Pourbaix [36], courtesy of Pergamon Press Ltd.

Reactions with Ammonia

McAlpine [64] reported some experiments with the reaction of I_2 with NH_3. This reaction is suggested:

$$I_2 + 2NH_3 = NH_2I + NH_4^+ + I^- \tag{50}$$

An average Keq of 1.8 was determined when the molar ratio of NH_3 to I_2 was the order of 4000:1. This is extremely unlikely in treated waters.

Other Chemical Disinfectants

Ozone

In addition to the physical properties cited above, ozone is frequently called a "powerful" oxidant. That is, it is unstable in water according to:

$$O_{3(g)} + 2H^+ + 2e = H_2O + O_{2(g)}; \quad V^0 = +2.076\,V \tag{51}$$

or

$$O_{3(g)} + 6H^+ + 6e = 3H_2O; \quad V^0 = +1.511\,V \tag{52}$$

The V^0 values for these two reactions exceed the one for $O_{2(g)}$, +1.228 V; consequently, its instability and its strong oxidative properties in water are noted. Ozone must be generated onsite by an electrical discharge in air or streams of oxygen. This produces in practice 1–2% O_3 when air is used, and 3–5% when pure $O_{2(g)}$ is used. It is moderately soluble in water; 0.494 cm^3/cm^3 at 0°C. Ozone has a half-life of about 40 min in pure distilled water at pH 7.6 and 14.6°C. At pH 8.5, the half-life is 10 min [65].

Ferrate

The ferrate ion, FeO_4^{2-}, has received some attention as a disinfectant. In this ion, iron has an oxidation state of VI, which is unstable in water as seen in these reactions [37]:

$$FeO_4^{2-} + 8H^+ + 3e = Fe^{3+} + 4H_2O, \quad V^0 = +1.7\,V \tag{53}$$

$$FeO_4^{2-} + 5H^+ + 4e = HFeO_2^- + 2H_2O, \quad V^0 = +1.001\,V \tag{54}$$

Here again, the instability of FeO_4^{2-} in water and the acceptance of 3 and 4 moles of electrons makes this ion an oxidizing disinfectant. It is prepared electrochemically from scrap iron. Aqueous ferrate solutions have a violet color similar to $KMnO_4$.

Chlorinated Isocyanurates

Cyanuric acid and various cyanurates have been employed for many years in outdoor swimming pools as stabilizers for active chlorine to prevent its photolytic decomposition. Chlorine residuals disappear rapidly (1–2 hr) on a sunny day. In the presence of 25 mg/l cyanuric acid, chlorine residuals may persist for several days from the formation of chlorinated cyanurates. The structure of cyanuric acid is:

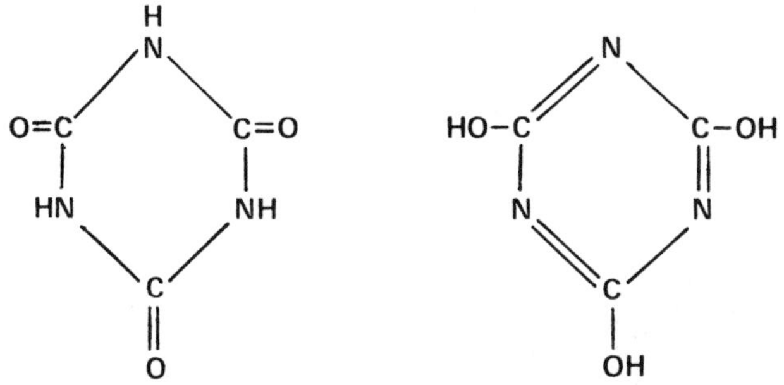

Keto Form **Enol Form**

Cyanuric acid (H_3Cy) is a weak, triprotic acid that ionizes, in sequence, to H_2Cy^-, HCy^{2-} and Cy^{3-}. The hydrogens are sequentially replaced by Cl^+ to yield H_2ClCy, HCl_2Cy and Cl_3Cy. All of this results in a rather complex equilibrium system (Figure 18). The equilibrium constants have been computed by O'Brien et al. [66] and are given in Table XVII, which also contains Equations 55–66.

EFFECTIVENESS OF DISINFECTION

The primary purpose of disinfecting drinking water is, of course, to destroy and eliminate pathogenic organisms responsible for waterborne diseases. Sterilization is not necessary, nor is it desirable to have killed all microorganisms. It is impractical to do so. An assessment of a reduction in the concentration of microbes is sufficient protection against the transmission of pathogens.

Biocidal efficacy of disinfectants is evaluated, usually, through laboratory studies using indicator organisms or the pathogens. As such, there is a need to control many variables that affect the effectiveness of the disinfection process. Among these variables are contact time, pH, temperature and halogen demand. (Note that the halogens, particularly chlorine, are used predominantly in drinking water. However, disinfection and chlorination are not necessarily synonymous.) In laboratory studies, these variables can be carefully controlled. In treatment practice, however, it is much more difficult to do so, especially in distribution systems.

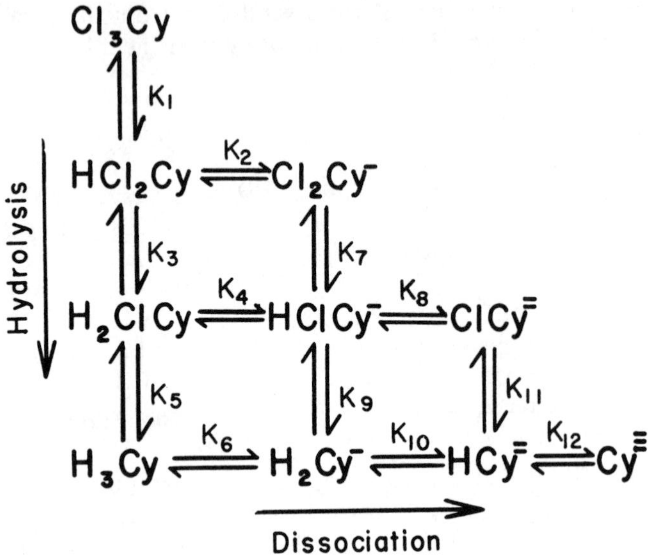

Figure 18. Equilibria among cyanuric acid and its chlorinated derivatives. Reproduced from O'Brien et al. [66], courtesy of Ann Arbor Science Publishers.

Table XVII. Equilibrium Constants for the Chlorinated Cyanuric Acids [66]

Reaction Number	Reaction	pK[a]
55	$Cl_3Cy + H_2O = HCl_2Cy + HOCl$	1.8 ± 0.2
56	$HCl_2Cy = H^+ + Cl_2Cy^-$	3.75 ± 0.03
57	$HCl_2Cy + H_2O = H_2ClCy + HOCl$	2.93 ± 0.07
58	$H_2ClCy = H^+ + HClCy^-$	5.33 ± 0.05
59	$H_2ClCy + H_2O = H_3Cy + HOCl$	4.07 ± 0.08
60	$H_3Cy = H^+ + H_2Cy^-$	6.88 ± 0.04
61	$Cl_2Cy^- + H_2O = HClCy^- + HOCl$	4.51 ± 0.09
62	$HClCy^- = H^+ + ClCy^-$	10.12 ± 0.02
63	$HClCy^- + H_2O = H_2Cy^- + HOCl$	5.62 ± 0.05
64	$H_2Cy^- = H^+ + HCy^-$	11.40 ± 0.10
65	$ClCy^{=} + H_2O = HCy^{=} + HOCl$	6.90 ± 0.11
66	$HCy^{=} + H^+ + Cy^{\equiv}$	13.5

[a]These values are expressed as alkaline hydrolysis constants. For comparison with the present study a factor of 6.5 should be added to these values (i.e., $pK_w - pK_{HOCl}$).

A major factor affecting an evaluation of the efficacy of a particular disinfectant is the test microorganism [59]. There is a wide variation in susceptibility among bacteria, viruses and protozoa (cyst stage) as well as among genera, species and strains of these microbes. Furthermore, the vast majority of the literature on water disinfection is concerned with use of model organisms instead of the pathogens. The hypothesis is that the reaction of the model to a disinfectant is the same as the pathogen. Defense of this approach is argued in Reference 59. A model and/or indicator microorganism should follow these criteria of Fair and Geyer [67]:

1. The indicator should always be present when fecal material is present and absent in clean, uncontaminated water.
2. The indicator should die away in the natural aquatic environment and respond to treatment processes in a manner that is similar to that of the pathogens of interest.
3. The indicator should be more numerous than the pathogens.
4. The indicator should be easy to isolate, identify and enumerate.

As mentioned earlier in this chapter, bacteria of the coliform group, especially *Escherichia coli,* have proven useful as indicator organisms and disinfection models for enteric bacterial pathogens. However, they are not very good indicators and models for nonbacterial pathogens. The bacterial viruses of *E. coli* are used as disinfection models for enteric viruses.

There is also the question of responses to disinfectants by laboratory cultures and their counterparts in natural waters. There is considerable evidence that "naturally occurring" bacteria in swimming pools, for example, are one to two times more resistant to disinfectants than cultured bacteria.

All of this points to two major areas of concern and interpretation of disinfection studies:

1. Is there a sufficient and an accurate correlation of the responses to disinfection by model microorganisms with those of pathogens?
2. Are the responses to disinfectants in treatment plant practice the same as, or, at least, similar to, those obtained in the laboratory?

A major factor in disinfection studies and practices is the relation between contact time and disinfectant dosage to effect a specified or equal destruction of microorganisms. An empirical equation has been developed for these two variables [29]:

$$t = aC^b \tag{67}$$

where
- t = contact time, usually in minutes
- C = disinfectant dosage, used in mg/l
- a = a proportionality constant for a given organism, pH value, and temperature
- b = the slope of a log-log plot of t vs C

Typical plots are seen in Figure 19 where most of the lines have a slope or b value of -1. Since b is negative, the curve is hyperbolic, whereupon:

$$t = a/C \tag{68}$$

or

$$a = Ct \tag{69}$$

where a is a constant that can be used conveniently to compare disinfectants, organism response, etc. so long as the percentage kill is 99.6–100% and the slope is -1. Frequently, b is called the coefficient of dilution [69]: When $b>1$, the efficiency of the disinfectant decreases rapidly as it is diluted, when $b<1$, time of contact is more important than dosage, and when $b=1$, C and t are weighted equally. The reports of disinfectant effectiveness that follow should be read with Equation 69 in mind.

Free Chlorine ($HOCl$ and OCl^-)

One of the early researches was conducted by Butterfield et al. [70] who studied percentages of inactivation as a function of time for *E. coli, Enterobacter aerogenes, Pseudomonas aeruginosa, Salmonella typhi* and *Shigella dysenteriae.* Various concentrations of free chlorine were utilized at pH values ranging from 7.0 to 10.7 and in two temperature ranges: 2–5°C and 20–25°C. This classical study dealt with the action of disinfectants on pathogens. In general, the primary factors affecting the bactericidal efficacy of free available chlorine are [59]:

1. the time of contact between the bacteria and the bactericidal agent, i.e., the longer the time, the more effective the chlorine disinfection process;
2. the temperature of the water in which contact is made, i.e., the lower the temperature, the less effective the chlorine disinfecting activity; and
3. the pH of the water in which contact is made, i.e., the higher the pH, the less effective chlorination.

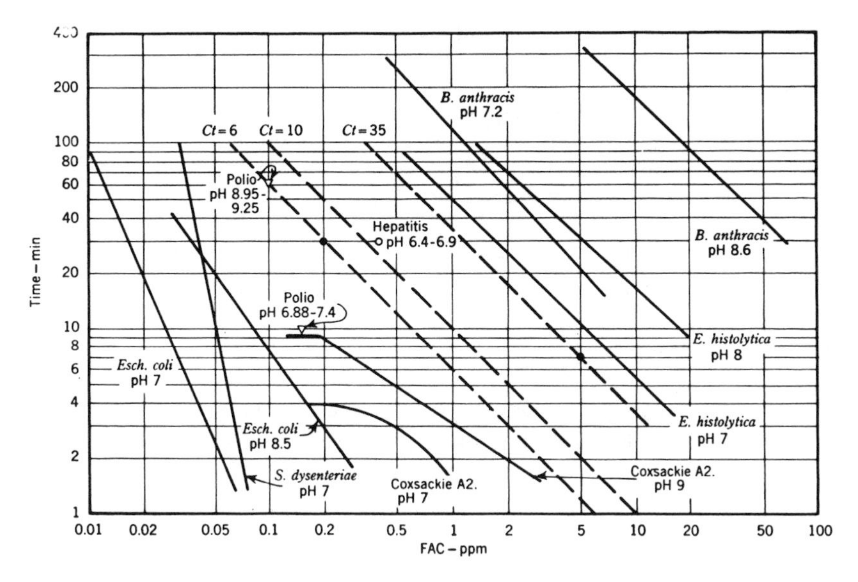

Figure 19. Disinfection vs free available chlorine residuals. Time scale is for 99.6–100% kill. Temperature was in the range 20–29°C, with pH as indicated. Reproduced from Baumann and Ludwig [68], courtesy of the American Water Works Association.

The results of Butterfield's free available chlorine research are summarized in Table XVIII. (It should be noted here that the 2–5°C temperature range is employed to slow the kinetics of disinfection and also to simulate the winter operation of treatment plants.) From these data, HOCl is more effective than OCl^- and the higher temperatures, 20–25°C, lowered the contact time and chlorine dosage for 99–100% kill of *E. coli*. This comparison is seen also in Figure 20. That there is some genera dependency is seen also in Table XVIII. *Salmonella typhi* required twice as much chlorine as *E. coli,* but the absolute quantities of 0.06 and 0.03 mg/l, respectively, are minute compared to the overall chlorine demand of a water.

Another early laboratory study of the effectiveness of chlorine was reported by Chang [71] for disinfection of the cyst of *Endamoeba histolytica*. The time-concentration effects are seen in Figure 21, where as long as 120 min are required for destruction of 30–75 cysts/cm^3 of water. It is interesting to note the convergence of the curves representing $Cl_{2(g)}$, HOCl and chloramines at 120 min. It appears that type of chlorine disinfectant is a significant factor at contact times of less than 60 min in the destruction of cysts of *E. histolytica*.

Clarke and Kabler [72] and Clarke et al. [73] reported the results of

Table XVIIIa. Effect of pH and Temperature on Contact Time (min) for 100% Kill of *E. coli* [70]

	Dosage (mg/l)		Contact Time (min)	
pH	2–5°C	20–25°C	2–5°C	20–25°C
7.0	0.04	0.04	3	1
8.5	0.14	0.07	10	10
9.8	0.72	0.06	10	20
10.7	0.30	0.40	60	10

Table XVIIIb. Effect of pH and Temperature for 100% Kill of *E. coli*, Contact Time = K, Cl_2 Dosage (mg/l) for 100% Kill, 10 min Exposure

pH	2–5°C	20–25°C
7.0	0.03	0.04
8.5	0.14	0.07
9.8	0.72	0.30
10.7		0.40

Table XVIIIc. Effect of Species, 5 min Exposure, pH 7, 20–25°C

Species	Cl_2 Dosage for 100% Kill (mg/l)
A. aerogenes	0.08
S. typhosa	0.06
Sh. dysenteriae	0.05
E. coli	0.03
Ps. pyocyaneus	0.03

laboratory studies on the inactivation of purified coxsackievirus and type 3 adenovirus, respectively, by chlorine. The time-concentration relationships are given in Figure 22 for two temperatures and two pH values. (It should be noted here that, at pH 7.0, HOCl is 80% of the $[Cl_2]_t$ (see Figure 5). Therefore, any disinfection at this pH value is not entirely accomplished by HOCl.) The b values for Equation 67 are:

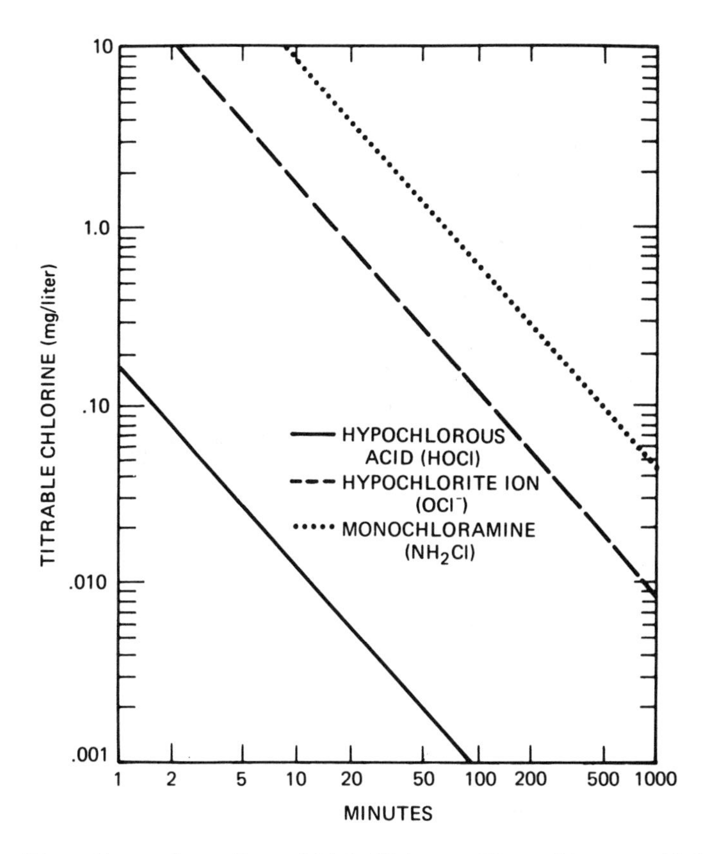

Figure 20. Comparison of germicidal efficiency of hypochlorous acid, hypochlorite ion and monochloramine for 99% destruction of *E. coli* at 2–6°C.

- pH 9, 3–6°C, $b = 1.07 \pm 0.26$
- pH 7, 3–6°C, $b = 0.72 \pm 0.07$
- pH 9, 27–29°C, $b = 0.63 \pm 0.10$

Somewhat similar results were reported for inactivation of the type 3 adenovirus in water [73]. The data in Table XIXa were plotted to be linear for the log-log relation of time and concentration of disinfectant. The b values were:

- pH 9, 25°C, $b = 1.46$
- pH 9, 4°C, $b = 1.04$
- pH 7, 4°C, $b = 0.66$

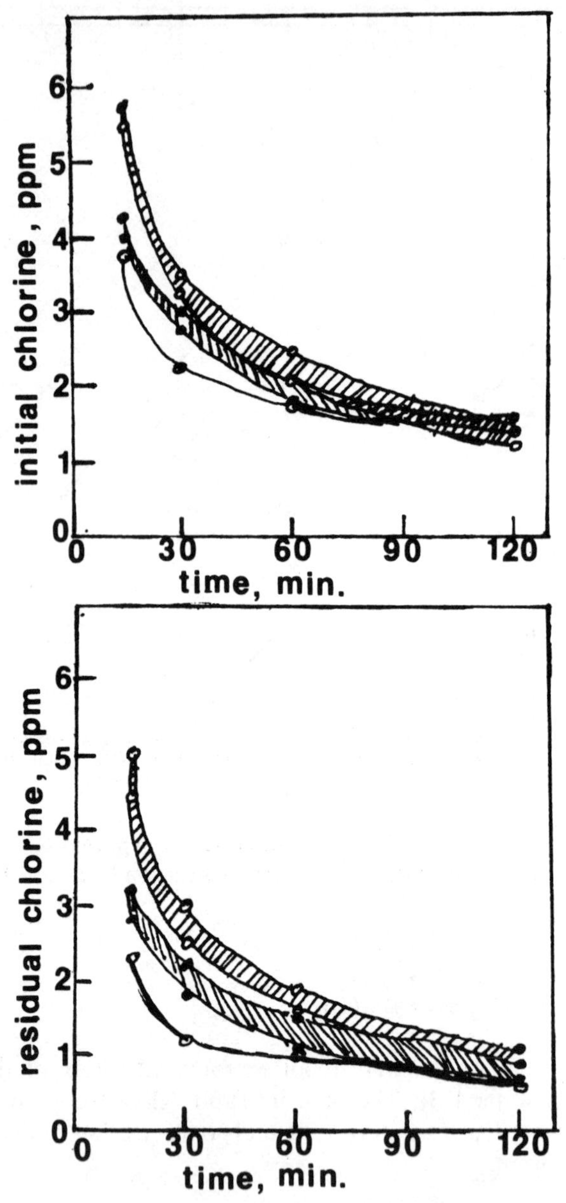

Figure 21. Comparison of lethal residual concentrations of the solution of hypochlorite, chloramines and gaseous chlorine for killing *E. histolytica* cysts in tapwater under similar conditions. T = 18°C, cyst density = 30–75/cm^3, total organic nitrogen = 0.1–0.2 ppm, pH 6.8–7.2. Solid circles = high test hypochlorite, open circles = chloramines, dotted circles = gaseous chlorine. Reproduced from Chang [70], courtesy of *War Medicine*.

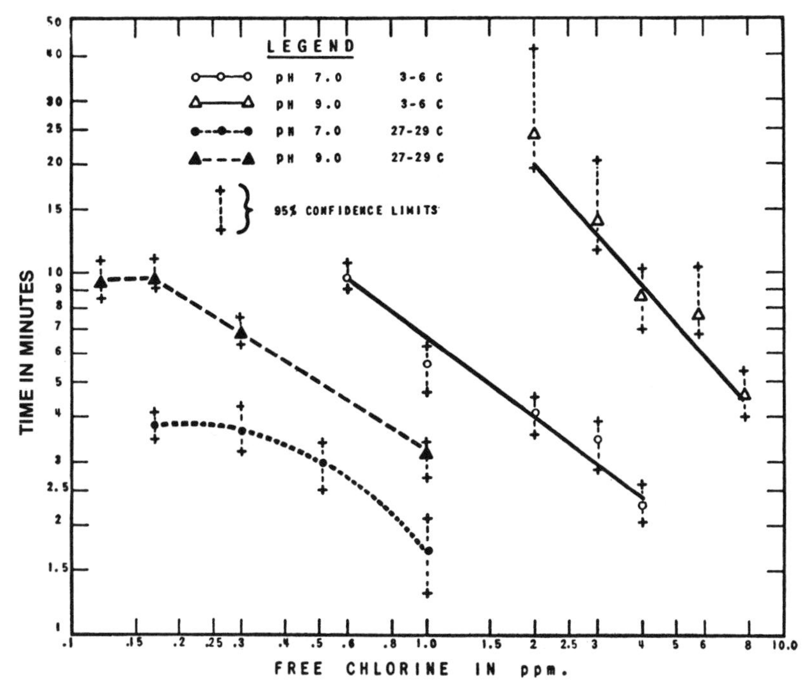

Figure 22. Time (min) required to inactivate 99.6% of coxsackievirus in water by chlorine. Reproduced from Clarke and Kabler [72], courtesy of the American Society of Hygiene.

Table XIXa. Time Required by Free Chlorine to Inactivate Approximately 99.8% of Purified Type 3 Adenovirus in Water[a]

pH	Temperature °C	Free Chlorine (ppm)	Time Required (sec)
8.8–9.0	25	0.1	120–140
		0.2	40–50
		0.5	8–16
8.8–9.0	4	0.1	180
		0.2	80–110
		0.5	30–40
6.9–7.1	25	0.1	8–16
		0.2	8–16
		0.5	<8
6.9–7.1	4	0.1	18–25
		0.2	8–20
		0.5	<10

Table XIXb. Approximate Time Required by Free Chlorine to Achieve >99% But <100% Kill of Various Microorganisms in Water[a]

Organism	pH	Temperature (°C)	Range of Free Chlorine (ppm)	Time
E. coli	7	25	0.08–0.12	<60 sec
A. aerogenes				60 sec
S. typhosa				60 sec
Type 3 Adenovirus				<16 sec
Coxsackievirus A2				240 sec[b]
E. histolytica Cysts				150 min[b]
B. anthrax Spores				360 min[b]
E. coli	7	4	0.08–0.10	<60 sec
S. typhosa				<60 sec
Type 3 Adenovirus				22 sec
Coxsackievirus A2				40 min[b]
E. histolytica Cysts				630 min[b]
B. anthrax Spores				24 hr[b]
E. coli	8.5	25	0.08–0.14	180 sec
S. typhosa	8.5	25		180 sec
Type 3 Adenovirus	9.0	25		130 sec
Coxsackievirus A2	9.0	25		10 min[b]
E. coli	8.5	4	0.14	300 sec
S. typhosa	9.8	4	0.40	10 min
Type 3 Adenovirus	9.0	4	0.14	130 sec
Coxsackievirus A2	9.0	4	1.0	45 min[b]

[a]Reproduced from Clarke et al. [73], courtesy of the American Society of Hygiene.
[b]Values obtained by extrapolation of interpolation.

The pH 7, 25°C data apparently were curvilinear and were similar to the pH 7, 27–29°C curve in Figure 20. Table XIXb shows the concentrations of HOCl and OCl^- required to destroy some pathogenic microorganisms. *E. coli* and *S. typhosa* require about the same $[Cl_2]_t$ and contact time for their demise.

Laboratory studies were conducted for the effect of chlorine over the pH range of 6.–10. on five types of polioviruses and two types of coxsackieviruses, all of which are enteric [74]. The results are summarized in Table XX. These data clearly show that HOCl, at pH 6., is at least four times as effective as OCl^-, at pH 9. Also, there are considerable differences in response of the various types of viruses to chlorine. This is especially the case at the lower temperatures of 1–5°C.

A comparative study of the inactivation of an animal virus and *E. coli*

Table XX. Time Required for 99.9% (or Greater) Inactivation by Free Residual Chlorine[a]

Temperature	Virus Strain	Concentration of Free Chlorine (ppm)	Time at pH (min)				
			6	7	8	9	10
25–28°C	Polio 1	0.01–0.10		16			
	(Mahoney)	0.11–0.20			5		
		0.21–0.30	2	3	4	8	30
	Polio 1 (487)	0.21–0.30		4			
	Polio 1 (MK 500)	0.21–0.30	4	6	12	16	>30
	Polio 2	0.11–0.20		2			
	Polio 3	0.11–0.20		2	8	16	
	Coxsackie B1	0.21–0.30		8	4	8	
		0.31–0.40		2			
	Coxsackie B5	0.21–0.30	2	1	2	8	16
1–5°C	Polio 1	0.11–0.20		8			
	(Mahoney)	0.21–0.30	4	8			16
	Polio 1 (MK 500)	0.21–0.30		30	120		
	Polio 2	0.01–0.10		60			
		0.11–0.20		30			
		0.21–0.30		16			
	Polio 3	0.11–0.20		30			
	Coxsackie B5	0.21–0.30		16	30		

[a]Reproduced from Kelley and Sanderson [74], courtesy of the American Public Health Association.

by HOCl and OCl^- was conducted through laboratory experiments [75]. Figure 23 shows the time-concentration relationships for a 99% inactivation of the enteric microorganisms. Hypochlorous acid was more effective for disinfection of *E. coli* than for the poliovirus I. However, this was reversed for OCl^- (Figure 23b). That OCl^- is more effective than HOCl for inactivation of the poliovirus I is seen in Figure 23c. This is contrary to the work of Clarke and Kabler [72]. Figure 23d shows that HOCl is about 50 times more effective than OCl^- against *E. coli*. This result has been reported by others [69] where the argument is in the arithmetic of the HOCl:OCl^- effectiveness ratio.

These studies were extended to two DNA phages, T2 and T5, two RNA phages, f2 and MS2, *E. coli* and two animal viruses, poliovirus I and coxsackievirus [76]. Their graphic results are seen in Figures 24 and 25. The two DNA phages were more responsive to HOCl than *E. coli*, whereas the other viruses require more free chlorine for their destruction.

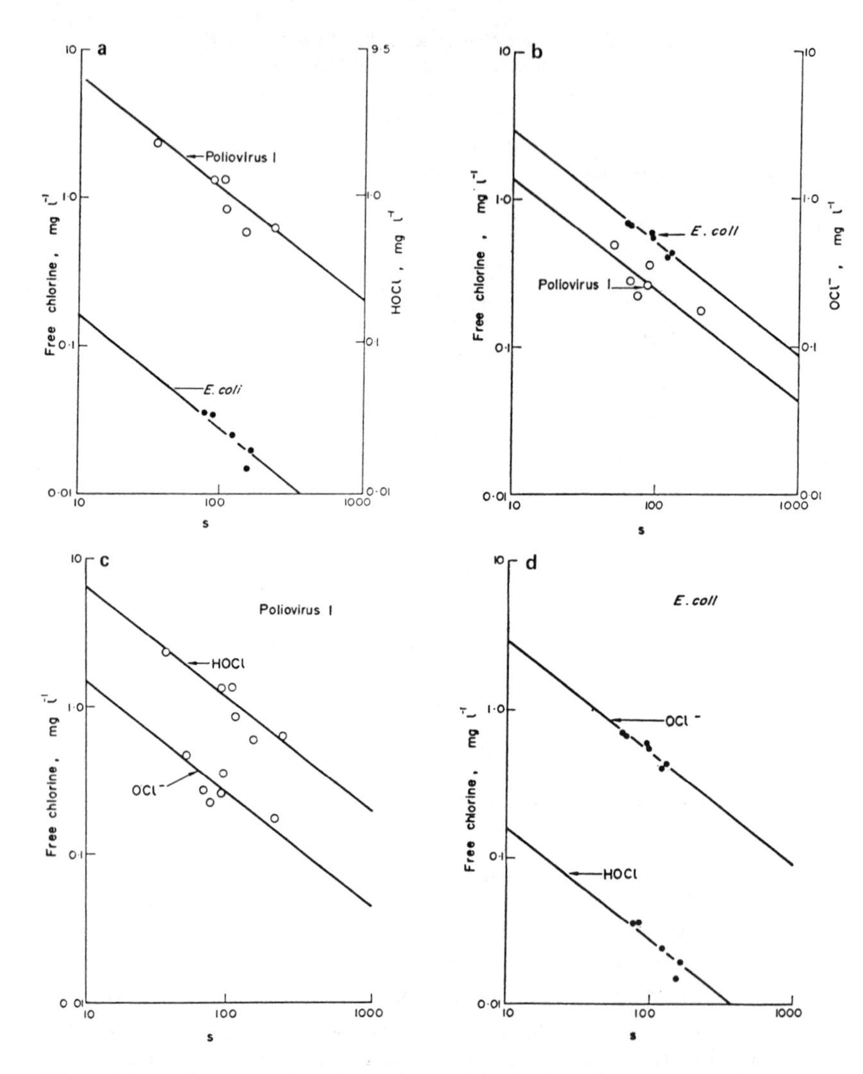

Figure 23. Concentration-time relationship for 99% inactivation of **(a)** poliovirus 1 and *E. coli* by HOCl at pH 6. and 5°C; **(b)** poliovirus and *E. coli* by OCl^- at pH 10. and 5°C; **(c)** poliovirus 1 by HOCl and OCl^- at 5°C; **(d)** *E. coli* by HOCl and OCl^- at 5°C. Reproduced from Scarpino et al. [75], courtesy of the International Association on Water Pollution Research.

All of the viruses and phages required less OCl^- than *E. coli.* Nearly all of the slopes or b values for Equation 67 were 1.0, which is interpreted as a first-order inactivation mechanism.

That chlorine inactivates various viruses and that $[H^+]$ is a factor was

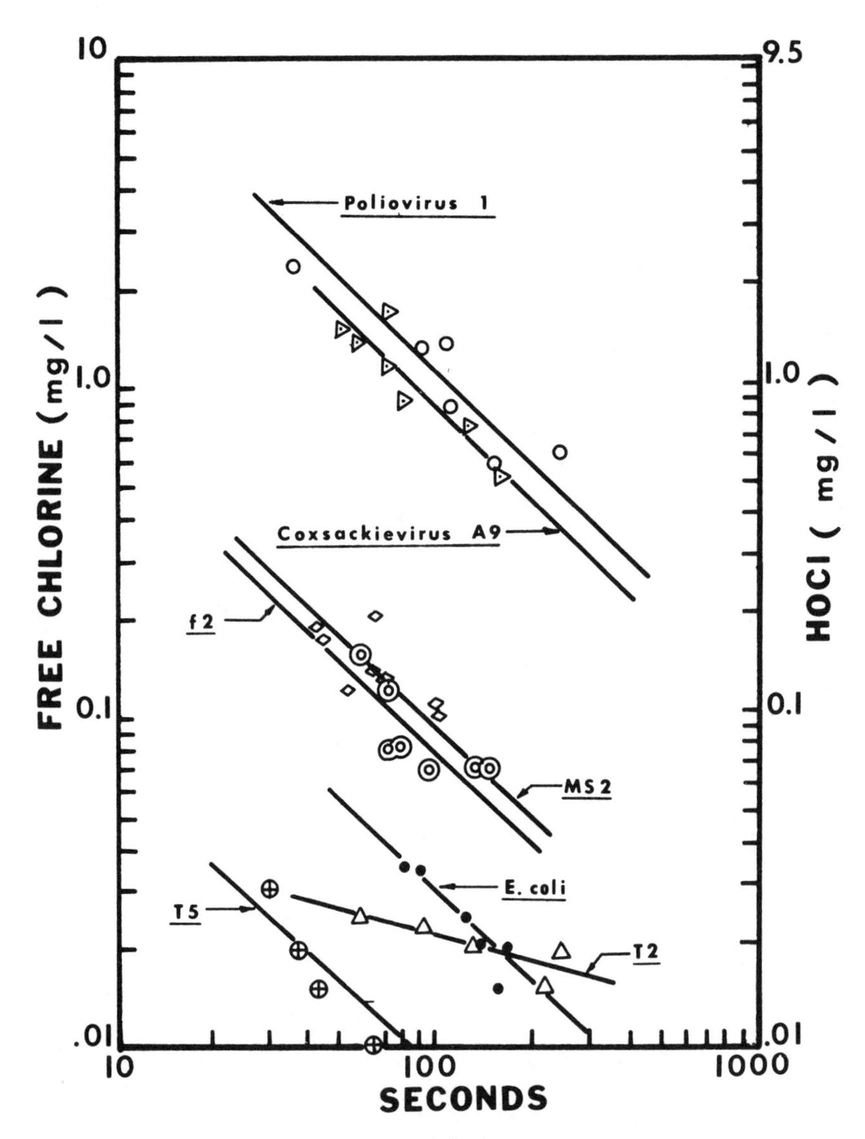

Figure 24. Concentration-time relationship for 99% inactivation of animal and bacterial viruses, and the reference bacterium *E. coli* ATCC 11229, by hypochlorous acid at 5°C and pH 6. Reproduced from Scarpino et al. [76], courtesy of Ann Arbor Science Publishers.

reported by Engelbrecht et al. [77]. Table XXI summarizes their data, giving the times and the chlorine concentrations needed for 99% inactivation of the viruses at pH values of 6.0, 7.8 and 10.0. These results indi-

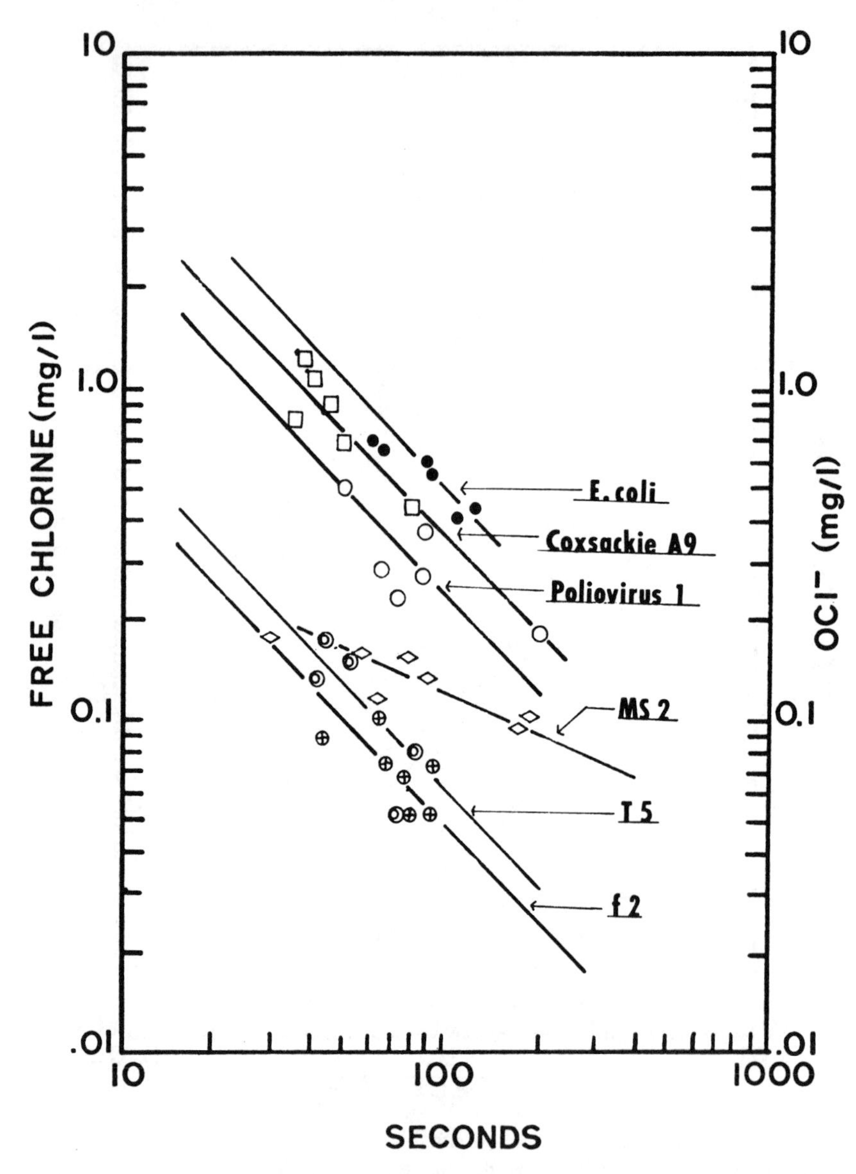

Figure 25. Concentration-time relationship for 99% inactivation of animal and bacterial viruses, and the reference bacterium *E. coli* ATCC 11229, by hypochlorite ion at 5°C and pH 10. Reproduced from Scarpino et al. [76], courtesy of Ann Arbor Science Publishers.

cate that there is a significant difference in the time required for the 99% inactivation at pH 6. (HOCl) and 10. (OCl^-). This is, of course, similar

Table XXI. Time Required for 99% Inactivation by Free Residual Chlorine at 5.0±0.2°C[a]

pH	Concentration of Free Chlorine (mg/l)[b]	Virus Strain	Minutes for 99% Inactivation	Rank Ordering
6.00	0.46–0.49	Coxsackie A9 (Griggs)	0.3	1
6.00	0.48–0.49	Echo 1 (Farouk)	0.5	2
6.00–6.02	0.48–0.51	Polio 2 (Lansing)	1.2	3
6.00–6.03	0.38–0.49	Echo 5 (Noyce)	1.3	4
6.00	0.47–0.49	Polio 1 (Mahoney)	2.1	5
6.00–6.06	0.51–0.52	Coxsackie B5 (Faulkner)	3.4	6
		Coxsackie A9 (Griggs)	ND[c]	
7.81–7.82	0.47–0.49	Echo 1 (Farouk)	1.2	1
		Polio 2 (Lansing)	ND[c]	
7.79–7.83	0.48–0.52	Echo 5 (Noyce)	1.8	3
7.80–7.84	0.46–0.51	Polio 1 (Mahoney)	1.3	2
7.81–7.82	0.48–0.50	Coxsackie B5 (Faulkner)	4.5	4
10.00–10.01	0.48–0.50	Coxsackie A9 (Griggs)	1.5	1
10.00–10.40	0.49–0.51	Echo 1 (Farouk)	96.0	6
9.89–10.03	0.48–0.50	Polio 2 (Lansing)	64.0	4
9.97–10.02	0.49–0.51	Echo 5 (Noyce)	27.0	3
9.99–10.40	0.50–0.52	Polio 1 (Mahoney)	21.0	2
9.93–10.05	0.50–0.51	Coxsackie B5 (Faulkner)	66.0	5

[a]Data from Engelbrecht et al. [77].
[b]Range of measured free chlorine residual in the "test" reactor at the termination of each of three separate experiments.
[c]ND = not determined.

to the results with *E. coli*. Engelbrecht et al. [77] agree with the results of Scarpino et al. [76] for the coxsackievirus A9 at pH 10. (OCl^-). However, their results disagree at pH 6. (HOCl). Engelbrecht reported that 0.3 min was required for 99% inactivation with about 0.5 mg/l free chlorine, whereas Scarpino reported that 3 min were required. In any event, there is an obvious species dependency for disinfection of bacteria and viruses by free chlorine.

Chloramines

Butterfield and Wattie [78] were among the first investigators to report the bactericidal properties of the chloramines at pH values ranging from 6.5 to 10.5 and in two temperature ranges, 2–5°C and 20–25°C. A summary of their results is given in Table XXII, where the chloramine dosage and contact times required for 100% kill of *E. coli* are noted. Dichlora-

Table XXIIa. Effect of pH and Temperature on Contact Time for 100% Kill of *E. coli* by Chloramines [78]

pH	NH_2Cl[a] (%)	Dosage (mg/l) 2-5°C	Dosage (mg/l) 20-25°C	Contact Time (min) 2-5°C	Contact Time (min) 20-25°C
6.5	35.		0.3		60.
7.0	51.	0.6	0.3	180.	90.
7.8	84.		0.3		180.
8.5	98.	1.8	0.6	20.	120.
9.5	100.	1.8	0.9	90.	180
10.5	100.		1.5		180

Table XXIIb. Effect of Species

Genera	Chloramine Dosage (mg/l)[b] pH 7.0	pH 8.5	pH 9.5
E. coli	1.2	1.8	1.8
A. aerogenes	1.5	1.8	1.8
S. typhosa	0.9	1.8	1.8
Sh. dysenteriae	0.9	1.8	1.8
Ps. pyocyaneus	0.9	1.8 (98)[c]	1.8 (99.2)[c]

[a]Estimated, balance is $NHCl_2$.
[b]For 100% kill at 20–25°C, 20 min contact time.
[c]Percent kill.

mine is a more effective disinfectant than monochloramine, since lesser dosages and lower contact times were required at pH values of 7.0 and 6.5. To obtain a 100% kill with the same contact time required about 25 times as much chloramine as free chlorine. For the same kill with equal quantities of chlorine and chloramine, approximately 100 times the exposure period is required for the latter disinfectant. That there is little genera difference is seen also in Table XXII. *E. coli* and *A. aerogenes* were slightly more resistant at pH 7.0.

Another early report was given by Kelly and Sanderson [79] who researched the effect of combined chlorine on polio- and coxsackieviruses. At 25°C and a pH value of 7.0, a concentration of at least 9.0 ppm was necessary for inactivation of polioviruses with a contact period of 30 min and of 6 ppm over 60 min; 0.5 ppm required seven hr of contact. No data were available from this study to distinguish the effectiveness of mono- and dichloramine.

At 15°C, a 99% kill of *E. coli* was effected in about 20 min using 1.0 ppm NH_2Cl at a pH value of 9.0 [80]. There was some evidence from this study also that *E. coli* was less resistant to NH_2Cl than several animal viruses.

The rates of disinfection of test organisms by $NHCl_2$ in demand-free phthalate buffer at pH 4.5 and 15°C was reported [81]. Figure 26 shows the comparisons that were made among two enteroviruses, poliovirus I and coxsackievirus A9, the bacteriophage ϕX-174, and *E. coli.* These data suggest that, at 15°C, poliovirus I was 17 times more resistant than the coxsackievirus A9, 83 times more resistant than ϕX-174 and 1700 times more resistant than *E. coli* to dichloramine. It was observed also that dichloramine was a more effective disinfectant than monochloramine in these experimental systems.

In other limited tests, Kelley and Sanderson [74] found that viruses were inactivated by chloramines, but considerably longer contact times and higher residuals of combined chlorine were required. For example, contact times of four hours, at least, with a residual of 0.72 ppm were required for 99.7% inactivation of a poliovirus (Mahoney) at pH 7.

Table XXIII summarizes the various types of chlorine dosages required for 99% inactivation of *E. coli* and poliovirus I. Johanneson [58] cites the individual studies in this table.

Chlorine Dioxide

There has been much interest and research into the efficacy of ClO_2 as a disinfectant for bacteria and viruses since the early 1940s [59]. Ridenour and Ingols [82] were among the early investigators to report that ClO_2 was at least as effective as Cl_2 against *E. coli* after 30 min with equivalent concentrations apparently over the pH range of 6.–10. The results of other studies in the 1940s and 1950s are questionable because their method of preparing ClO_2 invariably included the production of Cl_2 [59]. Analytical procedures employed in those years could not distinguish between ClO_2 and other oxychloro species.

The analytical problems apparently were overcome in the 1960s, enabling Benarde et al. [83] to compare the bactericidal effectiveness of Cl_2 with ClO_2 at pH values of 6.5 and 8.5. At the lower pH value, both compounds inactivated a fresh strain of *E. coli* in less than 60 sec, with Cl_2 (or HOCl) being slightly more effective. However, at the higher pH value, ClO_2 was much more effective than Cl_2. A 0.25 mg/l dosage of ClO_2 effected a 99% destruction of *E. coli* in 15 sec, whereas 5 min were required for Cl_2. Another study by Benarde et al. [84] showed the usual

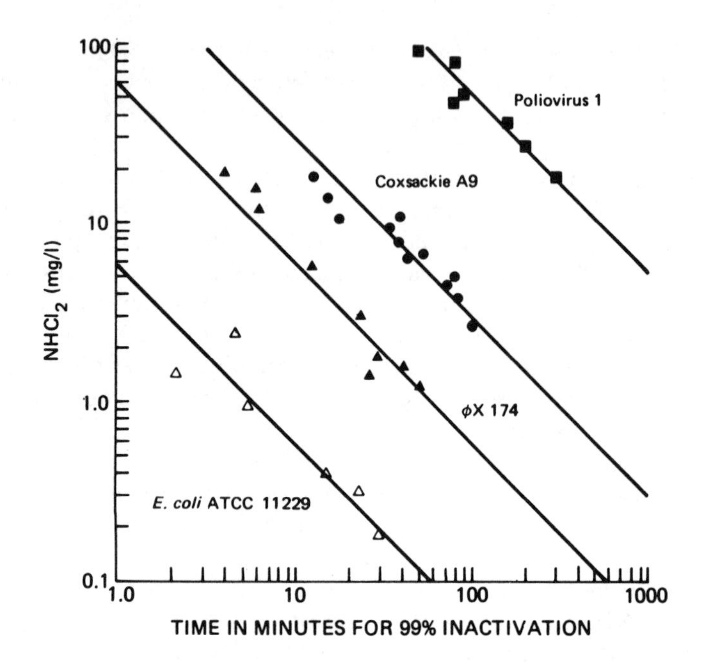

Figure 26. Inactivation of various microorganisms with dichloramine ($NHCl_2$) at pH 4.5 and 15°C. Reproduced from Esposito et al. [81], as reported in Reference 37.

temperature effects on the rate of disinfection in the 30–5°C range. For 99% inactivation of *E. coli* with 0.25 mg/l ClO_2, 190, 74, 41 and 16 sec were required at 5, 10, 20 and 30°C, respectively.

ClO_2 destruction of viruses and bacteria in water at 15°C and pH 7 was determined [85]. Figure 27 shows the concentration-time relationships for 99% disinfection of poliovirus I, coxsackievirus A9 and *E. coli.* From the relative positions of the curves in Figure 27a, poliovirus was 8.9 times and the coxsackievirus was 2.3 times more resistant than *E. coli* to ClO_2. Each curve had a slope of approximately 1, indicating a first-order kinetic disinfection reaction (see kinetics discussion below). The plots in Figure 27b show that ClO_2 is a more effective viricide at pH 9. than at pH 7. or 4.5.

Bromine and Bromine Chloride

Bromine or combined bromine compounds have been used rarely to disinfect drinking water. Their reasonably good potential as disinfectants

Table XXIII. Dosages of Various Chlorine Species Required for 99% Inactivation of *E. coli* and Poliovirus 1 [59]

Test Microorganism	Disinfecting Agent	Concentration (mg/l)	Contact Time (min)	Ct	pH	Temperature (°C)
E. coli	Hypochlorous acid (HOCl)	0.1	0.4	0.04	6.0	5
	Hypochlorite ion (OCl^-)	1.0	0.92	0.92	10.0	5
	Monochloramine (NH_2Cl)	1.0	175.0	175.0	9.0	5
		1.0	64	64.0	9.0	15
		1.2	33.5	40.2	9.0	25
	Dichloramine ($NHCl_2$)	1.0	5.5	5.5	4.5	15
Poliovirus I	Hypochlorous acid (HOCl)	1.0	1.0	1.0	6.0	0
		0.5	2.1	1.05	6.0	5
		1.0	2.1	2.1	6.0	5
		1.0	1.0	1.0	6.0	15
	Hypochlorite ion (OCl^-)	0.5	21	10.5	10.0	5
		1.0	3.5	3.5	10.0	15
	Monochloramine (NH_2Cl)	10	90	900	9.0	15
		10	32	320	9.0	25
	Dichloramine ($NHCl_2$)	100	140	14,000	4.5	5
		100	50	5,000	4.5	15

is seen below. These are confined, however, to small-scale or emergency situations and to swimming pools [86].

An early study by Tanner and Pitner [87] indicated that 0.15 mg/l HOBr was needed for a "complete kill" in 30 min, whereas 0.6 mg/l was needed for *S. typhi*. Spores of *Bacillus subtilis* required >150 mg/l of Br_2.

The effect of formation and decomposition of the bromamines on the efficacy of bacterial destruction was discussed by Johannesson [88]. A dosage of 0.28 mg/l of NH_2Br (as Br_2) effected a 99% inactivation in less than 1 min, whereas the same content of N-bromodimethylamine [$N(CH_3)_2Br$] required 12 min.

Wyss and Stockton [89] reported the effect of HOBr and bromamines on spores of *Bacillus subtilis* at pH 7. and 25°C. This was accom-

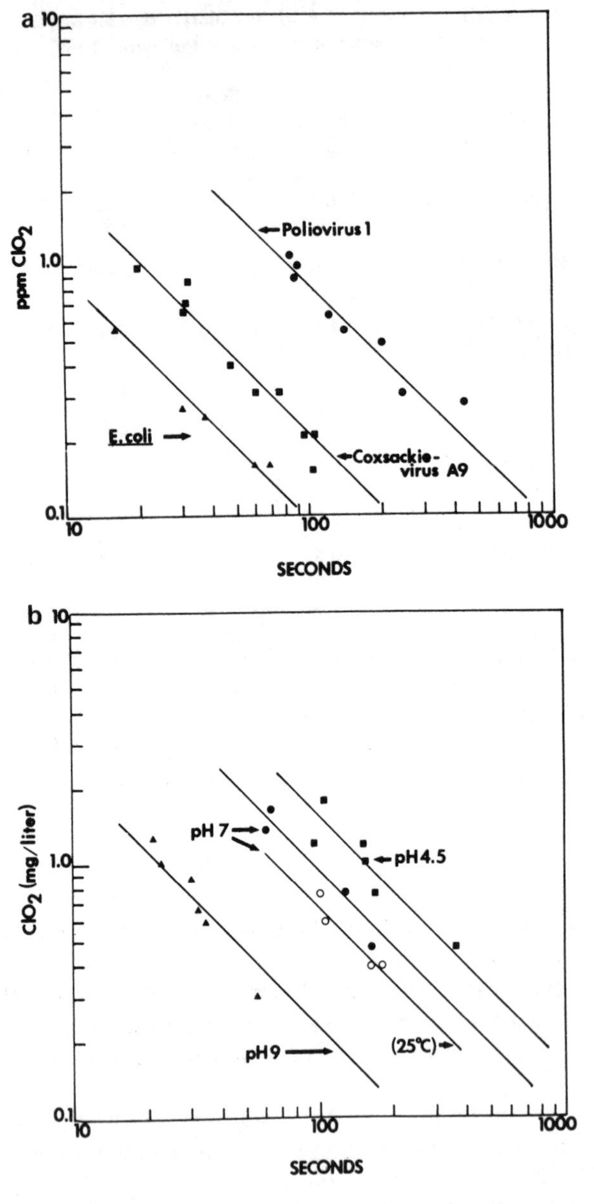

Figure 27. **(a)** Concentration-time relationship for 99% destruction or inactivation of poliovirus 1, coxsackievirus A9 and *E. coli* by chlorine dioxide at 15°C and pH 7.0. **(b)** Effect of pH on inactivation of poliovirus 1 at 21°C and pH 4.5, 7 and 9, and at 25°C and pH 7. Reproduced from Cronier et al. [85], courtesy of Ann Arbor Science Publishers.

plished simply by addition of increasing quantities of NH_3 to a 20-mg/l solution of bromine. Table XXIV shows the results of this experiment wherein a 99% kill was recorded for each $[NH_3]$. Hypobromous acid is the predominant compound when $[NH_3]=0$. At the 1 mg/l NH_3 level, NBr_3 is predominant. Between 10 and 30 mg/l, $NHBr_2$ is present, which is less stable and less effective than HOBr. At 100 and 1000 mg/l NH_3, NH_2Br predominates and is apparently as effective as NBr_3 and HOBr when the Ct values are compared.

Reoviruses and polioviruses have received considerable attention in bromine disinfection studies. The effect of HOBr on single vs aggregated particles at pH 7. and 2°C has been studied by several investigators [59]. The degree of aggregation had a marked effect on the observed inactivation rates. For example, reoviruses, as single particles, required only 1 sec for a 1000-fold decrease in their members at pH 7., 2°C and 0.46 mg/l HOBr as Br_2 [90]. For the same degree of inactivation, the time was doubled for aggregated particles.

The inactivation of single poliovirus particles in buffered, distilled water and constant residual contents has been studied by Floyd et al. [90] for six species of the bromines. Table XXV shows the calculated Ct values needed to yield the 99% inactivation by the six species whose concentrations were near 1 mg/l as Br_2. These researchers also demonstrated that $NHBr_2$, NBr_3 and HOBr were less efficient at higher contents, whereas the OBr^- became more effective as its concentration was increased. This observation is opposite to the OCl^-/HOCl system.

The chemistry and disinfectant properties of interhalogens and halogen mixtures—BrCl, ICl and IBr—were reported by Mills [91]. For

Table XXIV. Effect of Ammonia on Bromine Concentrations Required for 99% Inactivation of *Bacillus subtilis* Spores at 20 mg/l, pH 7, 25°C[a]

Ammonia, (mg NH_3-N/l)	10-min Residual (mg/l)[b]	Contact Time (min)	Ct[c]
0	20	14	280
1	16	19	304
2	2	>100	>200
10	8	85	680
30	9	70	630
100	14	18	252
1000	19	14	266

[a]Reproduced from Wyss and Stockton [89], as reported in Reference 37.
[b]Initial bromine concentration 20 mg/l.
[c]Concentration of bromine times contact time.

Table XXV. Exposure (Ct)[a] to Various Bromine Compounds Required for 99% Inactivation of Poliovirus 1, Mahoney[b]

Chemical Form	Ct	Temperature (°C)	pH
Dibromamine ($NHBr_2$)	1.2	4	7.0
Nitrogen Tribromide (NBr_3)	0.19	5	7.0
Bromine (Br_2)	0.03	4	5.0
Hypobromite Ion (OBr^-)	0.01	2	10.0
Hypobromous Acid (HOBr)	0.24	2	7.0
Hypobromous Acid	0.21	10	7.0
Hypobromous Acid	0.06	20	7.0

[a]Concentration of compound times contact time.
[b]Reproduced from Floyd et al. [90] as reported in [37].

example, Poliovirus II (Sabin), with 10 ppm NH_3 present, was sterilized in <5 min using 4.0 ppm BrCl (as Cl_2), whereas equivalent chlorine concentrations failed to kill all of the viruses within 60 min.

The comparative disinfection efficiency of BrCl and Cl_2 for poliovirus is seen in Figure 28 [92]. Experiments were conducted at pH 6. to enable a comparison between HOBr and HOCl. Several concentrations of BrCl and Cl_2 were prepared in a phosphate buffer and mixed for 4 and 10 min before 1 ml of purified poliovirus (10^5–10^6 PFU/ml) was added and allowed to react at 25°C for 15 min. Figure 28 shows that the minimum dose of BrCl required effectively to inactivate more than 99.99% of the virus was 0.075 mg/l, 4-min premix and 0.15 mg/l for the 10-min premix. The Cl_2 dosages were 0.25 and 0.30 mg/l for the same premix periods. These results suggested that BrCl (HOBr) is two to three times more effective than Cl_2 (HOCl) for inactivation of poliovirus.

That Br_2 and/or $HOBr/OBr^-$ is an effective bacteri- and viricide in laboratory studies has been discussed above. It is more effective than Cl_2 in the presence of NH_3 and is reasonably active over a wide range of pH values. However, a major disadvantage of Br_2 as a drinking water disinfectant is its reactivity with ammonia and other amines encountered in the treatment of drinking water [59].

Iodine

Iodine has had a long history as an antiseptic for skin wounds and mucous surfaces of the body and as a sanitizing agent in hospitals and laboratories [59]. The use of I_2 as a disinfectant of drinking and swimming pool waters has been limited due to a variety of reasons. It has been found useful for small-scale and emergency situations.

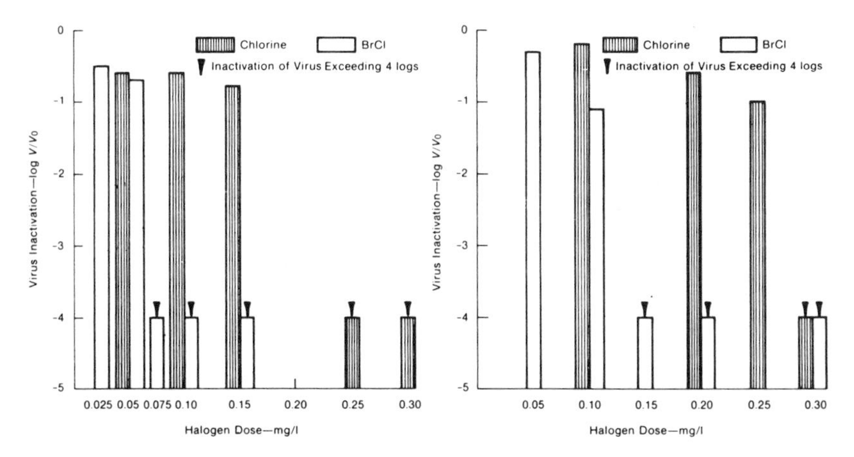

Figure 28. Stability and minimum concentration of BrCl and Cl_2 required to effectively inactivate poliovirus, with premixing times of 4. min (left) and 10. min (right). Reproduced from Keswick et al. [92], courtesy of the American Water Works Association.

Some of the early research into the efficacy of iodine as a drinking water disinfectant was conducted by Chang and Morris [93], Morris et al. [60] and Chang [61]. Chang and Morris [93] determined the usefulness of I_2 as an emergency disinfectant for drinking water. I_2 concentrations of 5–10 mg/l were found to be effective against all types of waterborne pathogenic organisms (enteric bacteria, amoebic cysts, cercariae, leptospira and viruses) within 10 min at room temperature. Germicidal activity is maintained over the pH range 3.–8. in the presence of a variety of natural water contaminants (organic color, loess, clay minerals, NH_3, urea, etc.). Typical results are given in Table XXVI [93]. Iodine was also reasonably effective against the cysts of *Endamoeba histolytica.* The concentration-time relation (Equation 69) was in the order of 200, 130 and 65 for 3, 10 and 23°C, respectively. The b value from the log-log plot was 1.5, which is the more or less typical Ct relationship. This work led to the development of the iodine tablets: Globaline, Potadine and Hexadine S&N, for use as field disinfectants in military situations [60]. In a later study, Chang [67] reported the Ct relationship for destruction of coxsackievirus B1 at 25°C by elemental iodine. In this case, $b = 0.96$ and $C^b t = 58.5$.

The effect of $[I_2]$, pH, contact time and temperature on 13 enteric bacteria in a laboratory study was reported [94]. The pH effect at 2–5°C was demonstrated when some bacterial species required 3–4 times as much I_2 for inactivation ($\approx$99.9% kill) at pH 9. than at pH 7.5. A summary of this work appears in Table XXVII where *E. coli* is the model organism.

Table XXVI. Destruction of Various Types of Bacteria by Iodine[a,b]

Organism	Geometric Mean No. of Survivors per 100 ml after: 5 min	10 min	20 min	30 min
E. coli (22 tests)	6.6	2.4	1.6	1.3
Sal. typhosa (8 tests)	1.4	<1	<1	<1
Sh. dysenteriae (9 tests)	1.3	1	1	<1
Sal. schöttmuelleri (9 tests)	21	5.1	2.7	2.1
Vibrio cholera (6 tests)	<1	<1	<1	<1
Mixed *coli aerogenes* flora of sewage, initial count 5000/ml (3 tests)	1.7	1.3	1.3	1.6

[a]Reproduced from Chang and Morris [93] courtesy of the American Chemical Society.
[b]Cambridge tapwater (10% seawater added for *V. cholera* tests; 10% filtered Cambridge sewage added for *coli-aerogenes* tests). Iodine dosage, 7–8 ppm; residuals after 10 min, 5–7 ppm; residuals after 30 min, 4–6.5 ppm. Temp, 25°C, pH values range from 4.5 to 8.1; low values obtained by addition of citric acid, high values by addition of $NaHCO_3$. Initial No. of organisms, 10^6/ml.

Table XXVII. Concentrations of Iodine and Contact Times Necessary for 99% Inactivation of *Escherichia coli*[a]

Iodine, (mg/l)	Contact Time (min)	Ct[b]	pH	Temperature (°C)
1.3	1	1.3	6.5	2–5
0.9	2	1.8	6.5	
1.3	1	1.3	7.5	
0.7	2	1.4	7.5	
0.8	1	0.8	7.5	
0.6	2	1.2	7.5	
0.8	1	0.8	8.5	
0.9	2	1.8	8.5	
1.8	1	1.8	9.1	
1.2	2	2.4	9.1	
0.35	1	0.35	6.5	20–25
0.20	2	0.40	6.5	
0.45	1	0.20	7.5	
0.30	2	0.60	7.5	
0.45	1	0.45	8.5	
0.40	2	0.80	8.5	
0.45	1	0.20	9.1	
0.30	2	0.60	9.1	

[a]Reproduced from Chambers et al. [94], as reported in Reference 37.
[b]Concentration of iodine times contact time.

This apparently is the best available information on the I_2 disinfection of bacteria [59].

Several studies have been conducted on the efficacy of I_2 destruction of viruses. These studies are summarized in Table XXVIII (see Reference 59 for original references). The viruses are, of course, more resistant to disinfection than are the negative cells of bacteria. This is reflected in the Ct values in Tables XXVII and XXVIII. In addition, at pH 6., I_2 is nearly as effective a viricide as HOCl.

Many studies have been reported on cyst inactivation by iodine [59]. Direct comparison of the results, however, is difficult because of a wide variety of experimental systems.

As a disinfectant in water, iodine is similar in many respects to Cl_2 and Br_2. Free I_2 is an effective bactericide over a relatively wide range of pH values. Field studies with small public water systems have shown that 0.5–1.0 mg/l of I_2 can be maintained in distribution systems, and that the magnitude of the residual is enough to produce safe drinking water with no apparent adverse effects on human health [59]. I_2 is effective against bacteria and cysts, although its effectiveness is reduced at "high" pH values ~9.–10. Its current use is restricted to emergency disinfection of field water supplies [95].

Ozone

In recent times, ozone (O_3) has been mentioned frequently as an alternative disinfectant for Cl_2 in drinking water in order to avoid formation of the low-molecular-weight halogenated hydrocarbons. However, it has been researched for efficacy of disinfection since the 1950s when Fetner and Ingols [96] reported the bactericidal activity of O_3 against *E. coli* at 1°C. Table XXIX shows the survival of *E. coli* in several O_3 contents and contact times, at pH 6.8 and 1°C. Ozone appears to have an "all or none" effect, as seen in the sharp demarcation between the 0.42 and 0.53 mg/l dosages. There is apparently a critical value above which O_3 is effective and below which it is not. Table XXIX also shows the survival of *E. coli* in chlorinated waters where the optimum dosage lies between 0.27 and 0.35 mg/l. The two disinfectants are nearly equal on a weight basis.

A somewhat different result was reported by Wuhrmann and Meyrath [97] who found that 99% of *E. coli* was inactivated in 21 sec with 0.0125 mg/l O_3, 62 sec with 0.0023 mg/l and 100 sec with 0.0006 mg/l at pH 7. and 12°C. The higher temperature used by Wuhrmann and Meyrath undoubtedly was responsible for the faster disinfection. Another "rapid inactivation" was reported by Katzenelson et al. [98] who observed that

Table XXVIII. Concentrations of Iodine and Contact Times Necessary for 99% Inactivation of Polio- and f_2 Viruses with Flash Mixing [59]

Test Microorganism	Iodine, (mg/l)	Contact Time (min)	Ct[a]	pH	Temperature (°C)
f_2 Virus	13	10	130	4.0	5
f_2 Virus	12	10	120	5.0	
f_2 Virus	7.5	10	75	6.0	
f_2 Virus	5	10	50	7.0	
f_2 Virus	3.3	10	33	8.0	
f_2 Virus	2.7	10	27	9.0	
f_2 Virus	2.5	10	25	10.0	
f_2 Virus	7.6	10	76	4.0	25–27
Poliovirus 1	30	3	90	4.0	
f_2 Virus	64	10	64	5.0	
f_2 Virus	4.0	10	40	6.0	
Poliovirus 1	1.25	39	49	6.0	
Poliovirus 1	6.35	9	57	6.0	
Poliovirus 1	12.7	5	63	6.0	
Poliovirus 1	38	1.6	60	6.0	
Poliovirus 1	30	2.0	60	6.0	
f_2 Virus	3.0	10	30	7.0	
Poliovirus 1	20	1.5	30	7.0	
f_2 Virus	2.5	10	25	8.0	
f_2 Virus	2.0	10	20	9.0	
f_2 Virus	1.5	10	15	10.0	
Poliovirus 1	30	0.5	15	10.0	

[a]Concentration of iodine times contact time.

an initial residual O_3 concentration of 0.04 mg/l, in demand-free water, disinfected approximately 3 logs (i.e., 99%) of *E. coli* in 50 sec, whereas a concentration of 1.3 mg/l achieved the same effect in 5 sec at 1°C and pH 7.2, as seen in Figure 29.

Broadwater et al. [99] also observed the "all or none" response of *E. coli, B. megaterium* and *B. cereus* to various concentrations of O_3 up to 0.71 mg/l at 28°C (pH not reported). With a constant contact time of 5 min, no inactivation of vegetative cells of *E. coli* or *B. megaterium* occurred until the initial concentration of O_3 was 0.19 mg/l, whereupon they were nearly destroyed. An initial O_3 content of 0.12 mg/l was needed for disinfection of *B. cereus.* These values of O_3 content are slightly lower than those of Fetner and Ingols at 1°C [96]. A greater $[O_3]$ was needed to inactivate the spores of *B. megaterium* and *B. cereus,* 2.29 mg/l within 5 min.

Table XXIX. Survival Percentages of *E. coli* in Different Concentrations of Ozone and Chlorine in Solution after Various Contact Times at pH 6.8 and 1°C[a]

Initial Concentration (mg/l)	Period of Exposure (min)						
	1	2	4	8	16	32	64
Ozone							
1.00	<1	<1	<1	<1			
0.75	<1	<1	<1	<1	<1	<1	
0.63	<1	<1	<1	<1	<1	<1	
0.53	<1	<1	<1	<1	<1	<1	
0.42	75	<1	71	<1	73	46	
0.31	96	97	74	70	92		
0.21	33	82	99	99	79	69	
Chlorine							
1.10	<1	<1	<1	<1	<1	<1	<1
0.52	<1	<1	<1	<1	<1	<1	<1
0.35	1.5	<1	<1	<1	<1	<1	<1
0.27	2.0	1.5	1.6	1.2	1.0	<1	<1
0.25	24	10	12	13	8	7	5
0.17	70	53	39	24	15	4	<1
0.10	87	66	64	46	46	45	40

[a]Reproduced from Fetner and Ingols [96], courtesy of the *Journal of General Microbiology*.

A dosage of 0.5 mg/l O_3 was required at 25°C for the inactivation of *E. coli, Staphylococcus aureus, S. typhimurium, Shigella flexmeri, Pseudomonas fluorescens* and *Vibrio cholerae* [100]. All were reduced by 7.5 logs in 15 sec.

The inactivation of poliovirus I and coliphage T_2 at 5°C and pH 7.2 by O_3 was investigated [98,101]. That the kinetics of this disinfection are rapid is seen in Figure 30. There is an apparent two-stage action of O_3. The first stage occurs in less than 8 sec with a virus kill of 99–99.5%. The second stage lasts from 1 to 5 min, during which residual viruses are disinfected. It is interesting to note that an increase of $[O_3]$ beyond 0.2 mg/l did not affect the rate or the percentage of kill. Some of these microorganisms apparently are resistant to ozonation. Later experiments suggested that single-particle viruses were destroyed in the first stage and the slower second stage was due to clumps of viruses. These experiments were conducted with O_3-demand free water.

Inactivation of >99.9% for vesicular stomatitis virus, >99.99% for encephalomyocarditis and 99.99% for GDV11 virus in 15 sec with an O_3 residual of approximately 0.5 mg/l were obtained by Burleson [100].

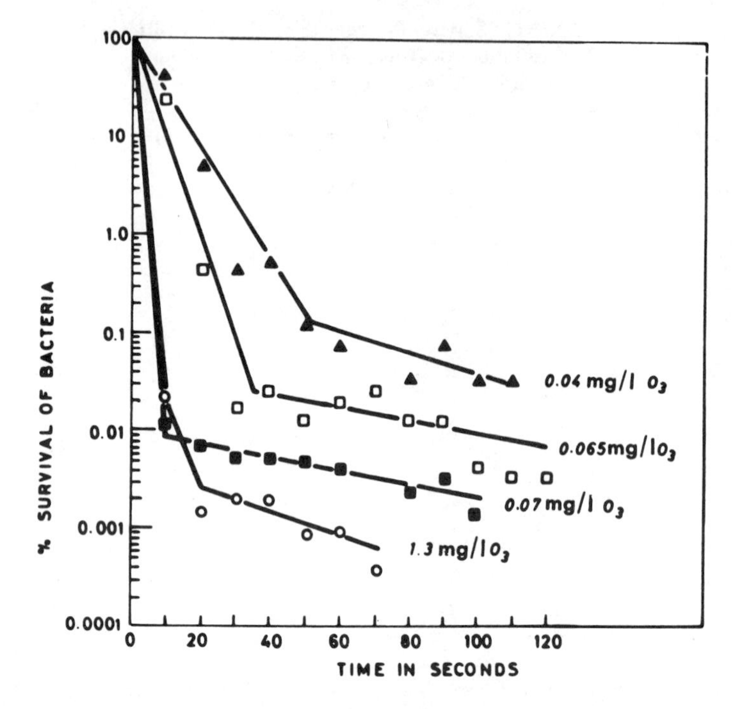

Figure 29. Inactivation kinetics of *E. coli* by various concentrations of ozone at 1 °C. Reproduced from Katzenelson et al. [98], courtesy of the American Water Works Association.

A wide range of ozone dosages, contact times and percentages of inactivation is noted above. Most of these studies were conducted in ozone-demand-free waters. However, this variable may account for some of the differences in the required ozone content. On the other hand, natural waters will have these demands. The pH value is apparently not a factor between 6. and 8.5, whereas temperature is with faster rates occurring at the higher values. Table XXX brings all of this together where Ct relationships are noted. It is seen that the Ct product varies over a broad range from 10^{-3} to 4.2×10^{-1}. In addition to the variables cited above, there are difficulties with the ozone analytical method and nonuniformity of microorganisms from one laboratory to another. Another difficulty with O_3 is its relatively short half-life in water, 10–40 min. Consequently, it is necessary to add another disinfectant to maintain an inactivation capability in a water distribution system. Additional kinetic studies of O_3 disinfection are discussed below.

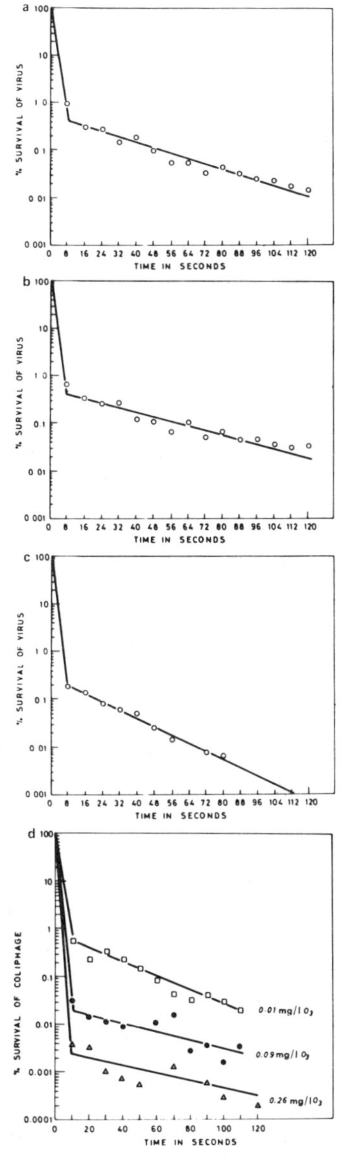

Figure 30. Inactivation kinetics of **(a)** poliovirus 1 by 0.3 mg/l ozone at 5°C; **(b)** poliovirus 1 by 0.8 mg/l ozone at 5°C; **(c)** poliovirus 1 by 1.5 mg/l ozone at 5°C; **(d)** coliphage T_2 by various concentrations of ozone at 1°C. Reproduced from Katzenelson et al. [101], courtesy of Ann Arbor Science Publishers.

Table XXX. Concentration of Ozone and Contact Time Necessary for 99% Inactivation of *E. coli* and Poliovirus 1 [59]

Test Microorganism	Ozone (mg/l)	Contact Time (min)	Ct[a]	pH	Temperature (°C)
E. coli	0.07	0.083	0.006	7.2	1
	0.065	0.33	0.022	7.2	1
	0.04	0.50	0.02	7.2	1
	0.01	0.275	0.027	6.0	11
	0.01	0.35	0.035	6.0	11
	0.0006	1.7	0.001	7.0	12
	0.0023	1.03	0.002	7.0	12
	0.0125	0.33	0.004	7.0	12
Polio 1	<0.3	0.13	<0.04	7.2	5
	0.245	0.50	0.12	7.0	24
	0.042	10	0.42	7.0	25
	<0.03	0.16	<0.005	7.0	20

[a]Concentration of ozone times contact time.

Comparative Studies of the Halogens and Ozone

In many instances, it is difficult to evaluate and compare the disinfection efficacy of the various agents cited above. Several studies are available, however, in which the experimental conditions were held constant, thereby allowing a direct comparison to be made.

An early study was reported by Krusé et al. [102] that compared the effectiveness of Cl_2, Br_2, I_2 and their amines as viri- and bactericides for water and wastewater. From the results of this study (Figures 31 and 32) it is seen that iodamines are more effective than bromamines and chloramines for both viruses and bacteria. Chloramine apparently is not viricidal within 30 min. That iodine is a more effective viricide than bromine and chlorine at pH 7. and 8. is seen in Figures 32a and b. At pH 6., chlorine is the most effective agent. This undoubtedly is due to the predominance of HOI at pH 7. and 8., whereas HOCl is predominant at pH 6.

To date, the only quantitative comparative study of the cysticidal properties of Cl_2, Br_2 and I_2 was published by Stringer et al. [103]. Table XXXI shows the contact times required for a 90% inactivation of amoebic cysts by the three halogens at pH values of 6.0, 7.0 and 8.0, at 30°C. At pH 8.0, for example, iodine is not as effective as bromine and chlorine. However, bromamine appears to be more effective at pH 8.0 than the iodamines and chloramines. Also, the halamines are generally less effective than the free halogens in cysticidal activity.

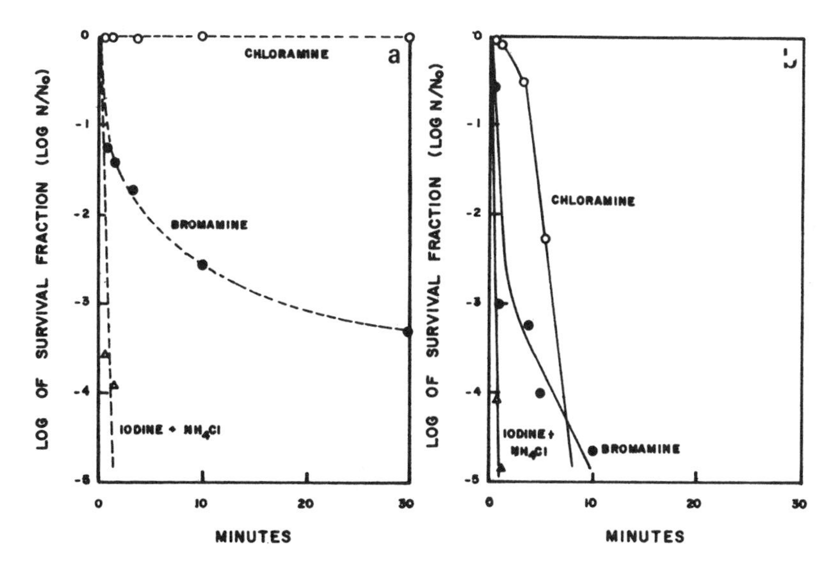

Figure 31. **(a)** Viricidal and **(b)** bactericidal activities of prereacted inorganic halamines at 0°C, pH 7.5, [halogen] = 8. mg/l. Reproduced from Krusé et al. [102], courtesy of the American Society of Civil Engineers.

From the studies of Cronier et al. [85], Figure 33 shows a comparison of the inactivation efficacies of ClO_2, HOCl, OCl^-, NH_2Cl and $NHCl_2$ for poliovirus at 15°C. On a weight basis, chlorine dioxide is as efficient a viricidal agent as HOCl.

The biocidal efficiencies of alternative disinfectants to chlorine were reported by Hoff and Geldreich [104]. Citing the research of others, the order of decreasing efficiency is: ozone, chlorine dioxide, free chlorine and chloramines, while the order of stability in water is the reverse of efficiency. An increase of pH value (in the range 6.–9.) has a beneficial effect on ClO_2, a detrimental effect on free chlorine and little effect on ozone or chloramine efficacies. As seen in Figures 33 and 34, neither order of efficiency nor degree of difference between the disinfectants is the same for both poliovirus and *E. coli*. These two figures summarize, in general, the state-of-the-art for the major disinfectants in water to date.

Other Disinfectants

Ferrate

That the oxidation state of Fe(VI) in the ferrate ion, FeO_4^{2-}, is easily reduced was discussed above. Several studies have been conducted on the

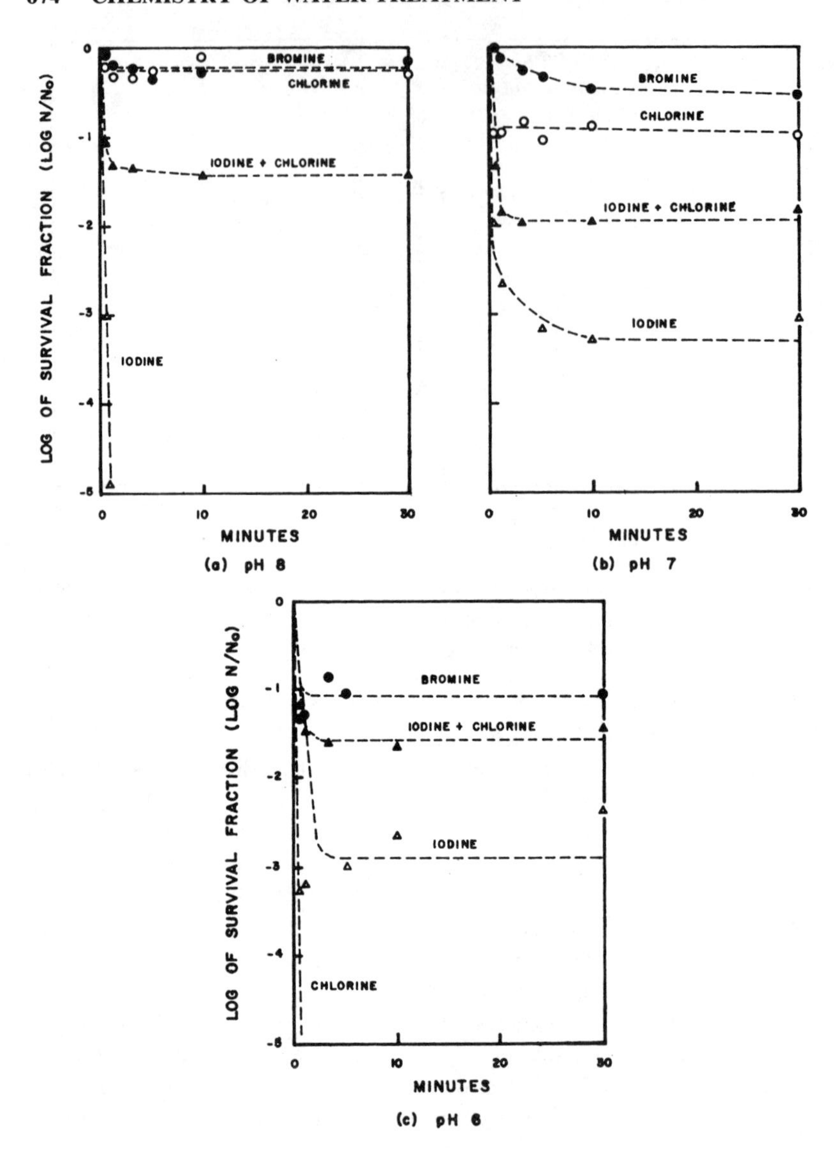

Figure 32. Activity of halogens (4-mg/l dosage) against f2 in filtered sewage effluent at 0°C. Reproduced from Krusé et al. [102], courtesy of the American Society of Civil Engineers.

efficacy of this oxidant as a disinfectant [59]. A pure culture of *E. coli* was inactivated at pH 8.0 with [FeO_4^{2-}] ranging from 1.2 to 6.0 mg/l [105]. A 99% disinfection was observed at 6.0 mg/l, pH 8.0 and 27°C

Table XXXI. Contact Times Necessary for Low Residual Bromine, Chlorine and Iodine in Water at 30°C to Effect 90% Inactivation of Cysts from Simian Stools[a,b]

		Contact Time Required (min) by 10-min Halogen Residual Concentration								
		0.5 mg/l			1.0 mg/l			2.0 mg/l		
	pH	Br_2	Cl_2	I_2	Br_2	Cl_2	I_2	Br_2	Cl_2	I_2
Buffered Water	6.0	10	10	20	4	4	10	3	3	5
	7.0	12	14	40	8	12	20	4	5	7
	8.0	15	20	ND[c]	10	15	80	5	10	20
Buffered Water in Presence of Excess Ammonia	6.0	10	65	20	8	35	10	4	22	5
	7.0	30	120	40	10	55	20	7	35	7
	8.0	35	ND	ND	13	80	80	9	50	20

[a]Reproduced from Stringer et al. [103], courtesy of Ann Arbor Science Publishers.
[b]The proportions of the molecular species present in the test system depend on the halogen, the pH and the presence or absence of ammonia.
[c]ND = not determined.

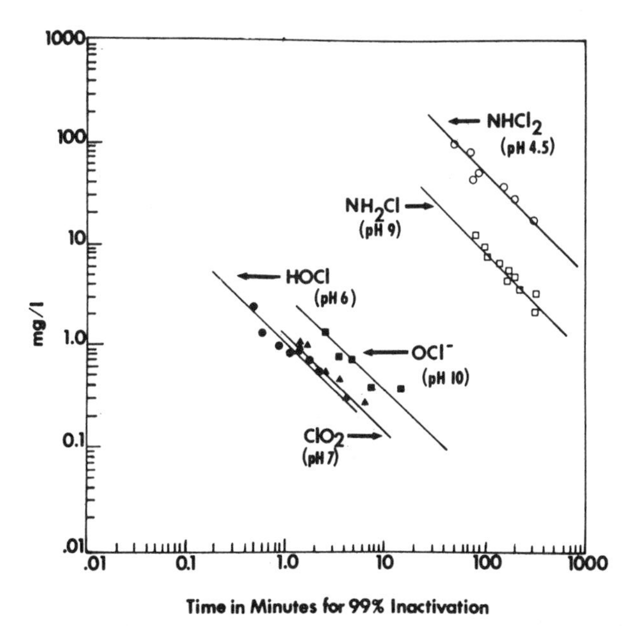

Figure 33. Comparison of the relative inactivation of poliovirus 1 by HOCl, OCl^-, NH_2Cl, $NHCl_2$ and ClO_2 at 15°C and different pH values. Reproduced from Cronier et al. [85], courtesy of Ann Arbor Science Publishers.

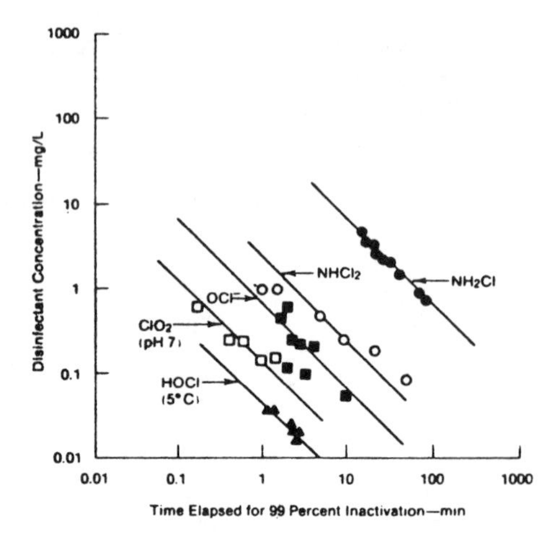

Figure 34. Inactivation of *E. coli* ATCC 11229 by free and combined chlorine species and chlorine dioxide at 15°C. Reproduced from Scarpino et al. [76] and Esposito et al. [81], courtesy of Ann Arbor Science Publishers and the American Society of Microbiology.

within 8.5 min. Waite [106] extended these studies to include enteric pathogens and gram-positive bacteria and also evaluated the effects of pH and temperature. The rate of *E. coli* inactivation by ferrate was increased as the pH value was decreased from 8.0 to 6.0. However, the benefit of this higher rate is offset by a greater rate of ferrate decomposition. *S. typhimurium* and *Shigella flexneri* were inactivated in a manner similar to *E. coli* (Table XXXII). However, considerably higher dosages (>12. mg/l) were needed to inactivate *Streptococcus faecalis.* Concentrations of 12 and 60 mg/l required 5 and 15 min, respectively, for 99% inactivation at pH 8.0.

The disinfection efficacy of ferrate is summarized in Table XXXII where dosage is given rather than a residual because of rapid decomposition in aqueous solution. The Ct values, therefore, should be interpreted as relative to the test microorganism and for ferrate, itself, when compared to other disinfectants. Rates of inactivation of *E. coli* with ferrate appear to be on the same order of magnitude as NH_2Cl [105]. Considerably more research is needed before ferrate is employed in water treatment [59].

Potassium Permanganate

That this oxidant has some disinfection capability is seen in Table XXXIII [107]. There are occasions when $KMnO_4$ is used for Fe and Mn

Table XXXII. Concentrations of Ferrate (FeO_4^{2-}) and Contact Times Necessary for 99% Inactivation of *E. coli*, *Streptococcus faecalis* and f_2 Virus[a]

Test Microorganism	Ferrate Dosage (mg/l)	Contact Time (min)	Ct[b]	pH	Temperature (°C)
E. coli	1.2	4.6	5.52	7.0	20
	6.0	2.5	15.0		
Streptococcus faecalis	1.2	365	438	7.0	20
	6.0	4.0	24		
f_2 Virus	1.2	3.7	4.4	6.0	20
	6.0	1.5	9		

[a]Reproduced from Waite [106], as reported in Reference 37.
[b]Concentration of ferrate vs contact time.

Table XXXIII. Concentrations and Contact Times Necessary for Potassium Permanganate to Effect 99% Inactivation of a 48-hr *E. coli* Lactose Broth Culture[a]

Temperature (°C)	pH	Contact Time					
		1 mg/l	2 mg/l	4 mg/l	8 mg/l	12 mg/l	16 mg/l
0	5.9	45	ND[b]	ND	ND	10	5
	7.4	c	ND	ND	ND	c	115
	9.2	ND	ND	ND	ND	ND	ND
20	5.9	95	45	15	15	10	5
	7.4	c	ND	c	80	80	25
	9.2	ND	ND	ND	ND	ND	c

[a]Reproduced from Cleasby et al. [107], as reported in Reference 37.
[b]ND = no data reported.
[c]Inactivation in 120 min was less than 90%.

removal, taste and odor control, etc. There will consequently be some disinfection incidental to the intended treatment. It is doubtful that $KMnO_4$ should be used as the principal disinfectant. The dosages cited in Table XXXIII are within those used in actual practice. According to Cleasby et al. [107], the major factor influencing the effectiveness of $KMnO_4$ is the $[H^+]$. The lowest contact times were obtained at pH 5.9 rather than at 7.4 or 9.2. Because the oxidation product is $MnO_{2(s)}$, a brown-black solid, any use of $KMnO_4$ would be limited to pretreatment.

Miscellaneous Compounds

Other disinfectants used in water treatments are: the chlorosulfamates, $NHClSO_3^-$ and $NCl_2SO_3^-$, for swimming pools [108], silver, Ag^+, for use in home treatment systems and swimming pools [59], the chloroisocyanurates for swimming pools [109], and ultraviolet radiation for disinfecting bacteria and other microorganisms in small water systems [59].

Summary

Tables XXXIV and XXXV bring together the major disinfection methods for drinking water and the comparative efficacy of the disinfectants. The biocidal activities of these disinfectants are compared through the Ct product that is required to achieve a particular degree of inactivation under similar conditions of temperature, $[H^+]$, etc. The lower the Ct value, the more effective is the disinfection by a particular agent.

KINETICS OF DISINFECTION

Chick's Law

Chemical destruction of the microorganisms cited above ideally should occur as a reaction between a single organism and the disinfectant. Also, the disinfectant should remain unchanged in chemical composition as well as in concentration throughout the period of contact. The water should contain no interfering substances. Under these conditions, the rate of disinfection is a function of contact time, disinfectant concentration and water temperature. The number of microorganisms may or may not affect this process.

Much of the kinetics of disinfection follows the time-rate of kill described by Chick's law of disinfection [110]. This "law" states that y, the number of organisms destroyed per unit time, is proportional to N, the number of organisms remaining, with the initial number being No (nomenclature follows Fair et al. [22]):

$$\frac{dy}{dt} = k(No - y) \tag{70}$$

where k is the rate constant ($time^{-1}$). Integration between the limits of $y = 0$, $t = 0$, and $y = y$ at time t yields:

$$ln\frac{N}{No} = -kt \tag{71}$$

or:

$$\frac{N}{No} = e^{-kt} \tag{72}$$

Consequently, a plot of log N/No vs time gives a straight line with a slope of $-k\log e = -k'$, and an intercept of 1 (or 100%) at time 0. For a reaction time of 1, $k't = 0.4343$, the surviving fraction is 0.368 or 36.8%. Other reaction times may be calculated from Equation 72. For example, a 99.995% kill is achieved with 10 reaction times. Chang [111] and others have indicated that disinfection is a first-order rate process. In other aspects of the quantitative assessment of germicidal efficiency, the reader is referred to the specific lethality coefficient presented by Morris [112] and the multi-Poisson distribution model, or the "multihit" concept described by Wei and Chang [113].

Departures from Chick's law are frequent and common in laboratory experiments and treatment plant practice. Rates of kill may increase or decrease rather than remain constant with time. Explanations are offered for departures from Chick's law, but no proof is given. In order to linearize these data, empirical coefficients are added to the t term in Equation 72 [22]:

$$\frac{N}{No} = e^{-kt^m} \tag{73}$$

An example is given in Figure 35, where linearization is achieved with $m = 2$. When $m > 1$, the rates of kill rise in time, whereas $m < 1$, the rates of kill decrease with time.

Temperature of Disinfection

It is well known that temperature affects the rate of disinfection. This was seen in many of the studies cited above. In accord with the rate laws, a diffusion-controlled disinfection reaction or, perhaps, an enzyme-disinfectant reaction will show temperature effects in conformance with the van't Hoff-Arrhenius relationship for which it is convenient to write:

$$\log\frac{t_1}{t_2} = \frac{Ea(T_2 - T_1)}{2.303RT_1T_2} \tag{74}$$

Table XXXIV. Summary of Major Possible Disinfection Methods for Drinking Water [59]

Disinfection Agent[a]	Technological Status	Efficacy in Demand-Free Systems[b]			Persistence of Residual in Distribution System
		Bacteria	Viruses	Protozoan Cysts	
Chlorine[c]	Widespread use in U.S. drinking water				Good
As hypochlorous acid (HOCl)		+ + + +	+ + + +	+ +	
As hypochlorite ion (OCl^-)		+ + +	+ +	NDR[d]	
Ozone[c]	Widespread use in drinking water outside United States, particularly in France, Switzerland and Quebec	+ + + +	+ + + +	+ + + +	No residual possible
Chlorine Dioxide[c,e]	Widespread use for disinfection (both primary and for distribution system residual) in Europe, limited use in United States to counteract taste and odor problems and to disinfect drinking water	+ + + +	+ + + +	NDR[d]	Fair to good (but possible health effects)
Iodine	No reports of large-scale use in drinking water				Good (but possible health effects)
As diatomic iodine (I_2)		+ + + +	+ + +	+ + +	
As hypoiodous acid (HOI)		+ + + +	+ + + +	+	
Bromine	Limited use for disinfection of drinking water	+ + + +[f]	+ + + +[f]	+ + +[f]	Fair
Chloramines	Limited present use on a large scale in U.S. drinking water	+ +	+	+	Excellent

[a]The sequence in which these agents are listed does not constitute a ranking.
[b]Ratings: + + + +, excellent biocidal activity; + + +, good biocidal activity; + +, moderate biocidal activity; +, low biocidal activity; ±, of little or questionable value.
[c]By-product production and disinfectant demand are reduced by removal of organics from raw water prior to disinfection.
[d]Either no data reported or only available data were not free from confounding factors, thus rendering them not amenable to comparison with other data.
[e]MCL = 1.0 mg/l because of health effects.
[f]Poor in the presence of organic material.

Table XXXV. Comparative Efficacy of Disinfectants in the Production of 99% Inactivation of Microorganisms in Demand-Free Systems [59][a]

Disinfection Agent	*E. coli*			Poliovirus 1			*Endamoeba histolytica* Cysts		
	pH	Temperature (°C)	Ct[b]	pH	Temperature (°C)	Ct[b]	pH	Temperature (°C)	Ct[b]
Hypochlorous Acid	6.0	5	0.04	6.0	0	1.0	7	30	20
				6.0	5	2.0			
				7.0	0	1.0			
Hypochlorite Ion	10.0	5	0.92	10.5	5	10.5		NDR[c]	
Ozone	6.0	11	0.031	7.0	20	0.005	7.5–8.0	19	1.5[d]
	7.0	12	0.002	7.0	25	0.42			
Chlorine Dioxide	6.5	20	0.18	7.0	15	1.32			
	6.5	15	0.38	7.0	25	1.90		NDR[c]	
	7.0	25	0.28						
Iodine	6.5	20–25	0.38	7.0	26	30	7.0	30	80
	7.5	20–25	0.40						
Bromine		NDR[c]		7.0	20	0.06	7.0	30	18
Chloramines									
Monochloramine	9.0	15	64	9.0	15	900		NDR[c]	
	9.0	25	40	9.0	25	320			
Dichloramine	4.5	15	5.5	4.5	15	5,000		NDR[c]	

[a]Conditions closest to pH 7.0 and 20°C were selected from studies discussed in the text. Values for other conditions and agents appear in the text along with discussions of the cited studies.

[b]Concentration of disinfectant (mg/l) times contact time (min).

[c]Either no data reported or only available data were not free from confounding factors, thus rendering them not amenable to comparison with other data.

[d]This value was derived primarily from experiments that were conducted with tap water; however, some parallel studies with distilled water showed essentially no differences in inactivation rates.

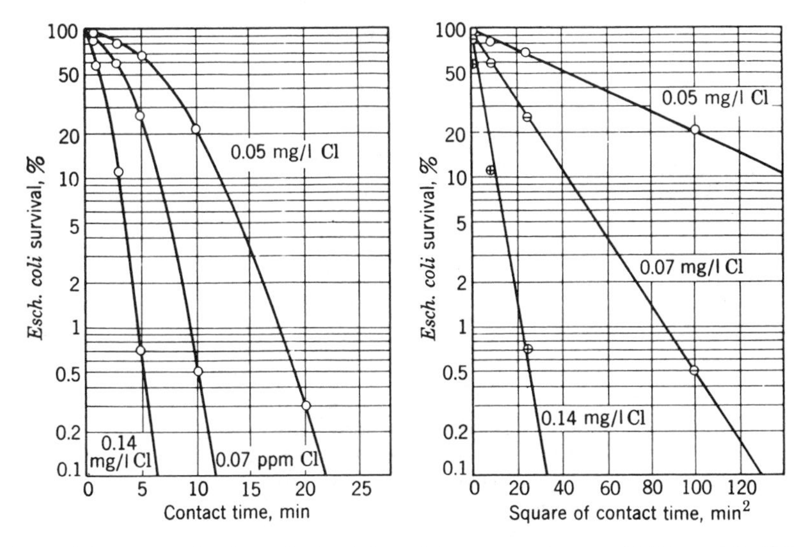

Figure 35. Length of survival of *E. coli* in pure water at pH 85. and 2–5°C. Reproduced from Fair et al. [22], courtesy of John Wiley & Sons, Inc.

where T_1, T_2 = absolute temperatures (°K)
t_1, t_2 = times required for equal percentages of kill at fixed contents of disinfectant
Ea = activation energy (cal)
R = the ideal gas law constant, 1.9872 cal/mol deg

When $T_2 - T_1 = 10$, the ratio of t_1/t_2 is called Q_{10}, which is helpful in comparing destruction rates of various disinfectants and microorganisms. Equation 74, at water temperatures near 20°C, becomes:

$$\log Q_{10} = \log \frac{t_1}{t_2} = \frac{Ea}{39{,}000.} \tag{75}$$

Fair et al. [22] fitted the data of Butterfield [70] into Equation 75 and calculated Ea and Q_{10} values (Table XXXVI).

Kinetic Studies

One of the early kinetic studies concerning disinfection of water was the iodine destruction of amoebic cysts by Chang and Morris [93]. The Ct products are given above for three temperatures, 3, 10 and 23°C. An activation energy of 9500 cal was obtained when the temperature was

Table XXXVI. Temperature Dependence of Disinfecting Concentrations of Aqueous Chlorine and Chloramines in the Destruction of *E. coli* in Clean Water[a]

Type of Chlorine	pH	Ea (cal)[b]	Q_{10}
Aqueous Chlorine	7.0	8,200[c]	1.65[c]
	8.5	6,400	1.42
	9.8	12,000	2.13
	10.7	15,000	2.50
Chloramines	7.0	12,000	2.08
	8.5	14,000	2.28
	9.5	20,000	3.35

[a]Reproduced from Fair et al. [22], courtesy of John Wiley & Sons, Inc.
[b]The higher the value of Ea, the slower is the reaction.
[c]The rate of reaction at pH 7.0 is relatively so fast that these values are probably unreliable. The magnitudes of Ea throw some light on the nature of the disinfecting process for the chlorine species released in water.

varied and the contact time was held constant. The cysticidal action of iodine was dependent less on temperature than HOCl (Ea = 13,000 cal) [93]. The Q_{10} value for the I_2-amoebic cyst system was 1.65.

The kinetics of viral (ϕX174am3, a DNA coliphage) inactivation by bromine was reported by Taylor and Johnson [114]. Figure 36 is a typical result where only 1.2 μM HOBr is required to obtain seven logs of disinfection in less than 3.15 min at 30°C and a pH of 7.0. Temperature results are given in Figure 37 where the rate constants are plotted against [HOBr] at 0, 15 and 30°C. The slopes of the 15 and 30°C curves are nearly equal, whereas the 0°C slope is obviously different. This suggests, to Taylor and Johnson [114] at least, that there was a change in the reaction mechanism between 0 and 15°C. The Arrhenius activation energy, 37,000 cal, was calculated from the plot seen in Figure 38. "Since this value is quite high, it supports the hypothesis that the overall process of ϕX174 inactivation by HOBr at pH 7 and 1.0 μM is chemically controlled" [114]. In order for a diffusion-controlled process to occur, the Ea value would have to be less than 10,000 cal [114].

An activation energy value of 18,300 cal was obtained in the continuous ozonation of *Mycobacterium fortuitum* at pH 7.0 [115]. These experiments were conducted for the feasible use of this organism as a microbial indicator for disinfection efficiency.

The kinetics of inactivation of single poliovirus particles were investigated in water with various species of bromine [90]. Figure 39a shows the usual semilogarithmic or first-order plot of the inactivation at pH 10,

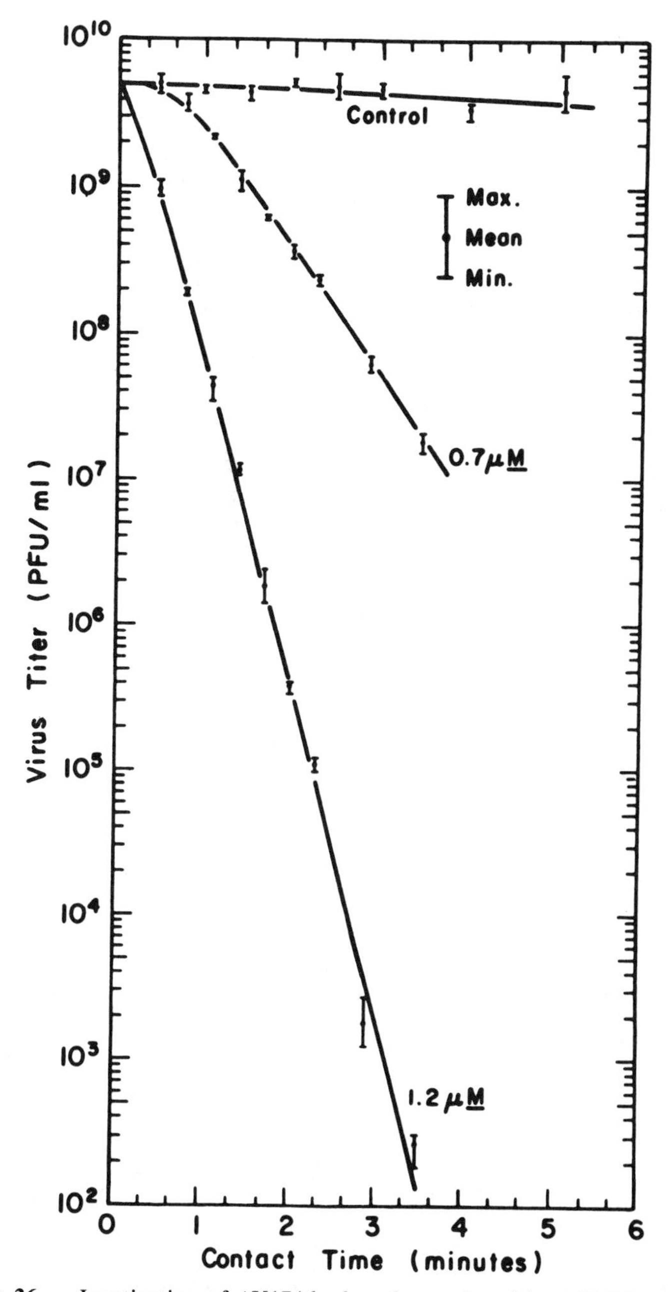

Figure 36. Inactivation of ϕX174 by hypobromous acid at pH 7.0 and 30°C. Reproduced from Taylor and Johnson [114], courtesy of Ann Arbor Science Publishers.

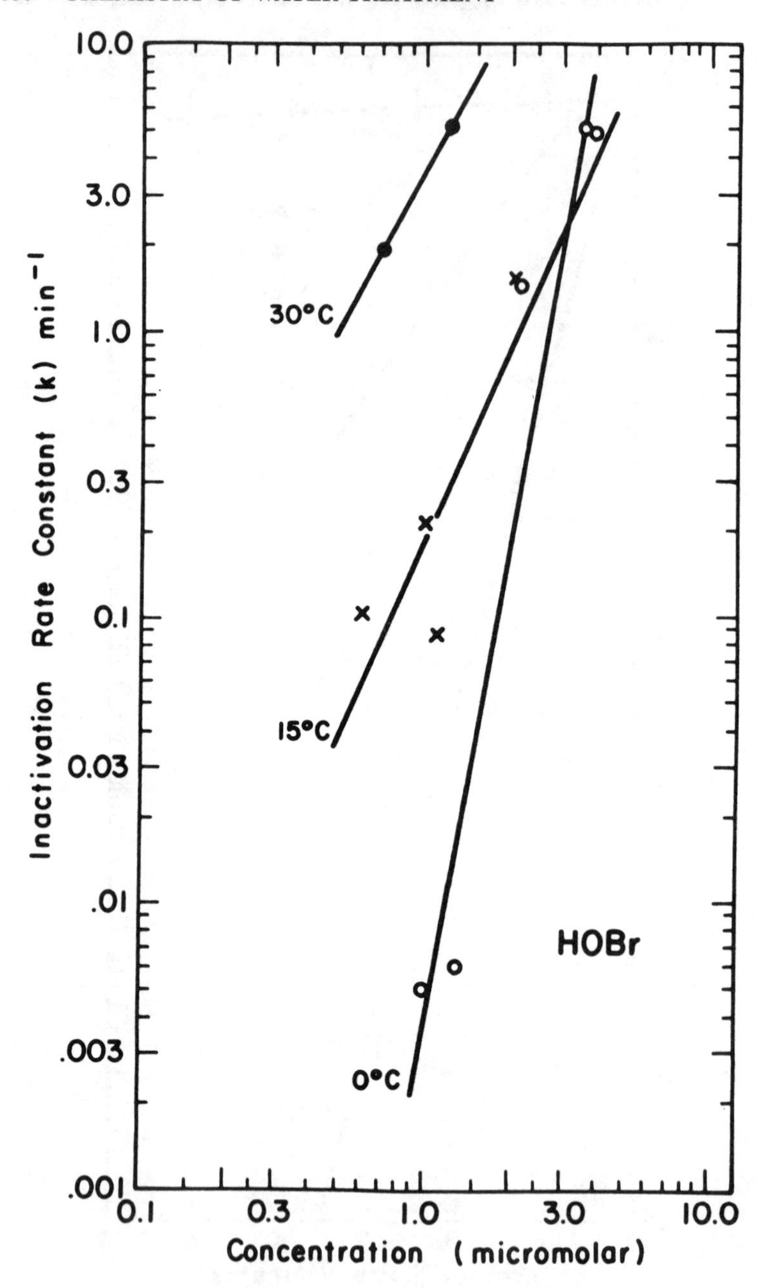

Figure 37. Variation in the inactivation rate constant (k) at three temperatures. Reproduced from Taylor and Johnson [114], courtesy of Ann Arbor Science Publishers.

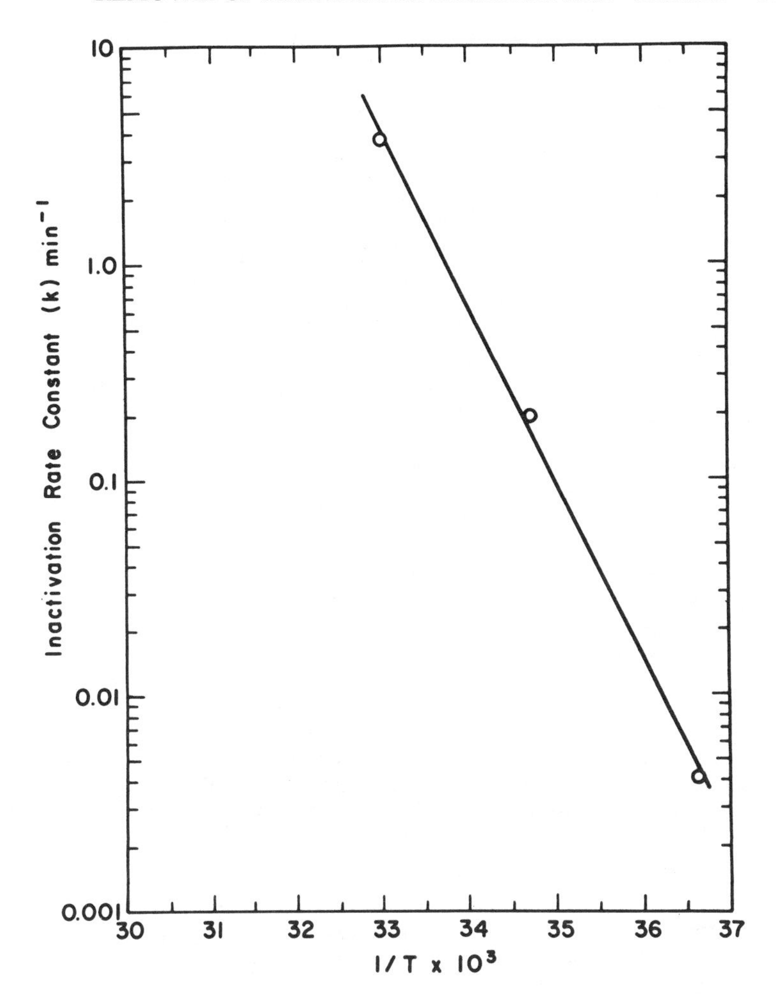

Figure 38. Arrhenius plot of the inactivation of ϕX174 by 1.0 μM hypobromous acid at pH 7.0. Reproduced from Taylor and Johnson [114], courtesy of Ann Arbor Science Publishers.

where OBr^- predominates. The inactivation rate typically was quite rapid initially, whereupon it decreased with time. Figure 39b gives the kinetics of inactivation of the poliovirus with Br_2 contents of 4.7, 12.9 and 21.6 μM (0.75, 2.06 and 3.46 mg/l). These results were linear and did not curve with time as did OBr^-. The reaction rates were −0.8, −1.81 and −2.5 $\log_{10}$/second, respectively, for the three Br_2 contents. These rates were slower than the initial ones for OBr^- but were approximately 10 times greater than for HOBr [116]. Figures 39c and d show the kinet-

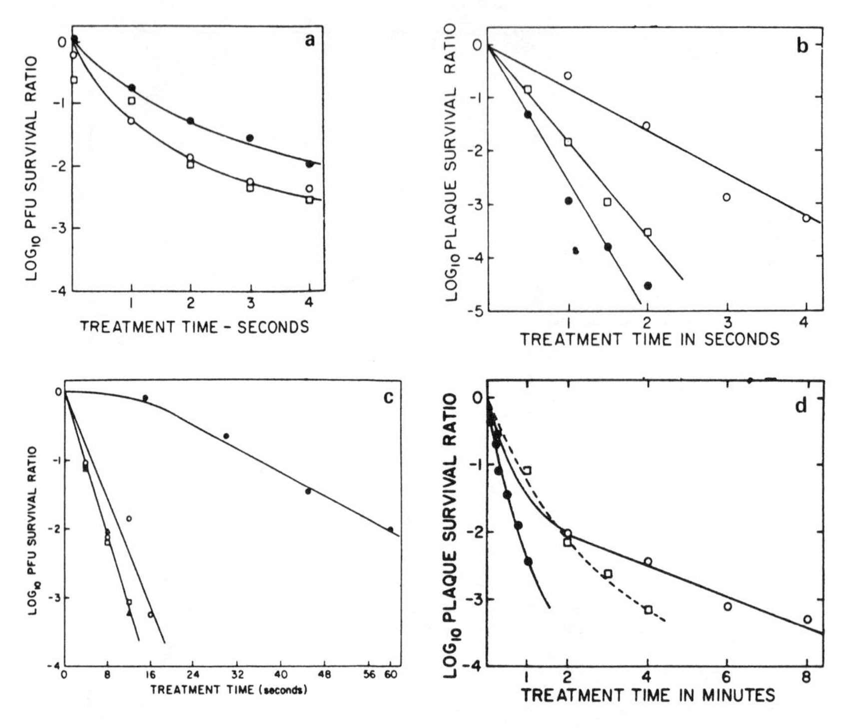

Figure 39. Kinetics of inaction of poliovirus at 4°C by **(a)** OBr^- at pH 10 (●—● = 2 μM, □—□ = 6 μM, ○—○ = 12 μM); **(b)** Br_2 at pH 5 in the presence of 0.3 M NaCl (○—○ = 4.7 μM, □—□ = 12.9 μM, ●—● = 21.6 μM); **(c)** NBr_3 (●—● = 3.2 μM, ○—○ = 12 μM, □—□ = 27 μM, △—△ = 49 μM); **(d)** $NHBr_2$ (○—○ = 2.9 μM, □—□ = 13.8 μM, ●—● = 39 μM; data points at 39 μM represent two experiments, one with samples taken up to 15 sec, and samples taken at 15–60 sec). Reproduced from Floyd et al. [90], courtesy of the American Chemical Society.

ics of inactivation of a poliovirus by NBr_3 and $NHBr_2$, respectively. The latter kinetics were definitely slower than NBr_3, Br_2 and OBr^-. Note that the time scale on Figure 39d is in minutes, whereas the others are in seconds. In summary, the initial rate for OBr^-, the linear rates of Br_2, NBr_3 and HOBr [116], and the initial rate of $NHBr_2$, all at 10 μM, are 3.85, 1.55, 0.15, 0.16 and 0.013 $\log_{10}$/second, respectively. The Ct values for these bromine species are given above in Table XXV.

Some kinetic aspects of the inactivation of poliovirus by HOCl and OCl^- were reported by Floyd et al. [117] who used virus preparations in which no less than 99% of the virus were free single particles. Figure 40 shows a reasonably consistent and similar mode of inactivation by HOCl

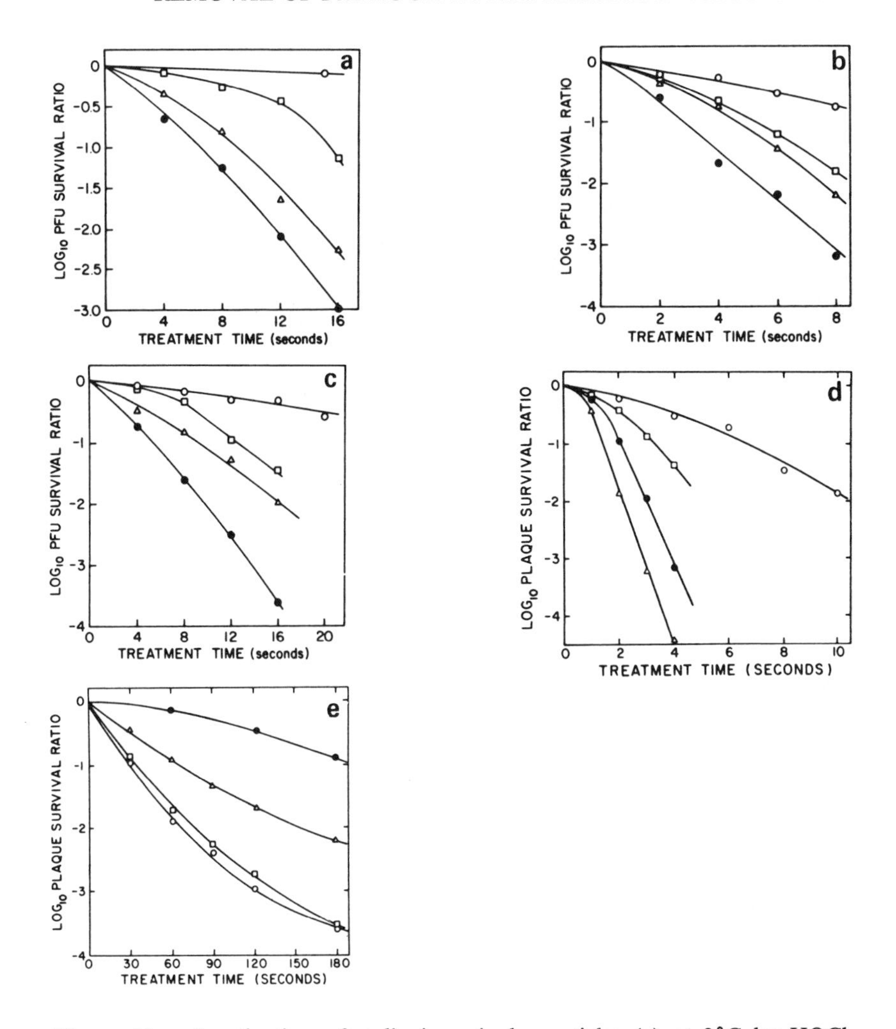

Figure 40. Inactivation of poliovirus single particles **(a)** at 2°C by HOCl, pH 6.0, ○ = 1.4 μM, □ = 10 μM, △ = 22 μM, ● = 40 μM; **(b)** at 20°C by HOCl, pH 6.0, ○ = 2.2 μM, □ = 11 μM, △ = 22 μM, ● = 35 μM; **(c)** at 10°C by HOCl, pH 6.0, ○ = 1.5 μM, □ = 11 μM, △ = 20 μM, ● = 41 μM; **(d)** at 30°C by HOCl, pH 6.0, ○ = 1.4 μM, □ = 10 μM, ● = 20 μM, △ = 42 μM; **(e)** at 20°C by OCl^-, pH 10.0, ● = 3.5 μM, △ = 7.5 μM, □ = 15 μM, ○ = 30 μM. Reproduced from Floyd et al. [117], courtesy of the American Chemical Society.

regardless of temperature. Note that the time scales on Figures 40a and b are twice the ones on Figures 40c and d. The results for virus inactivation with OCl^- is seen in Figure 40e. These reaction rates were considerably slower than with HOCl. All of these curves indicate that the inactivation rate is concentration dependent with exception of the 15 and 30 μM of OCl^-. Neither rate constants nor activation energies were calculated.

Comparisons may be made now between the rates of inactivation of the poliovirus by various bromine and chlorine species. The rates shown by HOBr and HOCl are quite similar: 22 μM HOBr at 2°C reduced the plaque survival level to 10^{-3} in 16 sec [117], whereas it required 40μM HOCl to produce essentially the same results (see Figure 40a). However, there are considerable differences when OCl^- and OBr^- are compared with each other and with HOCl and HOBr. The OBr^- is more effective than HOBr, whereas HOCl is far more efficient than OCl^-. The relative rates are in the order: $OBr^- > HOBr > HOCl > OCl^-$ [117].

Mechanism of Activity

The mechanism of disinfection by chlorine and other oxidants has received considerable discussion and speculation over the decades of research [59]. It seems reasonable to require disinfectants to penetrate the cell wall of the target organism and subsequently to react with one or more of the enzyme systems therein. An early reference to a mechanism of disinfection is an observation by Chang [71] that there was a greater uptake of chlorine by cysts of *E. histolytica* at a ''low'' pH value than at a ''high'' pH. It was hypothesized that uncharged HOCl was able to penetrate into this organism. There was considerable support for this hypothesis by several investigators [59]. Later, Kulikovsky et al. [118] implicated permeability damage to bacterial spores as a mechanism of chlorine inactivation. Studies with *E. coli* have shown that Cl_2 causes leakage of cytoplasmic material with protein existing first followed by RNA and DNA into the suspending menstruum.

Several studies have indicated that disinfection of bacteria was due to enzyme inhibition [59]. Green and Stumpf [119] and Knox et al. [120] suggested that the destruction of bacteria by Cl_2 was caused by inhibiting the mechanism for glucose oxidation. The specific reaction was an oxidation of the sulfhydryl group associated with the aldolase enzyme. Another study reported that Cl_2 significantly inhibits both O_2 uptake and oxidative phosphorylation [59]. Chang [111] concurred with the hypothesis that the quick disinfection of vegetative bacteria by Cl_2 was caused by extensive destruction of metabolic enzyme systems. Also, there was some speculation about the inactivation of viruses that are generally more resistant to Cl_2 than bacteria. Chang suggested that denaturation of the capsid protein of viruses was responsible. Venkobachar [121] reported some data that show the total dehydrogenase activity of intact *E. coli* cells was affected by Cl_2 doses which, in turn, correlated well with percent survivals. It was shown that the succinic dehydrogenase activity in crude extracts of *E. coli* was affected by Cl_2. Also, the effect of chlorine on thiol groups in the crude extract was studied. A reduction in these thiol groups suggested that they are vulnerable to oxidation. Haas and Engelbrecht [122] determined the uptake of radioactive chlorine by coliforms, acid-fast bacteria and yeasts. The Freundlich adsorption model was able to describe the uptake of chlorine. *Candida parapsilosis* and *Mycobacterium fortuitum* exhibited more uptake of Cl_2 at pH 7. than at pH 9.14. This study confirmed that the inactivation of vegetative microorganisms was from the cell-associated chlorine.

The reader is given reference [59] for additional speculations about mechanisms of disinfection by the chloramines, ozone, etc.

REMOVAL BY CHEMICAL COAGULATION

Bacteria and Viruses

Several studies have reported coagulation and subsequent removal of bacteria and viruses by polyelectrolytes, alum, etc. It is questionable whether or not a coagulation process alone would be effective enough for the required 100% removal of bacteria and viruses. On the other hand, these studies provide an insight into their incidental removal by coagulation.

A cationic polyelectrolyte was utilized as a primary coagulant and as a coagulant aid in the removal of a bacterial (bacteriophage T2 of *E. coli*) and an animal (Type 1 poliovirus) virus from artificially seeded water [123]. A summary of this laboratory study appears in Table XXXVII, where the polyelectrolyte alone was extremely ineffective for virus reduction. The clay mineral, montmorillonite, was added to simulate the effect of turbidity. It is significant to note that up to 82% virus removal was effected by the clay itself. When either alum or ferric sulfate was utilized as the primary coagulant, the virus removals were poor. However, when the clay was present and coagulated, virus removal as high as 86% was recorded. Presence of the polyelectrolyte did not improve this form of virus disinfection.

The alum coagulation of suspensions of *E. coli* was approached in a manner similar to removal of abiotic turbidity from water [124]. In this case, bacterial removal was measured by turbidity reduction. A typical result is seen in Figure 41. The initial turbidity represented by the 0.5

Table XXXVII. Removal of Poliovirus and Turbidity by Unaided Coagulation, and Cationic Polyelectrolyte Aided Coagulation with Aluminum and Ferric Sulfates in Waters with Varying Ionic Concentrations[a]

	$CaCl_2$ Concentration (M)									
	0.1		0.01		0.001		0.0001		0	
Substance	Virus Removal (%)	Turbidity Removal (%)	Virus Removal (%)	Turbidity Removal (%)	Virus Removal (%)	Turbidity Removal (%)	Virus Removal (%)	Turbidity Removal (%)	Virus Removal (%)	Turbidity Removal (%)
Aluminum Sulfate										
Polyelectrolyte only	6		36		25		10		0	
Clay only	79	91	82	91	13	26	0	2	0	0
$Al_2(SO_4)_3$ only	12		9		27		33		36	
$Al_2(SO_4)_3$ + clay	79	94	86	96	83	93	26	18	11	0
$Al_2(SO_4)_3$ + clay polyelectrolyte[b]	69	95	81	95	75	94	19	93	24	92
Ferric Sulfate										
Polyelectrolyte only	6		36		25		10		0	
Clay only	79	91	82	91	13	26	0	2	0	0
$Fe_2(SO_4)_3$ only	20		20		20		0		0	
$Fe_2(SO_4)_3$ + clay	85	93	83	93	4	0	0	0	0	0
$Fe_2(SO_4)_3$ + clay + polyelectrolyte	76	95	75	93	60	83	62	0	35	0

[a] Reproduced from Thorup et al. [123], courtesy of the American Water Works Association.

[b] pH—6.8, adjusted with 1.25–4.0 ml of 0.1 N NaOH. Dosages: polyelectrolyte (Purifloc C32) 1 mg/l; clay 50 mg/l; $Al_2(SO_4)_3$ 10 mg/l (0.29 mM); $Fe_2(SO_4)_3$ 11.6 mg/l (0.29 mM).

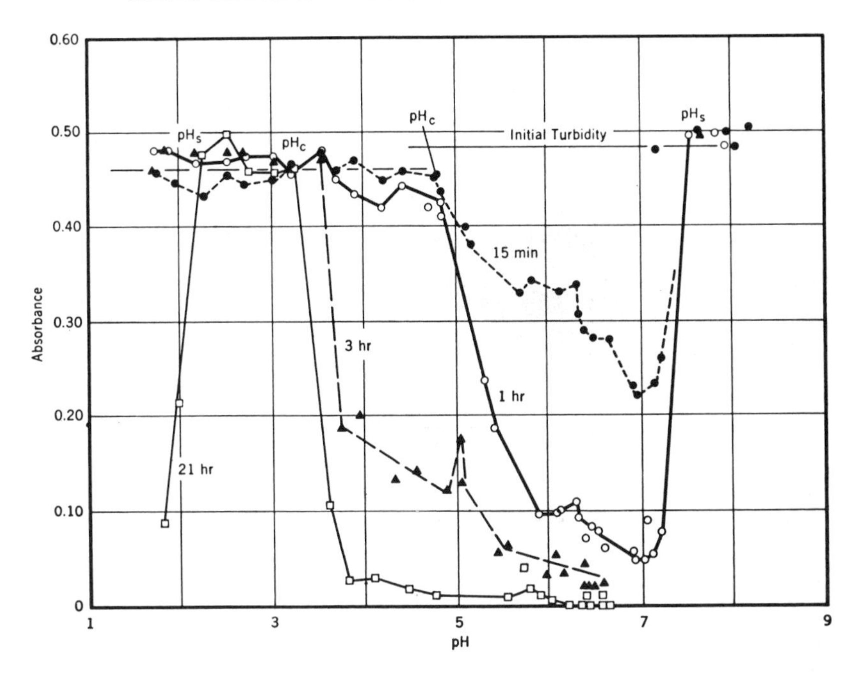

Figure 41. Coagulation with 4.5×10^{-5} *M* $Al_2(SO_4)_3$. Clarification of *E. coli* suspensions at different times as a function of pH. Reproduced from Hanna and Rubin [124], courtesy of the American Water Works Association.

absorbance apparently is the initial concentration of *E. coli* (2.5×10^8 cells/ml). In Figure 41, two distinct pH ranges of aggregation and settling are observed. Very slow coagulation, essentially complete within 21 hr, occurred in the lower pH range (<3.3). In the higher pH range (>4.75), clarification was essentially complete within one hr (sweep zone) and, apparently, was due to flocculation by hydrolyzed aluminum oxides. This "slow" and "rapid" coagulation was distinguished by the sedimentation data at 1 and 21 hr. The "critical" pH values, pHc and pHs, were obtained by extrapolating the steep portions of the sedimentation curves back to the turbidity of the control. Thus, pHc is that pH value above which coagulation occurs and, pHs is that pH value above which stabilization (no coagulation or restabilization due to charge reversal) is complete. These critical values at a given $[Al]_t$ are boundary points defining regions within the alum-pH stability domain (Figure 42). The authors did not indicate the bacterial contents represented by turbidities by less than 0.1 absorbance. It is nonetheless apparent that significant reductions of *E. coli* are accomplished in the coagulation process.

In another study, Manwaring et al. [125] employed ferric chloride for

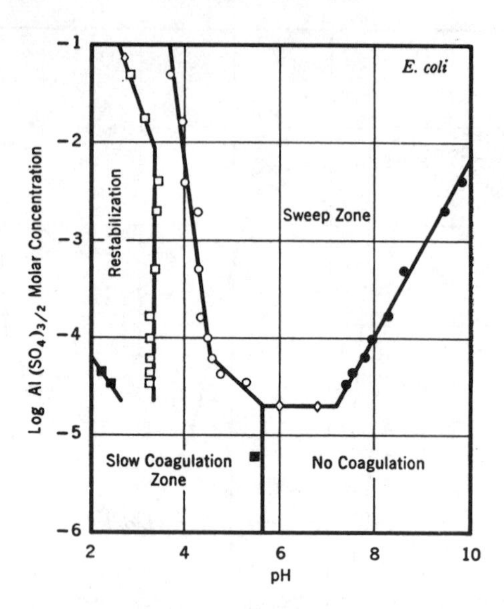

Figure 42. Aluminum sulfate domain of stability. Effect of any dose of alum on the stability and coagulation of *E. coli* at any pH. Circles (1 hr) and squares (21 hr) for turbidity-pH settling data, and diamonds (12 hr) for turbidity-concentration data; open symbols are coagulation values and solid symbols are stabilization values. Reproduced from Hanna and Rubin [124], courtesy of the American Water Works Association.

coagulating the bacterial virus, bacteriophage MS2 (against *E. coli*). Figure 43 shows the virus and clay turbidity removals at pH values ranging from 5.0 to 8.0 and coagulant contents up to 100 mg/l. Based on these removals, the optimum ferric chloride dosage was 50–60 mg/l with the optimum virus and turbidity removals occurring at a pH value of 5.0. Using an average input of 3.9×10^5 PFU/ml in each experimental system, the highest virus removal was 99.5%. These results with ferric chloride are as effective as with alum (Figure 44) [126].

A rather comprehensive study of virus removal from spiked raw surface water and wastewater was published by Shelton and Drewry [127] who utilized alum, ferric chloride, ferric sulfate, cationic, anionic and nonionic polyelectrolytes, and sodium aluminate ($Na_2OAl_2O_3$). The experimental virus was the bacteriophage f2 (host-specific to *E. coli*). Table XXXVIII shows the effectiveness of the various coagulants and combinations thereof on removal of this virus from raw surface water (Little Tennessee River, Tennessee). It appears that greater than 99% removals were obtained in all systems with alum considered to be the most effective coagulant by the authors. This consideration included tur-

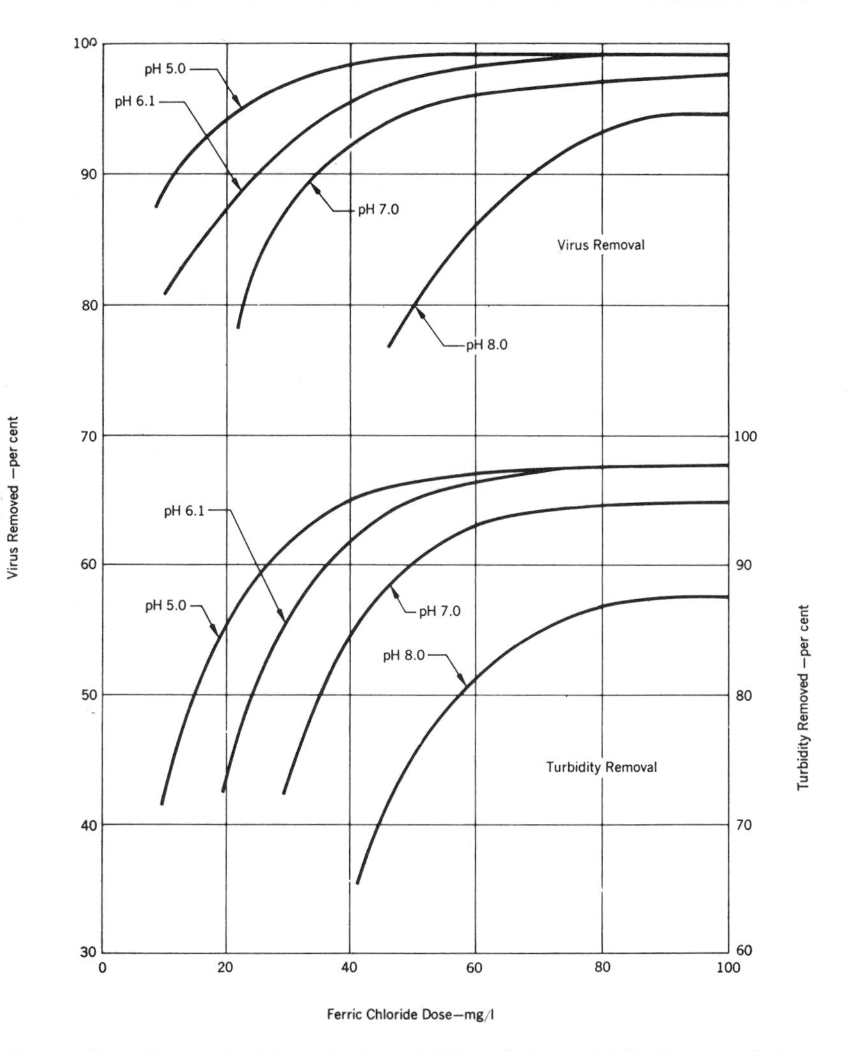

Figure 43. Removal of bacteriophage MS2 and clay turbidity by coagulation and flocculation. Average input virus concentration $= 3.9 \times 10^5$/ml. Reproduced from Manwaring et al. [125], courtesy of the American Water Works Association.

bidity and COD removals as well as the virus. The results were reported by York and Drewry [128] in a similar study for virus (same as above) removal from Fort Loudoum Lake, Knoxville, Tennessee, water by the same coagulants listed in Table XXXVIII. Alum and ferric chloride achieved >99% virus removal and the turbidity was reduced by 90%.

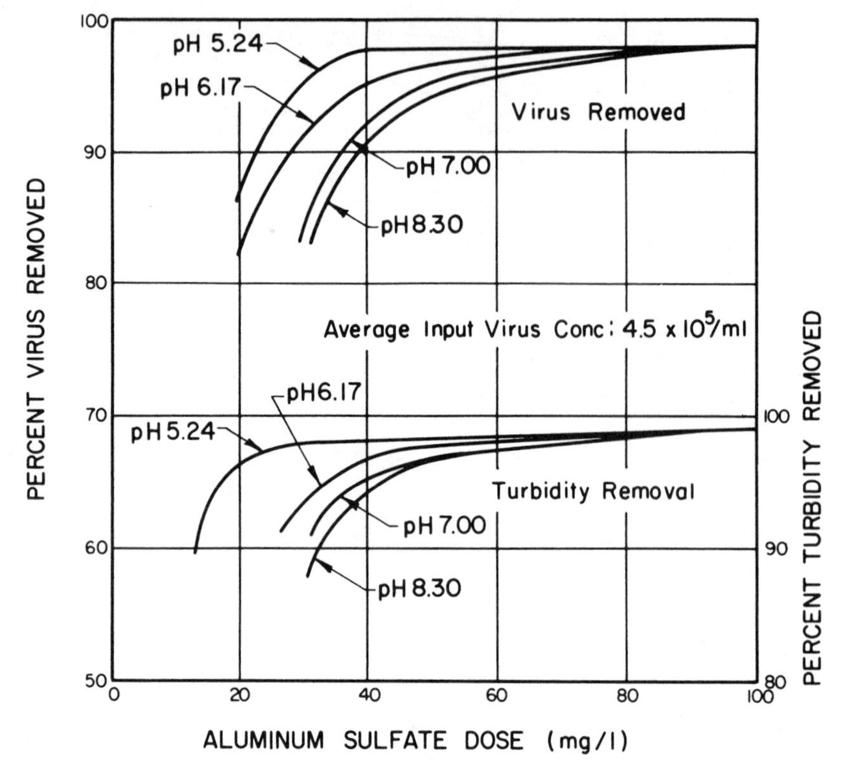

Figure 44. Removal of bacteriophage T_4 and clay turbidity by coagulation and flocculation. Reproduced from Chaudhuri and Engelbrecht [126], courtesy of the American Water Works Association.

That the removal of the bacterial virus (MS2 against *E. coli*) from distilled, deionized water by a laboratory bench-scale diatomaceous earth filter is feasible was demonstrated by Anurhor and Engelbrecht [129]. Filtration was aided through the addition of a cationic polyelectrolyte, Puriflor C-31, to the diatomaceous earth. The essential results are seen in Figure 45 at three concentrations of the polyelectrolyte where breakthrough curves describe the virus removal. Here Ce represents the virus content in the filtrate and Ci is the initial content (1.4×10^8 PFU/ml). It is interesting to note from Figure 45 that 0.5 and 2.0 mg/l of the polyelectrolyte caused rapid breakthrough of the virus, whereas there was "significant removal of the virus with 0.25 mg/l of the polyelectrolyte." The higher quantities of the Purifloc C-31 apparently completely coated the virus particles, thereby preventing their removal.

There was successful removal of a virus (bacteriophage MS2 for *E.*

Table XXXVIII. Comparison of the Effectiveness of the Coagulants on Raw Surface Water

Coagulant and Coagulant Aids	Coagulant Concentration (mg/l)	Maximum Virus Removal (%)	pH[b]	Coagulant Concentration (mg/l)	Maximum Turbidity Removal (%)	Coagulant Concentration (mg/l)	Maximum COD Removal (%)	Optimum Dosage[c]	
								A	B
$Al_2(SO_4)_3$	15	99.45	6.8	15	96.50	20	43.0	12	14
$FeCl_3$	40	99.10	6.8	34	94.80	48	73.0	37	32
$Fe_2(SO_4)_3 \cdot nH_2O$	63	99.91	7.2	35	90.50	38	88.50	65	42
Cationic Flocculant	2.2	92.10	7.5	0.56	46.0	0.51	26	[d]	1.5
$Al_2(SO_4)_3$ + Cationic Flocculant	8 + 0.25	99.75	6.8	8 + 0.25	97.4	8 + 2.25	49.5	9.2 0.25	7.2 0.25
$Al_2(SO_4)_3$ + Nonionic Aid 1	10 + 0.25	99.50	6.8	10 + 0.25	97.4	10 + 0.5	59.3	9 0.4	9 0.16
$Al_2(SO_4)_3$ + Anionic Aid 1	10 + 0.40	99.50	6.7	10 + 0.2	99.65	10 + 0.4	97.5	8.2 0.4	[e] [e]
$Al_2(SO_4)_3$ + Anionic Aid 2	9 + 0.2	99.93	6.8	9 + 0.2	98.6	10 + 0.2	79.5	7.1 0.2	[e] [e]
$Al_2(SO_4)_3$ + Nonionic Aid 2	10 + 0.4	98.60	6.8	10 + 0.2	97.20	7 + 0.4	85.00	6.4 0.4	5.4 0.4
$Al_2(SO_4(_3$ + $Na_2OAl_2O_3$	10 + 3	99.89	6.9	10 + 12	98.70	10 + 9	86.00	10 9	[e] [e]

[a]Reproduced from Shelton and Drewry [127], courtesy of the American Water Works Association.
[b]pH of supernatant immediately after sedimentation.
[c]Isoelectric point as indicated by (A) zeta potential; (B) colloidal titration.
[d]Isoelectric point was not reached with test dosage.
[e]More than one isoelectric point was indicated.

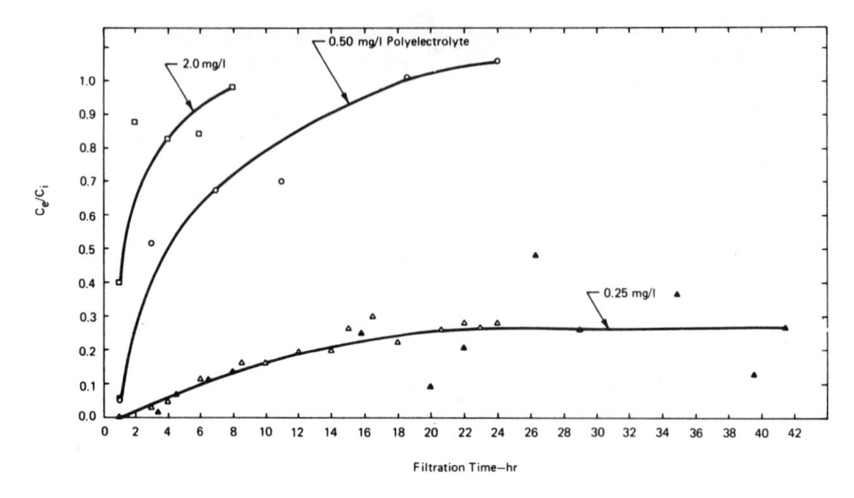

Figure 45. Removal of bacteriophage MS2 by uncoated diatomaceous earth with various levels of polyelectrolyte added to the suspending medium. pH 6.0, average virus input concentration = 1.40×10^8 PFU/ml; flowrate = 1 gpm/ft^2. Reproduced from Anurhor and Engelbrecht [129], courtesy of the American Water Works Association.

coli) from tapwater and simulated surface water (by addition of topsoil to tapwater) by alum and four cationic polyelectrolytes [130]. Table XXXIX shows the results of this study. Alum was not effective in removing the bacteriophage MS2 within the range of 5–10 mg/l. A removal of 99.7% was achieved with 50 mg/l alum when 14 NTU of turbidity were present. Nalco 8101 was able to reduce the virus content (initially 5.4×10^3 PFU/ml) by 96% at a dosage of 2 mg/l. Better removals of the virus were recorded when turbidity was present in the water.

Sproul [131] has reviewed critically the literature of virus removal by coagulation processes. Enteric virus removal should be in the 90–99.999% range by coagulation of water and wastewater containing poliovirus and coxsackievirus A2.

In a study of synthetic polymers for flocculation of *E. coli* in water, filtration rates, electrophoretic mobility and light transmission were measured to evaluate the effectiveness of removal [132]. *E. coli* concentrations of 50–2700 mg/l and pH values of 4.–9. were flocculated with nonionic, anionic and cationic polymers. Mol wts from 7000 to 3,000,000 for the nonionic and anionic polyacrylamides were not effective flocculants at concentrations of 0.1 to 50. mg/l. On the other hand, cationic polyethyleneimines with mol wts from 300 to 112,000 were effective at contents of 0.1–80 mg/l. These mol wts, per se, were not factors in the

Table XXXIX. Bacteriophage MS2 Removal from Water by Various Coagulants Without Sedimentation[a]

Coagulant	Dosage (mg/l)	MS2 Removal (%) In Water Without Turbidity	MS2 Removal (%) In Water With 14 ntu Turbidity
A	5	33	Insignificant
	6	27.5	Insignificant
	7	Insignificant	Insignificant
	8	Insignificant	Insignificant
	9	Insignificant	Insignificant
	10	Insignificant	Insignificant
	20		94
	30		99
	40		99.5
	50		99.7
B	2	Insignificant	75
	4	Insignificant	71.5
	6	Insignificant	64
	8	Insignificant	65.5
	10	Insignificant	45
C	2	96	96
	4	93.5	97
	6	93	95.5
	8	92	97
	10	90	96
D	2	48.5	63
	4	44.5	58
	6	30	28
	8	38	41
	10	39	38
E	2	56.5	57.5
	4	57	55.5
	6	39	23.5
	8	49.5	41
	10	54.5	43.5

[a]Reproduced from Malek et al. [130], courtesy of the American Water Works Association.

process. Redispersal of the bacteria was experienced with 10 mg/l of the higher-molecular-weight polymers.

Algae

Even algae are coagulated by alum at dosages from 10 to 40 mg/l. Lin et al. [133] researched the alum coagulation of 38 genera of algae in the Illinois River. A typical result was an overall algal reduction in excess of 85% at a coagulant content of 30 mg/l.

CASE HISTORIES

Disinfectants

The relatively recent concern over appearance of trihalomethanes (THM) in finished waters has prompted the use of alternative disinfectants and/or chlorination practices within the treatment plant. Brodtmann and Russo [134] present a summary of 30 years of experience with chloramines as primary disinfectants at Jefferson Parish, Louisiana, using water from the Mississippi River. An intensive study was conducted on the bacterial quality ("total" bacteria and coliforms) of the water (1) just before addition of chlorine and ammonia and sand filtration, (2) after filtration, and (3) in various parts of the distribution system. Table XL is a summary of the coliform tests for a period of five months. It was concluded:

> when properly applied at effective dosages (1.5 to 1.8 mg/l), chloramine produces 100% kills of pathogenic bacteria species and also effectively reduces total bacterial populations to an acceptable range. Moreover, from a large number of THM analyses, it is apparent that the use of chloramine as a sole disinfectant is effective in preventing the formation of THM during the treatment process and continued disinfectant control through the distribution system [134].

The Philadelphia Suburban Water Co. (PSWC) has reported essentially the same experience with chloramines as that cited above [135]. The principal sources of supply for this company are four rural streams that are tributaries to the Schuylkill and Delaware Rivers. Pre- and postchloramine treatment has been used for over 50 years with considerable success. Table XLI shows the average (total) bacteria and coliform counts for a 50-yr period at PSWC. There were obviously excellent results. The prechlorine to ammonia ratio averaged 3.3, the prechlorine dosage was 1.34 mg/l, and the average residual chlorine was .58 mg/l. It is significant to note that the residual chlorine concentration increased steadily from an average of 0.28 mg/l in the 1930s to greater than 1.0 mg/l in 1979.

The Louisville Water Co., Kentucky, investigated several disinfection processes using ClO_2, ClO_2 and NHxCly, and $HOCl$-OCl^- and NHxCly, in an effort to limit the formation of trihalomethanes and to attain adequate disinfection in the distribution system [136]. Figure 46a shows the treatment processes that were used prior to the research effort that began in 1974. Figure 46b depicts one of the experimental disinfection systems; namely, the prechlorination is eliminated, ClO_2 and NH_4^+ are added after

Table XL. Summary of Routine Microbiological Analyses for District 1 (East Jefferson) Operations[a]

Month	Filters		Inlets to Distribution		Field			
	Mean Coliform Density (Σ colonies/ 100 ml per no. of samples tested)	Number of Samples	Mean Coliform Density (Σ colonies/ 100 ml per no. of samples tested)	Number of Samples	Mean Coliform Density (Σ colonies/ 100 ml per no. of samples tested)	Number of Samples	pH	Temperature (°C)
January	0.0	307	0.0	400	0.0	248	8.70	3.3
February	0.39	257	0.0	324	0.0	199	8.58	3.3
March	0.32	307	0.0	400	0.0	248	8.34	7.2
April	0.01	301	0.0	262	0.0	228	8.44	15.0
May	0.0	310	0.02	382	0.0	240	8.50	18.9

[a]Reproduced from Brodtmann and Russo [134], courtesy of the American Water Works Association.
[b]Includes reservoirs, clearwells, and lines-to-mains samples.

Table XLI. Bacteria and Coliform Counts for a 50-year Period at PSWC[a,b]

	Average Bacteria (count/ml)				Coliforms (count/100 ml)			
Years	Raw Water	Water at Top of Filter	Filtered Water	Plant Effluent Water	Raw Water	Water at Top of Filter	Filtered Water	Plant Effluent Water
1930–1939	379	12.0	1.0	1.1	1,749	5.1	0.02	0.14
1940–1949	426	12.5	2.6	1.1	2,124	1.9	0.08	0.012
1950–1959	2,559	27.7	5.3	3.3	7,869	0.7	0.22	0.26
1960–1969	2,709	86.0	8.2	5.6	10,289	4.0	0.42	0.15
1970–1979	7,661	106.0	40.0	13.2	31,420	5.3	0.12	0.16
1979	10,265	51.0	13.9	4.5	45,000	0.6	0.01	0.00

[a]Reproduced from Shull [135], courtesy of the American Water Works Association.
[b]Each result based on approximately 2000 samples.

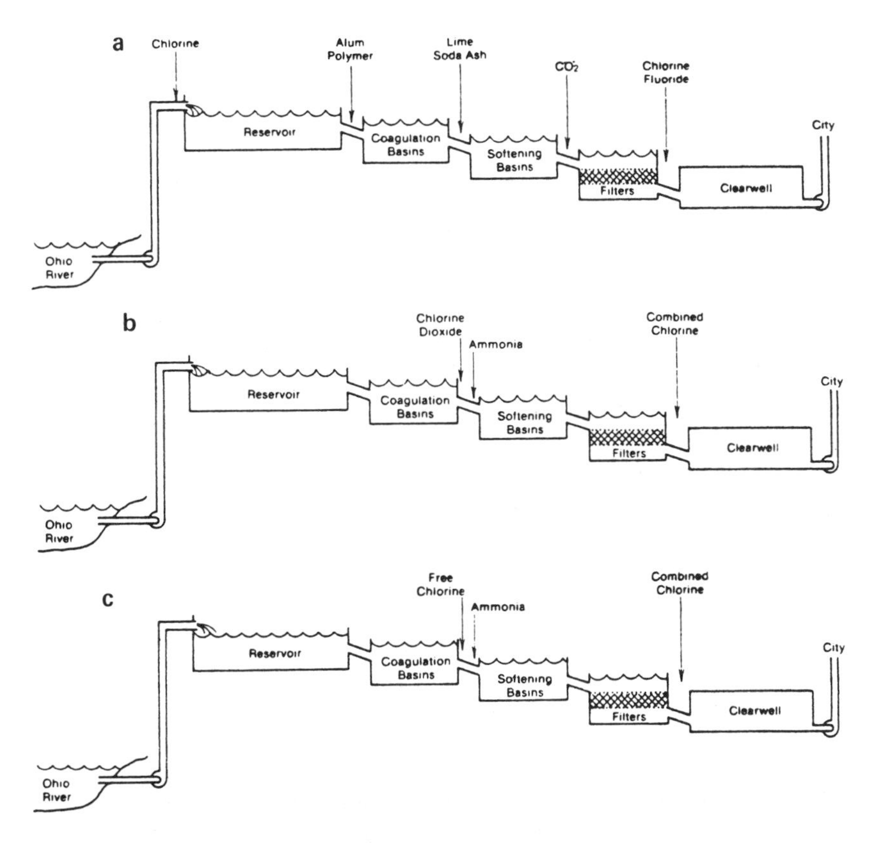

Figure 46. **(a)** Treatment processes used at the Louisville Water Company before 1974. **(b)** Chlorine dioxide-chlormaination disinfection system. **(c)** Short-term free chlorination disinfection system. Reproduced from Hubbs et al. [136], courtesy of the American Water Works Association.

the coagulation basin, and combined chlorine is added just before the clearwell. Figure 46c shows the disinfection system that eventually was chosen as the permanent water treatment process. The ClO_2-NH_2Cl system was investigated as an in-plant test from April 23 through May 8, 1979. Prior to this date, free chlorine was dosed at the influent of the coagulation basins, while ammonia was added to the influent of the softening basins. This was replaced by adding ClO_2 to the coagulation basin effluent at 0.6 and 0.8 mg/l followed by ammoniation about 10 min later. Figure 47a shows the standard plate count for the finished water through the test period. It would appear that the ClO_2-NH_2Cl treatment was essentially the same as the free chlorination practice. Figure 47b shows that coliform counts generally were less than 1/100 ml in the

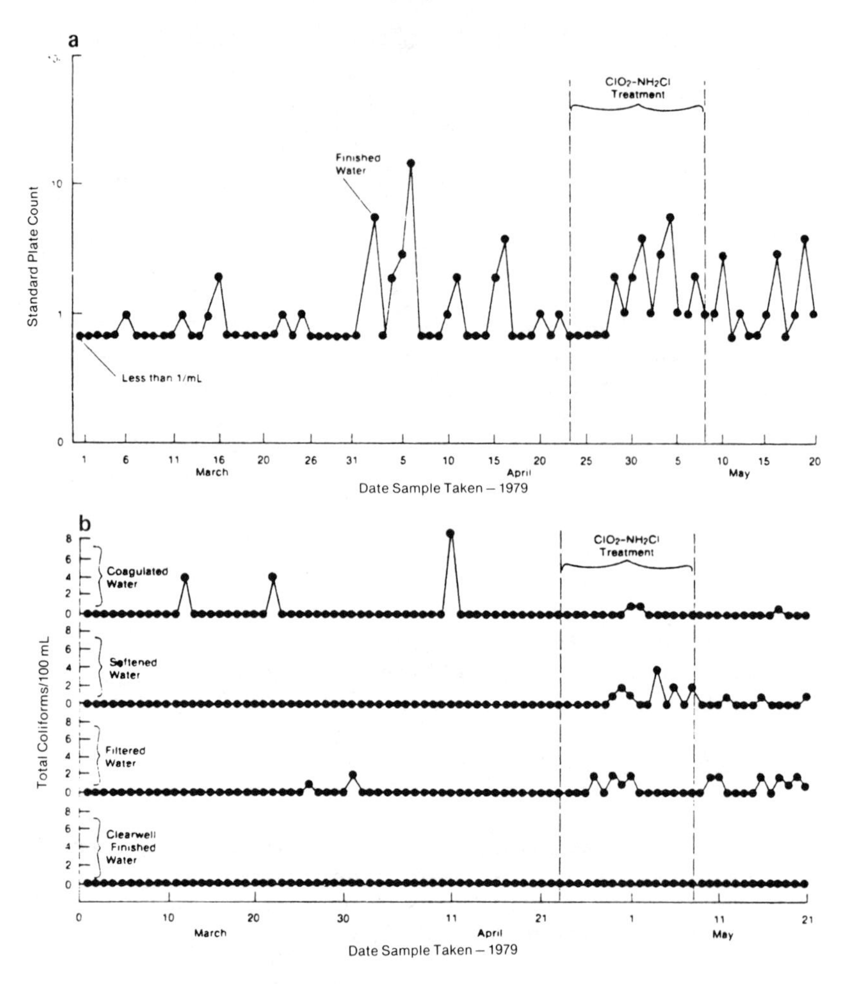

Figure 47. **(a)** Standard plate count data for finished water. **(b)** Total coliform data per 100 ml of coagulated, softened, filtered and finished water. Reproduced from Hubbs et al. [136], courtesy of the American Water Works Association.

coagulated, softened and filtered waters, whereas the finished water was always at the 0/100-ml level. Immediately after the investigation with the ClO_2-NH_2Cl combination, the disinfection system was modified to apply free chlorine at the coagulation basin with ammoniation about 10 min later (Figure 46c). The combined chlorine residual in the finished water varied between 0.8 and 1.8 mg/l. This treatment system of short-term free chlorination followed by conversion to monochloramine gave adequate disinfection through the treatment plant and distribution system. In addition, the concentration of trihalomethanes averaged less than 25 $\mu g/l$ for 1979. While the ClO_2-NH_2Cl system provided adequate disinfection in the test period, economic considerations and "established confidence in free chlorine as a disinfectant led to the decision to utilize the short-term free chlorination system followed by ammonification to produce monochloramine."

Pilot-plant (Figure 48) experiments, using O_3 for disinfection of the waters of Lake Winnipesaukee, the supply for Laconia, New Hampshire, were reported by Keller et al. [137]. This was an unusual case where the lakewater quality was such that complete treatment was not necessary. Instead, this study was conducted to determine the effectiveness of O_3 as a viricide following carbon adsorption. Two types of viruses were employed: type II poliovirus and coxsackievirus B-3 (Nancy strain). The essential results are seen in Tables XLII and XLIII. Sufficient disinfection was achieved at initial $[O_3]$s of 0.33–2.9 mg/l and with contact times of 5 min and less. One of the disadvantages of O_3 is its inability to carry a residual concentration into a distribution system. This could be overcome by addition of chlorine or chloramines prior to entry into the distribution system.

An important consideration in the application of chlorine or an alternative is the cost/benefit ratio. This is especially the case in view of the occurrence of THM in drinking water. Clark [138] evaluates the cost-effectiveness of chlorination relative to other types of disinfectants. According to this author, "Disinfection is one of the most cost-effective preventive practices in the field of public health, especially with filtration of surface waters." Clark gathered the cost data from the Cincinnati Water Works and the Hamilton County (Ohio) Health Department. Benefits were calculated as the decrease in costs of death and sickness due to introduction of disinfection and sand filtration. Because of this, the data base covers the years 1905–1927.

Detailed cost/benefit analyses were presented for chlorine, chlorine dioxide, ozone and chloramines. Only a summary is presented here. Figure 49 shows the overall benefits and costs for the years 1905–1927. It is obvious that the benefits of chlorination and filtration far exceed their operational costs. Table XLIV compares costs of the various disinfectants at presumed equivalent dosages. Chlorine is slightly less expensive than the others, although ClO_2 is a close second. Clark [138] concludes:

> The fact remains that disinfection and the combination of disinfection and filtration are highly cost-effective preventive public health practices. When various disinfectants are compared on the basis of

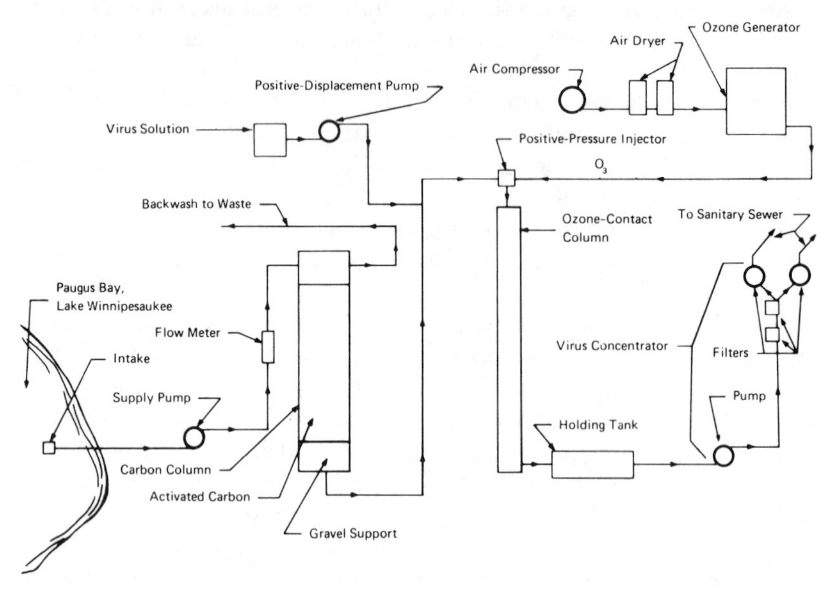

Figure 48. Schematic of pilot plant for treatment of Lake Winnipesaukee waters. Reproduced from Keller et al. [137], courtesy of the American Water Works Association.

Table XLII. Pilot-Plant Studies Results of Seeding Coxsackievirus B-3, Laconia Water Works[a]

Run	Date	Ozone Generated (ppm)	Ozone Dosage (ppm)	Ozone Residual (ppm)	Virus Application (PFU)	Virus Surv (PFU)	Inact (%)	Sample Coll. (gal)
1	8-2-74	4.03	2.9	1.2 after 1 min 0.47 after 4 min	1×10^5	0	>99.99	300
2	8-2-74	4.03	2.9	1.04 after 1 min 0.74 after 15 min	1×10^8	0	>99.999	150
3	9-12-74	2.0	1.53	0.59 after 1 min	7.6×10^7	0	>99.999	150
4	9-12-74	1.45	1.13	0.28 after 1 min	7.6×10^7	0	>99.999	150

[a]Reproduced from Keller et al. [137], courtesy of the American Water Works Association.

dosage and energy consumption, chlorine has the lowest marginal cost per unit increase in concentration. The costs of other disinfectants, however, are only slightly higher than that of chlorine, and disinfection on any basis shows a favorable cost-benefit relationship.

Table XLIII. Pilot-Plant Studies Results of Seeding Bacteria, Laconia Water Works[a]

Run	Date	Ozone Gen. (ppm)	Ozone Dosage (ppm)	Coliform Count Leaving Col. per 100 ml
1	9-20-74	3.16	2.36	<2
				<2
2	9-20-74	2.64	1.97	<2
				5
3	9-20-74	2.05	1.54	<2
				<2
4	9-20-74	1.48	1.12	<2
				8.8
5	9-25-74	2.03	1.57	<2
				<2
6	9-25-74	1.53	1.19	<2
				<2
7	9-25-74	1.02	0.82	>240
				<2
8	9-25-74	0.57	0.47	15
				15
9	10-2-74	1.56	1.22	<2
				<2
10	10-2-74	1.04	0.33	<2
				2.2
11	10-2-74	0.79	0.65	<2.0
				2.2

[a]Reproduced from Keller et al. [137], courtesy of the American Water Works Association.

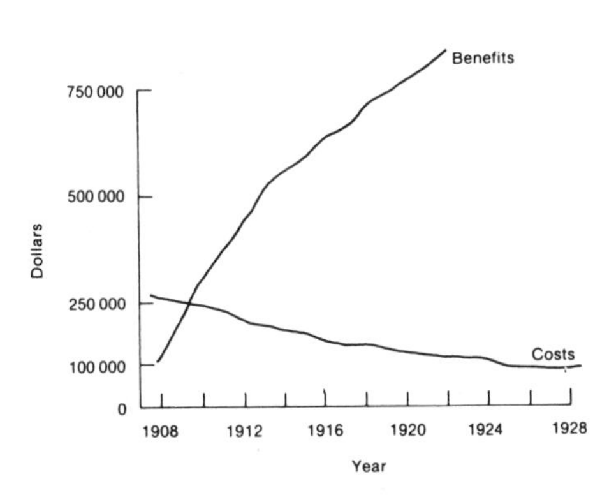

Figure 49. Costs vs benefits of chlorination and filtration. Reproduced from Clarke [138], courtesy of the American Water Works Association.

Table XLIV. Comparative Costs of Disinfection Practices[a]

Process	Concentration (mg/l)	Operating Costs for Plant Capacity of 38 ML/day (10 mdg)(¢/kl)			Operating Costs for Plant Capacity of 380 ML/day (100 mgd)(¢/kl)		
		O&M	Capital	Total	O&M	Capital	Total
Chlorination							
Without contactor	2	0.2	0.1	0.3	0.1	<0.1	>0.4
With contactor	2	0.2	0.2	0.4	0.1	0.1	0.2
Chlorine Dioxide							
Without contactor	1	0.2	0.2	0.4	0.1	<0.1	>0.1
With contactor	1	0.2	0.2	0.4	0.1	0.1	0.2
Ozone	1	0.2	0.2	0.4	0.1	0.1	0.2
Ammoniation							
Without contactor	3	0.5	0.2	0.7	0.3	0.1	0.4
With contactor	3	0.5	0.2	0.7	0.3	0.1	0.4

[a]Reproduced from Clark [138], courtesy of the American Water Works Association.

Operational Effectiveness

Many questions arise from disinfection of water by chlorine and other agents. As cited above, there are a multitude of laboratory studies on the effectiveness of the various disinfectants. On the other hand, are these agents effective in actual operation under field conditions? Does the presence of a chlorine residual assure disinfection?

The relationship between chlorine residual and the bacteriological quality in the distribution system was reported by Buelow and Walton [139]. Their study attempted to address two issues:

1. How much and what kind of chlorine residual should be carried throughout the distribution system;
2. Where and how many samples should be collected to maintain adequate surveillance to protect consumers?

The authors obtained their data from the Cincinnati Water Works and the National Community Water Supply Survey (CWSS) (969 water supply systems) conducted in 1969 by the Bureau of Water Hygiene, U.S. Public Health Service (PHS) [140]. The first issue was discussed by a consideration of the number of "repeat samples." At that time, the PHS standard required that daily samples be taken from the same location and be examined until two successive samples are free from coliform bacteria whenever a sample exceeds 4/100 ml. Table XLV shows the number of repeat samples for the Cincinnati Water Works during 1966 and 1967

Table XLV. Results of Repeat Samples for Cincinnati Waterworks[a]

	1966 Comb. Cl_2 Res.	1967 Comb. Cl_2 Res.	1969–70 Free Cl_2 Res.
No. of regular samples containing more than 4 coliform bacteria in 100 ml of water	19	20	8
No. of repeat samples required to obtain 2 successive 0/100-ml coliform counts	49	48	17
No. of repeat samples found to contain coliform bacteria	10	6	1

[a]Reproduced from Buelow and Walton [139], courtesy of the American Water Works Association.

when combined chlorination was practiced and during 1969–1970 when free chlorination was practiced. It appears, according to the authors, that free residual chlorine maintained in the distribution system has accomplished its purpose of providing control of local contamination. From the CWSS, it was observed that in chlorinated water supplies, a chlorine residual must be maintained throughout the distribution system in order to have the confidence that disinfection has been accomplished. Figure 50 is an example of the data to support this observation. It is interesting to note the percentage of unchlorinated drinking waters that existed in 1969.

A rigorous protective protocol must be followed during the installation of new mains and during the repair or replacement of existing mains in order to avoid bacteriological contamination of the distribution network. There are six areas of concern: (1) protecting new pipe sections at the construction site, (2) restricting the type of joint-packing materials used, (3) ensuring the preliminary flushing of pipe sections, (4) disinfecting pipes, (5) ensuring their final flushing, and (6) bacteriological testing. Buelow et al. [141] discuss the difficulties encountered in the disinfection of new mains.

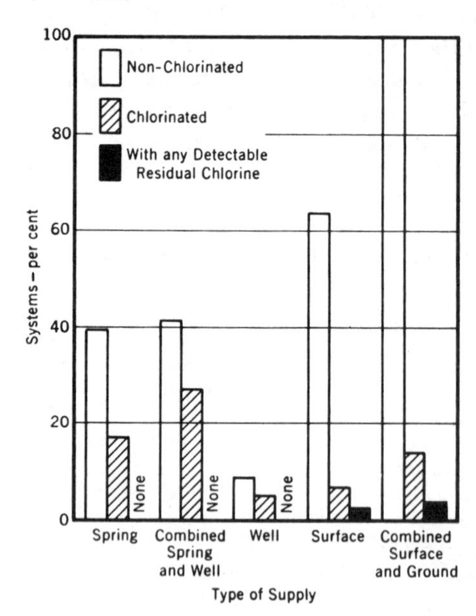

Figure 50. Percentage of water supply distribution systems found to have average total coliforms greater than 1/100 ml in various types of supplies. Reproduced from Buelow and Walton [139], courtesy of the American Water Works Association.

A number of techniques exist for new main disinfection in the field: Cl_2, $KMnO_4$, heavy metals and combinations thereof. Table XLVI shows the recurrence of failures after chlorinating various types of materials. The average failure of 18.5% is quite significant. Some of these failures apparently can be ascribed to the difficulty of disinfectants penetrating into the materials used to seal pipe joints. Other failures can be due to unnecessary contamination during construction. Strict adherence to the measures to meet the six areas of concern will provide a greater percentage of new mains becoming disinfected [141]. Pipe interiors, fittings, and valves must be protected against contamination. After completion, lines should be flushed at a minimum velocity of 76.2 cm/sec (2.5 ft/sec). After this, a free residual chlorine greater than 1 mg/l should be applied so that, at least, a 0.5-mg/l residual will remain in the system for 24 hr.

Dechlorination

There are occasions when it is necessary to reduce chlorine from the finished water prior to entry into the distribution system. This is usually the case following superchlorination to breakpoint. Dechlorination is accomplished by sulfur dioxide or other types of reduced sulfur compounds. There are occasions when activated carbon is employed. The following chemical reactions are applicable:

$$SO_2 + Cl_2 + 2\,H_2O = 2\,HCl + H_2SO_4 \tag{76}$$

$$NaHSO_3 + Cl_2 + H_2O = NaHSO_4 + 2\,HCl \tag{77}$$

$$Na_2SO_3 + Cl_2 + H_2O = Na_2SO_4 + 2\,HCl \tag{78}$$

Table XLVI. Mains Disinfection: Recurrence of Failures with Respect to Material[a]

Material	No. Laid	No. Passed First Time	Number Failing Chlorination					Failure (%)
			1X	2X	3X	4X	5+	
Asbestos Cement	391	329	48	11	2	1	0	16
Iron	70	54	11	3	1	0	1	23
PVC	282	222	23	12	8	1	5	21
All Materials	743	605	93	26	11	2	6	18.5

[a]Reproduced from Buelow et al. [141], courtesy of the American Water Works Association.

$$C + Cl_2 + 2H_2O = 4HCl + CO_2 \quad (79)$$

Since all four of these reactions produce acidic substances, the alkalinity of the water will be reduced. The stoichiometries of these reactions are:

Dechlorinating Agent	Parts Required per Parts Chlorine Removed (theoretical)	Alkalinity Consumed (as $CaCO_3$) (presumed to be mg/l) [21]
Sulfur Dioxide	0.90	2.8
Sodium Bisulfite	1.46	1.38
Sodium Sulfite	1.77	1.38
Activated Carbon	0.085	2.1

REFERENCES

1. Baker, M. N. *The Quest for Pure Water, Vol. I,* 2nd ed., (Denver, CO: American Water Works Association, 1981).
2. Lippy, E. C. *J. Am. Water Works Assoc.* 73:57 (1981).
3. Salvato, Jr., J. A. *Environmental Engineering and Sanitation,* 2nd ed. (New York: John Wiley & Sons, Inc., 1972).
4. Geldreich, E. E. In: *Water Pollution Microbiology,* R. Mitchell, Ed. (New York: John Wiley & Sons, Inc., 1972).
5. Taylor, F. B. *J. Am. Water Works Assoc.* 66:306 (1974).
6. Craun, G. F. *J. Am. Water Works Assoc.* 73:360 (1981).
7. Weibel, S. R. et al. *J. Am. Water Works Assoc.* 56:947 (1964).
8. Craun, G. F. et al. *J. Am. Water Works Assoc.* 68:420 (1976).
9. Craun, G. F., and R. H. Gunn. *J. Am. Water Works Assoc.* 71:422 (1979).
10. Craun, G. F. *Am. J. Public Health* 69:817 (1979).
11. Fenters, J. D., and J. M. Reed. *J. Am. Water Works Assoc.* 69:328 (1977).
12. Committee Report, "Viruses in Drinking Water," *J. Am. Water Works Assoc.* 71:441 (1979).
13. Mahdy, M. S. *J. Am. Water Works Assoc.* 71:445 (1979).
14. Geldreich, E. E. et al. *J. Am. Water Works Assoc.* 64:596 (1972).
15. Kehr, R. W., and C. T. Butterfield. *Public Health Rep.* 58:589 (1943).
16. Geldreich, E. E. *J. Am. Water Works Assoc.* 62:113 (1970).
17. "Water Quality Criteria, 1972." Committee on Water Quality Criteria, National Academies of Sciences and Engineering, Washington, DC (1972).
18. Geldreich, E. E., and B. A. Kenner. *J. Water Poll. Control Fed.* 41:R336 (1969).
19. Heukelekian, H., and H. B. Schulhoff. Bulletin 589, NJ Agricultural Experiment Station (1935).
20. *Bacterial Indicators/Health Hazards Associated with Water.* A. W.

Hoadley and B. J. Dutka, Eds., ASTM Special Tech. Pub. 635 (Philadelphia, PA: American Society for Testing and Materials, 1977).
21. Laubusch, E. J. In: *Water Quality and Treatment,* 3rd ed. (New York: McGraw-Hill Book Company, 1971).
22. Fair, G. M. et al. *Water Supply and Waste-Water Disposal* (New York: John Wiley & Sons, Inc., 1954).
23. Johannesson, J. K. *J. Am. Public Health Assoc.* 50:1737 (1960).
24. Moeller, T. *Inorganic Chemistry* (New York: John Wiley & Sons, Inc., 1958).
25. Ohmüller, O. *Arb. Kaiserl. Gesundl* 8:228 (1892).
26. Goldstein, M. L. et al. *J. Am. Water Works Assoc.* 52:247 (1960).
27. Gilcreas, F. W. et al. *J. New England Water Works Assoc.* 67:130 (1953).
28. *Standard Methods for the Examination of Water and Wastewater,* 15th ed. (Washington, DC: American Public Health Association, 1980).
29. White, G. C. *Handbook of Chlorination* (New York: Van Nostrand-Reinhold Company, 1972).
30. Winkler, L. "Math as Termeze Hudomany," *Ertisito,* Vol. 25 (1907).
31. Jakowkin, A. A. *Z. Phys. Chem.* 29:654 (1899).
32. Conmick, R. E., and Y. T. Chia. *J. Am. Chem. Soc.* 81:1280 (1959).
33. Morris, J. C. *J. Am. Chem. Soc.* 68:1692 (1946).
34. Shilov, E. A., and S. N. Solodushenkov. *Comp. Rend. Acad. Sci. l'URSS* 3(1):17 (1936).
35. Morris, J. C. *J. Phys. Chem.* 70:3798 (1966).
36. Pourbaix, M. *Atlas of Electrochemical Equilibria in Aqueous Solutions* (New York: Pergamon Press, Inc., 1966).
37. *Drinking Water and Health, Vol. 2* (Washington, DC: National Academy Press, 1980).
38. Weil, I., and J. C. Morris. *J. Am. Chem. Soc.* 71:1664 (1949).
39. Morris, J. C. In: *Principles and Applications of Water Chemistry,* S. D. Faust and J. V. Hunter, Eds. (New York: John Wiley & Sons, Inc., 1967).
40. Saguinsin, J. L. S., and J. C. Morris. In: *Disinfection: Water and Wastewater* (Ann Arbor, MI: Ann Arbor Science Publishers, Inc., 1975).
41. Morris, J. C. Lecture notes (1978).
42. Wei, I. W., and J. C. Morris. In: *Chemistry of Water Supply Treatment and Distribution,* A. J. Rubin, Ed. (Ann Arbor, MI: Ann Arbor Science Publishers, Inc., 1974).
43. Morris, J. C., and I. Wei. Preprint, American Chemical Society, Minneapolis, MN (April 15, 1969).
44. Griffin, A. E., and N. S. Chamberlin. *J. New England Water Works Assoc.* 55:371 (1941).
45. Griffin, A. E. *J. Am. Water Works Assoc.* 31:2121 (1939).
46. Palin, A. T. *Water and Water Eng.* 54:151, 189, 248 (1950).
47. Morris, J. C. et al. "Kinetic Studies on the Breakpoint Reaction with Ammonia and Glycine," Proc. Int. Cong. Pure Appl. Chem., New York (1952).
48. Sounier, B. M., and R. E. Selleck. *J. Am. Water Works Assoc.* 71:164 (1979).
49. Henderson, C. T. U.S. Patent No. 1,995,639 (1935).
50. Hildesheim, H. *Tech. Gemeind.* 39:56 (1936).
51. Mills, J. F. U.S. Patent No. 3,462,363 (1969).

52. Kanyaer, N. P., and E. A. Shilan. *Tr. Ivanor. Khim. Tekhvol. Inst.* (*USSR*) 3:69 (1940).
53. Latimer, W. M. *Oxidation Potentials,* 2nd ed. (Englewood Cliffs, NJ: Prentice-Hall, 1952).
54. Farkas, L., and M. Lewin. *J. Am. Chem. Soc.* 72:5766 (1950).
55. Galal-Gorchev, H., and J. C. Morris. *Inorg. Chem.* 4:899 (1965).
56. Johnson, J. D., and R. Overby. *J. San. Eng. Div., ASCE* 97:617 (1971).
57. Inman, Jr., G. W., et al. *Inorg. Chem.* 15:3037 (1976).
58. Johanneson, J. K. *J. Am. Public Health Assoc.* 50:1731 (1960).
59. *Drinking Water and Health, Vol. 2* (Washington, DC: National Academy Press, 1980).
60. Morris, J. C. et al. *Ind. Eng. Chem.* 45:1013 (1953).
61. Chang, S. L. *J. Am. Phara. Assoc.* 47:417 (1958).
62. Black, A. P. et al. *J. Am. Water Works Assoc.* 57:1401 (1965).
63. Marks, H. C., and F. B. Strandskov. *Ann. N.Y. Acad. Sci.* 53:163 (1950).
64. McAlpine, R. K. *J. Am. Water Works Assoc.* 74:725 (1952).
65. Stumm, W. *J. Boston Soc. Civ. Eng.* 45:68 (1958).
66. O'Brien, J. E. et al. In: *Chemistry of Water Supply Treatment and Distribution,* A. J. Rubin, Ed. (Ann Arbor, MI: Ann Arbor Science Publishers, Inc., 1974).
67. Fair, G. M., and J. C. Geyer. *Water Supply and Wastewater Disposal* (New York: John Wiley and Sons, Inc., 1954).
68. Baumann, E. R., and D. D. Ludwig. *J. Am. Water Works Assoc.* 54:1379 (1962).
69. Fair, G. M. et al. *Elements of Water Supply and Wastewater Disposal* (New York: John Wiley & Sons, Inc., 1971).
70. Butterfield, C. T. et al. *Public Health Rep.* 58:1837 (1943).
71. Chang, S. L. *War Medicine* 5:46 (1944).
72. Clarke, N. A., and P. W. Kabler. *Am. J. Hyg.* 59:119 (1954).
73. Clarke, N. A. et al. *Am. J. Hyg.* 64:314 (1956).
74. Kelly, S., and W. W. Sanderson. *Am. J. Public Health* 48:1323 (1958).
75. Scarpino, P. V. et al. *Water Res.* 6:959 (1972).
76. Scarpino, P. V. et al. In: *Chemistry of Water Supply Treatment and Distribution,* A. J. Rubin, Ed. (Ann Arbor, MI: Ann Arbor Science Publishers, Inc., 1974).
77. Engelbrecht, R. S. et al. "Virus Sensitivity to Chlorine Disinfection of Water Supplies," EPA Report 600/2-78-123, U.S. EPA, Washington, DC (1978).
78. Butterfield, C. I., and C. Wattie. *Public Health Rep.* 61:157 (1946).
79. Kelly, S. M., and W. W. Sanderson. *Amer. J. Public Health* 50:14 (1960).
80. Siders, D. L. et al. "Destruction of Viruses and Bacteria in Water by Monochloramine," Abstract E27, Amer. Microbiol., Washington, DC (1973).
81. Esposito, P. et al. "Destruction by Dichloramine of Viruses and Bacteria in Water," Abstract No. G99, American Society of Microbiologists, Washington, DC (1974).
82. Ridenour, G. M., and R. G. Ingols. *J. Am. Water Works Assoc.* 39:561 (1947).

83. Benarde, M. A. et al. *Appl. Microbiol.* 13:776 (1965).
84. Benarde, M. A. et al. *J. Appl. Microbiol.* 30:159 (1967).
85. Cronier, S. et al. In: *Water Chlorination, Vol. 2*, R. L. Jolley et al., Eds. (Ann Arbor, MI: Ann Arbor Science Publishers, Inc., 1978).
86. Johannesson, J. K. et al. *Am. J. Public Health* 50:1731 (1960).
87. Tanner, F. W., and G. Pitner. *Proc. Soc. Exp. Biol. Med.* 40:143 (1939).
88. Johannesson, J. K. *Nature* 181:1799 (1958).
89. Wyss, O., and J. R. Stockton. *Arch. Biochem.* 12:267 (1947).
90. Floyd, R. et al. *Environ. Sci. Technol.* 12:1031 (1978).
91. Mills, J. F. In: *Disinfection: Water and Wastewater* (Ann Arbor, MI: Ann Arbor Science Publishers, Inc., 1975).
92. Keswick, B. H. et al. *J. Am. Water Works Assoc.* 70:573 (1978).
93. Chang, S. L., and J. C. Morris. *Ind. Eng. Chem.* 45:1009 (1953).
94. Chambers, C. W. et al. *Soap. Sanit. Chem.* 28:149 (1952).
95. O'Connor, J. T., and S. K. Kapoor. *J. Am. Water Works Assoc.* 62:80 (1970).
96. Fetner, R. H., and R. S. Ingols. *J. Gen. Microbiol.* 15:381 (1956).
97. Wuhrmann, K., and A. Meyrath. *Schweiz Z. Path. Bakteriol.* 18:1060 (1955).
98. Katzenelson, E. et al. *J. Am. Water Works Assoc.* 66:725 (1974).
99. Broadwater, W. T. et al. *Appl. Microbiol.* 26:391 (1973).
100. Burleson, G. R. *Appl. Microbiol.* 29:340 (1975).
101. Katzenelson, E. et al. In: *Chemistry of Water Supply Treatment and Distribution,* A. J. Rubin, Ed. (Ann Arbor, MI: Ann Arbor Science Publishers, Inc., 1974).
102. Krusé, C. W. et al. In: *Proc. Nat. Spec. Conf. Disinf.,* American Society Civil Engineering, New York (1970).
103. Stringer, R. P. et al. In: *Disinfection: Water and Wastewater,* J. D. Johnson, Ed. (Ann Arbor, MI: Ann Arbor Science Publishers, Inc., 1975).
104. Hoff, J. C., and E. E. Geldreich. *J. Am. Water Works Assoc.* 73:40 (1981).
105. Gilbert, M. B. et al. *J. Am. Water Works Assoc.* 68:495 (1976).
106. Waite, T. D. Paper 33-4, *Proc. Am. Conf. AWWA,* Denver, CO (1978).
107. Cleasby, J. L. et al. *J. Am. Water Works Assoc.* 56:466 (1964).
108. Delaney, J. E., and J. C. Morris. In: *Proc. Nat. Spec. Conf. Disinf.,* American Society Civil Engineering, New York (1970).
109. Gardiner, J. *Water Res.* 7:823 (1973).
110. Chick, H. *J. Hyg.* 8:92 (1908).
111. Chang, S. L. *J. San. Eng. Div., ASCE* 97:689 (1971).
112. Morris, J. C. In: *Disinfection: Water and Wastewater,* J. D. Johnson, Ed. (Ann Arbor, MI: Ann Arbor Science Publishers, Inc., 1975).
113. Wei, J. H., and S. L. Chang. In: *Disinfection: Water and Wastewater,* J. D. Johnson, Ed. (Ann Arbor, MI: Ann Arbor Science Publishers, Inc., 1975).
114. Taylor, D. G., and J. D. Johnson. In: *Chemistry of Water Supply Treatment and Distribution,* A. J. Rubin, Ed. (Ann Arbor, MI: Ann Arbor Science Publishers, Inc., 1974).
115. Farooq, S. et al. *Water Res.* 11:737 (1977).
116. Floyd, R. et al. *Appl. Environ. Microbiol.* 31:298 (1976).

117. Floyd, R. et al. *Environ. Sci. Technol.* 13:438 (1979).
118. Kulikovsky, A. et al. *J. Appl. Bacteriol.* 38:39 (1975).
119. Green, D. E., and P. K. Stumpf. *J. Am. Water Works Assoc.* 38:1301 (1946).
120. Knox, W. E. et al. *J. Bacteriol.* 55:451 (1948).
121. Venkobachar, C. et al. *Water Res.* 9:119 (1975).
122. Haas, C. N., and R. S. Engelbrecht. *Water Res.* 14:1749 (1980).
123. Thorup, R. T. et al. *J. Am. Water Works Assoc.* 62:97 (1970).
124. Hanna, G. P., and A. J. Rubin. *J. Am. Water Works Assoc.* 62:315 (1970).
125. Manwaring, J. F. et al. *J. Am. Water Works Assoc.* 63:298 (1971).
126. Chaudhuri, M., and R. S. Engelbrecht. *J. Am. Water Works Assoc.* 62:563 (1970).
127. Shelton, S. P., and W. A. Drewry. *J. Am. Water Works Assoc.* 65:627 (1973).
128. York, D. W., and W. A. Drewry. *J. Am. Water Works Assoc.* 66:711 (1974).
129. Anurhor, P., and R. S. Engelbrecht. *J. Am. Water Works Assoc.* 67:187 (1975).
130. Malek, B. et al. *J. Am. Water Works Assoc.* 73:164 (1981).
131. Sproul, O. T. "Critical Review of Virus Removal by Coagulation Processes and pH Modifications," EPA Report 600/2-80-004, U.S. EPA, Cincinnati, OH (1980).
132. Dixon, J. K., and M. W. Zielyk. *Environ. Sci. Technol.* 3:551 (1969).
133. Lin, S. D. et al. "Algal Removal by Alum Coagulation," Report 68, Illinois State Water Survey, Urbana, IL (1971).
134. Brodtmann, Jr., N. Y., and P. J. Russo. *J. Am. Water Works Assoc.* 71:40 (1979).
135. Shull, K. E. *J. Am. Water Works Assoc.* 73:101 (1981).
136. Hubbs, S. A. et al. *J. Am. Water Works Assoc.* 73:97 (1981).
137. Keller, J. W. et al. *J. Am. Water Works Assoc.* 66:730 (1974).
138. Clark, R. M. *J. Am. Water Works Assoc.* 73:89 (1981).
139. Buelow, R. W., and G. Walton. *J. Am. Water Works Assoc.* 63:28 (1971).
140. McCabe, L. J. et al. *J. Am. Water Works Assoc.* 62:670 (1970).
141. Buelow, R. W. et al. *J. Am. Water Works Assoc.* 68:283 (1976).

INDEX